D0944610

INDUSTRIAL CHEMISTRY
Volume 1

ELLIS HORWOOD SERIES IN APPLIED SCIENCE AND INDUSTRIAL TECHNOLOGY

Series Editor: Dr D. H. SHARP, OBE, former General Secretary, Society of Chemical Industry; formerly General Secretary, Institution of Chemical Engineers; and former Technical Director, Confederation of British Industry.

This collection of books is designed to meet the needs of technologists already working in fields to be covered, and for those new to the industries concerned. The series comprises valuable works of reference for scientists and engineers in many fields, with special usefulness to technologists and entrepreneurs in developing countries.

Students of chemical engineering, industrial and applied chemistry, and related fields, will also find these books of great use, with their emphasis on the practical technology as well as theory. The authors are highly qualified chemical engineers and industrial chemists with extensive experience, who write with the authority gained from their years in industry.

Published and in active publication

PRACTICAL USES OF DIAMONDS
A. BAKON, Research Centre of Geological Technique, Warsaw, and A. SZYMANSKI, Institute of Electronic Materials Technology, Warsaw
NATURAL GLASSES
V. BOUSKA, *et al.*, Czechoslovak Society for Mineralogy & Geology, Czechoslovakia
POTTERY SCIENCE: Materials, Processes and Products
A. DINSDALE, lately Director of Research, British Ceramic Research Association
MATCHMAKING: Science, Technology and Manufacture
C. A. FINCH, Managing Director, Pentafin Associates, Chemical, Technical and Media Consultants, Stoke Mandeville, and S. RAMACHANDRAN, Senior Consultant, United Nations Industrial Development Organisation for the Match Industry
OFFSHORE PETROLEUM TECHNOLOGY AND DRILLING EQUIPMENT
R. HOSIE, formerly of Robert Gordon's Institute of Technology, Aberdeen
MEASURING COLOUR
R. W. G. HUNT, Visiting Professor, The City University, London
MODERN APPLIED ENERGY CONSERVATION
Editor: K. JACQUES, University of Stirling, Scotland
CHARACTERIZATION OF FOSSIL FUEL LIQUIDS
D. W. JONES, University of Bristol
PAINT AND SURFACE COATINGS: Theory and Practice
Editor: R. LAMBOURNE, Technical Manager, INDCOLLAG (Industrial Colloid Advisory Group), Department of Physical Chemistry, University of Bristol
CROP PROTECTION CHEMICALS
B. G. LEVER, Development Manager, ICI plc Plant Protection Division
HANDBOOK OF MATERIALS HANDLING
Translated by R. G. T. LINDKVIST, MTG, Translation Editor: R. ROBINSON, Editor, *Materials Handling News.* Technical Editor: G. LUNDESJO, Rolatruc Limited
FERTILIZER TECHNOLOGY
G. C. LOWRISON, Consultant, Bradford
NON-WOVEN BONDED FABRICS
Editor: J. LUNENSCHLOSS, Institute of Textile Technology of the Rhenish-Westphalian Technical University, Aachen, and W. ALBRECHT, Wuppertal
PROFIT BY QUALITY: The Essentials of Industrial Survival
P. W. MOIR, Consultant, West Sussex
EFFICIENT BEYOND IMAGINING: CIM and Its Applications for Today's Industry
P. W. MOIR, Consultant, West Sussex
TRANSIENT SIMULATION METHODS FOR GAS NETWORKS
A. J. OSIADACZ, UMIST, Manchester
MECHANICS OF WOOL STRUCTURES
R. POSTLE, University of New South Wales, Sydney, Australia, G. A. CARNABY, Wool Research Organization of New Zealand, Lincoln, New Zealand, and S. de JONG, CSIRO, New South Wales, Australia
MICROCOMPUTERS IN THE PROCESS INDUSTRY
E. R. ROBINSON, Head of Chemical Engineering, North East London Polytechnic
BIOPROTEIN MANUFACTURE: A Critical Assessment
D. H. SHARP, OBE, former General Secretary, Society of Chemical Industry; formerly General Secretary, Institution of Chemical Engineers; and former Technical Director, Confederation of British Industry
QUALITY ASSURANCE: The Route to Efficiency and Competitiveness, Second Edition
L. STEBBING, Quality Management Consultant
QUALITY MANAGEMENT IN THE SERVICE INDUSTRY
L. STEBBING, Quality Management Consultant
INDUSTRIAL CHEMISTRY: Volumes 1 and 2
E. STOCCHI, Milan, with additions by K. A. K. LOTT and E. L. SHORT, Brunel

INDUSTRIAL CHEMISTRY
Volume 1

E. STOCCHI

Translators:
K.A.K LOTT and E.L. SHORT
Department of Chemistry
Brunel, The University of West London
Uxbridge, Middlesex

ELLIS HORWOOD
NEW YORK LONDON TORONTO SYDNEY TOKYO SINGAPORE

This English edition first published in 1990 by
ELLIS HORWOOD LIMITED
Market Cross House, Cooper Street,
Chichester, West Sussex, PO19 1EB, England

A division of
Simon & Schuster International Group

This English edition is translated from the original Italian edition *Chimica Industriale*, Vol. 1 *Inorganica*, published by Edinco editrice Torino,

Printed and bound in Great Britain
by The Camelot Press, Southampton

British Library Cataloguing in Publication Data

Stocchi, E.
Industrial chemistry.
Vol. 1, E. Stocchi
1. Applied chemistry
660
ISBN 0-13-457318-8

Library of Congress Cataloging-in-Publication Data available

Table of contents

Second Part

Translators' Preface

Throughout the translation of this work we have attempted to follow strictly the intentions of the author while retaining clarity of expression in the English version. For the most part, a fairly literal translation has been possible but sometimes it has been necessary to translate idiomatic expressions somewhat more loosely.

Considerable effort has been put into checking many of the more specialized technical terms occurring in the text but, inevitably, there may be certain instances where a specialist in a given topic may have chosen an alternative and, here, the translators, of course, accept responsibility for any unconventional usage which may have crept in. If any such points are noted, we should be grateful if they could be brought to our attention.

Brunel University
August 1989

KAKL
ELS

1

The fundamental principles and problems of industrial chemistry

1.1 THE FIELD OF INDUSTRIAL CHEMISTRY

All of the laboratory research which has been carried out at an ever increasing rate by groups of scientists in well proven areas as well as on new themes over a period of one and a half centuries up to the present time, and the major advances which have followed from theoretical studies and practical experiments in thermodynamics, kinetics, mechanics, cybernetics, and many other sciences and technologies, would not have been justified on a practical plane if industrial chemistry had not evolved at the same pace; industrial chemistry being that branch of chemistry which formulates and solves the problems which are inherent in the exploitation of all raw materials, and which produces from them, on an appropriate scale, many other products which do not occur naturally, which are becoming scarce in nature, or whose methods of preparation are capable of being improved upon.

Industrial chemistry is the 'executive-technics' of chemistry, that is, the science which assists nature in unlocking its resources and in doing so is a stimulus to the pursuit of the optimal course, followed by natural creative evolution, thereby enhancing the natural resources. Consequently, being a synthesis of science and technology, industrial chemistry enables humanity to experience the benefits of the function performed by chemistry which may even improve upon certain natural processes.

Also, whilst the contemporary theoretical and practical structure of modern industrial chemistry tends to ignore the formal dichotomy in the subject, the emphasis in a seat of learning must still be put on the two traditional areas which are intimately related and indicate the nature of the materials and products which are dealt with. These two areas are *inorganic industrial chemistry* and *organic industrial chemistry*.

The scheme presented below summarizes the contents and aims of the abovementioned branches of industrial chemistry.

As indicated in each first section of the scheme, both inorganic and organic industrial chemistry have the primary aim (historically) of applying chemical

		Object	Specl aims
Industrial chemistry	*Inorganic*	Application of chemical transformation operations to natural inorganic materials	to iprove the quality and servs rendered by them (war, phosphates, uranium, etc.)
			to cvert them to technically usef products (glasses, ceraics, cements)
			to drive semimanufactured or maifactured components from thei (metals, synthesis gas, cok alloys, etc.)
		Large-scale production of inorganic substances	prie base materials (mineral aci, ammonia and soda, for exaple)
			inganic compositions of great indstrial interest (such as fertizers, salt products, and pigents)
	Organic	Application of chemical transformation operations to natural organic materials	to iprove the quality and serces rendered by them (e.g. ruber, natural fibres, skins, ancellulose)
			toonvert them to products ofarious use (such as soaps, fat gelatine, sugars, and fenentation products)
			toerive components or maufactured goods of great prtical interest from them (sphur, hydrogen, carbon, peol, diesel fuels, lubricants, etc
		Large-scale production or organic substances or derivatives based on organic substances	of clearly organic nature (abhols, acids, resins, dergents, solvents, dyes, elatomers, pharmaceuticals, and son)
			of mixed nature (explosives, peicides, optical whiteners, pltic materials, varnishes, an so on)

procedures to the transformation of natural materials. is in this way that the traditional distinction between *applied chemistry* and *indutrial chemistry* arises.

Nowadays, however, with the existence of so many subtitute materials which are hybrids of those belonging specifically to these two traitional areas, and, more importantly, with the rapid integration of the basic prcedures that lie at their foundation, a breakdown in such a formal distinction haensued.

1.2 INDUSTRIAL CHEMISTRY AND RAW MATERIALS

In order to understand the essential role played by chenistry in general and by industrial chemistry in particular, it may also be observed that any material starts off as a suitable product capable of exploitation by various other technologies by means of chemical processing. Consequently, the world economy is fundamentally dependent on products from the chemical industry.

As a further demonstration of this point it may be recalled that *commercial technology*, that is, the study of any really marketable object from the standpoint of the raw materials, derivatives, or intermediates from which it is formed, is a discipline which is twinned with *industrial chemistry*, so establishing, in the final analysis, the descriptive–appraisal–investment aspects of the products. The existence in the past of chairs of industrial chemistry and the commercial technology which were held jointly in Italian institutes by such illustrious professors as E. Molinari, D. Meneghini, and L. Cambi is proof of the essentially applied aspect which the whole of commercial technology addresses with relation to industrial chemistry.

From the natural environment, that is, from the lithosphere, the hydrosphere, the atmosphere, and the biosphere, industrial chemistry procures raw materials in order to convert them into 'intermediates' which are subsequently provided as base materials to every other kind of industry.

Raw materials of the lithosphere

Industrial chemistry takes the major part of the chemical elements, whether they are metals or non-metals, from the Earth's crust. Most of the elements occur in a chemically combined form from which they must be extracted.

The abundance of the elements is judged not so much on the effective distribution of the element but from the point of view of the ease with which its deposits can be exploited ('economic availability'). This is the case with tin and lead which are wrongly held to be more abundant than rubidium, while antimony which, contrary to what is generally believed, is more widely distributed than germanium.

Industrial chemistry also extracts coal, petroleum, and natural gases from the lithosphere. The activities of a whole branch of the subject are essentially founded on these materials.

Moreover, the salts forming the basis of many chemical products such as barium and calcium sulphates and fertilizers such as nitrates and phosphates are also derived from the Earth's crust. Finally, limestones, loams and silica in many varied forms and silicates for the ceramic, cement and glass industries are all obtained from the lithosphere.

Raw materials from the hydrosphere, the atmosphere, and the biosphere

From the marine and oceanic environment, that is, the *hydrosphere*, industrial chemistry extracts ever increasing amounts of the most important raw material, *water*, by means of desalination techniques.

On the other hand, sea salt is the best known raw material for the chemical industry which is furnished by the hydrosphere. In fact, it is not only the elements from which it is constituted that are obtained from sodium chloride but also all the compounds of sodium (soda, sulphates, and sulphides in various states of hydration, phosphates, polyphosphates, and so on). Moreover, the halogens iodine and, above all, bromine, originate for the greater part from the hydrosphere.

From the gaseous envelope which surrounds the Earth, that is, the *atmosphere*, industrial chemistry procures nitrogen, oxygen, and the noble gases. The fact should never be overlooked that the natural environment *air* is also systematically utilized in an integrated manner by many processes employed in industrial chemistry.

Finally, raw materials such as essences, cellulose, and other saccharides come to the chemical industry from the *biosphere*, principally via the plant kingdom. Fats, on the other hand, originate from both the plant and animal kingdoms, while enzymes for the biochemical industry are predominantly obtained from the animal kingdom.

1.3 THE PROBLEM OF ENERGY IN THE CHEMICAL INDUSTRY

Chemistry is one of the major industrial voices of the World and, among all man's activities, which create wealth from large-scale operations, the chemical industry is that which uses the greatest amounts of energy. In fact, in the more highly and harmoniously industrialized nations (USA, Japan, West Germany) the chemical industry consumes on average an amount of energy which is about 30% of the total requirement of the whole of industry. More specifically, the average annual consumption of energy by the chemical industry in the United States is about 6870 GJ ($1\ GJ = 10^9\ J$).

For this reason, when account is taken of the fact that many chemical processes furnish their own energy, it can be appreciated that the energy problem is simultaneously of equal concern to the chemical industry.

This is reflected in the acute crisis which affected the chemical industry at the same time as the grave energy crisis which occurred in the earlier part of the seventies and led to a recession which was in direct contrast to the great expansion witnessed by this industry between the fifties and seventies when energy had been 'cheap' everywhere and very readily available.

The basics of energy management for the improvement of economic budgeting

Energy must therefore be considered as an essential element for the chemical industry together with the land, the work force, the capital, and the raw materials. On the other hand, energy is not, as it may appear by a simple parallelism, a raw material since it lacks the property of mass. It is treated as an independent entity both as regards its provision and its application.

Today, the energy costs of the preparation of many products of the chemical industry tend to be about 35% of their sale price, which means that a great deal of attention has to be paid to 'energy management', that is, to a prearranged energy plan, to the control of energy use, to the improvement of yields, and to the recovery of energy in the best way. Erroneous estimates of energy costs and of the effective energy requirements constitute grave 'managerial' errors from which major crises in economic management follow.

However, while estimation of the cost of energy may result in data which are quite objective and furnished by the market economy, an energy requirement is, to a large extent, coupled with the capacity to recover energy in the productive process stages.

The operations which require more than half of the total heat which is used in industrial chemistry are those of distillation, desiccation, and evaporation. On the other hand, these are the processes which are most readily adapted to the saving of appreciable quantities of energy through the provision of good insulation, using more operational units (evaporators, for example) which are linked together so that units can successively utilize the heat produced by the preceding units, equipping the apparatus with automatic control systems in order to reduce any fluctuations by timely correction of the causes and the compression of steam in order to 'revitalize' it by raising its energy level.

In addition to the indirect systematic energy resources of the type mentioned above, there are also other 'occasional' sources which are related to the particular characteristics of the processes under consideration and the materials which are treated in these processes. These include a more thorough draining of the materials to be dried, the reduction of the drying medium to the bare minimum necessary, and improvement of the contact between the medium and the material to be dried.

Types of energy and energy sources for industrial chemistry

Basically, two types of energy are used in the chemical industry: thermal energy and electrical energy. The proportions in which these are used are about 58% and 42% respectively.

These percentage data refer to the amounts of energy which are fed *ex novo* into industrial chemical process plant from external energy sources. In practice, however, the amount of heat which is effectively employed is a higher percentage than that noted above, since the chemical processes themselves are a concentrated source of heat and also because electrical energy finishes up by being degraded into heat which can be utilized to a greater or lesser extent.

The commonest form of utilization of 'process' heat, that is, the heat which is generated for the two reasons which have already been stated in exothermic reactions and by operations which are carried out using electrical energy, is that destined to the production of steam with a certain energy content either as heat for the preheating of a steam boiler, as latent heat, or as heat for the superheating of the steam which has been produced.

The chemical industry directly exploits energy sources to procure the calorific energy which it requires while utilizing electrical energy produced by external conventional sources for the same purpose.

As usual, in terms of the amounts of each type which are actually used, the sources of electrical energy in order of importance are thermoelectric, hydroelectric, and, finally, nuclear electric. At the moment (and presumably at least for a long time in the future) the amounts of energy utilized in the chemical industry which are derived from biomass, winds, and other alternative energy sources in general are practically insignificant.

The energy sources which are directly exploited by the chemical industry in order to supply itself with calorific energy are shown statistically in Table 1.1, while, in the short term, it is not foreseen that there will be any appreciable contributions from alternative energy sources to the heat requirements of the chemical industry if one excludes regional utilization of biogas, the local exploitation of geothermal energy, and the widespread, but also always small, use of solar energy for the acceleration of evaporation and drying processes.

It can be seen from an analysis of the table that as with everything produced within the orbit of the chemical processes operated, the heat which is exploited in the chemical industry originates from the degradation of chemical energy, that is, the form of energy involved in the phenomenology of the bonding of atoms and molecules.

Table 1.1—Statistics of heat sources for the chemal industry

Energy sources	*Types*	*Contribution*
Natural gases	dry, with CH_4 predominating	54%
Medium-heavy petroleum fractions	(kerosenes), gas oil and combstion oils	10.8%
Coals	natural (excluding peat) and artificial (types of coke)	12.5%
Other combustibles	synthesis gas, refinery gas, gafrom various pyrolyses of petrols	18.6%
Steam	produced outside of the orbit which it serves†	4.1%

†This steam may originate, for example, from the execution of various petoleum refinery operations or from various operations involved in the distillation of coal.

1.4 RESEARCH IN THE CHEMICAL INDUSTRY

Another fundamental problem of the chemical industry is tat of 'research', which is principally directed to the improvement and the total rencation of plant and the manner in which processes are carried out. On the other hnd, the synthesis and application of new products also constitute areas of research inhe chemical industry.

The aims of research in the field of industrial chemistry ma be classified according to the following scheme:

Research { innovative
improvement and, in the limit, optiiization
application

In industrial chemistry, as well as in every other industria sector, the aims of the research which is pursued are distinctly different from the ims of research carried out in universities, scientific foundations, and so on. In the brmer case the aims are predominantly economic while in the latter case they are pursued in a research atmosphere which tends to increase knowledge, to formlate and validate new hypotheses in order that they may become theories. In accrdance with these aims it is therefore customary to say that research carried out in ndustry is of the nature of *applied research* while the more theoretical research is *pureresearch*. This, however, does not preclude the possibility that basic knowledge regading new processes and products may be obtained in a seat of learning through aplied research.

The effect of research costs in the chemical industry on a company's economy is considerable, varying from 2 to 5% of net turnover in variouly designated companies up to 15% in the avant-garde sectors of multinational chmical companies of the calibre of Hoechst.

To demonstrate the importance of the heed which is paid to research in the chemical industry it is convenient simply to take as an example the Hoechst company which, in terms of turnover, has leapt to a leading position among the World's chemical industries as a result of an extremely active attitude towards research.

It is of benefit to recall that Hoechst was the smallest of the three most important chemical companies (Bayer, BASF, and Hoechst) into which the Allies aspired to split the German chemical superholding IG. Farbenindustrie at the end of the second world war since it was held responsible for having given Hitler economic support in his assault on the rest of the World.

In fact, starting from an investment in research of around 5 million DM in 1947, Hoechst passed to 60 million DM in 1955 and to beyond 400 million DM in 1970 when it started to overtake rapidly all of its competitors worldwide.*

Innovative research

Any chemical industrial company which wishes to affirm its own identity by distinguishing itself from other competitors must study the introduction of new production processes and must carry out research on new products.

The aims of innovative research in the productive technologies and in products for sale are various: how to procure the esteem of prestige, to have the possibility of selling and exchanging licences for the construction and exploitation of plant, and the production and sale of new materials.

A research project which is cultivated in this field may be the fruit of the intuition of the design offices within an industry, or it may be suggested to these people by, among others, the economic management of the company, by those concerned with the setting up of the installations, by the plant managers, by those responsible for production, and by members of the sales offices in general and those concerned with customer service in particular. In fact, sensitivity to the exigencies of the market is the most stimulating guide to the implementation of new ideas which may lead to a reduction in the costs of matching the competition and in satisfying any needs which may be perceived.

Innovative research which is not based on an initial concrete fact because it involves the realization of an abstract project ('idea') requires the longest time for the initiative to be carried forward. There are great costs, above all, of risk taking insomuch as it is not infrequently the case that it may prove impossible to implement the basis of the idea within a framework which is technically sound and economically admissible. This leads to the cancellation of the project after considerable financial means have been expended upon it.

The onerous economic aspect of this kind of research is also witnessed by the fact that, whatever the objective may be (plant or product), it does not solely call for development in the pure study phase (theoretical basis) but also laboratory experiments, pilot scale studies, and, as we shall see (p. 91) semiscale studies with increasing costs due to the economic burdens arising from the retention of large research groups and highly specialized operatives and the preparation of the equipment on an ad hoc basis, using the most suitable materials in the most appropriate forms.

Product development research

Product development research is directed to the perfecting of the processes both with

---

* Karl Winnacker (ex-President of the Administration Council of Hoechst), 'Never lose hope' (Non perdersi mai d'animo), Casa Editrice *Il Ponte*, Milan.

regard to the adaptability of management as well as with resect to their suitability for increasing the purity of the products, their yields, and oppounities for application. Being characterized by relatively short fulfilment times, produ development research must not be interrupted by the management. This is to gurantee the systematic application of the results to the actual products concerned nd so as not to avoid the management being burdened with the onerous economic poblems of a postponed re-entry at a time when the concomitant costs are high.

Product development research compensates for the small dvantages which result from it in comparison with innovative research, with the low osts which are required to sustain it. Inter alia, this research, as a rule, does not reqire the development of specific technologies, and, in general, may be carried ot in the production departments in plants which already exist; it may be a queson of the modification of operating structures and the physical and/or chemicohysical conditions (of catalysts, for example).

Product development research initiatives are stimulated bythe economic management of a company and are generally proposed by the sales stf who react to requests made by customers.

Application research

This type of research has a three-fold aim: to extend the utization of the products, to increase their effectiveness, and to guarantee the absenc of any indications that they may be toxic or give rise to pollution problems when the are used. The extension of the uses and the increase in the efficacy of products mst permit, in order that innovative research also does not have to be carried out, thesetting up of application conditions which differ only in their environmental charater or in that there are new additives, carriers, synergizers, protectives, etc.

In its turn the guarantee of the absence of any indicatin that the product has deleterious properties is an ideal condition which must be prsued with the intention of maximizing the benefit from the use of the product and ninimizing any possible damage which may arise from its use. In practice, the actio (and also the use of a product) cannot be free from risk, and the benefits cannot aways be separated from the damage.

Application research is carried out in laboratories of diferent kinds (chemical, biological, technological) which are normally equipped, ad is carried out as the work of individual technicians or groups of workers close o the companies which make use of the products concerned, or it is the fruit of thework of teams who are dispatched onto the fields where the applications take pace. The latter is most frequently the case if the products to be applied are fertilizers phytopharmaceuticals, pesticides, and, in general, intended for the amelioration of human conditions of life, in particular climatic and environmental conditions.

The reasons behind the need to carry out application research o products of the chemical industry on site are many and complex to the point of imposing stict conditions on the costs (sale prices) of the finished products.

For example, it may be recalled that the surface of the Earth, icluded between the tropics (latitudes of 23 degrees 27 minutes North and South) in which te temperature is constantly high with frequent heavy downfalls of rain, poses a series of spedfic questions regarding the employment of phytoiatric products.

The temperature and humidity conditions have a decisive effect on the development of the vegetation, of the creatures which infest it, and on the relationships between the two. Among other things:

- cultivated plants and their parasites grow more rapidly between the tropics than in temperate climates,
- antiparasitic chemicals often provoke curative effects on plants which are unknown in our latitudes,
- the superficial adhesion of the products (generally exceedingly important in the struggle against parasites) is prejudiced by prolonged and frequent rains,
- sometimes there is a reinforcement of the initial action of pesticides with a concomitant reduction in the duration of their effect,
- the virulence of parasites and pathogenic agents tends to be synchronized with the optimal stage of plant development, that is, the physiological state when pesticide application is less favourable.

It is a combination of these factors relating to the environment which makes it necessary to check and evaluate on site the direction to be given to research carried out in order to guarantee 'custom-made' applications of industrial chemical products in the fight against plant pests.

Research of the kind which is conducted remotely from the parent company and from the industrial chemical plant which support the research, has a powerful influence on the financial balances of this industry. Plots of land must be acquired, buildings constructed, apparatus and equipment procured and qualified and specialized personnel employed (who are mentally flexible and endowed with initiative, among other things). These personnel must receive proper and orderly supplies of experimental materials, logistic support, and help of the kind which is likely to ameliorate their living conditions, which are frequently difficult to endure for those who are accustomed to the many commodities and services available in their home country.

1.5 TOXIC EFFECTS OF CHEMICAL PRODUCTS

Toxic effects constitute another serious problem which arises from activities in the field of industrial chemistry. Above all, field workers may become subject to an action, provoked by chemical products, which modifies or inhibits their biochemical and physiological systems. The treatment presented here will not be of the usual vague nature of 'avoiding accidents and industrial hygiene' which belongs more appropriately to the chemical plant sector, but it will be rather to define a set of fundamental notions regarding the toxicology of chemical products which is presented with the aim of providing the reader with a knowledge of how to defend himself, particularly with regard to potential poisons.

Environmental contamination by chemical products

Paracelcus, one of the major protagonists of Renaissance thinking, was to write: 'Everything is poisonous, nothing is poisonous, only the dose makes the poison' in order to indicate that *innocuous substances do not exist, because there is a toxic dose of every substance.*

On the other hand there is no doubt that chemical compounds, if for no other reason than that they are potential competitors in reactions involving substances which participate in cellular metabolism, are inherently toxic to living organisms and therefore to man.

Today, exposure to chemical products has reached a startling intensity so that man is exposed to an out and out pathological danger from chemical compounds.

In 1982 world production of organic compounds had touched 85 million tons. As a conspicuous amount of these compounds was dispersed, by many varied routes, into the environment where it may have been or may not have been readily degraded, the contamination of the planet became evident, exceeding 200 000 tons. However, organic compounds are not the only environmental chemical pollutants; others include coal and petroleum combustion products, sulphur compounds, metal powders, and ions of the medium-heavy and heavy metals.

In spite of the startling picture which has been painted of the objective situation regarding environmental damage caused by chemical products, cases of acute or immediate poisoning are comparatively rare. On the other hand, shorter and longer term cases of poisoning due to the additive effects of several chemical compounds or the effects of several compounds which mutually reinforce the action of one another on a toxicological plane (synergism) are frequently encountered.

Finally, products of the chemical industry* which are capable of producing the effects of poisoning are extremely numerous. Such compounds are injurious to organisms in a rather short term.

Classification of toxic effects

To be injurious to tissues a toxic chemical compound must come into contact with a part of the body which may be the skin, the eyes, the respiratory tracts, the digestive system, etc. Naturally, it is not only the part which comes into contact with the toxic agent which becomes damaged, but, depending on the substance, there may be repercussions on various parts of the body as a result of absorption and diffusion by contact or by transport by means of the physiological fluids. One may therefore have *local effects*, *generalized effects*, and *mixed effects*.

Furthermore, these toxic effects may be *very mild*, *light*, *serious*, or *extremely serious*, depending on the intrinsic toxicity of the agent responsible for provoking them.

Examples of different toxic effects classified on the basis of the mode in which they manifest themselves and which part of the body is affected, are given in Table 1.2.

Finally, the nature of a toxic effect and its gravity are determined by the degree of exposure, and it is found that *acute toxic effects* are a consequence of a single prolonged exposure to a certain chemical product, while chronic toxic effects tend to be manifested after repeated exposures for rather brief periods.

The nature of toxic effects

The biological reaction to poisoning by a chemical product is primarily dependent on the physicochemical properties of the agent in question and the exposure conditions.

The following is a review of the principal toxic effects.

Inflammation. Inflammation is characterized by an increase in the blood flow, by the infiltration of plasma into tissues, and by the afflux of particular blood cells into the affected zone; all of these being autodefence measures. Inflammations may be either acute or chronic. In the second of these two forms, which is the more serious, the processes of tissue replacement and replacement of the inflammatory agent continue at the same rate, thereby hindering curative remedies.

---

*The products of the chemical industry should not be confused with chemical compounds, since the latter are also derived from other sources (motor engines, industrial burners, metal–mechanical manufacture, etc.).

Table 1.2—Classification of toxic effects

Symptoms	*Location*	*Effects produced*	*Chemical agents*
acute	local	pulmonary damage	hydrogen chloride
	diffuse	haemolyses	arsine
	mixed	pulmonary damage	nitrogen oxides
		metahaemoglobinemia	
		sensitizations	ethylenediamine
		peripheral neuropathy	methyl n-butylketone
of short duration	local	irritation of the respiratory ways	pyridine
	diffuse		
	mixed	renal and hepatic disorders	
chronic	local	bronchitis	sulphur dioxide
	diffuse	hepatic angiosarcoma	vinyl chloride
	mixed	emphysema kidney damage	cadmium
latent	local	pulmonary oedema	phosgene
	diffuse	neuropathy and pulmonary fibrosis	organic phosphates and paraquat

Degeneration. When degeneration occurs the poisoned cells are affected by abnormal mutations in their tissue which process finally leads to death as a consequence of a long exposure to a toxic agent.

Necrosis. A term implying the death of certain tissues as a consequence of contamination by chemical agents.

Hypersensitization reactions. An uncontrolled response of the immunological system to an attack by external agents and substances leads to a state of hypersensitization. This trouble, in a toxicological context, occurs for the most part on the epidermis and in the lungs. 'Allergic forms' arise after a state of pronounced inflammation on the skin, while, in the case of the respiratory passages within the lungs, the process is classified as 'asthma'. Toluene diisocyanate is, for example, a cause of hypersensitivity of the respiratory tracts.

Destruction of the immunological system. The primary function of the immunological system is to provide protection against external pathogenic agents. The effect of substances which directly attack this defence system is therefore deleterious. On the other hand, some phenomena involving tumours are also responsible, since the immunological system plays an important role in the control of neoplastic cells.

Neoplasia. Neoplasms are abnormal masses of cells which 'have become deranged' in the sense that their growth and division parameters have leapt up so that growth and division are out of control.

Neoplasms are basically subdivided into benign and malignant neoplasms. Benign neoplasms are, in essence, confined, so that they do not diffuse and cause damage to other tissues, while malignant tumours spread, like 'a spot of oil', to every part of the body.

Mutagenesis. A chemically induced mutagenesis implies interference by external factors in the synthesis of desoxyribonucleic acid (DNA). It leads to very serious imbalances, not only at the somatic level. It has been proven experimentally that a chemical agent which stimulates mutagenesis is also carcinogenic.

Inhibition of enzymes. Several chemical agents produce toxic effects as a consequence of the inhibition of the biological enzyme system. For example, toxic organic phosphates inhibit the 'cholinesterase' group of enzymes among which the most important is acetylcholinesterase which is responsible for the transmission of nerve impulses. Inhibition leads to the interruption of such nerve impulses and thereby to paralysis.

Biochemical decoupling. The energy liberated in normal biochemical processes is stored up in phosphate molecules with a high concentration of chemical energy. These are adenosinetriphosphate molecules. Decoupling agents such as dinitrophenol impede the formation of these molecules, with the result that the energy is continually dispersed as heat.

Lethal syntheses. When a toxic substance is extremely similar to another substance which ordinarily enters into a biological process, it may happen that the former substance is metabolized to produce a toxic, extraneous, compound.

This is the case with fluoroacetic acid which is 'accepted' as if it were acetic acid in the Krebs cycle of tricarboxylic acids. The result is that fluorocitric acid is formed which is an inhibitor of aconitase, as a consequence of which energy production via the citric acid cycle is blocked.

Teratogenesis. Teratogenesis (production of 'monsters') is the abnormal development in the structural and functional profile of the foetus and embryo. According to the nature of the chemical agents which are responsible, teratogenesis is a consequence of diverse toxic effects such as mutagenesis and the inhibition of enzymes. The critical exposure period coincides with the early stage of pregnancy when cell differentiation and the definitive formation of the organs is taking place. Nevertheless, there is danger throughout the whole of the period of gestation.

Irritation of the sensory organs. Although this cannot be strictly considered as a toxic effect, more or less serious irritation of the sensory organs is a very common occurrence.

For example, the eyes may be troubled by an excess of tears and the involuntary closure of the eyelids (blepharospasmus), while the inhalation of irritants initially provokes disturbances of the olfactory passages, with the hypersecretion of mucus and then, when the irritant passes from the nose to the throat, with coughing.

Factors which influence toxicity

Many factors have shown themselves to be capable of influencing the nature, the gravity, and the probability of causing damage of a toxic nature.

The principal factors which influence the toxicity of a product are:

- *the number of exposures:* certain toxic effects are encountered which manifest themselves after a single exposure, others which manifest themselves after a few exposures, and yet others which only appear after a long series of exposures;
- *the intensity of the exposure;*
- *the race to which the subject belongs:* differences in metabolism and, more generally, physiology, determine whether the subject will exhibit a greater or smaller tolerance toward chemical agents;
- *the mode in which the toxic agents are taken up:* this leads, in general, to a different development of the effects;
- *the exposure period:* either with regard to the hour of the day (in the case of various degrees of sunstroke and, especially, of the night when there are different degrees of physiological activity) or the time of year (in relation to the seasons) which produce biological and physiological changes which can modify the extent of the toxicity of a product;
- *the impurities* in the toxic agents which, above all, may act as inhibitors or activators of the product itself.

Other factors are related to the general state of health of the affected person and his natural sensitivity to certain stimuli, etc.

Contamination routes

It has already been stated that damage caused by a chemical product requires the contact of the substance with a part of the body which changes in accordance with the mode in which the substance is taken up.

We now pass to a brief review of the principal channels of poisoning by chemical products.

Ingestion. The whole of the digestive system provides fertile ground for the reception of irritant and caustic effects or, even, carcinogenic effects. The parts which are usually of most interest are the mouth, the pharynx, the oesophagus, the stomach, and the gastro-intestinal tract.

The epidermic route. The skin is the principal contamination pathway in an industrial environment. The effects may be chronic or acute inflammations, allergic reactions, and neoplasia. The presence of wounds and the nature of the ambient environment together with the physicochemical characteristics of the pathogenic agent at the level of contamination attained are naturally the determining factors here.

Inhalation. Gases and vapours reach the pulmonary alveoli without any delay, and it is the solubility in water of a gas or a vapour which determines the level to which it penetrates into the respiratory system. For instance, the difference in the solubilities of chlorine and hydrogen chloride means that the extent of penetration and the irritant action of the two gases are different.

Particles with a size greater than 40 nm do not penetrate into the respiratory tracts: those with a size between 10 and 50 nm stay in the primary tract, while those with a size of the order of 2 to 10 nm progressively reach the trachea, the bronchia, and the bronchioli, and it is only those with sizes smaller than 1 to 2 nm which arrive at the alveoli. The aerodynamic behaviour of fibres is such that those with a diameter greater than 3 nm do not penetrate into the lungs, while those with a diameter of less than 3 nm and a length not exceeding 200 nm reach the lungs. Fibres with a length greater than 10 nm cannot be so readily removed by the pulmonary excretory system. Apart from the physicochemical properties of the pathogenic agent, the 'damage' naturally depends on the contact zone, that is, on the part of the respiratory tracts onto which the agent is deposited. Such damage may include acute or chronic inflammation, immunological hypersensitivity, and even neoplasia of a localized nature, or, most frequently, of diffuse neoplasia leading to loss of function of the affected organs.

Eye contact. Toxic effects relating to the eyes which originate from contact with solid, liquid, or gaseous chemical products range from acute inflammation to persistent and, sometimes, irreversible damage. Toxicologically, accumulations of materials absorbed by the peripheral blood vessels of the eyeball are the most dangerous.

Multiple exposures

In practice it may happen that exposure occurs to more than one toxic agent, and, consequently, interactions between the effects caused by the various toxic products are possible.

The effects are said to be 'independent' when each material manifests its own characteristic effects independently of the presence of any other agent.

The effects are said to be 'additive' when the total effect is equal to the sum of the individual effects.

On the other hand, the effect is said to be 'antagonistic' when the effects of two or more chemical agents cancel out or, at least, diminish the overall toxic effect.

'Reinforcement' is the term used when a low toxicity effect of a product is aggravated by the intervention of a second agent.

The toxic effect of a combination of two toxic agents which is greater than the sum of the partial effects of the two agents taken individually is said to be 'synergistic'. Synergism differs from reinforcement in that both of the products contribute to it.

However, exposure to combinations of chemical products does not always lead to effects attributable to an obvious interaction. Consequently, every situation requires an in-depth analysis of its implications.

Types of absorbed toxic chemical products

The general course which a chemical agent may follow once it has penetrated into an organism is illustrated schematically in Fig. 1.1.

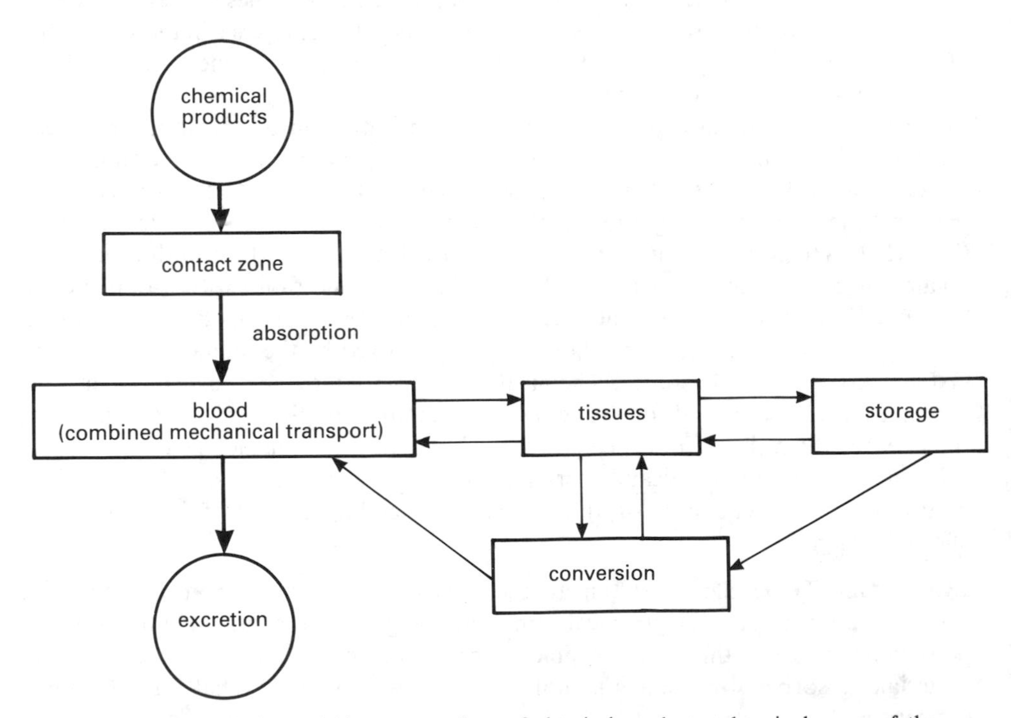

Fig. 1.1—Possible fate, within an organism, of chemical products taken in by one of the routes indicated in Fig. 1.2. The lungs and the liver are included among the 'tissues' referred to here. The latter is particularly employed in the conversion of toxic products into substances which can then be directly sent for excretion through the agency of the bile and sent to the kidneys via the blood stream.

According to the material and the part at which it enters, passive diffusion or diffusion facilitated up to a point where it is actively transported may occur. The cells then sometimes take it up by 'drinking' it ('pinocitosis').

As indicated in the scheme, if the blood takes care of the transport of the product, it does so mechanically or in a state of combination with its own components such as proteins.

After arriving at the tissues, the intrusive product may follow various pathways. It may be metabolized, for example, by biochemical conversion or else, by re-entering the blood, be excreted via the kidneys or by another route. However, the extent of stopping or retention of the product does influence its diffusion and limits either its metabolism or its excretion.

Metabolism. In certain cases the metabolic process reveals itself as being the optimal method of detoxification, but sometimes only makes matters worse.

Table 1.3 shows that the metabolic route can, in practice, be a 'double-edged weapon' with respect to the toxicity of chemical products.

Excretion. In addition to the kidneys, the principal excretory organs are the liver (which, in this case, acts by means of the bile) and the lungs in the case of products which are relatively volatile. On the other hand, perspiration is not very important from this point of view. This may be inferred by reference to Fig. 1.2.

It is now seen, in fact, that the liver turns out to be directly concerned in the most important route for the ingestion of toxic agents (the digestive system) and receives, as a matter of priority via the arterial circulation, those non-volatiles which the lungs have not been able to eliminate. Moreover, the liver, either because it partially transforms them chemically and allows them to be excreted by the lungs or the kidneys, or because it includes them by a different mechanism in the bile which undergoes a purging effect at the intestinal level, is the main organ for ridding the body of toxic products.

The skin, however, by virtue of the relatively low flow of blood through it and the lack of any substantial capacity to intervene in the chemical transformation of substances and the small size of the products which it is capable of expelling (above all into the countercurrent

Table 1.3—Deactivation and activation of toxic compounds by the metabolic route

Chemical agents	*Transformation*	*Conversion*
Cyanides (CN—)	detoxification	enzymatic to less toxic thiocyanate
Benzoic acid (C_6H_5COOH)	detoxification	into hippuric acid by combination with glycine
Isoniazide (pyridyl–C(=O)NHNH$_2$)	detoxification	*N*-acetylation leads to less toxic derivatives
Parathion ($(C_2H_5O)_2P(=S)$—O—C_6H_4—NO_2)	enhanced toxicity	into 'paraoxon': $C_{10}H_{14}NO_6P$ Powerful inhibitor of cholinesterase for oxidative desulphurization
Carbon tetrachloride (CCl_4)	enhanced toxicity	into products which are toxic to the liver
2-Acetylaminofluorene (fluorenyl–N(H)–C(=O)–CH_3)	enhanced toxicity	by *N*-hydroxylation into a more potent carcinogen

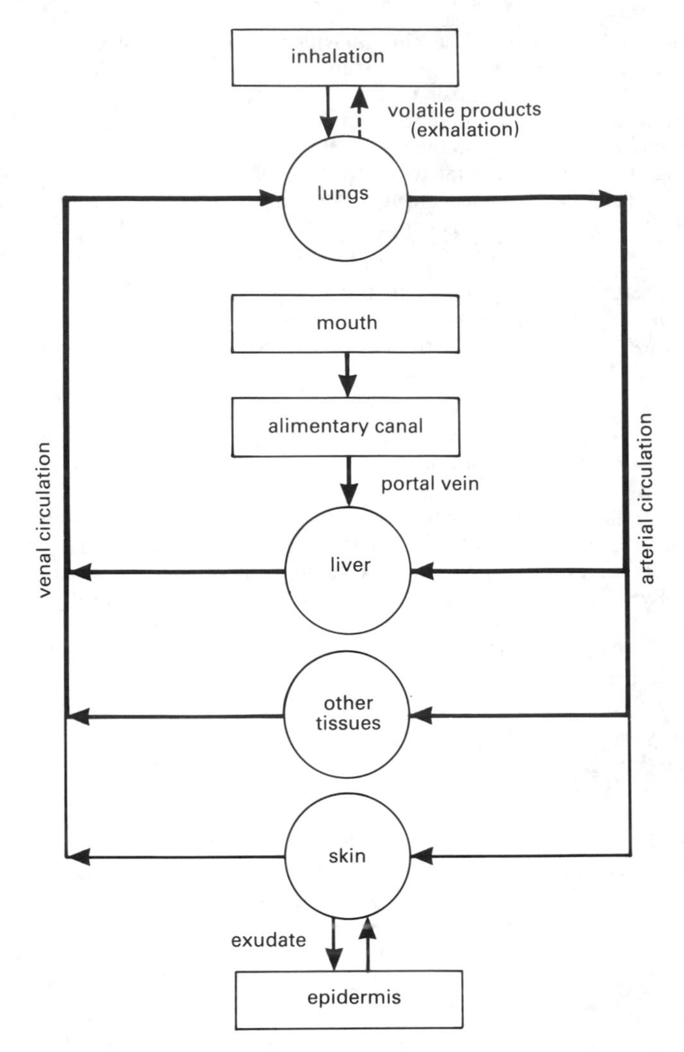

Fig. 1.2—Schematic representation of the pathways followed by toxic products which enter into an organism. The relative thickness of the lines indicates the magnitude of the flow of the product which is in the circulatory system.

with those which are entering into the organism from the skin), is not a prominent organ for the disposal of toxic substances.

1.6 THE RECORDS ACHIEVED AND THE DANGERS DERIVING FROM THE ACTIVITIES OF THE INORGANIC CHEMICAL INDUSTRY

For many years the invoices showing the quantities of products from inorganic chemical industry have been larger than those for the organic chemical industry. However, the reverse applies with regard to the total value of the invoices. This is due to the much higher value of the latter with respect to the former, above all with respect to the more sophisticated production technologies and research costs of the manufacture of the products of the organic chemical industry.

In the case of inorganic compounds, the level of direct or indirect use in the fields of chemistry, agriculture, and general industry, corresponds to the record for quantity. Compounds such as the mineral acids, soda and potash, ammonia, chlorine, and a gamut of salts constitute the framework of the entire chemical industry. Many of these inorganics are the compounds which are used as fertilizers, agents for the improvement and cultivation of soils, pesticides, and veterinary products. In addition, very many of these compounds are supplied in large quantities to serve the printing, textile, tanning, mechanical engineering, building, environmental protection, and many other industrial technologies.

From these simple facts it can readily be appreciated that the role of inorganic industrial chemistry is now essential in the biological and environmental economy and in planetary technology.

However, account should be taken just as readily of the fact that a number of problems of an ecological and health nature arising from pollution by inorganic chemical products has occurred at a global level at the same pace as the development of industrial inorganic chemistry.

One primarily thinks about the SO_2 emissions which lead to the production of sulphuric acid, the nitrogen oxide emissions from factories which produce and use nitric acid, and the 'poisonous exhalations of sulphides' from many chemical industries. The escapes of halogens and volatile halides, CO, cyanides, metal and/or mineral powders, and of assorted amounts of these substances from technologies which produce or exploit the chemical products of inorganic industrial chemistry, should be considered to be on the same level.

Passing now to the liquid effluents from laboratories, workshops, and industries which utilize inorganic chemical products, the problems become more complicated owing to the intervention of oxidizing agents, reducing agents, heavy metal ions, suspended metallic slimes, and, in general, by the intervention of all the above-mentioned pollutants. These are washed out by rain or controlled so as to satisfy emission standards, in order to protect the atmosphere, only to be subsequently discharged into sewers leading to the contamination of water courses.

In the course of this first volume we shall occupy ourselves more or less directly and extensively with the problems already mentioned which arise from the use and dispersion of the products of inorganic industrial chemistry; above all, in order to indicate radical remedies or effective antidotes for them.

2

Equilibrium constants and reaction yields

2.1 REAGENTS ARE CONVERTED TO REACTION PRODUCTS WHICH SUBSEQUENTLY ASSUME THE ROLE OF REAGENTS

On reacting methane with water at some fixed temperature the chemical reaction

$$CH_4 + H_2O \rightarrow CO + 3H_2 \tag{2.1}$$

occurs.

On the other hand, if carbon monoxide and hydrogen are reacted together at the same temperature as the last experiment, the reaction

$$CO + 3H_2 \rightarrow CH_4 + H_2O \tag{2.2}$$

occurs.

The difference between the two reactions (2.1) and (2.2) is that their rates are in general different.

On the basis of the law of mass action (Guldberg and Waage) these rates are *practically** proportional to the product of the concentrations of the reacting substances (the reagents) raised to the powers of their respective stoicheiometric coefficients.

Indicating concentrations in the conventional manner, the rate (v_1) of reaction (2.1) will be given by the equation:

$$v_1 = k_1 \cdot [CH_4] \cdot [H_2O]. \tag{2.3}$$

In the case of (2.2) the expression for the rate of reaction (v_2) will be:

$$v_2 = k_2 \cdot [CO] \cdot [H_2]^3. \tag{2.4}$$

The two expressions (2.3) and (2.4) clearly indicate that the rate of reaction at a

*Here and subsequently, the word *practically* signifies that this is true in practice even if matters are somewhat different in theory. In fact, in the latter case one should speak of *active masses* rather than concentrations. This concept is dealt with more fully within the confines of physical chemistry.

certain fixed temperature is greater, the greater the concentration of the reactants, and, consequently, this reaction rate will increase or decrease with a corresponding increase or decrease in the concentrations of the reactants.

Let us now react methane with water at the same temperature as before. We know that carbon monoxide and hydrogen will be formed. As soon as these products are formed, since the conditions apply under which process (2.2) considered above takes place, they will react in the sense indicated to reform CH_4 and H_2O.

The reaction products, therefore, in their turn, become reactants owing to the fact that as soon as they are formed the conditions are such that a reaction takes place between them.

2.2 CHEMICAL REACTIONS PROCEED TO ESTABLISH AN EQUILIBRIUM BETWEEN REACTANTS AND THE REACTION PRODUCTS

It might be easily concluded that, since (2.2), which is opposed to (2.1), occurs after CO and H_2 have been formed, that all reactivity would cease or, what is worse, that the reaction mixture would return to its initial state. This is, of course, absurd, since when (2.1) first commences, the concentrations of CH_4 and H_2O are high, while the concentrations of CO and H_2 are very small, and since the rates of reaction, as is indicated above, are proportional to the concentrations, it follows that v_1 is much greater than v_2.

However, as more and more CO and H_2O are transformed into CO and H_2, the concentrations of the products increase, while the concentrations of the reagents decrease with the result that (as is also indicated above) v_2 increases while v_1 continues to decrease.

A condition is therefore logically arrived at when the rates of the two opposing chemical processes (2.1) and (2.1) are equal ($v_1 = v_2$). This condition may be characterized by the equation:

$$k_1 \cdot [CH_4] \cdot [H_2O] = k_2 \cdot [CO] \cdot [H_2]^3.$$

At this point there is no further decrease in the amounts of CH_4 and H_2O present nor any increase in the amounts of CO and H_2 present because now, during each unit of time, the concentrations of CH_4 and H_2O (that is, the reaction rate!) decrease owing to their transformation into CO and H_2 by exactly the same amount as the concentrations of the same species increase owing to the reconversion of CO and H_2 into CH_4 and H_2O at the same rate.

It is therefore justly said that a state of equilibrium has been attained in which the probabilities of the existence either of the reactants or of the reaction products are equal.

Such a state is characterized by the rates of the two opposing reactions, that is, the rate of the reaction which leads from reagents to reaction products and the rate of the reverse reaction.

The simple, clear, and logical reasoning proposed here is valid for any chemical reaction, so that there are no chemical reactions which precisely proceed to completion in an exact sense, but there are only reactions which attain a state of equilibrium between the reagents and the reaction products.

2.3 THE THEORETICAL EQUILIBRIUM IS AN IDEAL CONDITION FOR A REACTION

Therefore, when a reaction has arrived at a point where it is statistically 'dead' because there are no further changes in the concentrations of its reagents and reaction products, it is at equilibrium.*

In practice, however, a reaction does not successfully reach the equilibrium which is calculated for it on the basis of the reagent concentrations which are initially allowed to react (the 'theoretical equilibrium') because such a reaction is generally accompanied by secondary or 'parasitic' reactions. Moreover, for various reasons connected with how the reaction is carried out (too briskly, with low activity catalysts, etc.), it does not effectively succeed in attaining the equilibrium state which is calculated for it at the experimental temperature even when there are no parasitic reactions.

The theoretical equilibrium condition is therefore an ideal condition for a reaction which is attainable only with difficulty, so that elaborate calculations based on experimental data should not be regarded so much as describing the theoretical equilibrium but rather as describing the effective equilibrium which is attained.

2.4 THE REACTION YIELD ATTAINED AT THE THEORETICAL EQUILIBRIUM

As was demonstrated earlier, reactions, even when they take place under conditions which are ideal for their development, never proceed precisely to completion because they reach a state of equilibrium between the reagents and products. It therefore follows that the products obtained from reactions are always smaller in amount than they would be if the reagents had been completely tranformed into products with a yield of 100%.

It therefore makes sense to speak of the existence of a 'yield of a certain product with respect to a certain reagent'. Such a yield is given as the ratio of the weight of the reaction product from which one calculates the return effectively obtained and the weight of the product which would have been obtained on the basis of a calculation assuming that the reagent had been completely transformed into the product.

Let us clarify this point with an example.

Calculate the theoretical yield of CO_2 ($M_r = 44$) produced with respect to $CaCO_3$ ($M_r = 100$) when 126 g of the latter is heated up to a temperature of 900°C (1173 K) in a closed vessel of volume 15 l.

This reaction involves the establishment of the dissociation equilibrium

$$CaCO_3 \rightleftharpoons CO_2 + CaO$$

for which $K_p = 1.30$ atm at the experimental temperature.

In the case that the 126 g of $CaCO_3$ had been completely transformed into CO_2:

$$\text{a yield } g_t = \frac{126}{100} \cdot 44 = 55.4 \text{ g of } CO_2 \text{ would have been obtained.}$$

*On the other hand, it may be noted that the fact that no further changes in the concentrations are found does not mean, in any absolute sense, that the concentrations do not change any more. They do change, but the changes compensate one another. In this sense the point of equilibrium is only statistically 'dead'.

Next, the number of grams of CO_2 which are actually obtained must be calculated on the basis of the equilibrium constant.

It is expedient to recall* that, when one is dealing with a heterogeneous equilibrium involving a single gaseous component, the equilibrium constant K_p is identical to the pressure of that component, that is:

$$K_p = p_{CO_2} = 1.30 \text{ atm}.$$

Upon substituting the available data into the ideal gas equation ($pV = nRT$), the number of moles of CO_2 which are present at equilibrium are found to be:

$$n_{CO_2} = \frac{1.30 \cdot 15}{0.082 \cdot 1173} = 0.2028 \text{ moles}$$

and, hence, the quantity of CO_2 obtained in practice is given by:

$$g_e = 0.2028 \cdot 44 = 8.92 \text{ g of } CO_2.$$

If the definition of the yield of a certain product with respect to a certain reagent is now applied, the reaction yield (η) is found to be:

$$\eta^{CO_2}_{CaCO_3} = \frac{8.92}{55.4} = 0.1615.$$

2.5 MAXIMUM YIELD AND GENERAL EXPRESSIONS FOR ITS CALCULATION

Any yield which is calculated by using the theoretical equilibrium conditions for a reaction is said to be the maximum yield, because, in the absence of parasitic reactions, a maximum conversion of reagents into products is attained at the point when the reaction is very advanced.

It is useful to derive general expressions for the calculation of the maximum yields of products with respect to the reagents.

For this purpose let us consider the general reaction:

$$rR \rightleftharpoons pP \tag{2.5}$$

where R and P indicate the chemical formulae of the reagent and of the product respectively and r and p are the respective stoicheiometric coefficients.

The maximum yield (η) of the product P with respect to the reagent $R(\eta^P_R)$ is, by definition, given by:

$$\eta^P_R = \frac{g \text{ (kg) of the product } P \text{ obtained under equilibrium conditions}}{g \text{ (kg) of the product } P \text{ obtained if completely converted}} = \frac{m^P_e}{m^P_t}. \tag{2.6}$$

In general, the data which are available are the relative molecular masses of $P(M^P_r)$ and $R(M^R_r)$ and the initial mass of R which we indicate by m^R_0.

A quantity which can always be determined with the data for the problem (frequently by means of the equilibrium constant) is the equilibrium mass of P. As in (2.6), this is denoted by m^P_e.

The other term (m^P_t) which appears in (2.6) can be formulated in terms of the

*This notion is employed in institutes of physical chemistry.

available data as follows:

$$m_t^P = \frac{m_0^R}{M_r^R} \cdot \frac{p}{r} \cdot M_r^P.$$

General expression for the maximum yield as a function of the masses

At this point, (2.6) may be reformulated as:

$$\eta_R^P = \frac{m_e^P}{\dfrac{m_0^R}{M_r^R} \cdot \dfrac{p}{r} \cdot M_r^P} = \frac{m_e^P \cdot r \cdot M_r^R}{m_0^R \cdot p \cdot M_r^P} \tag{2.7}$$

where the various symbols have the same meanings as above.

Equation (2.7) is the general expression which serves for the calculation of the maximum yield of P with respect to a reagent R in reaction (2.5).

It is convenient here to illustrate the validity of (2.7) by means of an example.

The partial pressure of sulphur trioxide which is in equilibrium, at a certain temperature and a total pressure of 1.2 *atmospheres, with sulphur dioxide and oxygen is* 0.20 *atm.*

Calculate the yield of SO_3 both with respect to SO_2 and oxygen in the case of a mixture which initially consists of 51.2 *g of SO_2 and* 35.2 *g of O_2. The data given are $M_r(SO_2) = 64$, $M_r(O_2) = 32$ and $M_r(SO_3) = 80$.*

The problem implies the occurrence of the reaction:

$$2SO_2 + O_2 \rightleftharpoons 2SO_3$$

which, in terms of the solution to be found, is not substantially different from (2.5) in that the separate yields of SO_3 with respect to the reagent SO_2 and of SO_3 with respect to the reagent O_2 are to be calculated.

Let us first solve the problem of determining, by normal chemical and physico-chemical calculations, the values of m_e^P and m_t^P to be substituted into (2.6) in order to answer the two questions which are posed in the problem.

The initial numbers of moles of the reactants are:

$$51.2/64 = 0.8 \text{ mole } SO_2 \text{ and } 35.2/32 = 1.1 \text{ mole } O_2.$$

Indicating the number of moles of SO_3 formed at equilibrium by x, the numbers of moles of the different species which are present under equilibrium conditions are:

$$n_{SO_2} = 0.8 - x; \qquad n_{O_2} = 1.1 - x/2; \qquad n_{SO_3} = x.$$

Hence the total number of moles present at equilibrium is

$$(0.8 - x) + (1.1 - x/2) + x = 1.9 - 0.5x.$$

To find x, use is made of the pressures taken from the data supplied in the problem, noting that the partial pressure of a gas which is a component of a gaseous mixture is equal to its mole fraction multiplied by the total pressure, that is, in the case of SO_3:

$$0.20 = 1.2 \cdot \frac{n_{SO_2}}{n_{total}} = 1.2 \cdot \frac{x}{1.9 - 0.5x}.$$

Upon solving this equation, it is found that $x = 0.292$.

The mass of SO_3 present under equilibrium conditions is therefore:

$$m_e^{SO_3} = 0.292 \cdot 80 = 23.36 \text{ g of } SO_3.$$

If the conversion of SO_2 had been complete, the calculated mass of SO_3 obtained would have been:

$$m_t^{SO_3} = 0.8 \cdot 80 = 64 \text{ g of } SO_3$$

since, when there is total conversion, the number of moles of SO_3 formed is equal to the initial number of moles of SO_2.

If, on the other hand, the conversion of O_2 had been complete, the resulting number of moles of SO_3 would have been:

$$m_t^{SO_3} = 2 \cdot 1.1 \cdot 80 = 176 \text{ g of } SO_3$$

since the attainable number of moles of SO_3 is double the number of moles of O_2 supplied.

Equation (2.6) is now applied twice to yield the two answers to the problem. Specifically:

$$\eta_{SO_2}^{SO_3} = \frac{23.36}{64} = 0.365 \quad \text{and} \quad \eta_{O_2}^{SO_3} = \frac{23.36}{176} = 0.133.$$

Let us now solve the problem more quickly by applying equation (2.7), the utility of which will subsequently become clear.

The data: $r = 2$, $p = 2$, $M_r(SO_2) = 64$, $M_r(SO_3) = 80$ and $m_0^{SO_2} = 51.2$ g are available for the application of equation (2.7) to find the yield with respect to SO_2, while, for the application of equation (2.7) to find the yield with respect to O_2, the only data available are $r_2 = 1$ and $m_0^{O_2} = 35.2$ g.

In anticipation, $m_e^{SO_2} = 23.36$ g is determined for both of the calculations.

At this point the values required are found by using equation (2.7):

$$\eta_{SO_2}^{SO_3} = \frac{23.36 \cdot 2.64}{51.2 \cdot 2 \cdot 80} = 0.365 \quad \text{and} \quad \eta_{O_2}^{SO_3} = \frac{23.36 \cdot 1.32}{35.2 \cdot 2 \cdot 80} = 0.133.$$

General expression for the yield as a function of the number of moles

Equation (2.7) may be algorithmically transformed to the following form:

$$\eta_R^P = \frac{(m_e^P/M_r^P) \cdot r}{(m_0^R/M_r^R) \cdot p} = \frac{n_e^P \cdot r}{n_0^R \cdot p} \tag{2.8}$$

where n_e^P is the number of moles of the product with respect to which the equilibrium yield is calculated and n_0^R is the initial number of moles of the reagent with respect to which the yield is calculated. It is obvious that if the weights are given in kg, n_e^P and n_0^R are expressed in units of kmole.

Let us now see an example of the application of equation (2.8).

In the recovery of chlorine from hydrogen chloride which is carried out in the dehydrochlorination process 1752 kg of HCl is treated with 6622.32 m³ of air at 400°C and 1 atmosphere. The equilibrium constant K_p for the conversion reaction:

$$4HCl + O_2 \rightleftharpoons 2Cl_2 + 2H_2O$$

is $K_p = 42$ atm^{-1} under the given conditions. Calculate the yield of chlorine with respect to HCl. The approximate relative molecular mass of HCl is 36.5.

The number of kmoles of HCl is $1752/36.5 = 48$.

The number of kmoles of air, found by applying the equation $PV = nRT$, is 120.

The number of kmoles of O_2 is then $120 \cdot 0.2 = 24$.

In order to apply equation (2.8) it is sufficient to find $n_e^{Cl_2}$, that is, the number of kmoles of Cl_2 which is present at equilibrium, since we already have all the other quantities which are required.

Let there be x kmoles of Cl_2 at equilibrium. Then, the number of kmoles of each component at equilibrium may be written (on the basis of the reaction) as:

$$n_{Cl_2} = n_{H_2O} = x, \qquad n_{HCl} = 48 - 2x, \qquad n_{O_2} = 24 - x/2$$

while the total number $(\sum_n)$* of moles will be given by:

$$\sum n = 72 - 0.5x.$$

The following expressions are obtained for the partial pressures from the total number of kmoles at equilibrium and the fact that the total pressure is 1 atmosphere:

$$p_{Cl_2} = p_{H_2O} = x/(72-0.5x); \qquad p_{HCl} = (48-2x)/(72-0.5x)$$

and

$$p_{O_2} = (24-0.5x)/(72-0.5x).$$

The equation for K_p is then constructed using these values for the partial pressures:

$$K_p = \frac{[x/(72-0.5x)]^2 \cdot [x/(72-0.5x)]^2}{[(48-2x)/(72-0.5x)]^4 \cdot [(24-0.5x)/(72-0.5x)]} = 42.25 \text{ atm}^{-1}$$

which is subsequently solved to yield the value: $x = 18.72$ kmole.

Having thus found $n_e^{Cl_2}$, equation (2.8) is applied to give the final answer to the problems:

$$\eta_{HCl}^{Cl_2} = \frac{18.72 \cdot 4}{48 \cdot 2} = 0.78.$$

2.6 CALCULATION OF THE MAXIMUM YIELD FROM THE EQUILIBRIUM CONSTANT

The case of a reaction which takes place without any change in the number of moles

Consider the equilibrium for a general reaction in which there is no change in the number of moles

$$A + B \rightleftharpoons C + D \tag{2.9}$$

and suppose that equimolar initial quantities of A and B have been mixed together at a total pressure P.

By means of equation (2.8), we express the yield of C with respect to A† as

$$\eta_A^C = n_e^C / n_0^A \tag{2.8a}$$

since, in this case: $p = r = 1$.

Because a proportionality exists between the number of moles and the partial pressures of a gas in a gaseous mixture, (2.8a) may assume the form:

$$\eta_A^C = p_e^C / p_0^A, \tag{2.8b}$$

*The sign $\sum$ indicates summation, that is, the total sum. The subscript represents the kind (moles in this case) of quantities which are being added together, while additional symbols which may be put above and below the summation sign indicate the limits over which the summation is carried out.

†One would proceed in an analogous manner in calculating the yield of D with respect to A, the yield of C with respect to B and the yield of D with respect to B.

or, given that at the beginning of the reaction $p_0^A = P/2$:

$$\eta_A^C = \frac{p_e^C}{P/2}. \tag{2.8c}$$

It can be shown that the expression for the equilibrium constant which is to be exploited in order to recover the yield is as follows*:

$$K = \frac{(p_e^C)^2}{\frac{1}{4}(P - 2p_e^C)^2}. \tag{2.10}$$

In fact, the expression for the equilibrium constant (K) of equation (2.9) is:

$$K = \frac{p_C \cdot p_D}{p_A \cdot p_B} \tag{2.10a}$$

and the relationship

$$p_A + p_B + p_C + p_D = P$$

is always valid.

Given that C and D are produced in equimolecular quantities $p_C = p_D = p_e^C$ and since the pressure varies in the same manner during the reaction in the case of the two initial reactants, $p_A = p_B = p$.

Consequently, equation (2.10a) assumes the form:

$$K = \frac{(p_e^C)^2}{p^2}. \tag{2.10b}$$

In turn, the expression originally proposed for the total pressure may be written as:

$$P = p + p + p_e^C + p_e^C = 2p + 2p_e^C$$

which allows the partial pressure p to be expressed in the form:

$$p = (P - 2p_e^C)/2.$$

When this expression is substituted into (2.10b), we obtain (2.10) as required.

By obtaining p_e^C from (2.8c), substituting it into (2.10), and carrying out some manipulation, we finally arrive at the expression:

$$K = \frac{(\eta_A^C)^2}{(1 - \eta_A^C)^2}. \tag{2.10c}$$

The expression for the maximum yield as a function of equilibrium constant that is derived from (2.10c) is:

$$\eta_A^C = \frac{\sqrt{K}}{1 + \sqrt{K}}. \tag{2.11}$$

Equation (2.11) shows that the maximum yield for a reaction which occurs with no change in the number of moles (and therefore no change in the volume of the system) is independent of the pressure.

---

*The generality of the equilibrium constant K, which has only been formulated as a function of the pressures, is due to the fact that, in the case in question, $\sum v_i = 0$, and therefore $K_p = K_c = K_n = K_x = K$.

Let us now consider an example of the application of (2.11) and thereby demonstrate its validity.

Calculate the maximum yield of hydrogen with respect to water and carbon monoxide for the reaction:

$$H_2O_{(g)} + CO_{(g)} \rightleftharpoons H_{2(g)} + CO_{2(g)}$$

at a temperature at which the equilibrium constant for the dissociation of water vapour is $K_{p_1} = 1.16 \cdot 10^{-10}$ *atm*$^{1/2}$ *and of gaseous* CO_2 *is* $K_{p_2} = 1.09 \cdot 10^{-11}$ *atm*$^{1/2}$. *The initial reactants, which must undergo reaction (conversion to water gas), are an equimolar mixture of* H_2O *and* CO.

The reactions for which the K_p values are given are:

$$H_2O_{(g)} \rightleftharpoons H_{2(g)} + 1/2 O_{2(g)}; \qquad K_{p_1} = 1.16 \cdot 10^{-10}\ \text{atm}^{1/2}$$

$$CO_{2(g)} \rightleftharpoons CO_{(g)} + 1/2 O_{2(g)}; \qquad K_{p_2} = 1.09 \cdot 10^{-11}\ \text{atm}^{1/2}.$$

The reaction indicated in the question can be obtained in terms of the two latter reactions by subtracting the second of these two reactions from the first.

Thus:

$$\begin{array}{rl} & H_2O_{(g)} \rightleftharpoons H_{2(g)} + \tfrac{1}{2}O_{2(g)} \\ - & CO_{2(g)} \rightleftharpoons CO_{(g)} + \tfrac{1}{2}O_{2(g)} \\ \hline & H_2O_{(g)} + CO_{(g)} \rightleftharpoons CO_{2(g)} + H_{2(g)} \end{array}$$

It is known from physical chemistry that, if an equilibrium reaction can be expressed as the difference between two other reactions, its equilibrium constant may be obtained as the ratio of the equilibrium constants for the first reaction, and that for the reaction which is subtracted from it. In this case, it therefore follows that:

$$K^* = \frac{K_{p_1}}{K_{p_2}} = \frac{1.16 \cdot 10^{-10}\ \text{atm}^{1/2}}{1.09 \cdot 10^{-11}\ \text{atm}^{1/2}} = 10.6.$$

The value for the equilibrium constant found in this manner may now be employed in (2.11) to determine the (identical) yields of hydrogen with respect to water and carbon monoxide; that is,

$$\eta_{H_2O}^{H_2} = \eta_{CO}^{H_2} = \frac{\sqrt{10.6}}{1 + \sqrt{10.6}} = 0.765.$$

It is obvious that the same yields may be calculated with the aid of (2.8). Assuming, for this purpose, that the initial reactants consisted of a mixture of 1 kmole of H_2O and 1 kmole of CO, the numbers of kmoles of the various species at equilibrium will be:

$$\begin{array}{cccc} CO_{(g)} + & H_2O_{(g)} \rightleftharpoons & CO_{2(g)} + & H_{2(g)} \\ 1-x & 1-x & x & x \end{array}$$

The equilibrium constant (with a value of 10.6) is given by the expression:

$$10.6 = \frac{x^2}{(1-x)^2}$$

whence we obtain $x = 0.765$.

The value of x found in this manner is none other than the n_e^P which occurs in (2.8) which

---

*K is general here for the reason indicated in the footnote pertaining to equation (2.10). Moreover, as can be seen, it is dimensionless.

when applied, as follows, using the available data ($n_0^R = 1$ kmole), gives:

$$\eta_{H_2O}^{H_2} = \eta_{CO}^{H_2} = \frac{0.765 \cdot 1}{1 \cdot 1} = 0.765.$$

As can be seen, the results are in perfect agreement.

The case of a reaction occurring at a pressure *P* in which the number of moles varies

Let us consider the generic reaction

$$aA + bB \rightleftharpoons cC + dD. \tag{2.9a}$$

We intend

(1) to obtain an expression for the equilibrium constant K_p of this reaction in terms of the pressure P at which the reaction takes place and the stoicheiometric coefficients a, b, c, and d which occur in it and as a function of the mole fraction of the reaction product C, x_C.
(2) to show that the mole fraction x_C, together with the coefficients for the reaction, are sufficient to express the yield of C with respect to the reactant A.

Since the problem of finding the values of the stoicheiometric coefficients and the pressure at which the reaction occurs has not been posed here, we are in a position of having to calculate once again (albeit indirectly) maximum yields as a function of the equilibrium constant.

N.B. It follows, of course, that the procedures being proposed here can be employed in an analogous manner to calculate the maximum yield of C with respect to B, of D with respect to A, and of D with respect to B.

The equilibrium constant as a function of a single mole fraction

The chosen mole fraction, as has already been said, is x_C which is related to the mole fraction x_D by the equation:

$$x_D = \frac{d}{c} \cdot x_C,$$

in so far as it is obvious that the occurrence of D in the mixture (x_D is the mole fraction of D) with respect to the mole fraction of C present depends on the ratio of the relative numbers of moles which are formed.

In order to express the remaining mole fractions, that is, x_A and x_B, as a function of just x_C, we start off by noting the relationship:

$$x_A + x_B + x_C + x_D = 1$$

from which, after some simple manipulation and making a substitution, we obtain:

$$x_A + x_B = 1 - (x_C + x_D) = 1 - \left(x_C + \frac{d}{c} \cdot x_C\right) = 1 - x_C \cdot \frac{c+d}{c}.$$

By multiplying and dividing the second term by $a + b$ and then separating the ratio which has been introduced, the following expression results:

$$x_A + x_b = \frac{a+b}{a+b}\left(1 - x_C \cdot \frac{c+d}{c}\right) = \left(\frac{a}{a+b} + \frac{b}{a+b}\right)\left(1 - x_C \cdot \frac{c+d}{c}\right)$$

There are therefore two possible individual formulations of x_A and x_B as a function of x_C:

$$x_A = \frac{a}{a+b}\left(1 - x_C \cdot \frac{c+d}{c}\right); \qquad x_B = \frac{b}{a+b}\left(1 - x_C \cdot \frac{c+d}{c}\right)$$

as can be proved by the fact that when these expressions are added together they yield the total value of $x_A + x_B$.

The equilibrium constant K_p is related to the analogous constant K_x by the relationship: $K_p = K_x \cdot P \sum v_i$. For our purposes, however, K_x is written explicitly in terms of mole fractions to yield:

$$K_p = \frac{x_C^c \cdot x_D^d}{x_A^a \cdot x_B^b} \cdot P \sum v_i.$$

By substituting in this expression the values for the various mole fractions which have been derived above in terms of x_C, one arrives at the expression for the equilibrium constant as a function of just a single mole fraction. After carrying out some simplifying manipulations, the generally accepted formulation of this expression is obtained:

$$K_p = \frac{(a+b)^{a+b} \cdot \left(\frac{d}{c}\right)^d \cdot x_C^{c+d}}{a^a \cdot b^b \cdot \left(1 - \frac{c+d}{c} \cdot x_C\right)^{a+b}} \cdot P \sum v_i \tag{2.12}$$

as further manipulation does not lead to any appreciable simplification.

Maximum yield as a function of a single mole fraction

Let us retain (2.9a) as the reference reaction and x_C as the mole fraction in question and reconfirm the validity of (2.8) by suitably modifying it firstly to correspond to the yield of C with respect to A and then rearranging it into the form:

$$\eta_A^C = \frac{n_e^C \cdot a}{n_0^A \cdot c} = \frac{n_e^C / c}{n_0^A / a}. \tag{2.8d}$$

By putting $n_0^A = a$ and expressing n_e^C as a function of the mole fraction x_C ($x_C = n_e^C / n_{total} \Rightarrow n_e^C = x_C \cdot n_{total}$), the equation for the yield is transformed to the following form:

$$\eta_A^C = \frac{n_e^C}{c} = \frac{x_C \cdot n_{total}}{c}. \tag{2.8e}$$

Starting from b, the initial number of moles of B, and denoting the 'degree of conversion' of A (that is, the fraction of the moles of A which has reacted) by g, the total number of moles present at equilibrium is calculated as*:

$$n_{total} = a - g \cdot a + b - \frac{b}{a} \cdot g \cdot a + \frac{c}{a} \cdot g \cdot a + \frac{d}{a} \cdot g \cdot a$$

*This calculation is based on the fact that the moles of B which decompose and those of C and D which are formed are functions of the number of moles of A which react by virtue of the ratio between the coefficients of B, C and D and the coefficient of A.

$$= a - g \cdot a + b - b \cdot g + \underline{c \cdot g} + d \cdot g$$

$$= a + b - g \cdot (a + b - c - d). \tag{2.13}$$

One may convince oneself of the practical validity of formula (2.13) derived above by considering, as an example, the reaction involving the conversion of hydrogen chloride into chlorine by reaction with oxygen:

$$4HCl + O_2 \rightleftharpoons 2Cl_2 + 2H_2O.$$

Let us assume that there are initially 4 moles of HCl and 1 mole of O_2 and suppose that the degree of conversion of HCl is 0.25. Then, at equilibrium, there will be the following balance between the number of moles of the various species:

$4HCl$ +	O_2 →	$2Cl_2$ +	$2H_2O$
the initial number of moles of this reagent less the number of moles transformed	an amount (moles) equal to one quarter of the amount of HCl transformed	an amount equal to one half of the moles of HCl transformed is produced	an amount equal to one half of the moles of HCl transformed is produced
$4 - 0.25 \times 4$	$1 - 1/4 \times 0.25 \times 4$	$1/2 \times 0.25 \times 4$	$1/2 \times 0.25 \times 4$

Hence, the total number of moles present is:

$$\sum \text{moles} = 4 - 1 + 1 - 0.25 + 0.5 + 0.5 = 4.75.$$

Now, by applying the formula derived above, it is found that:

$$n_{total} = 4 + 1 - 0.25(4 + 1 - 2 - 2) = 4.75$$

in perfect agreement with the value deduced by using a normal balance.

The degree of conversion may be extracted from the following equation:

$$x_C \cdot [a + b - g(a + b - c - d)] = c \cdot g, \tag{2.14}$$

the terms of which are equal because they both express the number of moles of C which are present at equilibrium.

Actually, the first term is no more than the application of the fact, which has previously been recalled in the derivation of (2.8e), that $n_e^C = x_C \cdot n_{total}$.

As regards the second term, it may be noted that this occurs as an underlined addend in the deduction of (2.13) where it has the significance of the number of moles of C at equilibrium.

Therefore, by solving (2.14) for g, it is found that:

$$g = \frac{x_C \cdot \dfrac{a+b}{c}}{1 + x_C \cdot \dfrac{a+b-c-d}{c}}. \tag{2.14a}$$

Upon substituting (2.14a) into (2.13) and simplifying, the expression for the total number of moles becomes:

$$n_{total} = \frac{a+b}{1 + x_C \cdot \dfrac{a+b-c-d}{c}} \tag{2.13a}$$

The formula for the maximum yield as a function of a single mole fraction is obtained at this point by introducing (2.13a) into (2.8e):

$$\eta_{\mathrm{A}}^{\mathrm{C}}=\frac{x_{\mathrm{C}}\cdot\dfrac{a+b}{c}}{1+x_{\mathrm{C}}\cdot\dfrac{a+b-c-d}{c}}. \qquad (2.15)$$

In order to demonstrate that (2.9) is a particular case of (2.9a) and that, consequently, expressions (2.12) and (2.15) also apply to reactions of the form of (2.9), let us calculate, using these expressions, the maximum yield of hydrogen with respect to H_2O and CO under the conditions of the problem on page 48.

The use of (2.12) firstly enables one to determine K (it is also valid for K_p) for reaction (2.9) (as has already been done on page 48). It is found that $K=10.6$. In this case, since $(\sum v_i)=0$, the factor $P^{\sum v_i}$, which appears in (2.12), is unity. Then, by applying this relationship in full (for didactic reasons), we have:

$$10.6=\frac{(1+1)^{1+1}\cdot\left(\dfrac{1}{1}\right)^{1}\cdot x_{\mathrm{C}}^{1+1}}{1^{1}\cdot 1^{1}\cdot\left(1-\dfrac{1+1}{1}\cdot x_{\mathrm{C}}\right)^{1+1}}=\frac{4x_{\mathrm{C}}^{2}}{(1-2x_{\mathrm{C}})}.$$

This equation is solved in the following steps:

$$2.65=\frac{x_{\mathrm{C}}^{2}}{(1-2x_{\mathrm{C}})^{2}}\Rightarrow\sqrt{2.65}=\frac{x_{\mathrm{C}}}{1-2x_{\mathrm{C}}}\Rightarrow x_{\mathrm{C}}=0.3825.$$

Finally, the required yields are found by applying equation (2.15):

$$\eta_{\mathrm{H_2O}}^{\mathrm{H_2}}=\eta_{\mathrm{CO}}^{\mathrm{H_2}}=\frac{\dfrac{1+1}{1}}{1+\dfrac{1+1-1-1}{1}}=2\cdot 0.3825=0.765.$$

The value which has been found here agrees with that found on page 48 by using equation (2.11), that is, by using a process which is operatively more convenient than that applied here.

Let us now solve two other problems which put into relief some of the principles of the application of (2.12).

The equilibrium constant K_p for the synthesis of phosgene

$$Cl_{2(g)}+CO_{(g)}\rightleftharpoons COCl_{2(g)}$$

at a certain temperature and a pressure of 1 atm is 0.645 atm^{-1}. Determine the maximum yield of phosgene with respect to the reagents under these conditions.

Here, when (2.12) is applied for the determination of x_C we have $(\sum v_i=1)$

$$0.645=\frac{(1+1)^{1+1}\cdot x_{\mathrm{C}}^{1}}{1^{1}\cdot 1^{1}\cdot\left(1-\dfrac{1}{1}\cdot x_{\mathrm{C}}\right)^{1+1}}\cdot 1^{1}=\frac{4x_{\mathrm{C}}}{(1-x_{\mathrm{C}})^{2}}.$$

It is seen that, if a term is missing from (2.9a), its coefficient is not put equal to zero when applying (2.12), but is simply disregarded.

Solution of the equation first set up on the basis of (2.12) leads to the recovery of the value: $x_C=0.124$.

This value of x_C is now substituted into (2.15) to yield the answer to the problem, that is:

$$\eta_{Cl_2}^{COCl_2} = \eta_{CO}^{COCl_2} = \frac{0.124 \cdot 2}{1 + 0.124} = 0.22.$$

* * *

The equilibrium constant K_p for the synthesis of ammonia at a certain temperature and a total pressure of 200 atm is $5.78 \cdot 10^{-4}$ atm^{-2}. The reagents are mixed in a proportion of 1:3 (N_2/H_2). Calculate the maximum yield of ammonia with respect to hydrogen and nitrogen.

The reaction in question is:

$$1N_{2(g)} + 3H_{2(g)} \rightleftharpoons 2NH_{3(g)}.$$

To solve this problem, we shall first attempt to apply (2.12):

$$5.78 \cdot 10^{-4} = \frac{(1+3)^{1+3} \cdot x_C^2}{1^1 \cdot 3^3 \cdot \left(1 - \frac{2}{2} \cdot x_C\right)^4} \cdot 200^{-2} = \frac{256 \cdot x_C^2}{27(1 - x_C)^4} \cdot \frac{1}{200^2}.$$

As can be seen, this is a case when the application of (2.12) turns out to be particularly difficult. For this reason, in such cases, one looks for a different route to obtain x_C.

On the other hand, it would be advisable for those with access to a computer (since the alternative is by trial solution) to use it to solve the above equation and compare the result with that obtained by the other route thereby obtaining experience in the use of computer solutions to problems in industrial chemistry.

To determine x_C in the simplest manner, the partial pressure of nitrogen may be denoted by y, the partial pressure of hydrogen by $3y$, and the partial pressure of ammonia by $200 - 4y$. The equation leading to the determination of y is the normal one for K_p expressed in terms of partial pressures:

$$5.78 \cdot 10^{-4} = \frac{(200 - 4y)^2}{y \cdot 3^3 y^3} = \frac{(200 - 4y)^2}{27y^4}$$

$$\sqrt{(27 \cdot 5.78 \cdot 10^{-4})} = \frac{(200 - 4y)}{y^2} \Rightarrow 0.125y^2 + 4y - 200 = 0$$

and solving the quadratic equation for the positive value of y gives

$$y = 27.04 = p_{N_2}.$$

The partial pressure of ammonia will therefore be:

$$p_{NH_3} = 200 - 4 \cdot 27.04 = 91.84 \text{ atm}$$

and its mole fraction at equilibrium (i.e. the value of x_C which is being sought) is found to be:

$$x_{NH_3} = \frac{91.84}{200} = 0.459.$$

At this point it is necessary to apply (2.15) to obtain the maximum yield:

$$\eta_{max} = \frac{0.459 \cdot \frac{1+3}{2}}{1 + \frac{1+3-2}{2} \cdot 0.459} = 0.629.$$

The maximum yield which has been found is that of NH_3 with respect to nitrogen, to hydrogen, and to the stoicheiometric mixture composed of these two.

In fact, (2.15) allows one to calculate, for any reaction having reactants A and B, the maximum yield of the product C with respect to A or with respect to B or with respect to $A + B$ when they are used in their stoicheiometric ratio.

2.7 REACTION YIELDS—TIME AND TEMPERATURE FACTORS

What are reaction isotherms?

The yields for conversions of reactants into products increase with the passage of time until the maximum yield has been achieved. However, such increases are not constant with respect to time. They become progressively smaller because, on the one hand, as the concentration of reaction products increases there will be an increase in the rate of the reverse reaction, and on the other hand, the rate of the forward reaction will decrease as the concentration of the reactants falls off.

When operating at a constant temperature, the change in the yield as a function of time is therefore represented as a curve (*b*) rather than a straight line (*a*) in the yield/time diagram (Fig. 2.1).

Since the yield of a reaction is proportional to the degree of advancement (or development) of the reaction, curves such as (*b*), representing the course of reaction at a constant temperature, are known as *isothermal reaction curves* or simply 'reaction isotherms'.

Reaction isotherms and enthalpy values

For a given endothermic reaction the reaction isotherms at several different temperatures tend asymptotically to the thermodynamic yield without ever crossing one another. In this case, as Fig. 2.2 shows, the reaction yields and reaction rates increase as the temperature characterizing the isotherm increases (while reducing the contact time).

For kinetic and thermodynamic reasons, that is to increase the equilibrium constant and hence the maximum yields obtained, *it is necessary in the case of endothermic reactions to work at the highest temperature possible.*

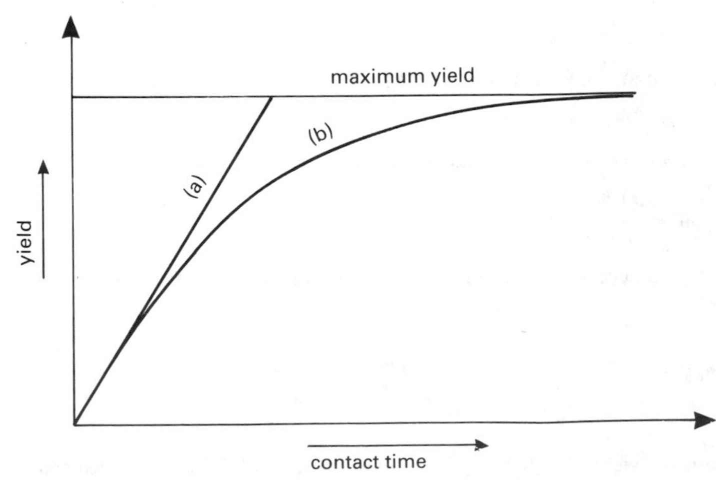

Fig. 2.1—Theoretical line (a) and practical curve (b) showing the time/yield course of a reaction.

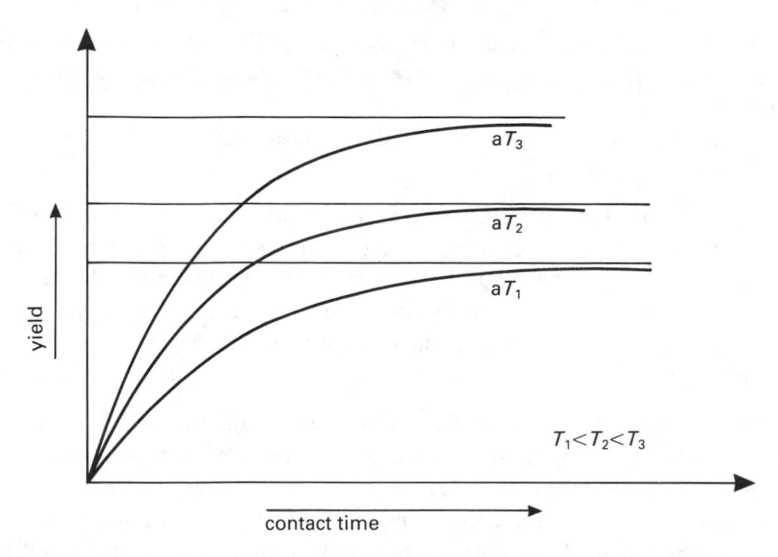

Fig. 2.2—Reaction isotherm for the same endothermic reaction carried out at constant but different increasing temperatures.

Apart from the consumption of thermal energy in starting up the reaction and losses from radiation, if insufficient attention has been paid to achieving suitable insulation, the main drawback to operating at a very high temperature is that of wear and tear on the reactors and accessories.

The reaction isotherms for exothermic reactions still tend asymptotically to the thermodynamic yield with the passage of time.

However, as is shown in Fig. 2.3, which depicts three reaction isotherms for the same reaction:

- the reaction isotherm for the highest temperature (T_{max}) reaches the lowest thermodynamic yield (r_{min}) most rapidly;

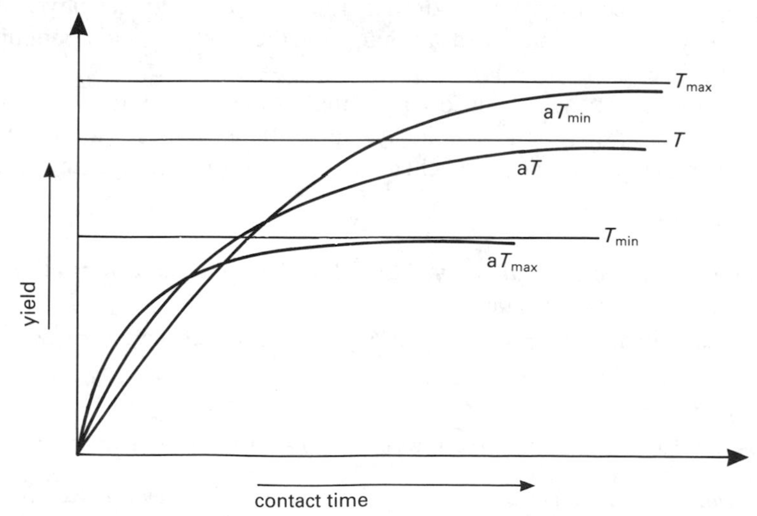

Fig. 2.3—Reaction isotherms for an exothermic reaction carried out at decreasing temperature to increase the yield.

- the isotherm for the intermediate temperature (T) rises less rapidly than the preceeding isotherm, indicating a lower reaction rate, but proceeds to a higher yield (r);
- the isotherm for the lowest experimental temperature (T_{min}) rises more slowly to the highest yield (r_{max}).

Given the way that they progress, the reaction isotherms cross each other in this case. This is because the pronounced fall-off in the yields as time passes, due to the progressive decrease in the equilibrium constant with increasing temperature, outweighs the opposing effect whereby these yields are enhanced by the speeding up of the rate of reaction accompanying such a temperature rise.

The shapes of these curves explain the two fundamental sets of operating conditions:

(1) the reaction is carried out initially at a high temperature, so as to favour a high rate of reaction, and then subsequently at a reduced temperature in order to increase the thermodynamic yield;
(2) suspension of the reaction conditions (usually by interrupting contact with a catalyst), when to prolong the reaction time would lead to only small increases in yield, or, if management of the process is costly (due to pressure of business and/or corrosion problems and/or working practices etc.).

The second type of reaction isotherm shape, that is the one exhibited by exothermic reactions, has been of great interest in industrial practice for many years, because the best known *synthetic reactions* (including among others, sulphur trioxide, ammonia and methanol), *oxidative conversions* and *reforming isomerisations* are of this type.

2.8 THE INFLUENCE OF RESPECTIVE CONCENTRATIONS OF REAGENTS AND OF THE WORKING PRESSURE UPON YIELDS FROM REACTIONS BETWEEN GASES

The chemical factor 'composition of the reaction mixture' and the physical factor 'pressure of the system' can lead to drastic changes in the reaction yields from processes involving gas phase reactions.

We shall treat the two themes by solving actual problems that are of considerable interest both theoretically and practically; in particular, these are examples of mathematical methods applied to chemical calculations concerning industrial processes.

The problem of variation in synthetic yields with a change in the composition of the reagents in a gas phase reaction

When air is heated up to a temperature of 2360°C, nitrogen and oxygen combine to form nitric oxide:

$$N_2 + O_2 \rightleftharpoons 2NO,$$

and, when equilibrium is attained, 2.3% (*by volume*) *of the reaction mixture present in the reactor consists of nitric oxide.*

We wish to find: (1) *the percentage composition of the reactants which leads to the maximum yield of the reaction product*; (2) *the percentage, by volume, of NO that is present at equilibrium at the temperature of* 2360°C *starting from the optimal percentage composition of the reactants.*

Part 1

Since there is no change in the number of moles of reactants and products in this reaction, $K_p = K_c = K_n = K_x = K$.

Denoting the initial mole fractions of nitrogen and oxygen by x_{i,N_2} and x_{i,O_2} and that of nitric oxide at equilibrium by $x_{e,NO}$, the mole fractions of nitrogen and oxygen at equilibrium may be written as $(x_{i,N_2} - \frac{1}{2}x_{e,NO})$ and $(x_{i,O_2} - \frac{1}{2}x_{e,NO})$ respectively.

The expression for the equilibrium constant will be therefore:

$$K = \frac{x_{e,NO}^2}{(x_{i,N_2} - \frac{1}{2}x_{e,NO})(x_{i,O_2} - \frac{1}{2}x_{e,NO})}.$$

It is obvious that $x_{i,N_2} + x_{i,O_2} = 1$, whereupon

$$x_{i,O_2} = 1 - x_{i,N_2}$$

and the equilibrium constant may assume the form:

$$K = \frac{4x_{e,NO}^2}{x_{e,NO}^2 - 2x_{e,NO} + 4x_{i,N_2} - 4x_{i,N_2}^2}.$$

By transforming this equation into a quadratic in $x_{e,NO}$ one has:

$$(4 - K)x_{e,NO}^2 + 2Kx_{e,NO} - 4Kx_{i,N_2} + 4Kx_{i,N_2}^2 = 0$$

and applying the standard formula for the solution of a quadratic equation gives:

$$x_{e,NO} = \frac{-K + \sqrt{(K^2 - (4-K)\cdot 4Kx_{i,N_2}^2 + (4-K)\cdot 4Kx_{i,N_2})}}{4 - K}.$$

The value of x_{i,N_2} which leads to a maximum value of $x_{e,NO}$ is calculated by setting the derivative of $x_{e,NO}$ with respect to x_{i,N_2} equal to zero. This derivative is:

$$\frac{dx_{e,NO}}{du_{i,N_2}} = \frac{1}{4-K}\cdot\frac{-8K(4-K)x_{i,N_2} + 4K(4-K)}{2\sqrt{(K^2 - (4-K)\cdot 4Kx_{i,N_2}^2 + (4-K)\cdot 4Kx_{i,N_2})}}.$$

It is known that the derivative will be zero when the numerator in the fraction representing it is zero. It is therefore required that:

$$-8K(4-K)x_{i,N_2} + 4K(4-K) = 0$$

which upon solution gives:

$$x_{i,N_2} = \tfrac{1}{2} = 0.5.$$

This result leads to the conclusion that the maximum yield of NO with respect to the percentage composition of the reactants is achieved when equimolar quantities of nitrogen and oxygen are initially present; that is, the initial mixture consists of 50% N_2 and 50% O_2.

Part 2

To answer the second part of this problem one must first calculate the value of K taking into account the fact that in air there is 79% N_2:

$$K = \frac{4\cdot 0.023^2}{0.023^2 - 2\cdot 0.023 + 4\cdot 0.79 - 4\cdot 0.79^2} = 3.44\cdot 10^{-3}.$$

At a constant temperature the value of K does not vary just by changing the initial reactant composition. However, the composition of the gas will have to change at equilibrium, and, in particular, $x_{e,NO}$ which can be found by solving the equation:

$$3.44\cdot 10^{-3} = \frac{x_{e,NO}^2}{(0.5 - \frac{1}{2}x_{e,NO})(0.5 - \frac{1}{2}x_{e,NO})}$$

which is done by taking the square root of both sides of the equation:

$$5.86\cdot 10^{-2} = \frac{x_{e,NO}}{0.5 - \frac{1}{2}x_{e,NO}}$$

from which is obtained:

$$x_{e,NO} = 2.85 \cdot 10^{-2}.$$

It is seen that the yield is increased by starting out from a mixture consisting of 50% of N_2 and 50% of O_2 because there is 2.85% of NO present in the gas at equilibrium instead of the 2.3% which is present when the reaction mixture initially consisted of 79% of N_2 and 21% of O_2.

The problem of the variation of yields from synthesis reactions when the pressure on the system is varied in a gas phase reaction

A mixture of nitrogen and hydrogen in stoicheiometric proportions is brought into contact with a normal catalyst for the synthesis of ammonia at a temperature of 573°C *and at a pressure of* 10 *atm. The equilibrium*:

$$N_2 + 3H_2 \rightleftharpoons 2NH_3$$

is attained, and it is found that the amount of ammonia formed is 14.73% *by volume of the gaseous mixture.*

Calculate what the percentage of ammonia formed at the same temperature would be if the pressure had been raised to 300 atm.

Given the proportionality between moles and volumes (*Avogadro's hypothesis*), if the system consists of 1 mole of gas at equilibrium, there is 0.1473 mole of ammonia and 0.8527 mole of hydrogen and nitrogen. These latter two gases, which allow variations in the reaction in accordance with the ratio in which they are initially present, make up $\frac{3}{4}$ and $\frac{1}{4}$ of the equilibrium mixture of H_2 and N_2 respectively. The number of moles of the two gases per mole of the mixture present at equilibrium will therefore be:

$$n_{H_2} = \tfrac{3}{4} \cdot 0.8527 = 0.6396,$$

$$n_{N_2} = \tfrac{1}{4} \cdot 0.8527 = 0.2132.$$

Consequently, the partial pressures of the three components of the gaseous mixture at equilibrium under a total pressure of 10 atmospheres will be*:

$$p_{NH_3} = 0.1473 \cdot 10 = 1.473 \text{ atm},$$

$$p_{H_2} = 0.6936 \cdot 10 = 6.936 \text{ atm},$$

$$p_{N_2} = 0.2132 \cdot 10 = 2.132 \text{ atm}.$$

Under these conditions the expression for the equilibrium constant (K_p) for the ammonia synthesis reaction will be:

$$K_p = \frac{1.437^2}{6.396^3 \cdot 2.132} = 3.89 \cdot 10^{-3} \text{ atm}^{-2}.$$

The value of the equilibrium constant which has just been found is fixed at 300 atm provided that the temperature remains constant.

However, the new expressions for the partial pressures will now be:

$$p_{NH_3} = x_{NH_3} \cdot 300 \text{ atm}$$

$$p_{H_2} = (1 - x_{NH_3}) \cdot \tfrac{3}{4} \cdot 300 = (1 - x_{NH_3}) \cdot 225 \text{ atm}$$

$$p_{N_2} = (1 - x_{NH_3}) \cdot \tfrac{1}{4} \cdot 300 = (1 - x_{NH_3}) \cdot 75 \text{ atm}$$

and the new expression for the equilibrium constant will be:

$$3.89 \cdot 10^{-3} = \frac{x_{NH_3}^2 \cdot 300^2}{[(1 - x_{NH_3}) \cdot 225]^3 \cdot [(1 - x_{NH_3}) \cdot 75]}.$$

*Here, the number of moles of each component is identical to the corresponding mole fraction since there is just one mole of the mixture present overall.

By simple mathematical manipulations one obtains:

$$3.89 \cdot 10^{-3} = \frac{x_{NH_3}^2}{(1 - x_{NH_3})^4} \cdot 1.05 \cdot 10^{-4}$$

that is,

$$\frac{x_{NH_3}}{(1 - x_{NH_3})^2} = \sqrt{\frac{3.89 \cdot 10^{-3}}{1.05 \cdot 10^{-4}}} = 6.08.$$

When solved, this equation leads to the result $x_{NH_3} = 0.69$. On the basis of the expressions for the reaction yield presented in this chapter it is evident that, since the value of x_{e,NH_3} changes from 0.1473 to 0.69 when the pressure is raised from 10 atm to 300 atm, the yield increases sharply as a function of the pressure applied to the system in reactions which involve gases.

3

Theoretical and practical principles in the management of industrial chemical processes

3.1 THE OBJECTS OF INDUSTRIAL CHEMISTRY

From the standpoint of development, that is from their occurrence under fixed conditions, all possible chemical reactions are always regulated by two factors:

- by the probability of occurrence (the 'thermodynamic factor'),
- by the rate at which they occur (the 'kinetic factor').

Practically, these factors are extremely important: the first because the absolute value of the yield of a chemical process ('reaction') depends on it, and the second because it controls the time within which such a yield is attained.

It is a special task of physical chemistry both to establish whether a chemical reaction is possible under certain conditions and, also, to determine, if the reaction in question is found to be possible, the degree of probability and the kinetics. Meanwhile, *industrial chemistry is principally concerned with the optimization of the operational conditions which will allow feasible reactions of practical interest* (here we encounter the 'industrial' aspect) *to occur to the greatest extent in the shortest possible time*. Everything is obviously carried out on the basis of physicochemical forecasts.

The tasks of physical chemistry are therefore preeminently theoretical, while the main function of industrial chemistry turns out to be essentially practical.

From what has been said it is evident that it is an error to confuse, as is often done, the functions of physical chemistry with those of industrial chemistry, even if the fulfilment of the former turns out to be a prerequisite for a rational foundation for the discharge of the functions of industrial chemistry.

Other important complementary functions relevant to industrial chemistry are the prearrangement of chemical processes which are of interest to it in a form adapted to the reagents, the separation of by-products from reaction products, the eventual refining and/or concentration of the latter products, and the installation of all the ancillary and control plant for all the functions which are carried out.

3.2 INDUSTRIAL CHEMISTRY AND PLANT CHEMISTRY

It is logical that industrial chemistry should pay direct attention to its basic activity of optimizing yields from reactions so that maximum amounts of products are obtained in the shortest possible times. On the other hand, the study and execution of its ancillary activities are part of chemical plant theory and technology. Such ancillary activities are generally limited to taking action within the framework of the description of the chemical processes which are of interest and the solutions adopted by chemical plant technologists to sustain such processes.

It is therefore clear that it is difficult to distinguish between the limits of industrial chemistry and industrial plant technology. Industrial chemistry is the 'raison d'etre' for the latter, for which it forms an essential requirement.

In the second part of this chapter, as a demonstration of the close relationship between the activities of chemical plant technology and industrial chemistry, examples of operations are reported which are commonly implemented in the field of industrial chemistry but with theoretical and applicative foundations typical of plant technology.

FIRST PART

THEORETICAL PROPOSITIONS CONCERNING THE MANAGEMENT OF THE MAIN ACTIVITIES OF INDUSTRIAL CHEMISTRY

3.3 PHYSICOCHEMICAL PHENOMENA EXPLOITED IN THE PRACTICE OF INDUSTRIAL CHEMISTRY

In further expanding on the concepts which have already been mentioned, it may be said that industrial chemistry has as its principal objectives:

(1) to displace a reaction in a direction which leads to the maximum yield of a certain product;
(2) to minimize production times.

To achieve these goals, industrial chemistry has identified, among physicochemical phenomena, *the dynamic nature of chemical equilibrium* and the *ability of molecules to become activated* as being of particular importance. The discussion of these topics which is subsequently presented is an essential prerequisite for the understanding of the management of industrial chemical processes.

Chemical equilibrium is a dynamic state: implications and consequences

The dynamic nature of chemical equilibrium and its significance

The passivity characterizing a state of physical (electrical, thermal, acoustic, etc.) or physicochemical (fusion, evaporation, etc.) equilibrium is always apparent only as a result of the constancy of just the macroscopic properties (that is to say, the properties which are experimentally observable) of the system in question. The same applies in the case of a chemical equilibrium attained by a reaction in so far as the molecules taking part in it are always in a state of chaotic motion, colliding with one another and reacting, during which time the reactants continue to form products while the latter react to reform the reactants. The macroscopic properties (concentration, colour, pressure, etc.) remain constant because the rates of the two opposing processes involved in the system are equal.

Consequently, *a state of chemical equilibrium is to be understood in a dynamic sense, that is, as a state of a chemical process in which the rates of the opposing reactions involved in the process are equal* (the kinetic aspect of chemical equilibrium).

From an interpretative point of view it may be noted that *the dynamic nature of a state of chemical equilibrium is a symptom of an equal probability for the formation of reactants and products which is achieved at equilibrium as a consequence of the probabilities for their existence being the same* (the thermodynamic aspect of chemical equilibrium).

The physical effect of 'action and reaction' applies to chemical equilibrium

Chemical equilibrium is included among those natural states which are governed in accordance with the general laws of physical dynamics. In particular, the law of 'action and reaction' applies to a system in chemical equilibrium. This law is invariably stated as:

> *Every change, however small, in the factors determining an equilibrium will produce a reactive response which tends to eliminate the effect of the change.*

In substance, this is Le Châtelier's law which became such (that is, a *law*) after the indisputable validity of the empirical principle enunciated by H. L. Le Châtelier a century ago (1884) had been recognized and, also, proven by a thermodynamic route.

The importance of Le Châtelier's law, which is a variant of the law of action and reaction in physical dynamics, is very great in chemical working practices in so far as it logically describes (since it is related to the general laws of physics), in a readily comprehensible qualitative form, the behaviour of systems governed by the laws of thermodynamics. The latter are expressed in terms of formulae whose general application is offset by the difficulty in their deduction, comprehension, and use.

The dynamic nature of chemical equilibrium leads to its inertia. Alternative use of the two aspects in application

Physical systems react to stimuli ('actions') through inertia, that is, by a tendency not to change the state by which they are characterized. On the basis of this, the law of action and reaction may be considered as a corollary to the law of inertia.

The prospect of unifying the sciences with the support of objective practical verification induces one to apply the same considerations to the physicochemical system of a 'chemical equilibrium', whose experimentally proven inertia is demonstrated by the existence of *the equilibrium constant*, a quantity which is identically formulated either by considering the kinetic or thermodynamic aspect of an equilibrium.

The equilibrium constant is of great importance in practice because, on the basis of it, one can easily arrive at an understanding of how certain properties of a chemical equilibrium can be exploited in order to modify the equilibrium.

The use of an equilibrium constant is no less a rigorous forecasting method because, as already indicated, the expression for this quantity, apart from being derived from kinetic considerations, results from mathematical, but less intuitively obvious, arguments based on the laws of thermodynamics.

The application of Le Châtelier's law, which validates the use of the equilibrium constant, for the prediction of the modifications which must be made to chemical processes with a view to optimizing the yields from them is also made more justifiable for the same reasons.

Molecular activation and reactivity

Every chemical species with a finite probability of existence (a 'molecule') is inherently endowed with chemical inertia which is an obstacle to its reactivity.

'Probability of existence', which forms part of the very definition of a molecule is, in fact, inherent in this *inertia*.

If the chemical inertia is minimal or small, the molecule is characterized by a (more or less) high reactivity, while, on the other hand, as the molecular chemical inertia increases, the reactivity progressively falls off until a point is arrived at when the molecules are extremely stable and exhibit little tendency to react.

It is obvious that the time required for the transformation of reactants into reaction products is determined in accordance with the chemical inertia of the molecules of such reactants. A reduction in the molecular chemical inertia, that is, 'molecular activation', is therefore a prerequisite if molecules which are naturally unreactive are to become reactive.

3.4 HOW INFLUENCING THE CHEMICAL EQUILIBRIUM IMPROVES YIELDS

It is stated in the preceding paragraph that it is necessary to exploit the dynamic nature of the chemical equilibrium for a reaction of interest in order to influence the production from an industrial chemical process.

Such exploitation implies that changes are made in the physicochemical factors which influence the equilibrium by their differential action and exert an effect on the chemical behaviour of the reagents. Such factors are:

(i) the *concentration* of the reagents in question at equilibrium,
(ii) the *temperature* at which the equilibrium is established,
(iii) the equilibrium *pressure*.

Here, we subsequently intend to study, both on the basis of Le Châtelier's law and on the basis of the chemical equilibrium expressed in terms of the relative constant, how the changes can be predicted in these factors which are required in order to enhance reaction yields.

The influence of changes in concentration on an equilibrium

Let us consider the usual generalized equilibrium:

$$aA + bB \rightleftharpoons cC + dD \qquad (3.1)$$

and assume that one of the components (product D, for example) has been partially removed from the system.

As the system, on the *basis of Le Châtelier's law*, will tend to neutralize the perturbation to which it has been subjected, the equilibrium is displaced to the right, thereby restoring the substance which had been removed. Since there is always a reaction to a perceived perturbation, if there is an increase in the concentration of one of the components on the left hand side of (3.1) (component A, for example), the equilibrium will be displaced to the right, thereby decreasing the concentration of the component which has been added.

With reference to the expression for the *equilibrium constant* of reaction (3.1):

$$K_c = \frac{[C]^c \cdot [D]^d}{[A]^a \cdot [B]^b}, \tag{3.2}$$

the changes to the system proposed above respectively bring about a decrease in the numerator and an increase in the denominator of the fraction representing K_c. It follows from this that, since K_c remains constant, there must be in both cases an increase in the value of the numerator of this fraction, leading to an increase in the reaction products C and D or, in other words, a displacement of the reaction to the right as already has been said.

Predictions based on the Le Châtelier law are therefore in complete agreement with the results obtained from the equilibrium constant.

The influence of pressure changes on an equilibrium

Changes in pressure influence only the equilibria of reactions involving gases and/or vapours, since pressure does not have any appreciable effect on condensed phase reactions.

Variations in pressure also have no effect on an equilibrium when the reaction does not lead to any change in the total number of moles; that is, if $c + d - a - b = \sum_{v_i} = 0$, where a, b, c, and d are the reaction coefficients.

In other cases, on the basis of *Le Châtelier's law*, a change in pressure will, in general, displace the equilibrium in such a way as to attempt to neutralize the action to which it has been subjected; that is, so as to reduce the pressure if it has been compressed and vice versa.

It is necessary to distinguish two cases:

(1) $a + b > c + d$ (for example, $4HCl + O_2 \rightleftharpoons 2Cl_2 + 2H_2O$)

(2) $a + b < c + d$ (for example, $2CH_4 + O_2 \rightleftharpoons 2CO + 4H_2$).

In the first case ($\sum_{v_i} < 1$), an increase in pressure displaces the reaction to the right, while a decrease in pressure drives it in the other direction.

If, indeed, the reaction goes to the right the number of moles in the system decreases and the increase in the applied pressure is (at least in part) thereby neutralized. If, on the other hand, the reaction goes to the left owing to a decrease in pressure, there is an increase in the number of moles in the system which at least in part neutralizes the diminution in pressure. In the second case ($\sum_{v_i} > 1$) compression displaces the reaction to the left, while a reduction in pressure drives it to the right.

This is due to the fact that if the reaction goes to the right the number of moles in the system is reduced, thereby neutralizing, at least in part, the effect of the compression. If, however, the reaction is displaced toward the right by lowering the pressure, the number of moles in the system increases, thereby compensating, at least in part, for the reduction in pressure.

This behaviour is determined by the fact that the pressure of a system in equilibrium does not tend to change. Actually, since the pressure is a consequence of the density of the molecules in the surrounding, it increases upon either decreasing the volume or increasing the absolute number of molecules. Therefore, for there to be no changes, either the volume must decrease or there must be an increase in the number of molecules and vice versa.

Being gases, if predictions are to be made based on the equilibrium constant, it is first necessary to express (3.2) as a function of the partial pressures.

For this purpose it is necessary to extract from $pV = nRT$ an expression for the concentration of a general chemical species i as follows:

$$\frac{n_i}{V} = \frac{p_i}{RT}$$

and, subsequently, to substitute this expression, suitably manipulated, into (3.2) giving:

$$K_c = \frac{\dfrac{p_C^c}{(RT)^c} \cdot \dfrac{p_D^d}{(RT)^d}}{\dfrac{p_A^a}{(RT)^a} \cdot \dfrac{p_B^b}{(RT)^b}}.$$

By means of simple manipulation, one finally arrives at:

$$K_c = \frac{p_C^c \cdot p_D^d}{p_A^a \cdot p_B^b} \cdot (RT)^{a+b-c-d} = K_p \cdot (RT)^{-\sum v_i}.$$

Since the temperature is constant, the factor $(RT)^{-\sum v_i}$ is constant and, consequently, so is the factor K_p. It is thereby proven that, at a constant temperature, not only K_c but also K_p is constant. This means that variations in temperature not only permit adjustments in the concentrations of the components in the system but, likewise, permit adjustments in the partial pressures for the reasons just explained.

Having found the expression:

$$K_p = \frac{p_C^c \cdot p_D^d}{p_A^a \cdot p_B^b} \tag{3.3}$$

we shall first apply (3.3) to the reaction cited in case 1 on page 64 to demonstrate that, also in this case, predictions based on Le Châtelier's law are in complete agreement with the results obtained from the equilibrium constant. One has:

$$K_p = \frac{p_{Cl_2}^2 \cdot p_{H_2O}^2}{p_{HCl}^4 \cdot p_{O_2}}. \tag{3.3a}$$

At this point, if the hypothetical increase and decrease in the pressure acting on the system is quantified by multiplying the partial pressures by n and $1/n$ respectively, equation (3.3a) assumes the form*:

$$K_p = \frac{n^2 \cdot p_{Cl_2}^2 \cdot n^2 \cdot p_{H_2O}^2}{n^4 \cdot p_{HCl}^4 \cdot n \cdot p_{O_2}} \quad \text{and} \quad K_p = \frac{1/n^2 \cdot p_{Cl_2}^2 \cdot 1/n^2 \cdot p_{H_2O}^2}{1/n^4 \cdot p_{HCl}^4 \cdot 1/n \cdot p_{O_2}} \tag{3.3b}$$

*It may be noted that, according to Dalton's law, multiplying the total pressure by n or $1/n$ leads to the partial pressures being changed by the same factors.

whence:

$$K_p = \frac{p_{Cl_2}^2 \cdot p_{H_2O}^2}{n \cdot p_{HCl}^4 \cdot p_{O_2}} \quad (3.3b) \qquad \text{and} \qquad K_p = \frac{n \cdot p_{Cl_2}^2 \cdot p_{H_2O}^2}{p_{HCl}^4 \cdot p_{O_2}} \quad (3.3c)$$

In the case of (3.3b), the denominator of the fraction increases in correspondence with an increase in pressure on the system, while, in the case of (3.3c), the numerator of the fraction increases in correspondence with a decrease in the pressure. To ensure that K_p remains constant in the case of (3.3b), there must be an increase in the numerator of the fraction, while, in the case of (3.3c), there must be a decrease in the denominator.

This is equivalent to saying that the reaction is displaced to the right in the first eventuality and to the left in the second, exactly as predicted on the basis of Le Châtelier's law (page 62).

* * *

The agreement between the conclusions drawn on the basis of Le Châtelier's law and those obtained by using the equilibrium constant may be similarly demonstrated by referring to the example of the reaction presented as case 2 on page 64.

The influence of temperature changes on an equilibrium

As we shall see, the temperature affects both the rate at which equilibrium is attained as well as the position of equilibrium.

However, while the rate of reaction is always increased by an increase in temperature, the equilibrium position may be changed in either of the two opposing directions when the temperature is raised. Analogously, a lowering of the temperature leads to a diminution of the reaction rate, but can likewise change the equilibrium position in two directions.

To make predictions concerning the direction in which the equilibrium of a reaction will be displaced as a consequence of a change in temperature, one must always (that is, when reference is made to Le Châtelier's law or the equilibrium constant) have regard to whether the reaction is exothermic or endothermic.

In the very rare case (of certain electrochemical reactions, for example) when the reaction in question is *athermic*, temperature will have no effect on the position of relative equilibrium.

While this observation is not so important per se, it does allow one to intuitively perceive, what is actually encountered experimentally, that the effect of changes in temperature on reaction equilibria is greater, the greater the heats of reaction of the processes in question, and that this effect decreases and finally vanishes as the heats of reaction become smaller and tend to zero.

Predictions of the evolution of equilibria as the temperature is varied are based on the fact that raising the temperature leads to the addition of heat, while reductions in temperature are due to the removal of heat.

On the basis of these results it may be understood by the Le Châtelier law *that a chemical equilibrium is displaced in the direction of the development of heat when the temperature at which the equilibrium is established is lowered and displaced in a direction leading to the absorption of heat when the temperature is increased.* The equilibrium will therefore lead to the development of heat if heat is removed from the system, and to the absorption of heat if heat is added to it. In both cases there is a tendency to recreate the initial thermal conditions of the system.

With regard to the exothermic ammonia synthesis reaction, for example:

$$N_{2(g)} + 3H_{2(g)} \rightleftharpoons 2NH_{3(g)} + 22 \text{ kcal}$$

a reduction in the synthesis temperature drives the reaction to the right (the direction of the development of heat), while an increase in temperature drives it in the opposite direction (the direction in which heat is absorbed*).

On the other hand, the reaction:

$$CH_{4(g)} + H_2O_{(g)} \rightleftharpoons CO_{(g)} + 3H_{2(g)} - 49.3 \text{ kcal}$$

goes to the left when the temperature is lowered and to the right when the temperature is increased.

In industry there are no reactions which may be realistically treated as being independent of temperature.

Predictions concerning displacements of the equilibrium position of a reaction due to variations in the temperature using the *equilibrium constant* are made by utilizing the van't Hoff equation which, for qualitative purposes, may be formulated at once as:

$$\log \frac{K_2}{K_1} = \frac{Q}{4.55} \cdot \frac{T_1 - T_2}{T_1 \cdot T_2} \qquad (3.4)$$

where Q is the enthalpy of the reaction and T_1 and T_2 are the temperatures at which the equilibrium constant has the values of K_1 and K_2 respectively.

By applying (3.4) to an exothermic reaction, it can be readily understood that, by making the reaction take place at a temperature T_2 which is greater than T_1, the second part of the right hand term becomes negative, whereas this term will be positive when the reaction takes place at a temperature T_2 which is less than T_1.

Since the sign of the left hand side of the equation must be the same as that of the right hand side, one has in the first case that:

$$K_2 < K_1$$

and, in the second case, that:

$$K_2 > K_1.$$

By lowering the equilibrium constant, as occurs in the first case, the reaction yields are reduced, that is, the reaction is reversed (displaced to the left). However, when the equilibrium constant is increased, as in the second case, the reaction yields go up, that is, the reaction is promoted (displaced to the right).

It is therefore concluded, in returning to the Le Châtelier law, *that an exothermic reaction is favoured by a lowering of the experimental temperature and disfavoured by an increase in the temperature.*

* * *

When (3.4) is applied to an endothermic reaction, it is noted that, when the experimental temperature is raised ($T_2 > T_1$), the right hand term is positive while a

*The reversibility of reactions implies that reactions which are exothermic in one direction are endothermic in the opposite direction (The 'first principle of thermodynamics').

reduction in the temperature at which the reaction occurs ($T_2 < T_1$) leads to the right hand side of the equation becoming negative. Therefore, in the case of an endothermic reaction:

$K_2 > K_1$ when the experimental temperature is increased and

$K_2 < K_1$ when the experimental temperature is decreased.

It is concluded, on the basis of Le Châtelier's law, *that an endothermic reaction is favoured* (displaced to the right) *by a reduction in the experimental temperature.*

3.5 THE PROBLEM OF MOLECULAR ACTIVATION

As we have already had occasion to state, it is not sufficient that a proposed reaction should be carried out under conditions which would give high yields which can be forecast on the basis of the equilibrium constant ('thermodynamic' yields) thereby giving results that are acceptable to industry, but it must also take place at a rate which ensures a satisfactory hourly production of the product.

To understand how this 'kinetic' aspect, which is a prerequisite for the industrial scale manufacture of chemical products, can be suitably satisfied we must deal with the modes in which the molecules partitipating in a reaction can be activated.

Nature of the reaction phenomenon

A reaction implies a combinative rearrangement of atoms, that is, a break-up of the molecules into atoms followed by the recombination of atoms to form a molecular species which is more stable than the reagents (at least, under the reaction conditions).

It is important to make clear that the reaction products are not the most stable products in an absolute sense because those species which have a high probability of existence, under certain conditions, cannot convert into other forms. Also, they tend to persist owing to 'locking' of the formation equilibrium attained under suitable conditions.

Thus, ammonia dissociates at a high temperature to form nitrogen and hydrogen which persist at ordinary temperatures in spite of the high probability of them recombining to form ammonia under normal conditions.

Reactions of interest to the chemical industry generally permit various chemical species (such as NH_3 and CO_2 or CH_4 and SO_2) to enter into reaction, and obviously require a collision mechanism between such molecules under conditions which energetically favour breaking of the existing bonds between their atoms with the subsequent spontaneous formation of new products.

On the other hand this is also the mechanism of industrial reactions involving just a single type of reagent such as polymerizations and dissociations. In this case collisions between the molecules contribute to their degradation and provide opportunities for the products of bond rupture to couple together in new molecular aggregates. For example, this is how one explains the formation of molecular hydrogen and nitrogen by the degradation of ammonia molecules, of crystalline aggregates of graphitic carbon as well as molecules of hydrogen by the pyrolysis of hydrocarbons.

In conclusion, the kinetics of chemical reactions, in general, and those of interest in industrial chemistry, in particular, are determined by two factors: *by the number of collisions between the molecules of the reactants and the energy state of the molecules participating in such collisions*; the 'nature of the reagents' (the other kinetic factor sometimes specified at a speculative physicochemical level) being considered as none other than an essential indicator of the molecular energy state.

How collisions between molecules or between activated surfaces are increased

The case of liquid and gaseous reactants

The collisions between molecules of liquid or gaseous reagents are influenced above all by *concentration* because the greater the number of molecules per unit volume the greater the probability of a collision between them.

The temperature is also a favourable factor in increasing the number of collisions, but its influence on the rate of reaction is, for another cause as we shall see, so great as to make it practically impossible to propose a connection between the increase in thermally induced collisions and the consequent increase in the rate of reaction.

In fact, it has been demonstrated experimentally that, upon a certain increase in temperature, the intermolecular collisions increase by 1–2% while, at the same time, the rate of reaction increases by 200–300%.

For kinetic purposes, the possible collisions between molecules are classified as *effective* and *non-effective*: the first type of collision occurs in a mode (molecular orientations, interactions upon collision with decisive effects on the bonds, etc.) which favours reaction, while the second type of collision occurs with steric characteristics which are unfavourable or indifferent towards reaction.

It is observed that the atoms in a molecule undergo vibrations causing the bonds between them to stretch to some extent. When such atoms are at the maximum distance from one another they have the greatest probability of being removed from the molecule. If two molecules, as shown in Fig. 3.1, collide while in a state of maximum elongation, the collision which is effective because of bond resonance effects leads to the formation of new chemical species which are more stable than those which collided.

The case of solid reactants

In the case when solid reagents are involved in reactions, intermolecular collisions cannot take place because, even in the highest states of subdivision, the particles are

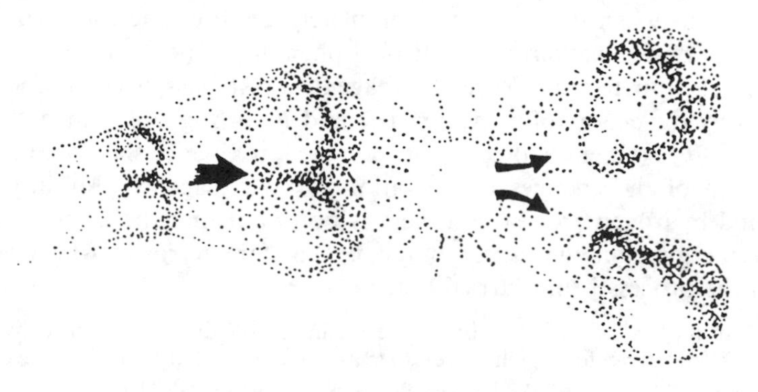

Fig. 3.1—Collision of molecules.

still extremely large compared with molecular dimensions. If one is dealing with molecular solids or when dealing with atomic and ionic solids one is always concerned with aggregates consisting of an enormous number of components. Moreover, solids are so inert that it is difficult for active ('effective') kinetic collisions to occur between the particles.

Even though it is impossible to realize the optimal conditions which appertain in the case of intermolecular collisions, conditions can always be established which are more favourable for solid reactions by, on the one hand, increasing their state of subdivision and, on the other hand, maintaining them in an agitated suspension.

When solid reagents are subdivided, new surfaces are developed to a proportional extent on which there are active atoms due to the instability arising from the anisotropic interactions at a surface. In its turn, agitation leads to an increase in the number of collisions and in the factors favouring reaction (noticeable increases in bond lengths, increases in kinetic energy, etc.).

This is reflected in the great importance which is always paid to the milling and mixing of solid reagents which are provided by industrial plant technology with the aim ofincreasing the rates of reactions involving such reagents.

The kinetic influence and changing the molecular energy state

The problem of activation energy

Chemical reactions require that chemical bonds in the reagents are broken, that is, they necessitate the breaking of bonds involving orbitals which, as has already been noted, correspond to low energy levels.

The physical reason for the formation of chemical bonds lies in the fact that there is a lowering of the energy upon their formation and the need to raise the molecular energy level (and thereby the energy of the pertinent bonding orbitals) to bring the molecules to a condition where such bonds can be broken (to bring about reactions).

The energy which must be supplied to molecules to bring them to a state when they are capable of reacting is called 'the energy of activation'.

The energy of activation is not a molecular constant but a variable which is dependent on the state of the molecules, being smaller when the molecular energy content is already high.

To truly comprehend the kinetics of a reaction, which is the outcome of an objective evaluation of experimental results, it must be recalled that molecules can also react under conditions when they have not completely acquired the energy of activation, since bond rupture (reaction) is a statistical phenomenon.

For any reaction to take place the reagents must pass over an energy barrier represented by the energy of activation (E_1 and E_2 in Fig. 3.2) which is specific for each of them. It is therefore logical that the reaction kinetics should be governed by the maximum of the energies of activation for the reagents. For instance, when nitrogen and hydrogen are reacted, there is no doubt that the former has a higher activation energy than the latter, and it is therefore nitrogen which governs the kinetics of the process being carried out.

The entry of molecules into reaction is one of those phenomena governed by the laws of large numbers (statistics) for which there exists a probability that a pertinent event may also take place before conditions have been attained which, *with respect to a single event*, would need to be achieved before it could occur.

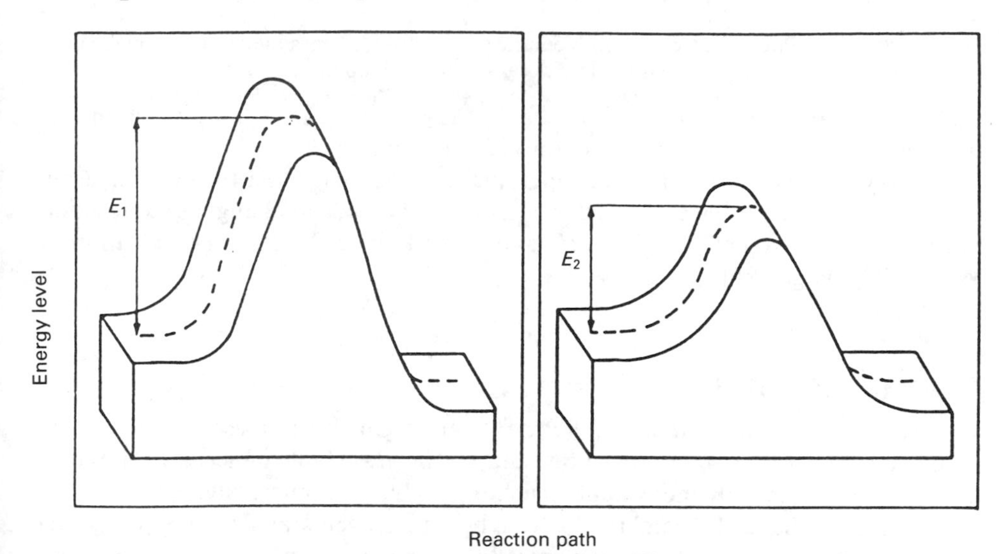

Fig. 3.2—Energy barriers.

The probability of a reactive event, defined as the ratio between the number of cases of bond rupture and the number of molecules in the reaction system is obviously greater or smaller, depending on how close to, or far from, acquiring the whole of the activation energy the molecules may be. This probability must become unity at that moment when all the molecules in the system react effectively (a ratio of one between favourable cases and possible cases.

Reactivity and temperature

The activation energy can be reduced and, in the limit, made to vanish by furnishing energy to the molecules in various forms: electrical energy, electromagnetic (light) energy, and, above all, in the form of thermal energy which raises the temperature.

But it is not necessary that the increase in temperature of the reagents is such as to reduce the molecular activation energy to zero by the absorption of heat in order that an increase in the rate of reaction may be observed. Actually, any temperature increase is successful in bringing about such an increase in the rate, owing, as was stated above, to the statistical nature of the reaction phenomenon.

The reason for this, apart from the strictly scientific considerations already mentioned, can be readily understood by recourse to an intuitive comparison.

A six metre pole vault is an exceedingly difficult athletic achievement. To be certain (probability $=1$) that an individual athlete attempting this will succeed in getting over the bar it would be necessary to give to the champion a surplus of strength and high grade training (energy of activation) compared to the amounts of these responsible for his usual performances.

However, if the number of athletes (reactant molecules) attempting to overcome this obstacle is allowed to increase excessively,* it is certainly not necessary to train such a crowd to any large extent in order that someone succeeds in getting the better of this simply compelling contest. Furthermore, the number of those who are victorious increases rapidly even with a small increase in the strength and quality of training of the athletes.

---

*It is of benefit to note that there are always more than 10^{20} molecules per cm^3 in any reaction carried out practically.

Therefore, that which is extremely problematic (commonly 'impossible') at the level of single individuals offers certainties when the law of large numbers applies.

Therefore: an *increase in temperature* by any amount *will always favour the development of reactions*, that is, the 'rate of reaction'.

It is only extremely rarely that a temperature increase acts other than in increasing the rates of kinetic processes. In such rare cases it succeeds in changing the reaction mechanism in a manner which is very favourable kinetically as in the oxidation of NO to NO_2 (page 294).

Reactivity and catalysis

The nature and action of catalysts. Reactants are often susceptible to combining in a labile manner with certain substances, and, in this less stable molecular state, they yield thermodynamically more stable reaction products more readily.

On the other hand, by combining strongly with reagents and thereby rendering the formation of the more thermodynamically stable reaction products more difficult, other substances exhibit a 'negative' action on the reaction kinetics.

Substances capable of speeding up or slowing down the rates of reactions are called 'catalysts'.

Schematically, a reaction of the type $A + B \rightarrow C$ which is slow becomes fast in the presence of a positive catalyst (Cat) which is capable of acting in the following manner:

$$A + B \rightarrow ACat + B \rightarrow C + Cat.$$

While this mechanism assumes that A has an affinity for Cat, the strength of this interaction must be sufficiently low as to allow the subsequent formation of C.

The way in which a negative catalyst acts is analogous to that for a positive catalyst, but, in this case, the intermediate product of type $ACat$ in the preceding example is a relatively stable compound.

It is implicit in this mechanism for the action of catalysts that the catalyst does not form part of the reaction products except when it acts as a simple initiator (in certain polymerizations) or when it is produced from a reagent (acids in esterifications).

The action of catalysts had already been perceived intuitively by J.J. Berzelius who, in 1836, defined them, using a Greek term which means 'break up' to indicate precisely that they decrease the stability of substances, making it more likely that they will react.

Energy of activation and catalysts. The reactive act in a non-catalyzed reaction is generally unique, while there are many combinative stages of catalyzed reactions. At each stage in a catalyzed reaction the activation energy is smaller than for the corresponding act in the non-catalyzed reaction if the catalysis brings about an increase in the rate of reaction, while if the catalysis leads to a diminution in the reaction rate, at least one stage involves an activation energy greater than that for the corresponding act in the uncatalyzed reaction.

Figure 3.3 presents diagrams of the possible activation energy profiles during the course of a non-catalyzed reaction, a reaction which is speeded up by a catalyst and a reaction retarded by a catalyst.

It is seen that, in every case, the catalyst provides an alternative possible pathway

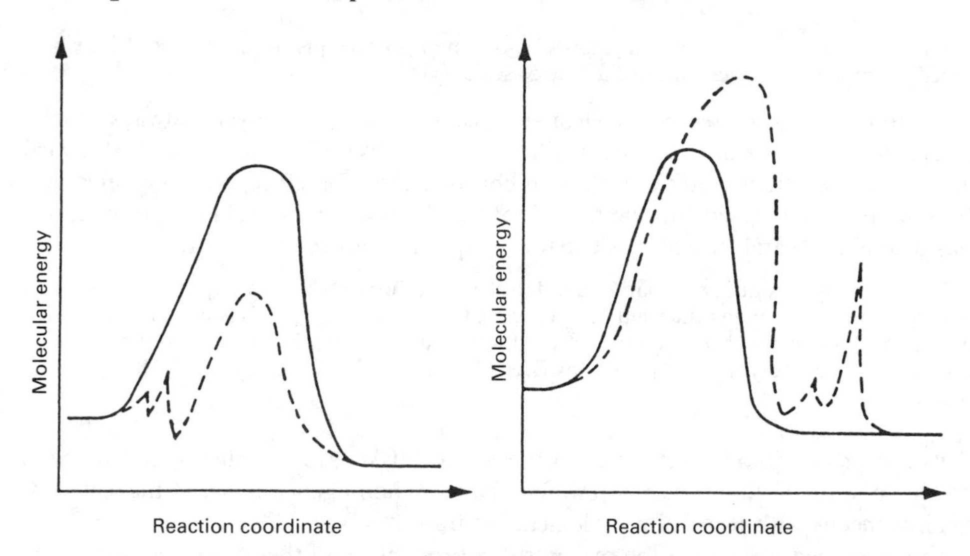

Fig. 3.3—On the left: the solid line depicts the normal energy surface of a reaction while the dotted line depicts the energy profile for the reaction when catalyzed. On the right: the analogous profiles of another reaction are shown where the solid line applies to the uncatalyzed reaction whereas the dotted line now refers to a negatively catalyzed reaction. These examples clearly demonstrate the different actions of the two types of catalyst.

which is energetically more or less preferable to the pathway followed by the uncatalyzed reaction.

Usefulness of catalysts. Catalysts do not change the thermodynamic characteristics of the reactions in which they are active, that is, the equilibrium conditions. However, they do accelerate the attainment of the equilibrium state, providing alternative, more or less facile, pathways for the reaction.

The properties used to characterize a catalyst are:

(i) the *specificity*, which is the exhibition of catalytic properties with respect only to a certain type of reaction (hydrogenation, oxidation, hydrolysis, etc.);
(ii) the *activity*, which is measured by the number of moles transformed or produced per unit time per unit weight of the catalyst;
(iii) the *selectivity*, that is, the capacity to increase the rate of formation of just one of the possible products of a certain type of reaction (for example, of a carboxylic compound rather than an epoxide in the oxidation of alkenes).

The specificity and selectivity clearly distinguish the kinetic promotional action of catalysts as compared with the action of temperature which is non-specific and non-selective. For this reason, catalysts are much better than temperature as agents for modifying reaction kinetics.

Moreover, while temperature is exclusively a positive kinetic agent (page 72), catalysts may be systematically classified as positive or negative agents because, in addition to those which speed up the kinetics, there are others among them which retard the rate of reaction (antioxidants, antipolymerization agents, antidetonation agents, etc.).

On the other hand, as a simplification, it is common practice just to discuss catalysts which speed up processes on account of their large-scale use.

It should also be remembered when using catalysts that a catalyst is always equally active (for reasons which can be justified theoretically) with respect to the forward and reverse reactions leading to the equilibrium state. Therefore, once a catalyst has been found which speeds up the rate of formation of a certain reaction product, one has available a catalyst which is capable of speeding up the reverse reaction.

Inter alia, upon putting a platinum catalyst into benzene which is kept in contact with hydrogen, the thermodynamic equilibrium state of the process: $C_6H_6 + 3H_2 \rightleftharpoons C_6H_{12}$ is rapidly arrived at. Vice versa, using the same catalyst under identical conditions, but starting off with cyclohexane, the very same equilibrium is established at the same rate as in the previous reaction.

Types of catalysts. The fundamental types of catalysts are established on the basis of the nature of the substrate, which may be homogeneous with the catalyst, heterogeneous with it, or of a biological nature.

In *homogeneous catalysis* the reagents, the products, and the catalyst all form part of a single phase—which is usually liquid.

Typical examples of homogeneous catalysis are the oxidation of SO_2 with NO_2, the dehydration of alcohols and, for that matter, the hydration of alkenes and the esterifications $\rightleftharpoons$ hydrolyses (of esters).

Heterogeneous catalysis is the most common both in the field of inorganic chemistry and industrial inorganic chemistry. In this type of catalysis the reactants are usually liquids or gases and the catalysts are solids.

As regards their nature, the substances exhibiting catalytic activity which are most important in this field are the metals, metal oxides, and salts which behave as Lewis acids.

Enzymatic catalysis is promoted by protein-like substances (frequently they are simple proteins) produced by living cells* which are known as *enzymes.*

The action of enzymes governs the chemical kinetics of a myriad of chemical reactions on which life in all its manifestations depends. This action is explained in terms of a combination between an enzyme and its substrate (the substance whose transformation is catalyzed by the enzyme) to form a complex within which the substrate becomes activated in the sense that its conversion into reaction products is facilitated.

The product having been formed detaches itself from the complex, thereby liberating the enzyme which can then react in a cyclic manner with other substrate molecules.

The characteristic which fundamentally distinguishes enzymes from non-biological catalysts is their lability when heated, non-biological catalysts generally being very stable to heat. This is because increasing the temperature destroys or, at least, denatures† enzymes while changes in temperature over wide ranges cause no damage to inorganic catalysts.

---

*Moreover, the age of industrial production using syntheses by means of biocatalysts has begun. This is one of the main fields of bioengineering.

†The denaturation of enzymes involves the disruption of the secondary, tertiary, and quaternary structures which are present in proteins with a loss of the stereospecific characteristics which are essential if the enzymes are to exhibit enzymatic activity.

However, both enzymes and biological catalysts can be *poisoned*, that is, deactivated by substances with which they preferentially combine to yield compounds which are relatively more stable than those which are similar but subject to catalytic action. On the other hand, enzyme poisons understandably involve polyfunctional organic compounds which are far more numerous and varied than inorganic catalyst poisons.

Nowadays, enzymes are also prepared industrially by separating them from their producers (for the most part bacteria) and crystallizing them. Enzymes are sold and used in this crystalline form for the catalysis of chemical processes which form the foundation of a large number of technological industries concerned with cleansing, softening, energy production from biomass, the modification of the organoleptic properties of foodstuffs, and so on.

Classification of industrial catalysts. The principal industrial catalysts are listed in Table 3.1 on the basis of their structural characteristics and the catalytic functions which they display.

The uses made of the catalysts indicated in the table will be described in detail and illustrated together with the relevant mechanism in the remainder of this book where specific notes on their preparation will also be given.

SECOND PART

RECYCLING AND CLEANSING OPERATIONS IN INDUSTRIAL CHEMISTRY

3.6 THE CONCEPT AND FUNCTION OF RECYCLING

In many industrial chemical processes an aliquot of the substances which have already participated in the process is returned to an earlier stage in the cycle. This operation is known as *recycling*.

The aims of recycling are numerous and include:

- increasing the global percentage yields,
- concentrating solutions,
- increasing the degree of separation of the components of vapours,
- the reutilization of solvents,
- saving energy.

Let us now pass to their detailed description, giving examples of applications of the functions displayed by recycling.

Increasing global percentage yields

The manner in which many industrial chemical processes are carried out is often, either for economic or other operational reasons, far from the conditions which can guarantee the attainment of the best yields.

In other cases yields are unsatisfactory because of the low intrinsic value of the equilibrium constant of the reaction being exploited.

Table 3.1—Classification of industrial catalysts

			Type of catalytic reaction	*Formulations*
Industrial catalysts	Structural (that is, in relation to their intrinsic nature)	Acid–base	—Cracking	SiO_2–Al_2O_3, Ce, Zeolites
			—Dealkylation–alkylation	$AlCl_3$, HF, BF_3
			—Dimerization	H_2SO_4, Na, Li and K
			—Esterification and trans-esterification	H_2SO_4
			—Hydration–dehydration	P_2O_5, Al_2O_3
			—Isomerization about double bonds	H_2SO_4, Al_2O_3, H_3PO_4, Na, K, SiO_2–Al_2O_3
			—Isomerization of molecular framework	P_2O_5, BF_3, $AlCl_3$
			—Oligomerizations	H_2SO_4, H_3PO_4
			—Polymerization	Na, Li, BF_3, $AlCl_3$
			—Hydrolysis of fats	$ZnCl_2$, $BaCl_2$, $AlCl_3$
		Metal-alkyls, transition metal compounds and complexes	—Dimerization	$RhCl_3$, $RuCl_3$, $CoCl_2$–$AlEt_3$
			—Oligomerization	$AlEt_3$, $CoCl_2$–$AlEt_3$
			—Generic stereospecific polymerization	allyl-NiBr, $AlBrEt_2$, LiR, $MgEt_2$, $RhCl_3$
			—Stereospecific polymerization Ziegler–Natta	$TiCl_3$–$AlEt_3$, $TiCl_4$–$AlEt_3$ $VCl_4 + AlEt_3$, $CoCl_2$–$AlClEt_2$
	Functional (in relation to function)	Oxidation–reduction	—Chlorination	$FeCl_3$, $SbCl_3$, $CuCl_2$
			—Dehydrogenation	ZnO–Cr_2O_3, Fe_2O_3, Pt–C, MoO_3–Al_2O_3
			—Hydrogenation	Pt, Pd, Ni, Co, Fe, Rh, Ru, MoS_2 ZnO–Cr_2O_3, $Co_2(CO)_8$, CuO–Cr_2O_3, $RhCl_3$
			—Oxychlorination	$CuCl_2 + KCl$
			—Oxidation	Pt, Ag, V_2O_5, N_2O_5, Cu_2O, molybdates, antimonates
		Polyfunctional	—Hydrocracking–Hydro-dealkylation	CoO–MoO_3–γ-Al_2O_3
			—Hydrodesulfurization	CoO–MoO_3–SiO_2–Al_2O_3
			—Reforming	Pt–Al_2O_3 (+HF or HCl)

The first aim of recycling is rightly that of improving the *global percentage yields* of a process, which is expressed as the ratio, multiplied by 100, between the amount of one of the components (usually the most valuable) or the amount of the stoicheiometric mixture of the reactants which are transformed into reaction products and the corresponding amounts used in the manufacture, that is:

$$\eta\% = \frac{\text{mass (or number of moles) of a component or of a mixture transformed}}{\text{mass (or number of moles) of initial reaction mixture}} \cdot 100 \tag{3.5}$$

A diagram of the apparatus for this type of recycling is shown in Fig. 3.4.

The most well known cases of recycling for the purpose of increasing the yields are those employed in the synthesis of ammonia and urea.

Concentration of solutions

A feasible recycling operation is that of concentrating solutions emerging from reaction vessels or solutions recovered from the reaction products. The concentration of solutions derived from reaction vessels by recycling is carried out, for example, in the synthesis of sulphuric acid, using homogeneous catalysis and the concentration by recycling of the solutions emerging from the reaction product recovery vessels in the process of the preparation of many of the rather soluble salts, including $CuSO_4 \cdot 5H_2O$ (page 222). The return to the boiling vats of sugar solutions obtained by the centrifugation of sugars which have already been purified and concentrated once in the sugar industry also belongs to this type of recycling.

Recycling in rectification processes

In rectification columns it is essential that there should be liquid present to ensure exchange between the liquid and vapour phases by means of which the separation of the components in the mixture being distilled is realized. The liquid becomes scarce on the higher plates of rectification columns and may be completely absent on the last plate of a column, so that, on the one hand, a part of the rectification column then becomes unused while, on the other hand, a pure component is not obtained at the head of the column.

To prevent this, liquid is recycled into the column as a 'reflux', and this may be

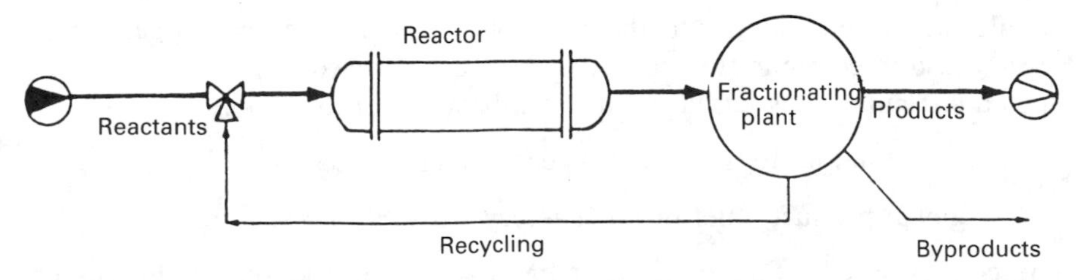

Fig. 3.4—Recycling scheme for the purpose of increasing the global percentage yields of a process. It is essentially that part which has not reacted, after being separated from the reaction products, which is recycled. The by-products, instead of being recycled, often find special uses in other processes.

considered as the best known recycling process in the field of chemical plant technology.

Examples of this type of recycling in rectification are extremely numerous over the whole domain of industrial chemistry, since rectification is extensively used in the reaction product separation field.

Recycling for the re-use of solvents

In the crystallization of slightly soluble molecular and ionic solids, the successive centrifugations or filtrations make available, at each cycle, solvent which is practically pure or of a (constant) high grade of purity. Recycling of the solvent is then regularly built into the industrial processes.

The solvent is also systematically recycled after the desorption of adsorbed products as is seen, for example, in the process scheme considered in section 3.7.

Finally, at the end of extractive distillations, including those of reforming effluents, the solvent, which is separated by means of a suitable column from the extract released into it, is selectively recycled into the first column.

The recovery and dispersion of heat by recycling

After many reactive exothermic processes the products in question (such as synthesis gas, steam, and air) are at least partially recycled in succession to the evaporation, distillation, and drying operations with the aim of *recovering heat*.

The fraction of the products recycled is dependent on plant and energy costs such as those of electricity for recirculation, and thermal energy for reactive preheating. The criterion of plant cost in the programming of recycling is fundamental because, as the amount of recycling is increased, there is a need to operate, if parity of production is to be maintained, plants of larger and larger size.

Sometimes, recycling is carried out in order to *get rid of heat* rather than to recover it. In this case the recycled heat does not have the capacity to do what is wanted of it at the outlet of the apparatus which produces it (unlike recycling for the recovery of heat), and the required removal of heat is achieved by external refrigeration. An example of this will be seen in the illustrative plant which forms the basis of a numerical calculation in the next section.

3.7 CALCULATIONS ON THE OPERATION OF PLANT USING RECYCLING

In a calculation appertaining to the running of a plant it often happens that it is necessary also to compute the amounts which are recycled.

Such a computation is based on material balances of the type:

$$\text{total feed} = \text{fresh feed} + \text{recycled substances}$$

$$\text{global product} = \text{net product} + \text{recycled substances.}$$

The correct specification of these material balances and the analogous energy balances ensures that a plant is run under more favourable economic conditions when the relative amount of recycling has been established, using data obtained in this manner.

A plant calculation which illustrates the three above-mentioned types of recycling (firstly, increasing global percentage yields, then solvent reutilization and then removal of heat) is now developed, only the principal recycle being calculated.

The reaction

$$4HCl + O_2 \xrightarrow[500-550^\circ C]{CuCl_2} 2Cl_2 + 2H_2O$$

is involved in the production of chlorine from hydrogen chloride.

It is found in laboratory trials that 60% of the HCl can be oxidized when pure HCl and O_2 as a gaseous mixture in a molar ratio of 2:1 are introduced into the reactor. The operational conditions are such as to guarantee the same degree of conversion.

Knowing that the plant is fed with 2000 kg/h of HCl, that the desiccant is initially 96% H_2SO_4 but becomes diluted to 70% during the process and that 95% of the water produced condenses in the scrubbing tower carrying hydrochloric acid with it to an extent that it contains 20 mol% HCl, calculate:

(i) the composition of the overall feedstock,
(ii) the percentage ratios for the recycling of HCl and O_2,
(iii) the hourly production of Cl_2 (in kg),
(iv) the global percentage yield of the process with respect to hydrogen chloride,
(v) the amount of sulphuric acid required to dry the gases coming out of the water removal tower.

The values of the data required to solve the problem which are presented here take account of the effects on the system of the subsidiary operations (recycling to remove the heat generated upon the dissolution of hydrogen chloride in water by means of a water-cooled condenser and, particularly, variations in the feedstock due to the oxidation of traces of the absorber of chlorine, S_2Cl_2, to H_2SO_4) which are depicted in the plant scheme shown in Fig. 3.5.

Denoting the number of kmoles of recycled HCl by x, the quantities needed to solve this problem may be calculated on the basis of the data as follows:

- the number of kmoles of HCl in the fresh feedstock is: $2000/36.47 = 54.8$,
- the number of kmoles of HCl participating in the process is: $54.8 + x$,
- the number of kmoles of HCl converted into chlorine is: $(54.8 + x)\cdot 0.60$,*
- the number of kmoles of either water or chlorine produced is: $\frac{1}{2}\cdot(54.8 + x)\cdot 0.60$,
- the number of kmoles of water separated in the water removal tower is: $\frac{1}{2}\cdot(54.8 + x)\cdot 0\cdot 60\cdot 0\cdot 95$,
- the number of kmoles of HCl dissolved in the water removed at the water separation stage is: $\frac{1}{2}\cdot(54.8 + x)\cdot 0.60\cdot 0.95\cdot 0.20$.

Consequently, the material balance which can be established with respect to the HCl feedstock is:

$$54.8 + x = x + \frac{(54.8 + x)\cdot 0.60}{2}\cdot 0.95\cdot 0.20 + (54.8 + x)\cdot 0.60.$$

By solving this equation, it is found that $x = 28.6$ kmoles.

It immediately follows from this, on the basis of the quantities already calculated, that:

- the number of kmoles of HCl participating in the process is: 83.4,
- the number of kmoles converted into chlorine (60% conversion) is: 50.04,
- the number of kmoles of water and chlorine produced is: 25.02,
- the number of kmoles of water which is removed at the low temperature condensation stage is: 23.77,
- the number of kmoles of HCl dissolved in the water condensed out is: 4.75.

---

*This and the following decimal numbers are the various percentages taken from the text of the problem and divided by one hundred.

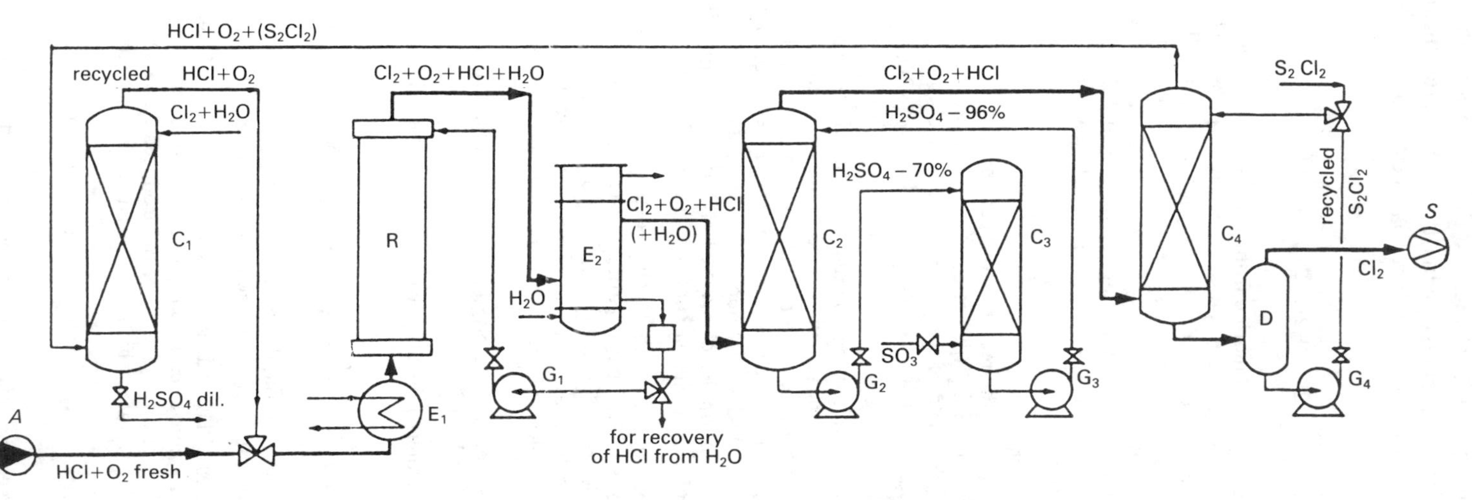

Fig. 3.5—Normalized scheme for the maximum recovery of chlorine from hydrogen chloride. The flow of the mainstream is indicated by heavy black lines. The course of the process can be wholly followed on the basis of the text of the problem, by observing the substances in circulation, the type of plant apparatus, and the auxiliary equipment shown in the scheme.

If y is the number of kmoles of fresh oxygen feed and z is the number of kmoles of recycled oxygen, then, allowing for the fact that the feedstock is supplied in such a way that there is a 2:1 ratio of HCl to O_2, the following equation results:

$$y + z = \frac{83.4}{2} = 41.7$$

which can be solved by taking account of the fact that the amount of fresh oxygen feedstock consumed in the reaction is equal (in kmoles) to half of the chlorine produced, that is, $y = 25.02/2 = 12.51$.

The amount of recycled oxygen, in kmoles, is therefore: $41.7 - 12.51 = 29.19$.

All the remaining questions posed in the problem can now be answered on the basis of the values which are now available:

- the composition of the feedstock is:

$$125.1 \text{ kmoles total} \begin{cases} 83.4 \text{ kmoles HCl} & \begin{cases} 54.8 \text{ kmoles fresh HCl} \\ 28.6 \text{ kmoles recycled HCl} \end{cases} \\ 41.7 \text{ kmoles } O_2 & \begin{cases} 12.51 \text{ kmoles fresh } O_2 \\ 29.19 \text{ kmoles recycled } O_2 \end{cases} \end{cases}$$

- the percentage recycling ratio of HCl is:

$$R_{HCl} = \frac{HCl_{recycled}}{HCl_{total}} \cdot 100 = \frac{28.6}{83.4} \cdot 100 = 34.29,$$

- the percentage recycling ratio of O_2 is:

$$R_{O_2} = \frac{O_{2\,recycled}}{O_{2\,total}} \cdot 100 = \frac{29.19}{41.7} \cdot 100 = 70.0,$$

- the hourly production of Cl_2 ($M_r = 70.92$) in kg is:

$$70.92 \cdot 25.02 = 1774.4,$$

- the percentage global yield required is calculated, using equation (3.5), as:

$$\eta\% = \frac{50.04}{83.4} \cdot 100 = 60.0\%,$$

- the amount (w) of sulphuric acid required for drying is calculated by solving the balance equation:

$$w \cdot 0.96 = [(25.02 - 23.77) \cdot 18 + w] \cdot 0.7,$$

where $w \cdot 0.96$ is the amount of pure sulphuric acid used, $(25.02 - 23.77)$ is the number of kmoles of water not removed at the condensation stage, 18 is the relative molecular mass of water and w is the mass of sulphuric acid used which, when added to the mass of the uncondensed water, yields the total final mass of 70% sulphuric acid. It is therefore clear that the second term also represents the amount of pure sulphuric acid used.

The value of w in kg is: $w = 86.5$.

Figure 3.6 shows the flow scheme for the process being studied, together with the quantities of materials constituting the mass balance. Where no units are given, the quantities are in kmole.

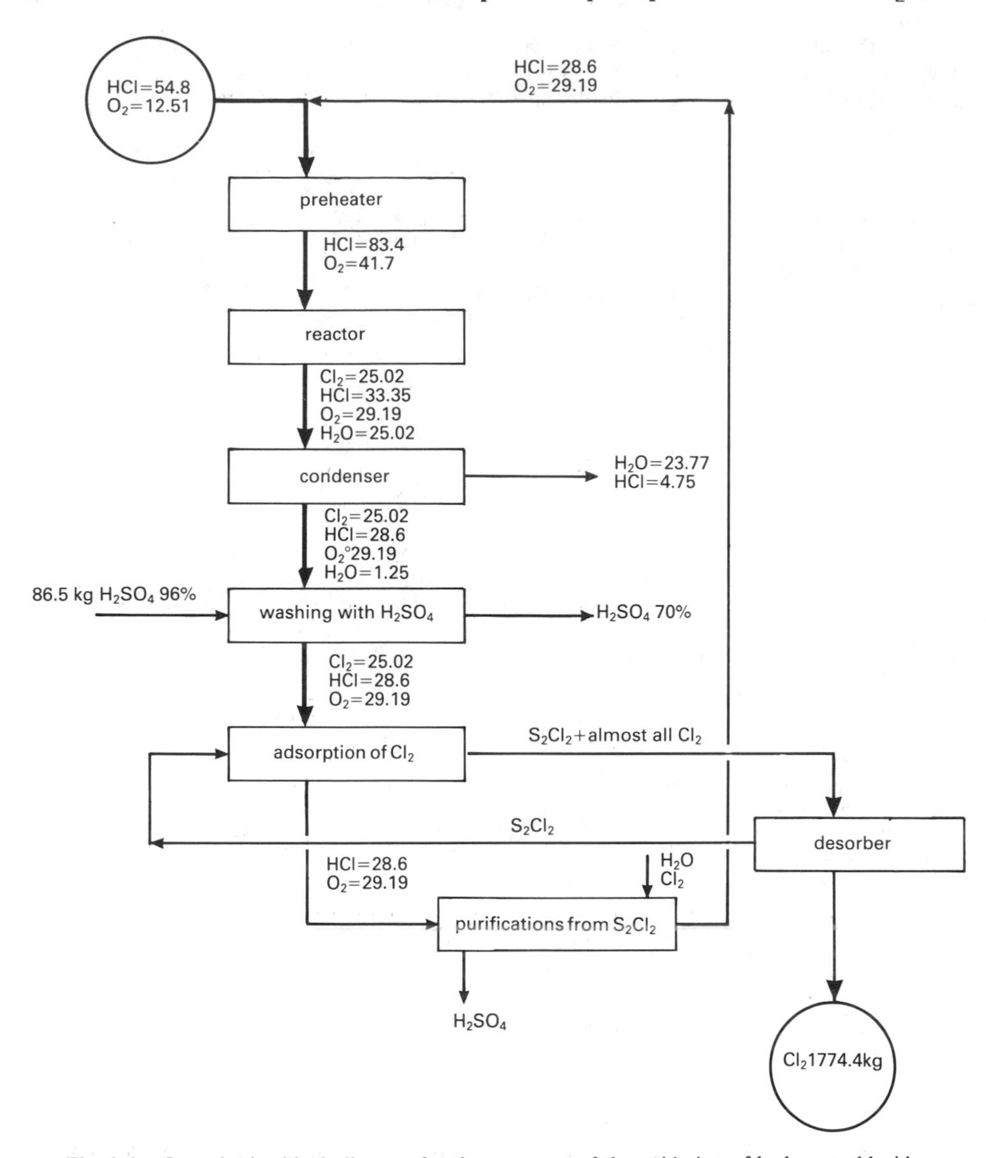

Fig. 3.6—Quantitative block diagram for the processes of the oxidation of hydrogen chloride to chlorine, together with the data and results derived during the preceding management exercise.

3.8 THE ACCUMULATION OF INERT SUBSTANCES AND THE CONTAMINATION OF REAGENTS IN RECYCLING

When unreacted materials are recycled, inert substances which have remained unchanged at the end of a cycle are added to the fresh feedstock which likewise already contains some of them. As the process is continued, the inert substances contaminate the reactants to an ever increasing extent, with the result that the yields from the process become progressively lower. This fall-off in the yields of the process is in line with the fact that, as the number of inert molecules present in the system progressively increases, they replace the molecules which can react in the collisions implied by the reaction mechanism, thereby leading to an ever increasing number of collisions which are ineffective.

For instance, if oxygen contaminated with nitrogen or, worse still, contaminated with air, were used as the oxidant in the industrial recovery of hydrogen chloride illustrated in the preceding section, recycling would rapidly lead to an unacceptable fall-off in the degree of oxidation of HCl.

The conventional remedy to the drawback of excessive contamination of reactants by the accumulation of inert substances is the removal ('escape', in the case where a gas is concerned) of a fraction of the substances destined to be recycled. This operation is known as *purging*, and is carried out at a suitable stage in the functioning of the plant.

Thus, in the conventional process for the recovery of chlorine from HCl, which is still taken as the example, purging is carried out, where requested, after purification of the recycling gas from S_2Cl_2.
Calculations involving purging are treated in a subsequent numerical exercise.

The best known examples of purging operations, other than those used in recycling to increase yields, are the re-use of solvents in distillation or extractive crystallization processes and those concerned with the conditioning of the electrolytic solutions used in galvanic baths for the electrochemical refining of metals. In these cases purging consists of substitution of a fraction of the contaminated solutions by pure solvents or new solutions.

In every case the purged systems can be subjected to special treatments to remove the impurities either by extraction of the inert components or recovering the useful ingredients from them. On the other hand, when these purification processes are too costly, the purged systems are discarded as a whole after suitable decontamination in order to protect the environment.

The recycles in the re-oxidation of HCl to Cl_2 are subjected to purging with recovery of hydrogen chloride by repeated thorough washing with water, that which remains being discharged subsequently into the atmosphere.
In the purging carried out in the recycling stages in the synthesis of ammonia, it tends, instead, to be the inert substances (the noble gases) which are extracted.
Finally, the inert substances removed by the various cyclic treatment operations for the softening and decontamination of waters are discharged directly into the atmosphere.

3.9 EFFECT OF THE ADDITION OF INERT GASES ON GAS PHASE EQUILIBRIA

When gaseous substances, contaminated with inert gases, are recycled in systems where there are equilibria between the gases, these equilibria, at a constant pressure, may or may not be affected by the addition of the inert gases, according to the type of stoicheiometry involved in the equilibrium. On the other hand, there is never any effect on an equilibrium at constant volume.

It is desirable to be more precise about these statements, as they refer to situations which may be encountered in industrial chemical processes which may or may not permit corrective steps to be taken on the relevant systems.

To study the behaviour of gas phase equilibria in the case when inert gases have been added, it is necessary to start again with the expression for the equilibrium constant in terms of the partial pressures (K_p) as a function of the equilibrium constant in terms of the numbers of moles of the various gases involved (K_n). This is:

$$K_p = K_n \left(\frac{P}{n_{tot}} \right)^{\sum v_i} \tag{3.6}$$

where P is the total pressure, n_{tot} is the total number of moles of the gases (including inert gases) making up the system, and ($\sum v_i$) is the algebraic sum of the reaction coefficients when the coefficients of the products are taken as being positive and the coefficients of the reactants are taken as being negative ($\sum v_i = +c+d-a-b$).

It may be noted that the numbers of moles of the active gaseous components in the system appear in the expression for K_n:

$$K_n = \frac{n_C^c \cdot n_D^d}{n_A^a \cdot n_B^b},$$

while the numbers of moles of the inert gases are also included in n_{tot}.

It is then noted that K_p always remains constant at a constant temperature.

Therefore, when inert gases are added to the system at constant pressure, it is K_n which must vary as a function of the reaction characteristics, expressed by the exponent ($\sum v_i$) in order to compensate for the change which n_{tot} undergoes as the result of such an addition.

To be precise:

- if $\sum v_i = 0$, that is, if the reaction does not change the number of moles in passing from the reactants to the products, the addition of inert gases has no effect on the equilibrium position since, in this case, $(P/n_{tot})^0 = 1$ and K_n remains constant like K_p;
- if $\sum v_i > 0$, that is, when the reaction is accompanied by an increase in the number of moles, the addition of inert gases leads to a decrease (an increase in the denominator) of the exponential ratio $(P/n_{tot})^{\sum v_i}$ and, according to equation (3.6), since K_p remains constant, it must increase K_n by driving the reaction to the right to yield a greater amount of reaction products;
- if $\sum v_i < 0$, K_n must decrease, according to equation (3.6), since the addition of inert gases increases both n_{tot} and the exponential ratio $(P/n_{tot})^{\sum v_i}$: that is, the reaction goes backwards to the left with a decrease in the yields.

Examples of the first of the reaction types mentioned above are the synthesis of nitric oxide from the elements, the reaction for the conversion of water gas: $CO + H_2O \rightleftharpoons CO_2 + H_2$, and the condensation of nitrogen dioxide into dinitrogen tetroxide.

Examples of the second type of reaction are, in general, all dissociation reactions, the reforming of methane $CH_4 + H_2O \rightleftharpoons CO + 3H_2$ and the catalytic oxidation of H_2S with SO_2 which leads to the formation of gaseous water and sulphur vapour.

Finally: syntheses in general (for example of ammonia, of SO_3 from SO_2), the oxidative process for the recovery of chlorine from HCl, the production of CS_2 and H_2S from methane and sulphur, and many other reactions forming the basis of industrial processes, are examples of reactions of the third type.

It is concluded that, at *constant pressure*, one may neglect the effect of the addition of inert gases either to systems with reactions where there is no change in the number of moles or to systems with reactions in which there is an increase in the number of moles, as, in such cases, there is no reduction in the yields. However, inert substances must be purged from reaction mixtures in which there is a decrease in the number of moles if reduced yields are to be avoided.

At *constant volume*, as there is no change in the ratio P/n_{tot} because no simultaneous, proportional increase occurs either in the numerator or the denominator, such an addition has no effect of any kind on the system of interest.

3.10 EXAMPLES OF CALCULATIONS ON PURGING OPERATIONS

In the preceding sections it has been seen that purging is a process of considerable importance in industrial chemical process operations.

On the other hand, purging needs to be used with caution in order not to lose too much expensive or useful substances while not foregoing the benefits of recycling in general.

The following problems will provide an introduction to the times and modes of purging operations with respect to the results which are expected from them.

Problem 3.1 Purging of a solvent contaminated during an extractive distillation process.

In an extractive distillation process the amount of solvent in circulation is 1000 kg/h. Calculate the weight of solvent to be purged every hour, when it is replaced by fresh solvent, assuming that the solvent is contaminated to an extent of 0.15% every hour and the tolerance limit for contamination is 3%.

The calculation of the amount of contaminated solvent to be purged is carried out by acknowledging that the amount of impurities absorbed by the solvent is equal to the amount of impurities extracted with the purge. Moreover, it is assumed that the purge is carried out up to an impurity concentration at the tolerance limit.

Therefore, indicating the amount of solvent to be purged each hour by x, the equality

$$1000 \cdot 0.0015 = x \cdot 0.03$$

must be solved, whereupon it is found that

$$x = 50 \text{ kg/h}.$$

Therefore, every hour, 50 kg of contaminated solvent must be replaced by the same amount of fresh solvent.

Problem 3.2 Calculation based on the principle of the oxidation of hydrogen chloride to chlorine, using a purging operation.

In a process of the above type the system is fed with oxygen containing 1.8% of nitrogen and the maximum tolerable nitrogen/oxygen ratio in the gases which emerge from the reactor is 40%. The feed of 2000 kg/h of HCl is supplied, using an HCl/O_2 ratio of 2:1, and the degree of conversion achieved is 60%. The water passed down which is condensed to an extent of 95% carries 20% of the weight of the HCl with it. Calculate:

- *the amounts of oxygen, nitrogen, and hydrogen chloride purged;*
- *the reflux ratio in the process with respect to oxygen and hydrogen chloride;*
- *the amount of chlorine produced per hour;*
- *the global percentage yield of chlorine with respect to hydrogen chloride.*

The purging must be such as to guarantee that the amount of nitrogen emerging from the process is equal to the amount which enters.

Let us assume that 1 kmole of HCl (fresh + recycled) participates in the process. Then,

- the number of kmoles of HCl which react is: 0.6,
- the number of kmoles washed out by the water (which, in kmoles, is half of the HCl which has reacted) is: $\frac{1}{2} \cdot 0.6 \cdot 0.95 \cdot 0.2 = 0.057$,
- the number of kmoles of HCl present in the purging gases is: $1 - (0.6 + 0.057) = 0.343$,

* the number of kmoles of O_2 (fresh + recycled) participating in the process is: 0.5 (it is recalled that the HCl/O_2 feed ratio is 2:1),
* the number of kmoles of O_2 reacted is: $0.6/4 = 0.15$,
* the number of kmoles of O_2 present in the purging gases is: $0.5 - 0.15 = 0.35$.

Indicating the amounts of oxygen, nitrogen, and hydrogen chloride present in the purging gases by Q_{O_2}, Q_{N_2} and Q_{HCl}, one then has:

* from the data given: $Q_{N_2}/Q_{O_2} = 0.40$,
* from what has been deduced above: $Q_{HCl}/Q_{O_2} = 0.343/0.35 = 0.98$.

The quantity Q_{O_2} can be expressed as a function of the number of kmoles of O_2 in the feed on the basis of the requirement stated at the end of the formulation of the problem, that is,

$$\underset{\text{nitrogen in feed}}{y \cdot 0.018} = \underset{\text{nitrogen purged}}{Q_{O_2} \cdot 0.40}$$

It follows from this that:

$$Q_{O_2} = \frac{y \cdot 0.018}{0.40}$$

and, consequently that the amount of HCl present in the purging gases is given by the expression:

$$H_{HCl} = \frac{y \cdot 0.018}{0.40} \cdot 0.98.$$

The number of kmoles of fresh HCl fed into the process is $2000/36.47 = 54.8$ and if x, y and z indicate, respectively, the number of kmoles of recycled HCl, the number of kmoles of O_2 fed in and the number of kmoles of recycled O_2, a material balance can be set up with respect to HCl:

$$\underset{\text{HCl total}}{54.8 + x} = \underset{\text{HCl recycled}}{x} + \underset{\text{HCl reacted}}{(54.8 + x) \cdot 0.6} + \underset{\text{HCl dissolved in the water}}{\frac{(54.8 + x) \cdot 0.95 \cdot 0.6 \cdot 0.2}{2}} + \underset{\text{HCl purged}}{y \frac{0.018 \cdot 0.98}{0.4}}$$

and one obtains a material balance with respect to O_2:

$$\underset{\text{oxygen fed in}}{y} = \underset{\text{oxygen reacted}}{\frac{(54.8 + x) \cdot 0.6}{4}} + \underset{\text{oxygen purged}}{y \cdot \frac{0.018}{0.4}}$$

Taking account of the HCl/O_2 feed ratio (2:1), one likewise arrives at the balance equation:

$$\underset{\text{total oxygen}}{y + z} = \underset{\text{one half total HCl}}{\frac{54.8 + x}{2}}$$

By solving the three balance equations which have just been formulated, it is found that:

$$x = 27.4 \text{ kmoles}; \qquad y = 12.35 \text{ kmoles}; \qquad z = 28.75 \text{ kmoles}.$$

The amount of O_2 purged is therefore

$$Q_{O_2} = 12.35 \cdot \frac{0.018}{0.4} = 0.56 \text{ kmoles}$$

for HCl the amount purged is:

$$0.56 \cdot 0.98 = 0.55 \text{ kmoles}$$

and for nitrogen the amount purged is:

$$0.56 \cdot 0.4 = 0.224 \text{ kmoles}.$$

The total amount purged is therefore:

$$0.56 + 0.55 + 0.224 = 1.334 \text{ kmoles}.$$

The oxygen reflux ratio is:

$$\frac{\text{oxygen refluxed}}{\text{oxygen input}} \cdot 100 = \frac{28.75}{12.35 + 28.75} \cdot 100 = 69.9$$

and the HCl reflux ratio is:

$$\frac{\text{HCl refluxed}}{\text{HCl input}} \cdot 100 = \frac{27.4}{27.4 + 54.8} \cdot 100 = 33.3.$$

The amount of chlorine ($M_r = 70.92$) produced per hour is:

$$82.2 \cdot 0.6 \cdot \tfrac{1}{2} \cdot 70.92 = 1749 \text{ kg}.$$

The global percentage yield is:

$$\eta\% = \frac{49.32}{82.2 + 0.55} \cdot 100 = 59.6,$$

on the basis of equation (3.5), taking account that, in calculating the total number of kmoles of HCl, the number of moles of purged HCl must also be included.

Problem 3.3 Calculation on the principle of inert gas purging in an ammonia synthesis process.

An ammonia synthesis plant is supplied with nitrogen and hydrogen in a stoicheiometric ratio. 30% of the mixture of the two gases is converted, and the resulting gaseous mixture is separated by condensation as the gases leaving the reactor, are cooled. The remaining nitrogen and hydrogen are recycled.

Given that the nitrogen feedstock contains 0.8% of argon and that the tolerance limit for argon in the overall nitrogen feedstock is 14%, calculate the recycling ratio and the fraction of the gas which must be continually purged.

The material balance for the system, under the assumption that it is fed with 100 kmoles of nitrogen and 300 kmoles of hydrogen, is presented graphically in Fig. 3.7.

By indicating the amounts of nitrogen which are recycled and purged by x and y respectively, a series of relationships can be established which are useful in obtaining expressions for the solution of the problem. More precisely:

- ∗ the total number of kmoles of nitrogen fed in is: $100 + x$;
- ∗ the number of kmoles of recycled hydrogen is: $3x$;
- ∗ the total number of kmoles of hydrogen fed in is: $300 + 3x$;
- ∗ the number of kmoles of ammonia formed is: $(100 + x) \cdot 0.3 \cdot 2$;
- ∗ the maximum number of kmoles of argon (Ar) which may be present in the total feedstock is: $0.14 \cdot (100 + x)$;
- ∗ the number of kmoles of Ar, per kmole of nitrogen, which leaves the condenser is: $14/70 = 0.2$*;
- ∗ the number of kmoles of Ar purged is: $0.2 \cdot y$.

Since the number of kmoles of Ar purged is equal to the 0.8 kmoles of Ar present in the fresh feedstock (as indicated in the text of the problem), the first equation to be solved is:

$$0.2 \cdot y = 0.8$$

whence,

$$y = 4 \text{ kmoles of } N_2 \textit{ purged}.$$

---

*Since, in fact, 30% of the nitrogen entering the system is transformed into ammonia, the argon–nitrogen ratio at the end will be 14/70 rather than 14/100 which it was at the beginning.

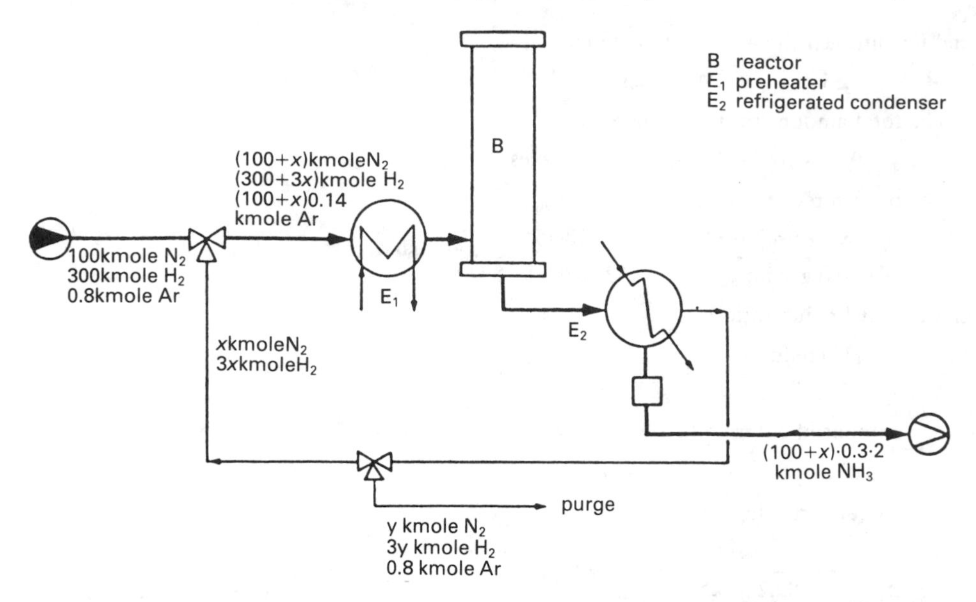

Fig. 3.7—Graphical representation of the material balance in the system relating to the calculation in Problem 3.3. The amounts of substances x and y must be calculated in order to answer the question.

The number of kmoles of hydrogen purged is then immediately calculated, by taking account the fact that the H_2/N_2 ratio is always 3/1, yielding a value of 12 kmoles.

The total amount of gas purged is therefore:

$$4 + 12 + 0.8 = 16.8 \text{ kmoles}.$$

The second equation to be solved for x is obtained from the nitrogen balance:

$$\underset{\text{nitrogen fed in}}{100} = \underset{\text{nitrogen which is converted into } NH_3}{(100 + x)\cdot 0.3} + \underset{\text{nitrogen purged}}{4}$$

By solving this equation it is found that $x = 220$ kmoles of recycled nitrogen.

The recycling ratio for nitrogen, which is the same as the recycling ratio for hydrogen by virtue of the fact that the two gases are always fed in stoicheiometric proportions, is given by:

$$\frac{220}{100 + 220} = 0.67.$$

The amount of recycled argon is equal to:

$$220 \cdot 0.2 = 44 \text{ kmoles},$$

and the amount of recycled hydrogen is:

$$3 \cdot 220 = 660 \text{ kmoles}.$$

Finally, the purging fraction is equal to:

$$\frac{16.8}{220 + 44 + 660} = 0.017.$$

Given the smallness of the purge (1.7%), it may not always be carried out, since the losses which occur in the plant may be sufficient to maintain an argon concentration which is below the tolerance limit.

4

Industrial chemical process and industrial chemical plant technology

4.1 THE INDUSTRIAL CHEMICAL PROCESS AND INDUSTRIAL CHEMICAL PLANT TECHNOLOGY

A 'chemical process' is composed of a combination of *chemical reactions* such as syntheses, oxidation, hydration, and salt formation, for example, and *operations based on physical phenomena*, such as evaporation, crystallization, distillation, and extraction, which lead to the transformation of raw materials (such as petroleum or coal, air, and water) into finished products (such as nitric acid and nitrogenous fertilizers).

'Process technology' is the complex set of characteristics of the apparatus in which a chemical process is carried out and which determine its specification. More precisely, a chemical process is said to be on a 'laboratory scale' if the apparatus, for the greater part, is made out of glass with reduced dimensions and the process is carried out under rather mild conditions. If, however, the operational conditions are rigorous and the apparatus is constructed from metals or ceramics with dimensions such that quantities of substances of the order of tons per day can be treated (barring exceptions),* then one is dealing with what is defined as an *industrial chemical process*.

The assemblage of apparatus by means of which an industrial chemical process is realized in an *industrial chemical plant* and the technology used in the setting up of conditions of such a plant, in a special way, is the result of the enterprise.

An industrial chemical plant which is intended to carry out a certain industrial chemical process comprises all the apparatus necessary to satisfy requirements regarding temperature, pressure, filtration, drying, etc. which may be requested in order to carry out the *basic reactions* or *secondary reactions*.

---

*It is the whole collection of plant parameters and, above all, whether the experimental application is provisional or definitive, which decides whether a process is an industrial process, a laboratory process, or a pilot process. A single characteristic says little. Among other things, an industrial process in the 'fine' chemical sector may produce amounts of the order of kilograms or less per day.

'Basic' reactions are those involving the transformation of prime materials into finished products, that is, into the products which the plant was intended to produce, while 'secondary' reactions are those which lead to the formation of side products and means of recovering and eliminating them.

4.2 CHOICE OF AN INDUSTRIAL CHEMICAL PRODUCTION SYSTEM AND THE PROCESSES WHICH SERVE IT

The factors which lead to the setting up of a certain industrial chemical manufacturing plant are many and varied.

The fundamental factor is the economic advantage to be gained from the choice made with regard to the availability, cost, and logistic facilities for the provision of the required raw materials as well as with regard to the existence of markets which are capable of ensuring an adequate sales volume for the finished products at profitable prices.

Other factors which, in certain cases, may even determine the choice are, inter alia, the compatibility of the projected production with the conditions appertaining in the territory in which it is set up and the opportunities for integrating it with other activities of the district in which it is located. Such activities are not necessarily chemical but may be mechanical engineering, agricultural and food industries, manufacturing industries, and so on.

By the statement 'compatibility with the conditions appertaining in the territory' one understands all the local conditions which in the long run may impose burdens on the manufacture of the product such as the quality of water required with regard to local water resources, the danger of liquid or gaseous effluents in densely populated areas where ecological problems may arise, and the legislative requirements which decree safety and pollution limits which must be guaranteed locally.

It is, for example, this incompatibility with the conditions appertaining within territories in advanced and ecologically conscious countries with advanced environmental protection schemes in which they institutionally participate which urges, above all, some multinational companies to set up certain production facilities without adequate guarantees of being able to cope with either the dangers which are inherent in their activities or operational anomalies in countries where, to the contrary, there is a dearth of satisfactory legislative provisions regarding environmental protection, and this makes such operations formally compatible with the laws appertaining in such a territory.

The present discourse may be considered to apply to the Seveso event, which was brought about by 'delegation' from a company in a country which rigorously upholds proper environmental protection safeguards to another country with legislative shortcomings in matters of environmental protection, or, equally, to the Bhopal tragedy in India which was incomparably more serious in every respect.

Closely connected with the decision to start up an industrial chemical process, at the point when the reasons for the choice are frequently identified, is the adoption of a process (or processes) which must be used in its implementation.

Also, in the case of the processes at the foundation of the choice, there are the economic considerations which are on a level with the evaluation of their technical suitability with regard to the characteristics of the plant apparatus (the 'process technology').

However, the best technical suitability is not always the decisive element in choosing how to carry out the processes. This is the case, above all, when the production includes several processes, because the motives of an agreed company policy of coordinating the various processes, making them served by technologies which are integrated to the greatest extent, may take the upper hand. Also, the possession of an exhaustive amount of reliable information on the operating conditions, on the technologies used in the construction and regulation of the apparatus and control instruments, of the structural and maintenance requirements, and the protection of the catalyst used, and so on,* may be yet more determining than the optimization of the efficiency from the point of view of the technical profile when selecting a process.

An example of this is furnished, in the case of companies who have spent considerable effort in the overhauling of their own processes for the synthesis of ammonia, by the adoption of some of these processes even if they turn out to be technically less capable than more recent developments or subsequently come to the forefront for their acknowledged technical worth.

The choice of a process may depend on:

- the raw material to be exploited in it,
- the physicochemical conditions to be observed in it,
- whether the productive activity is to be continuous or not.

A case where the *raw material* to be used in the process is of interest in the discussion is that of the production of acetaldehyde, as it is possible to start out from ethylene, acetylene, or ethanol. Choices of this kind are mainly made on the basis of the availability and the cost of the raw material.

There is a choice to be made regarding the *physical and chemical conditions* to be employed in the process, for example, when it is possible to produce a polymer, starting out from the same monomer by different methods. In this and in similar cases, the process to be employed is decided upon, firstly, by paying regard to the characteristics which the product must possess and, secondly, on the basis of production costs, among which plant ammortization costs play a primary role.

Finally, it happens extremely frequently that a choice must be made between *continuous and discontinuous processes* for the productive activity. The criteria on which a preference is based which arise in this case include the quantities which are to be produced and whether there is a continuous demand or just a periodical demand for the resulting products. Other factors to be considered in making the choice between a continuous process and a discontinuous process are, for example, the reduction in the plant and automatic control costs arising from discontinuous processes and the reduction in manual labour costs and the volume of the apparatus, for the same production levels, which are attainable when using continuous processes.

4.3 PLANT REALIZATION

Once a choice has been made regarding an industrial process, it has to be implemented in the form of a plant.

This operation is relatively simple if the process is of an already known type, that is, it is a copy of plants which have already been constructed. In such a case it is only a matter of making any necessary change in scale in relation to the changed

*The entirety of the reliable knowledge concerning every facet of the construction and functioning of such installations, leading to the highest level of competence in their realization, is described by the term 'process and plant know-how'.

dimensions of the new plant with respect to those which have already been put into operation and taking account of any technological improvements to be made to the plant to correct anomalies which have been encountered in earlier plants or to bring about an absolute improvement in the services.

True innovations may be requested in the planning whenever a new plant comes on stream in a productive unit which is notably or completely different in terms of its location and/or structural features and/or management from those which have shared similar plant which has been constructed earlier. In such a case, in fact, new and different sources of energy, water, steam, and means of utilizing intermediate products, etc. could also suggest that the whole plant design should be radically changed such that a fundamentally new project must be envisaged. The implementation of plant which involves some new planning project, that is, in the absence of the know-how concerning the plant to be constructed and run, always carries a greater burden of responsibility and is, of course, the most laborious. This is demonstrated by the fact that the realization of such an installation is always the work of a highly qualified technical staff, and we shall therefore confine the discussion to a delineation of the contents of the most important operative stages.

In this case there is the need, as a rule:

- to understand the chemistry of the process by means of theoretical studies supported by laboratory experimentation in order to anticipate the stages of the process and, as a rule, the operations based on physical phenomena which the projected plant must handle;
- to confirm and eventually integrate or modify the forecasts which have been made experimentally on the basis of pilot scale studies. Here, it is also necessary to review the quality, shape, and dimensions of the apparatus, the constructional materials, and the operational conditions which allow, in the final analysis, the minimization of the costs;
- to pass, if it is necessary, to 'semi-scale' industrial production to confirm or suitably modify the projected executive functions of the plant while putting into practice process technology which utilizes the prototypes of the real apparatus to be used in the industrial plant;
- to put the technological and constructional planning of the plant into hand: the former, before the definitive refining of the flow schemes and quantities handled, requires the final fixing of variables such as the type, dimensions, and specific characteristics of all the apparatus and process controls, while the latter has as its objective the dimensioning of the area and height of the plant, of the fabrications, and of the 'services', such as tubing and other transport mechanisms, separation, milling, mixing, and so on.

Finally, when the costs of the plant and its operation have been defined and the decision has been taken to construct the plant, it passes to the stage of putting the work in hand which is followed by the set-up of the plant and the making of trials.

4.4 CRITERIA RELATING TO THE DIMENSIONING OF INSTALLATIONS

The choice of the dimensions of an industrial chemical plant is fundamentally made on the basis of the forecast sales of the product over the years for which the plant is to operate (a decade). The criterion to be strictly observed in connection with this is that of minimizing under-utilization of the plant, because operation at reduced capacity means that the capital invested is not bearing fruit.

Technologically, one should not immediately exhaust the full productive capacity of a plant, because one will benefit from a more favourable situation when, in the first years of operation of the plant, there is a growth in the volume of sales, and thereby a substantial correspondence with the full productive capacity of the plant.

It is necessary to proceed with care in selecting the dimensions of a plant in the sense that, if the forecasts concerning the sales of the product are uncertain or it is not possible to predict them absolutely, one must foresee, first of all in the defined area of the market which the plant serves, the possibility of an elastic enlargement in the time for which the plant operates. Another sensible decision is that of installing the basic pieces of apparatus (reactors, evaporators, distillation and extraction columns, various types of furnaces and ovens) with the greatest forseeable capacity, and the secondary pieces of apparatus (such as filters, centrifuges, mills, and storage tanks) with a capacity which is lower than the maximum forecast, in a plant context when there is spare space for the acquisition of similar auxiliary secondary pieces of apparatus or replacements of greater capacity or productivity.

The solution which has just been put forward is particularly advantageous when the plant functioning at full capacity requires the use of primary pieces of apparatus with the smallest dimensions which are technically and economically acceptable, where, in such a case, there is no opportunity of reducing the capacity of this apparatus.

Finally, it must be kept in mind that the analyses both of the costs of a variable productive capacity plant which has been put into operation ('erected'), defined as the *physical costs of the plant* and of the costs associated with subsequent enlargements also play an important role in the choice of the various possible solutions regarding the size of installations.

4.5 THE BASIC PLANT APPARATUS. ELEMENTS OF REACTOR CHEMISTRY

Reactors: types, functions, and criteria of choice

Reactors are the apparatus in which reactions take place in industrial chemical plants. They are therefore the plant structures which necessitate or justify all the other equipment on the site. They are thus justifiably looked upon as the 'heart' of chemical plants.

There are many types of reactors, because there are many types of reactions, of conditions under which these reactions proceed, and of reagent and reaction product characteristics.

However, two fundamental functions are common to all reactors:

- that of providing optimal contact between the components taking part in the reactions (reactants and catalysts) which are frequently present in reactors in heterogeneous phases;
- that of ensuring that the reaction is carried out under the required physical conditions (temperature, pressure, and contact time).

The choice of reactor type is made on the basis of a guarantee that it will respond to all of the fundamental functions which a reactor offers (the technical factor) while always, on the other hand, paying comparative regard to the economic factor.

Classification of reactors

The criteria on the basis of which reactors are classified are various, but basically:

- in relation to the manner in which the process is to be carried out,
- as a consequence of the heat exchange system which is adopted,
- in accordance with the physical characteristics of the substances to be treated.

The manner in which the process is carried out gives rise to the classes (A):

(a) discontinuous reactors,
(a′) continuous reactors,
(a″) semicontinuous reactors.

The heat exchange system employed leads to class (B):

(b) adiabatic reactors,
(b′) non-adiabatic reactors.

The physical characteristics of the substances in question lead to class (C):

(c) reactors for homogeneous reactions,
(c′) reactors for heterogeneous reactions.

Other classifications of reactors are based on the modes of feeding in and/or of the systematic operation and/or the discharge of the materials participating in the reaction which, naturally, includes catalysts as well as solvents and diluents. However, no matter what class they belong to, reactors can be adapted for different purposes by means of suitable constructional modifications.

Review and examples of reactors belonging to the various classes

Class A reactors

Discontinuous reactors can be initially charged with the materials participating in the reaction which remain in the reactor for the period which is necessary for the programmed extent of reaction to be attained. The reaction can then be interrupted and, finally, the reaction products can be discharged.

The very commonly encountered ‘vats’ in their different forms, some of which are shown in Fig. 4.1, belong to the class of discontinuous reactors. Here, one is concerned with vessels of cylindrical profile which frequently have a hemispherical base and open or closed plates at the other extremity.

In these reactors, mixing of the charge is accomplished by means of stirring and external recycling mechanisms. Finally, it should be noted that closed vats built to withstand high pressures are known as autoclaves. Such systems are always equipped with adequate safety devices.

Continuous reactors, some typical examples of which are shown in Fig. 4.2, are subdivided into ‘homogeneous’ and ‘heterogeneous’ reactors according to whether the reactions which take place in them involve one or more phases, including catalysts.

The common characteristic of continuous reactors is that of being continuously

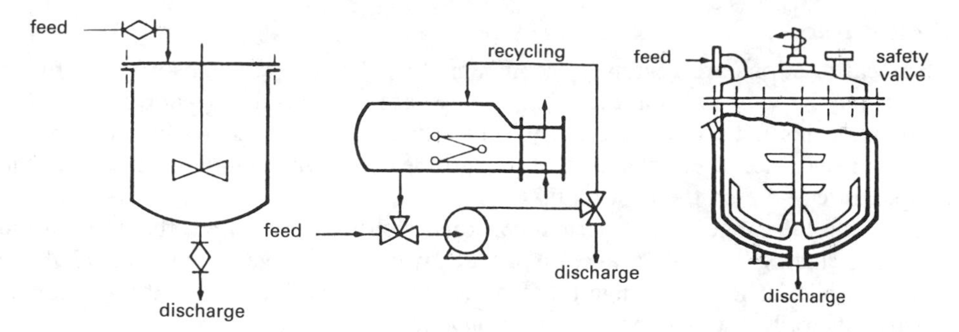

Fig. 4.1—Discontinuous reactors. From the left: closed vat with stirrer, vat with external recycling ('kettle'), autoclave with a stirrer and an external heat exchange jacket.

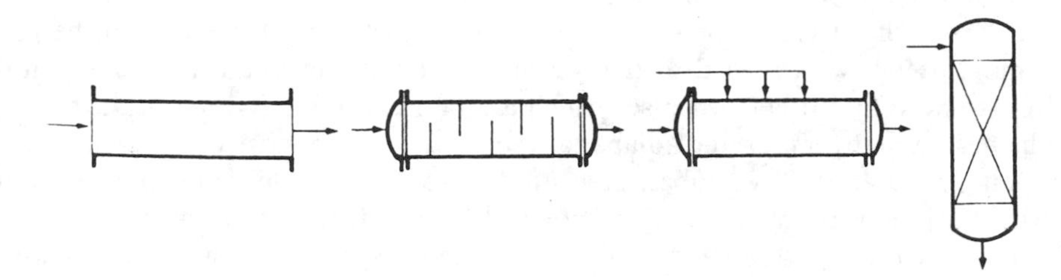

Fig. 4.2—Continuous reactors. From the left: tubular piston-type flow reactor, tubular diaphragm reactor, multi-inlet tubular reactor, tower reactor with filling. Here, in every case, one is concerned with reactors with a high ratio of length to diameter, the optimization of which enables one both to balance the reaction time with that taken by the system of reagents to flow through the reactor and to control the ambient temperature in the best way for kinetic purposes while avoiding denaturing and destructive reactions.

charged with reactants as the reaction products are discharged. This takes place without a bed in the case of homogeneous reactors and with a fixed or mobile ('fluid') bed in heterogeneous reactors.

There are many versions of continuous reactors, but these may be subdivided into *vats* with mixing or recycling, *tubular* reactors with a piston-like flow or with multiple inlets for the reagents, and *towers* which may be empty, fitted with plates or diaphragms, or equipped with a filling. Figure 4.2 schematically illustrates many of these variants.

Semicontinuous reactors are characterized by a continuous inflow of materials but a continuous outflow of certain reagents only. Among the reagents which are most frequently fed into semicontinuous reactors are hydrogen and oxygen when they are employed in reduction and oxidation reactions respectively. Water, in the form of steam, and hydrogen chloride are, on the other hand, reagents which are often continuously removed from semicontinuous reactors used for esterifications and various condensations.

From a structural point of view semicontinuous reactors are of the vat type, manufactured in such a way so as to permit the easy introduction of a reagent which is continuously fed in and the easy, uninterrupted, removal of a product.

Class B reactors

Chemical reactions are generally accompanied by the release or absorption of heat which may be very pronounced; that is, they are exothermic or endothermic. On account of this it becomes necessary, other than in those cases when the thermal effect is negligible, to equip the reactors with heat exchange systems to control the temperature at which the process takes place.

If, within the range of the thermal control which has to be used, the body of the reactor, including the walls,* takes no part in any heat exchange, then it is an *adiabatic reactor*, while, if heat exchange takes place towards the interior of the reactor or towards its walls, the reactors are *non-adiabatic*.

When adiabatic reactors are employed, thermal control is applied by means of independent external heat exchangers.

Several specimen models of the two types of reactors with the corresponding heat exchange equipment are shown in Fig. 4.3. The structural aspects and functional particulars of the two reactor systems which have just been discussed will be gone into more deeply with regard to the synthesis of ammonia, the catalytic oxidation of SO_2 to SO_3, and other processes which are of great industrial importance.

The heat exchangers serving adiabatic reactors are of the most varied types (using air, rain, finned and coated tubes, pipe-still furnaces, etc.), while the choice is more restricted in the case of the exchangers intended to operate with non-adiabatic reactors which, for the most part, are served by heat exchange systems consisting of an 'outer jacket', an 'internal coil', or 'external half-tubes'.

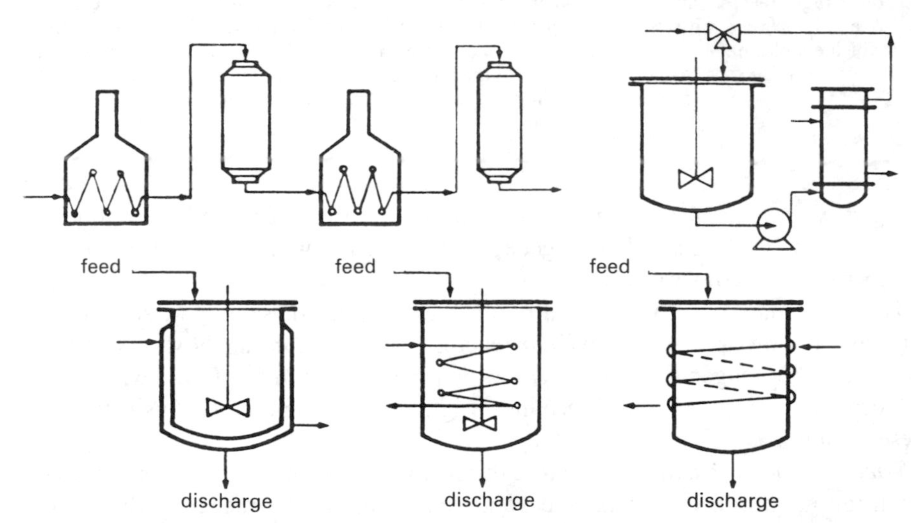

Fig. 4.3—Above. Adiabatic reactors: alternated with pipe-stills (left) and with a vertical multitube heat exchanger (right). Below. Non-adiabatic reactors: vat equipped with mixer and external jacket (left), an internal coil (çentre), and a simple vat with external half tubes flattened onto the surface of the reactor which they serve (right).

*All reactors which are not equipped in the interior or towards the walls with specific heat exchange systems, even if, it should be noted, heat is always effectively exchanged in a more or less spontaneous manner with the exterior via the structure of the reactor, are considered as adiabatic reactors to a high degree of approximation. This is because the reaction enthalpies are of an order of magnitude, or more, different from that of normal heat exchanges with the exterior.

The choice of the type of heat exchanger is essentially conditioned by the fluids employed for the heat exchange in combination with the temperature at which the system is operated.

Class C reactors

When a reaction must be carried out in a *homogeneous phase*, discontinuous reactors of class A or continuous reactors also of class A may be used, according to the type of process.

However, to handle reactions which have more than a single phase, that is, *heterogeneous reactions*, the range of reactors which can be used is the most extensive, allowing, above all, the use of semicontinuous reactors which, in principle, are concerned with heterogeneity in so far as the phases that are fed into them or continuously removed from them are different from that (or those) of the reaction. In addition, homogeneous phase reactors may also lend themselves to the fulfilment of these reactions, provided that one is dealing with dispersions (emulsions, suspensions, etc.) which have a liquid phase and in which good mixing between the phase can be conveniently guaranteed.

Moreover, almost all common reactors may be used for heterogeneous reactions, and there are many reactors with special constructional characteristics.

Firstly, there are the inclined, rotating cylindrical furnaces (or the *Lurgi* type) and the *rotating*, *shelved* furnaces (of the *Herreschoff* type). The first kind are all continuous, while the others are also continuous but in stages, enabling one to maintain the temperature at determined values during the course of the reaction. Furnaces of the *gazogene* type also exist which are characterized by a bed of reactive solid through which a gaseous reactant passes.

However, in the field of heterogeneous phase functions fixed bed and moving bed reactors stand out, nowadays, in importance. Such reactions are defined in accordance with how the catalyst in them is put into operation. The following scheme indicates how, as a rule, these reactors are constructed.

The scheme clearly indicates the different consistency (threads or granules) of the catalysts in the 'fixed bed' variant, while, in the other variant, the catalyst is present as a powder.

Several solutions with respect to the disposition of the catalyst within the reactor are shown in Fig. 4.4.

Reactors of the type listed in the schemes presented above, and, in particular, fluid bed reactors, are also used to carry out uncatalyzed heterogeneous reactions as in the roasting of pyrites.

4.6 WHERE TO SITE PLANTS

The choice as to where an industrial chemical plant should be sited is nowadays a typical problem which is resolved by means of the processing of data with computers, because so many safeguards are necessary and they are not always concordant.

Special necessities in the management of this type of plant are the consistent availability of water and electrical energy and the attitude of the authorities toward getting rid of aerial and effluent emissions apart from that which the treatment of

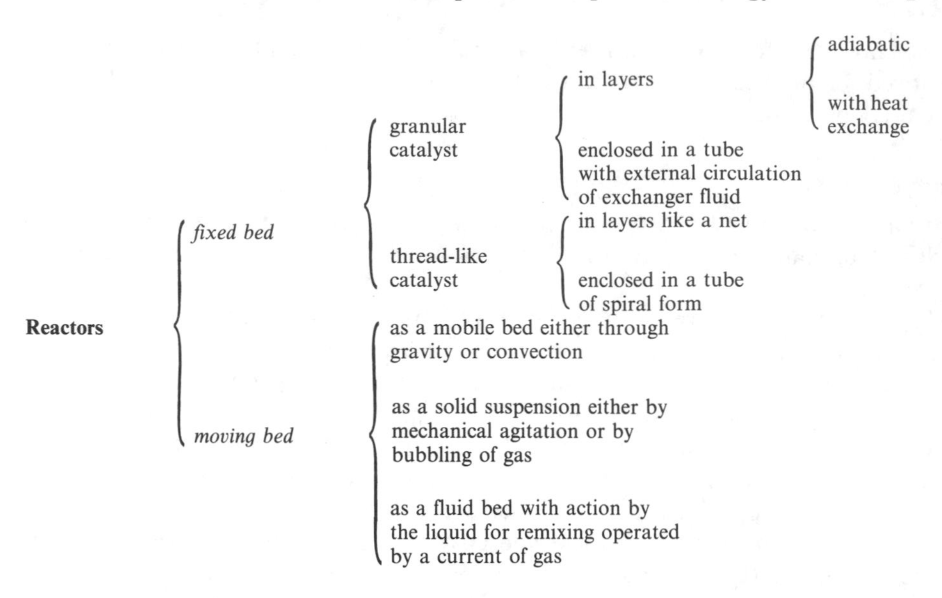

such emissions removes. As in every industrial activity, easy links are also required for the resupply of raw materials and for the distribution of the products (roads, motorways, railways, and finally, ports) besides the availability of a local labour force with a suitable degree of professionalism. The site therefore needs to be close to large centres and, moreover, centres providing social services and schools which, in general, is in contrast to the particular needs of the plant management (ready availability of water and the possibility of getting rid of various waste products and effluents).

Even from the brief analysis presented here, it can be understood that the situation is complex in itself, and is even more so because of the contradictions which arise in it. It is therefore expedient to single out the zones of preselected limited areas which, upon a preliminary examination, show themselves to be suitable in some way for the siting of the plant, and then to proceed with the collection of pertinent information as to the suitability of those zones without overlooking anything.* This provides the data which are to be computerized in order to pick out with certainty where it is expedient to site the new plant.

*Sometimes particulars which at the time appear irrelevant may, in fact, be of considerable importance. Such is the case when one is dealing with the siting of a chemical plant with gas emissions which cannot be readily deodorized. Then, in fact, the study of the prevalent wind directions, which is not very significant in its own right, can make it ecologically inconvenient to situate the plant in the vicinity of a populated centre on account of inadequate dispersion and the changing of the direction of the emissions by the winds.

Fig. 4.4—Above. Fixed bed reactors: with layers of granular catalysts and interspaced heat exchange (left); with granular catalyst enclosed in a tube and external circulation of the exchanger fluid (centre); with a thread-like catalyst arranged spirally in tubes (right). Below. Moving bed reactors: moving bed falling under gravity and lifted by convection (left), doubly agitated solid suspension (centre), and fluid bed with recovery cyclone (right). The common characteristic of such reactors is the provision of contact between the reagents and the catalyst.

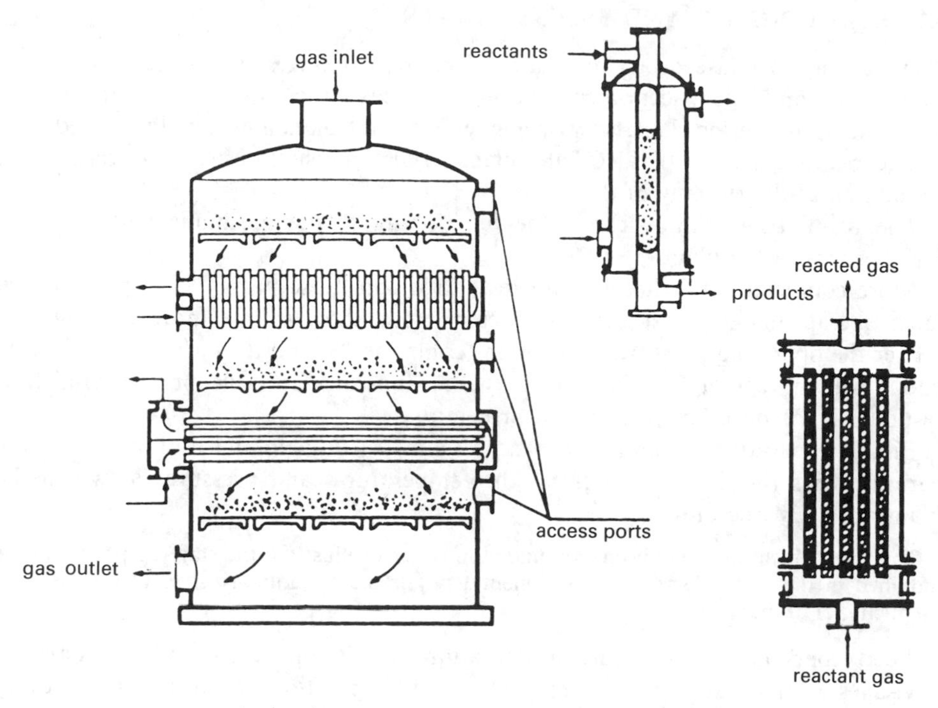
gas inlet
reactants
products
reacted gas
access ports
gas outlet
reactant gas

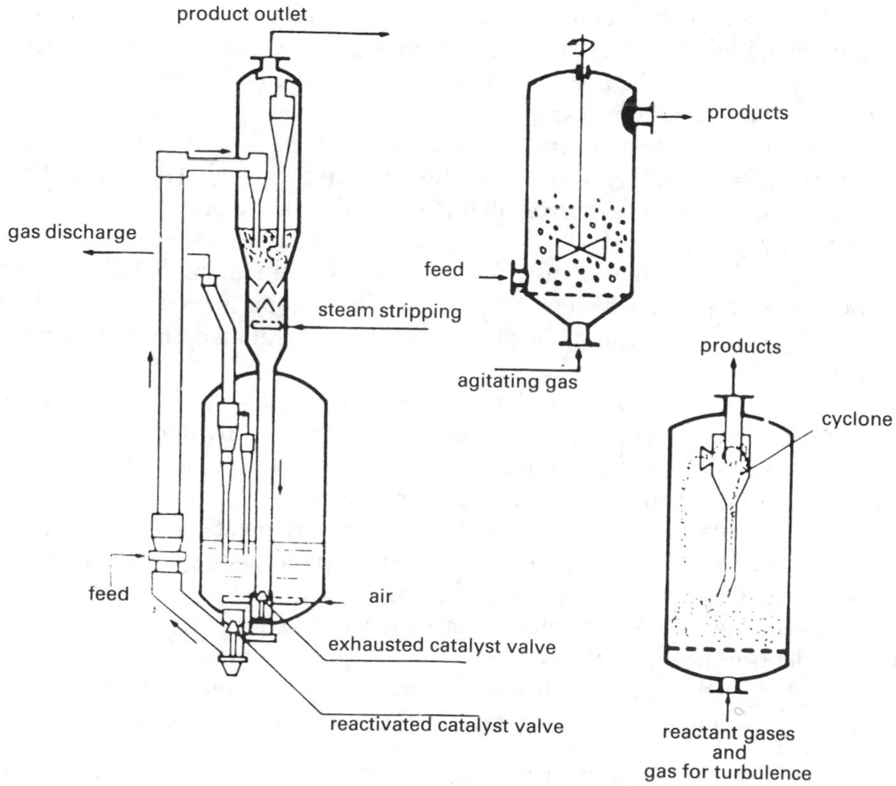
product outlet
products
gas discharge
feed
steam stripping
agitating gas
products
cyclone
feed
air
exhausted catalyst valve
reactivated catalyst valve
reactant gases
and
gas for turbulence

4.7 STARTING UP AND PLANT TRIALS

The starting-up phase of a plant takes place in two stages: first, 'plant trials' are carried out, and, then, section by section in the case of complex plants or, usually, all at once for a routine plant, the working regime of the plant is gradually realized.

The plant trials are divided into blank trials, pressure tests, and trials using simulation of the running of the plant.

The blank trials simply check the functioning of the mechanical assemblages forming part of the plant.

More compelling are the pressure tests to which those parts of the plant which must operate under pressure are subjected. Here, the tests are more rigorous the higher the operating pressure in the plant components, and they are those precisely laid down in regulations prescribed by the apposite Councils or the Ministries responsible for both health and civil protection.

Finally, operational simulation tests require starting up the flow cycles of the various fluids (liquids and gases) at the temperatures and pressures at which the plant normally operates.

The rules which have just been expounded in order to illustrate the start-up phase can be modified in the case of special environmental requirements (anticorrosive, of inertness, of asepticity, of dryness, etc.).

The second phase of the start-up operation, that is, the actual and true start-up, envisages an integral, even gradual, putting into operation of parts or the entirety of the plant until, in the section relevant to the exercise (or the whole of the plant), the operational conditions for the normal functioning of the plant have been attained and maintained. Once the operation has been regularized, the running of that part of the plant which has been started up is maintained for sufficient time in which to make a sound judgement regarding the mode of operation by which it must be characterized. The checking of yields and the characteristics of the products obtained provides a picture of the course of what is happening in the plant.

The plant acceptance trials follow the starting-up phase.

In the case of continuous plants, these trials consist of making the plant function precisely in accordance with the planned conditions for a period of several days, while, in the case of discontinuous plants, the plant operating cycles are carried out repeatedly.

In both kinds of plant, both at the start-up stage and, above all, at the plant trial stage, extremely frequent analytical controls must be carried out not only on everything that comes out of the plant (from gas exhausts, purges, and as products and by-products) but also on intermediate products.

Both in the case of continuous and discontinuous plants, the maximum and minimum levels at which the plant can operate without altering the quality of the product* are ascertained during the plant trials.

The tested plant has an *operational elasticity* which is proportional to the spread between the two measured levels.

At the plant trial stage an outline is also produced of the 'plant operating scheme' which provides guidance on how to start the plant up, the control sequences, methods

---

*It is, of course, clear that stagnation in the apparatus or passing through it too rapidly, outside certain limited routines, lead *per se* to a check on the negative consequences on the characteristics of the products.

of management, and how to close it down, as well as the routes followed by all the reagents from the collection site to the reactor and the routes of the products from the reactor to the storage facilities, and of every type of fluid present in the process.

In order that this large amount of data may be recorded in an orderly and rational manner a guide has been elaborated for this purpose which anticipates the use of conventional symbols and numbers. The reader is referred to the UNICHEM manuals where examples are provided for a useful and complete first-hand knowledge of this guide.

4.8 PLANT OPERATIONAL CONTROL

The operational control of plants, that is, ensuring that the plants are functioning as laid down at the planning stage with a view to the desiredproducts, both qualitatively and quantitatively, is carried out at the plant trial stage, and then in a continuous manner during the normal operation of the plant.

The operational control of plants therefore tends to stabilize the operating conditions, and is accomplished by means of a number of pieces of equipment which may be indicators, that is direct reading devices or registers which are respectively continuous (producing a diagram) and periodic (at suitable intervals of time). Registers, especially those producing diagrams, have been favoured for many years as they are automatic and provide a more distinct representation of the course of processes.

Operational controls are employed both to reveal anomalies in the functioning of a plant—providing an opportunity to change the operating variables in such a manner as to remove the anomalies, and with the aim of establishing the dependence between variation in the operating conditions and the characteristics of the products.

Operational control of normal plants

The scheme given also shows the manner in which small and intermediate size plants are controlled.

		Types of apparatus	*Examples*
Control	*in-line* (that is, on the plant)	physical	thermometers, photometers, polarimeters
		physicomechanical	manometers, densitometers, flow-meters
		chemicophysical	viscometers, thickness meters, conductivity meters
	in the laboratory	physical	X-rays, mass spectrometers, NMR
		chemicophysical	chromatographs, colorimeters, coulometers
		chemical	pipettes, burettes, crucibles

Operationally, the in-line controls are continuous, while those performed in the laboratory are of an intermittent nature.

Today, in-line controls are by far preferred to laboratory controls, and, if manual chemical analytical instrumentation is excluded, it is completely artificial to distinguish between the devices used for in-line control and laboratory apparatus, because the application of all of them in the two fields is very much the same.

It is then clear that, by the various instrumental terms, one understands the most diverse types of analogous devices. Thus, the term 'thermometers' is understood to include not just the classical mercury thermometers but thermocouples, thermoresistances, and pyrometers, while 'colorimeters' is understood to include, inter alia, pieces of apparatus such as fluorimeters and nephelometers.

Operational control in large plants

From the time when the plant is planned, all large plants are equipped with a room containing a centralized display panel which has the object of carrying out 80–90% of the plant measurements. It is only for the solution of specific problems which, for example, demand speedy intervention that decentralized control panels exist at appropriate locations.

Certain operators superintend such panels by modifying, when required, the different sets of variables such as temperature, pressure, flow rate, differential flow rate, and density.

However, the most modern solution to the control of the operating conditions appertaining in a plant and, especially, in a large plant, use computers. The values of the sets of variables which are detected by the sensors are constantly checked by microprocessors which are connected to one or more computers.

By supervising the data furnished by the microprocessors, the computers are in a position to process and visualize the plant indices to be managed (material balance, specific consumptions, etc.) and to execute checks and optimizations on the basis of which the microprocessors themselves can be made to intervene and modify certain variables of the operating process.

A consequence of the use of a centralized control panel is the almost total disappearance of the control panel operative working as an individual. He has been replaced by the operator who has the task of supervising the video display and keyboard channel to the computer operator.

The sophisticated modern operational controls in large plants not only ensure the instantaneous restoration of situations or normality in the control systems by means of suitable corrections of the operating variables, but they also promote the precise specification of fluctuations in the operating conditions and the characteristics of the products. Such controls, together with any systematic profile concerning the feeding, production, and operational behaviour of plants, permit one, on the basis of the data gathered, to carry out the preparation of *statistical process controls*. This latter term indicates the possibility of detecting variations in the functioning or in the quantity and quality of the products, and, hence, in the limit, to control the production costs of selected materials possessing rigorously preset properties.

5

The problematics of water

Industrial chemical technology is responsible both for an adequate supply of water for every purpose and for ensuring that the ecology of the hydrosphere is safeguarded.

This is because, on the one hand, the chemical aspect of the treatments which the supply and the protection of waters require is collectively more important by far than any other aspect (mechanical, physical, and biological), and, on the other hand, the interdisciplineary preparation which is necessary for handling these tasks is certainly more appropriate to chemists than to experts in other technologies.

Moreover, industrial chemical technology is collectively in the avant-garde in the creation of serious problems concerning the condition of the terrestrial water supplies which will be handed down to later generations.

For these reasons the subject of water will be dealt with here, as suggested by the title to this chapter, from the point of view of all the complex problems which can arise.

5.1 THE TERRESTRIAL WATER CYCLE AND THE ORIGINS OF WATER PROBLEMS

All of the water present on Earth is cyclically linked in the sense that water is continually leaving the surface of the Earth by evaporation, only to fall back again in the form of various meteorological products.

It is calculated that, on average, 3.6×10^{14} tons of water evaporate from the Earth, the lithosphere, and the hydrosphere each year, and that as much falls back there in the form of snow, rain, and hail stones. About one quarter of the water which is redeposited runs over the land masses ('lithospheric water'), while the remainder enters the hydrosphere. Part of the lithospheric water finishes up in servicing the biosphere, while much of it is lost by returning to the seas, where on account of the greater subaerial area, evaporation is far more intense.

Nowadays man, at least in the more advanced nations, harnesses an appreciable fraction of the lithospheric and other water which he extracts from subterranean folds, in order to put it to use in agriculture ('irrigation waters'), for civil purposes ('drinking waters'), and industrial ('industrial waters'). The estimates of the percentages of water going to these destinations are: 39% for agricultural purposes, 12% for civil purposes, and 49% for industrial purposes.

As a consequence of this, more than 60% of the water exploited by man is subject, in a country such as Italy, to a deterioration in its purity as a result of the uses which are made of it. It is both this systematic deterioration of the quality of a large part of the available water, and, given the increasing demands, the ever growing shortage of water which is not too saline, which contribute to the creation of water problems.

Furthermore, there are also other reasons why problems with waters arise. Among other things, waters effect a solvent action on rocks, thereby becoming loaded with 'hardness salts', and, in doing so, acquire a composition which makes them unsuitable for many industrial and civil purposes.

5.2 REQUIREMENTS REGARDING WATERS IN RELATION TO THEIR USE

The destinations of lithospheric waters sought by man are therefore agrarian, civil, and industrial.

Irrigation waters must be:

- transparent on the surface and clear in the bulk,
- free from phytotoxins,
- not too saline.

Surface *non-transparency*, which is harmful in irrigation waters, is a consequence of pollution by hydrocarbons, foams, and powders, while varying *bulk opacity*, which is also harmful, is for the most part due to suspensions of powders, mineral oil emulsions, or hydrophobic solvents and oil-based mud dispersions.

Above all, cyanides, phenols, aromatic compounds in general, sulphites and heavy metals are phytotoxic.

The salts in irrigation water must not exceed a limit of 2.5 g/litre and the chlorides, in particular, must not exceed a limit of 0.5 g/litre.

Irrigation waters must also be adequately aerated.

Drinking water (or 'water for civil purposes') must be:

- bacteriologically harmless,
- colourless, odourless, and clear,
- free from poisons and from the symptoms of poisons (such as ammonia and nitrites),
- moderately endowed with certain dissolved salts and well aerated so that they have a pleasant taste.

From a qualitative point of view, the admissible anions in good drinking waters are bicarbonates, chlorides, sulphates, and nitrates, while the cations Na^+, K^+, and Li^+ are, of course, allowed. Moderate amounts of Ca^{2+} and Mg^{2+} are also permissible, and small amounts of aluminium, iron, and manganese cations are tolerated. Heavy metals such as lead, mercury, copper, and chromium must be completely absent.

Quantitatively, the weight of the dry residue of good drinking waters must not exceed 0.5 g/litre, and, in this range, the calcium and magnesium salts must not be greater than 0.2 g expressed as $CaO + MgO$, sulphates must not contain more than 0.1 g of SO_3, chlorides not more than 0.035 g of Cl, and the nitrates must not correspond to more than 0.027 g of N_2O_5.

Finally, waters which are adjudged as being potable must have a temperature lying within the limits 7 to 15°C.

Industrial waters must, as a rule:

- not exhibit turbidity due to true suspensions (of earth, powders, etc.) or colloidal suspensions (of clays, silica, etc.),
- not contain dissolved corrosive gases (O_2, CO_2, H_2S, and H_2) at all or any other gas (e.g. N_2) in amounts such as to constitute potential danger upon entry under high pressure,
- not contain dissolved salts which are corrosive (e.g. $MgCl_2$, Na_2CO_3, and nitrates) or salts which are deposited in the form of scale (for the most part, calcium and magnesium salts).

Such salts turn out to be harmful to industrial water for several other reasons. For instance, calcium and magnesium bicarbonates, apart from producing CO_2 which is aggressive towards heated metallic structures, encourages the formation of insoluble precipitates (of $CaCO_3$, for example), and nitrates are damaging because of the oxidation phenomena which they bring about and also because they cause the formation of caustic hydroxides which are capable, for example, of yielding precipitates upon reaction with magnesium ions:

$$8NaNO_3 + 9Fe + 4H_2O \rightarrow 3Fe_3O_4 + 8NO + 8NaOH$$

$$Mg^{2+} + 2OH^- \rightarrow Mg(OH)_2 .$$

Of course, the requirements regarding industrial waters are extremely varied and specific for each type of industry, in which they must be used, to a point where they cannot be predetermined in a general manner.

Among other things: waters for use in sugar refineries must not contain anions which hinder the crystallization of the sugar leading to an increase in the yield of molasses (e.g. CO_3^{2-}, NO_3^-, and SO_4^{2-}); waters to be used in food industries must, inter alia, be bacteriologically pure and free from ammonia and nitrites, and water for use in refrigeration plants must be sterilized, de-aerated, and stripped of every scale-forming agent in order to prevent, above all, through the development of sticky microorganisms (algae), good heat transfer.

5.3 THE STATE OF WATERS WHICH ARE PRACTICALLY AVAILABLE

A preliminary review of how waters are made available is presented.

Waters used for the first time and re-used waters

Raw untreated waters are available in nature. These are (principally*) represented by:

- hard marine waters;
- sweet or slightly hard waters: from mountain rivers, rivers, lakes, and ponds;
- sweet or hard subterranean waters from springs or artesian wells.

The terms *sweet* and *hard* are used to characterize waters in relation to the amounts of certain salts in these waters.

Raw, natural, non-marine waters may find some uses in one or more fields of the practical utilization of water: agrarian, civil, or industrial. All of these untreated

*Other natural waters are rain waters which are, in any case, collected to be exploited, and the waters from snowfields and glaciers. Here, one is concerned with very pure but rarely-used waters.

natural waters are used in every field after being subjected to procedures which remove the components from them which are undesirable as regards the uses to which they are to be put; that is, they are used in the form of treated waters.

When they are used, natural waters, whether they are raw or treated, are to be considered as waters which are used for the first time.

Once they have been used, such waters become 'reflow' or 'discharge' waters. Reflow waters are sent into water reservoirs after having been acclimatized, and are re-used after suitable treatment or without any further treatment, according to whether there is any need to restore certain of their characteristics. For instance, before recycling water which has been used for the scrubbing of smoke particles, it is filtered, while one tends to use water in a cleansing process (for example) which has been heated up in a cooling operation.

Waters which are recycled with or without treatment are referred to as 're-used waters'.

The hardness of waters

Water running off the soil and in the subsoil is laden with salts, and is more or less enriched in these salts. The process of the dissolution of limestone brought about by water which loosely contains carbon dioxide due to atmospheric washing and running through vegetated terrain is of particular importance:

$$Ca(CO)_3 + CO_2 + H_2O \rightarrow Ca(HCO_3)_2 \tag{5.1}$$

The salts which have significant consequences in civil and industrial uses, may be distinguished as:

- alkaline salts which are usually considered to be innocuous;
- calcium and magnesium salts (and, possibly, those of other alkaline-earth and earth metals such as iron and manganese) which are always harmful if they exceed certain amounts.

Hard waters are defined as those which contain more than a certain amount of calcium and magnesium salts, etc. ('hardness salts').

At the civil level, extreme hardness renders the waters undrinkable, hinders cooking and baking and impedes washing with soaps due to the precipitation of salts as scums. The following reaction shows how such precipitates are formed.

$$\underset{\text{hardness cation}}{Me^{2+}} + \underset{\text{anion of soap}}{2RCOO^-} \rightarrow \underset{\text{insoluble salt}}{Me(RCOO)_2}$$

Hard waters are industrially inadmissible for a number of reasons. Let us point out some of them.

By obstructing tubes and depositing scale on kettles, hard waters lower the coefficient of heat exchange, thereby making it necessary to apply costly and dangerous superheating. Inter alia, possible flaws and cracks in the incrustation bring water into contact with the superheated metal plates, thereby leading to unexpected and abnormal steam generation which can cause explosions or serious anomalies in the mechanical strength required of the plant.

There are also dangers of ruptures due to the corrosion of the structures under the incrustations as a consequence of the phenomenon of *differential aeration*. It so happens that, owing to the lack of oxygen, there is no demand for reducing electrons under the scale, while, on the other hand, such a demand exists at every uncovered point of the structure which is accentuated by the 'electrophilic' lack of balance between the two zones. Therefore, electrons required at the aerated points migrate to these areas via the metal from non-aerated regions which, as a consequence, undergo oxidation (electron loss), that is, corrosion.

Moreover, electrical units (e.g. batteries), paper mills (where, among other things, the incorporation of the paper antiexpandant is impeded), tanneries (the swelling of the hides is impeded), dyeing plants (the colours tarnish), etc. cannot be fed with hard waters.

The hardness of water is traditionally expressed in terms of *degrees of hardness* which are understood as the amounts of standard substances which are potentially contained in a certain amount of water. The standards are:

- calcium carbonate for the French degrees of hardness (°F), for the English degrees of hardness (°Br), and for the American degrees of hardness (°USA);
- calcium oxide for the German degrees of hardness (°D).

We shall specifically discuss only what applies in the case of the French degrees of hardness and the German degrees of hardness.

A water has a hardness of 1°F if it contains hardness salts in an amount corresponding to 1 g of $CaCO_3$ in 100 litres (10 mg/litre), and has a hardness of 1°D if its content of hardness salts corresponds to 1 g of CaO in 100 litres (10 mg/litre).

Nowadays, however, hardness is frequently expressed in meq/litre* and ppm of $CaCO_3$ and, less commonly, in meq/l or in ppm ('parts per million') of CaO.

Table 5.1 shows the conversions between the various degrees of hardness and the correspondences in meq/l and in ppm of $CaCO_3$.

Hardness may serve as a criterion in the classification of waters. This is briefly shown in Table 5.2.

In Table 5.3 the distinctions which are usually made between the different types of hardness are shown.

Salt waters

While hardness is the typical characteristic of raw natural surface and subterranean waters, the distinguishing quality of natural marine waters and some surface and subterranean waters ('brackish waters') and certain reflow waters is saltiness.

There is no limit of saltiness which would enable one to define a water as being salted (salty), and whether waters which are not pronouncedly salty are considered as such or not depends on the uses to which they are to be put.

Table 5.1—Comparison of water hardness

Type of degree	*°F*	*°D*	*°Br*	*°USA*	*meq/l*	ppm_{CaCO_3}
French	1	0.56	0.7	0.58	0.2	10
German	1.79	1	1.25	1.05	0.36	17.85
English	1.43	0.8	1	0.84	0.29	14.3
USA	1.71	0.96	1.2	1	0.34	17.1

†meg is milliequivalents

Table 5.2—Classification of waters on the basis of hardness

Water	*Hardness (°F)*	*Examples*
very sweet	0–4	rain
sweet	4–8	oligomineral
medium hard	8–12	potable waters according to the WHO*
fairly hard	12–18	treated river and lake
hard	18–30	permeating limestones
very hard	30 . . .	from chalk downlands

*World Health Organization.

Table 5.3—Types of water hardness

Classification	*Definition*	*Principal salts present*
temporary hardness	disappears when water is boiled	hardness metal bicarbonates
permanent hardness	boiling has no effect	hardness metal-sulphates, chlorides, and nitrates
total hardness	sum of the above	all the hardness salts

In general, salt waters are also hard, but this is not necessarily so, as all the salts present in it (including alkali metal salts) contribute to the saltiness.

Apart from the inconveniences arising from the hardness, if it is present, salt waters are:

* inadmissible for agricultural use as they contain appreciable amounts of salts which are harmful to plants, especially through the dehydrating osmotic effects which their tissues are subjected to;
* in civil use, they are quite unsuitable for quenching thirst and for all purposes which require the absence of ionic solutes;
* industrially, they are harmful in very many different processes owing to corrosion phenomena, damaging ebullioscopic effects, incompatibility with chemical processes, and so on.

The return, upon desalination, to the use of raw salt waters, especially sea waters or recovered waters, is today becoming a common practice, owing, above all, to the improved methods which we shall see are available for this purpose.

Polluted waters

The EEC directives indicate what is considered as the artificial pollution of waters in the following words: ‘Hydric pollution is the effect of the discharge into aqueous surroundings of substances or of energies such as to compromise human health, to damage living resources, and, more generally, the hydric ecological system and to constitute an obstacle to any legitimate use of the waters, including environmental attractions'.

Artificial pollution is today by far the most widespread and serious, but it is clear that all the waters containing substances or heat energy of the same type as those caused by man-made pollution are to be considered as being polluted. Examples are: the natural waters which have become chalky owing to the washing away of argillaceous sites in the Appennines, and stagnant waters where the development of certain fauna and flora has first led to an excessive reduction in the amount of oxygen present and then to them becoming very rich in the anaerobic microorganisms responsible for putrefaction.

Levels of pollution of waters and their causes

Pollution of waters can develop at three different levels:

- changes in the qualities of waters due to natural causes, which cannot be suppressed in that rain water comes into contact with airborne substances or materials (in volcanic districts, for example) and, more generally, substances of the mineral and biological worlds are brought into solution or into suspension when carried from the soil (natural pollution);
- a higher state of pollution than that which has just been mentioned derives from the presence, due to non-natural causes, of different kinds of pollutants including energy (heat) in amounts which, on the other hand, do not exceed the auto-purifying capacity of the water concerned (temporary induced pollution);
- a third and more serious state of pollution when the pollutants are such, both qualitatively and quantitatively, as to inhibit the autopurification capacity of the water and, thereby, to provoke the permanent aesthetic or functional degradation of the water body (permanent induced pollution).

* * *

Natural pollution of water (including the saltiness of sea waters) is always due in the first place to the terrestrial litho-biospheric structure and, secondly, to activity outside the Earth and to certain meteorological phenomena.

Artificial factors leading to the pollution of waters are essentially due to discharges of domestic sewage, industrial effluents, and effluents associated with arable farming and livestock husbandry from which sometimes follows, so to speak, the 'fuse' for a process which is the inverse of autopurification, that is, in essence, 'eutrophication'.

Discharges of domestic sewage carry into waters, above all, the products of human metabolism with a relatively high bacterial content, together with contamination from domestic activities (food and washing) and refuse carried away by the drainage of streets, squares, and workshops.

Discharges of industrial effluents containing residues of raw materials and the intermediate and final products produced in the operation are of variable composition, depending on the nature of the industry which produces them.

For example, it may be recalled that industries such as food, textiles, tanning, and paper add to waters both residues which are characteristic of urban discharges as well as their own specific products such as colorants and finishing additives. The printing and electroplating industries are furnishers of acid and basic waters which are contaminated by, inter alia, heavy metal cations, chromates, reducing agents, and sometimes cyanides, while petroleum refineries return (in particular) ammonia, sulphurs, and mineral oils to water.

The *discharges which are associated with arable farming and livestock husbandry* contaminate waters, above all, with metabolic fluids and fluids from slaughterhouses and the processing of dairy products, pesticides, and leached-out fertilizers.

The term *eutrophication* indicates the excessive growth and disordered multiplication of aquatic plants, principally algae, due to the presence of extremely high quantities of nutrients in the waters.

When used in the above manner 'eutrophication' indicates that an environment is able to supply optimal nutrition to certain living species. The fact that there are, in practice, various possible degrees of trophication, as can be seen from Table 5.4, is not further dealt with here.

The nitrogen and phosphorus compounds coming from domestic and industrial discharges and from the leaching-out of agricultural fertilizers are principally responsible for eutrophication.

Lakes in which the exchange of water proceeds at only a modest rate, and pools in general, are particularly prone to this phenomenon which, when it takes place, renders the environment hostile to fish and other aerobic living creatures even if the supply of nutrients to the water is subsequently interrupted.

It happens, in fact, that the excessive development of plantlife with a short life, such as algae, leads to large quantities of organic material in putrefaction. The latter requires the consumption of large amounts of oxygen on account of the tendency of organic residues to be transformed into CO_2 and H_2O. In this very transformation nitrogen and phosphorus are released and become available again, which, in turn, favours, in so far as is possible, the redevelopment of plant life, thereby making the environment permanently unsuitable for aerobic living organisms.

Table 5.4 summarizes the contemporary terminology used to describe the waters affected by trophism. It also describes the state of such waters and gives the amounts of oxygen, phosphorus, and nitrogen present in such waters.

Oxygen demand as an index of the pollution of water

Not only when there is eutrophication but also when a body of water (sea, river, or lake water) holds, either in suspension or free, oxidizable substances such as sulphides and sulphurous compounds, nitrites and ferrous salts, the products of animal metabolism, hydrocarbons, and the enormous majority of organic compounds of industrial origin, the body becomes inhospitable to all types of fish and the aerobic microorganisms which must promote the autopurification processes in it. Since, in general, the pollutants of bodies of water are substances which consume oxygen, the extent to which such bodies are polluted may be considered to be proportional to the amount of oxygen which is required to eliminate the action of the reducing agents which are contaminating them.

The *oxygen demand of a water body* is that amount of oxygen, expressed in mg/l (which is the same as g/m^3), required to free it from the action of the agents present in it which consume oxygen.

The oxygen demand of a body of water can be expressed either in terms of the chemical oxygen demand (COD) or the biological oxygen demand (BOD) which are

Table 5.4—Grades of water trophication and their characteristics

Types of waters	*Appearance*	*Oxygen present and phosphorus and nitrogen content*
Eutrophic	Infested with algae, turbid, greenish, plainly malodorous, and of bad taste.	*Oxygen*: <2 mg/l (at 20°C and 1 atm) *Phosphorus* (as P): >0.01 mg/l *Nitrogen* (as N): >0.3 mg/l
Mesotrophic	Unclear and incipient greenish coloration, distinctly unpleasant odour, and taste readily discernible.	*Oxygen*: from 2 to 20 mg/l (at 20°C and 1 atm) *Phosphorus*: ca. 0.002 mg/l *Nitrogen*: ca. 0.06 mg/l
Oligotrophic	Clear and colourless to the naked eye. Practically odourless and tasteless. Very few algae.	*Oxygen*: up to 40mg/l (saturation limit of oxygen in water at 20°C and 1 atm, 44.34 mg/ml) *Phosphorus*: >0.001 mg/l *Nitrogen*: <0.03 mg/l

respectively defined as:

- the oxygen consumed in the oxidation by chemical means of all substances exhibiting a reducing action which are present in it under specified conditions, apart from chlorides (COD);
- the oxygen consumed for the biological oxidation, also under fixed conditions, of a fraction of the substances with a reducing action which are present in it (BOD).

The techniques for the determination of these two parameters which are, respectively, of an analytical inorganic chemical character in the case of COD and of a biochemical nature in the case of BOD are described in texts dealing with analytical chemical techniques and in treatises on water treatment.

Sometimes, but not always, the BOD is specified as BOD_5 to show that the determination has been carried out over a period of five days. The BOD_5 represents only a fraction of the total oxygen necessary for the complete transformation of the biodegradable organic substances because the entire process takes place over a period which is much longer than five days (about one month). For example, in the case of normal domestic sewage fluids, the BOD_5 is slightly more than 68% of the total BOD.

While the BOD shows, by means of how it is measured experimentally, the content of biodegradable substances, the COD expresses the amount of oxygen which is necessary in order to obtain the effects produced by the chemical oxidation energy due to potassium dichromate in a sulphuric acid solution with which it is determined. Therefore, for the purposes either of determining the purifying action of biological treatments or of following the course of pollution of bodies of water by biodegradable substances, systematic reference is made to the BOD.

The Royal Commission on Sewage Disposal, the British organization which has perfected the method for the determination of the BOD, considers waters with a BOD of 1 ppm* as being very clean, those containing substances which pollute the water to yield a BOD value of 5 ppm as suspect, and a water with a BOD of 10 ppm as certainly polluted.

According to existing legal provisions, companies which already exist in Italy can discharge effluents with a BOD up to 250 ppm into public waters, while new companies must ensure that the bod of their effluents does not exceed 40 ppm.

5.4 THE ELIMINATION OF SUSPENSIONS FROM WATERS

Waters derived from the most widely varying sources may contain suspended bodies (suspensions) covering a wide range of sizes. In fact, they vary from bulky solids with dimensions of the order of one metre to colloidal dispersions with a size of the order of one micrometre ($1\ \mu m = 10^{-6}$ m).

The treatments used to free waters from suspensions are:

- *screening* to remove solids with dimensions† greater than 1 cm;
- *sedimentation* which can be used for the elimination of suspensions with sizes from 1 cm to 5×10^{-3} cm;
- *mechanical filtration*‡ which can be used to eliminate suspensions with sizes from 5×10^{-3} cm to 10^{-5} cm;
- *coagulation–flocculation* followed by sedimentation and/or filtration for the removal of colloidal dispersions.

It is noted that mechanical filtration can in its own right replace sedimentation, since it is undoubtedly possible to separate by filtration every particle that undergoes sedimentation. Since, however, sedimentation is less costly than filtration and this difference increases the greater the amount of suspended material which has to be separated, sedimentation is carried out before filtration in industrial practice in order to make the task of filtration easier and, thereby, to lower the costs.

Screening

The passage of the water through screens is normally the first operation which is encountered in a plant for the purification of mining and discharged waters. By means of this operation, larger solids, such as mineral fibres, pieces of wood, rags, and paper are eliminated. The operation is carried out by making the water pass through a grid in the form of bars, a mesh, or a perforated plate of flat, convex, or drum shape. The grids with the largest apertures from 0.5 cm upward are normally of the bar type, while those with small apertures are either in the form of a mesh or perforated

*It is recalled that 1 ppm is equal to 1 mg/litre or to 1 g/m^3.

†The sizes given here are to be looked upon as 'orders of magnitude'. In fact, in apparent contrast to what will shortly be said there are grids with apertures from 0.5 cm, and it is found that there are also particles which attain sizes of scarcely 10^{-3} cm among the sedimentation products. In spite of the lower limit given here for particles which can be removed by filtration, particles with a diameter slightly larger than 10^{-5} cm are found to have passed through filters. For this reason, in practice, when a large fraction of the suspended particles have dimensions on the limit requiring the use of one method of separation or another, these methods are applied successively.

‡As we shall see (page 153), a 'biological filtration' does also, in fact, exist.

plates. The amount of material removed by the grid depends, of course, not only on the dimensions but also on the shape of the apertures.

Fig. 5.1 illustrates the principle by which a certain type of grid, used to intercept waters and remove the coarser pieces of suspended material from them, operates.

Sedimentation

The theory of the sedimentation process ('decantation') is developed in courses on industrial chemical plant and is only approximately applicable to the field of water purification where the suspended particles have different shapes, dimensions, and non-uniform specific weights. This is at variance with the assumptions of the abovementioned theory, and, for this reason, the theory will not be presented here.

In practice, if a water which contains suspended materials of differing densities which are held in suspension by a turbulent motion finds itself in calmer conditions, sedimentation of the heavier materials in it occurs. Suspended materials of a certain range of sizes (and weights) which are held in suspension by turbulence are therefore separated by means of sedimentation from a body of water under the action of gravity.

Both the values regarding the dimensioning of the sedimentation reservoirs in which the waters must remain for a given time in order to guarantee that certain particles suspended in them may have the possibility of undergoing sedimentation, and the other operating conditions which ensure that the liquid is clarified to the

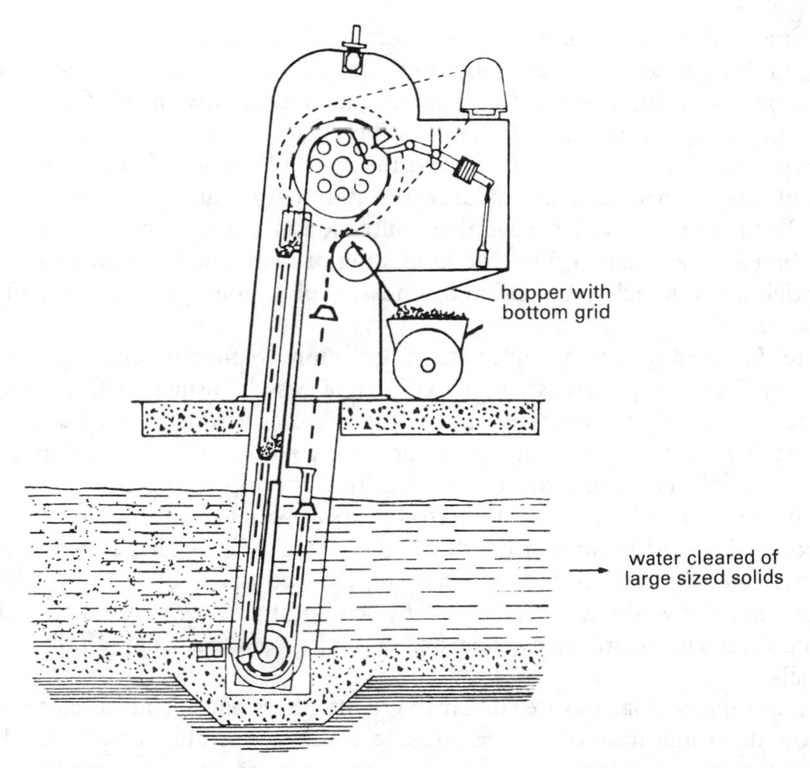

Fig. 5.1—The screens intercept solids of appreciable dimensions, gathering them into buckets which are raised and then upturned, so loading a small wagon with the discharge.

greatest extent by means of this procedure, are derived in practice from laboratory experiments.

These conditions primarily concern the structural characteristics of the solids to be made to undergo sedimentation. These may be *granular solids* such as sands and certain powders or *flocs* such as some hydrated oxides and many components of biological slimes which impose conditions on the shape, dimensions, and the operating conditions of the apparatus in order to instigate a high throughput and to guarantee residence times for the water being treated which are favourable to its clarification by sedimentation.

Sedimentation devices normally consist of circular or rectangular vessels made out of cement or iron with the bottom gently inclined toward a zone where the sludge which has settled out is carried away by using suitable scraping mechanisms for its subsequent discharge. As an alternative, the sludge can be directly removed from the vessels with mobile suction tubes.

A sedimentation apparatus is reproduced in plan and section in Fig. 5.2. Water entering this apparatus passes into a distributor situated at the centre of the vessel, while the effluent flows from the device via coarse blankets of material arranged in the periphery. The distributor has the main task of ensuring the entry of the liquid without creating turbulence.

The sludges are taken away to the discharge point by the sludge scraper blades while substances floating on the surface are gathered using foam removal blades. Both the sludge scraping blades and the foam removal blades are driven by a single revolving mechanism.

Filtration

Filtration exploits a pressure difference between the upper and the lower face of a filter bed. It operates on suspensions which cannot be removed by sedimentation in reasonably short times, and serves primarily for the improvement of the quality of water bodies in accordance with the use which is to be made of them.

For the development of the theory pertinent to filtration it is necessary to refer, as has already been done in the case of sedimentation, to courses in industrial chemical plant. With regard to water technology, filtration is generally carried out by using filters based on a granular filter bed consisting of inert silicates with an assortment of particle sizes (pebbles, gravels, and sands) or of another type such as granulated anthracites.

Water filters may operate under gravity or, more frequently, under pressure.

Gravity filters (Fig. 5.3) essentially consist of vessels, usually of concrete, which operate open to the atmosphere, and on the bottom of which a drainage system is installed. The filtering medium (sands, gravels, coke, etc.) is arranged in layers of particles of ever increasing size on passing from the top to the bottom. The liquid to be filtered enters from above the filtration bed, permeates through it, and passes out from it into the bottom of the drainage system. After a certain period of usage the filter must be regenerated by removing the entrained solids from it. For this purpose counter-washing is carried out by sending the washing water to below the filtering strata via the drainage system and making it come out from suitable collection channels.

Pressure filters (Fig. 5.4) are similar to gravity filters but operate as closed systems to allow the application of an overpressure which accelerates the process. In these filters, as in gravity filters, the filtering elements are arranged in order of increasing weight towards the bottom, a position which they retain even after counter washing.

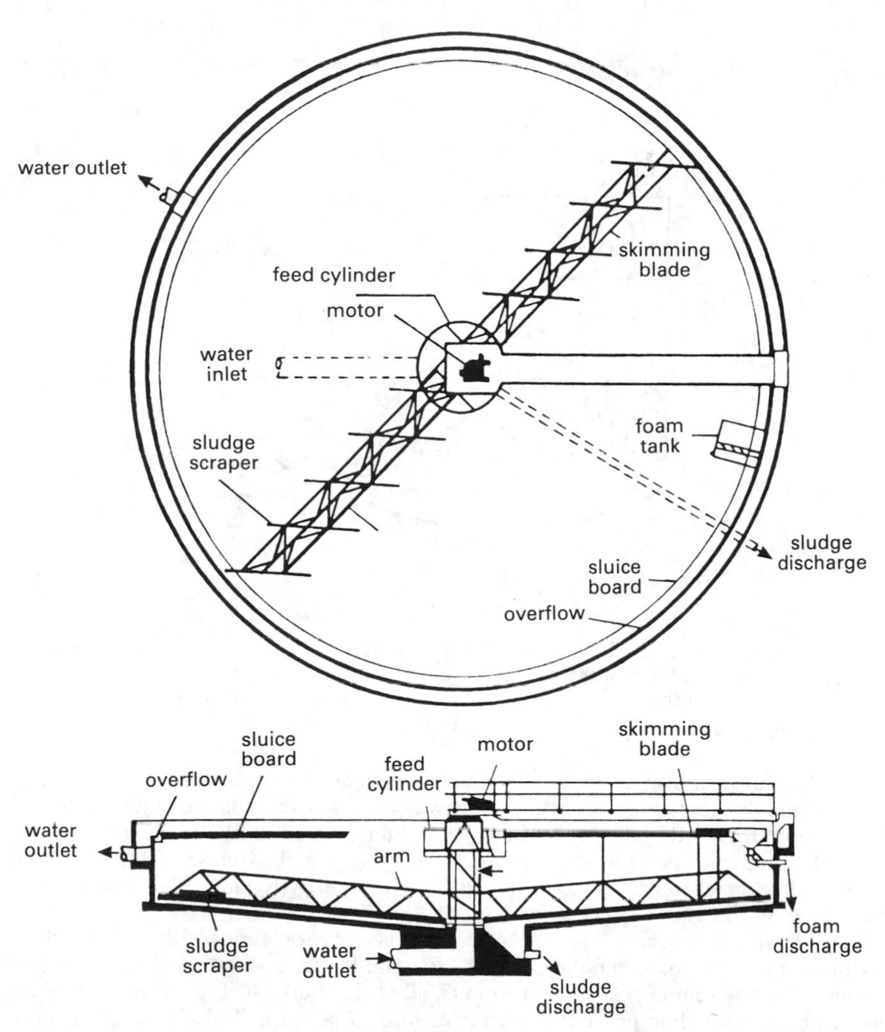

Fig. 5.2—Plan and section of a circular sedimentation unit with two-armed sludge scrapers. In circular sedimentation units the inlet distributor, sited at the centre of the vessel, is sometimes incorporated in the column which drives the sludge-scraping mechanism.

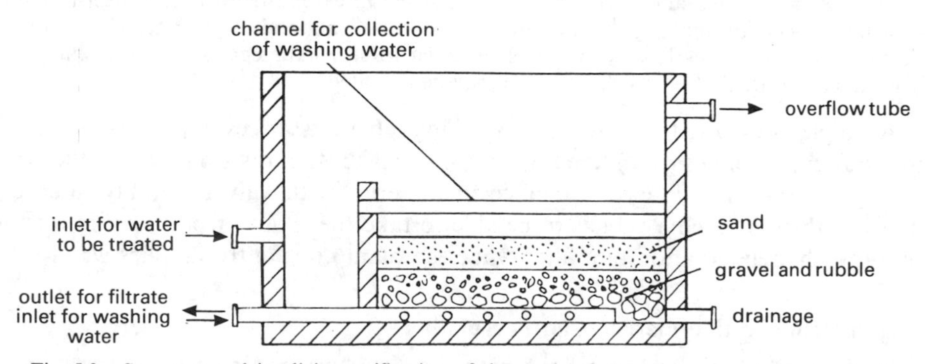

Fig. 5.3—Structure and implicit specification of the mode of operation of a simple 'gravity' type sand filter.

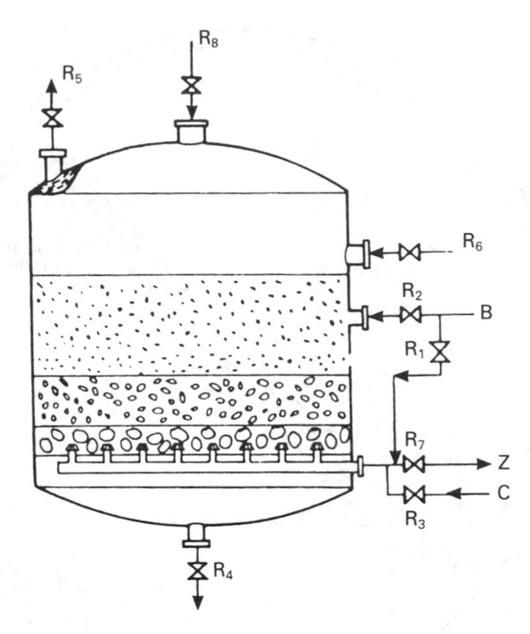

Fig. 5.4—Pressure filter with layers of silicaceous filtering elements of assorted granular dimensions. A description of the manner in which the filter works by alternation of the state of the valves with which it is equipped is given in the text.

The filter shown in Fig. 5.4 is equipped with a system of valves the action of which enables one to gain a detailed comprehension of the way in which the apparatus functions.

When only valves R_6 and R_7 are open, the water enters the filter, runs through the filter bed, and comes out as clear water from Z. To speed up this phase of the operation, it is possible, after the entrance of the water via R_6, temporarily to close this valve and to admit compressed air via R_8.

When the filtering capacity of the device is exhausted (or almost), the valves which were previously open are closed and the valves R_5, R_3, and R_2 are opened. Compressed air is blown in from B, and by sending clear water from C, all of the mud which has formed over the filter bed during the preceding phase is carried back into suspension. The air–water–small particle disperse system which is formed in this manner passes out from the vent, high on the left, which is controlled by the valve R_5. The outlet from the vent is covered by a fine metallic mesh to prevent any of the filtering elements from escaping.

Valve R_2 is reclosed and R_1 is opened successively to extend the washing effect to the less obstructed mass by making the mixture of compressed air and clear water enter from the bottom. Finally, valves R_4 and R_5 are kept open to allow the egress of the washing water from the apparatus, after which the cycle may be repeated.

A single pressure filter, when a washing phase is necessary, guarantees only intermittent supplies of filtered water. To make the distribution of the water continuous, several filters are connected in series with valves suitably interposed between them in such a way as to be able to take them out or put them into action to allow the regeneration of a filter while the plant constantly supplies water.

Coagulation and flocculation. Clariflocculation

If the suspended materials in the water to be treated are of dimensions corresponding to the colloidal state, neither any reduction in turbulence associated with sedimentation nor the capillary phenomena of filtration are sufficient to separate out such particles. In this case it is necessary to resort to techniques for the destabilization of colloids by annulling, in a compensatory manner, the surface charges carried by their particles in such a way that the repulsion between them is removed, and their aggregation to form particles with dimensions suitable for sedimentation and/or filtration becomes possible.

This destabilization of the charges on colloids which has just been mentioned is known as 'coagulation'.

The coagulation effect is achieved by the addition of electrolytes such as aluminium and iron sulphates and ferric chloride or polyelectrolytes, that is, natural organic polymers (such as alginates), artificial organic polymers (such as certain amide derivatives and derivatives of cellulose), or synthetic organic polymers (such as the quaternary salts of polyvinylpyridine) to the water. Shown below is the chemical structure of a synthetic polyelectrolyte of the 'quaternary ammonium salt' type.

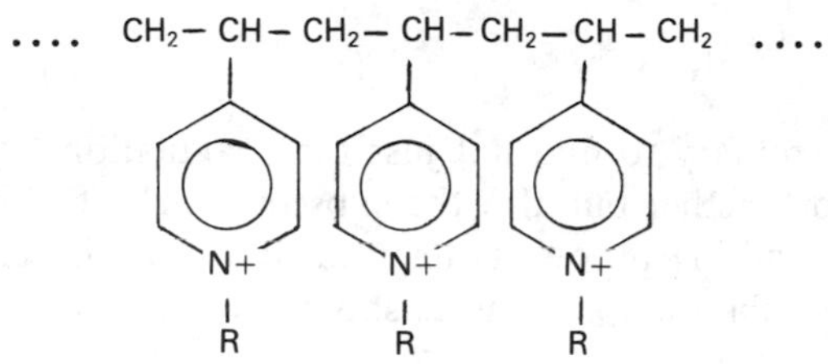

.... CH_2– CH – CH_2– CH – CH_2 – CH – CH_2 ....

N+ N+ N+

R R R

N-alkylpolyvinylpyridine

The polyelectrolytes are soluble in water because they carry ionizable groups along the length of the chain.

Since the colloidal particles of the suspensions to be separated normally carry negative charges, organic or inorganic coagulants with a cationic effect are customarily employed. However, the use of inorganic electrolytes (such as sodium aluminate: $Na_2Al_2O_4$) or organic polyelectrolytes (which arise from the polyacrylic acids which can be formulated as: ...–$CH(CH_3)$–$CH(COO^-)$–$CH(CH_3)$–$CH(COO^-)$–$CH(CH_3)$–...) which exhibit an anionic effect, cannot, of course, be excluded when it is necessary to separate carbon powders dispersed in the form of positively-charged particles.

In every case it is necessary to adjust the ambient pH value in order to increase the effect of the coagulant, as it is only over certain intervals of pH that electrolytes and polyelectrolytes exhibit their coagulating action. It is also necessary to maintain a high degree of agitation during coagulation in order to promote the dispersion of the coagulating agent.

* * *

The choice of coagulating agents is always intended to create conditions which are favourable to the aggregation of the discharged colloids into particles with dimensions of several millimetres ('flocs'). The mechanism of the formation of such flocs is known as 'flocculation' and adsorption phenomena lie at the foundations of this mechanism.

The pH value which is adopted on the one hand to favour the coagulating action and, on the other hand, the formation of flocculating agents is sometimes imposed by the solutes in the water. For instance, when water with appreciable temporary hardness is treated with aluminium sulphate, CO_2 is evolved and aluminium hydroxide is formed:

$$Al_2(SO_4)_3 + 3Ca(HCO_3)_2 \rightarrow 3CaSO_4 + 2Al(OH)_3 + 6CO_2.$$

The carbon dioxide carries the pH value of the water being treated to between 5 and 7, favouring coagulation*, while the voluminous cationic complex formed from the hydroxide provides the optimal conditions for flocculation.

Because flocculation may be favoured it is necessary that encounters between the particles which must form the flocs which, once they have been formed, do not break up. It can then be understood that flocculation requires conditions which contrast with those required for coagulation with the result that one, unlike the other, is favoured by gentle agitation. For this reason, the two operations are carried out in separate apparatus or in different compartments of the same apparatus.

* * *

In a plant it is customary to find not just the coagulation and flocculation of a turbid water coupled together but also the subsequent clarification of the water by sedimentation. The resulting process is referred to as 'clariflocculation'. A labelled diagram of a clariflocculation apparatus is shown in Fig. 5.5.

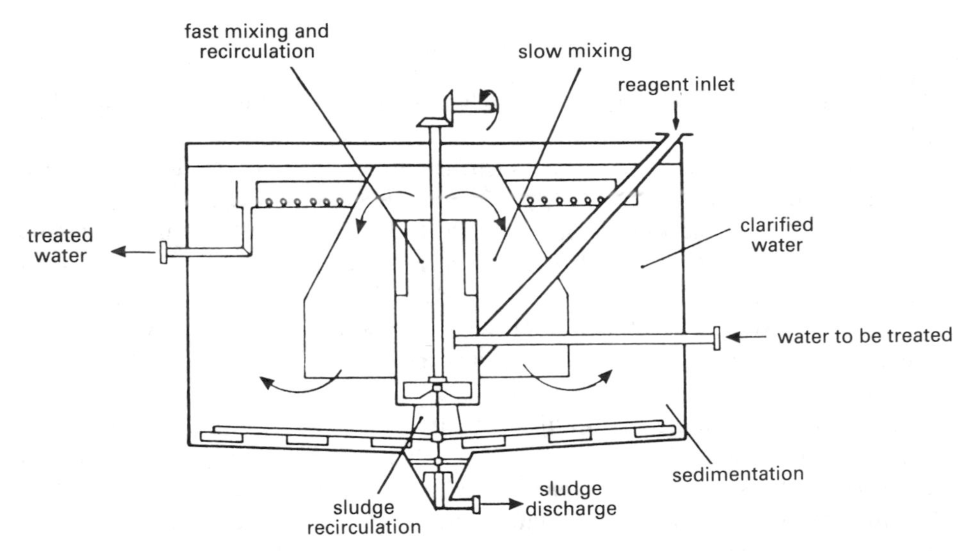

Fig. 5.5—Apparatus for clarifocculation. The water to be treated is mixed with the reagents first vigorously and subsequently slowly. During the fast mixing the sludges are, in part, recycled by eddies, while maximum flocculation takes place during the slow mixing. Sedimentation carries the sludges onto thc bottom of the apparatus from where they are sent to the sludge outlet.

---

*Actually, at this pH value aluminium hydroxide forms a 'polynuclear complex' with Al^{3+} cations:

$$Al^{3+} + 5Al(OH)_3 \rightarrow [Al_6(OH)_{15}]^{3+}$$

derived from the dissociation of aluminium sulphate. Complexation hinders the hydrolysis of such cations. The cationic polynuclear complex which is formed is firstly an active coagulant and secondly an excellent flocculating agent.

Clarification is practised both in the field of untreated waters and recycled waters, and filtration may also be exploited as a complement to it.

5.5 THE SOFTENING OF WATERS

Rarely in the field of rendering waters potable, but frequently within the compass of civil use (such as laundries and heating plants), and almost always for waters for industrial use, the hardness of the waters must be lowered, and ultimately eliminated. The treatments necessary for this purpose are softening processes or 'processes for the sweetening' of water.

The soda–lime process

Chemistry and plant

By treating the water to be softened, in a cylindrical apparatus with a conical base for the collection of the sludges, with a solution prepared by mixing lime water, powdered lime, and Solvay soda in water, reactions of the following type takes place:

$$\text{temporary hardness} \begin{cases} Ca(HCO_3)_2 + Ca(OH)_2 \rightarrow 2CaCO_3 + 2H_2O \\ Mg(HCO_3)_2 + 2Ca(OH)_2 \rightarrow Mg(OH)_2 + 2CaCO_3 + 2H_2O \end{cases}$$

$$\text{permanent hardness} \begin{cases} CaSO_4 + Na_2CO_3 \rightarrow CaCO_3 + Na_2SO_4 \\ \begin{cases} MgCl_2 + Ca(OH)_2 \rightarrow Mg(OH)_2 + CaCl_2 \\ CaCl_2 + Na_2CO_3 \rightarrow CaCO_3 + 2NaCl \end{cases} \end{cases}$$

with the intervention of just lime in the case of temporary hardness and both lime and soda in the case of permanent hardness.

The process is carried out at 60–70°C with preheating of the water to be treated and the blowing of live steam into the reaction mixture.

The sludge, consisting of the whole variety of the many precipitates which are formed, gathers at the base of the reactor (Fig. 5.6). The water overflows from the top of the reactor and is pumped to the filtration unit after having been treated with a solution of trisodium phosphate and with polyphosphates both to refine the process and to counteract, when it is

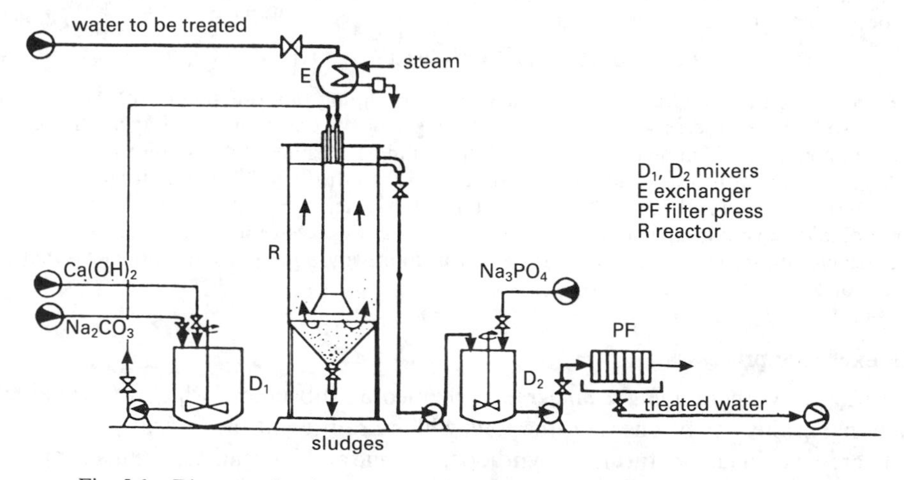

Fig. 5.6—Diagram of a plant for the softening of water by the soda-lime process.

necessary, the consequences of the caustic action of the effluent which will be explained in detail below.

The soda-lime process reduces the hardness of the water which is treated to, at least, 5°F, and, after the addition of phosphates, it is practically removed altogether.

Consequences and remedies

Normally, iron alloys corrode only slowly in a basic environment, as they tend to become covered with a layer of the oxide of their base metal (iron) which passivates them. However, at high temperatures and pressures, serious corrosion phenomena at the expense of iron alloys occur in a basic environment, and this takes place primarily in zones affected by heat treatments and mechanical working, that is, in welded and rivetted areas.

The process:

$$2Fe + 2OH^- + 2H_2O \rightarrow 2FeO_2^- + 3H_2$$

is basically responsible for corrosion by a basic ('caustic') environment under drastic conditions of temperature and pressure. The ferrite anion (FeO_2^-) is hydrolyzed with the regeneration of alkali:

$$2FeO_2^- + H_2O \rightarrow Fe_2O_3 + 2OH^-$$

so that the phenomenon can continue.

The damage caused to iron alloys by the hydrogen which is developed under pressure is still more serious than the corrosion due to the abovementioned causes, since this hydrogen enters into combination with the carbon component of such alloys to form methane ($C + 2H_2 \rightarrow CH_4$), leading to a distinct embrittlement of the iron alloys.

The weakening of the mechanical properties of iron alloys which is brought about in this manner is technically known as *caustic embrittlement*.

Waters which have been softened using the soda–lime process primarily cause caustic embrittlement due to the excess of Na_2CO_3 which is used. In fact, sodium carbonate, as a result of the hydrolysis which it undergoes, produces a volatile acid at high temperatures and renders the environment caustic:

$$CO_3^{2-} + 2H_2O \rightarrow 2OH^- + H_2CO_3 \rightarrow H_2O + CO_2\uparrow$$

By avoiding the use of an excess of Na_2CO_3 and refining the softening of the water treated by the soda–lime process with Na_3PO_4, the danger of the development of aggressive CO_2 is avoided and an environment is created which is almost harmless to iron alloys.

Actually, the hydrolysis of sodium phosphate leads to the formation of hydrogen phosphate anions ($H_2PO_4^-$ and HPO_4^{2-}) as well as phosphoric acid (H_3PO_4), which, since they are non-volatile, have the capacity to counter efficiently any excessive increase in the ambient pH, and which, in addition, can also passivate iron alloys by adherent and compact superficial layers of phosphates.

Ion exchange processes

Natural or synthetic *ion exchangers* are defined as substances which are capable of exchanging their own ions with the ions which are present in water.

There exist, in fact, natural and synthetic, organic and inorganic, systems containing cations (Na^+, H^+, etc.) which compensate for the electrostatic charges on their anionic radicals. When such systems come into contact with hard waters containing Ca^{2+} or Mg^{2+} ions, they release their own cations in exchange for the latter cations.

Synthetic substances which are entirely of an organic nature and which are capable of exchanging certain ions (OH^-, Cl^-) with other anions (HCO_3^-, CrO_4^{2-}, NO_3^-, SO_4^{2-}) are also available. They function by a mechanism which is analogous to that which has just been described.

Ion exchangers are therefore classified as follows:

$$\textbf{Ion exchangers}\begin{cases}\textit{cationic} & \begin{cases}\text{natural}\\ \text{synthetic}\end{cases}\\ \textit{anionic} \longrightarrow & \text{only synthetic}\end{cases}$$

The action and regeneration of an ion exchanger

The exchange capacity of an ion exchanger represents the quantity of ions, suitably expressed, which such an exchanger is, on the one hand, capable of releasing and, on the other hand, capable of reacquiring, that is, capable of exchanging. Such a capacity may be expressed either as an absolute value, that is, as the *total exchange capacity* or as an operational value, that is, the *practical exchange capacity*.

The *total exchange capacity* is that which is measured from when the exchanger starts to function until it has completely exhausted its exchange capacity, while the *practical exchange capacity* is that which is measured from when the exchanger starts to function to when a tolerable lower limit in its ability to exchange ions has been reached.

To gain a better understanding let us consider the effective mode of operation of an ion exchanger, that is, of a porous mass which is capable of being permeated by a solution containing certain ions to yield an effluent which contains other ions.

At the beginning of the process, the number of ions which are not exchanged and pass out into the effluent is analytically insignificant, but, at a certain point, more of these ions appear in the effluent as the exchanger starts to become saturated with them. If the experiment is continued it is found that the concentration of the ions which are not exchanged increases until the concentrations of the ions in the solution entering the ion exchanger and the solution passing from it are the same. It is then said that the ion exchanger is exhausted. In practice, the operation of an ion exchanger is not considered as being finished at the point at which the ion exchanger is exhausted, but rather when the amounts of undesirable ions in the effluent which have been retained by the exchanger exceeds a limit which is no longer tolerable in so far as the use to which the water is to be put is concerned.

Then: the 'total exchange capacity' is measured at the absolute limit of exhaustion of the exchanger, while the 'practical exchange capacity' is measured at the limit of exhaustion imposed in practice.

Both the total as well as the practical exchange capacity can be expressed in milliequivalents of ions fixed by exchange:

- per cm^3 of exchanger (meq/cm^3)
- per litre of exchanger (meq/l)
- per gram of exchanger (meq/g)

or, less commonly, in grams of a standard compound (usually $CaCO_3$) per cubic centimetre of exchanger (g/cm^3), per liter of exchanger (g/l), or per gram of exchanger (g/g).

* * *

However, the most interesting aspect of the behaviour of ion exchangers is their ability to be regenerated, that is, in water softening technology, for example, the capacity to take up alkali metal cations again by a reverse process in which the cations responsible for hardness are released from the exchanger.

This regeneration of an exchanger is achieved by washing the exchanger with a concentrated solution of a soluble compound containing the ion to be restored into the exchanger. So, an exchanger for the softening of water which initially contains Na^+ ions is regenerated, after it has been used to fix Ca^{2+} and Mg^{2+} ions, by washing it with a concentrated solution of Na^+Cl^-, while an exchanger which is initially in the hydrogen form due to the presence of the radicals $-SO_3^-H^+/-COO^-H^+$ is regenerated after it has been used to fix metallic cations by washing it with suitably concentrated solutions of hydrochloric or sulphuric acids.

It is therefore concluded that ion exchangers operate reversibly, and the direction of the exchange process is determined by the difference in the nature and the concentrations of the ions occupying the exchangers and those in the effluent flowing through them.

Mechanisms of ion exchange

An ion exchanger, into which an aqueous solution containing other ions diffuses, is electrolytically dissociated, thereby creating fixed electrically charged sites on the exchanger and the free counter-ions of the exchanger*. With the flow of the aqueous solution, other conditions being equal, there is a degree of repartitioning of the ions to be fixed between the exchanger and the aqueous solution; the extent of this repartitioning being determined by the affinity of the ions themselves towards the exchanger as well as by the concentrations of the ions ('mass action').

The conditions which must be held constant because the exchange is solely a function of the affinity of the ions towards the exchanger and of mass action, are generally:

- the *contact time*, which is determined by the amount of water passing with respect to the volume of the exchanger and, hence, to the height of the ion-exchanging layer;
- the *temperature*, which influences both the amount of water percolating through the exchanger owing to the dilatation of the pores as well as the velocity of the exchange reaction;
- the *pH value of the water* which is treated if the exchanger is acid or basic since, in such a case, the degree of dissociation of the exchanger can be influenced by it.

Under operational ion exchange conditions, that is, when the ion exchange and regeneration are carried out alternately, the sole decisive condition as regards the exchange activity is the *level* (or 'degree') *of regeneration* of the exchanger which is expressed as the ratio of the mass of the exchanger which is completely restored to its full exchange capacity to the total mass of the exchanger.

The combination of the two factors, affinity and mass action, which is proportional to the difference in the concentration of the exchanging ions between the solution and the exchanger, constitutes the motive force for the exchange which lies at the foundation of the phenomenon. This force has its maximum effect on first coming

*The term 'counter-ions' is employed to indicate the mobile ions of an ion exchanger.

into contact with the ion exchange bed, but subsequently diminishes owing to the reduction in the concentration difference which is responsible for it. So, the maximum exchange force is attained at a site located just beyond the point where the water enters into the ion exchanger bed, which, by analogy with the mechanism which has just been described, rapidly gives up this role to the next successive layer in the bed. Thus, the continuing displacement of the maximum exchange force along the length of the bed proceeds until a point is reached when there is no longer such a maximum force within the system, and, at this point, the symptoms of a loss in the ability to exchange ions will become evident and ever more pronounced. This loss in the ability of the system to exchange ions will continue until it becomes intolerable upon reaching the 'practical exchange capacity' which was defined above.

The kinetics of exchange reactions

From the point of view of the rates at which they proceed, ion exchange reactions are slow since the solution containing the ions to be exchanged must penetrate into the structure of the exchanger. It is therefore this diffusion of the solubilized ions into the structure of the exchanger which is the rate-limiting step in exchange reactions. This step is less significant, and, thereby, less kinetically decisive the greater the porosity of the exchanger, as measured by its specific surface area.

The action of ion exchangers can therefore be related, not only distantly, to the action of catalysts in so far as, like catalysts, these are active in one direction or the other of the exchange process ('reaction'), but, more correctly, because ion exchange is a surface phenomenon which takes place by a mechanism which is analogous to that of heterogeneous catalysis with which it has, in common, the ability to be increased by increasing the specific surface area of the substrate.

Deactivation ('poisoning') phenomena are also evidence of the analogy between catalysts and ion exchangers, both of them being sensitive to 'thermal shock' and to other causes whereby the relative physical structure becomes modified.

Among other things, swelling and weakening brought about by variations in the osmotic pressure during the contrasting conditions to which the exchanger is subjected in the exchange and regeneration phases leads to variations in its mechanical resistance and porosity.

Moreover, exchangers, like catalysts, are sensitive to poisons which alter their chemical structure with negative consequences as regards their exchange properties just as poisons reduce catalytic activity.

The nature and action of mineral cation exchangers

The first substances to be used with the ability to act as ion exchangers were the *zeolites*. These are natural aluminosilicates which contain both alkali metal cations which can be exchanged for the cations responsible for the hardness of water (Ca^{2+} and Mg^{2+}) as well as intercrystalline water which can be reversibly lost and reacquired.

In the natural economy the zeolitic component of clays is of great importance. Inter alia, the zeolites which are disseminated throughout the argillaceous strata of the Earth's crust soften hard surface waters provided that they penetrate downwards to some extent.

The scarcity in nature of zeolite concentrates has, however, led to the search for their synthetic substitutes which are known as *permutits* and can be prepared, for example, by the fusion of mixtures of quartz, kaolin, and sodium carbonate in a ratio of 1:2:4 by weight. Moreover, by taking their constituents in suitable proportions, it is possible to produce certain permutits with mechanical properties and, thereby, resistance and ion exchange capacities which are qualitatively better than those of zeolites.

The possible chemical composition of permutits is, in fact, variable, permitting opportune changes in the raw materials used in their production. Among other things, permutits have been prepared containing Rb^+ and Cs^+ instead of Na^+ and K^+, and others with Ga^{+++} and Ge^{++++} substituted, at least in part, for aluminium and silicon.

The reactions which are of general interest in the softening of waters by means of mineral exchangers are of the type shown in Fig. 5.7.

As can be seen from the reaction for the elimination of temporary hardness shown in the caption to Fig. 5.7, natural ion exchangers cannot eliminate carbon dioxide from waters. For this reason, it is necessary to carry out mixed treatments with lime, and an exchanger when decarbonation is requested.

Manufacture, state of supply, and the useful lifetime of organic ion exchangers

Certain limitations associated with the cation of mineral ion-exchangers have promoted research into synthetic organic products of a porous nature which are permeable to aqueous solutions and possess ion exchange capability.

Among other things, mineral exchangers remain active over only relatively short periods, with a tendency for their exchange properties to fall off more rapidly, the more active they are at the beginning. Moreover, permutits and zeolites exchange only cations, whereas the need frequently arises also to remove certain anions from waters.

Synthetic organic ion exchangers ('ion exchange resins') are structurally amorphous, macromolecular polymers. The manner of how and the conditions under which they exchange will be dealt with thoroughly in the second volume of this work.

Here, some general remarks are made concerning organic ion exchangers, leaving it to the illustrative brochures on the commercial products from various companies to provide details of the particular characteristics (density, mechanical resistance, exchange capacity, maximum operating temperature, etc.) of the different kinds of resins.

Here is, first of all, a classification of such resins according to the active groups which they contain.

Ion-exchange resins	cationic	*strongly acidic*	with sulphonic acid groups $-SO_3H$ capable of forming salts (e.g. $-SO_3^-Na^+$)
		weakly acidic	with carboxylic acid groups $-COOH$ capable of forming salts (e.g. $-COO^-Na^+$)
	anionic	*strongly basic*	with hydroxytetraalkylammonium groups $-\overset{+}{N}R_3OH^-$ capable of forming salts ($-\overset{+}{N}R_3X^-$)
		weakly basic	with hydroxyalkylammonium groups $-\overset{+}{N}H_3OH^-$ ($-\overset{+}{N}H_2ROH^-$) capable of forming salts ($-\overset{+}{N}H_3X^-$ or $-\overset{+}{N}H_2RX^-$)

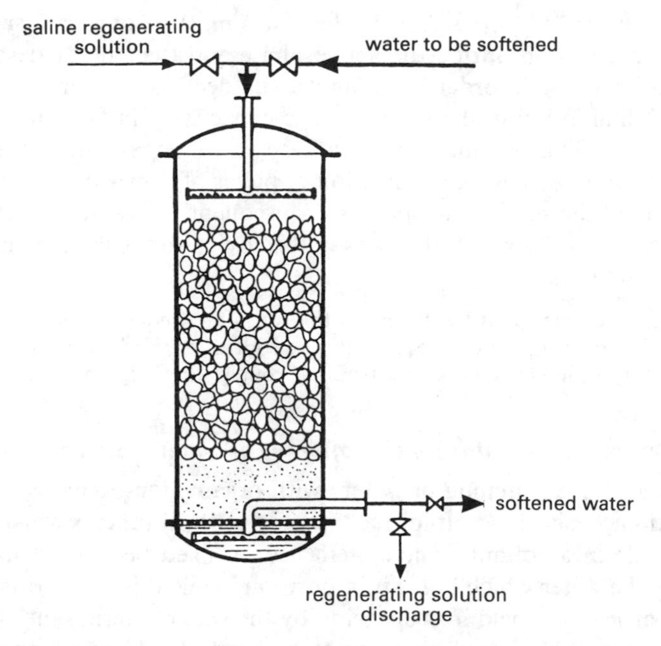

Fig. 5.7—Water softening by means of natural ion exchangers is carried out in cylindrical apparatus similar to fast filters and controlled by an appropriate series of valves. In practice, a number of such units are arranged in parallel so that one of them can be regenerated while the others are operating. At the base of these softening units there is about 30 cm of sand and a mass of exchanger which is about 1.20 to 1.30 metres in height. The characteristic exchange reactions, where Pe indicates a permutit or zeolite anion, are respectively:

* in the case of temporary hardness:

$$Na_2Pe + Ca(HCO_3)_2 \rightarrow CaPe + 2NaHCO_3$$

* and, in the case of permanent hardness:

$$Na_2Pe + CaSO_4 \rightarrow CaPe + Na_2SO_4$$

Once it has reached its practical exchange capacity, the exchanger is regenerated by washing it with a concentrated solution of NaCl which brings about the reaction:

$$CaPe + 2NaCl \rightarrow Na_2Pe + Ca_2Cl_2,$$

thereby producing a very hard aqueous purge which is discarded.

Cationic resins are able to exchange cations and, in particular, Ca^{2+} and Mg^{2+}, while anionic resins exchange the widest range of anions (CO_3^{2-}, Cl^-, SO_4^{2-}, etc.).

The resins are supplied by the producers either in a dry state or a moist state after having been immersed in water or a saline solution. The dry resins must be allowed to swell slowly before use so that they can take up solution. The details of this preliminary treatment are suggested by the suppliers on the basis of studies and tests.

The porous structure of ion exchange resins, which is raised to a point where the voids constitute 30–40% of the total volume, is necessary to guarantee a high specific surface area which favours the diffusion of ions and facilitates exchange. Also, when there is equivalence between the matrix and the active groups, it is possible to achieve the most diverse exchange behaviour both with regard to ions, non-ionic substances, and the small pieces of debris present in the liquids treated. As regards the non-ionic constituents, the empty pore spaces in the resins must be of such a size as to permit readily their adsorption during the active phase and their ready desorption during the regenerative phase in order to minimize any decrease in porosity with use.

A gradual diminution in the ion exchange capacity is, in fact, inevitable with ion exchange resins. This is due, above all, to small errors in the operating and maintenance conditions, to the deposition of pieces of debris in the pores, and to the fragmentation of the small resin spheres. Consequently, it is necessary to change the resin regularly. This is done on the basis of the advice and recommendations supplied by the manufacturers.

It should also be noted that treatments with a resin are frequently preceded by operations such as filtration or coagulation, decantation and filtration, and so on, of the waters to be treated, to diminish the dangers of inactivation brought about by any cause.

The use of ion exchange resins for the softening of waters with various end results

The pure and simple softening of waters with organic ion-exchangers is carried out by using cationic resins in a salt form consisting of small spheres which may be either fixed or mobile in a column. The system uses a 'fixed bed' if the mass of resin is traversed by the water which holds it in position, while it is a question of a 'fluidised bed' if the particles are held in suspension by the water entering the column and the spheres are prevented from escaping by means of perforated plates with small holes.

The water softening reactions, as they apply to calcium salts and an ion exchanger in the sodium form which is depicted as *Re*Na, are:

- in the case of temporary hardness:

$$2Re\mathrm{Na} + \mathrm{Ca(HCO_3)_2} \rightarrow Re_2\mathrm{Ca} + 2\mathrm{NaHCO_3}$$

- in the case of permanent hardness:

$$2Re\mathrm{Na} + \mathrm{CaSO_4(CaCl_2)} \rightarrow Re_2\mathrm{Ca} + \mathrm{Na_2SO_4(2NaCl)}.$$

In this way the ions which are responsible for the hardness (Ca^{2+} in this case) become fixed, while Na^+ ions pass into solution as illustrated in Fig. 5.8.

The regeneration operations which have to be carried out after the practical

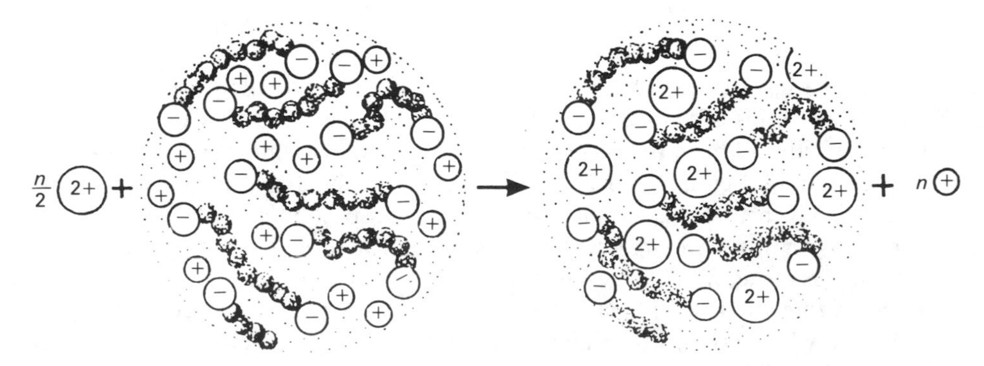

Fig. 5.8—The bivalent cations responsible for hardness displace twice their number of alkali metal cations from the resins which they replace in order statistically to neutralize the anionic groups on the resin molecules.

exchange capacity has been attained are:

- washing with other water in the opposite direction to the flow of the water which has been treated in order to remove any solid particles (pieces of debris of various shapes and consistency) which were retained by the resin;
- putting in the regenerating solution which, in the case considered above, would be sodium chloride under the operating conditions specified by the suppliers of the resin;
- washing with deionized water* using a flow in the same direction to eliminate the excess of the regenerating solution.

Fig. 5.9 shows a schematic diagram of a water treatment plant with a cationic resin in the sodium form put into columns which are arranged in parallel.

When a water softening process is adopted which solely makes use of an alkaline cation exchange resin, effluent waters from the softener are obtained which, especially when the inflowing water contains much temporary hardness, can potentially become very alkaline when heated, owing to the decomposition of the alkaline bicarbonates:

$$HCO_3^- \rightarrow OH^- + CO_2.$$

When the alkalinity of the water makes its use inadmissible for the purposes for which it was intended, combined treatments are carried out which use 'weak' cation exchange resins in the sodium form and 'strong' cation exchange resins in the acid form loaded into separate columns. The following reactions then take place:

* in the column containing the weak cation exchange resin:

$$2\mathit{Re}\mathrm{Na} + \mathrm{Ca(HCO_3)_2} \rightarrow \mathit{Re}_2\mathrm{Ca} + 2\mathrm{NaHCO_3}$$

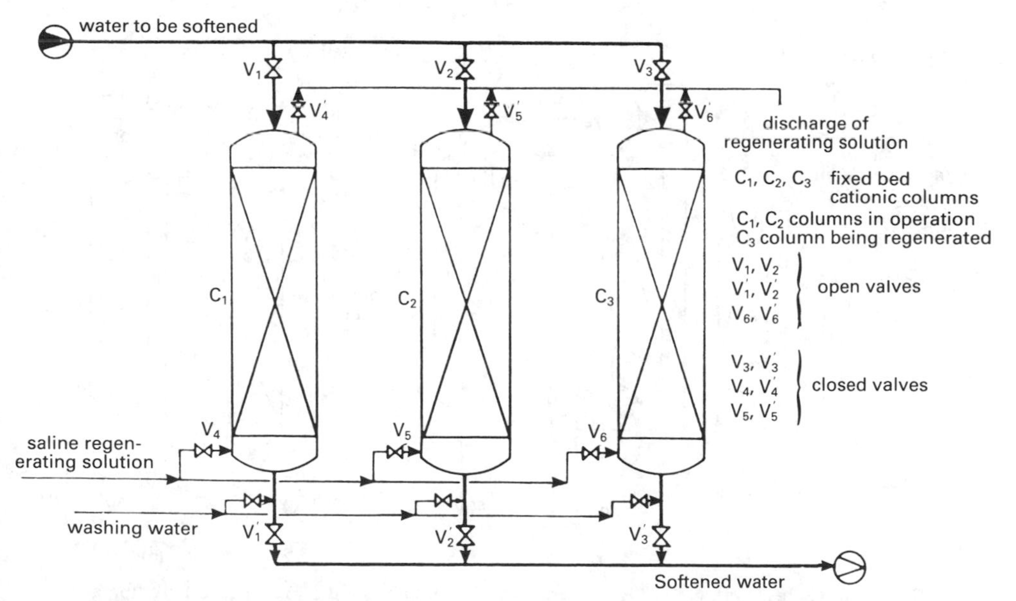

Fig. 5.9—A three-column plant for the softening of water. At any time two columns are in operation and one is being regenerated so as to ensure a continuous supply of softened water.

*If raw water were to be employed in this phase there would be a loss in the exchange capacity of the resin.

* in the column containing the strong cation exchange resin:

$$2ReH + CaCl_2(CaSO_4) \rightarrow Re_2Ca + 2HCl(H_2SO_4)$$

Contact between bicarbonate ions and the mineral acid, for example, then leads to:

$$HCO_3^- + HCl \rightarrow H_2O + Cl^- + CO_2.$$

The CO_2 is finally removed from the water flowing out of the plant by blowing in air.

The method would be optimal if there were to be stoichiometric correspondence between the temporary and the permanent hardness.

Figs. 5.10(a) and (b), respectively, schematically show two plant schemes for the implementation of the method of water softening which has just been described.

However, it is not easy to compensate for the ratio of alkaline bicarbonate arising from the temporary hardness and the acids arising from the permanent hardness of the waters which are treated.

For this reason, where waters of minimum hardness and which are free from potential alkalies are requested, a plant which is more complex than that shown in Fig. 5.10 is adopted. This plant consists of:

* a weak acid cation exchange resin column in which the following reaction occurs:

$$2ReH + Ca(HCO_3)_2 \rightarrow Re_2Ca + 2H_2O + CO_2,$$

* a tank which is used for the partial removal of the CO_2 by purging,
* a column packed with a strong acid cation exchange resin in which strong acids are liberated:

$$2ReH + CaCl_2(CaSO_4 \text{ etc.}) \rightarrow Re_2Ca + 2HCl(H_2SO_4 \text{ etc.})$$

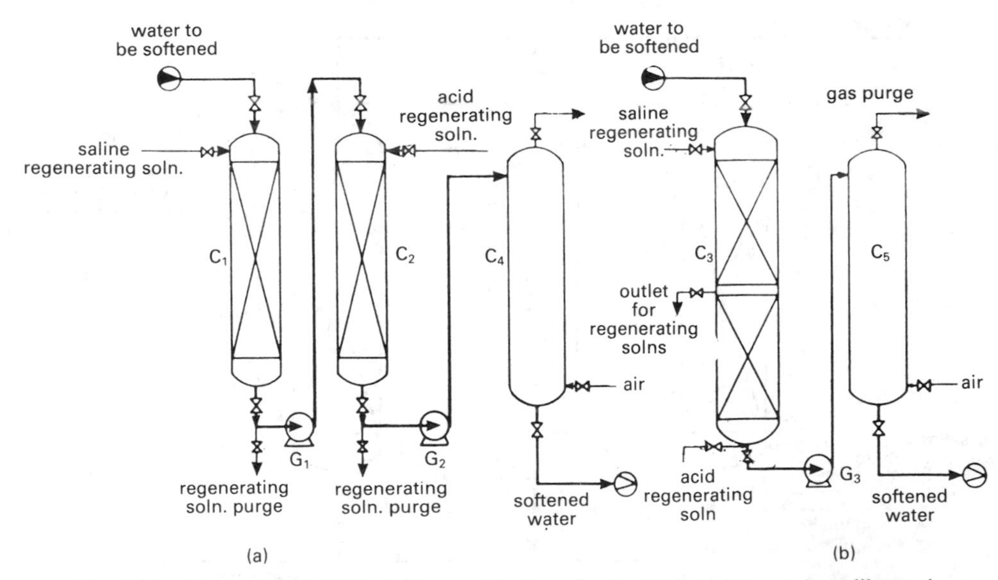

Fig. 5.10—Scheme (a). $NaHCO_3$ is liberated in C_1 and acids (HCl, H_2SO_4, etc.) are liberated in C_2. Contact between the bicarbonate and the acids leads to the development of CO_2 which is removed in column C_4 by stripping with air. Scheme (b). An analogous process to that described in (a), which uses the upper part of column C_3 as the entire column C_1 and the lower part as the whole of column C_2. Column C_5 is then identical in its structure and mode of operation to C_4.

where further CO_2 is released from bicarbonates as in the preceding column,
* a column for the complete purging of CO_2,
* a column containing an anion exchange resin which neutralizes all the acidity:

$$Re\overset{+}{N}H_3OH^- + HCl(HNO_3, \text{etc.}) \rightarrow Re\overset{+}{N}H_3Cl^-(NO_3^-, \text{etc.}) + H_2O.$$

A plant for the softening of water with the production of a neutral effluent is shown in Fig. 5.11.

In this type of plant the columns are regenerated with acid (HCl, HNO_3, H_2SO_4) in the case of cation exchange resins and with alkali (NaOH) in the case of anion exchange resins.

5.6 THE REMOVAL OF CATIONS FROM WATERS AND THE DEMINERALIZATION OF WATERS

Frequently, the complete removal of cations from water is the first stage of deionization, that is, the complete removal of all solute ions from the waters. For this reason it is expedient to consider the two processes together.

The removal of cations

This process is an end in itself when it is a question of recovering cations which are of interest (Cu^{2+} and Ni^{2+}, for example) or the elimination of noxious ions (Cd^{2+}, Zn^{2+}, Ca^{2+}, for example) from waters. The removal of cations is carried out by using cation exchange resins in the acid form. The reactions are of the type:

$$2ReH + MeX_2 \rightarrow Re_2Me + H_2X_2$$

where X indicates either $1/2SO_4^{2-}$ or Cl^- or another salt anion.

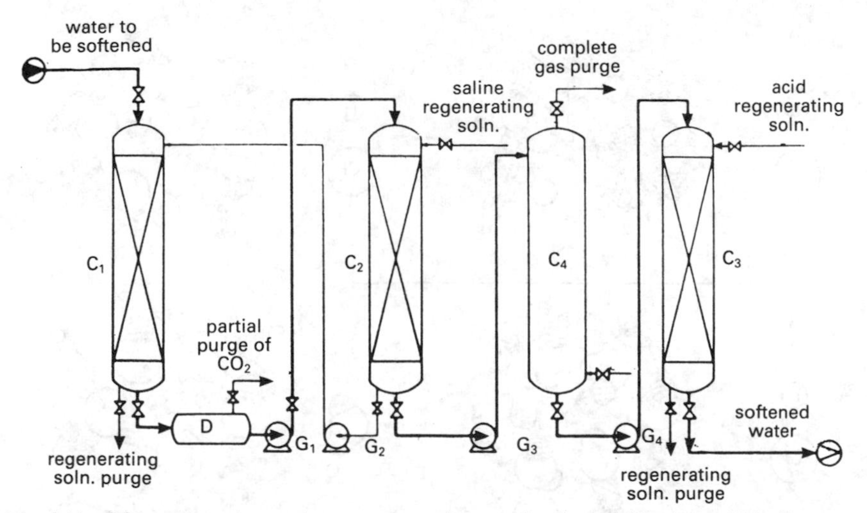

Fig. 5.11—The weak cation exchange resin exchanges the Ca and Mg ions of bicarbonates, the strong cation exchange resin exchanges the Ca and Mg ions of sulphates and chlorides, and the CO_2 is subjected firstly to partial and then to complete purging. Finally, the acidity is neutralized by using an anion exchange resin.

After it has attained its practical exchange capacity, the ion exchange resin is regenerated by using acids (HCl, H_2SO_4).

Deionization ('Demineralization')

The mechanism of the demineralization of water, using the simple example of the removal of NaCl, is shown in Fig. 5.12.

Ion exchange resins for demineralization are both costly in themselves and costly to operate. For this reason, when it is necessary to carry out the demineralization of waters with an appreciable degree of hardness, it is customary for the process to be preceded by softening treatments using mineral exchangers or calcium hydroxide as well as by pretreatments using coagulation, filtration etc. All of this is done to make the task of the resins easier.

The actual resin plant is composed of columns of cation exchange resins in the acid form, columns of weakly basic anion exchange resins, air degasifiers, columns of strongly basic anion exchange resins, and, finally, mixed bed columns containing a cation exchange resin in the acid form and a basic anion exchange resin. A diagram of the complete plant is shown in Fig. 5.13.

The plural which has been used above in listing the component parts of the demineralization plant which has just been described does not indicate that it is necessary to use more than one unit at each stage, but simply indicates the convenience of systematically having some beds in operation while others are being regenerated to maintain the continuous operation of the process.

In the case of mixed bed columns the regeneration is carried out in two stages: first, a solution of acid is passed over the cation exchange resin from the bottom and is made to come out, by means of aspiration, at a point corresponding to the interface between the two resin strata in the central zone of the column. A solution of alkali hydroxide is then passed into the bed from above and removed by aspiration at the point corresponding to the interface between the two resin strata.

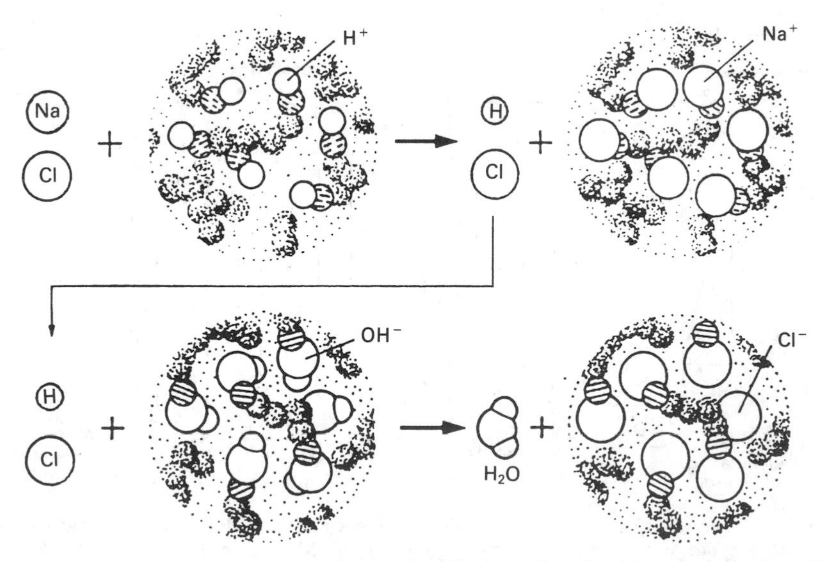

Fig. 5.12—The cations in the water to be demineralized are first fixed by a resin in the acid form and the anions are then fixed by means of a basic resin.

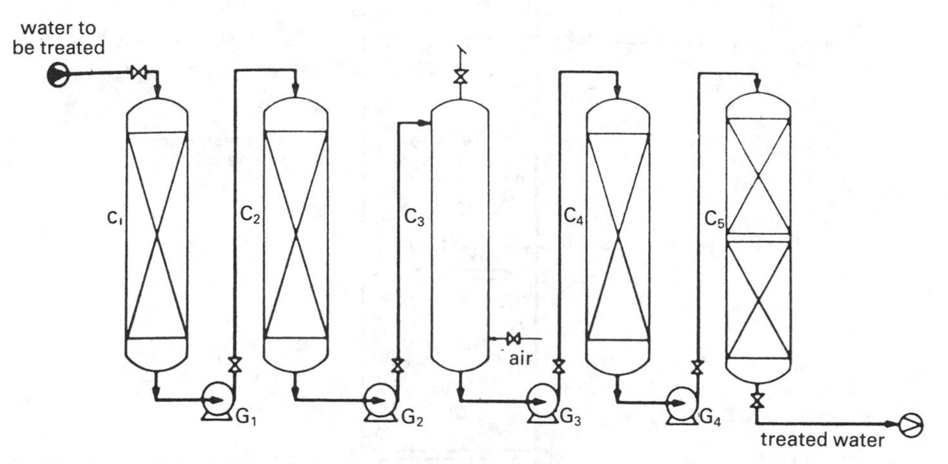

Fig. 5.13—Columns C_1 and C_2 eliminate the greater part of the cations and anions present in the water. Column C_3 is used to blow out the CO_2. Column C_4 is able to eliminate the silica. Finally, the mixed bed column C_5 removes the last traces of ions.

The two resins are subsequently washed with pure water, while air is blown in, to bring them into a state of suspension. Finally, the resins are washed in a counter current of water in such a way that the spheres of the resin of different densities stratify by carrying the cation exchange resins upward and the anion exchange resins downward. The countercurrent flow of washing water for the restratification of the resins is not shown in Fig. 5.14.

5.7 THE DESALINATION OF WATERS

The deionization of waters which has already been discussed is, in itself, a very refined method for the desalination of water. However, the term *desalination of water* is correctly used to indicate the process of the more or less complete elimination of salts from 'salt waters'. Nowadays, the desalination of water is of great practical importance owing to the ever-increasing general need for water.

Technological processes for the desalination of water consist of four operations which are of a predominantly physical nature: two (*distillation* and *freezing*) exploit changes of state, one (*reverse osmosis*) depends on osmotic phenomena, and one (*electrodialysis*) fully exploits the phenomenon of electrostatic attraction.

In the following sections we shall widen our knowledge of the four processes.

Distillation

'Distilled water' is a very widely known form of particularly pure water which is used in practice. It is traditionally obtained, as its name suggests, by distillation. Nowadays, the distilled water used in industry is produced by the deionization of drinking water or softened water, while production by true distillation is possibly confined to water which must have particular aseptic characteristics and be free from any residues of chemical reagents and/or traces of pathogenic germs.

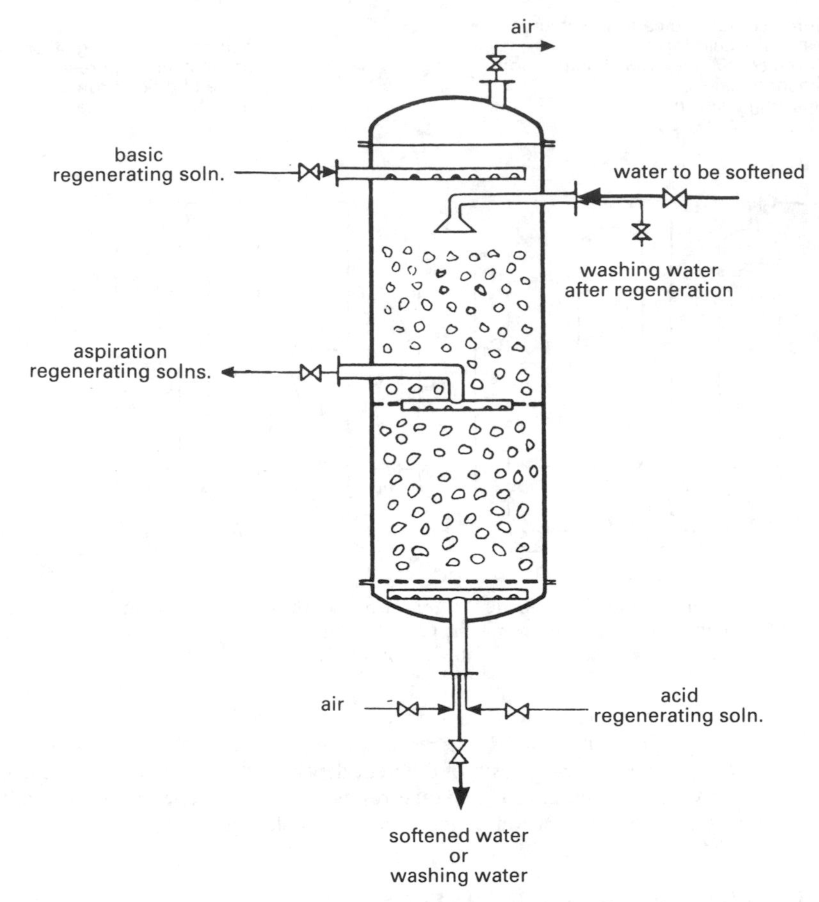

Fig. 5.14—Detailed scheme showing the functioning of a mixed bed column with indication of the air flow which displaces the beds which have been washed with water after regeneration.

The major obstacle to the systematic exploitation of the distillation of water lies in the large amount of heat which is required to evaporate water. In fact, when account is taken of the fact that it is necessary to heat the water up to 100°C and, subsequently, to supply the latent heat of evaporation, the heat required for the distillation of water exceeds 6×10^5 kcal/m^3, and, even if it is possible to make some economies by the recovery of heat, the costs are still always relatively high.

Today, distillation plays an extremely important role in the desalination of marine and brackish waters owing to the lack of operational systems which are clearly more advantageous in design. Among other things, the water obtained can be of extremely high purity (with a residual salt content lower than 5 ppm) and is therefore suitable both for industrial use and for civil purposes after suitable treatment to make it drinkable.

When distillation is adopted as a process for desalination, it is always the case that only a fraction of the aqueous solution treated is evaporated, thereby leading, in every case, to the production of desalinated water and brine. Another common characteristic of desalination methods which employ distillation is the use of a primary stage in which water is evaporated from a saline solution and a subsequent stage at which the vapour which has been produced is condensed with the recovery of the latent heat which is then used either in the preheating of the feed water and for the generation of other steam or for both of these purposes.

From the point of view of the energetics of the process, the less the energy needs to be supplied by external sources, the greater the amount of energy recovered internally during the process. The amount of heat recovered is a function of the efficiency of the heat transfer process by means of which it takes place. Such heat transfer is directly proportional, for equal times, to the area through which the exchange takes place, to the conductivities of the materials out of which the heat exchangers (coils, tubes, etc.) are constructed, to the temperature difference which, by suitable means, can be successfully exploited, and inversely proportional to the thickness of the walls through which exchange occurs and the extent to which it is encrusted with deposits.

In terms of the energy consumed, desalination by distillation is characterized by the evaluation of the *specific production* or 'yield' which is understood as the weight of water (in kg) which is desalinated per unit (kg) of reheating steam employed.

High-yield plants consume less energy, but investment costs are higher as it is necessary to increase the area of the heat exchange surface and/or to employ materials with a high thermal conductivity and/or to adopt anti-incrustation strategems and so on. The range of variability in the yield from desalination plants is rather wide, and may assume values from 6 to 12.

The environmental and operational chemicophysical conditions also reflect on the economics of the process. Above all, this is due to problems associated with incrustation and corrosion which make it necessary to close the plant down in order to clean it and to replace components. Incrustation problems can, to some extent, be avoided by means of pretreatments of the feed water and the adoption of proper values of the operating parameters (rate of flow, thermal gradients, etc.). The corrosion which is brought about by the high conductivity of the saline solutions which are treated and hydrolytic phenomena which become pronounced under the thermal conditions used in practice, are attenuated by a preliminary elimination of dissolved gases (oxygen and carbon dioxide, in particular) and are forestalled by the use of corrosion-resistant metallic materials possibly protected by resins and rubbers. Among other things, the heat exchange surfaces are made of copper and the pumps of stainless steel.

From a plant technology point of view, the systems which are employed for the concentration of other solutions by evaporation, that is, systems which were already in vogue for the production of distilled water from soft waters, today find only extremely limited use in the desalination of water by distillation. There are 'simple effect', 'multiple effect' and 'thermal compression' processes for details of which the reader is referred to courses on chemical plant design and operation.

In this field, today, evaporation processes based on multiple expansions (multiflash), which are shown schematically in Fig. 5.15, have become very successful. Multiflash processes are mainly used for the desalination of seawater to attain very high purities (residual salts at about 5 ppm) for industrial and civil use.

The water to be desalinated first passes through the bundle of tubes in the 'cooling section' E_1 of the plant and is then subjected both to anti-incrustation pretreatments in D and to degassing (the removal of O_2 and CO_2) in C_1. A portion of the brine is successively recycled and then passes through the bundle of tubes in the 'recovery section' E_2 where there is an increase in temperature at the expense of the heat of condensation of the steam which has been produced. Finally, the highest temperature in the cycle is attained in E_3, always by heat

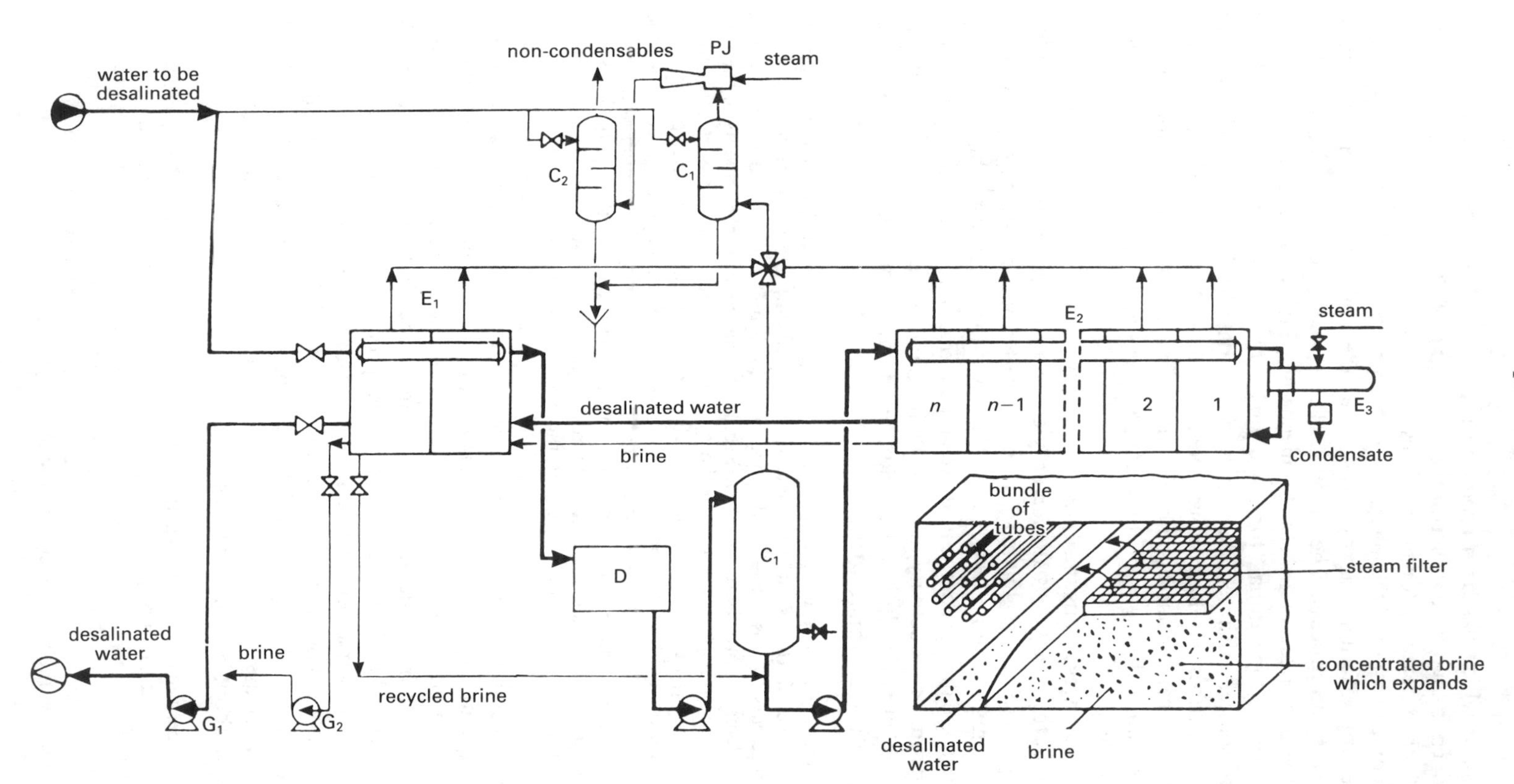

Fig. 5.15—The desalination of water by the 'multiflash' process which is described in detail in the text. On the right, under the recovery section E_2, the details of one 'stage' of such a multistage section is shown from which the mechanisms of the exchange, evaporation, and condensation processes can be seen.

exchange. After that, the saline solution enters into the first stage of the recovery section as a fluid which evaporates under reduced pressure. The steam produced yields heat by condensing on the bundle of tubes through which the water to be desalinated arrives, and the condensate formed is collected in a suitable duct as desalinated water.

The solution then passes into the following stages which operate in an analogous manner to that which has just been described but under an ever-increasing degree of vacuum and decreasing temperature. The mechanism by which every evaporator functions ('stage') is shown by a label on the right hand side under the figure.

The vacuum for the whole of the system is provided by the ejector PJ connected to the barometric condensers C_2 and C_3 which are fed from above with a fraction of the desalinated water.

The pumps G_1 and G_2 respectively extract the desalinated water and the brine from the last stage. The latter, as has already been stated, is partially recycled.

Freezing

This desalination process also includes the separation of water from a saline solution by the exploitation of the virtual existence of the barrier formed by the surface of separation between the liquid and the solid which is permeable to water molecules but not to the salts in solution. In fact, at a temperature below a certain value which is a function of the amount of salt in the water being treated, a solid phase separates out upon cooling which consists solely of water molecules (ice).

Freezing is a desalination process which, by operating at a low temperature, minimizes corrosion problems and is particularly favourable from the point of view of the energetics of the process in countries with a cold climate which, however, are generally less interested in the problem of the desalination of waters.

The energy required for the realization of freezing processes is essentially mechanical, and it is necessary to bring about large reductions in pressure to make the water evaporate (with a cooling effect) at the lowest possible temperature. From an energy consumption point of view, the rest of the process essentially consists of the recovery of refrigerating units.

The scheme shown in Fig. 5.16 lists, in a coordinated manner, the operations used in the total transformation of salt waters into desalinated waters by means of freezing.

The freezing of water with the separation of brine can also be carried out indirectly by exploiting hydrocarbons (propane and butanes) which have vapour pressures which are much higher than that of water and can therefore be evaporated at relatively high pressures. The heat of evaporation is, in this case, also (but indirectly) removed from the water which subsequently freezes. In their turn, the hydrocarbon vapours are recompressed and subsequently made to condense by exploiting the cooling due to the fusion of the ice which is produced.

Reverse osmosis

By putting into contact with a 'semipermeable membrane', that is, a porous membrane which allows the passage of solvent but not of solutes, the water tends to be carried to the other side of the membrane if there is a solution (salt species) present there. A hydrostatic pressure is thereby generated in the compartment containing the solution, the value of which, when equilibrium is reached, provides a measure (in atmospheres, for example) of the *osmotic pressure* of such a solution (Fig. 5.17(a)).

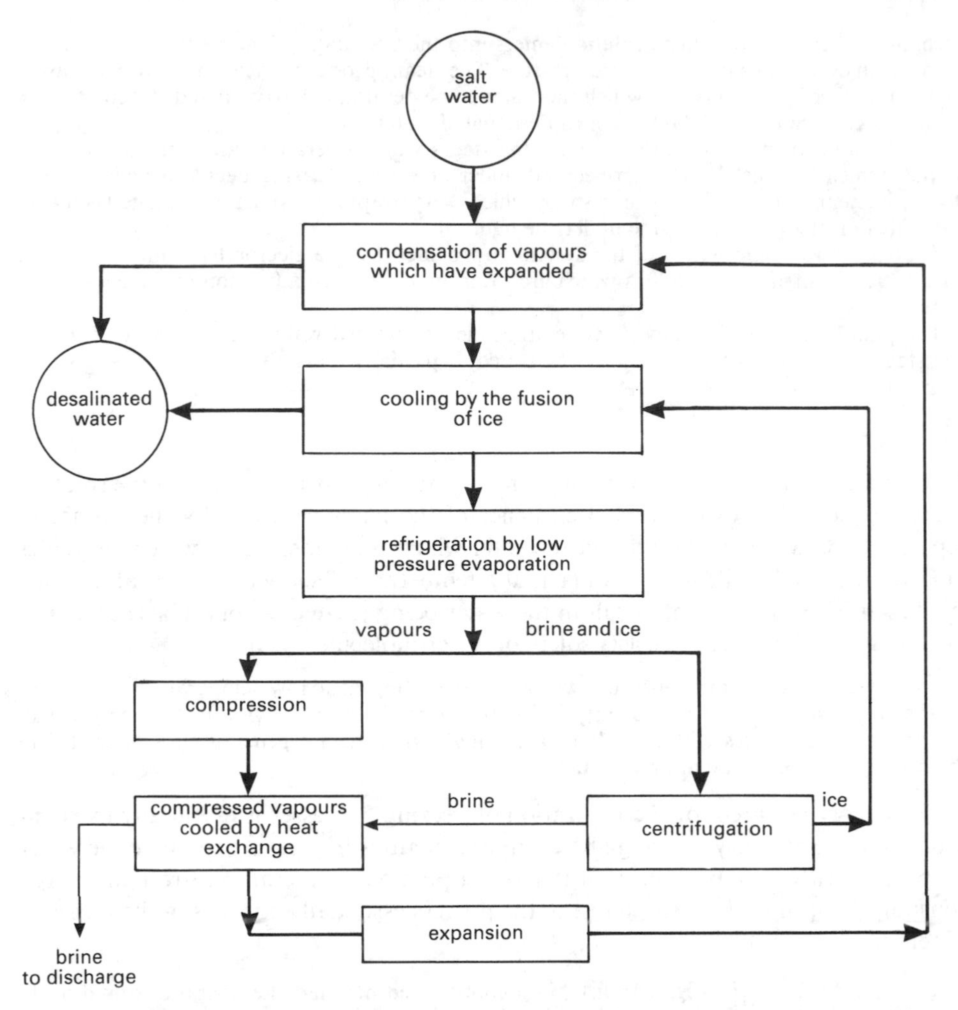

Fig. 5.16—Block diagram of the production of sweet water from salt waters by freezing, with recovery of the refrigerating units to make the process economically admissible.

From a physicochemical point of view osmotic pressure* is interpreted as the tendency of a solute to become more dilute, that is, to pull in water if the solute is entrapped by a semipermeable membrane as a consequence of the tendency of solutes to occupy, by diffusion, the greatest possible volume and to attain thereby the maximum possible degree of disorder ('entropy').

It is therefore understandable that, as a result of this natural tendency, the pure solvent (or the solvent in the less concentrated solution) should pass from one side of the membrane to the other in opposition to the hydrostatic pressure which, as a consequence, is set up on the side where the liquid arrives. Since this hydrostatic pressure continues to increase until there is an equal tendency of the solutes to diffuse, its value, measured under equilibrium conditions, directly expresses the value of the osmotic pressure of the solution.

The osmotic pressure is independent of the type of solute and of the number of 'molecules' dissolved, but is solely dependent on the number of particles contained

*The term 'pressure' therefore indicates only the manner in which this tendency is measured.

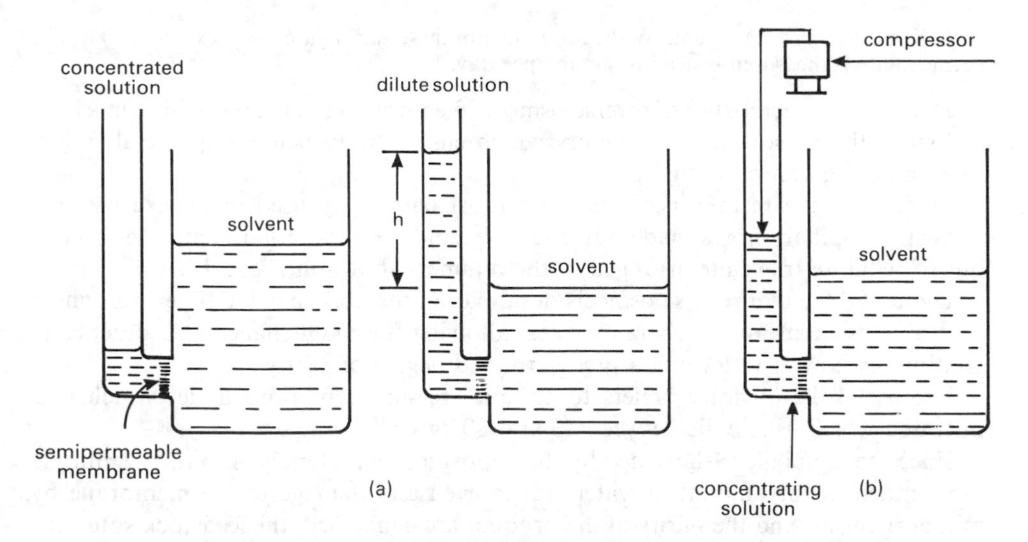

Fig. 5.17—In (a) it is seen that a semipermeable membrane allows the spontaneous establishment of a hydrostatic pressure on the side containing the solution. In (b) it is shown schematically that a mechanical pressure exerted on a solution in contact with a semipermeable membrane drives the solvent towards the solvent reservoir.

in the solution. The values of the osmotic pressure are therefore proportional to the number of gram-ions of solutes (not moles) in solutions which are electrolytically dissociated or ionized.

So, when 58.5 g (1 mole) of NaCl in water is taken, this does not produce an osmotic pressure proportional to one mole of solute, but, since the salt is dissociated into Na^+ and Cl^- ions, the resulting pressure has the value doubled by the number of gram-ions which are formed and is equal to that due to one mole of KI or 2/3 of a mole of $CaBr_2$.

For this reason, the osmotic pressures of salt waters are inherently high.

The phenomenon of osmosis is reversible in the sense that, if mechanical pressure which is higher than the osmotic pressure is applied to a solution in contact with a semipermeable membrane, the passage of solvent through the membrane is brought about, that is, 'reverse osmosis' is occurring (Fig. 5.17(b)).

Desalination by reverse osmosis therefore exploits the compression of saline waters, under a mechanical pressure which is greater than their osmotic pressure, against semipermeable membranes. As the process proceeds it is obvious that the mechanical pressure must be raised because of the increasing osmotic pressure resulting from the progressive rise in the concentration of the water undergoing desalination.

The decisive factor in the exploitation of this extremely interesting process for the desalination of water is the quality of the selective membranes which constitute the basic element of such desalination apparatus. These must be semipermeable, allow appreciable rates of flow, and be able to withstand both the high mechanical pressures employed in the process and chemical attack by solutes.

The semipermeability of the membranes is expressed as a *rejection*, that is, on the basis of their effective capacity to oppose the passage of solutes (to repel them) while the solvent flow rates are related to the *permeability* of the membranes to the solvent molecules. The rejection is measured as the percentage of the solutes initially present in the solutions treated which are still there after the treatment. With good membranes, this figure may exceed 95%. The permeability is measured instead in m^3/m^2 per day.

The main characteristics of reverse osmosis membranes are therefore high mechanical strength and solvent permeability which ensure from them an appreciable flow rate and a high rejection figure.

At present, the membranes which are most commonly used in reverse osmosis consist of capillary fibres made out of a polyamide resin and flat membranes made out of cellulose triacetate arranged in the manners shown in Fig. 5.18.

The useful life of a reverse osmosis desalination membrane is the time over which it retains its permeability characteristics, allowing the maintenance of a practically continuous flow of water and a predetermined degree of purity.

The useful life obviously refers to certain pressure conditions under which it is operated which usually lie between 20 and 80 kg/cm^2.

Since the useful life is shortened by the deposition of materials onto the membrane in so much as the amount of water transported and damage to the membrane by microorganisms and the purity of the product are concerned, the feedstock solutions are pretreated with hypochlorites to render them aseptic and with acids and sequestering agents to prevent incrustation and they are filtered to eliminate any turbidity.

Electrodialysis

Principle of the method

It is known that dialysis is the operation of separating certain solutes from other solutes owing to the fact that the first can pass through permeable membranes whilst the others cannot. In principle, this occurs because the membranes used for dialysis (dialyzing membranes) have apertures ('holes') of dimensions which allow certain solutes to pass but not others.

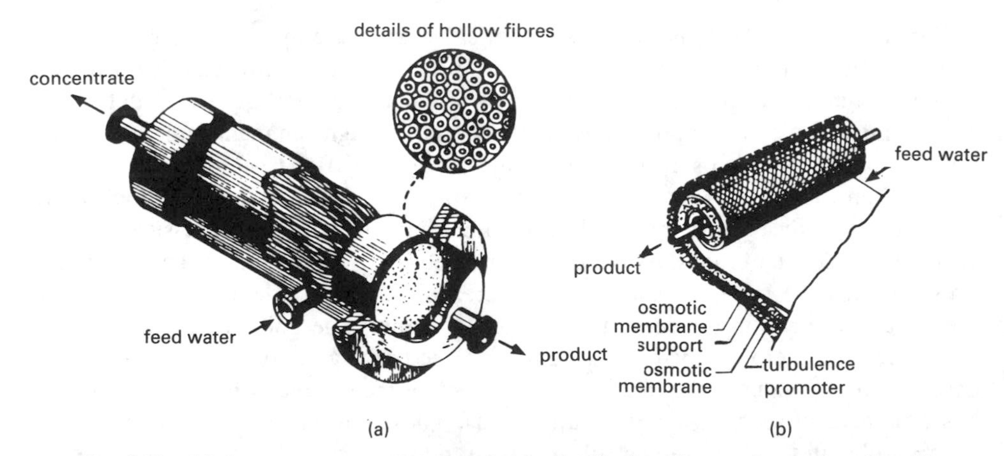

Fig. 5.18—(a) shows a structure consisting of a bundle of hollow fibres made out of a polyamide resin which, when surrounded along their external surfaces by salt water under pressure, permit desalinated water to pass through them. This emerges at one end as the 'product', while the concentrate emerges from the other end. (b) shows a structure formed from flat cellulose acetate membranes which have been wrapped around a central tube to form a roll. This central tube acts as a collector of the water which has been produced from the salt water which has been compressed around the membrane system which is placed in a cylindrical steel container.

To speed up the phenomenon, which is inherently slow, because the apertures do not allow the spontaneous passage of masses of ions and the membranes constitute a general obstacle to diffusion, a continuous difference in potential can be established by the application of electrodes of opposite sign. Then, the ions (which are generally the solutes with dimensions which are favourable for passing through dialyzing membranes) are prompted to migrate towards the electrode of the opposite sign to that on the ion.

Industrial electrodialysis cells

In industrial practice electrodialysis is carried out in cells with multiple compartments formed by special membranes which select the solutes which they will allow to pass. Dividing membranes are employed which can either be traversed by just positive ions or only negative ions. Since the membranes are arranged in an alternating manner with respect to the type of ionic migration which they permit, electrolyte drainage compartments ('drainage traps') and desalinated water compartments are set up. Fig. 5.19, which is appropriately labelled, illustrates the structure of such cells.

On the one hand, ions enter the drainage traps because they are attracted by the electric charge of the opposite sign to their own, and they encounter membranes

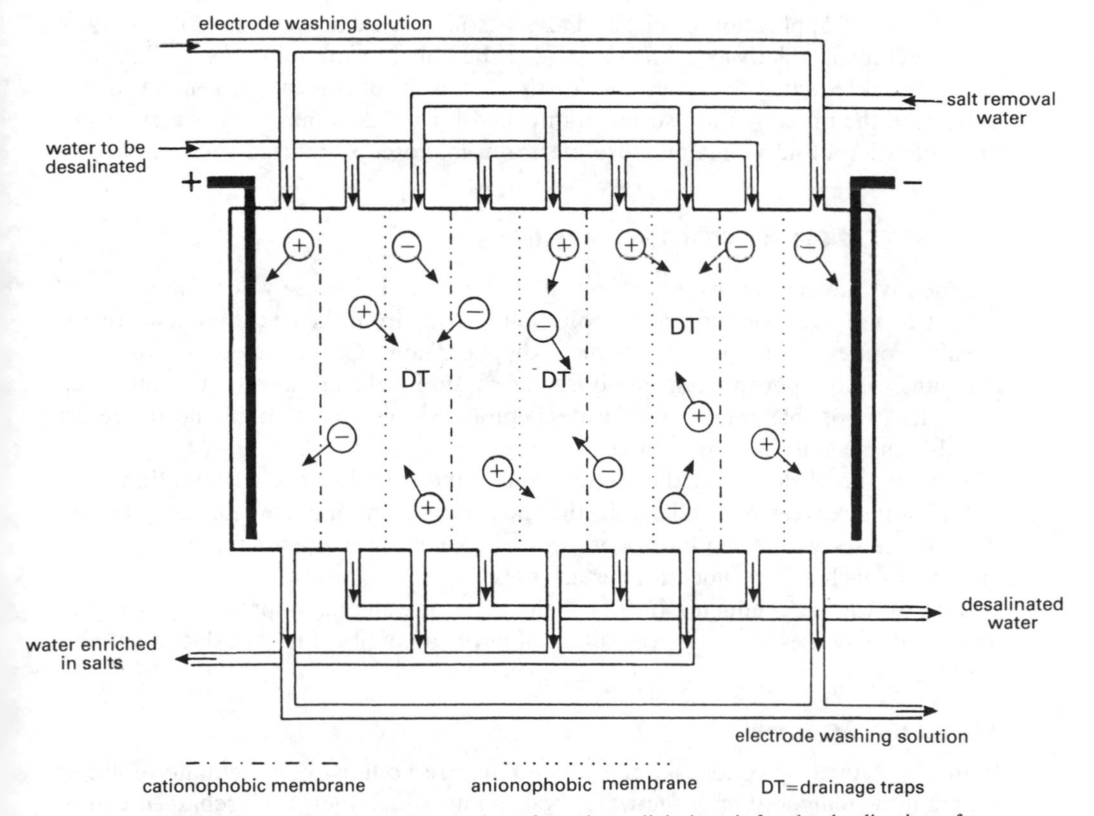

Fig. 5.19—Scheme showing the functioning of an electrodialysis unit for the desalination of water, indicating the possibilities which anions and cations have of crossing the membranes which form the partitions in the apparatus.

which are permeable to them and, on the other hand, they cannot get out of such traps because of the impermeability of the membrane on the other side although the electric field draws them towards it.

The structure of membranes and the applications of electrodialysis

Industrial electrodialysis membranes are 'cationophobic' if they are permeable to just anions, and 'anionophobic' if they allow only cations to pass.

Cationophobic membranes have tetralkylammonium cations $-\overset{+}{N}R_3$ attached to them, while the anionophobic membranes have sulfonate anions $-SO_3^-$ attached to them. Less commonly, such membranes may contain alkylammonium groups $-\overset{+}{N}RH_2$ or carboxylate groups $-COO^-$ respectively.

The preparation of these membranes consists of a first phase in which the polymer network is constructed with sites at which the respective $-\overset{+}{N}R_3$ or $-SO_3^-$ functions can be incorporated (benzene nuclei are usually the sites which lend themselves best to this) and a second phase during which such active groups are introduced by means of known synthetic routes (sulphonation, etc.). The frequency of the electrostatically repelling groups within the resin molecules must be high enough to ensure a strong rejection field.

The resins prepared in this manner are 'hooked', using special techniques, to mechanically resistant and chemically inert fibres (PTFE, for example). The pores with which the system must finally be furnished should be narrow and of an order of size of 25–35 Å.*

The field of application of electrodialysis is almost exclusively that of waters of low salt content: relatively saline rivers (e.g. the Nile), or the solutions produced by industries in recycling processes and for the treatment of effluents in general. In the latter case the process is not so much intended for the desalination of waters as for the separation of non-ionic macrosolutes from microscopic ionized solutes.

5.8 WATER CONDITIONING PROCESSES

Frequently, raw natural waters cannot be taken directly for use while reflow waters cannot be re-used owing to energy pollution (arising from their heat content) or the specific solutes or suspensions which they contain. Gases, solvents, iron and manganese salts, phosphates, microorganisms, and reducing agents are but a few examples of possible pollutants of waters employed for the first time and (there are possibly more still) in reflow waters.

Here, we shall deal with the elimination of individual causes of pollution from waters while successively indicating the most opportune moment for carrying out these processes within the framework of more general treatments for the cleansing of waters which are intended for certain uses.

A study will be conducted into how the state of waters is improved with regard to individual causes which lead to a fall-off in the amounts of it available of suitable quality.

The cooling of waters

Waters obtained by condensation and cooling are polluted or potentially polluted waters if the temperature of the water bodies into which they are discharged can be

*1 Ångstrom unit = 10^{-10} m.

significantly raised by the quantity and the heat content of the effluent waters. Moreover, it is frequently necessary to recycle such waters within the framework of the economy of the process.

In these cases, the need arises to remove the excess heat from the waters concerned, which is generally achieved by means of cooling towers (Fig. 5.20) where the water is distributed onto subdividing grills and subjected to currents of fresh air which are blown in. A particle separator prevents the entrainment of water by the air, which, having been drawn upwards by the fan, emerges from the top while the water which has been cooled is collected at the bottom.

It is inevitable that there is an appreciable loss of water due to the humidification of the cooling gas, and, for this reason, the effluent is reintegrated when it serves for recycling.

Degassing

With waters which are to be decontaminated by the removal of gases there is a need to distinguish cases in which such gases are 'living', that is, free gases which are dissolved in the water, and those in which the gases are only potentially present in them. An example of the second kind is presented by water which contains ammonia as NH_4OH or as an ammonium salt.

In the first case, which is typified by CO_2 and hydrocarbons, one resorts to blowing in air if, as is often true, the purified water can tolerate, as regards the destinations for which it is intended, a certain oxygen content. In other cases, a current of nitrogen

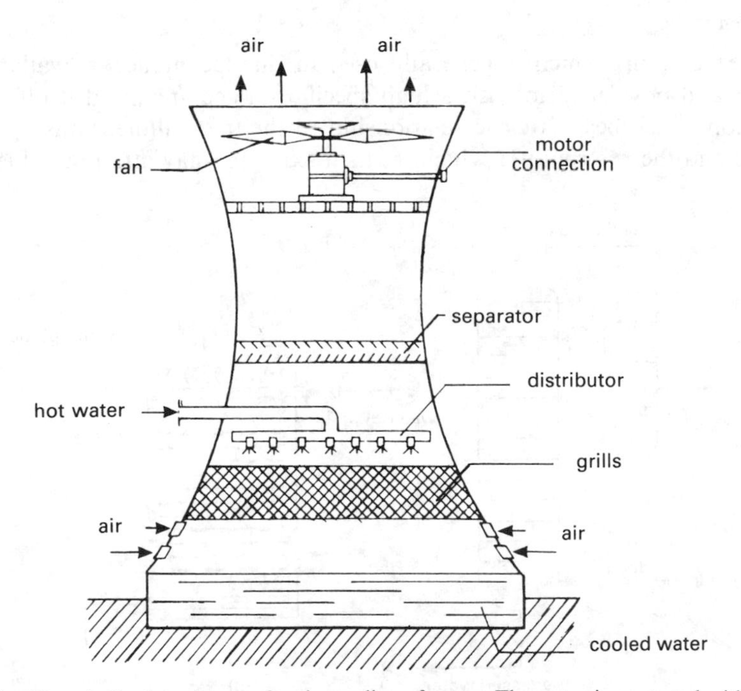

Fig. 5.20—Hyperbolic-shaped tower for the cooling of water. The water is contacted with air in the lower part of the tower, while the upper part provides a pathway for the aspiration of the air which carries away the heat.

is used. When nitrogen is employed it is necessary to recover the volatile gaseous materials which have been removed in order that the stripping gas may be recycled.

It is also possible to degas water with steam, using a countercurrent of steam in a tower fitted with plates (Fig. 5.21(a)). This exploits both the entrainment effect of the steam as well as the lowering of the solubility of the gases with increasing temperature.

In the case when the gases are only potentially contained in the water it is first necessary to liberate them by means of a chemical reaction. For instance, the addition of alkali to waters containing ammoniacal contaminants transforms them to gaseous ammonia which can then be stripped by using one of the methods described above. In Fig. 5.21(b) the stripping of ammonia from water at pH ≈ 12 is shown.

Adsorption

To remove certain substances from waters, recourse is made to adsorption on granular porous materials which, in this operation, are generally charcoals or *macroporous polymers*.

For the greater part, the mechanism of adsorption brings into play intermolecular links of the van der Waals' force type in the case of *physical adsorption* and true chemical bonds in the case of *chemisorption*. Physical adsorption is preferred both because it allows one to re-use the adsorbents by means of alternate adsorption and desorption cycles and because, when it is required, the adsorbate may also be recovered.

Use of carbons

'Activated carbon', which is generally used in this technique, is available in both granular and powder form with a high specific surface area, and it offers a rate of adsorption which bears a close relationship to the mean dimensions of the carbon granules and the external pH which, at its maximum, may attain a value of 8 to 9.

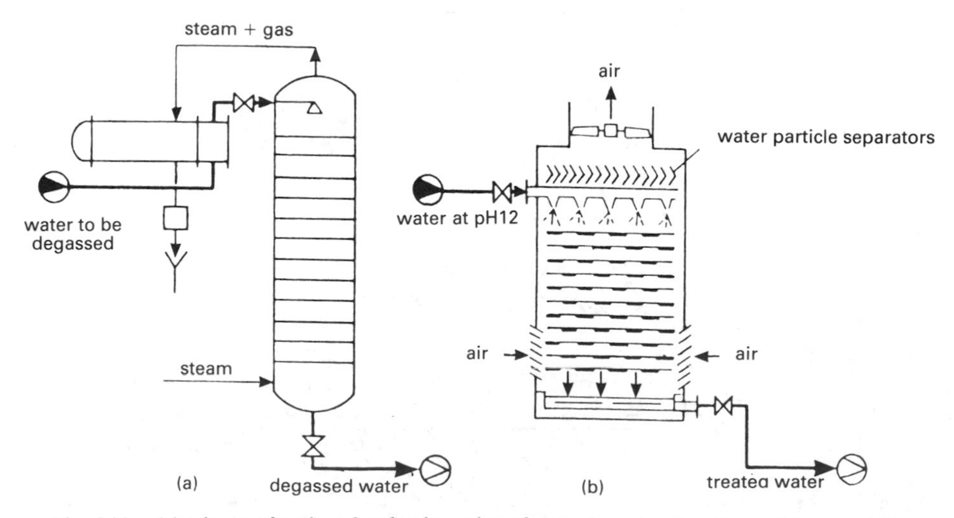

Fig. 5.21—(a) scheme of a plant for the degassing of water by using live steam, (b) method for stripping ammonia from water by means of air.

While powdered carbon is put into operation by its agitation within vessels, granular carbon is employed in columns arranged in series or in parallel. In Fig. 5.22 there are three columns, two of which are in an operating state while one is closed off to discharge the carbon so that it may be sent to the regeneration plant.

To avoid a rapid and even permanent reduction in the adsorption capacity of the carbon, the waters are subjected to a preventive treatment in which they are freed from turbidity and any suspended solids by clarifying them and, possibly, filtering them.

The granular carbon adsorbent is regenerated, according to the actual cases, using four distinct techniques:

- by washing with solvents,
- by washing with alkali or acid,
- by steam regeneration,
- by thermal regeneration.

Most frequently, one of the first three of these techniques is combined with the fourth. Regeneration is carried out in furnaces of the Herreshof type (page 192) at 900–930°C in an atmosphere with a low oxygen content in such a manner as to ensure the complete combustion or vaporization of the organic substances while confining the losses of carbon to less than 10%. Nevertheless, the adsorption capacity of the carbon falls off with every regeneration cycle.

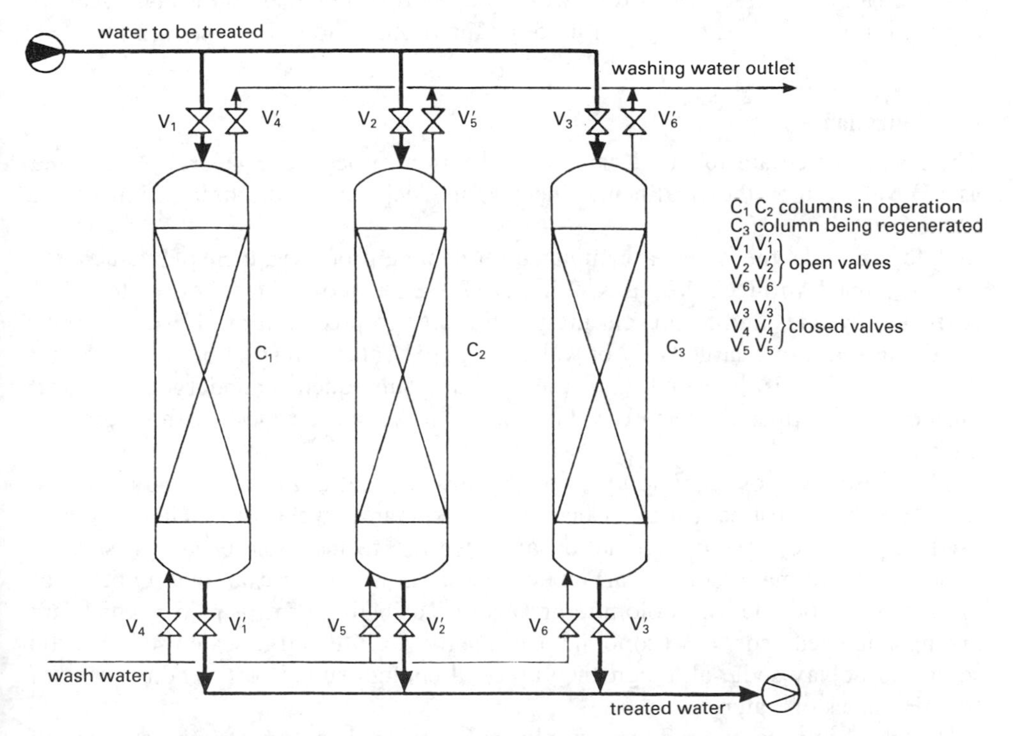

Fig. 5.22—Plant consisting of columns in parallel for the adsorption of, for example, non-polar high molecular weight organic substances on granular activated carbon. To understand how it functions, attention is drawn to the system of valves with which it is equipped.

Powdered carbon cannot be regenerated, and for this reason it is disposed of in some manner after it has been recovered by mechanical separation in apparatus reminiscent of the type used in clariflocculation (page 118).

The use of macroporous polymers

'Macroporous polymers' for adsorption vary in their constitution according to the nature of the compounds which are to be adsorbed. The polymers which are basically used are:

- styrene copolymers for the adsorption of apolar compounds,
- acrylic copolymers for the adsorption of rather polar compounds.

All such copolymers are resistant to the action of not too concentrated chemical reagents such as those which are always present in treated waters.

Then, by attaching particular groups to the copolymers, selective adsorption towards particular types of compounds such as phenols, amines, chlorinated compounds, etc. is guaranteed.

The macroporous polymers are prepared from the corresponding monomers by suspension polymerization in the presence of agents which induce porosity and which can be removed afterwards. These polymers are regenerated by steam stripping and elution with solvents. In every case there is the possibility of recovering the substances which have been adsorbed.

The treatment of recycled waters with macroporous polymers on a plant scale is carried out in the types of column used for the regeneration of active carbons.

Neutralization

The results which are required of a 'neutralization' process are generally evaluated as pH values after the treatment. These values depend on the destination of the treated waters.

If, for example, the waters are intended for cooling processes, their pH value must be maintained within quite narrow limits (7.3–8.3), above all, in order to avoid corrosion by acids and to prevent incrustation due to precipitation. If waters are to be discharged into a large body of water, their pH values must lie between 5.5 and 9.5, while, if it is intended that they should subsequently undergo biological purification treatments, their pH values must also lie, more or less, within the same limits.

The substance most commonly employed for the neutralization of acidity is lime water or, the almost equivalent, calcium hydroxide in powder form. These are also used because they favour coagulation and give rise to insoluble products (such as the calcium and magnesium bicarbonates present in water, for example). Quite a lot of calcium carbonate and dolomite are also used in the form of pieces which are arranged in fixed beds. For economic reasons, there is little large-scale use of caustic soda and Solvay soda, although they are well enough suited to the adjustment of the pH values of water.

When calcium compounds are employed as neutralizing agents, the amount of acidity due to sulphuric acid must be contained within limits so as not to produce appreciable amounts of calcium sulphate precipitates.

Sulphuric and hydrochloric acids are employed for the neutralization of alkalis, but it is still a common practice to reduce the alkalinity of recycled waters by mixing them with various acidic effluents.

Oxidations and reductions

Changing the oxidation number of several compounds of varying toxicity which are present, above all, in recycled waters has the aim of transforming these compounds into others which are less damaging or not damaging at all, or to promote their removal from the waters. For instance, sulphides can be oxidized to sulphates, and cyanides and cyanates can be oxidized to a mixture of nitrogen and carbonates.

Oxidations also often exhibit a biocidal action, that is, an action leading to the destruction of various microorganisms.

Among reductions which, in general, are not so extensively used as oxidations, that involving the transformation of Cr(VI) compounds into Cr(III) compounds stands out in importance.

Oxidations and reductions are carried out by a chemical route, that is, by the addition of suitable oxidizing and reducing agents to waters.

The commonest oxidizing agents are atmospheric oxygen, chlorine, various compounds of chlorine, and ozone. Less commonly, nitrates and oxygenated water are employed. The normal reducing agents are sulphur dioxide and certain of its derivatives, ferrous sulphate, hydrazine, and metallic iron.

It should also be mentioned that reductions are often complementary to oxidations in the treatment of waters. For instance, the excess of chlorine after an oxidation in which it has participated can be removed by means of hydrazine or SO_2:

$$N_2H_4 + 2Cl_2 \rightarrow N_2 + 4HCl$$

$$SO_2 + Cl_2 + 2H_2O \rightarrow H_2SO_4 + 2HCl.$$

The complementary nature of oxidation–reduction processes is also illustrated by the dechlorinations which are carried out by percolating the treated waters over calcium sulphite.

The nature and action of oxidizing agents

For a long time aerial oxygen has undoubtedly been the most important oxidant in the case of bodies of water, though this gas finds relatively little application in rapid and biocidal oxidations.

The most important fast-acting and biocidal oxidizing agents are certain chlorine compounds and, above all, chlorine itself.

Chlorine reacts with water in accordance with the equilibrium disproportionation:

$$Cl_2 + H_2O \rightarrow HCl + HOCl$$

which is displaced to the right under conditions which are favourable to the removal of HOCl.

The latter occurs:

(1) by the decomposition at a pH value of 5 to 6:

$$HOCl \rightarrow HCl + O$$

(2) by the formation of hypochlorite anions:

$$HOCl + OH^- \rightarrow H_2O + ClO^-$$

when the pH value of the treated water lies between 6 and 9.5.

The amount of chlorine employed varies according to the oxidation processes used: from a maximum for the oxidation of organic substances to a minimum to make water drinkable.

It is often necessary to study the operational conditions in the case of the destructive oxidation of organic substances with chlorine. For example, the chlorination of phenols is carried out in the presence of an excess of chlorine and at pH values lower than 7 in order to avoid the inherently high reactivity of chlorine with respect to electrophilic substitution reactions in an aromatic ring.

Among the inorganic substances which are the most readily oxidized by chlorine in water are, as has already been stated, sulphurous acid and cyanides. The latter are readily oxidized in an alkaline environment (pH of 10–11):

$$CN^- + 2OH^- + Cl_2 \rightarrow CNO^- + 2Cl^- + H_2O$$

$$2CN^- + 5Cl_2 + 12OH^- \rightarrow N_2 + 2CO_3^{2-} + 10Cl^- + 6H_2O.$$

About 3 g Cl_2 per g of CN^- is consumed in the first of these reactions, while about 7 g Cl_2 per g of CN^- is required to guarantee that the second reaction occurs. The alkalis must then be systematically added to the reaction mixture to ensure that the correct pH for the process is maintained.

Chlorine also exhibits a powerful antiseptic action.

This antiseptic action of chlorine is traditionally interpreted on the basis of the oxidation of the protoplasm of a bacterial cell brought about by the oxygen liberated from the hypochlorous acid which is first formed. More recently, it is attributed both to the denaturation of the proteins which are constituents of algal and bacterial structures and to postulated chemical transformations of dietetic and reproductory factors in the bacteria which are brought about by hypochlorite ions, ClO^-.

Of the oxygenated compounds of chlorine which are used as oxidants for aqueous effluents, there is interest in:

- *bleaching powder*, Ca(OCl)Cl, which is available as a powder and acts by releasing chlorine;
- granular *calcium hypochlorite* and *sodium hypochlorite* in solution which act, via the hypochlorite ions ClO^- which are strongly oxidizing and as has already been said, antiseptics;
- *sodium chlorite*, which is stored as a solid and when dissolved in water, furnishes oxygen;
- *chlorine dioxide*, produced at the moment of use by the reaction:

$$2NaClO_2 + Cl_2 \rightarrow 2NaCl + 2ClO_2$$

which is strongly oxidizing because it yields both chlorine and oxygen.

From a technical point of view, it is necessary to take precautions of a general nature regarding the storage, transport, and handling of these compounds. Thermal isolation is necessary for the storage of these compounds to prevent decomposition, and they are transferred by means of tubes of hardened rubbers and reinforced plastics while, for their metering, volumetric apparatus lined with suitable plastic materials are used to avoid having to use expensive metals and alloys.

Ozone. Produced, as is well known, by a silent electric discharge in a flow of air, the gas stream containing the ozone is directly injected into the effluents to be treated where it exhibits powerful antiseptic, anti-odour, and strongly oxidizing properties, especially, with respect to cyanides, phenols and aromatic amines.

Reducing agents and reduction processes

The reducing agents most commonly used in the treatment of waters are *sulphur dioxide* and *derivatives of* SO_2. In order of their importance, they are:

- sodium metabisulphite, $Na_2S_2O_5$;
- sodium bisulphite, $NaHSO_3$;
- sodium sulphite, Na_2SO_3.

All of these compounds act in such a way that they are oxidized to sulphates and are active in one of the reactions schematically depicted below for two of them which involve the reduction of dichromates*:

$$Cr_2O_7^{2-} + 3SO_2(3SO_3^{2-}) + 2H^+(8H^+) \rightarrow 2Cr^{3+} + 3SO_4^{2-} + H_2O(4H_2O).$$

When the reduction is carried out with sodium metabisulphite $Na_2S_2O_5$ ($Na_2SO_3 + SO_2$), the process involves the complete coupling of the reactions which have just been schematically balanced.

As the reduction potential of chromates is a function of the ambient hydrogen ion concentration, the operation is easier, the lower the pH at which it takes place.

For reasons of cost, hydrazine is, in general, a reducing agent which complements SO_2 and its derivatives, in the sense that it is used to complete reductions which, for the greater part, have been brought about by sulphite reducing agents. It acts in the following manner:

$$2Cr_2O_7^{2-} + 3N_2H_4 + 16H^+ \rightarrow 4Cr^{3+} + 3N_2 + 14H_2O.$$

The relatively high cost of hydrazine is therefore partly compensated for by the fact that it is a reducing agent which leaves no traces.

Owing to the large amounts of sludges which are formed, iron and ferrous salts are less convenient reductants. Their use is generally connected with their ready availability. This is particularly pronounced in certain iron-working industries where they are produced both during machining as well as during fettling operations.

The removal of iron and manganese

Iron and manganese are present with an oxidation number of two, both in waters being used for the first time and in recycled waters. The elimination of these ions, which is required in the process of rendering the waters drinkable and in the case of industrial water supplies, is based on simple treatments in which these two metals are oxidized to an oxidation state of three by means of atmospheric oxygen. The maximum rate of reaction is attained at pH 7–7.5.

---

*One often speaks of the reduction of 'chromates', but only dichromates exist in an acidic environment.

The reactions are:

$$2Fe^{2+} + 1/2O_2 + 5H_2O \rightarrow 2Fe(OH)_3\downarrow + 4H^+$$

$$2Mn^{2+} + 1/2O_2 + 3H_2O \rightarrow 2MnO(OH)\downarrow + 4H^+.$$

Temporary hardness in water is favourable to the removal of iron and manganese because it exhibits a buffering effect:

$$HCO_3^- + H^+ \rightarrow H_2O + CO_2.$$

In other cases it is necessary to operate on carbonate beds or with added doses of alkali.

Where there is a need for water containing the lowest possible amounts of iron and manganese, the operation which has just been described is refined by using ion exchange on resins or, more specifically, on 'manganese permutits', in which the manganese is present in a tetravalent state.

Manganese permutits are compounds which are prepared from normal sodium permutits (page 124) by washing them with a solution of a manganous salt and subsequently oxidizing them by percolating a permanganate solution through them.

When the molecular formula of the exchanger is indicated by Pe_2MnO, the reactions leading to the fixation of Fe^{2+} and Mn^{2+} ions are:

$$Pe_2MnO + 2Fe^{2+} + 5H_2O \rightarrow 2Fe(OH)_3\downarrow + 4H^+ + Pe_2Mn$$

$$Pe_2MnO + 2Mn^{2+} + 3H_2O \rightarrow 2MnO(OH)\downarrow + 4H^+ + Pe_2Mn.$$

The permutit beds are regenerated by washing them in an aqueous countercurrent which carries away the $Fe(OH)_3$ and $MnO(OH)$ floc. They are subsequently reoxidized with permanganate.

In a plant for the complete removal of iron and manganese (Fig. 5.23) compressed air is injected into the water which then passes to a carbon filter coupled in parallel to a second filter which is regenerated as the first filter operates, and vice versa. $Fe(OH)_3$ and $MnO(OH)$ are deposited on the carbon which catalytically favours the oxidation while the flow of water decreases owing to clogging. More rarely, it is necessary to renew the carbon which has been subjected to mechanical degradation.

After percolating through the carbon, the water is de-aerated and then passed alternately, because here also an active phase and a regenerative phase are necessary, through columns filled with permutits.

The removal of phosphates

As has already been stated, the amounts of phosphates in waters which are discharged is one of the major causes of eutrophication. On account of this, among other treatments, it is necessary to remove the phosphates from the sewage consisting of the discharges from various domestic and mechanical engineering cleansing processes, from the washing and various treatments of fibres and cloths, from fermentation and cheese-making activities, and where the cations in the waters have been sequestered.

As a result of many of the processes which have just been mentioned, the phosphate is present in urban discharge waters in the form of orthophosphates, polyphosphates, and organically combined phosphate. Of these three forms, orthophosphate is the most readily removed by chemical precipitation. Nevertheless, biological treatments carry away the 'organic' phosphate and part of the phosphorus present as polyphos-

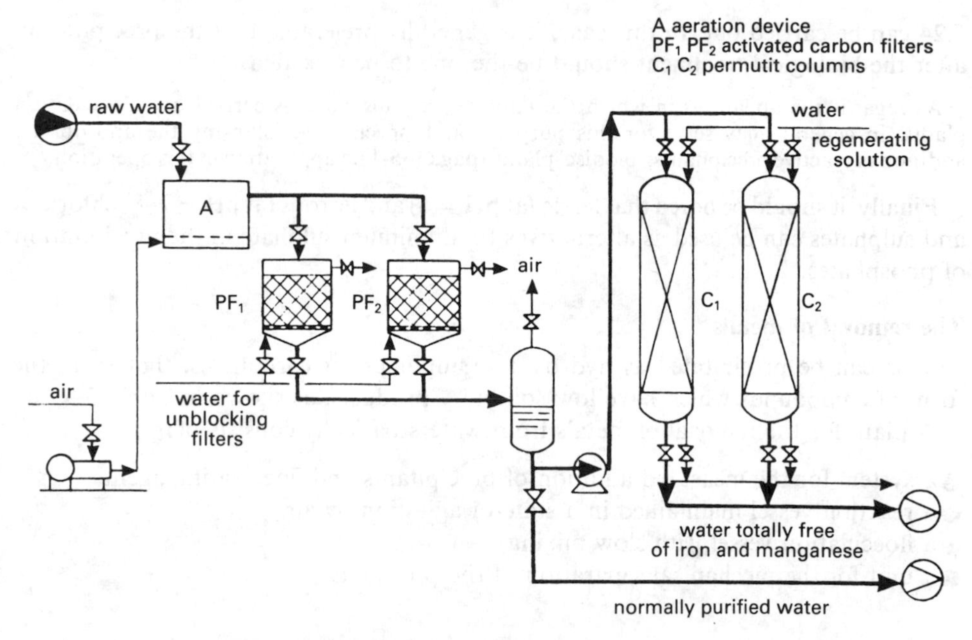

Fig. 5.23—Plant for the normal and refined purification of waters by the removal of iron and manganese. By suitable operation of the system of valves the two carbon filters PF_1 and PF_2 and the two columns C_1 and C_2 may be alternately put into operation or regenerated.

phate into the sludges, while almost all the remaining phosphorus present as polyphosphate is converted into orthophosphate.

Allowing for this, the outline scheme in Fig. 5.24 shows the way in which all the phosphates can be removed from water in a practical manner.

When such a complete removal of phosphates from aqueous effluents is not required, only one of the three precipitations indicated in the scheme depicted in Fig.

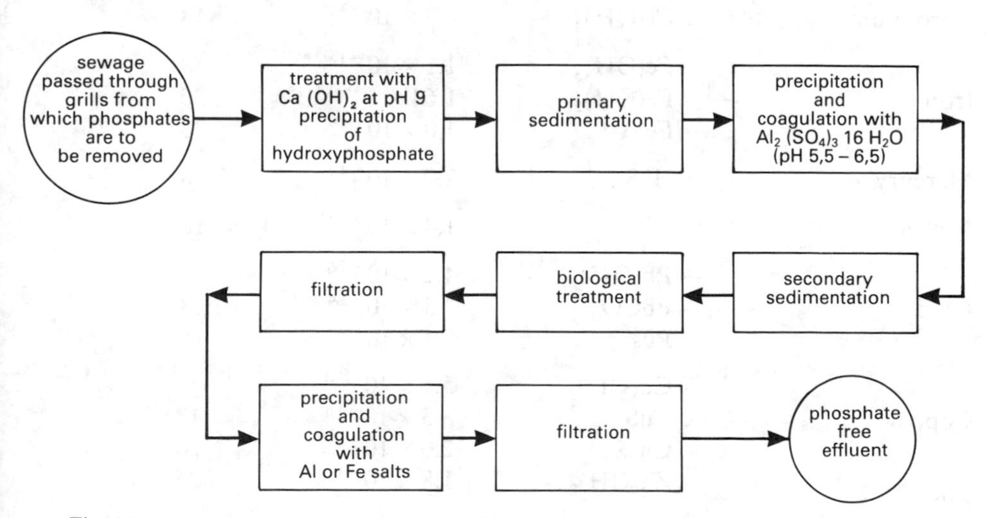

Fig. 5.24—Outline of the treatments in an integral scheme for the removal of phosphates from sewage which is initially highly contaminated with phosphates.

5.24 can be carried out. In this case, it is generally preferable that the precipitation after the biological treatment should be the one to be executed.

As regards the apparatus in which the removal of phosphates is carried out, flocculation plants, in general, may serve for this purpose, and, specifically, following the appropriate addition of reactive precipitants, 'biodisc' plants (page 155) find application in this operation.

Finally, it should be noted that ferric (at pH = 5) and ferrous (at pH = 7–8) chlorides and sulphates can be used as alternatives to aluminium sulphate in the precipitation of phosphates.

The removal of metals

Metals can be precipitated as hydroxides, sulphides, or carbonates, that is, in the form of compounds which have low solubility products, as shown in Table 5.5.

A plant for the removal of metals from waters basically consists of:

- a system for the measured addition of precipitants and flocculating agents,
- a reaction vessel maintained in a state of agitation by air,
- a flocculation vessel with slow mixing,
- a unit for the mechanical separation of the precipitate.

Table 5.5—Solubility products of compounds used for the removal of metals from solutions

Metals	*Formula of compound*	*Solubility product*	*Temperature/°C*
Silver	$AgOH$	1.52×10^{-8}	20
	Ag_2CO_3	6.15×10^{-12}	25
	Ag_2S	1.6×10^{-49}	18
Cadmium	CdS	3.6×10^{-29}	18
Chromium	$Cr(OH)_3$	6.7×10^{-31}	18
Iron	$Fe(OH)_3$	1.1×10^{-36}	18
	$Fe(OH)_2$	1.64×10^{-14}	20
	FeS	1.0×10^{-19}	25
Mercury	HgS	2.0×10^{-49}	18
Nickel	NiS	1.4×10^{-24}	18
Lead	$Pb(OH)_2$	4.2×10^{-16}	25
	$PbCO_3$	3.3×10^{-14}	18
	Pbs	3.4×10^{-28}	18
Copper	$Cu(OH)_2$	5.6×10^{-20}	
	CuS	8.5×10^{-45}	18
	Cu_2S	2.0×10^{-47}	20
Zinc	$Zn(OH)_2$	1.8×10^{-14}	25
	ZnS	1.2×10^{-22}	18

The precipitate may be separated by sedimentation and/or by filtration, and sometimes even flotation can be successfully employed.

It is also necessary to provide a means of efficiently controlling the pH which must assume the specific values according to the precipitate which is being obtained.

* * *

Methods for the removal of specific metals from water are summarized in Table 5.6. Such procedures are also applicable, however, to other metals which have to be separated.

5.9 THE BIOLOGICAL TREATMENT OF WATERS

Microbiological activity in treatments for the purification of waters

Nowadays, certain biological processes which destroy organic substances by mechanisms which are analogous to those operating in the autopurification of a body of water are employed on a vast scale to eliminate such substances from effluent waters.

The difference lies in the fact that, when urban and industrial effluents are purified by the removal of organic substances from them by purpose-built apparatus with very high concentrations of the agents promoting the destructive action, there is an increase in the velocity and yield of the transformation in comparison with the corresponding natural processes.

As in nature, the technological conditions under which the biological water purging processes are carried out may be aerobic or anaerobic, that is to say, characterized

Table 5.6—Examples of summarized procedures for the removal of metals from waters

Metals	*Precipitate formulae*	*Reagents*	*pH*	*Separation*	*Precautions*
Chromium	$Cr(OH)_3$	$Ca(OH)_2$	9–10	clariflocculation, filtration	precipitation only after complete reduction
Mercury	HgS	H_2S	8.5	sedimentation, filtration, adsorption	final separation of excess H_2S
Lead	$\begin{cases} PbCO_3 \\ Pb(OH)_3 \end{cases}$	$\begin{cases} Na_2CO_3 \\ Ca(OH)_2 \end{cases}$	6–10	sedimentation and filtration	optimal pH value determined during the exercise
Copper	$Cu(OH)_2$	$Ca(OH)_2$*	9–10.3	flocculation filtration	preventative removal of CN^- and NH_3
Zinc	$\begin{cases} Zn(OH)_2 \\ (ZnS) \end{cases}$	$\begin{cases} Ca(OH)_2 \\ (H_2S) \end{cases}$	9–10 (8.5)	sedimentation filtration on sand	preventative removal of CN^- and chromates

*Caustic soda is used as the precipitant only in the presence of a large quantity of sulphate ions because the calcium sulphate, which would otherwise be precipitated, occludes $Cu(OH)_2$ with serious economic losses of metal.

by the intervention of air or by the absence of air. Heterotrophic microorganisms are present in both cases. These are microorganisms which require organic substances as plastic cellular material and as a substrate for energy production.

The scheme shown in Fig. 5.25 illustrates the mechanisms of action for the two types of heterotrophic microorganisms generally used in the processes of biological water purification.

It can be seen from this scheme that the two types of microorganisms differ in the mechanism of their action, since the former take energy from oxidation reactions while the latter obtain energy from the rupture of chemical bonds.

However, in artificial biological treatments intended for water purification, as in the rest of nature, microorganisms also operate which may be active both in an aerobic sense as well as in an anaerobic sense. The kind of activity which they exhibit is determined by the environmental conditions under which they happen to find themselves.

This should not surprise us if it is recalled that even higher organisms such as man can obtain the energy required to sustain life both from respirative (oxidative) processes as well as from fermentations (degradative processes).

The reagents which can potentially act in two ways are referred to as *optional microorganisms*.

Operating conditions and practical results

Artificial biopurification treatments never require complex environmental conditions but take place, generally, under the same conditions as in nature.

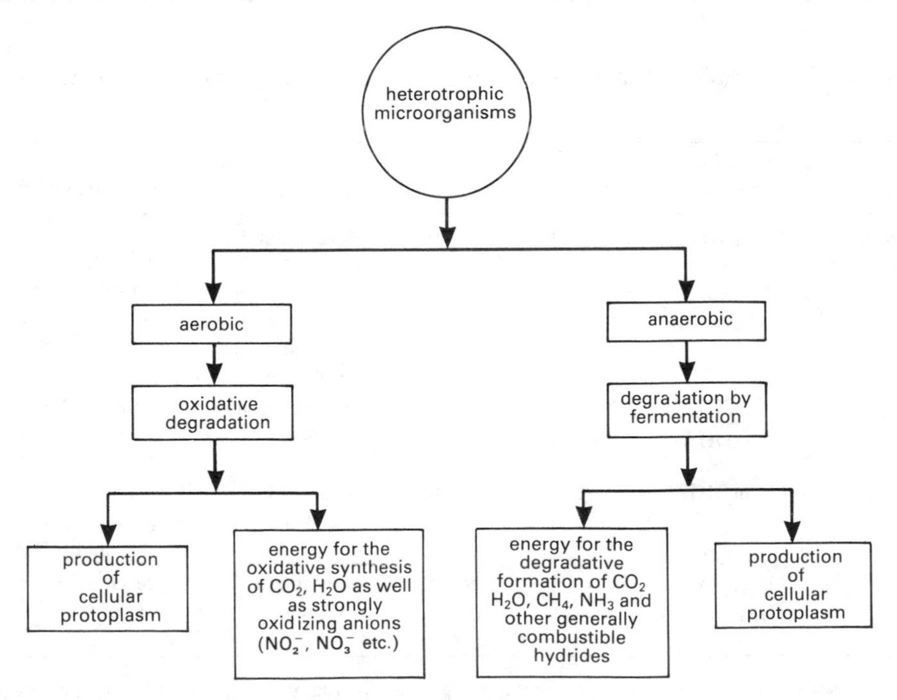

Fig. 5.25—Mechanism and the products of reaction of heterotrophic aerobic and anaerobic microorganisms.

The complexity of the situations arising in nature with regard to the development of microorganisms is evident. In fact, three zones can be recognized in every body of water (pond, lake, etc.): an upper zone, a central zone, and a lower zone. Aerobic microorganisms develop in the upper zone, microorganisms exhibiting both aerobic and anaerobic behaviour develop in the central zone (where there may be a lack of oxygen), while anaerobic microorganisms develop in the lower zone where oxygen is systematically absent.

In technological biopurification processes the type of microorganisms to be made to act is singled out and the conditions most favourable to their cleansing action are adopted, as this increases the efficiency of the treatment.

The choice of the type of microorganisms and the conditions is made on the basis of the results which one wishes to obtain; microbiological aerobic processes which proceed at a high rate are adopted when the purification of the water is the main aim, but anaerobic microbiological processes are used when alternative energy in the form of biogas is to be produced by a slower route requiring little waste of energy. The fundamental characteristic of aerobic processes is their utilization of the oxygen dissolved in the water under conditions which are favourable to the maintenance of the activity of the microorganisms.

The result is the production of a large amount of flocculant biological material which remains attached to the surfaces of the treatment apparatus in certain types of plant and remains dispersed in the mass of liquid in other types of plant. In every case the flocs of biological material pick up fine colloidal particles and adsorb other dissolved substances.

As the microorganisms remain active provided that the concentration of oxygen in solution never drops below a certain level, it is necessary to continuously resupply oxygen by means of the appropriate devices.

In anaerobic processes, the growth of the microorganisms, which are supplied with energy by the rupture of chemical bonds, leading to 80–90% of the organic substances present in the water being converted into methane (as well as into NH_3, H_2S, and PH_3) and CO_2, is low. The sludges are therefore of small volume and are readily disposed of. In these processes no energy is required for the supply of oxygen to the system but only to warm up the surroundings somewhat and to stir the mass. However, the methane ('biogas') obtained can be used to produce an amount of energy which is greater than that required to maintain the process in which it is formed.

Methods for the exploitation of aerobic microorganisms

Technologies and structure of equipment

The principal technological processes in which aerobic microorganisms are made use of are:

- the use of percolating and biodisc filters,
- treatments with active sludge systems,
- the use of oxidation basins ('aerated lagoons'),

in conjunction with other auxiliary and complementary plant structures.

While differing in the structure of the equipment required and the manner in which the processes are carried out, the three methods are analogous in that the mechanisms underlying the biochemical transformations of the organic substances are the same.

Percolating filters consist of a porous bed, formed from quite large randomly arranged pieces of material (stones, mineral coal) or materials of preformed shape (bricks, sheets of plastic) which are normally moulded to produce a honeycomb-like structure. These are structures with an enormous area per unit volume, and this large surface is covered with a biologically active film.

Passage of the liquid over the biological film permits the biochemical transformation of the organic substances carried in the water, and distribution of the liquid is entrusted to a rotating arm system fitted with spraying nozzles located over the whole of the active section of the bed.

The growing mass of cells is continually renewed because any excess becomes detached from the biological film in the form of flocs and passes out of the apparatus with the effluent. On the bottom of the apparatus there is a drainage system for the treated water which is subsequently sent to a secondary sedimentation vessel in which the biological slime is separated out. Such a system is illustrated in Fig. 5.26.

The small fraction of the oxygen used up in the biochemical reactions is that ordinarily present in the water being treated, while more comes from that which is picked up as the water is distributed onto the bed and, above all, from that which is dissolved in the liquid film of the porous bed both naturally as well as a consequence of the drawing in of air brought about by the drainage of water through the perforated layer. To raise the amount of oxygen present still further, the rate of circulation of air in the porous bed can be increased over that resulting from natural draughts.

Artificial forced and induced aeration is practised in medium-capacity and, especially, high-capacity* percolating filters and, according to the ventilation mechanism, may be arranged at the bottom or the top of the bed. Artificial aeration, apart from permitting the use of higher capacity percolating filters, also prevents the blockage of their drainage holes.

As far as the effect produced by the filter is concerned, it is observed that the

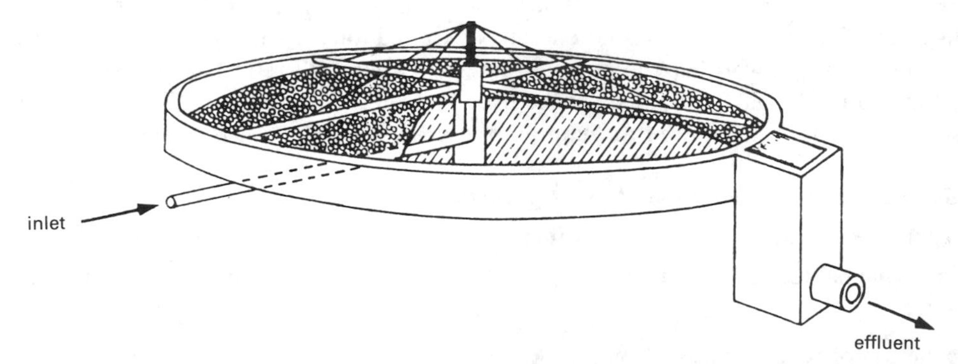

Fig. 5.26—Scheme showing the arrangement of the input, liquid sewage distribution, and outlet in a Dorr–Oliver percolating filter. This section shows the meshed structures covered with rubble which provide the enormous surface area which is favourable to the formation of a biological film. This is active owing to the high degree of contact with air which is ensured by such a filter.

*This classification may refer to the amount of water transported or to the amount of organic substances in the water arriving for treatment.

efficiency with which the 'organic load' is removed from an effluent depends on, apart from the temperature and the initial concentration of organics present, the parameters which guarantee a high oxygen concentration in the liquid film of the porous bed. These are the rate of oxygen transport, the contact area per unit volume, and the volume of the filter per unit of water to be treated which passes through the filter.

Moreover, since the concentration of oxygen in the liquid film of the porous bed is not solely dependent on its integration into the liquid film but also on the rate at which it is consumed in biological processes, its concentration can also be increased by lowering the amounts of organic substances present by recourse to recycling of the treated water.

It can be seen from Fig. 5.27 that the employment of percolating filters always implies the concomitant use of a sedimentation unit with recycling of the effluent from the latter.

Biodiscs are made of a battery of disc-shaped structures which are fixed to a central shaft and placed in a horizontal semi-cylindrical vessel.

The shaft is slowly rotated (2–5 rpm) so that the discs alternately have half of their surface exposed to the atmosphere and half of their surface exposed to the sewage which fills the vessel.

The discs with diameters of up to 2–3 metres are made out of a plastic material (expanded polystyrene or urea resins). Sometimes, metal mesh cylindrical casings containing various filling materials (Raschig rings, small plastic spheres, etc.) are used. Cylindrical structures consisting of alternate flat and corrugated sheets of polyethylene connected together in such a way as to create a 'bees' nest' structure are also employed. Such structures have a high specific surface area which attains values of 35–40 m^2/m^3. A schematic diagram is shown in Fig. 5.28.

Only very limited maintenance of biodiscs is necessary owing to the simplicity of these units. It essentially consists, apart from the routine removal of any blockages, of checking and lubricating the moving parts in accordance with the manufacturer's instructions.

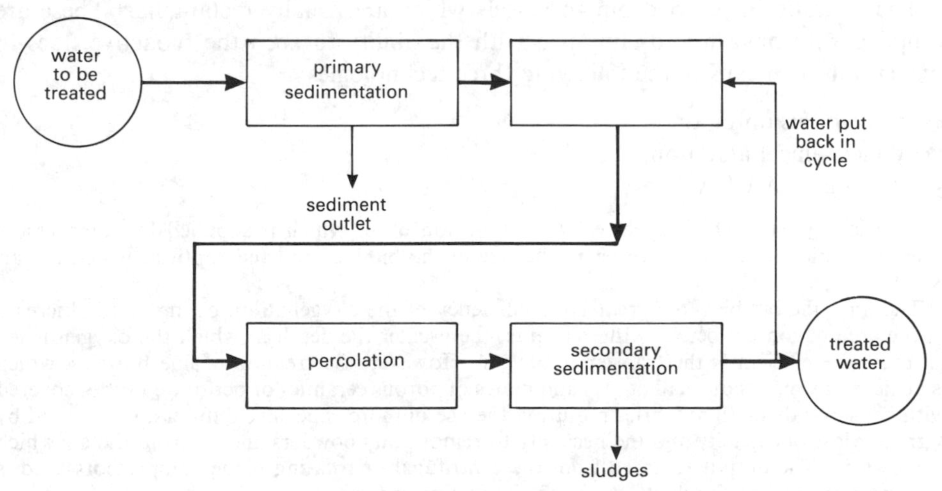

Fig. 5.27—Block diagram of the biological treatment of water in percolating filters with recirculation of the treated water.

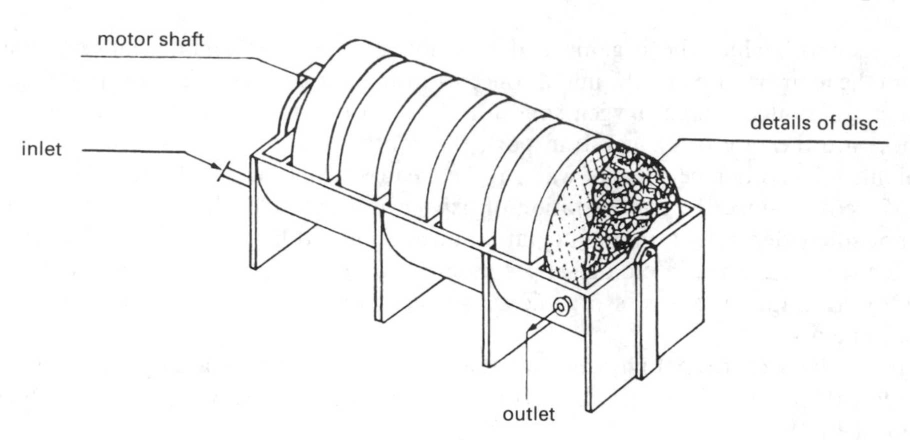

Fig. 5.28—Schematic diagram of a biodisc oxidation plant, showing in detail one of the discs filled with Raschig rings. It should be noted that every vessel consists of two drums (a two-stage plant). However, in practice, there are batteries of single drum vessels (single stage plants) and of vessels with more than two drums (multistage plants).

'Active sludges' are nothing more than sludges which have been produced in an oxidative biological water treatment and matured by repeated recycling. In this way, the enzymatic catalytic agents in the sludges are, in fact, increased with the double effect of:

- accelerating the oxidative phenomena,
- promoting the more complete utilization of the oxygen which is available to them.

The main steps in this treatment are the production of cellular material which aggregates into flocs and the adsorption on these flocs of the substances which are to be removed by biocatalytic oxidation. To promote absorption and subsequent oxidation of the substances to be removed, the flocs must be kept in suspension and supplied with oxygen.

The operation is carried out in vessels which are usually rectangular. These are supplied with oxygen and equipped with the ability to keep the bioactive flocs in suspension by means of the following three technologies:

- by the diffusion of air,
- by mechanical agitation,
- by using a mixed system.

Aeration systems which operate by the diffusion of air which is supplied by compressors (Fig. 5.29) differ from one another in the size of the bubbles and the depth at which the air is fed into the vessel.

The finer the bubbles, the greater the efficiency of the oxygenation, owing to the increase in the area of contact between the reagents; the greater the depth at which the oxygen is fed in, the more efficient is the transfer of oxygen. However, the creation of fine bubbles, which is achieved by means of small domes and plates of porous ceramics or perforated tubes covered with dense synthetic fibre fabrics requiring the use of more expensive diffusers, is attended by a greater loss of capacity and the necessity to remove any powders and oil from the air which is blown in. The output required from the centrifugal or rotating blade compressors used is then proportional to the depth at which the oxygen is fed in.

Aeration systems with mechanical agitation employ surface aerators which act by promoting

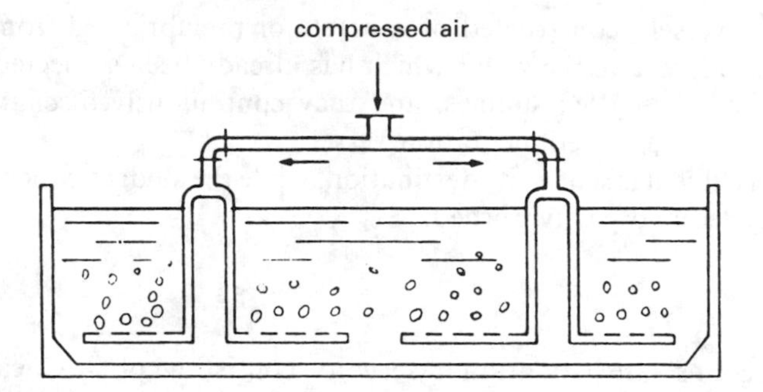

Fig. 5.29—Sectionof a typical sludge vessel activated by an air diffusion system in which air is blown in under pressure. The oxygen supplied leads to a luxuriant growth of microorganisms which impregnate the sludges and degrade the organic substances which pollute the waters.

contact between atmospheric air and the sewage in the basin. These aerators may be of two types:

- turbines floating on a vertical axle which create an umbrella-shaped jet of fluid with a large area of contact between the air and the sewage in the surface layer of the reservoir,
- rotors on a horizontal axle consisting of horizontal, rotating drums in which there are radial blades which promote recirculation of the liquid mass, thereby giving rise to a spray of fine liquid droplets which present a high surface area for the absorption of oxygen.

The essential features of the structures of the surface aerators which have just been classified are shown in Fig. 5.30.

Mixed aeration systems bring about aeration both by the injection of air and by mixing promoted by agitators which are placed above the points where the oxygen is introduced.

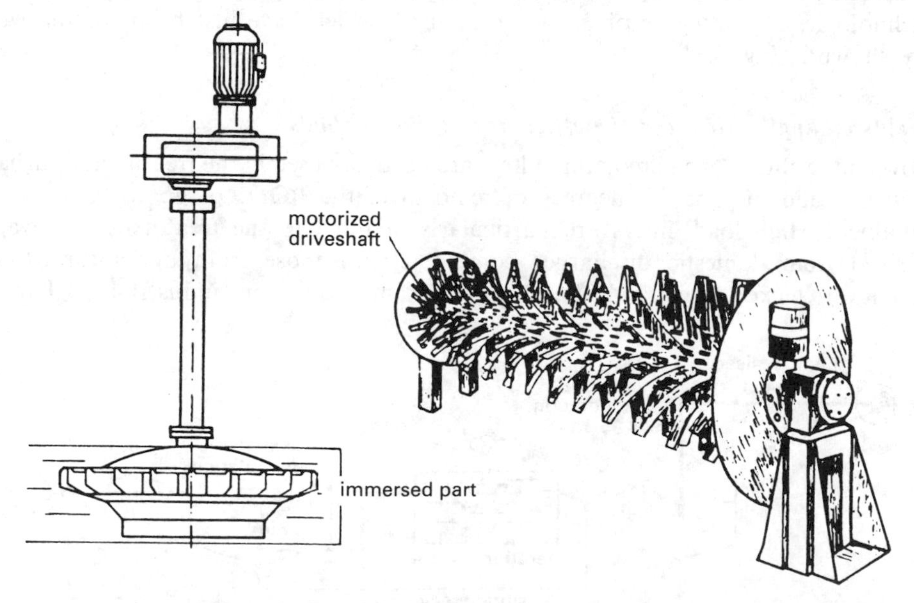

Fig. 5.30—Mechanical aeration systems for efficient liquid/gas mixing in active sludge treatments. Left: vertical shaft turbine. Right: brushes on a horizontal shaft. The turbine forms an umbrella-shaped jet of liquid which produces a large area of contact between the air and the liquid. The brushes produce minute droplets which present a high surface area thereby promoting the absorption of oxygen from the air into which they are nebulized.

The aeration vessels, constructed of concrete or prefabricated from iron which is suitably protected, receive the water which has already been subjected to sedimentation and the active recycled sludges, and they continuously discharge the aerated mixture into a secondary sedimentation vessel.

The purified effluent is sent ot its destination, while the sludges which are discharged are partly recycled and partly purged.

* * *

'Aerated lagoons' are oxidation reservoirs consisting of large-capacity vessels, formed by the excavation and banking of earth, which are made impermeable if the nature of the site requires it, and normally have a depth of no greater than 2 m. The residence times of the water treated in aerated lagoons are long, and oxidation without any recycling of sludges is therefore possible.

Oxygen is supplied by means of floating turbines (Fig. 5.30) or by means of an air diffusion system in the case of lagoons which are deeper than 1.5 m. In oxygen reservoirs of modest depth (1–1.5 m), the light reaching the bottom promotes the development of photosynthesizing algae which contribute to the ambient oxygen supply.

If the depth of the reservoirs exceeds 2 m, a non-aerated zone functioning under an anaerobic regime is established on the bottom. We shall consider this later.

* * *

Sections of the support and complementary plant making use of all the purification technologies exploiting aerobic microorganisms which have just been summarized are shown in Fig. 5.31.

Fields of application and the efficiency of the methods

Low and medium load percolating filters are normally used in the treatment of urban effluents, and they permit degrees of removal of the BOD_5 of up to 95% to be obtained. 'High load' filters are particularly adapted to the treatment of sewage derived from domestic discharges combined with those from agricultural/food activities (chicken farms and abattoirs, for example) or from industrial discharges

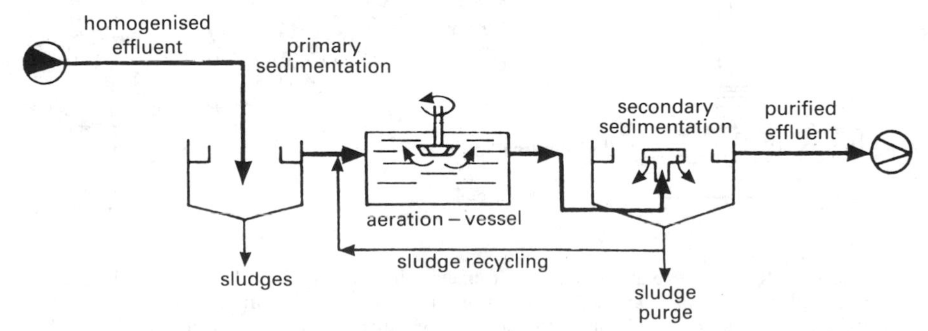

Fig. 5.31—Panoramic schematic diagram of plant for the appropriate treatment of effluent waters with an active sludge system, which also refers to other aerobic biopurification processes.

(breweries, petroleum refineries, etc.) and are capable of removing, on average, 60–70% of the BOD_5 of the initial effluent.

In biodisc systems it is possible to treat the discharges coming from tanning works, abattoirs, the dairy industry, and pharmaceutical establishments as well as from towns with populations of up to 100 000. The efficiency of the treatment depends, above all, on the dimensions of the surface of the discs. In the case of urban discharges, 80% of the BOD_5 is typically removed by employing 1 m^2 of disc surface per inhabitant, while about 95% of the initial BOD_5 is removed by the use of 3 m^2 of surface area per inhabitant.

Of all the methods based on the exploitation of microorganisms for the treatment of effluents containing organic substances, active sludge treatment is by far the most preferable. In fact, by varying the amount of air supplied and the aeration system, the flow rate, the age of the sludge, the sludge recycling ratio and its residence time, it lends itself to the treatment of every kind of urban and industrial effluent. It is frequently operated by combining the purification processes for the two groups of effluents and, when necessary, carrying out preventative purification procedures on the industrial effluents by means of chemical treatment. This is schematized in Fig. 5.32 within the framework of a rational synergetic coupling of the two types of microorganisms.

The efficiency of the BOD_5 removal by active sludge systems which have been modified in different ways varied over quitc wide limits, going from 75% in the case of high load normally treated effluents to 95% upon prolonged aeration and loading the aeration vessel progressively, or, if an economic source of oxygen is at hand, 'aerating' it with pure oxygen.

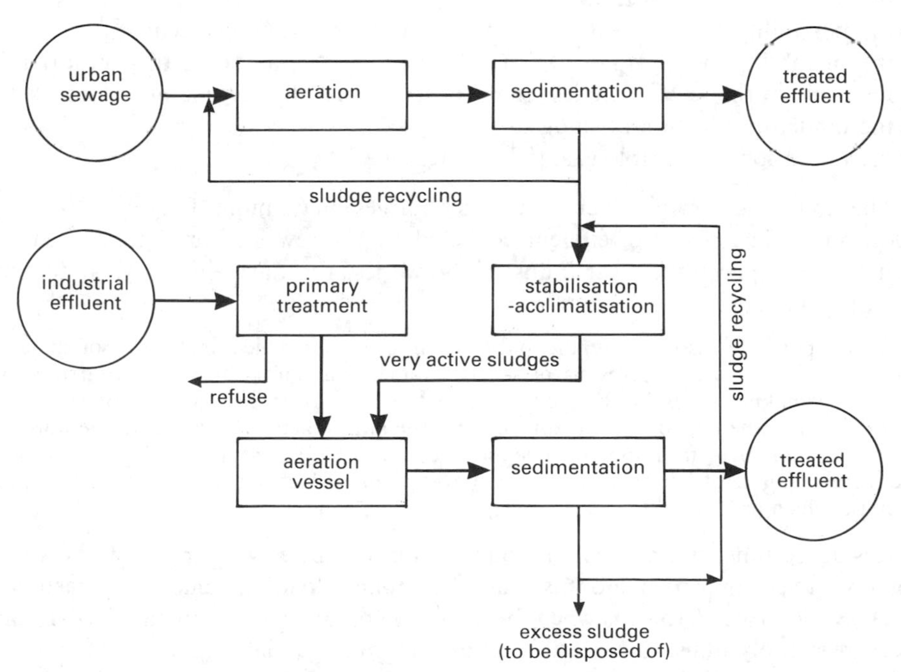

Fig. 5.32—Scheme for the treatment of industrial effluents using particularly active sludges derived from the treatment of urban sewage.

Table 5.7—Operational characteristics of aerated lagoons

Type of lagoon	*Depth* (m)	*Conc. of algae* (mg/l)	*pH*	*Residence time* (*in days*)	*Suspended solids* (mg/l)	BOD_5 *removal efficiency* (%)
Artificially aerated	1.5–4	—	6.5–8	3–10	80–250	80–95
Naturally aerated	1–1.5	40–100	6.5–10.5	10.40	80–140	80–95

Aerated lagoons, the use of which is favoured by the availability of ample space, the tolerance of poisonous emissions, and sunlight, lend themselves, in particular, to the treatment of industrial effluents polluted by organic substances such as hydrocarbons, phenols, amines and their derivatives, aldehydes, and micropolymers.

Several operational characteristics of these oxidative reservoirs are presented in Table 5.7.

Methods of exploiting anaerobic microorganisms

Anaerobic biopurification by means of enzymatic processes brought about by bacterial activity and normally used for the treatment of liquid discharges with a high BOD_5 (>2000 mg/l), presents, as has already been mentioned, several advantages over aerobic treatments:

- the low cost of the operation,
- the possibility of treatment also in the case of discontinuous feeding,
- the possibility of carrying out the process when the amounts of microorganism nutrients in the sewage are less than those required by aerobic processes,
- the limited production of sludges,
- the formation of a usable gas, rich in methane.

The industries from which liquid discharges may more frequently arise, in addition to the volume of effluent collected by the sewer system, are the canning industry, the pharmaceutical industry, breweries, the dairy industry, and animal breeding industries.

'Septic pits' are also among the purification processes which make use of anaerobic microorganisms. These are vessels of reduced dimensions into which sewage arrives. This sewage, by making use of the abundant local bacterial flora, forms a scum of protective agents on the surface where it comes into contact with the atmosphere. The sludge at the bottom of the vessel is removed from time to time and must be rendered aseptic by chlorination, given the large amounts of bacteria which it contains, and the fact that it also contains types of bacteria which are very pathogenic.

Residence times in anaerobic biopurification processes vary from ten days up to somewhat less than two months, and the organic loading removed, measured in COD, varies from 70 to 98% when the operating parameters (temperature, pH, times, etc.) are suitably adjusted. Nevertheless, given the high initial concentration of organic substances, the effluents must generally be subjected to further purification processes when the efficiency of removal is less than 90%.

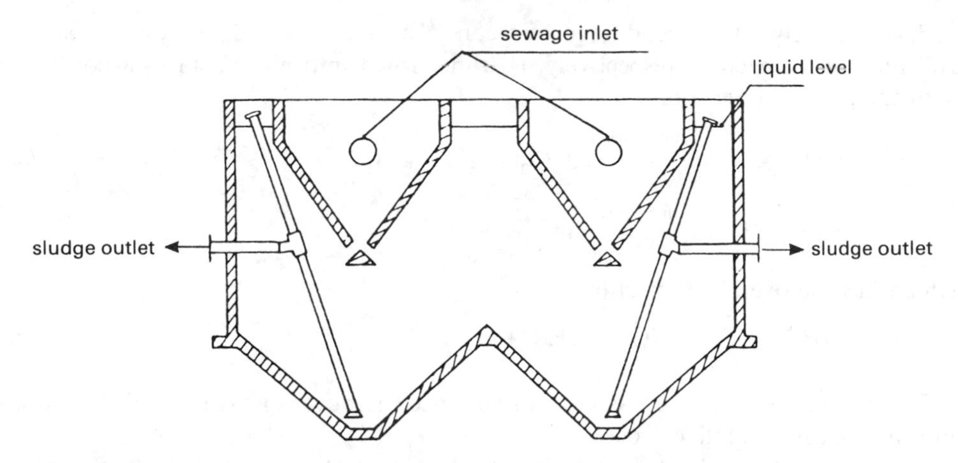

Fig. 5.33—Section of an 'Imhoff vessel'. It may be noted that the 'sludges' are made up of everything which, having entered the vessel, is thickened there. Inter alia, the sludges contain 90–95% of water and, after drying, they are easily-handled fertilizers, that is, such as may be supplied to operators who undertake to make them into a commercial product by means of chemicomechanical refinement techniques. These methods make them primarily suitable for agrarian use.

Fig. 5.33 is a schematic diagram of a classical anaerobic biopurification apparatus.

'Imhoff vessels' are typical devices which are used in the anaerobic biological purification of waters. They have been extremely widely used, but their deployment has fallen considerably and it is now almost restricted to small plants. They represent a very simple type of apparatus whose use and supervision do not call for any particular technical expertise, especially on account of the absence of any particular moving mechanical parts. In fact, Imhoff vessels probably require no more than fitting with an articulated device, similar to that used in sedimentation units, for the removal of the sludges.

Imhoff vessels (Fig. 5.33) consist of two compartments. The sewage to be purified is let into the upper compartment where sedimentation takes place, while the lower compartment receives the sludges formed by sedimentation which pass down to undergo digestion there. This second compartment is fitted with tubing for the extraction of the sludges. The gas produced tends to rise to the surface and comes out of the unit through suitable vents. Both the upper and lower parts of the apparatus are inclined, the former in order to favour the conveyance of the sedimentation from the top to the bottom and the latter in order to facilitate sludge extraction.

Bionitrification and biodenitrification processes

At present, biological nitrification–denitrification is adopted in the majority of plants for the removal of nitrogen from industrial process effluents.

Usually, denitrification is therefore the process which follows nitrification, but, sometimes, it is treated as a process in its own right as in the treatment of effluents which are very rich in nitrates which arrive from plants concerned with the purification of uranium.

Biological nitrification uses the biochemical oxidation of ammoniacal nitrogen by means of autotrophic nitrifying bacteria.

The process can be carried out in biological film reactors (percolating filters and biodiscs) or in reactors containing biomass in a dispersed phase (active sludge plants).

The chemistry of the process consists of two stages: during the first stage the organic nitrogen in the effluent is rapidly transformed into an ammonium compound, and, successively in the second stage, by means of catalyses carried out by *nitrosomonas* and *nitrobacter* bacteria respectively, is transformed into nitric acid in accordance with the following reactions:

$$NH_4^+ + 3/2O_2 \xrightarrow{\text{nitrosomonas}} NO_2^- + 2H^+ + H_2O \tag{a}$$

$$NO_2^- + 1/2O_2 \xrightarrow{\text{nitrobacter}} NO_3^- \tag{b}$$

which has the overall effect of (a) + (b):

$$NH_4^+ + 2O_2 \rightarrow NO_3^- + 2H^+ + H_2O.$$

During nitrification, there is also an enhanced removal of organic substances as in a normal biooxidative process.

On the other hand, denitrification uses the transformation of the nitrogen in nitrates into molecular nitrogen in an anaerobic environment. In practice, however, recourse is made either to slow mechanical agitation or the injection of a controlled amount of air to create sufficient agitation to avoid sedimentation of the biomass, but not so much as to endow the surroundings with an appreciable oxygen content.

If denitrification follows nitrification it is necessary to reintroduce a carbonaceous substrate into the system which acts as an electron donor in the reduction of nitrates. Methanol is the most commonly employed source of carbon, in which case the enzymatically catalyzed oxidation–reduction reaction may be represented by:

$$NO_3^- + 0.833CH_3OH + 0.167H_2CO_3 \rightarrow 0.5N_2 + 1.33H_2O + HCO_3^-.$$

As can be seen, in this way the nitrogen present in nitrates becomes an integral part of the nitrogen present in the atmosphere.

Fig. 5.34 shows a block diagram of the course of the nitrification–denitrification process in the context of an active sludge plant.

It is seen that, after a primary treatment in the sedimentation unit TP for the effluent I, sludge recycling takes place followed by biooxidation and nitrification in the aerated vessel (a = air) B + N. The sludges produced are separated out in the sedimentation unit S_1 which yields the effluent which is to be denitrified. A fraction of these sludges is recycled, while another fraction (W_1) is purged from the system. Recycled sludges and methanol are now added to the new effluent, and it is anaerobically denitrified in the vessel D. Aerobic stabilization is carried out in the percolating filter A, and, finally, the effluent E in the sedimentation unit S_2 is separated from the sludges for recycling and discarding (W_2).

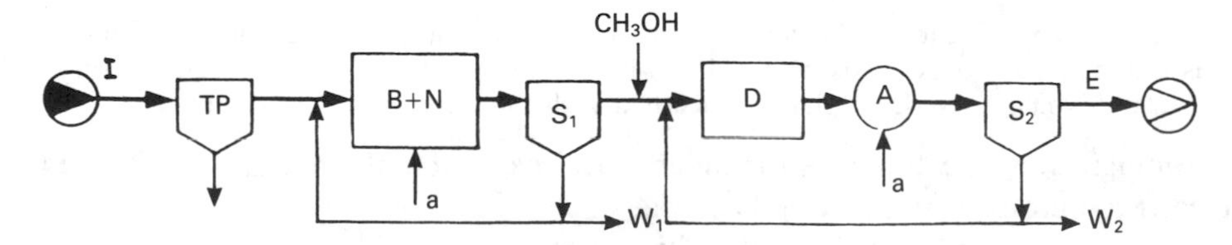

Fig. 5.34—Block diagram of a plant for the elimination of organic nitrogen from industrial process effluents. Only the second part can serve for the elimination of nitrates from effluents (such as those from the nuclear industry) by means of an initial seeding with the appropriate bacteria.

5.10 THE TREATMENT OF SLUDGES

Water purification processes produce sludges which are derived from the separation of suspended materials, from the precipitation of dissolved materials, from the addition of chemical products, and, above all, from the transformation of organic substances into a microbial cellular mass. The sludges, both in their own right and because they ferment, in general emit unpleasant odours and are often infective. Moreover, they disfigure the environment owing to their large volume and appearance. It is therefore necessary to thicken them, to deactivate them, and to dispose of them directly or after drying.

The thickening of sludges

Sludges can be subjected to *thickening by gravity* in 'thickening' devices which are similar to sedimentation units and are fitted with rotating mechanisms which break up the sludge flocs in different ways and thereby remove the gases and occluded water from them. The material to be treated is fed in centrally, and the thickened sludge is extracted from a small central well while the supernatant liquid which has separated out is collected in peripheral channels. *Thickening by flotation* is also possible. It employs the addition of chemical additives which modify the surface of the particles in such a way as to make them adhere to bubbles of air. In this way the particles are carried up to the surface where they form a dense layer which favours their removal.

The stabilization of sludges

Sludges are biologically active systems. If they are to be disposed of by means of drying at high temperature (350–400°C) or incineration, they must be deactivated. Sludges may be rendered biologically inactive by digestion or by storing them in biological lagoons.

There is *aerobic digestion* in which the sludge is oxidized in open vessels to ensure aeration. This process is economical in plant terms but leads to the loss of the biogas and requires powerful means of agitation and aeration because the concentration of the medium being treated is about ten times greater than that in the waters which are similarly treated in active sludge processes.

Anaerobic digestion is more useful (and widely used). This is a complex biochemical procedure in which numerous groups of aerobic and optional microorganisms assimilate and degrade the organic material. At the first stage acid fermentations take place, while, at the second stage, the acids which have been formed are degraded to CO_2 and methane (biogas).

The operational conditions and the important consequences of anaerobic digestion processes are summarized in Table 5.8.

The apparatus used in the two stages of a 'high load' anaerobic sludge digestion process is shown in Fig. 5.35.

In the first stage the digester on the left is heated by the circulation of water, and the sludge itself is mixed by a special agitation system.

In the second stage the digester on the right is reserved for the separation of the digested sludge from the supernatant fluid and for the complete release of the gases produced.

In certain plants recycling of the sludges passing out from the second stage back into the

Table 5.8—Characteristic parameters and the microorganisms removed

Type of sludge	*Temperature*	*Retention times (days)*	*No. of stages and type of stage*		*Microorganisms removed*
low load	ambient	30–60	1		
medium load	30–40°C	10–20	2	without sludge recycling	100% *Entamoeba hystolica* and *Escherichia coli*, 92% *Salmonella typhosa*, 90% *Mycobacterium tuberculosis*
high load	35–56°C			with sludge recycling	

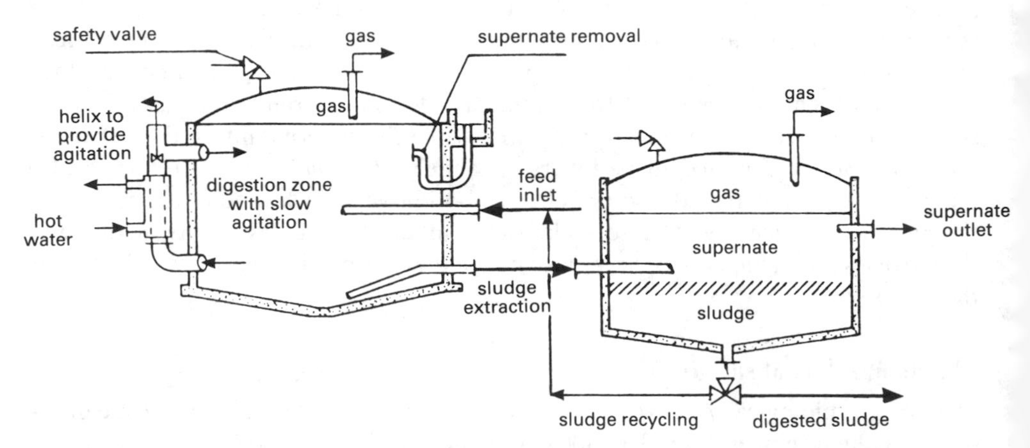

Fig. 5.35—Modern anaerobic sludge digester (left) and the sorting of the final products (right). The details of the agitation system and the heating by means of hot water circulation in a tubular sleeve through which the sludge is made to pass slowly are noteworthy.

first stage is also carried out in order to seed the feedstock and thereby accelerate the degradation time. This procedure is shown in Fig. 5.35.

Storage in lagoons, the third method for the stabilization of sludges, is practised when sites are available which are not too remote from the treatment plant but which are sufficiently isolated and inexpensive. Storage in lagoons consists of treating the sludges in reservoirs, with earth bases and embankments, and about 1.5 m deep with controllable accesses. The method is simply a lengthy storage of the sludges (for a period of the order of 24–36 months). The supernatant sewage which is formed partly evaporates and is partly retreated biologically like that which has already been through digestion processes.

The drying of sludges

After they have been stabilized, the water content of sludges may be reduced to about 70% (dehydration) and, subsequently, to about 10% (drying). Dehydration can be carried out, after a possible further thickening phase, by centrifugation but also by filtration under vacuum in Dorr–Oliver filters or by treating it in a filter press.

For details regarding these technologies the reader is referred to institutions concerned with chemical plant engineering.

The drying of dehydrated sludges is brought about in special 'flash furnaces' and in the rotating furnaces which are shown diagramatically in Fig. 5.36.

Whether they are of one or the other type, these furnaces operate at 350–400°C on account of the large amount of water which the feedstock contains. Even at these temperatures, however, there is a reduction in every case of unpleasant odour, infection, and so on. For this reason, if this type of drying is carried out, the preventive stabilization of the sludges primarily has the aim of reducing them in volume and exploiting them for energy production purposes. Nevertheless, solely by thickening and dehydrating the sludges, they can be transported to the point where they can be conveniently dried by this technique.

An alternative method of drying which necessarily requires preventive stabilization is that carried out on the draining surfaces of sand and gravel (drying beds) which permit the removal of the liquid which separates out from the sludge under gravity and is exposed to the atmosphere in such a way that the mass dries out by natural evaporation.

Depending on the climatic conditions appertaining at the installation site, the beds may be covered with glass or with plastic sheets, always ensuring, however, that adequate ventilation is maintained. This, apart from providing shelter when the water is bad, is very advantageous from an aesthetic point of view, even if there is some extra cost of maintenance and the removal of the dried sludges.

The disposal of sludges

Once they have been stabilized, sludges may be disposed of by spreading them onto

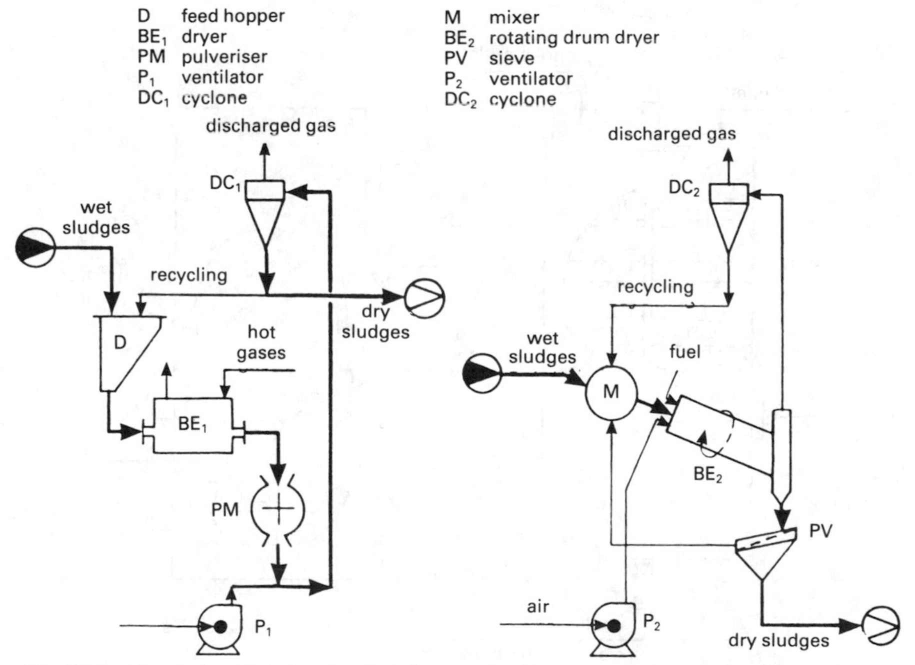

Fig. 5.36—The drying of sludges in 'flash furnaces' (left) and the drying of sludges in rotating furnaces (right). In both systems dried sludges are recycled to thicken the mass before it enters the drying furnace.

the soil, with the double advantage of utilizing the organic content as a humus and part of the mineral content as a fertilizer. The same thing can be done, depending on the type of terrain, only after having dehydrated the sludge apart from having stabilized it. Controlled amounts of sludges, including those which are still more or less active, are sometimes simply buried or confined.

Another way of disposing of sludges is spreading of the sludge, which has been stabilized and dried on the 'drying beds' previously discussed, onto agricultural soil.

Nowadays, however, burning ('incineration') of the sludges after they have been thickened and dehydrated is becoming ever more common. Recourse is made to this mode of sludge disposal, in particular when there is no demand for stabilized or dried sludges (for agricultural use, for example).

The incineration is carried out in 'multiple-hearth' furnaces or in 'fluidized bed' furnaces: the former being operated at 550°C to 1000°C and the latter at 760°C to 820°C. These furnaces are illustrated in Fig. 5.37.

Here, it is a question of promoting solid–gas reactions, and it is therefore essential to establish optimal contact between the phases participating in the process.

In the 'hearth-type' furnaces this is achieved by flowing the reactants in opposite directions, and, in 'fluidized-bed' furnaces, by stimulating the subdivision (which is favoured by the intervention of sand as a third body) and turbulence which fluidizes the reacting system.

The ashes produced by every kind of sludge serve as filling materials and find use in the ceramic industry and, above all, are usefully converted into ingredients for the preparation of constructional conglomerates including the moulding of artificial load-bearing stones for civil and industrial buildings, which are cemented together by suitable binders and procedures.

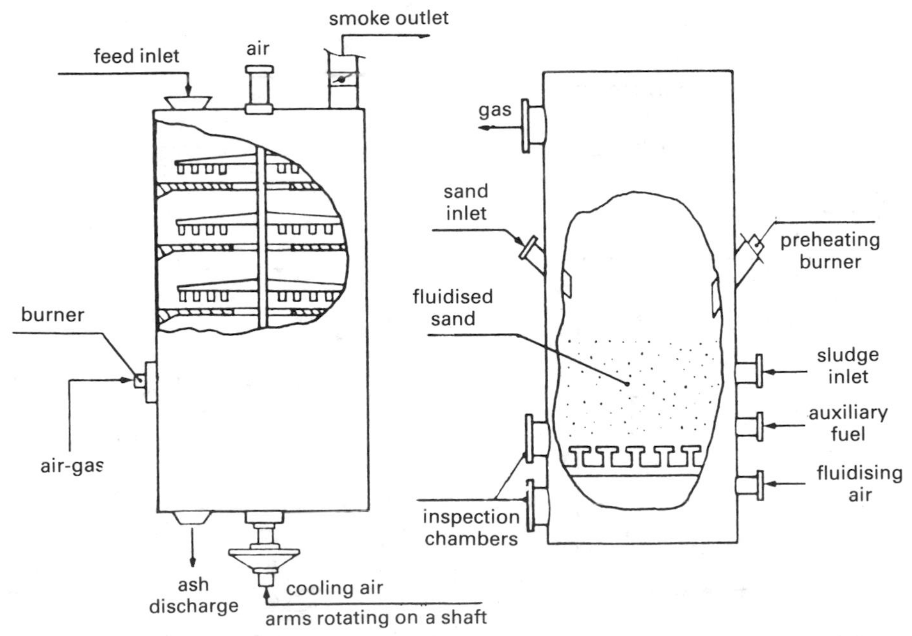

Fig. 5.37—On the left: Multiple-hearth sludge incinerator. On the right: Sludge incinerator operating under less severe thermal conditions to provide the most favourable contact between air and sludge.

5.11 TREATMENTS OF WATERS INTENDED FOR SPECIFIC PURPOSES

Everything which has been seen up to now regarding the clarification, the demineralization, the conditioning, etc. of waters is not, in most cases, carried out as a number of independent processes but is executed within the context of a programmed series of treatments on a body of water with the view to it being subsequently used for a certain specific purpose.

The number and the quality of the treatments required vary according to the state of the waters and their destination.

A complete review of the treatment cycles which may be necessary is impossible, and even an extensive review of the matter is out of the question. Nevertheless, after what has been said up to now and on the basis of the examples to be presented, which refer to typical and practically very important cases of the usage-oriented treatment of waters, the following material will allow some headway to be made in comprehending the problems associated with the provision of an adequate supply of water in any field when the water bodies to be exploited are in a condition which is totally different from that requested.

* * *

Here, we shall consider the following treatments of raw waters being used for the first time or processes for the regeneration of typical re-use waters which are required in various areas:

- making surface waters fit for drinking,
- the treatment of effluents from a petrochemical establishment,
- the capacity to supply boilers with drinking waters,
- the purification of urban effluents.

Finally, we shall consider the subject of the purification of a body of water within a chemical factory by illustrating the treatment of reflux water containing aldehydes.

The conversion of raw waters into drinking water

Nowadays, water supplies to urban areas are taken from rivers and lakes to an ever increasing extent.

It is not possible to give a standard list of treatments required in order to render river and lake waters drinkable, on account of the variability of these waters and of the quantities of material carried in them which make such waters raw and even polluted.

In principle, the processes which have to be carried out in order to make surface waters drinkable are:

- clarification and clariflocculation,
- the elimination of reducing agents, compounds of ammonia, and, above all, pathogenic microorganisms,
- the removal of medium to heavy metals (iron and manganese in particular) and the correction of excessive hardness.

It should also be noted that cases are becoming ever more frequent in which the chromates in the waters to be rendered suitable for drinking must be reduced to Cr^{3+} and eliminated together with zinc, cadmium, lead, and other heavy metals. Among the non-metallic pollutants which are frequently present, there are phosphates, surface-active compounds, phenols, hydrocarbons, and their chlorinated derivatives which have to be removed.

On the basis of analytical results, requests are made for particular treatments which have already been illustrated in detail or for their inclusion in the overall framework of the biological treatments.

Operations of a chemical, mechanical, and physicochemical nature for making water acceptable for drinking are therefore commonly required. These consist of:

- the addition of oxidizing–sterilizing reagents and of flocculating agents,
- clarification by sedimentation,
- adsorption on beds of active carbon.

Fig. 5.38 illustrates a standard plant for rendering common raw river or lake waters suitable for drinking.

Chlorine (Cl_2) and chlorine dioxide (ClO_2) are employed as the oxidizing and sterilizing agents, while $FeCl_3$ and, ever more commonly, aluminium polychloride $[Al(OH)Cl_2]_5$ are used as flocculants. This latter compound is active over a wide pH range and has a high coagulating capability toward all kinds of colloidal suspensions (clays, organic substances, etc.).

After oxidation, a wide range of modifications are carried out on the contents of the effluent being treated: the metals are converted to the oxidation state required for their precipitation (by possibly correcting the pH value with lime to favour such precipitation) and the hardness is reduced by the pH changes which occur, ammonium compounds are oxidized to nitrogen, and the microorganisms become dead organic material which flocculates or is subsequently treated with granular active carbon which also picks up traces of surface active compounds, phenols, and other organic pollutants.

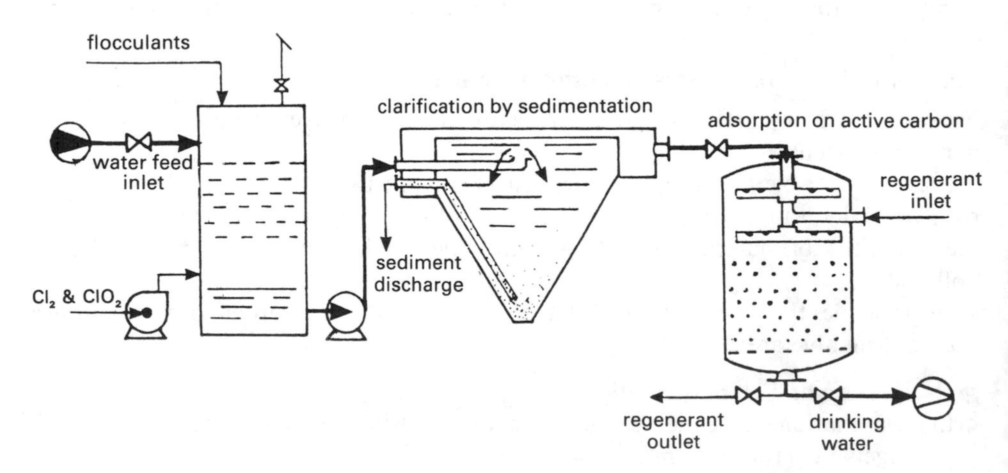

Fig. 5.38—Diagram of a plant for rendering river and lake waters, without particular pollution problems, suitable for drinking.

The treatment of effluents from petrochemical companies

The waters used in great quantities in petrochemical plants both for quenching and, more generally, cooling, either as components in processes (as in steam cracking, for example) or as a vehicle for acids and bases and so on, emerge from such processes having undergone pronounced changes. These changes are the amounts of organic substances which they contain and the extent to which they are loaded with reducing agents (that is, with high COD and BOD values), in the concentrations of ions present (NaCl alone may be present in amounts of 10–15 g/l), and in the amounts of the various and often oily dispersions and suspensions which are now present.

A standard treatment for such waters, which also lends itself to the cleaning up of effluents from many chemical industries, is shown in Fig. 5.39.

When there are appreciable amounts of oily substances floating on and emulsified with the polluted effluent, the treatment must be preceded by a slow-winding downflow of the liquid in the tanks where, for the greater part, the oily component becomes detached from the water, rises to the surface, and overflows.

The adaptation of drinking waters for use in boilers

The feeding of waters to boilers, that is, plant units employed in the production of steam, constitutes an important and delicate operation which, generally, makes use of drinking water as raw material.

These waters cannot be used directly for this purpose because their hardness and the amounts of gases dissolved in them are inadmissible for this process, which

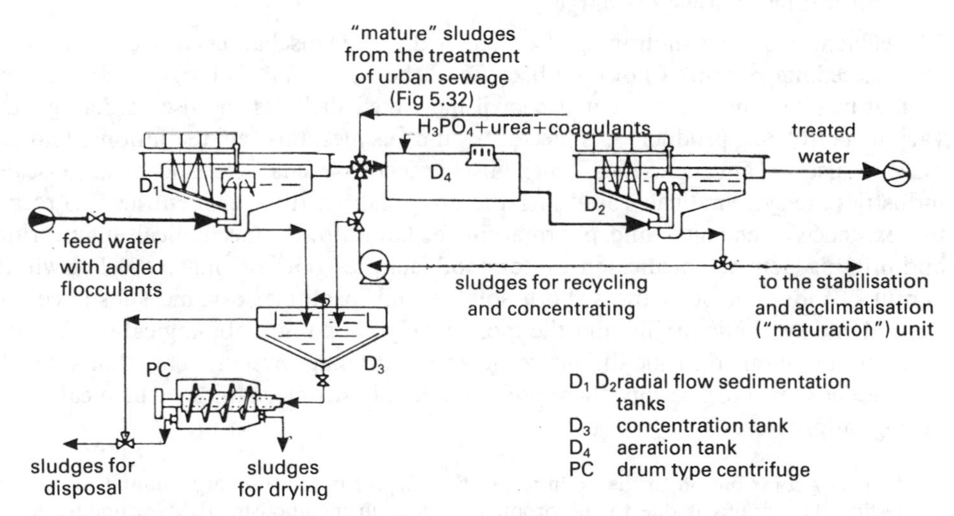

Fig. 5.39—Diagram of a plant for the treatment of the effluent from a petrochemical works. After the removal of any oil, the effluent is fed, having added flocculants, into the primary sedimentation unit from where it progresses to undergo biological oxidation. This is carried out by using sludges which are very active either owing to their nature (a mixture of directly recycled sudges and mature sludges after they have been blended with other sludges from the treatment of urban sewage and a fraction of the sludges recycled from this process), or to the nutrients specifically administered to the sludge-producing microorganisms, or to the efficient flotation–aeration put into operation. Secondary sedimentation follows with the production of sludges, which go to three destinations (as can be seen), together with treated water. The fate of the various sludges produced is evident.

requires:

- the practical absence of silica (<0.02 ppm),
- a salt content so low as to produce an electrical conductivity of less than 0.2 μS/cm at 20°C,
- a maximum hardness of 0.05°F,
- a maximum oxygen content of 0.02 mg/l.

To convert drinking water into boiler feed water it is necessary to carry out:

- preliminary treatments using lime and soda, to lower the hardness, with the addition of $FeSO_4.7H_2O$ and polyelectrolytes which produce flocs which are capable of occluding the fine precipitates to which the process gives rise;
- subsequent demineralization by ion exchange;
- efficient degassing of the feed water.

The plant which is required for this purpose is shown in Fig. 5.40.

If raw water from a normal water course were to be used as the starting material rather than drinking water, the processes of sand removal, sedimentation, and/or clarifying filtration and antialgal oxidation would have to be carried out, since boiler feed waters need to be limpid, colourless, and to be guaranteed to be free from any materials which might give rise to the formation of incrustations.

The purification of urban discharges

The effluents from urban drainage (sewage or drainage discharges) are, constitutionally, exceedingly complex not only because of the variety of discharges produced by animal metabolism and present day civil life which includes the use of detergents, various petroleum products, disinfectants, and disinfestants, acid solutions, and so on, but also because activities which discharge refuse characteristic of large-scale industries are systematically built into the urban fabric. It is pertinent here to recall the existence of chemical and pharmaceutical laboratories, mechanical engineering and printing factories, medical institutes, food factories, dairies, and so on. The whole gamut of inorganic substances, both soluble and insoluble, organic substances of natural and synthetic origin, and the most widely varied microbiological species are therefore represented in the discharges from the drainage system. The plants which are needed for purification therefore include physicomechanical, chemical, and biological processes.

The major contribution to the pollution of the effluent in the discharges found in normal domestic urban drains is due to the products of human metabolism. At a secondary level, there are both the residues of domestic activities connected with nutrition, personal hygiene, and the washing of clothes, and the dregs and refuse carried by waters which are washed off the streets.

The plants are more or less complex with regard to their type, dimensions, and the numbers of units employed in them according to the urban agglomerate which they serve and, hence, the volume of sewage which has to be purified. They may include small-scale, medium-scale, or large-scale operating processes.

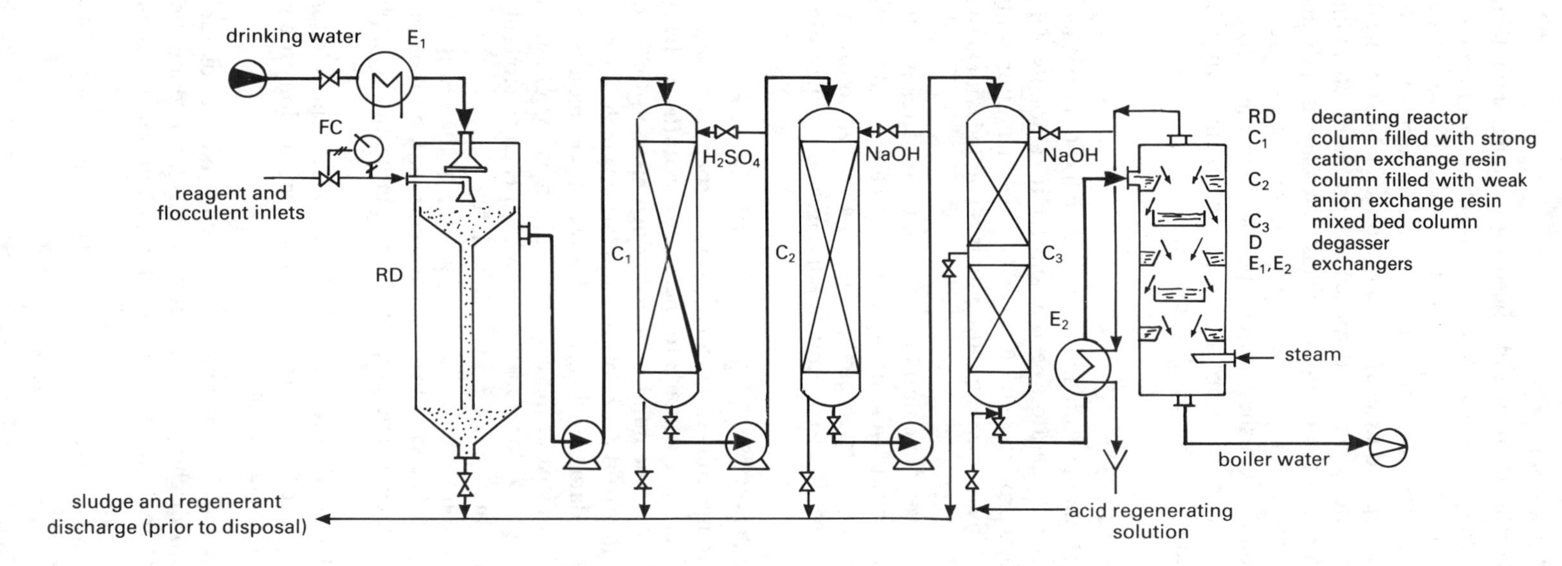

Fig. 5.40—The production of boiler feed water from drinking water. This involves, firstly, treatment of the water with $Ca(OH)_2$, Na_2CO_3, and flocculants, then the removal of cations in C_1 and the removal of anions in C_2, and, finally, removal of the last traces of cations and, above all, anions, in a mixed-bed column filled with a strong anion resin which even removes silica. After stripping the gases with live steam, the water is sent to the boilers.

Small-scale purification processes

Simple plants of the type which has already been shown diagrammatically in Fig. 5.31 serve in the case of small urban centres. The aeration tanks in these plants are always relatively spacious in order to guarantee the 'total oxidation of the sludges'. Thus, by allowing the discharges to remain in the oxidation tanks for a long period, a high level of sludge recycling and extensive and prolonged oxidation lead to a large reduction in the sludges which have to be disposed of, and the practical inactivation, by oxidative autodestruction, of the microorganisms which are already active in the process. By operating in this manner, it is easy to dispose of the sludges.

In this area of work rotating brushes of the type shown in Fig. 5.30 can be used as aerators, thereby saving energy in comparison with systems employing the injection of compressed air or aeration by means of a turbine.

Medium-scale purification processes

As the amounts of sewage from the drainage system which has to be treated increases, the feasible residence times in the aerators become shorter, the amounts of unstabilized or slightly stabilized sludges produced in the secondary sedimentation units increase, and the supernatant liquid in such sedimentation units is infested with active bacterial flora.

In this case the products of sedimentation carried out downstream of the aerators must be suitably treated. The supernatant liquid is subjected to chlorination before being discharged into the water reservoir, and the sludges which are destined for disposal are subjected to digestion, dehydration, drying, and incineration.

Large-scale purification processes

To purify the sewage from medium-scale to large-scale urban communities (250 000 to 500 000 inhabitants)*, recourse is made to the plant apparatus of the type shown diagrammatically in Fig. 5.41. The number of operating units and/or the dimensions of these units vary according to the size of the population which has to be served.

There are two pieces of data which are indicative of the requirements of the dimensions of the plant in this case: the aeration tanks must be of such a capacity as to put a volume of about 100 litres at the disposal of each inhabitant, while the entire plant is developed in an area of 3.5 to 5.5 hectares.

Various modifications are introduced into the scheme of the sewage purification plant design shown in Fig. 5.41 in the form of supplementary apparatus, especially in relation to where the effluent emerging from the purification complex will be discharged.

For example, a final chlorination is indispensable if the effluent is fed into bodies of water used for swimming or from where water is drawn for domestic purposes.

Nowadays, care is always taken to ensure:

- the specific elimination, for the protection of the ichthyofauna, of surface active compounds poured into discharge drains and arising from 'domestic' washing activities, textile workshops, and laundries sited within urban communities,

---

* Larger urban communities generally treat their sewage in several distinct purification plant units, primarily so as to simplify collection problems.

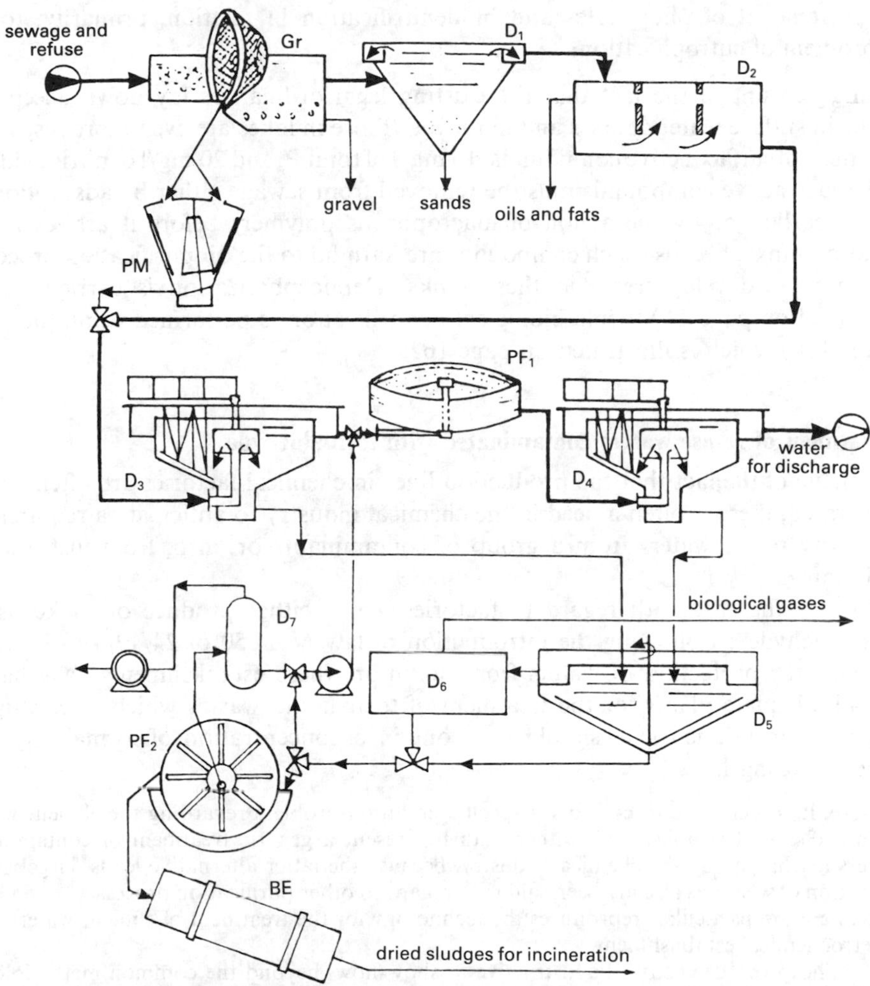

Fig. 5.41—The large-scale purification of urban sewage. The indicated numbers of operational units refer to a city of 220 000 to 270 000 inhabitants. The filthy effluent is mechanically purged and oil is taken off by skimming in tanks (not shown in the figure) where partial sedimentation of the particles to be removed occurs. After a primary sedimentation, biological oxidation with active sludges is carried out in the percolating filters followed by a secondary sedimentation which furnishes purified water. This is followed by the application of anaerobic digestion (which produces biogas) and the final disposal of the various kinds of sludges (including the biomass) which are not recycled to the biological oxidation units. If the sewage being treated contains conspicuous amounts of substances which poison the microorganisms responsible for the biological oxidation, these poisons must be eliminated upstream of the percolating filters.

- Gr screening and the removal of broken rubble (2 or more units of a capacity of 700 m^3);
- D$_1$ removal of sand (number and capacity as for Gr);
- D$_2$ oil removal (2 or more units of a capacity of $1–1.2 \times 10^3$ m^3);
- PM mill;
- D$_3$ primary sedimentation unit (2 or more units with a capacity of $10–15 \times 10^3$ m^3);
- PF$_1$ percolating filter (4 or more units with a capacity of $3.5–5 \times 10^3$ m^3);
- D$_4$ secondary sedimentation unit (4 or more units of $6–8 \times 10^3$ m^3);
- D$_5$ concentrator (4 or more units with a capacity of $6–8 \times 10^2$ m^3);
- D$_6$ anaerobic digester (2 or more units with a capacity of 10^3 m^3);
- PF$_2$ rotating drum filter;
- BE drier;
- D$_7$ storage tank.

- the removal of phosphates and biodenitrification in relation, primarily, to the problem of eutrophication,

taking account of the fact that the existing legal ordinances lay down acceptable limits in surface water courses and in the sea (those in lakes are even more restricted) of 2 mg/l of surface active compounds, 10 mg/l of total P, and 20 mg/l of nitric acid.

Surface active compounds must be removed from sewage either by adsorption on active carbon or by the action of macroporous polymers before it arrives in the aeration tanks, because such compounds are harmful to the biopurification processes which must develop freely in these tanks. Dephosphatization is carried out as described on page 149, while biological denitrification is performed using the plant technology which is illustrated on page 162.

Treatment of re-use waters contaminated with formaldehyde

By virtue of the fact that the production lines in chemical factories are often closely integrated, there is often a need in the chemical industry to anticipate a requirement to purify re-use waters from a group of contaminants or, also, from just a single pollutant.

This is the case with regard to factories which either produce or make use of formaldehyde*. Following the introduction of law No. 650 of 24/12/79 (the Merli law) on the protection of waters from pollution, these establishments have had to install adequate plants for the treatment of their liquid wastes which, according to the abovementioned law, should not contain a concentration of formaldehyde in excess of 2 mg/l.

Here, it is preferable to consider a specific pollution problem regarding the effluent waters from a chemical establishment rather than to present a generic treatment of contaminated waters discharged by the chemical industry, because the latter alternative leads, largely, to a repetition of what has already been said with regard to other purification processes for polluted waters and, in particular, reproduces the technology for the treatment of effluent waters from a petrochemical establishment.

It is, however, deemed to be instructive to show how, beyond the common methodologies for the purification of aqueous effluents from the chemical industry, the need often arises for the introduction of processes intended for the removal and, possibly, recovery of specific contaminants.

Waters for re-use containing formaldehyde are generally collected in two distinct storage tanks, depending on the concentration of the aldehyde, and, as a result, are classified as:

- strong waters containing from 0.1 to 0.2% by weight of formaldehyde,
- weak waters which contain two solutes in amounts varying from 0.01 to 0.1% by weight.

Also, there are possibly effluents due to overflows, losses, and drainage which have collected in tanks within departments during the various phases of the operations.

---

*Formaldehyde finds use in many, many different fields. We recall, besides the direct uses to which it is put in various manufacturing processes, those of phenolic plastics, amine plastics, acetalresins, pentaerythritol, and urotropine. It also finds use over a very wide area including the dyeing, paper, printing, and tanning industries and in explosives, pesticides, pharmaceuticals and so on.

These effluents are generically referred to as 'dregs'.

The dregs in the collection tanks allow the suspensions which they contain to form sediments, so that it is only the clarified supernatant solution which has to be treated. It is not considered to be worthwhile to recover the formaldehyde, but its content must be lowered to below the limit which the law permits regarding the effluent discharged from the factory.

The weak waters and the dregs are subjected solely to an oxidative decontamination treatment, while the strong waters are subjected to a preventative operation in which the formaldehyde is recovered.

As can be seen from Fig. 5.42, the strong waters are fed into the rectification column C_1 after having been preheated in the heat exchanger E_1 with the vapours from the head of the column itself which operates in the current of steam.

The vapours at the head of the column are subsequently condensed in E_2 and sent to a storage tank D from where a fraction of the liquid is returned to the column as a reflux while the fraction cooled in E_3 constitutes the distillate which is returned to the production plant or to plants utilizing formaldehyde. Those substances in the system which cannot be condensed out are sent from E_2 and from D and are scrubbed with water in the column C_2, thereby obtaining solutions which may be discharged into the drainage system as their formaldehyde content is less than 2 mg/l.

The product from the end of the column C_1, having been united with the weak waters and

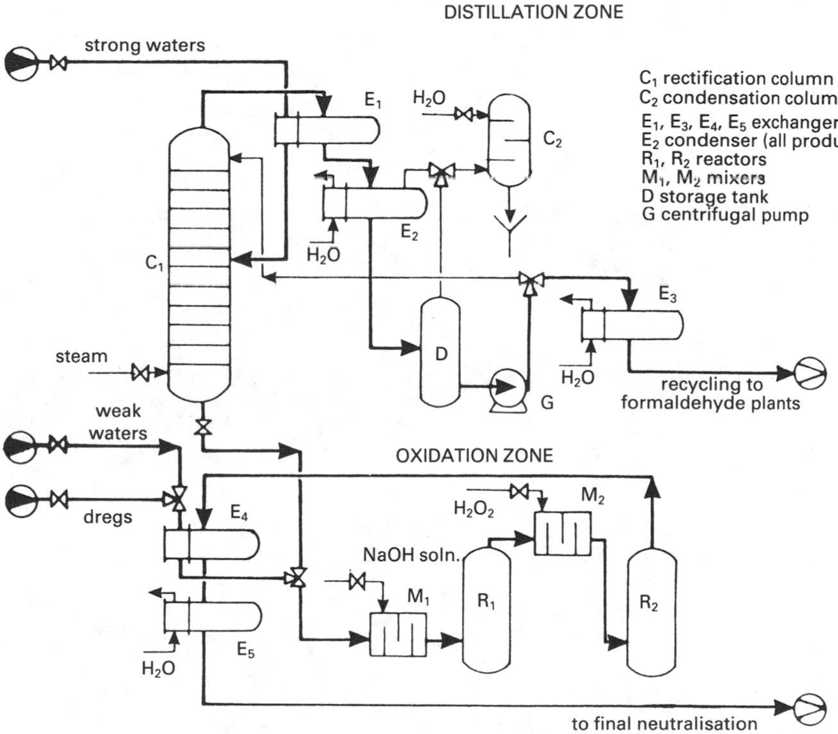

Fig. 5.42—The treatment of re-use waters contaminated with formaldehyde. The first part of the plant ('the distillation zone') corresponds to the rational structural development of an efficient process while the second part is solely to be considered as indicative of the treatments which have to be carried out on less contaminated effluents and the apparatus which is used for this purpose.

the dregs, is preheated and sent to the 'oxidation zone' where the two reactors R_1 and R_2 are operating (they are depicted only in outline in the scheme).

In the first reactor, following the addition of alkali, true and proper oxidation does not take place but rather a Cannizzaro disproportionation occurs which transforms part of the formaldehyde into formate and methanol. In the same reactor another fraction of the formaldehyde condenses to form higher carbonyl compounds (glycolaldehyde, glyceraldehyde, dimethylolketone) and sugars. On the other hand, true oxidation takes place with oxygenated water in the reactor R_2.

Finally, the oxidation product gives up its heat of reaction in the heat exchangers E_4 and E_5 before progressing to undergo preventative neutralization and being discharged as part of the effluents from the factory.

6

Sulphur and its principal compounds

6.1 GENERALITIES REGARDING SULPHUR

Natural state and sources

Sulphur is diffuse in nature both as a free element and, especially, in the form of sulphides and sulphates. Overall, sulphur constitutes a little less than $5 \times 10^{-2}\%$ of the lithosphere and almost $9 \times 10^{-2}\%$ of the hydrosphere which is particularly attributable to its compounds which are water soluble.

When in the pure state, the sulphur in the lithosphere is sometimes well crystallized, but, for the greater part, it forms compact and powdery microcrystalline masses which are accompanied by calcaerous, chalky, and argillaceous gangue.

The commonest crystal habit of sulphur is rhombic. The crystals are generally yellow, more or less modified by the gangue; and the presence of selenium, as in the case of certain Chinese and Japanese sulphurs, makes it red.

Native sulphur is generally of geothermal origin, that is, it is produced from hot currents of gaseous H_2S and H_2S in water either by aerial oxidation or by oxidation by means of SO_2:

$$2H_2S + O_2 \rightarrow 2S + 2H_2O$$

$$2H_2S + SO_2 \rightarrow 3S + 2H_2O.$$

It has been geologically verified that hydrogen sulphide is the product of the action of water vapour on sulphide materials:

$$CaS + H_2O \rightarrow CaO + H_2S,$$

and that sulphur dioxide and sulphur trioxide originate, for the most part, from combined equilibria of the type:

$$MeSO_4 + H_2O \rightleftharpoons MeO + H_2SO_4 \rightleftharpoons H_2O + SO_3 \rightleftharpoons SO_2 + 1/2O_2$$

(where Me = a bivalent metal) which are possible under the conditions existing for the igneous rocks in the subsoil.

On the surface, however, some of the sulphates which constitute the gangue of sulphide deposits are formed by the action of sulphur dioxide and air on the carbonates which are present.

The principal natural sulphides are, in the order of their widespread occurrence: FeS_2 (pyrites), FeAsS (arsenopyrites), ZnS (blende), PbS (galena), $CuFeS_2$ (chalcopyrites), and Cu_2S (chalcocite), while the most abundant natural sulphates are: $CaSO_4$ (anhydrite), $CaSO_4.SH_2O$ (gypsum), $MgSO_4.H_2O$ (kieserite), $BaSO_4$ (barytes), $Na_2SO_4.10H_2O$ (glauberite), and $FeSO_4.7H_2O$ (green vitriol).

Industry traditionally recovers sulphur from all of its natural sources and, in particular, from native sulphur deposits, from FeS_2, from ZnS, and from both anhydrous and hydrated calcium sulphates. Nowadays, however, it is common practice to systematically exploit the sulphur-containing gases which occur both naturally and, especially, as by-products of typical operations in applied chemistry such as the hydrodesulphurization of petroleum derivatives and the recovery of hydrogen sulphide from the gases obtained during the distillation of fossil fuels.

Properties of sulphur

Elementary sulphur may occur variously as crystalline sulphur, fused sulphur, sulphur vapour, plastic sulphur, and colloidal sulphur.

Crystalline sulphur

While possessing a variety of structural crystalline habits, solid sulphur exists in two main modifications. These are:

* *rhombic sulphur* which is stable at a pressure of 1 atmosphere up to 95.5°C in the form of yellow octahedra which, upon rapid heating, melt at 113°C under normal pressure and at 151°C under a pressure of about 1290 atmospheres;
* *monoclinic sulphur* which is stable at a pressure of 1 atmosphere between 95.5 and 119.25°C when it is very pure and can also be prepared at a normal temperature by the rapid cooling of molten sulphur.

These characteristics of crystalline sulphur may be deduced from the well known phase diagram of sulphur.

Fused sulphur

Just above its fusion point sulphur is a yellow, mobile liquid formed of octa-atomic molecular rings. Upon further heating, liquid sulphur becomes noticeably more fluid and changes from a reddish colour, taking on an even darker colouration up to 160°C owing to the progressive degradation of the molecules. In this process more and more sulphur atoms with an unpaired electron, which are notoriously chromomorphic, are liberated . . . S–S . . . → . . .S˙ ˙S . . .

Above 160°C the viscosity of the fused mass increases as a consequence of the formation of long chains, owing to a coupling of the sulphur atoms by bond formation involving the unpaired electrons. At 188°C the viscosity attains a maximum value at which point the product does not pour out when the container holding it is tipped up.

Upon raising the temperature still further, the viscosity decreases and the colour, which had become dark red, slowly changes to black until, at normal pressure, it boils at 444.6°C.

Sulphur vapours

Over a small range of temperatures sulphur vapours have a yellow–orange colouration (the S_8 molecular form). They then take on a red coloration which first darkens and then becomes brighter. This is because the sulphur progressively changes to the S_6, S_4 and S_2 molecular forms. It finally forms sulphur atoms (S).

Plastic sulphur and colloidal sulphur

When sulphur is heated up to its boiling point and then suddenly cooled, for example, by pouring a liquid thread of it into water, it forms a yellow–brown, transparent, structurally amorphous mass of so-called plastic sulphur because it can be drawn into threads. Plastic sulphur becomes hard, and when treated with carbon disulphide only part of it dissolves, and the more the original liquid sulphur was heated before being 'tempered', the lower is the amount which dissolves.

Colloidal sulphur is prepared either by the acidification of solutions of thiosulphate or alkaline/alkaline-earth polysulphides or by bubbling H_2S into an aqueous solution of SO_2.

Both plastic sulphur and colloidal sulphur are structurally disordered; that is, they are, in essence, amorphous.

6.2 THE EXTRACTION OF NATURAL SULPHUR

The processes employed in the extraction of natural sulphur vary, depending largely on the nature of the deposit. Sulphur mines are of two types: mines capped by compact rocks and mines capped with friable rocks. The first type of mine permits the opening of galleries from which the mineral can be brought to the surface; but this is not possible in the other case.

The Sicilian deposits, which today have become unworkable economically, on account of their low sulphur content and because alternative sources of sulphur are more convenient, belong to the first type. On the other hand, the central and north American sulphur-bearing districts (the Gulf of Mexico, Louisiana, and Texas), which today supply the greater part of the naturally extracted sulphur, belong to the second type.

Deposits of the first type are worked by autofusion or by the indirect fusion of sulphur.

The most rational autofusion method is that of the 'Gill furnace' which consists of chambers which can alternately function as hot beds, preheaters, and as melting furnaces. Such a furnace is constructed of 20–40 m^3 masonry chambers ('cells') internally lined with acid refractories, and with each of them fitted with apertures for the loading of the mineral, the removal of the exhausted gangue, for the collection of the fused sulphur, for the circulation of the hot vapours in the system, for the injection of air, and for the discharge of the combustion products to the chimney. The cells are grouped together in fours or sixes.

A diagram of the manner in which a Gill furnace operates is shown in Fig. 6.1, where F_1, F_2, F_3, and F_4 are the cells making up a set of four, a, b, c, . . . , n are control valves which

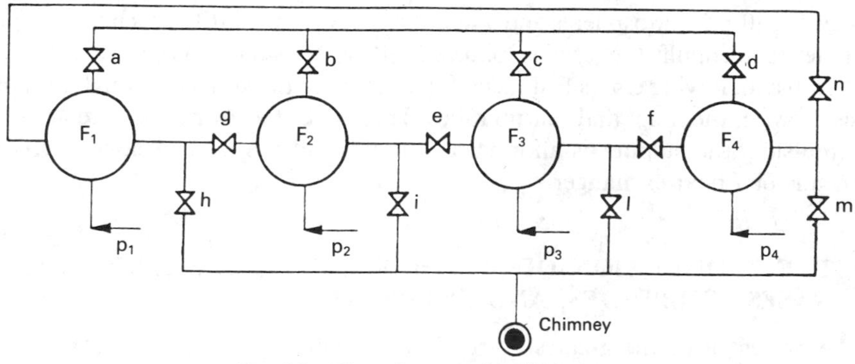

Fig. 6.1—Plant diagram of a Gill furnace.

are appropriately opened or closed at each phase of the process, and p_1, p_2, p_3, and pp_4 are air holders.

The system operates by making one of the four cells (the one which has acted as a melting furnace in the preceding phase and from which all sulphur recoverable from the mineral has been taken off at the tap hole) act as a hearth for the taking in of air from the base. Very hot vapours pass out of this cell which, upon entering the next cell, make it into the fusion cell, while the vapours emerging from this second cell are still quite hot and can be used, when they enter a third cell, to preheat the fresh mineral there. The fourth cell from which all the sulphur has been removed after it has functioned as a hearth in the preceding phase remains, in the meantime, shut off from the rest of the system and is duly refilled with the mineral.

The four phases in which this system operates can be identified as a consequence of successively keeping open the four valves shown below while all the other valves are closed:

a b e l *c d n h*

b c f m *d a g i*

In the first phase it is F_1, having received air from the base, which acts as a hearth, and F_2 which, having been preheated in the preceding phase, which now constitutes the fusion furnace of the system owing to the effect of the hot vapours originating from F_1. F_3 is then preheated by the vapours emerging from F_2 which subsequently pass into the chimney via the valve 1. In the meantime, the exhausted material is removed from F_4 and fresh material is loaded into it.

In the second phase F_2 acts as the hearth, F_3 is the fusion furnace, F_4 is the preheating cell, and the contents of F_1 are discharged and it is reloaded. The successive phases proceed in an analogous manner so as to provide continuous operation.

Among the methods for the indirect fusion of sulphur contained in rock which has been brought to the surface and crushed, the Japanese method excels. This makes use of superheated steam which is passed into containers riddled with small holes and filled with the sulphur-containing rock. These containers, in turn, are placed in large autoclaves. The yield of this process is greater than 90%, and up to 96% pure sulphur is obtained.

To exploit sulphur deposits which are capped by loose and friable rocks, that is, typically, the North American deposits, the 'Frasch' method is adopted. This technique entails the underground fusion of sulphur by means of superheated steam and the forcing of a fused sulphur–water emulsion resulting from the condensation of the steam to the surface by means of hot, compressed air.

The essential part of the apparatus for this process is a borer consisting of three concentric steel tubes: an external tube (15–20 cm diameter) with holes in the lower part, a central tube of 7–8 cm diameter and an inner tube of 3–4 cm diameter. As Fig. 6.2 shows, the steam is forced down through the annular space between the outer and central tubes. The superheated fluid melts the sulphur by diffusing into the earth surrounding the bottom part of the borer. The hot, compressed air at a pressure of 20–25 atmospheres and a temperature of 120–140°C which is forced down the inner tube engulfs the fused sulphur and the water which has condensed out from the steam and, having emulsified them, forces them to the surface through the annular space between the inner and central tubes. Then, after the air has been removed from the emulsion, the sulphur is allowed to solidify, and the heat released is recovered by means of a heat exchanger.

6.3 THE PRODUCTION OF SULPHUR FROM SULPHURIFEROUS GASES, SULPHIDES, AND SULPHATES

At the present time the greatest part of the sulphur produced in the World comes from certain natural gases (Canada and the USSR are at the head of the league for

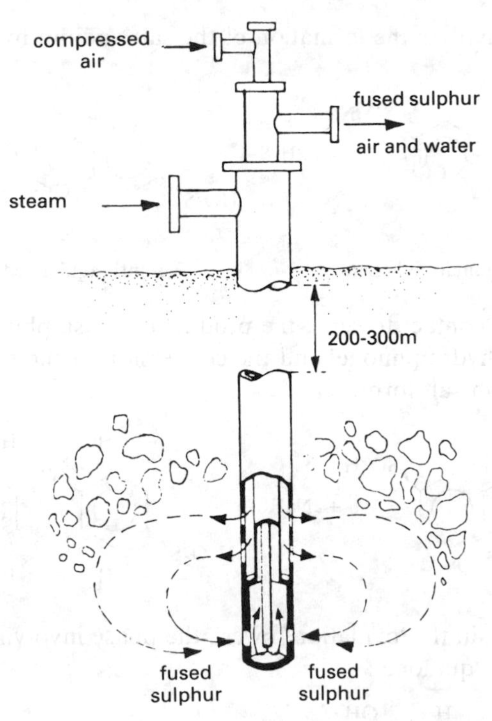

Fig. 6.2—Schematic diagram of the Frasch method for the production of sulphur by extraction from its natural deposits. Apart from the non-compactness of the capping rocks, the use of this procedure is favoured by the high concentration of sulphur and the impermeability of the rocks surrounding the diapiric sulphur-bearing domes ('cap rocks').

producers of sulphur using this route), from petroleum refinery gases (which are characteristic sources in petroleum producing and refining countries, including Italy), and from various coal distillation processes (a source peculiar to continental Europe). A significant fraction of the sulphur is still produced (Spain, Scandinavia, Tennessee and Virginia, USSR) from sulphide minerals and, in particular, from pyrites (FeS_2).

As always, natural sulphates may also constitute a notable source of sulphur.

Sulphidic gases and their exploitation

The characteristic of all the gases which are made use of as a source of sulphur is their appreciable content of hydrogen sulphide (H_2S).

Sulphur can be recovered from such gases by fixing the H_2S component in them and oxidizing it, thereby liberating sulphur in the same plant from the form in which it was fixed or by oxidation of the hydrogen sulphide after it has been released from the product, in which it was fixed, using a separate plant.

Processes which lead directly to sulphur

Among the processes which lead directly from gases containing H_2S to elementary sulphur, the following stand out:

- the *anthraquinone disulphonate method* which, in the plant depicted schematically in Fig. 6.3, first involves the formation of the salt sodium hydrogen sulphide:

$$\text{NaO}_3\text{S}\text{–(anthraquinone)–SO}_3\text{Na} + 2H_2S \rightarrow \text{HO}_3\text{S}\text{–(anthraquinone)–SO}_3\text{H} + 2NaHS$$

2,6-anthraquinonesodiumdisulphonate 2,6-anthraquinonedisulphonic acid

and, then, in the 'contact tower'®, the production of sulphur by the reduction of the 'quinone' to 'hydroquinone' and the conversion of the sulphonic acid groups back to the sodium salt form:

$$\text{HO}_3\text{S}\text{–(anthraquinone)–SO}_3\text{H} + 2NaHS \rightarrow \text{NaO}_3\text{S}\text{–(C(HO)(H))}_2\text{–SO}_3\text{Na} + 2S$$

and, subsequently, in the 'oxidation tower', the phase involving the reoxidation of 'hydroquinone' to 'quinone':

$$\text{(H)(OH)C} \;\; \text{C(H)(OH)} \xrightarrow{O_2} \text{O=C} \;\; \text{C=O} + 2H_2O$$

regenerated reagent

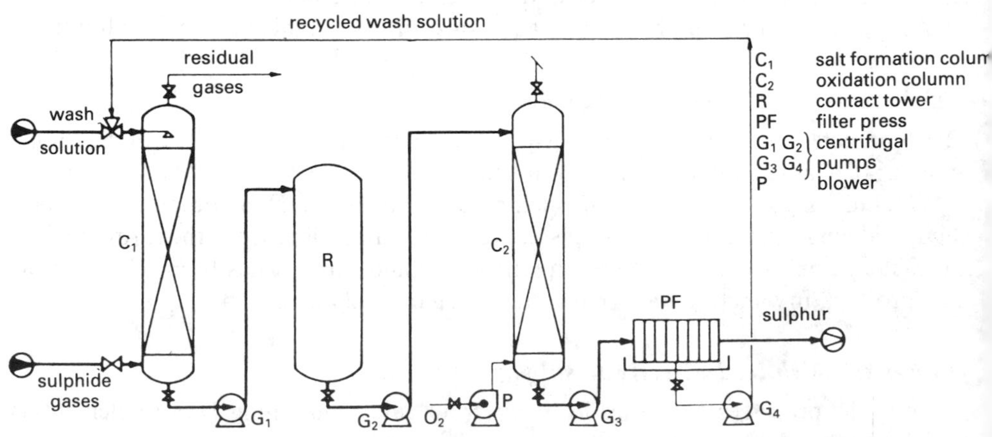

Fig. 6.3—The anthraquinone disulphonate method for the recovery of sulphur. A sufficiently long contact time in R allows the quinone → hydroquinone reduction to take place which does not occur in the preceding salt-formation column.

which is finally followed by the recovery of the sulphur which has been produced by means of filtration and recycling of the reagent solution;

- The *Thylox process*, which consists of firstly washing the gases with a mixture of sodium carbonate (or ammonium carbonate) and arsenious oxide which, in water, form disodium arsenite. This reacts with hydrogen sulphide as follows:

$$\underset{\text{disodium arsenite}}{Na_2HAsO_3} + 2H_2S \rightarrow \underset{\text{disodium oxodithioarsenite}}{Na_2HAsOS_2} + 2H_2O$$

followed by the regeneration of the reagent with air:

$$Na_2HAsOS_2 + O_2 \rightarrow Na_2HAsO_3 + 2S$$

and, finally by filtration, recovery of the sulphur which has been liberated;

- the *Koppers–Seabord process*, which involves washing the sulphidic gases with a dilute solution of soda:

$$Na_2CO_3 + H_2S \rightarrow NaHCO_3 + NaHS$$

followed by the catalyzed reoxidation of the bisulphide at an elevated temperature with air:

$$NaHCO_3 + NaHS + 1/2O_2 \xrightarrow[\text{compounds}]{\text{Fe(II)}} Na_2CO_3 + H_2O + S$$

which leads to the regeneration of the reagent and the production of sulphur;

- the *Laming mass process* which is mainly utilized in the production of sulphur from gases obtained during the distillation of coals. The gases to be desulphurized are made to pass through masses formed of hydrated iron oxides mixed with sawdust and other inert materials which increase the surface area of contact and render the masses porous. These masses which are arranged on a mesh in suitable containers are known as ‘Laming masses’.

Reactions of the type:

$$Fe_2O_3.3H_2O + 3H_2S \rightarrow 2FeS + S + 6H_2O$$

occur to a considerable extent within these mixtures.

After a certain amount of hydrogen sulphide has been fixed, the mass is reactivated with moist air:

$$2FeS + 3H_2O + 3/2O_2 \rightarrow Fe_2O_3.3H_2O + 2S.$$

When the sulphur has attained a level of about 50% it is either recovered in the form in which it exists or it is utilized, using combustion which converts it into SO_2, or it is utilized in both of these ways in plants of the Gill furnace type.

Solvent extraction is a leading method in the integral recovery of sulphur from Laming masses which are rich in it. Another method involves the fusion of the sulphur after the masses have been loaded, using small wagons with a mesh bottom, into wide boilers in which they are acted upon by a flux of superheated steam.

Processes which lead to high levels of H_2S

Among the processes which involve the fixation of H_2S from gaseous mixtures and its recovery as pure H_2S within a first plant and the production of sulphur from hydrogen sulphide within a second plant, there are:

- treatment with aminoalcohols (the *Alcazid process*) which is achieved by passing, after care has been taken to get rid of the heat of reaction, the gaseous mixture through successive washing towers with a countercurrent of ever fresher ethanolamines so that one finally has, on the one hand, a gas which is devoid of H_2S, and, on the other hand, ethanolamines saturated with H_2S is accordance with the reaction, when diethanolamine is used as the absorbent:

$$\underset{\text{diethanolamine}}{H_2S + NH(CH_2CH_2OH)_2} \rightleftharpoons \underset{\text{diethanolamine hydrogen sulphide}}{H\overset{-}{S}\overset{+}{N}H_2(CH_2CH_2OH)_2}$$

  The reagent is regenerated simply by heating which leads to the recovery of hydrogen sulphide owing to the reversibility of the fixation reaction;
- the use of phenoxides (the *Koppers process*) which is technologically well established and, as with the aminoalcohols, involves exothermic equilibrium reactions of the type:

$$\underset{\text{sodium phenoxide}}{C_6H_5ONa} \rightleftharpoons \underset{\text{phenol}}{C_6H_5OH} + NaHS$$

  so that high purity H_2S can be recovered from the reaction products at the end of the process by heating.

General method for the oxidation of H_2S to sulphur

The gases obtained from the Alcazid and Koppers processes can be converted to sulphur in plants (Fig. 6.4) which are essentially reproductions of the classical *Claus method* for the oxidation of H_2S to sulphur with air.

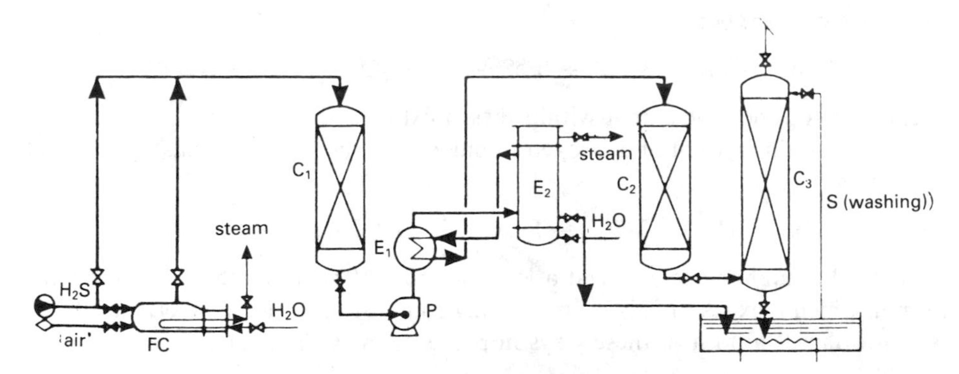

Fig. 6.4—The Claus method brought up to date. FC = furnace-boiler; C_1 = primary reactor which raises the temperature of the reactants from 320 to 390°C; C_2 = secondary reactor which requires the hot gases at 250–270°C with a low level of free sulphur if it is to function well. The reaction products from C_1 are therefore cooled by sending them, by means of the blower P, at 140–160°C into the exchanger E_1 and then further cooled by water in the sulphur condenser E_2 and successively brought back up to around 260°C by recycling in E_1. C_3 is a packed column which is fed with fused sulphur at 120°C which overflows into the tank of newly formed sulphur located under the column. The gaseous products which are purged consist of N_2, H_2O, CO_2, and small residues of H_2S and SO_2.

New plants have been perfected which also permit the recovery of sulphur from the tail gases arising from various processes which contain low levels of H_2S. While, in the original Claus method, the oxidation was carried out with good yields in a single heat-dispersing tower containing bauxite as the catalyst and exploited the overall reaction:

$$3H_2S + 3/2O_2 \xrightarrow[200\text{–}300°C]{\text{bauxite}} 3H_2O + 3S$$

modern plants use more active catalysts with longer useful lifetimes and consist of a furnace-boiler in which a third of the H_2S is oxidized to SO_2 with air:

$$2H_2S + 3O_2 \rightarrow 2H_2O + 2SO_2$$

in a primary reactor containing activated alumina or a mixture of cobalt–molybdenum oxides as well as more sophisticated catalysts of unrevealed composition where the reaction:

$$2SO_2 + 4H_2S \rightarrow 6S + 4H_2O$$

involving about $\frac{3}{4}$ of the H_2S still present, takes place, and, in a second reactor, at a lower temperature, where the remaining H_2S (about $\frac{1}{4}$) is oxidized by means of SO_2. More precise details of the process are given in Fig. 6.4.

Any hydrocarbons which may be present in the charge are transformed in the furnace-boiler into CO_2, CO, and H_2O, and, moreover, reactions leading to the production of COS and CS_2 may take place there. These reactions are of the type:

$$2H_2S \rightarrow 2HS^{\cdot} + 2H^{\cdot} \rightarrow 2S + 2H_2$$

$$CH_4 + 4S \rightarrow CS_2 + 2H_2S$$

$$CO_2 + H_2S \rightarrow COS + H_2O.$$

The higher temperatures appertaining in the first reactor tend to suppress the exothermic reaction involving the production of sulphur from H_2S and SO_2, but favour the reactions involving the other sulphur-containing gases which are formed in accordance with the reactions shown above or are present in the initial charge:

$$COS + H_2O \rightarrow CO_2 + H_2S; \qquad 2COS + SO_2 \rightarrow 3S + 2CO_2$$

$$CS_2 + 2H_2O \rightarrow CO_2 + 2H_2S; \qquad CS_2 + SO_2 \rightarrow 3S + CO_2$$

and so on.

The reaction mixture arrives in this state at the second reactor where the strongly exothermic sulphur oxidation–reduction reaction involving the H_2S/SO_2 couple proceeds to completion.

Production of sulphur from pyrites

There are a number of methods for the preparation of elementary sulphur from pyrites. The most important of these are:

- the thermal decomposition of the mineral FeS_2 at temperatures above 1000°C in the absence of air;
- treating pyrites at 600–700°C with water vapour;
- reacting pyrites with sulphur dioxide at 800–900°C.

All of these methods result in only the partial transformation of the sulphur contained in the raw material into elementary sulphur, and the residues are therefore suitable for use in the preparation of sulphur dioxide by roasting them with air. In other words, the exploitation of pyrites for sulphur is always only partial.

Production of sulphur from anhydrite and gypsum

Sulphur can be prepared, in suitable furnaces, from minerals based on $CaSO_4$, using the following reactions:

$$3CaSO_4 + 6C \xrightarrow{900^\circ C} 3CaS + 6CO_2$$

$$3CaS + CaSO_4 \xrightarrow{1200\text{–}1300^\circ C} 4CaO + 4S.$$

There is considerable interest in this method for the production of sulphur in Western European countries, and particularly in Germany, where there is a shortage of other local sources of sulphur.

6.4 REFINING, THE COMMERCIAL TYPES OF SULPHUR, AND THEIR USES

The sulphur extracted from natural deposits or as a product from sulphuriferous gases, sulphides, and sulphates is raw sulphur. That is, it is contaminated with earth, dross, and tars, etc. It is refined, whenever necessary, by making use of the fact that it can be readily melted and vaporized.

The commonest type of plant for the refining of sulphur consists of a battery of retorts (Fig. 6.5). In this plant the sulphur is fused, using superheated steam in a boiler F_1 from which it descends into a tank S which distributes it into the individual components of a battery of 8–12 retorts sited within a single furnace which is heated by the burning of naphtha or combustible gases. The sulphur sublimes in these retorts and thence passes to the air condenser F_2. Now, having become liquid again, it can be cast in moulds of various shapes, solidifying to form 'roll sulphur', 'sulphur bricks', 'sulphur pipes', etc.

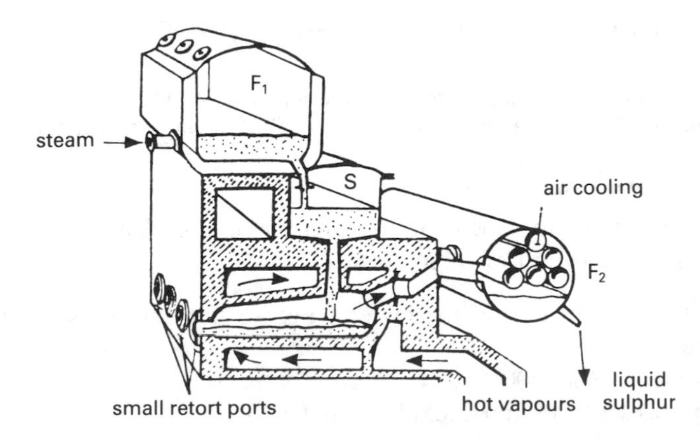

Fig. 6.5—Battery of retorts for the refining of sulphur. The liquid sulphur produced can be shaped in large moulds or passed to a system of nebulizers which turn it into powder of a varying degree of fineness.

The sulphur vapours emanating from the retorts can also be made to condense in large chambers (with a volume of 700–900 m^3 for every retort serviced) internally lined with argillaceous refractories.

Raw sulphur, as well as refined sulphur, are sometimes crushed, and the product of the milling can be variously classified by sieving or by its removal by a current of air. Different degress of fineness of the sulphur are obtained in this manner.

The fineness of the sulphur is estimated in *degrees Chancel*. Sulphur which is so subdivided that 45 g which settles out from a suspension in ether occupies 25 ml at 17.5°C has a fineness of 100 degrees Chancel. Good agricultural sulphurs are of 75 degrees Chancel, while the best are of 95 degrees Chancel. In the case of these latter sulphurs, 97% of their weight adheres to the leaves of the plants on which they are applied.

The various commercial forms of sulphur depend on how it has been obtained originally, on how it has been refined, and, eventually, treated. The principal forms include:

- *Frasch sulphur* of the 'dark' (raw) type, of the 'bright' type (purified from bituminous substances by fusion with the addition of kaolin followed by filtration), and of the 'super-bright' type (distilled);
- *raw sulphur* from any source whatsoever containing limestone, chalk, bitumens, etc. which is sold commercially in the form of rolls, bricks, or in pieces;
- *refined sulphur*, moulded in various forms or milled or blown (that carried away by a current of air from the products of the crushing of very pure sulphur in centrifugal mills);
- *sublimed sulphur* ('flowers of sulphur'), produced by the condensation in large chambers of the vapours coming from the retorts of the purification process of sulphur;
- *colloidal sulphur*, the preparation of which has already been described and from which 'precipitated sulphur' is recovered. It is available as an extremely fine-grained powder obtained by the careful drying of colloidal sulphur.

There are many, many uses of sulphur, and some of these are extremely important. Besides the $\frac{3}{4}$ of the sulphur which is consumed in the sulphuric acid industry, 7% is used for the production of SO_2 and its derivatives (page 196) and 8% is used for the synthesis of chemical products among which carbon disulphide, sulphur chloride, and the polysulphides stand out. The remainder is used in matches, plastic materials, pigments (ultramarine, for example), sulphur colorants, and detonation powders, as well as in rubber vulcanization works, synthetic elastomers, the production of veterinary and pharmaceutical preparations, and fungicide formulations for agricultural use.

6.5 SULPHUR DIOXIDE

Properties and uses of SO_2

Under normal conditions sulphur dioxide is a gas which is colourless itself but produces white fumes in air. It has an irritating odour and a suffocating effect. It is soluble in water and organic solvents which are more or less polar (alcohols, acids, and ethers) and bases (amines), forming salts which are thermally labile in the latter case.

Structurally, SO_2 is a resonance hybrid, the canonical forms of which are mostly polar:

$$\left\{ {}^{(+)}|S\begin{matrix}\nearrow \bar{O}|^{(-)} \\ \searrow\!\!\searrow \bar{O}\end{matrix} \leftrightarrow {}^{(+)}|S\begin{matrix}/\!\!/ \bar{O} \\ \searrow \bar{O}|_{(-)}\end{matrix} \leftrightarrow |S\begin{matrix}/\!\!/ \bar{O} \\ \searrow\!\!\searrow \bar{O}\end{matrix} \leftrightarrow |S\begin{matrix}/\!\!/\!\!\swarrow O|^{(+)} \\ \searrow \bar{O}|_{(-)}\end{matrix} \leftrightarrow |S\begin{matrix}\nearrow \bar{O}|^{(-)} \\ \nwarrow\!\!\searrow\!\!\searrow O|^{(+)}\end{matrix} \right\}$$

while the most symmetric, apolar form makes the largest contribution to the hybrid which is constructed from all five of the canonical structures shown above.

Table 6.1 lists the properties of SO_2.

Sulphur dioxide, as such, has been used in the past as a refrigerant fluid and is currently used as a disinfectant, conserving agent, and decolorant in food technology and in the paper, textile, conserving, and wine industries. It is also used as an absorbent for compounds with olefinic characteristics.

Sulphur dioxide lies at the base of the production of various types of salts and, above all, sulphuric acid. It is estimated that more than 88% of the total amount of sulphur dioxide is at the present time destined for the production of acid derivatives (SO_3, H_2SO_4, H_2SO_3, $NaHSO_3$, etc.) and salts (sulphites, thiosulphates, meta-bisulphites, sulphoxylates, and others).

Preparation of sulphur dioxide

The raw materials for the production of sulphur dioxide are, for the major part, sulphur, pyrites, and hydrogen sulphide. However, sulphides and certain sulphates, especially that of calcium, may also be used for this purpose.

Notable amounts of SO_2 are also recovered from the effluent gases from the burning of combustible oils, coals, natural gases, and gas oils by fixing it selectively with reagents or by making use of its solubility in suitable solvents.

Research in this field has been directed toward the perfection of the most selective fixing reagents and to the extent to which they can be regenerated unchanged after the fixation of SO_2.

Sulphur dioxide from sulphur

The production of sulphur dioxide from sulphur is based on the exothermic reaction:

$$S + O_2 \rightarrow SO_2 + 71\ \text{kcal/mole}.$$

Table 6.1—Physicochemical properties of sulphur dioxide

Property	*Values*	*Property*	*Values*
Melting point	-72.7°C	Solubility at 20°C	$39.4\ l_{SO_2}/l_{H_2O}$
Boiling point	-10°C	Solubility at 0°C	$79.8\ l_{SO_2}/l_{H_2O}$
Heat of evaporation	82 kcal/kg	Temperature at which solubility = zero	~80°C
Critical temperature	+157°C	Liquefaction pressure at 20°C	~2.5 atm
Critical pressure	77.7 atm	Reducing power in water	high, with oxidation to SO_4^{2-}
Density	2.264 kg/dm^3	Compounds	H_2SO_3 with water and HSO_3^-, SO_3^{2-}, $S_2O_5^{2-}$ with alkalis
Relative dielectric constant	17.3	Solvent power	high for olefins
Dipole moment	1.18 D		

It is necessary to ensure the intimate mixing of the air and the sulphur to avoid losses of the element by sublimation. The amount of air to be used in this reaction, above and beyond the theoretical amount, is established on the basis of the use to which the SO_2 produced is subsequently to be put.

The combustion may be carried out in the first place in rotating furnaces covered externally with sheet steel and internally lined with acid refractories. Such furnaces differ from one another essentially in the manner in which the reagents are fed in: this may be done by injection to a fused sulphur ejector (Fig. 6.6) or by the mixing of sulphur powder with 'primary air' and by the subsequent addition of secondary air after a first, partial, combustion in order to complete the combustion in a suitable chamber.

Cascade furnaces (or 'film' furnaces) can also be used (according to the state of the sulphur being burnt in them) with continuous operation over the whole lifetime of the apparatus (four years on average). These furnaces are shown in Fig. 6.7.

Finally, horizontal or vertical spray furnaces (Fig. 6.7) may also be used. These are fed with fused sulphur and are externally enclosed in steel but lined internally with acid refractories.

Sulphur dioxide from metal sulphides

The metal sulphide which is most widely used in the production of SO_2, is pyrites (FeS_2). However, blende (ZnS) is also employed, and, to a much lesser extent than this, other sulphides by virtue of their diffuseness in nature and the yield of sulphur obtained.

The reaction, which forms the overall basis for the production of sulphur dioxide by using pyrites, is:

$$4FeS_2 + 11O_2 \rightarrow 2Fe_2O_3 + 8SO_2 + 411 \text{ kcal/mole}.$$

Actually, the overall reaction shown here takes place in stages which involve the processes:

$$4FeS_2 \rightarrow 4FeS + 4S \quad \textit{thermal dissociation of pyrites}$$

$$4S + 4O_2 \rightarrow 4SO_2 \quad \textit{combustion of sulphur}$$

$$4FeS + 7O_2 \rightarrow 2Fe_2O_3 + 4SO_2 \quad \textit{combustion of iron(II) sulphide.}$$

Owing to the catalytic effect of Fe_2O_3, the reaction:

$$SO_2 + 1/2O_2 \rightleftharpoons SO_3$$

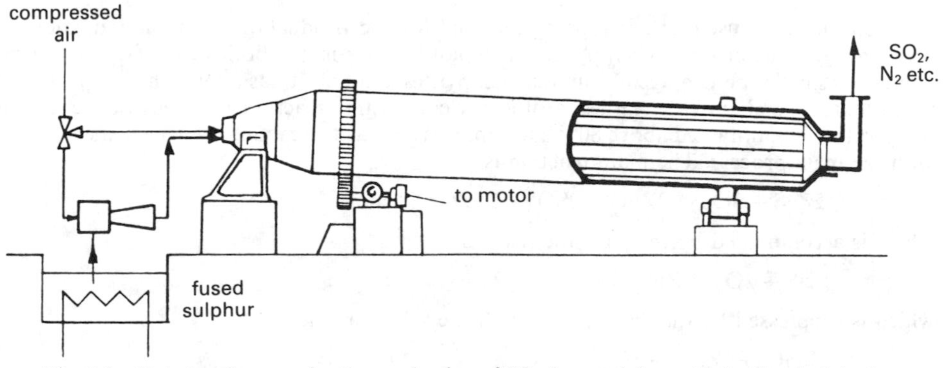

Fig. 6.6—Rotating furnaces for the production of SO_2 from sulphur with feeding by injection to an ejector.

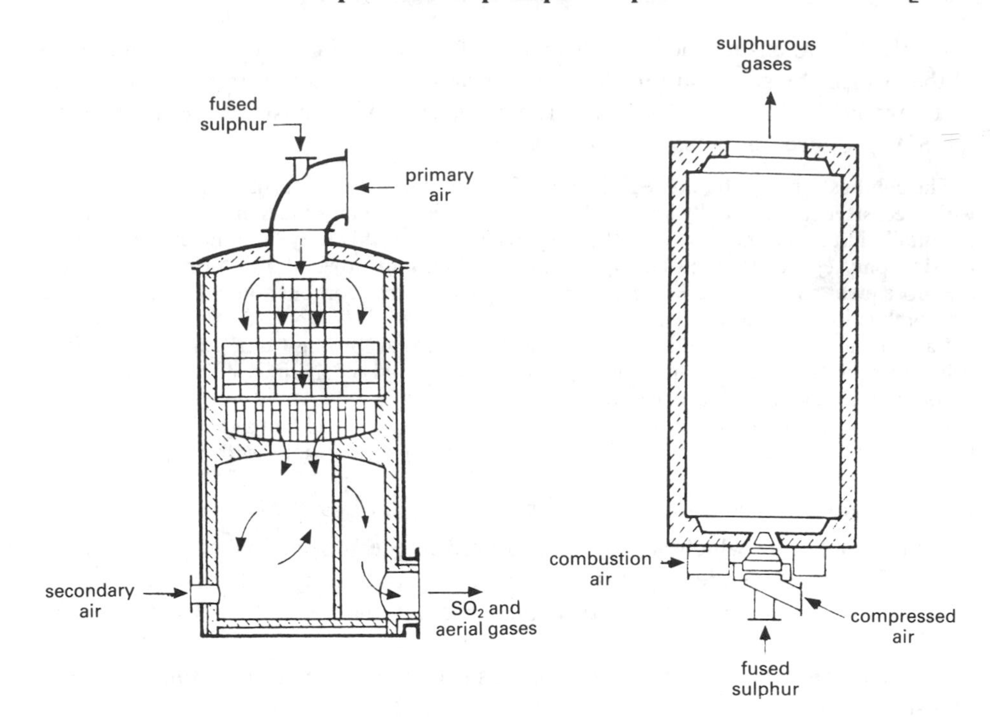

Fig. 6.7—(Left). Cascade furnace for the production of SO_2 from S. The upper part of the furnace consists of a 'honeycomb' of refractory bricks. The primary air burns the bulk of the sulphur, while the secondary air burns the last traces. Ashes accumulate in the bottom owing to the burning of impure sulphur. (Right). A spray furnace for the production of SO_2 from S. The injection nozzle is intended for the atomization of the sulphur. The insulation of the layer of steel which encloses the refractory kiln of the furnace can be seen. Inside the kiln there are perforated bricks which promote the combustion while slowing down the gas flows and encouraging contact.

occurs as a side reaction and, as a consequence:

$$Fe_2O_3 + 3SO_3 \rightleftharpoons Fe_2(SO_4)_3.$$

The higher the temperature of the process, the more these equilibrium reactions are driven to the left, and this is already very pronounced at 500–600°C.

When blende is used as the starting material for the production of sulphur dioxide, the problems are greater than when pyrites is employed. This is due both to the fact that blende contains a smaller percentage of sulphur than pyrites and that it has a lower heat of formation for the oxide, which implies a far less exothermic combustion reaction, and because the roasting temperature is maintained above 800°C to suppress a parasitic reaction which occurs in parallel with the main process. The main reaction is:

$$2ZnS + 3O_2 \rightarrow 2ZnO + 2SO_2 + 115.6 \text{ kcal/mole}$$

which is accompanied by the parasitic reaction:

$$ZnS + 2O_2 \rightarrow ZnSO_4$$

which is suppressed by the decomposition of the salt that is formed:

$$ZnSO_4 \rightarrow ZnO + SO_2 + 1/2O_2$$

which starts to take place at above 760°C and is complete at temperatures from 900–1000°C.

To avoid the occurrence of this parasitic reaction, therefore, the process is carried out at a temperature higher than 1000°C.

If blende is used, measures must therefore be taken to recover the heat and to use it in heating the air to be used in the combustion, and provision must be made, possibly, for supplementary ambient heating.

Furnaces for the production of sulphur dioxide from metal sulphides are structurally and operationally different, depending mainly on the dimensions of the particles of the mineral which is treated.

There are:

- *hearth furnaces*, *kiln furnaces* and *rotating furnaces* for crushed sulphides with particle sizes from 6 to 15 mm;
- *fluid bed furnaces* for particle sizes between 2 and 8 mm;
- *flash furnaces* for flotation concentrates and products of milling with a diameter of 0.05 to 0.20 mm.

In hearth furnaces one uses a greater excess of air, on account of the larger sizes of the pieces of mineral and the lower operating temperature. For this reason, the gas produced in terraced furnaces is less rich (up to 9.6% on average) in SO_2 than either the gas produced in flash furnaces (11.2% on average) or the gas produced in fluid bed furnaces (13.6% on average).

The optimal conditions for these latter furnaces are: operating temperature 880–900°C and an amount of air which is 1.05 to 1.15 times the theoretical amount.

Vertical hearth furnaces. Among the furnaces which are adaptable to the treatment of crushed pyrites the most useful and widespread are those in the form of a vertical cylindrical structure made out of plate steel and internally lined with aluminosilicate refractories.

The central metallic shaft, on which combs with cast iron teeth are fixed, is rotated to ensure both the removal of the material which is being roasted on the various shelves to prevent it from clustering and to transfer the material itself to the lower shelves. The shaft and the combs are cooled by a stream of air blown in from the bottom in furnaces of the Herreshoff–Kauffmann type (Fig. 6.8) and by the circulation of water in the Wedge types. In this way the wear and tear on the system is reduced and its operational lifetime lengthened.

However, when roasting blendes and certain other sulphides not only is it essential not to provide cooling but the soles of the furnaces must be heated. The reasons, which have already been indicated, for this operation and the preheating of the combustion air using the heat recovered from the combusted gases are essentially connected with the requirement that the salts (sulphates) which first tend to be formed by oxidation should be decomposed.

The feedstock is preheated on the top of the furnace and is ignited on the first shelf on which it is deposited by means of suitable loading mechanisms. The material being combusted is then made to descend toward the base through the apertures which are provided, thereby undergoing a zig-zag movement. Finally, the solid residue ('pyrites ashes') is removed and carried away. The air for combustion, which is in excess, rises from the base of the furnace to meet the combustible material, and what remains of it after the combustion process constitutes the 'vapours' of the furnace together with the SO_2 produced.

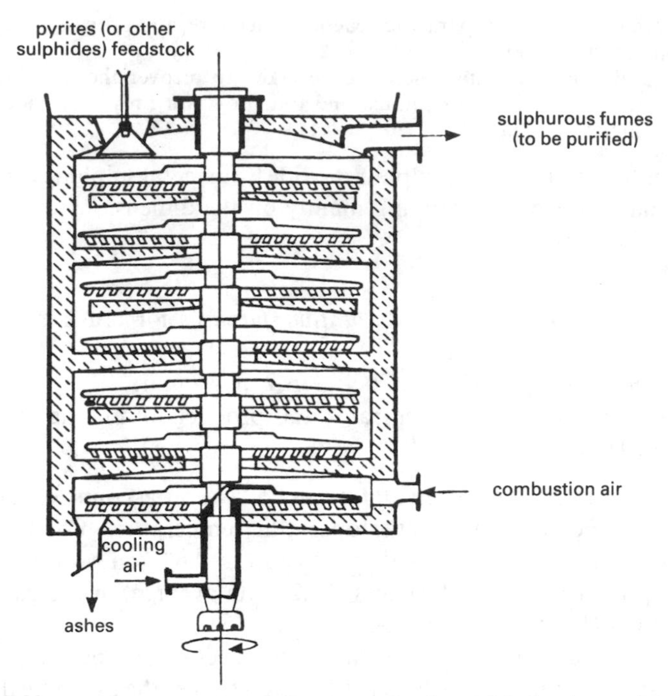

Fig. 6.8—Herreshoff–Kauffmann furnace for the production of SO_2 by the roasting of mineral sulphides.

When using operating conditions such as those which have just been described it is inevitable that the temperatures on the upper shelves will be higher than those on the lower shelves in the furnace. This can be circumvented by using a multiple feed system which involves a repartitioning of the mineral, which has reached the first shelf, between the second and fourth shelves in furnaces with seven levels, and between the second, fourth, and sixth shelves in furnaces with nine levels.

Rotating furnaces. Here, one is concerned with furnaces of the 'Lurgi' type which consist of an inclined rotating cylinder as shown in Fig. 6.9.

These furnaces are supplied with air which is distributed along the whole of the body of the apparatus to prevent local superheating leading to sintering of the material and to avoid, at the best, the non-uniform desulphurization which takes place when all the air is blown in from one end. These furnaces have great potential, but the degree of desulphurization of the mineral is lower than that achieved in vertical hearth furnaces.

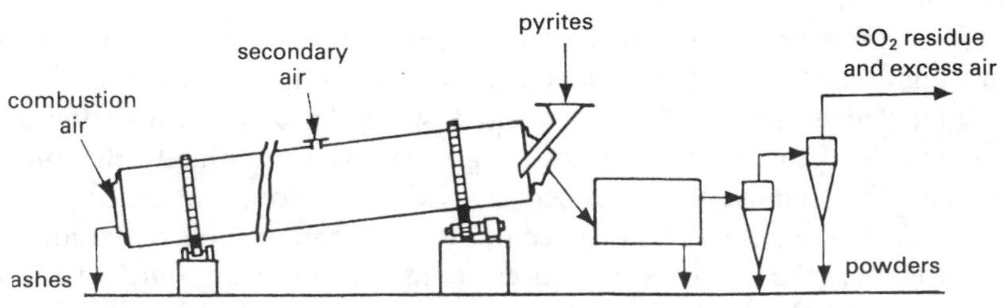

Fig. 6.9—Lurgi type rotating furnaces for the roasting of pyrites.

Fluid bed furnaces. These usually consist of steel cylinders internally lined with aluminosilicate acid refractories as shown diagrammatically together with their mode of operation in Fig. 6.10.

The mineral feedstock, suitably subdivided, arrives at a perforated plate through which a stream of combustion air is passed. A vortex consisting of an intimate mixture of the mineral and air is produced in this way, which behaves as a fluid and which is subjected to high-speed combustion owing to the great increase of the contact interface between the reagents.

The temperature rises to very high values and to a point when it is necessary to transport some of the heat away by spraying water into the furnace and/or providing the apparatus with forced water circulation heat exchangers to prevent the clustering of the pyrites ashes. The vapours leave through a port high on the side of the apparatus, while the ashes, which have fallen down with a parabolic motion, are removed and taken away to be reunited opportunely with the powders recovered in the separators where removal from the vapours has been effected.

Flash furnaces. The characteristics of these furnaces are: the removal of the heat of combustion by the circulation of air in the cavity between the two walls of the furnace, the sudden ignition (from which the process takes its name) of the preheated air, and powdered pyrites mixture which is forced into the furnace by means of a suitable nozzle and the admission of a secondary stream of air into the furnace to slow down the descent of the ashes and thereby to remove the last traces of sulphur from the mineral being treated by combustion.

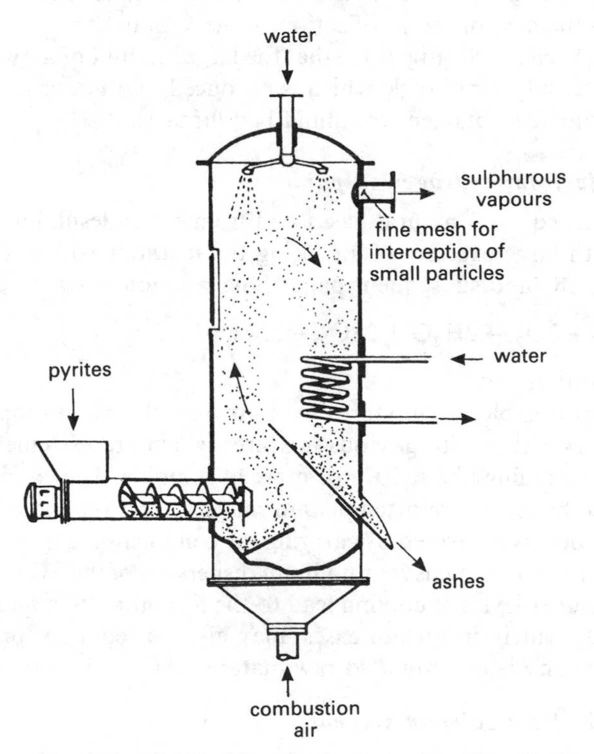

Fig. 6.10—Fluid bed furnaces for the roasting of pyrites.

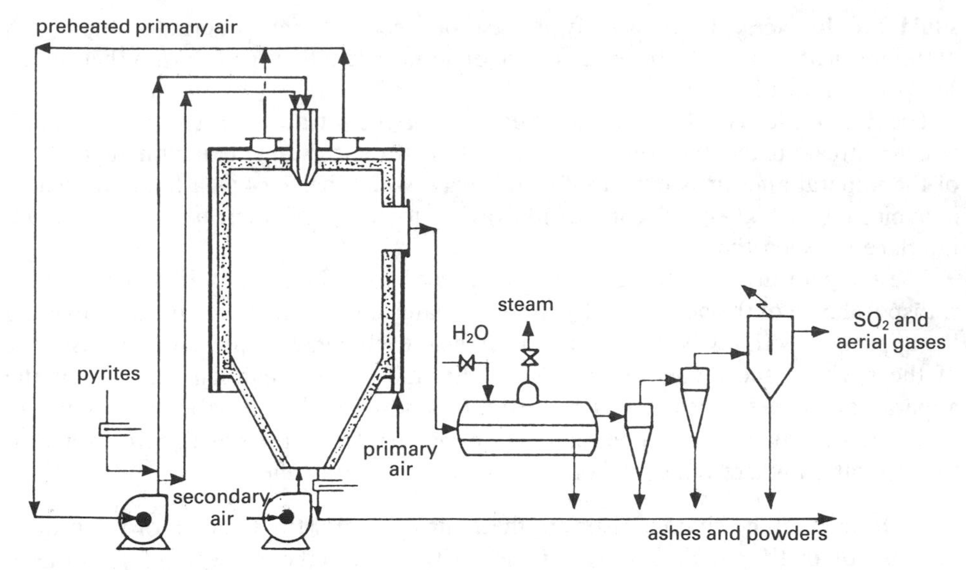

Fig. 6.11—Flash furnaces for the roasting of pyrites. The extent of removal of the sulphur from the mineral is high, but the notable amounts of air used dilutes the gas to a greater degree than in fluid bed furnaces.

The manner in which the system functions can be seen in detail from Fig. 6.11. Quite large amounts of powders are carried over in this process, thereby adding considerably to the cost of the purification of the vapours.

It has already been noted (Fig. 6.11) that the large amount of air which is circulated further dilutes the sulphur dioxide which is produced, but not as much as in the case of the sulphur dioxide obtained from fluid bed furnaces.

Sulphur dioxide from hydrogen sulphide

When H_2S obtained from natural gases and from hydrodesulphurization processes etc. is burnt with large excesses of air, using the methods which have already been described (page 182 and subsequent pages), the reaction:

$$2H_2S + 2O_2 \rightarrow 2H_2O + 2SO_2 + 210\ \text{kcal}$$

takes place quantitatively.

Since H_2S forms explosive mixtures with air over the range from 4 to 46%, it is necessary to work either with gaseous mixtures which are extremely low in H_2S or with mixtures containing about 50% or more of sulphide. The combustion is carried out in a furnace-boiler where attempts are made to recover the maximum amount of heat. Further heat is recovered by carrying the combustion gases out of the reaction zone at about 60°C so as to ensure minimum dispersion of the SO_2 in the condensing water. In fact, higher levels of cooling lead to the formation of concentrated aqueous solutions of SO_2 which, in certain cases, may also be required, provided that they are derived from an uncontaminated raw material (H_2S).

Sulphur dioxide from calcium sulphate

By fusing calcium sulphate with clay or sand in the presence of coke (or a coke–lignite mixture) in proportions by weight of approximately 6.4:1.6:1:2 in a distinctly oxidizing atmosphere to prevent the liberation of sulphur, as in the process described on page 186, SO_2 is produced by the reaction:

$$CaS + 3CaSO_4 \rightarrow 4CaO + 4SO_2$$

together with a good cement clinker. This is formed from the lime which is produced and from the acid components of the clay and the sand which were introduced.

The process is carried out in a rotating steel furnace internally lined with an acid china clay refractory in which a current of air which has been preheated by using recovered heat is passed in the opposite direction to that in which the solid is loaded in.

Purification of sulphurous vapours

Generally, sulphur dioxide forms a part of the smokes* emerging from production furnaces. The degree of purification which is required depends on the source from which the SO_2 has been derived and on the uses to which the sulphurous gas is to be put.

As far as the origin of the raw material is concerned, it should be noted with reference to the four principal sources of sulphur dioxide that:

- SO_2 originating from high purity sulphur burnt with prehumidified air does not require any further purification treatment,
- SO_2 obtained by the combustion of pure H_2S forms part of a gas from which the water is removed by cooling, from which the finest mists of sulphuric acid are removed and, when necessary, is subsequently dried,
- SO_2 originating either from furnaces for the roasting of pyrites or from furnaces which produce SO_2 together with clinker from $CaSO_4$ must have any powders and mists of fine particles removed from it and eventually be dried.

According to the uses to which it is to be put, SO_2 may be purified up to a point where it is a chemically pure compound.

Before any use, it is always absolutely necessary to free the sulphur dioxide from any suspended particles and mists. The primary removal of powders is carried out by using simple 'bag' or 'mesh' type filter systems.

Suspensions of larger particles can also be removed by passing the gas into 'powder chambers' which are fitted internally with diaphragms in such a way as to make the gas undergo multiple deviations. The smoke thus deposits owing to collisions leading to retardation of the motion of the solid particles. These are then taken away to the spiral screw of a discharge system at the centre of the floors of the chambers which are suitably shaped with a depression.

The commonest purification methods are:

- using *cyclones* for particles with dimensions greater than 20 μm, bearing in mind that the apparatus employed here is quite cheap both from the point of view of setting up and operation;
- using *scrubbers* for particles with dimensions up to 2 μm;
- using *electrostatic precipitators* which allow the gases to be freed from any type of suspension whatsoever, provided that they are pretreated by using one of the purification techniques with a more concise action.

*It is recalled that a 'smoke' is a dispersion of solid particles in a gas.

The reader is referred to treatises on chemical plant technology with regard to both the principles and mode of operation of such apparatus which are of general interest.

The production of pure sulphur dioxide

The gases which are contained together with SO_2 in the smokes from the production furnaces are essentially nitrogen and oxygen. When it is necessary to produce pure SO_2, after any powders have been removed from the smokes and they have been denebulized, the high solubility of SO_2 in water (about 113 kg per m^3 of solution at 20°C) is exploited. This solubility may be contrasted with that of oxygen (0.0414 kg/m^3 at 20°C) and, above all, with that of the nitrogen in the air (0.0282 kg/m^3 at 20°C).

For this purpose the gases to be purified are sent into a washing tower where they meet with a spray of cold water delivered from above. Nitrogen and oxygen escape from this tower*. The aqueous solution containing the SO_2 emerges from the bottom of the tower and is sent to be degassed in a plate or a packed column. The SO_2 which is liberated by the steam passes out from the top of this tower and is sent to be expanded, using a technique such that water is formed at a temperature which does not permit any appreciable redissolution of the sulphur dioxide. Finally, the moist SO_2 is dried in a countercurrent of concentrated sulphuric acid in a filled column and subsequently stored.

The plant used for the process which has just been described is shown in Fig. 6.12.

Plants also exist for the absorption of SO_2 in water under a pressure of 4 to 5 atmospheres, followed by decompression and direct heating, while, if the gases being treated are very low in sulphur dioxide, they can be utilized by washing them with a mixture consisting of equal parts of toluidine and xylidine (the *Sulphidin process*) which are capable of readily forming salts with SO_2 from which the SO_2 can be freed by heating them up to 100°C.

Principal compounds of sulphur dioxide

Sulphur dioxide yields a series of compounds of technological importance.

Bisulphites and sulphites

By bubbling SO_2 into an aqueous solution of alkaline carbonates or into suspensions of alkaline earth carbonates in water, bisulphites are obtained when the solutions are saturated:

$$\underset{\text{solution of density 1.308 kg/l}}{Na_2CO_3 + H_2O + 2SO_2} \rightarrow \underset{\text{solution of density 1.375 kg/l}}{2NaHSO_3 + CO_2}$$

The point at which saturation is attained is determined by measuring the density.

Sulphites are produced by the neutralization of bisulphites with alkalis:

$$NaHSO_3 + NaOH \rightarrow Na_2SO_3 + H_2O$$

*This solution does not contain more than 10–15 kg/m^3 of SO_2 either because saturation is not attained under the dynamic conditions, or because it is in the presence of a lot of extraneous gas, or because the temperature gradually rises as the dissolution of SO_2 proceeds, thereby reducing its solubility.

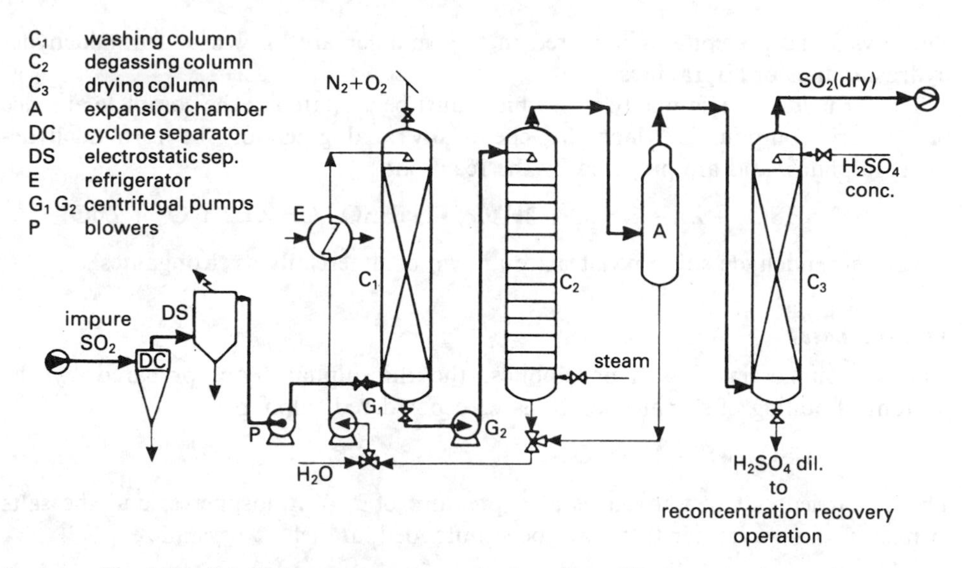

Fig. 6.12—Principal method for the production of pure SO_2 from roasting smokes.

Sulphites tend to crystallize as their heptahydrates from unsaturated solutions, while anhydrous sulphites crystallize from saturated solutions at temperatures above 40°C.

Pyrosulphites

Pyrosulphites (metabisulphites or disulphites) are prepared by the action of SO_2 on oxides, hydroxides, and anhydrous carbonates. For example:

$$2NaOH + 2SO_2 \rightarrow Na_2S_2O_5 + H_2O$$

In practice, this is carried out in rotating drums where the gas comes into contact with the granular alkaline or alkaline-earth compounds.

Hydrosulphites

A suspension of zinc powder in water is transformed into zinc hydrosulphite (dithionite) by SO_2:

$$2SO_2 + Zn \xrightarrow{\text{under } 30^\circ C} ZnS_2O_4$$

from which the alkali metal counterparts can be prepared by double exchange:

$$ZnS_2O_4 + Na_2CO_3 \rightarrow ZnCO_3\downarrow + Na_2S_2O_4$$

When solutions of hydrosulphites are saturated with sodium chloride, they precipitate the dihydrate: $Na_2S_2O_4.2H_2O$. The anhydrous compounds are prepared from the dihydrates by heating them up to 50–60°C.

The reducing power of hydrosulphites is increased and they are stabilized by reacting them with an aqueous solution of formaldehyde:

$$Na_2S_2O_4 + 2CH_2O + 4H_2O \rightarrow Na_2S_2O_4.2CH_2O.4H_2O\downarrow$$

The crystalline precipitates prepared in this manner are known as formaldehyde-hydrosulphites or 'Hyraldites'.

The formaldehyde-sulphoxylates, which must be prepared in the complete absence of air, are compounds which are chemically analogous for the formaldehyde-hydrosulphites, and are prepared by the reaction:

$$NaHSO_3 + Zn + CH_2O + 2H_2O \rightarrow NaHSO_2.CH_2O.2H_2O\downarrow + ZnO$$

The formaldehyde sulphoxylates are known commercially as 'Rongalites'.

Thiosulphates

Known commercially as hyposulphites, the thiosulphates are prepared by the prolonged boiling of sulphite solutions with powdered sulphur:

$$Na_2SO_3 + S \rightarrow Na_2S_2O_3$$

This is carried out in autoclaves at a pressure of 2–4 atmospheres, and the salts formed in this way, after they have been unloaded, are left to crystallize.

* * *

Bisulphites are sulphites are principally used as antiseptics, for the bleaching of fibres, and as solubilizers of the lignin which encrusts the cellulose in wood. Heavy metal (barium) sulphites are employed as fillers and bleachers in the paper industry. Metabisulphites are used more than sulphites and bisulphites in the photographic field and, in general, as reducing agents.

The three types of compounds which have just been mentioned are also used as conserving agents and decolorants in the food and drink industries in various commercial forms (powders, crystals) and with different EEC food additive symbols (E221, E223, etc.). These compounds are also very extensively used in the fields of food preservation, sugar manufacture, and the treatment of wines.

All of the hydrosulphites are put to various uses in the colorant industry and, even more so, in the servicing of the dyeing industries.

The thiosulphates are largely used in analytical chemistry and in the photographic and pharmaceutical industries. They are produced with various grades of purity and in various commercial forms.

6.6 SULPHUR TRIOXIDE

At normal temperatures sulphur trioxide is a vapour consisting of monomeric SO_3 and dimeric S_2O_6 which are in dynamic equilibrium with one another. In the liquid state sulphur trioxide is a cyclic trimer or an open chain trimer or tetramer. It boils at 44.8°C.

At normal temperatures, when the process is catalyzed by moisture, SO_3 tends, however, to polymerize (Fig. 6.13) to form solid chains of variable lengths. The white mass of solid sulphur trioxide is fibrous, like asbestos. Boron halides and B_2O_3 inhibit this process.

Sulphur trioxide, which dissolves violently in water and controllably in sulphuric acid, dissociates into SO_2 and O_2 with the result that it is a strong oxidizing agent.

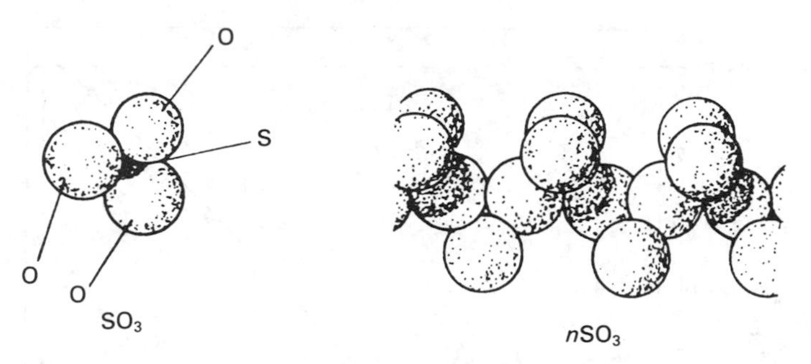

Fig. 6.13—The monomer of SO_3 (left) and the solid state polymer of this substance, S_nO_{3n} (right).

It oxidizes metals to sulphates, sulphides and oxides to sulphates, KCl and analogous compounds to chlorine, and sulphates and persulphates and phosphorus and sulphur to their respective anhydrides P_2O_5 and SO_2.

Sulphur trioxide is principally used in the production of 'oleum' by dissolving it in the monohydrate of sulphuric acid, as well as a sulphonating agent in a number of syntheses in industrial organic chemistry and, above all, in the dyeing industry.

6.7 SULPHURIC ACID

Properties, compounds with H_2O and SO_3, and commercial forms

At normal temperature above its melting point (+10.36°C) the compound H_2SO_4 is an oily, colourless liquid. It has a density of 1.8393 kg/dm^3 and fumes in air with the loss of a mist of SO_3.

When heated above 300°C, it systematically loses SO_3 until it arrives at the azeotropic composition of 98.54% H_2SO_4 and 1.46% of H_2O. It is in this form that it boils at 317°C with partial decomposition. If the undecomposed material is heated further, it continues to dissociate until this process is complete (at 450°C).

The anion formed from sulphuric acid (SO_4^{2-}) is a resonance hybrid with canonical structures of the form:

$$\left\{ \begin{array}{ccc} |\overline{O} & & \overline{O}| \\ & S & \\ |\underline{O} & & \underline{O}| \end{array} \leftrightarrow \begin{array}{ccc} |\overline{O} & & \overline{O}| \\ & S & \\ |\underline{O} & & \underline{O}| \end{array} \leftrightarrow \begin{array}{ccc} |\overline{O} & & \overline{O}| \\ & S & \\ |\underline{O} & & \underline{O}| \end{array} \leftrightarrow \begin{array}{ccc} |\overline{O} & & \overline{O} \\ & S & \\ |\underline{O} & & \underline{O} \end{array} \leftrightarrow \cdots \right\}^{2}$$

Commercial sulphuric acid (the 'monohydrate') has the composition 95.5 to 98.7% H_2SO_4 and a density of 1.84; that is, it is slightly more dilute or concentrated than the acid, which is prepared by distillation which contains 97.35% of the acid owing to the formation of the azeotrope.

Pure sulphuric acid (100%) can be prepared by cooling the acid of azeotropic composition until it crystallizes, followed by centrifugation to remove the crystals of pure H_2SO_4 which, as has already been stated, melts at 10.36°C. This is, of course, the 'laboratory' method of preparation. In practice, it is done by mixing sulphuric acid monohydrate with calculated amounts of oleums of known composition.

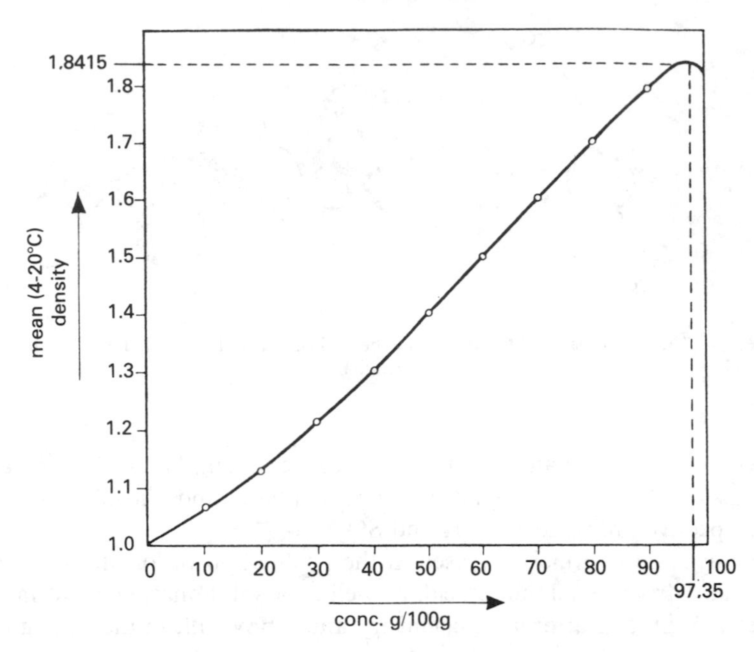

Fig. 6.14—The relationship between the concentration and mean density of sulphuric acid over a temperature range from 4–20°C. It is seen that the maximum density (the density of the azeotrope) is achieved by distillation when there is 97.35% of the acid present. The small circles on the curve indicate the densities characteristic of H_2SO_4–H_2O mixtures. The commercial acid with a density of 1.84 contains either 95.5% or 98.7% of H_2SO_4; that is, it is subazeotropic or superazeotropic.

The mean density of sulphuric acid over a temperature range from 4 to 20°C is shown in Fig. 6.14 as a function of concentration.

However, commercially, the density of sulphuric acid is usually expressed in *degrees Baumé* (°Bé) which are determined from the conventional scale by means of a density nomograph.

The scale for the measurement of °Bé is established by putting the Baumé hydrometer in pure water (at 15°C) and marking the point, at which the tube above the ballasted bulb of the device emerges from the water, as zero. The point where the tube on the same hydrometer emerges from a 10% aqueous solution of NaCl is marked as 10. The interval between the two marks is divided into 10 parts (degrees Baumé). Notches with the same spacing are put along the whole length of the tube.

Degrees Baumé are converted to a density in g/cm^3 at a certain temperature (and, in practice, to the density over the whole of the normal temperature range) by means of the formulae which are valid, respectively, in the two large zones of the World which are the most concerned with commercial use of sulphuric acid:

in the EEC — *in the USA*

$$d_{15^\circ\mathrm{C}} = \frac{144.30}{144.30 - {}^\circ\mathrm{B\acute{e}}_{(\mathrm{EEC})}} \qquad d_{60^\circ\mathrm{F}} = \frac{145}{145 - {}^\circ\mathrm{B\acute{e}}_{(\mathrm{USA})}}$$

The °Bé$_{(USA)}$ are slightly higher in value than °Bé$_{(EEC)}$.

*   *   *

The characteristics of the compounds which sulphuric acid can form with H_2O and sulphur trioxide are shown in Table 6.2.

The compounds listed in Table 6.2 constitute the crystalline phases which are obtained by cooling sulphuric acid containing various amounts of free water and free sulphur trioxide.

Table 6.2—Compounds formed by sulphuric acid with H_2O and SO_3

with water			*with sulphur trioxide*		
Formula	*% H_2SO_4*	*mp (°C)*	*Formula*	*% free SO_3*	*mp (°C)*
$H_2SO_4.H_2O$	84.5	+8.5	$4H_2SO_4.SO_3$	17	−11
$H_2SO_4.2H_2O$	73	−38	$H_2SO_4.SO_3$	45	+34.8
$H_2SO_4.4H_2O$	58	−27	$H_2SO_4.3SO_3$	71	+0.7

*   *   *

Compositionally, the types of sulphuric acid which are needed for industrial techniques, and are therefore marketed, are principally:

- *sulphuric acid monohydrate* which has already been considered (page 199);
- *sulphuric acid of various concentrations* which are mixtures of sulphuric acid and water so as to yield the mono-, di-, and tetra-hydrated species;
- *oleums*, mixtures of sulphuric acid compounds with sulphur trioxide usually containing 25, 30, or 60% of free SO_3.

Later on, when their uses are considered, the types of sulphuric acid required for different purposes will be specified in greater detail.

The preparation of sulphuric acid by heterogeneous catalysis

The method of production of sulphuric acid which is of the greatest industrial importance involves the aerial oxidation of sulphur dioxide to sulphur trioxide, using a solid catalyst consisting of vanadium pentoxide (V_2O_5) or, much less commonly, platinum.

The following equilibrium reaction is involved in this process:

$$2SO_2 + O_2 \rightleftharpoons 2SO_3 + 46.3 \text{ kcal}.$$

The expressions for the equilibrium constant (K_p) as a function of the partial pressure (p_i) and the mole fractions (x_i) are:

$$K_p = \frac{p_{SO_3}^2}{p_{SO_2}^2 \cdot p_{O_2}} \quad \text{and} \quad K_p = \frac{x_{SO_3}^2 \cdot P^2}{x_{SO_2}^2 \cdot P^2 \cdot x_{O_2} \cdot P}$$

where P is the total pressure of the reaction system.

The values of K_p at various temperatures are shown in Table 6.3.

Table 6.3—Values of K_p for the reaction $2SO_2 + O_2 \rightleftharpoons 2SO_3$

Temperature (K)	K_p	*Temperature (K)*	K_p	*Temperature (K)*	K_p
600	4180	852	14	1052	1
700	257	900	6.55	1100	0.64
800	32	1000	1.80	1200	0.26

The effects of temperature, pressure, and the concentration of O_2 on the $2SO_2 + O_2 \rightleftharpoons 2SO_3$ equilibrium

As can be seen from the K_p values shown in Table 6.3, which could have been forecast on the basis of the fact that the reaction in question is highly exothermic, the value of the equilibrium constant drops sharply as the temperature at which the reaction is carried out is increased. It is also obvious from this table that the reaction changes direction at a temperature of 1052 K (779°C) so that it is necessary to operate slightly below this temperature.

The fact that the reaction attains an equilibrium state where $\Delta G = 0$ at 1052 K can be deduced from the relationship:

$$\Delta G = \Delta H - T\Delta S$$

where: $\Delta H = -23.15$ kcal/mole^{-1}, $\Delta S = -22 \times 10^{-3}$ kcal/mole^{-1}/K^{-1}, and $T = 1052$ K.

The reason for the negative entropy change for the sulphur trioxide synthesis is to be found in the fact that this occurs with a reduction in the number of moles and, therefore, with a lowering of the 'disorder factor' in the system.

As far as the pressure is concerned, it can be demonstrated that the mole fraction of SO_3 present at equilibrium increases to a relatively small extent if the pressure is raised. For this purpose it suffices to recover x_{SO_3} from the expression for K_p as a function of the mole fractions. When this is done:

$$x_{SO_3} = x_{SO_2} \cdot x_{O_2}^{1/2} \cdot K_p^{1/2} \cdot P^{1/2},$$

from which it can be seen that the mole fraction of SO_3 is proportional to just the square root of the pressure acting on the system.

There is therefore no need to carry out the synthesis under pressure, which simplifies the construction of the plant and lowers operating costs.

Finally, it remains to discuss the role of the third variable which can affect the equilibrium, namely, that of the concentration. This is done with reference to the chemical species participating economically in the equilibrium whose concentration can be most readily changed, that is, with reference to oxygen.

On the basis of the expression for K_p, as a function of the partial pressures, it can readily be understood that the higher the concentration of O_2 and p_{O_2} thereby is increased, the more the partial pressure of the sulphur trioxide increases since K_p remains constant provided that the temperature is not varied. This is equivalent to saying that the yields from the process increase.

At first sight it might be thought that the yield of SO_3 can be increased without limit simply by continually increasing the amount of oxygen in the reagent gases. In effect, however, using an increasing excess of air leads to the ever-increasing influence of the inert gas (nitrogen) on the synthetic yield. This has a negative influence in this case because here one is dealing with the reaction which involves a contraction in volume.

Nevertheless, the sulphurous gases which arrive at the reactor have the mean composition: 7% SO_2, 10% O_2, and 83% N_2.

The kinetic problem in the synthesis of sulphur trioxide

From a kinetic point of view it is primarily the kinetic inertia of molecular oxygen which is the cause of the low rate of synthesis of sulphur trioxide.

This inertia, which corresponds to a high energy of activation, must be lowered if the reaction is to be kinetically favourable, and, as usual, this can be achieved by increasing the temperature and by the use of catalysts.

The use of temperature alone would make it necessary to operate under conditions which would render the reaction thermodynamically impossible. On the other hand, the catalyst must also be maintained at a suitable temperature if it is to act as such, in so far as that it only lowers the energy of activation but does not eliminate it altogether.

The temperature range which best meets the kinetic and thermodynamic requirements for the production of high yields in the synthesis of sulphur trioxide is located between 400°C and 500°C, with an optimal temperature at about 450°C.

The synthesis of SO_3 from SO_2 and O_2 is therefore carried out in practice under catalytic conditions at a temperature of 450°C.

Synthesis catalysts

Platinum and vanadium pentoxide, when suitably prepared, have been identified as the best-adapted catalysts for the synthesis of sulphur trioxide.

However, after the perfection of particularly active forms of V_2O_5, the use of platinum has almost been abandoned on account of its high cost and its susceptibility to poisons*.

The catalytic action of platinum is attributable to its ability to adsorb both molecular oxygen and, most importantly, atomic oxygen. It is the coming into play of this affinity which allows, under suitable conditions in practice, the occurrence of processes such as:

$$Pt_{O_{2ads}} \underset{-SO_3}{\xrightarrow{+SO_2}} Pt_{O_{ads}} \underset{-SO_3}{\xrightarrow{+SO_2+O_2}} Pt_{O_{2ads}}$$

where the states in which the molecular and atomic forms of oxygen from labile ('adsorption') complexes with the platinum are indicated by means of the conventional symbols O_{2ads} and O_{ads}.

The first stage in the catalysis is favoured by the tendency for platinum to preferentially adsorb atomic oxygen, while the second stage is favoured by the great affinity of SO_2 for this adsorbed atomic oxygen†.

It is therefore concluded that the platinum acts by transferring oxygen from air to the SO_2.

The catalysis being considered is a typical surface reaction, and the catalytic activity is therefore proportional to the surface area of the catalyst in the sense that increasing one also increases the other.

---

*Among other substances, the halogens, phosphorus, arsenic, selenium, tellurium, and their compounds are serious poisons for platinum in contrast to certain oxides (e.g. Fe_2O_3) and sulphates (e.g. $Al_2(SO_4)_3$).

†Some authors interpret this affinity as a tendency for SO_2 to reduce the oxide of platinum PtO to which $Pt_{O_{ads}}$, in effect, is seen to correspond.

For this reason, the platinum catalyst is prepared in a highly dispersed state by impregnating a support consisting of asbestos, magnesium sulphate, or silica gel with chloroplatinic acid, and heating the preparation after the addition of suitable reductants such as hydroxylamine and hydrazine.

Catalysis by vanadium pentoxide essentially involves the following steps:

$$\underset{\text{vanadium pentoxide}}{V_2O_5} + SO_2 \rightarrow \underset{\text{vanadium dioxide}}{VO_2} + \underset{\text{vanadyl-sulphate}}{VOSO_4} \xrightarrow{-SO} \underset{\text{divanadium tetroxide}}{V_2O_4} \xrightarrow{+1/2O} V_2O_5$$

which has been deduced by studying the separate steps and comparison with the chemistry of vanadium compounds.

The catalytic phenomenon is activated in the presence of alkali metal oxides which, by forming molten sulphates (e.g. K_2SO_4) and pyrosulphates (e.g. $K_2S_2O_7$) in the reaction zone lead to the process taking place in the liquid phase, giving a distinct increase in the kinetic rate. The mechanisms of this process remain unknown. Since one is dealing with surface catalysts and it is necessary to ensure that the V_2O_5 promoter is dispersed, it is formed on pumice stone or on other porous silicate materials by the thermal decomposition of ammonium vanadate which is precipitated onto the support from solution:

$$2(NH_4)_3VO_4 \rightarrow 6NH_3 + 3H_2O + V_2O_5.$$

This is a pyrolytic process which leads to the further subdivision of the catalyst brought about by the gases which are formed.

Sulphur, hydrogen sulphide, iron compounds, and arsenious oxide are among the commonest chemicals which bring about a reduction in catalytic activity and therefore poison V_2O_5 which, on the other hand, is less sensitive than platinum to catalyst inhibitors. Powders and moisture act as mechanical reducers of catalytic activity by reducing the active surface area of the catalyst.

The operation of a reactor with multiple catalytic layers for the conversion of SO_2 into SO_3

The apparatus in which the reaction leading to the conversion of SO_2 into SO_3 is carried out (converters) is designed so as firstly to achieve a high rate for the conversion process and then the highest possible thermodynamic yields. With this aim, the converters, which assume the form of vertical cylinders, are subdivided into several compartments in which there are separate layers of the catalytic mass supported by meshes.

In principle, the process develops as follows in a four compartment reactor (Fig. 6.15).

Upon entering the reactor from the top, the sulphurous gases, possibly with the addition of more pure, dry air and having been heated to about 400°C by heat exchange carried out earlier on the sulphurous gases themselves, the added air, or the mixture of them, are heated up to about 600°C whereupon they react. The rate of reaction is high but the yield does not exceed 75%.

Upon leaving the first compartment the temperature of the partially converted gases is lowered by 100° in the gas–gas heat exchanger E_1, and they are returned to the converter where, in correspondence with the temperature of the catalytic bed in

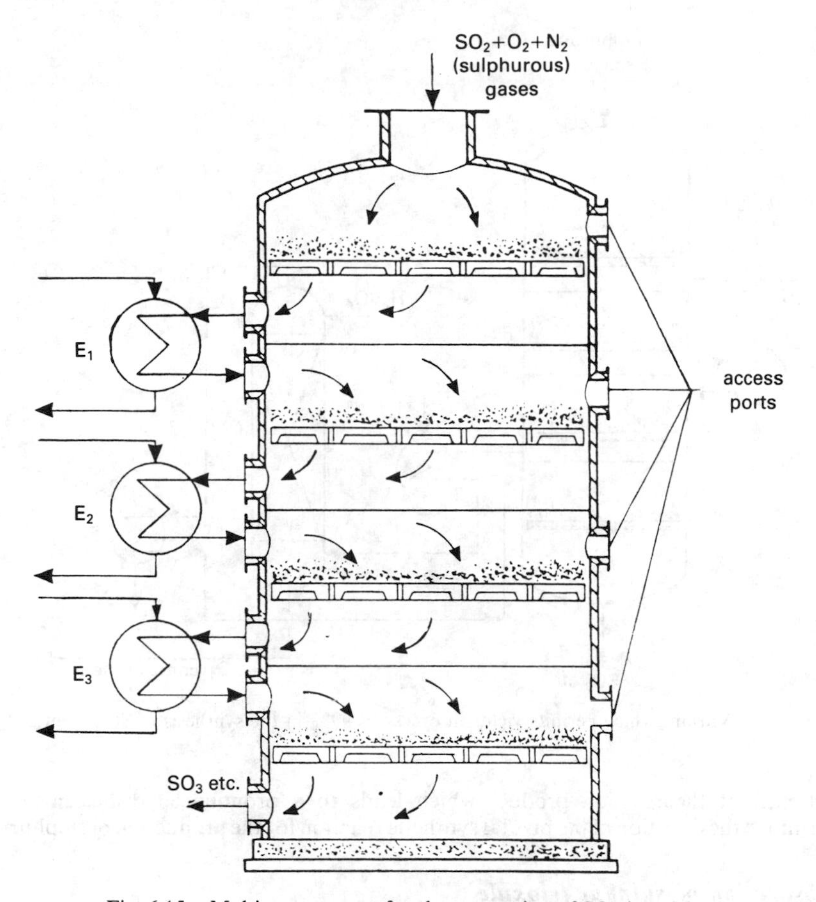

Fig. 6.15—Multistage reactor for the conversion of SO_2 into SO_3.

the second compartment, they are brought up to about 550°C and react to form further SO_3 from the SO_2. The rate of reaction is lower but the yields go up to 85%.

The gases are again sent out of the reactor and their temperature is reduced yet again by 100° by means of the heat exchanger E_2. They are then returned to the third reactor compartment where the conversion yields are raised up to 95% by passing through the catalytic bed at a temperature of about 480°C. The rate of conversion is further lowered, but now there is only a small quantity of gas to be converted.

Finally, after the temperature has been lowered for a third time in the external heat exchanger (E_3), the gases are passed back into the reactor where they undergo, on the catalytic bed in the fourth compartment, final conversion at around 450°C under conditions which give yields of 98–99%.

A variant (Fig. 6.16) of the last stage of this operation involves cooling the gases emerging from the third compartment by 100°C in E_1, fixing the SO_3 contained in them in the absorption tower C_1, adjusting their temperature to about 430°C in the exchanger E_2, and then sending them back into the fourth compartment of the reactor R for the completion of their conversion. This modification enables the final conversion yield to be increased also by removing the

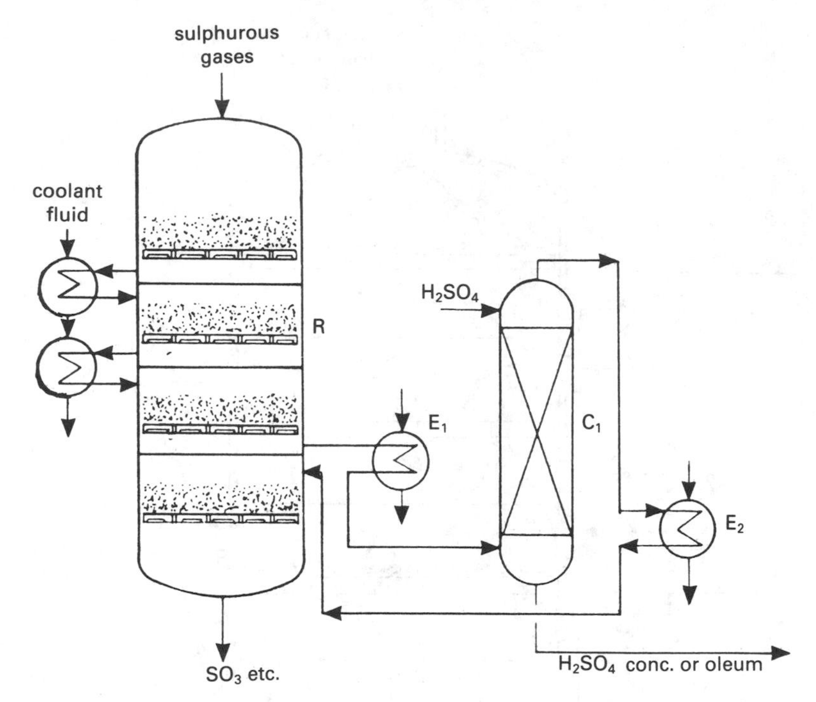

Fig. 6.16—Variant which permits yields in excess of 99% in the synthesis of SO_3 from SO_2.

greatest part of the reaction product which leads to a pronounced displacement of the equilibrium in the direction of the pivotal synthetic reaction for the production of sulphuric acid.

The absorption of sulphur trioxide

The gases, containing, on average, 8% of SO_3 mixed with O_2, N_2, and traces of SO_2 which come out of the converters, are suitably cooled and then sent to SO_3 absorbers; that is, they are passed into packed towers.

They are absorbed by 98–98.3% H_2SO_4 if 100% H_2SO_4 is to be prepared or by 100% H_2SO_4 if they are to be converted to oleum. In the latter case, however, it is necessary to carry out a double absorption to fix the appreciable quantities of SO_3 which escape during the first absorption stage.

The absorption temperature should be as low as possible in order to minimize fog formation and the losses of SO_3. Nevertheless, if oleum is being produced, the temperature must not be allowed to drop below 40°C (in practice, it is kept at 50–70°C) to avoid the dangers arising from the crystallization of the products.

It is also necessary to take care that there is a good and uniform distribution of the absorbent over the whole of the column packing material because the efficiency of the operation is equally as dependent on this as it is on the other conditions.

The variety of plant technologies for the industrial production of sulphuric acid by heterogeneous catalysis

Once the technologies for the production of sulphur trioxide from sulphur, hydrogen sulphide, and metallic sulphides together with the systems for the purification of the resulting gases, the reactors for the conversion of SO_2 into SO_3 and the absorbers of the SO_3 which is produced have been developed, as have in fact, been described, it remains for the various parts to be assembled in order to arrive at a point where information can be offered regarding the construction of comprehensive industrial plants for the production of sulphuric acid.

In different cases individual plant solutions are adopted from among the many which have been described at some time or other in order to achieve the conversion of raw materials into sulphuric acid. This, in particular, will differ according to the methods used for the production of SO_2 and its purification, while it remains understood in every case that the other solutions which have already been described are still valid.

It should be emphasized that the simplified, overall plant designs which are schematically depicted here reflect, in an average way, while arising from the syntheses which have been studied, all those employed in the largest modern processes used by BASF, Monsanto, Titlestad, etc., who are leaders in this field. The differences in the processes employed between them with regard to the constructional details of individual sections of the apparatus (especially with regard to the converters) are greater than for the apparatus as a whole.

Industrial plants using sulphur as the raw material. This commences with purified air in DS which is dried in C_1, preheated in the exchangers E_1 and E_2 and by the fused and sufficiently pure sulphur since the combustion gases do not need to be purified. See Fig. 6.17.

The raw material is combusted in the furnace H_1 which furnishes combusted gases which can be used for the production of water vapour in the boiler H_2 in such a way that, after receiving a further possible supplement of preheated air, they are at a temperature of about 400°C. In this condition they enter into the reactor–converter A_1 which operates with external refrigeration of the gases as the conversion proceeds and allows preabsorption of the sulphur trioxide which has been produced before the final conversion stage. This preabsorption involves the steam-producing boiler H_3, the absorption column C_2, and the exchanger–reheater E_3 through which the recycled gases pass.

The use to which the 100% H_2SO_4 which gathers at the bottom of the column C_2 is put is evident from the scheme. The gas emerging from the bottom of the reactor–converter A_1 is cooled by producing steam in H_4, and the SO_3 is then absorbed in the column C_3, using

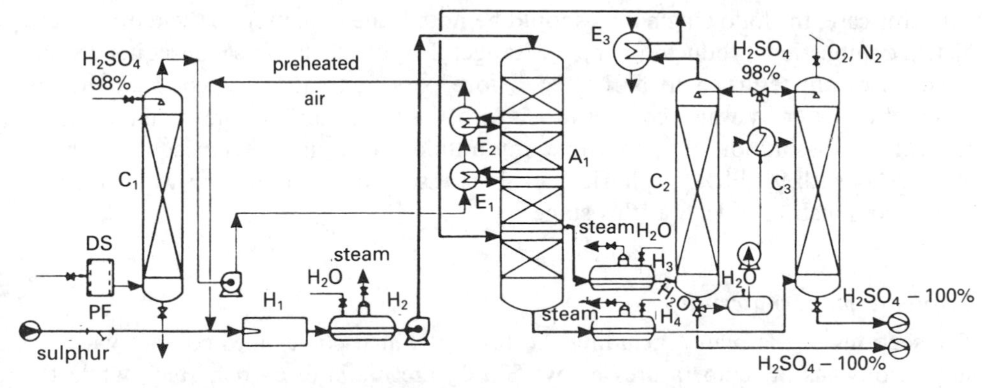

Fig. 6.17—Overall view of the process for the synthesis of 100% H_2SO_4 starting out from sulphur. Operations of very great industrial importance, both now and in the foreseeable future, are involved. In fact, in addition to the use of natural sulphur in this process, it is possible to exploit the sulphur prepared by the petrochemical and coal processing industries and by technologies, which uses the reduction of natural sulphates (of calcium, in particular).

the 98% H_2SO_4 derived, by suitable dilution, from the H_2SO_4 at the bottom of the column C_2. Similar acid is made in parallel by the absorption of SO_3 in C_2. The residual gases are exhausted from the top of C_1, while 100% H_2SO_4 collects at the bottom.

Industrial plant using pyrites as the raw material. It can be seen from the diagram presented in Fig. 6.18 that the raw material is roasted, having been introduced by the screw conveyer system into the fluid bed furnace H_1, with preheated air which is taken directly from the heat exchanger E_3 which operates with the gases removed from the third stage of the conversion reactor. The combustion gases are cooled in a boiler H_2 where steam is produced and where the coarsest suspended particles are deposited; the smaller particles being thrown down in the cyclone DC.

The ashes from the reactor and the powders which accumulate are gathered in a single collector, cooled and stored. Next, there is a cooling and washing column (C_1) which operates in a closed circuit with a dilute solution of sulphuric acid and purging of the small increase in the volume of the acid which occurs due to the condensation of moisture and the reaction of traces of SO_3 present in the gases.

In the following electrostatic filter DS the gases are denebulized, thoroughly purified and, after the addition of a measured quantity of clean air, pass to the drying stage in the column C_2, filled with 98% H_2SO_4. The gases are then first preheated in E_4 at the expense of the gases emerging from the base of the reactor TS and, subsequently, by heat exchange in E_1 and E_2 with the gases from the first and second compartments of the same reactor. The gases which have been prepared in this manner now enter into the converter–reactor where the oxidation of SO_2 to SO_3 commences. The process continues with heat exchange between the incoming gases and the gases descending through the column in E_1, E_2, and E_3, and is concluded in the fourth compartment of the reactor from where the gas emerges. This gas is all cooled in E_4 and, in part, refrigerated in E_5 before being absorbed in 100% H_2SO_4 in the column C_3 to yield oleum which is withdrawn from the bottom of this column.

The other part of the gas emerging from TS, after cooling in E_4, is absorbed in the column C_4 from the bottom of which 100% H_2SO_4 is withdrawn. Both the absorbent in column C_3 and the absorbents in columns C_2 and C_4 (after dilution with water) are derived from this product.

Industrial plant using hydrogen sulphide as the raw material. Here, one is concerned with a collection of plant units (Fig. 6.19) which are analogous to those employed in the process which starts out from sulphur as the raw material except that in the section where the combustion gases from the raw material (SO_2, H_2O, O_2 and N_2) are brought to the converter–reactor.

In this case, the following points should be noted: the reduction in the temperature of the combustion products in the exchanger E_1 to which, as a consequence, the dried gases are returned to be heated before passing to the catalytic system; the removal of water by water cooling in E_2 with the formation of a hot condensate to prevent the dissolution of appreciable amounts of SO_2, and the subsequent drying of the SO_2 with 93–98% sulphuric acid. It is essential to remove the water formed by the combustion of H_2S at this stage.

Requisite plant materials

The sections of the plant, including the tubing, which comes into contact with the sulphurous gas at temperatures below 500°C, are made of carbon steel, while the structures handling the gases involved in the synthesis of H_2SO_4 at temperatures above 500°C must be produced from alloy steels.

The grids in the reactors supporting the layers of catalyst are constructed from high quality cast iron.

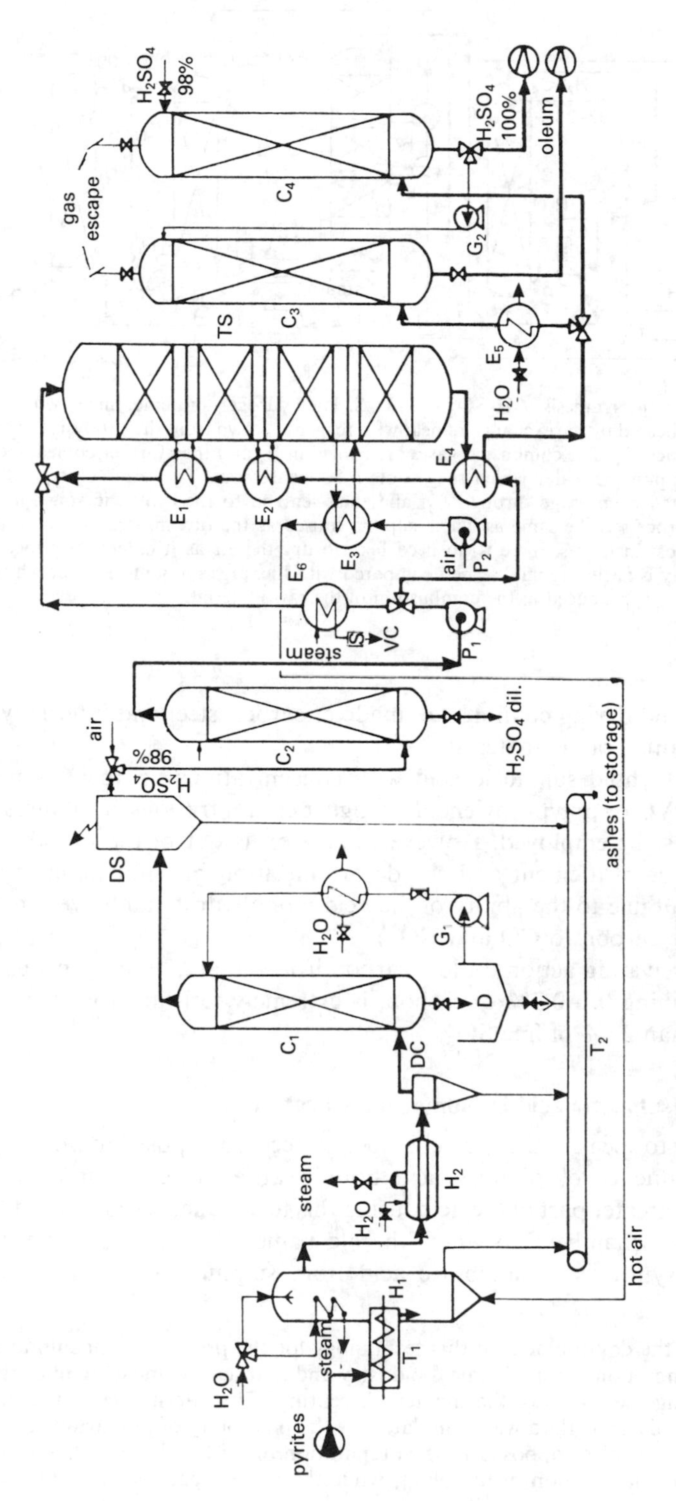

Fig. 6.18—Production of sulphuric acid using a process for the synthesis of SO_3 which starts out from pyrites (FeS_2) and other sulphides. The plant is generally well described in the text. It need only be added that the exchanger E_6 serves to bring the reagent gases which are dried in C_2 to a suitable priming temperature for the synthesis both when the plant is first put into operation and after any interruptions of its activity for any reason whatsoever.

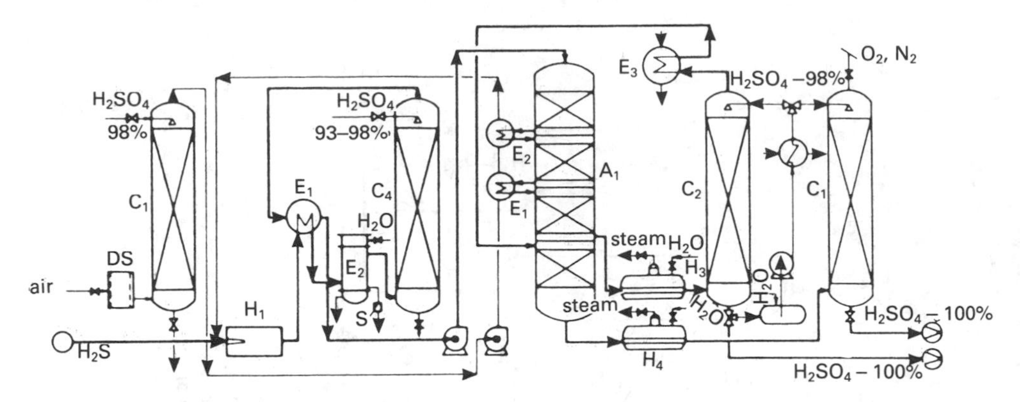

Fig. 6.19—Plant for the synthesis of H_2SO_4 from H_2S. The hydrogen sulphide, mixed with air which has been preheated by the exchange of heat with reagents drawn from the catalysis tower, is burnt in the furnace H_1. The combusted gases are cooled in E_1 and then further cooled with water in E_2 whereupon the water which they contain condenses out. They are subsequently dried in C_4, reheated by passage through E_1, and then sent on to the synthetic stage. The remainder of the process is the same as when sulphur is used as the raw material apart from the fact that, in most instances, there is no need here to dry the air as it enters the process because its humidity is fairly negligible when compared with the large amount of water which is produced in the combustion of the raw material.

The absorption and drying columns are made of carbon steel and internally lined with an acid-resistant ceramic material.

The tanks used to hold sulphuric acid with concentrations up to 96% may be constructed from PVC or polypropylene. For higher concentrations, structures made of carbon steel must be employed. However, cast iron is out of the question both on account of the embirttlement and the decarbonization brought about by such acids (this is attributable to the ability of the acid monohydrate and, even more so, of oleum to oxidize carbon to CO and CO_2).

Oleum has a passivating action on low-carbon iron. In fact, 'homogeneous iron', that is, steel containing 0.1–0.2% of carbon, is extremely resistant towards oleums containing more than 27% of free SO_3.

The preparation of sulphuric acid by homogeneous catalysis

It is incorrect today to speak of the 'lead chamber' process as opposed to the 'catalytic' method for the production of sulphuric acid, both because the vessels, in the modern plants in which the greater part of the acid, using this technique, is synthesized today, are not the classical 'chambers' to which the old name refers, and because, in this case as well, the synthesis of sulphuric acid from sulphur dioxide is promoted catalytically.

The adoption and the development of this technology for the production of sulphuric acid signalled the beginning of our great chemical industry and marked the most significant stages in its evolution during the whole of the nineteenth century. The alternative to it offered by the heterogeneous catalysis method was stimulated by the possibility of preparing acid of any concentration by this means as opposed to the preparation of acid with a maximum possible strength of 78% in a small section of the plant, while the greater part of the acid produced was of a concentration not exceeding 64%.

The steps in the production of sulphuric acid by homogeneous catalysis

The homogeneous catalysis which characterizes this method is carried out by using nitrous fumes, that is, by a mixture of nitric oxide and nitrogen dioxide which dissolve in sulphuric acid and behave as nitrous anhydride (N_2O_3). It is for this reason that the solution of nitrous ($NO + NO_2$) fumes in sulphuric acid is referred to technically as a nitrous solution.

The 'lead chamber' process has the following five steps if metallic sulphides (and, in particular, FeS_2) are used as the raw materials:

(1) the production of sulphur dioxide (page 188 and the following pages);
(2) purification of the combustion products at the 'powder chamber' level for the removal of solid suspended material (page 195), and, possibly, further purification steps depending on the quality of the sulphuric acid to be produced*;
(3) the mixing of the gases with the 'nitrous' solution;
(4) the catalytic reactions in which SO_2 is oxidized to SO_3 in the presence of air and water;
(5) the recovery of the 'nitrous' fumes.

Step 2 is omitted when sulphur is used as the raw material, and is replaced by a dehydration (not 'drying') of the gases if one starts out from hydrogen sulphide.

The technique and the vessels for the mixing of the nitrous gas

The reagents and the catalyst in this process are intimately mixed together in a column (a *Glover tower*) packed with Raschig rings, encased by lead plates which are fixed by autogenous welding to the outside, and with an internal wall lining made of a trachytic type of stone derived from lava, the 'Volvic' type from the Pyrenees being particularly suitable.

The combusted gases mixed with a suitable excess of air arrive at the bottom of the Glover tower, while both the nitrous mixture, which has been reformed in the final section of the plant, and more dilute sulphuric acid, which has been produced in the chambers, are fed in from the top of the tower.

This liquid feedstock makes the tower act as a thermal regulator in the process by lowering the temperature of the gases to the values which are the most favourable (60–70°C) for the reactions leading to the synthesis of the acid.

These reactions partly take place in the Glover tower itself, by utilizing the water content present in the dilute sulphuric acid feedstock which is thereby concentrated to a point where the most concentrated sulphuric acid obtained from the process (78% by weight) is produced in this section of the plant.

The variously localized chemical reactions in the process

The dilution of the nitrosylsulphuric acid (an essential component of the nitrous solution) within the Glover tower leads to its decomposition:

$$\underset{\text{nitrosyl-sulphuric acid}}{2HSO_4.NO} + H_2O \rightarrow 2H_2SO_4 + \underset{\text{'nitrous fumes'}}{NO + NO_2}$$

*Catalyst poisoning problems do not arise in this process, and the degree of purification of the synthesis gases is established in accordance with the purity required for the acid produced.

and the nitrous fumes produced catalyze the synthesis of sulphuric acid when they come into contact with sulphur dioxide and water:

$$NO + NO_2 + SO_2 + H_2O \rightarrow H_2SO_4 + 2NO \quad (1)$$

The following chain shows the reactions involved in (1) in detail:

$$NO + NO_2 \rightarrow N_2O_3 \xrightarrow{H_2O} 2HNO_2 \xrightarrow{SO_2} H_2SO_4 + 2NO$$

Moreover, the hypotheses regarding how, in effect, this catalytic oxidation takes place via a series of partial reactions characterized by low energies of activation are all supported by more or less convincing proofs.

Reaction (1) can be repeated cyclically by the partial reoxidation of the nitric oxide produced in the process by the excess air which forms part of the 'sulphurous' gas coming from the combustion section:

$$2NO + 1/2O_2 \rightarrow NO + NO_2 \quad (2)$$

Reactions (1) and (2) mainly occur in the chambers following the Glover tower until the sulphur dioxide has been exhausted, after which there is the problem of recovering the nitrous gases which are always present since they act only as catalysts.

It is the task of the *Gay Lussac towers*, as we shall see, to ensure that these gases are recovered. A reaction which is the reverse of the decomposition of the nitrous mixture, taking place in the Glover tower, takes place in these towers:

$$2H_2SO_4 + NO + NO_2 \rightarrow 2HSO_4.NO + 2H_2O \quad (3)$$

This is, in effect, a typical equilibrium reaction which is particularly sensitive to the 'mass action' effect exercised by water.

In fact, one may write:

$$2HSO_4.NO + 2H_2O \rightleftharpoons 2H_2SO_4 + NO + NO_2,$$

and this reaction is displaced to the right in the Glover tower where water is relatively abundant and to the left in the Gay Lussac tower which is supplied with sulphuric acid which is naturally transformed into nitrosylsulphuric acid by absorbing an equimolecular mixture of NO and NO_2 on account of its remarkably high concentration (78%).

High-efficiency catalytic plant equipment

Since process (1), indicated above, is an overall reaction consisting of a number of partial reactions which take place in the liquid phase, the development of the surfaces which are covered in this liquid is a factor of fundamental importance in promoting the synthesis of sulphuric acid.

The problem of increasing the surfaces which are covered by the liquid in this process has been successively solved in two ways: firstly, if nebulized water is brought into contact with the sulphurous gases in voluminous reaction zones using ever more highly refined techniques then, with packing systems, the areas of contact between the gas and the liquid are increased.

In the classical plants the reaction zones which were used to ensure the maximum reaction yields had the shape of large parallelepipeds with a square base ('chambers') made of lead which is resistant even to hot sulphuric acid so long as its concentration does not exceed 70%.

In addition to the feedstock, the water required for the synthesis was fed into the chambers in the form of a fine spray.

Later, 'tangential' lead chambers were adopted. This name is attributable to the fact that the sulphurous gases, which have been premixed with the nebulized water within the ducts, move tangentially ('spirally') in them leading to continuous remixing (Fig. 6.20).

Finally, in Petersen plants, use is made of the type of packing employed in Glover towers but in a group of three towers. Two of these are fed from above with the nitrous mixture and dilute sulphuric acid, coming from the third tower which is only supplied with water. The

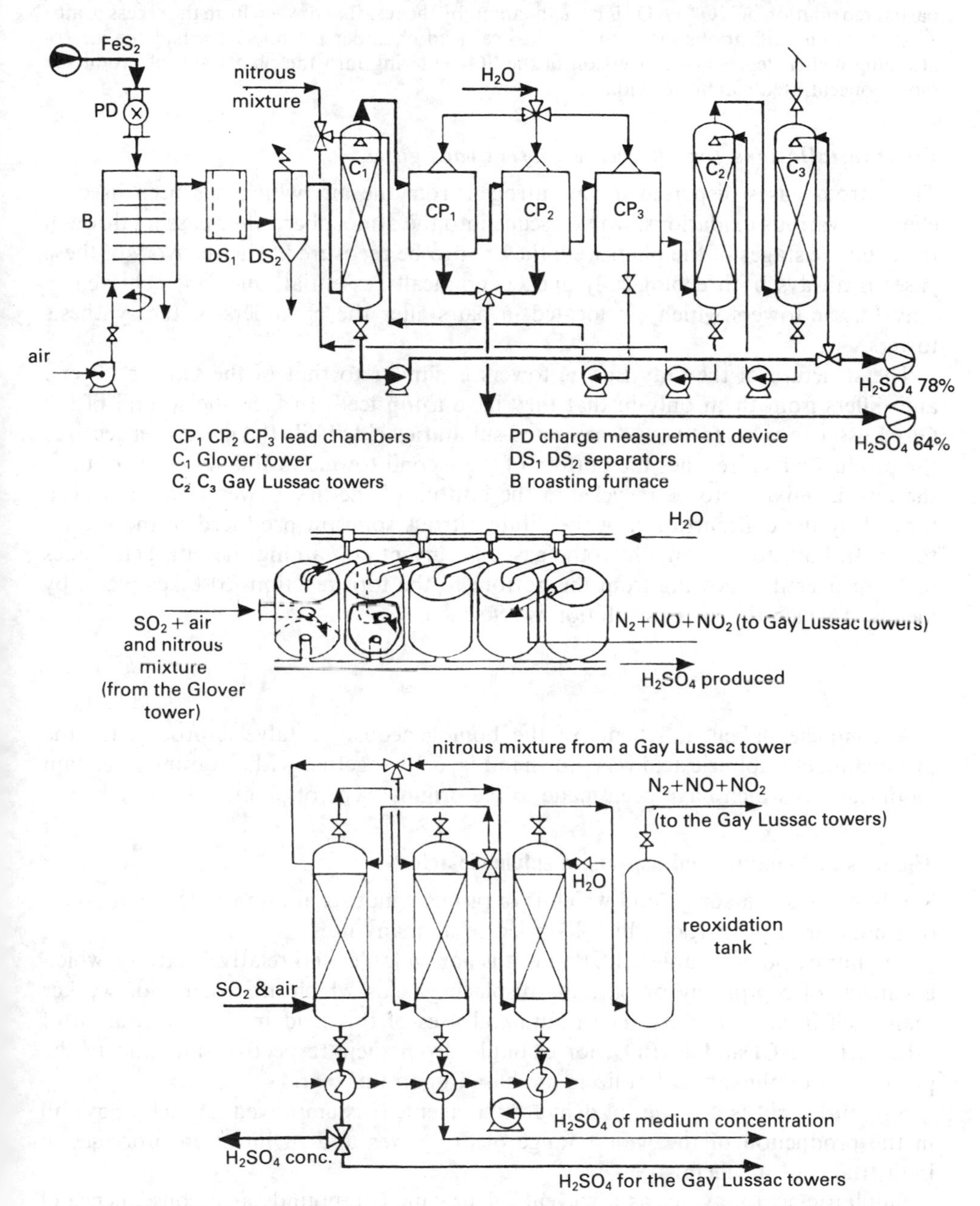

Fig. 6.20—Above: Diagram of a plant for the production of sulphuric acid by the 'lead chamber' process. Centre: 'Tangential chambers' which, in certain plants, have replaced lead chambers. Below: Plant using 'Petersen towers' which replace the Glover tower and the other chambers.

most concentrated acid is obtained from the first tower (Fig. 6.20), while fairly concentrated acid is obtained from the second tower and dilute acid from the third tower which, as has been noted above, is recycled to supply the water required for the production of the acid in the first two towers.

Plants of the Petersen type, that is, with packed towers replacing the chambers, do not have a Glover tower. A fourth tower (a 'reoxidation tank') is simply inserted in these plants where partial reoxidation of NO to NO_2 is brought about by the residual oxygen from the excess of air.

All of the modifications made to the classical lead chamber method have had the aim of attaining higher degrees of conversion of the SO_2 entering into the plants and of producing more concentrated sulphuric acid.

Plant installations for the recovery of nitrous gases

The nitrous gases dispersed in the nitrogen from the air, which has been used to effect the various oxidations, would escape into the atmosphere after passing through the catalytic stages of the plant were they not to be recovered. The recovery of these gases is always both ecologically and economically essential, and it is achieved in Gay Lussac towers which are located in pairs after the chambers or the synthesis towers.

The structure of the Gay Lussac towers is similar to that of the Glover towers, and differs from them only in that they have a top feed. In fact, the second of the Gay Lussac towers receives concentrated sulphuric acid, while the first tower receives the product which reaches the bottom of the second tower. In this way, one obtains the nitrous mixture to be recycled at the bottom of the first tower which has been formed by the concentration of the dilute nitrous solution produced in the second tower. In both towers, in which the gaseous current containing the nitrogen oxides to be recovered is flowing from the bottom to the top, reaction (3) takes place, by means of which the nitrous solution is reformed.

* * *

A complete, labelled, scheme of the homogeneously catalyzed process for the production of sulphuric acid was shown in Fig. 6.20, together with diagrams of certain modifications which have been made to the original type of plant.

The uses and commercial aspects of sulphuric acid

Sulphuric acid is a compound which is of primary industrial importance on account of a number of properties which have led to its many uses.

Sulphuric acid is a stable acid, that is it is non-volatile, and relatively strong, which is capacle of completely or partially displacing acids which are volatile or weaker than itself from their salts. The traditional uses of this acid in the preparation of other acids (HCl and H_3PO_4, for example) from their respective salts and in the production of phosphate fertilizers are based upon this fact.

Sulphuric acid is a powerful dehydrating agent. It is employed as such above all in the production of the widest range of explosives and in nitration processes in industrial organic chemistry.

Sulphuric acid may act as a solvent for organic compounds as a consequence of sulphonation reactions and as an oxidant of metals (Cu and Hg, for example) and organic compounds at elevated temperatures.

Sulphuric acid acts as a salt former while solubilizing oxides and metals in pickling processes and in producing, sometimes also by reacting with bases, the whole range of industrial sulphates which are either directly used or employed as raw materials for the manufacture of other compounds.

Finally, the combination of the different properties of this compound (mainly its acidic and dehydrating properties) make it a highly valued catalyst which is used in petroleum chemistry and in the acceleration of synthetic–hydrolytic equilibria.

* * *

Commercially, sulphuric acid is classified according to its concentration. It is available as:

- *dilute sulphuric acid*, 60–65% (about 50° Bé),
- *concentrated sulphuric acid*, 76–80% (up to 60° Bé),
- *sulphuric acid monohydrate*, 98–100% (about 66° Bé),
- *oleums* with 25, 30, 40, or 60% free SO_3.

Acid properties are characteristic of the first type, the sulphonating and more or less dehydrating properties are characteristic of the intermediate types, while the oleums are characterized by strongly dehydrating and oxidizing capabilities.

6.8 THE PRINCIPAL SULPHATES OF INDUSTRIAL IMPORTANCE

Various salts of sulphuric acid are of considerable or very great industrial importance. Before passing to a review of the industrial production of several of these, we present in Table 6.4 a synoptic picture which lists some of the properties and uses of these compounds.

A review of the preparation of sulphates involving the use of sulphuric acid is next presented. The preparation of sulphates derived from other sources is postponed until their uses are considered.

The production of sodium bisulphate and sulphate

Sodium bisulphate is produced solely by a chemical route, while sodium sulphate can also be produced by the fractional crystallization of natural saline mixtures.

For instance, 'Glauber's salt' crystallizes out from the leached products of the saline residues from the Stassfurt deposits when they are allowed to stand at 10–15°C. This occurs according to the reaction:

$$2NaCl + MgSO_4 + 10H_2O \rightarrow Na_2SO_4.10H_2O + MgCl_2$$

The anhydrous salt, when it is required, can be prepared from Glauber's salt by dehydration. Additionally, it should be mentioned that, although it is more convenient to transport Na_2SO_4, it is the decahydrate which is the most used industrially, especially in the textile, dyeing, and tanning industries.

Sodium bisulphate is produced chemically in suitable ovens (Fig. 6.21) ('bisulphate ovens') moulded in the form of a capsule and fed with 78% sulphuric acid and sea salt. The following reaction occurs at 200°C:

$$NaCl + H_2SO_4 \rightarrow NaHSO_4 + HCl$$

Table 6.4—Information on the principal industrial sulphates

Formulae	*RMM*	*Density*	*Appearance*	*Solubility in water*	*Uses*
$NaHSO_4$ and Na_2SO_4	120 142	2.65 2.67	white and crystalline	very high	dyes, glasses, pickling agents etc.
$Al_2(SO_4)_3$ $Al_2(SO_4)_3.18H_2O$	342.16 666.45	2.71 1.62	white and crystalline	low high	dyeing, tanning, paper
$K_2SO_4.Al_2(SO_4)_3.24H_2O$	948.76	1.75	white and well crystallized	1p:10p_{H_2O} at 10°C, 20 times this at 100°C	dyes, textiles, tanning and pharmaceuticals
$K_2SO_4.Cr_2(SO_4)_3.24H_2O$	998.86	1.83	violet and well crystallized	32.3 g/100 g at 25°C	dyeing, tanning
K_2SO_4	174.27	2.66	white, crystalline	12.5 g/100 g at 20°C	fertilizers and preparations of other salts
$(NH_4)_2SO_4$	132.15	1.78	white, crystalline	very high	fertilizers and preparations of other salts
$CuSO_4.5H_2O$	249.69	2.28	blue crystals	1p:3.5p_{H_2O} at 15°C 1p:1p_{H_2O} at 100°C	preparation of Cu salts, electro plating, and as a fungicide

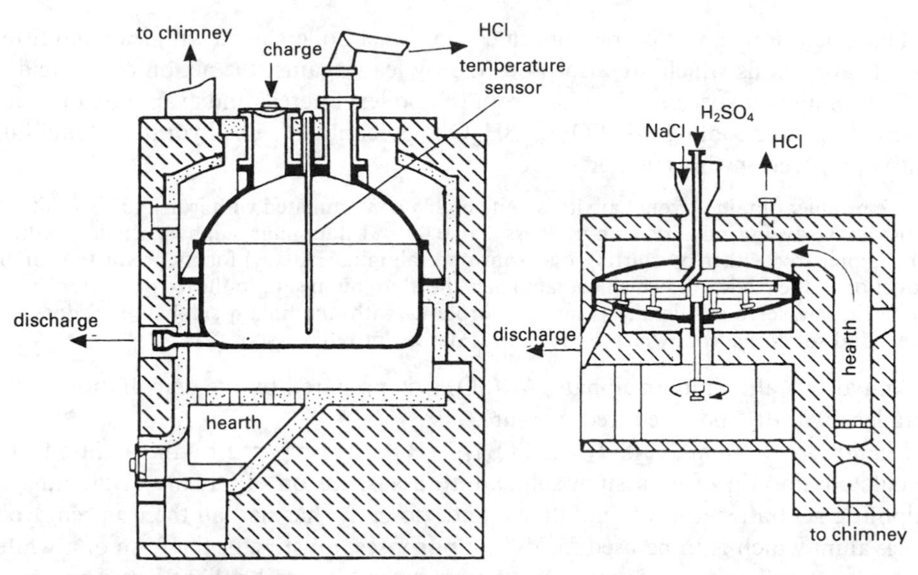

Fig. 6.21—Capsule furnace for the production of sodium bisulphate (left) and lenticular furnace for the production of sodium sulphate (right). The temperature and the stoicheiometric ratios of the reagents are selected according to whether the one salt or the other is to be produced.

Gaseous hydrogen chloride is liberated, and the fused bisulphate is discharged through a suitable channel.

Sodium sulphate is prepared in 'lenticular muffle furnaces' (Fig. 6.21) which are heated by the smoke originating from an external hearth and fed from above with a mixture of sea salt and sulphuric acid or sea salt and bisulphate. The reactions are:

$$2NaCl + H_2SO_4 \rightarrow Na_2SO_4 + 2HCl$$

$$NaCl + NaHSO_4 \rightarrow Na_2SO_4 + HCl$$

The development of gas (HCl) is favoured by the fairly high temperature, and the reactions are also displaced to the right by the small excess of sodium chloride (the less valuable reagent) which is used.

The furnace is continuously charged with the two reactants, while the hydrogen chloride leaves the system via a suitable exhaust port.

The rather high temperature of this process (350–500°C) and the friction caused by the remixing in the furnace contribute to the contamination of the product with iron sulphate, as the former factor leads to corrosion and the latter to scraping. To purify the product it is dissolved in water, treated with an aerated aqueous solution of lime, and the precipitated ferric hydroxide is then filtered off. Finally, the salt is recrystallized.

The industrial production of aluminium sulphates

When bauxite or crushed alumina are maintained in suspension by means of suitable agitation in 62–65% sulphuric acid for several hours at a temperature of 90–95°C, reactions of the following type occur:

$$\underset{\text{diaspore}}{2AlO(OH)} + 3H_2SO_4 \rightarrow Al_2(SO_4)_3 + 4H_2O$$

This operation can be carried out in carbon steel boilers lined with lead and fitted with heating coils which are also protected by lead against the action of the acid.

The solution is successively taken from the boiler, filtered, and evaporated in such a way that, after cooling, $Al_2(SO_4)_3.18H_2O$ is precipitated. The hydrated aluminium salt is prepared in this manner.

The product obtained from bauxite is quite highly contaminated with iron and is unsuitable for use in the dyeing industry. Iron-free (less than 0.005%) aluminium sulphate octadecahydrate can be prepared either by starting out from pure alumina (Al_2O_3) for its production or by reduction of the ferric salt, which contaminates the aluminium salt, to the ferrous state. Unlike the ferric salt, ferrous sulphate is not isomorphous with aluminium sulphate and does not form solid solutions from which it crystallizes out as 'mixed' crystals.

Anhydrous aluminium sulphate, $Al_2(SO_4)_3$ is prepared by the dehydration of the octadecahydrate under reduced pressure.

Finally, 'rock alum', $K_2SO_4.Al_2(SO_4)_3.24H_2O$, is prepared by the addition of a calculated amount of potassium sulphate to a suspension of bauxite or alumina in sulphuric acid after it has been filtered, and, possibly, has had all the iron removed. Rock alum which is to be used for dyeing purposes, must be in the form of a white, transparent, or opaque crystals. For other purposes, it may be slightly grey or faintly yellow.

Industrial preparation of chrome alum

The salt $K_2SO_4.Cr_2(SO_4)_3.24H_2O$ crystallizes from sulphate solutions in which the oxidation of organic compounds, such as anthracene and anthraquinone, with potassium dichromate is carried out.

It can also be prepared from solutions in which potassium dichromate has been reduced with sulphur dioxide:

$$3SO_2 + K_2Cr_2O_7 + H_2SO_4 \rightarrow K_2SO_4 + Cr_2(SO_4)_3 + H_2O$$

The commercial name of this salt is derived from its isomorphism with the better known rock alum.

Industrial methods for the production of ammonium sulphate

Here, we shall meet all the main industrial methods for the production of $(NH_4)_2SO_4$ because, among these, those which make use of sulphuric acid are outstanding.

Sulphuric acid processes

The common reaction in these methods is:

$$2NH_3 + H_2SO_4 \rightarrow (NH_4)_2SO_4 + 220\ \text{kcal/kg}$$

In the first method, which is depicted diagrammatically in Fig. 6.22, the starting material is 65% sulphuric acid, and, to concentrate the solution and thereby bring about the crystallization of the salt, use is made of the large heat of reaction.

In a second method the acid is first nebulized before coming into contact with the gaseous ammonia in order to speed up still further the evaporation of the water contained in the sulphuric acid used as the starting material. A diagram of the apparatus used in this process is shown in Fig. 6.23.

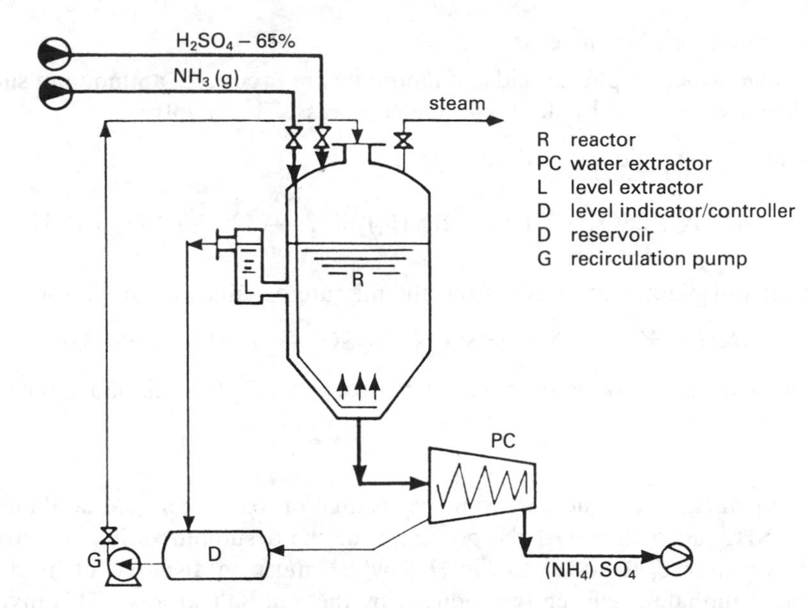

Fig. 6.22—Production of ammonium sulphate by the bubbling of gaseous ammonia into sulphuric acid of medium to high concentration.

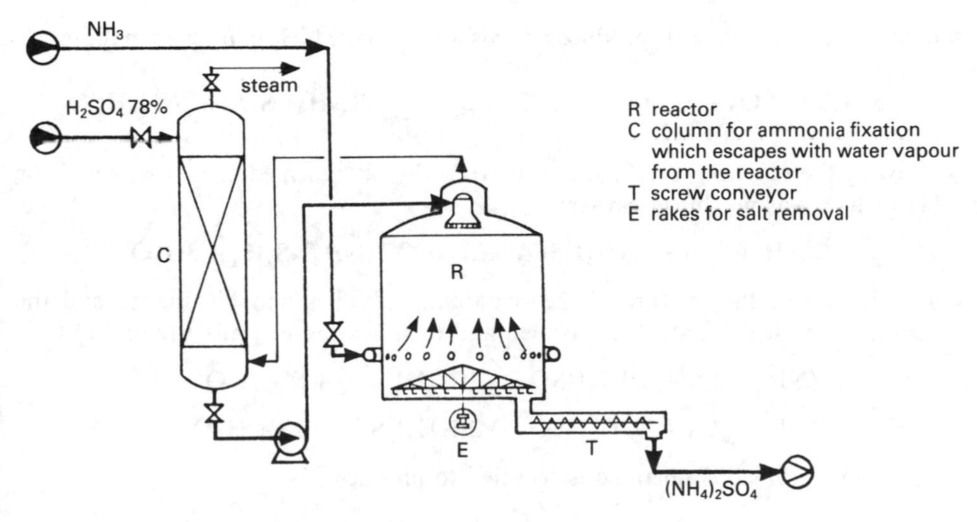

Fig. 6.23—The production of ammonium sulphate by contacting NH_3 with nebulized sulphuric acid. The large heat of reaction ensures the 'stripping' of the water in the sulphuric acid and, as a consequence, the direct crystallization of the ammonium salt.

The mixture produced by washing the gases from the distillation of cannel coal, after the removal of tar, with aqueous sulphuric acid (35%) must be kept to a specific concentration by evaporation of the saline solution, otherwise the salt may crystallize out. The salt is crystallized out only after the H_2SO_4 has been completely saturated with ammonia. This process will be described in Volume 2 of this work.

Processes using sulphurous gases

In factories in which sulphuric acid and ammonia are produced, ammonium sulphate is conveniently prepared by the *Guggenheim process*. This involves:

- fixing the sulphur dioxide with ammonia:

$$4NH_3 + 2SO_2 + 2H_2O \rightarrow 2(NH_4)_2SO_3 \xrightarrow[H_2O]{SO_2} 2NH_4HSO_3 + (NH_4)_2SO_3$$

- the partial displacement of SO_2 from the mixture of salts with sulphuric acid:

$$(NH_4)_2SO_3 + 2(NH_4)HSO_3 + 2H_2SO_4 \rightarrow 2(NH_4)_2SO_4 + 3SO_2 + 3H_2O$$

This method can also be used for the recovery of pure SO_2 from sulphurous gases.

* * *

In coke plants, where the gas from the distillation of cannel coal contains both H_2S and NH_3, downstream of the processes for the desulphurization of petroleum products when NH_3 is added to the H_2S which must be disposed of from there, ammonium sulphate is efficiently produced by the *Katasulf process*. This involves:

- burning a large part of the H_2S present:

$$2H_2S + O_2 \rightarrow 2H_2O + 2SO_2$$

- fixing, in the cold, the SO_2 produced in this way with the NH_3 contained in the gas:

$$6NH_3 + 3SO_2 + 3H_2O \longrightarrow 3(NH_4)_2SO_3 \xrightarrow[H_2O]{SO_2} 2(NH_4)_2SO_3 + 2NH_4HSO_3$$

- allowing the reaction to take place between the salts formed in this way and the H_2S which was not burnt earlier:

$$2(NH_4)_2SO_3 + 2NH_4HSO_3 + 2H_2S \rightarrow 3(NH_4)_2S_2O_3 + 3H_2O$$

- acidification of the mixture of the remaining sulphites and bisulphites and the thiosulphates which have been formed and then heating this mixture to 130°C:

$$(NH_4)_2SO_3 + 2NH_4HSO_3 \rightarrow 2(NH_4)_2SO_4 + S + H_2O$$

$$(NH_4)_2S_2O_3 + 2NH_4HSO_3 \rightarrow 2(NH_4)_2SO_4 + 2S + H_2O$$

The sulphur which is liberated is recycled to produce SO_2.

Processes using calcium sulphate

By crushing gypsum ($CaSO_4.2H_2O$) or anhydrite ($CaSO_4$), suspending them in water at 50–55°C and adding ammonia and carbon dioxide, the following reaction is brought about:

$$CaSO_4 + 2NH_3 + H_2O + CO_2 \rightarrow (NH_4)_2SO_4 + CaCO_3 + 984\ \text{kcal/kg}_{(NH_4)_2SO_4}$$

which goes to completion after a few hours.

As an alternative, to avoid problems concerned with the dissipation of large amounts of heat in the reactor, it is often preferred to proceed as shown in Fig. 6.24 where the CO_2 is

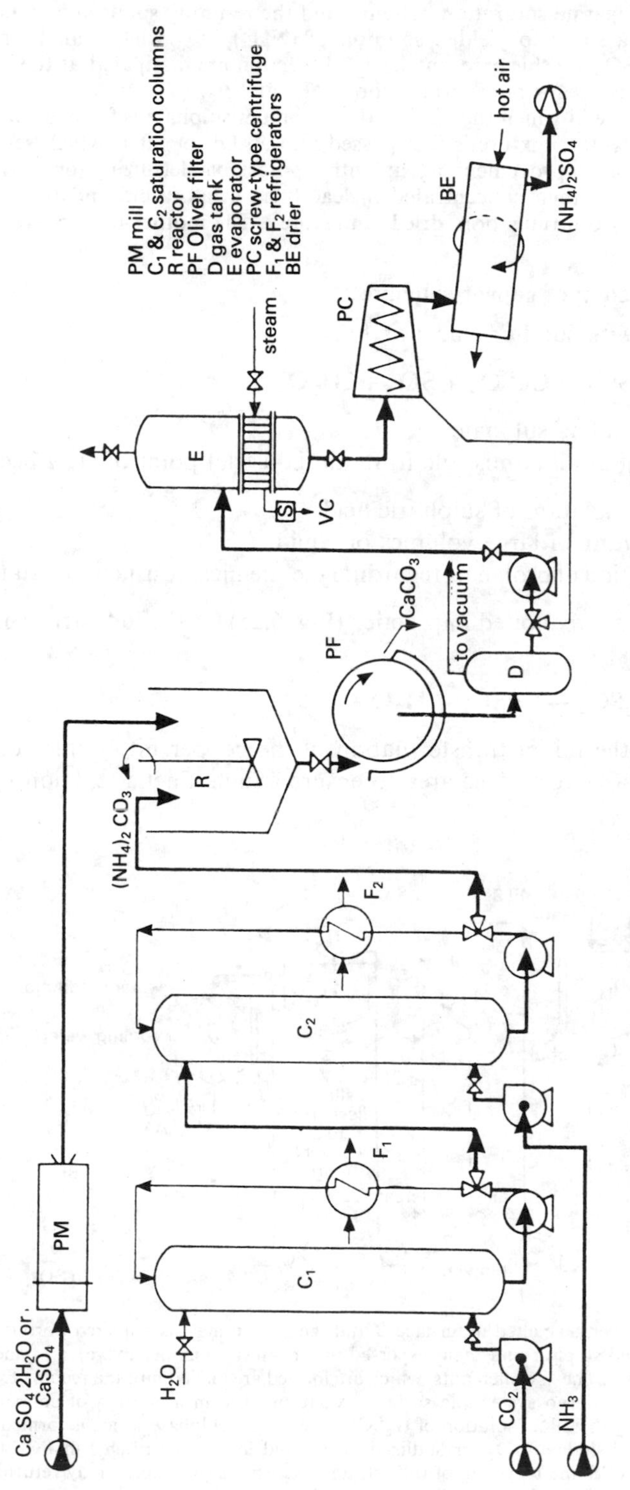

Fig. 6.24—Process for the production of ammonium sulphate from gypsum or anhydrite employing the reaction for the stepwise formation of the intermediate $(NH_4)_2CO_3$. By partial recyclings, with cooling, to every stage, the products are made more concentrated and the heat of reaction is dissipated.

dissolved in water in separate saturation columns and the resulting solution is made to react with gaseous ammonia so as to yield a solution of $(NH_4)_2CO_3$ which can be reacted by exchange with the $CaSO_4$. In this case, the heats of reaction are dissipated at the level of the individual operations by means of the water coolers F_1 and F_2.

From the reaction zone in which the dissolved ammonium sulphate is formed together with the precipitated $CaCO_3$, the mixture is first passed into an Oliver filter, which separates the $CaCO_3$: a valuable soil improvement agent with appreciable fertilizing properties (0.4% nitrogen). The solution is then concentrated in lead-lined evaporators and the ammonium sulphate is recovered by centrifugation, dried, and stored. The mother liquors are recycled.

The industrial production of copper sulphate

Copper reacts well with sulphuric acid at 130°C:

$$Cu + 2H_2SO_4 \rightarrow CuSO_4 + SO_2 + 2H_2O$$

to yield the corresponding sulphate.

The method is not at all admissible from an industrial point of view because:

- it involves the degradation of sulphuric acid,
- it requires the heating of large volumes of liquid,
- it requires the solution of problems regarding the chemical resistance of materials.

The reaction which is exploited in practice (Fig. 6.25) for the industrial production of copper sulphate is:

$$CuO + H_2SO_4 \rightarrow CuSO_4 + H_2O$$

This necessitates the initial transformation of the copper into cupric oxide with the development of a large surface area to ensure a high kinetic reaction rate.

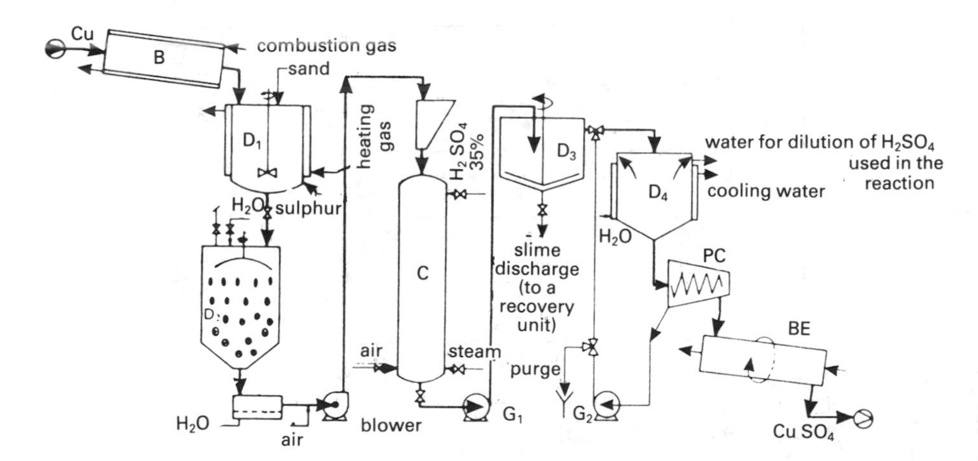

Fig. 6.25—The copper is smelted in furnace B and the molten mass is rendered porous by the addition of sand and sulphur and then dispersed in a dropwise manner in water by means of a suitable mechanism. The "swollen nuts' which are formed are loaded into the reactor C where, when heated, they undergo systematic surface oxidation to form a coating of copper oxide which is removed by the dilute solution of H_2SO_4. The copper sulphate which is formed is first freed from sand and sludge in D_3, crystallized in the cold in D_4, centrifuged in PC, dried in BE, and then stored. Some recycling of the salt and the acid is provided for by returning the mother liquors from the crystallization stage back into the process as indicated.

Conditions are provided which favour the reaction taking place in the different phases of the overall process. These are, more precisely:

- smelting of the copper in furnaces heated with natural gases or with Gazogene;
- the addition of sand to the fused mass which gives rise to porosity in it, and of sulphur which partially burns to yield SO_2 which is soluble in the melt;
- pouring the fused mass dropwise into cold water because the steam which is formed and the SO_2 which tends to be liberated may expand the metallic mass;
- loading the spherules, which have previously been formed in an expanded state, into suitable reaction columns which are fed from above with a dilute solution of sulphuric acid (35%) and fed from the bottom with an ascending stream of air and steam.

As the reaction bed sinks, the tower is systematically loaded with more spherules.

The oxidizing mixture (air + steam) is particularly efficient in the bottom of the column where, for this reason, the reaction between the copper oxide and the sulphuric acid is more active.

The resulting solution may be decanted into vessels because the sand and sludges, which possibly contain precious metals which are recovered, accumulate on the bottom. Once it has been clarified, the solution passes into other steel or lead-coated or plastic-coated vessels where copper sulphate slowly crystallizes. It is then centrifuged, washed in a centrifuge, dried, and stored.

6.9 THE PRODUCTION OF SODIUM SULPHIDES

When pure, crystalline anhydrous sodium sulphide, its hydrate ($Na_2S.9H_2O$), and the bisulphide are white. In commerce, however, these salts are more or less yellow. The bisulphide (NaHS) melts at 350°C, while the sulphide melts at 950°C.

All three salts are very soluble in water. The maximum density is possessed by the hydrated form (2.47), followed by the sulphide (1.85) and the bisulphide (1.79). They are primarily used in dyeing and colouring. Nevertheless, they also find uses in tanning techniques, flotation, and appreciable quantities of these compounds are employed for the reduction of nitro-compounds.

The classical method for the production of disodium sulphide uses the reduction of disodium sulphate with carbon in pit furnaces lined with alumina refractories. The reactions are:

$$Na_2SO_4 + 2C \rightarrow Na_2S + 2CO_2 \quad \text{at 700–800°C}$$

$$Na_2SO_4 + 4C \rightarrow 4CO + Na_2S \quad \text{at around 1000°C}$$

The batch nature of the process and its labour intensiveness have led to a decrease in its importance.

The replacement processes which have become of considerable importance are primarily:

- production from disodium sulphate by hydrogen reduction or, less commonly, other reducing gases such as CO and CH_4;
- the preparation from caustic soda and hydrogen sulphide.

The modern process of reduction with hydrogen has become dominant because it allows the synthesis to be carried out in a continuously rotating furnace, which is heated by the transmission of heat, by making the gas and the disodium sulphate reagents move in opposite directions under conditions where the reaction is catalyzed by the presence of iron oxides. The reduction reaction is:

$$Na_2SO_4 + 2H_2 \rightarrow Na_2S + 2H_2O$$

At temperatures below 600°C, the salts do not stick together, and the process can be carried out rapidly and continuously. Furthermore, the product is much, much purer than that obtained using the old thermal method.

The production of sodium sulphide from hydrogen sulphide and caustic soda is primarily based on the ready availability of the raw materials which nowadays are obtained from the chlorine process which furnishes sodium hydroxide and the desulphurization of petroleum products which yields the hydrogen sulphide required for the reaction. The pertinent reaction is:

$$2NaOH + H_2S \rightarrow Na_2S + 2H_2O$$

In this case the sulphide solutions are ready for concentration, while, after the thermal processes, it is still necessary to dissolve the product obtained in water to form a suspension, which is filtered and then concentrated to a point when the salt contents are those required for commercial purposes.

The most common commercial form is that which is obtained by concentrating the solutions to about 65% and then allowing them to cool. The salt $Na_2S.9H_2O$ is formed in this way which is either reduced to the form of flakes which can be put into bags or sealed ('fused') into steel drums.

Sodium bisulphide (NaHS) is prepared either by saturating concentrated solutions of Na_2S with H_2S, supersaturating NaOH solutions with H_2S, or reacting flakes of Na_2S with H_2S at temperatures slightly below 400°C. Under the conditions applying in the latter case the bisulphide is formed in a molten state while the Na_2S remains unfused.

Another method for the production of sodium bisulphide ('sodium hydrogen sulphide') is based on the reaction of gaseous H_2S with metallic sodium suspended in inert organic solvents.

In commerce, sodium bisulphide is sealed in iron barrels or put into sacks made out of a resistant plastic material.

6.10 THE PRODUCTION OF BARIUM SULPHIDE AND ITS MAIN DERIVATIVES

Disodium sulphide is also a by-product of the barium sulphide industry: this salt is obtained by reducing a mixture of barytes with finely ground carbon in a rotating furnace which is directly heated by naphtha or methane:

$$BaSO_4 + 4C \rightarrow 4CO + BaS$$

Barium sulphide is transformed into disodium sulphide during the production of barium carbonate which is the raw material for a very wide range of barium compounds:

$$Na_2CO_3 + BaS \rightarrow BaCO_3 + Na_2S$$

After filtration, the disodium sulphide solutions are concentrated up to the commercially requested concentrations or concentrated up to beyond 60% of the salt, sealed in drums, and left for the various hydrates ($Na_2S.9H_2O$, $Na_2S.6H_2O$, etc.) to crystallize.

However, the industrial preparation of barium sulphide is primarily connected with the production of the white pigment 'lithopone', a mixture of salts which is prepared by double decomposition with a solution of zinc sulphate:

$$BaS + ZnSO_4 \rightarrow \underbrace{BaSO_4 + ZnS}_{\text{lithopone}}$$

The precipitate, which is continuously filtered, is dried and then sent to rotating furnaces which operate under a highly reducing atmosphere. When roasted in this manner, lithopone acquires a high covering power which it retains after quenching, when cold.

6.11 THE CARBON DISULPHIDE INDUSTRY

Carbon disulphide is a volatile liquid (d_4^{20} 1.26 and b.p. 46.3°C) with a smell which is not unpleasant when the liquid is pure, but, in practice, it is evil smelling owing to the presence of traces of nauseous mercaptans. Carbon disulphide vapours are highly flammable and show a tendency, within the limits from 1 to 52%, to form explosive mixtures with air. Carbon disulphide (CS_2) is stored under a layer of water to prevent its evaporation. It has a low flammability point (236°C).

The major uses of CS_2 are in the production of xanthates which are employed in the synthesis of important cellulose derivatives (rayon and cellophane) and in flotation techniques. This compound also has very important applications as a solvent, fumigant, antiparasitic agent, pesticide carrier, and a reagent for the production of vulcanizing agents.

Carbon disulphide, which has been produced previously by synthesis from charcoal or coke with a low ash content (<2%) and sulphur, is mostly produced today from methane and sulphur.

The reaction in which CS_2 is synthesized from carbon and sulphur is endothermic:

$$C + 2S \rightarrow CS_2 - 142.1 \text{ kcal/kg}_{CS_2},$$

which means that the reagent mixture must be suitably heated in retorts.

The synthesis of CS_2 from carbon and sulphur involves bringing the solids and the sulphur vapours into contact at about 900°C. The heating which melts the sulphur and makes the carbons red hot can be produced either by the circulation of combustion gases or electrothermally, In the first case both the retort into which the piled-up carbon descends and the duct which carries the sulphur to the base of the pile of carbon are constructed of cast iron or chrome steel. In the second case, the apparatus consists of two concentric cylinders of a refractory material. When an electric current is forced to pass through this, the large resistance which it encounters leads to powerful heating across the heaped carbon in the internal cylinder. The electrical current is brought to the pile of carbon by means of electrodes, fixed in the base, which produce a large amount of heat where they come into contact with the carbon which vaporizes the sulphur arriving there via the cavity between the two refractory walls.

The mixture of gases coming out from the reaction retorts is formed from sulphur,

CS_2, H_2S, and COS, and is separated into fractions as follows:

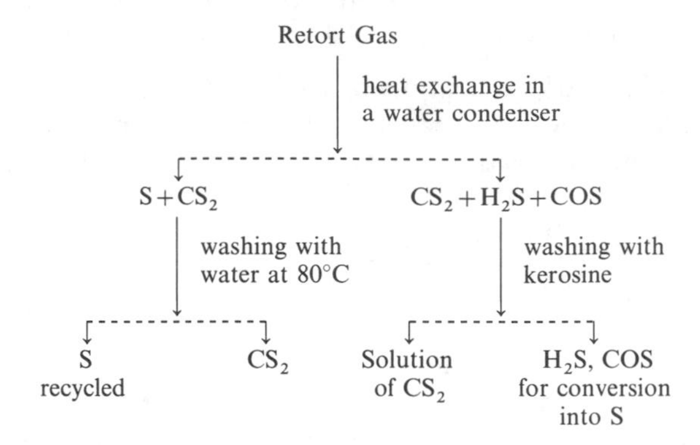

The CS_2 which is recovered by vaporization from the kerosene solution is reunited with the other product, and the kerosene is recycled.

Finally, the carbon disulphide is washed with alkalis and, after separation by decantation, it is distilled.

The reaction lying at the foundation of the production of CS_2 from methane and sulphur is also endothermic:

$$CH_4 + 4S \rightarrow CS_2 + 2H_2S - 38.1 \text{ kcal/kg}_{CS_2}$$

This reaction is carried out at about 650°C and under a pressure of 3–4 atmospheres and is catalyzed by silica gel or alumina activated with Cr_2O_3.

A diagram of the plant for the production of CS_2 from methane and sulphur is shown in Fig. 6.26.

The methane which has been cleaned up in the line filter PF_1 and heated in E_1 is mixed with vaporized sulphur in E_2. This mixture, after compression to a pressure of 3–5 atmospheres, passes into the superheater B_1 where its temperature is raised to about 650°C before entering into the reactor H_1 where, for the greater part, the endothermic reaction of the formation of CS_2 takes place. At the outlet from the reactor the mixture is once again brought up to a temperature of 600°C and then passed into H_2 to increase the conversion of the reagents into CS_2. The resulting products pass to the sulphur condenser E_3 where steam is produced. The sulphur separated in SV is recycled. The greater part of the crude CS_2 condenses out in the next exchanger E_4 and is collected in D. The rest of the CS_2 is dissolved in mineral oils in C_1 from the top of which about 95% H_2S is taken off and sent for the sulphur to be recovered. The CS_2 is stripped from the mineral oil in C_2: the oil being recycled in C_1 while the CS_2, having been reunited with that separated in D, is sent to the distillation column C_3. Both C_2 and C_3 are fitted underneath with standard reboiling facilities and furnish H_2S at the top which is sent off for the sulphur to be recovered. The CS_2 produced is already sufficiently pure for many uses and therefore goes into storage. When necessary, further distillation is carried out.

6.12 PRODUCTION AND USES OF HALIDES AND OXYHALIDES OF SULPHUR

Sulphur yields compounds primarily with chlorine which may or may not contain oxygen. The most important data on the compounds in question have been gathered together in Table 6.5.

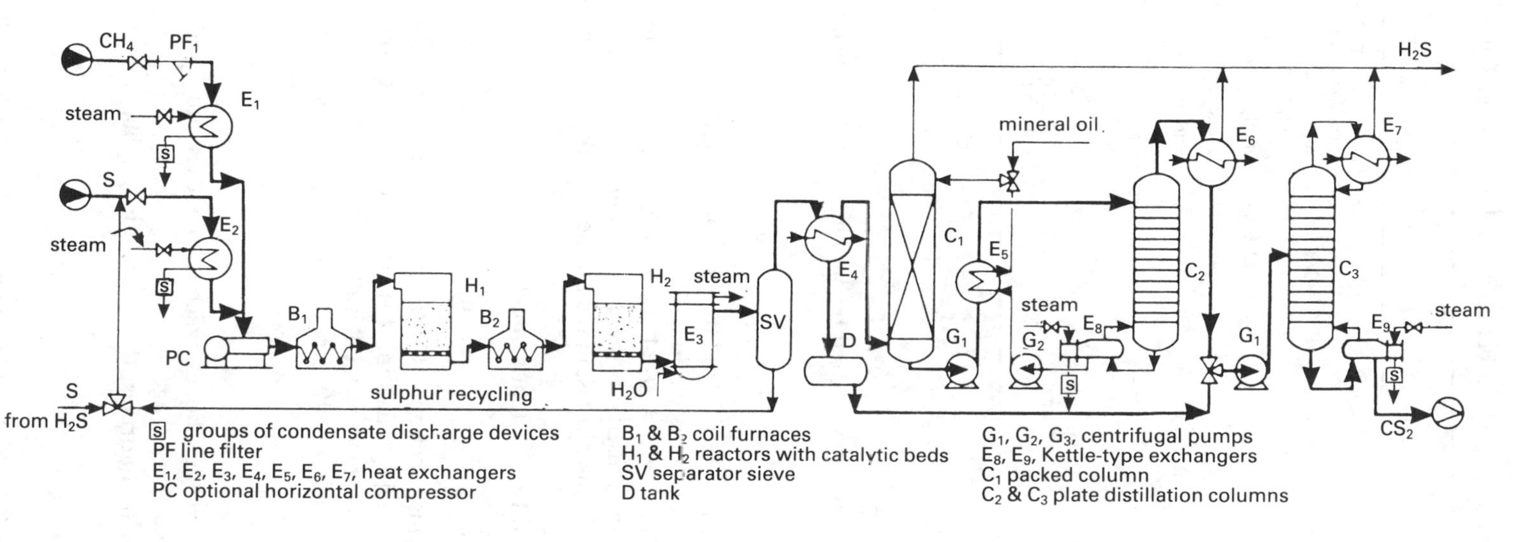

Fig. 6.26—Diagram of a plant for the production of CS_2 from methane and sulphur with the simultaneous recovery of sulphur from the H_2S produced. Up to the present, when use is made of carboniferous raw materials with a hydrogen/carbon ratio lower than that of methane (such as gasoils and naphthas), which are readily available at low prices on many international markets, the costs for the production of even just 'technically pure' carbon disulphide are unfavourable. This is because of the great difficulties which are encountered in making such raw materials compatible with the catalysts and the purification of the products, including hydrogen sulphide which must be sent to a Claus unit to recover the sulphur. This alternative to the use of the methane process is at the moment therefore to be considered as being only in the investigatory stage.

Table 6.5—Principal compounds formed between sulphur and the halogens

Names	*Formulae*	*Appear-ance*	*Colour*	*m.p. (°C)*	*b.p. (°C)*	*Special characteristics*
Sulphur chloride	S_2Cl_2	oily	dark yellow	−80	138	irritating vapours, liquid
Sulphur dichloride	SCl_2	liquid	dark red	−78	60	corrosive to the skin
Sulphur tetrachloride	SCl_4	liquid	dark red	−30	dec.	does not exist at normal temperatures
Sulphur hexafluoride	SF_6	gas	colourless	sub. at −64		highly toxic
Sulphonyl chloride	SO_2Cl_2	fuming liquids	colourless	−54	69	pungent taste ($SO_2 + Cl_2$)
Thionyl chloride	$SOCl_2$	fuming liquids	colourless	−104	78	irritating and toxic
Thionyl bromide	$SOBr_2$	fuming liquids	orange	−52	dec.	decomposes to yield S_2, Br_2, and S_2Br_2
Chlorosulphonic acid	HSO_3Cl	fuming liquids	colourless	−80	152	decomposes into SO_3, SO_2, SO_2Cl_2, Cl_2, SO_2 and HCl

dec. = decomposes, sub. = sublimes.

Sulphur monochloride is synthetically prepared by bubbling dry chlorine into fused sulphur:

$$2S + Cl_2 \rightarrow S_2Cl_2$$

This reaction is carried out at 125–130°C, and the reaction product mixture (S_4Cl_2, S_6Cl_4, etc.) is separated by fractional distillation.

Sulphur monochloride is, from an industrial point of view, the most important chloroderivative of sulphur. It is used in the production of reactive gases, in insecticides, in rubber and mastic substitutes, in the purification of juices and sugars, and as a synthetic intermediate.

Sulphur dichloride is prepared from sulphur monochloride:

$$S_2Cl_2 + Cl_2 \xrightarrow{(PCl_3)} 2SCl_2$$

The PCl_3 which catalyzes this reaction also stabilizes the sulphur dichloride so that it can be purified by fractional distillation. It is also used in insecticides, for the vulcanization of rubbers, as a solvent, and a chlorinating agent.

Direct chlorination of sulphur at −(40–50°C) leads to the formation of solid *sulphur tetrachloride*:

$$S + 2Cl_2 \rightarrow SCl_4,$$

which, after melting at −30°C, yields a liquid which decomposes into S_2Cl_2 and Cl_2 at −15°C acting as a hyperchlorinating agent.

When reacted with alkali metal fluorides, the chlorine in sulphur tetrachloride is replaced by fluorine:

$$4KF + SCl_4 \rightarrow 4KCl + SF_4$$

The sulphur tetrafluoride prepared in this manner, is the intermediate in the synthesis of *sulphur hexafluoride*:

$$SF_4 + F_2 \rightarrow SF_6$$

which is an important dielectric used in electronics. Both sulphur hexafluoride and, even more so, sulphur tetrafluoride are extensively used as fluorinating agents both in industrial and preparative organic chemistry.

Sulphonyl chloride is prepared by the catalyzed reaction between sulphur dioxide and dry chlorine at temperatures below 30°C:

$$SO_2 + Cl_2 \xrightarrow[\text{activated with camphor}]{\text{porous carbon}} SO_2Cl_2$$

Since this reaction proceeds in the reverse direction when sulphonyl chloride is heated, it is the optimal process for chlorination in synthetic organic operations.

Sulphonyl chloride is also used in the finishing of fabrics, as a solvent, and even as a poisonous gas.

Thionyl chloride is prepared by the reaction between sulphur monochloride or sulphur dichloride and sulphur trioxide:

$$S_2Cl_2 + SO_3 \rightarrow SOCl_2 + SO_2 + S$$

$$SCl_2 + SO_3 \rightarrow SOCl_2 + SO_2$$

This important reagent can also be prepared by direct synthesis from the elements:

$$2S + O_2 + 2Cl_2 \xrightarrow[\text{carbon}]{\text{activated}} 2SOCl_2$$

Thionyl chloride is used in the preparation of alkyl, aryl, and acyl halides, in the synthesis of pharmaceuticals, in the making of detergents, dyes, and printing products.

Thionyl chloride is also an intermediate in the preparation of *thionyl bromide*:

$$SOCl_2 + 2HBr \rightarrow SOBr_2 + 2HCl$$

which energetically donates bromine due to the products which arise when it is decomposed.

Chlorosulphonic acid is produced when gas containing 7–8% of SO_3 reacts with anhydrous HCl:

$$SO_3 + HCl \xrightarrow{\sim 170°C} HSO_3Cl$$

In organic chemistry, it is both a sulphonating agent and a chlorinating agent, depending on the substrate with which it reacts.

Chlorosulphonic acid decomposes upon contact with water to form H_2SO_4 and HCl, and it reacts with red phosphorus to yield phosphoryl chloride:

$$2P + 5HSO_3Cl \xrightarrow{\text{heat}} POCl_3 + H_3PO_4 + 5SO_2 + 2HCl$$

Chlorosulphonic acid is used in the synthesis of pharmaceuticals, pesticides, sweetening agents (saccharin), and dyes as well as the production of "Caro's acid", H_2SO_5, by allowing it to react with oxygenated water.

In industrial dyeing chemistry this important acid is produced by the characteristic reaction:

$$HSO_3Cl + H_2O_2 \rightarrow H_2SO_5 + HCl$$

Finally, it should be observed, more generally, that chlorosulphonic acid is a synthetic intermediate which is both a sulphonating agent and a chlorinating agent according to its mode of decomposition (Table 6.5).

6.13 SULPHAMIC ACID AND SULPHAMATES

Nowadays, many technologies and numerous chemical and galvanic processes make use of appreciable quantities of sulphamic acid NH_2SO_3H and sulphamates NH_2SO_3Me.

As examples of the technologies in which it is used, mention might be made of pickling processes, weed killers, flotation, and fireproofing, while sulphonation and the electrolytic refining of lead are included among the chemical and galvanic processes.

Very pure sulphamic acid can be prepared by the reaction between urea and oleum:

$$(NH_2)_2CO + H_2SO_4.SO_3 \xrightarrow{50-70^\circ C} 2NH_2SO_3H + CO_2,$$

while ammonium sulphamate can be produced by the reaction between ammonia and chlorosulphonic acid:

$$2NH_3 + ClSO_3H \rightarrow NH_2SO_3NH_4 + NH_4Cl$$

However, the industrial production of sulphamic acid is carried out in two stages:

- by synthesizing a white powder, which is a mixture of the acid NH_2SO_3H and the salts $NH_2SO_3NH_4$ and $NH(SO_3NH_4)_2$, from a mixture of $SO_{3(g)}$ and $NH_{3(g)}$ in a ratio of 2:3 at 200–300°C;
- by the treatment of this white powder at 30°C with 60–65% H_2SO_4.

The acid and salt components of the powder are present in a ratio of 1:1.1:2. By its action on them, sulphuric, on the one hand, yields sulphamic acid if it is present there and, on the other hand hydrolyses the ammonium iminosulphonate $NH(SO_3NH_4)_2$:

$$NH(SO_3NH_4)_2 + H_2O \rightarrow NH_2SO_3NH_4 + NH_4HSO_4$$

and, finally, liberates sulphamic acid from the ammonium sulphamate initially present in the mixture and from that which is formed as the result of hydrolysis:

$$NH_2SO_3NH_4 + H_2SO_4 \rightarrow NH_2SO_3H + NH_4HSO_4$$

As a result, an aqueous solution of sulphamic acid and ammonium bisulphate is obtained, and the two salts can be separated by crystallization. The sulphamic acid is finally purified by fractional crystallization.

Some of its unusual properties such as its high solubility in anhydrous SO_2 and sulphuric acid monohydrate and, on the other hand, its insolubility in dilute acids in more (alcohols) or less (ethers and acetone) polar organic solvents may further lead to an understanding of the techniques used in the separation of sulphamic acid from the most widely varying mixtures.

6.14 TOXICOLOGICAL AND ACCIDENT PREVENTION NOTES ON THE PRINCIPAL INDUSTRIAL COMPOUNDS OF SULPHUR

The aeriform (gaseous and vaporous) compounds of sulphur are, more or less, toxic to living organisms in general and to man in particular and among the liquid compounds of sulphur or compounds which are soluble in water, there are some originating from sulphuric acid, which are strongly aggressive agents towards every kind of living organism. To sum up: it may be said that, among the compounds of sulphur, there are some which count as the most common industrial poisons.

Table 6.6 shows the nature and the degree of the dangers to man from the principal aeriform compounds of sulphur*.

On the basis of the data given in Table 6.6 the extent of the danger arising from the poisons considered with regard to living organisms in general can be deduced.

Among the toxic compounds of sulphur which can form microdispersions in air (mostly as 'mists'), sulphuric acid stands out on account of its large and widespread use†. Here the tolerability limit is 1 mg/m^3.

While sulphuric acid is not flammable, it may cause other combustible material to burst into flames (carbon, H_2S, etc.) especially in the presence of oxidants (nitrates, chlorates, etc.).

Explosions brought about indirectly by sulphuric acid arise from the presence of hydrogen in the air which is formed upon contact of the acid with metallic materials. From this point of view, persistent mists of sulphuric acid stagnating in the sections of production and manufacturing units which involve its use are often the cause of danger.

Table 6.6—Degrees of toxicity and the toxic effects of gaseous sulphur compounds

Compounds	*Maximum tolerable limits*	*Levels at which they can be smelt*	*Toxic effects ascribable to them*
Hydrogen sulphide	20 ppm (27 mg/m^3)	Variable, depends on the contaminants. Always before tolerance limit is attained as it is perceived well before becoming harmful.	Within 1 hour, irritation of eyes and respiratory ducts at 109 ppm (152 mg/m^3). It is dangerous at 218 ppm (1 h) and extremely dangerous after $\frac{1}{4}$ h when present in amounts of 544 ppm.
Carbon disulphide	20 ppm (60 mg/m^3)	Perceptible even in doses of 0.01 ppm	Malaise and sleeplessness within 3–5 days at 33–66 ppm (99–198 mg/m^3). Bad headache and mental sluggishness within 2–3 h for doses of 333 ppm and the most serious damage (coma) in half an hour at doses above 2000 ppm.
Sulphur dioxide	5 ppm (13 mg/m^3)	3 ppm.	Tolerable for 1 minute at 200 ppm, causes coughing at 8–12 ppm, irritation at 20 ppm, and is rapidly lethal at above 1000 ppm.

* In fact, the tolerance limits of many gaseous sulphur compounds ($SOCl_2$, SO_2Cl_2, COS, etc.) are still to be defined while those for sulphur trioxide refer to sulphuric acid.

† Microdispersions of sulphuric acid in air generally arise from escapes of SO_3 from environments in which sulphur and its compounds are burnt since SO_3 forms mists of sulphuric acid when it comes into contact with the moisture in air.

Concentrated sulphuric acid destroys tissues and is very likely to cause blindness when it comes into contact with the eyes. Dermatitis is caused by prolonged contact of sulphuric acid (including dilute sulphuric acid) with the skin.

Finally, it should be remembered that inhalation of the vapours emanating from sulphuric acid first causes pulmonary lesions and loss of consciousness before leading to death if the causes are not radically removed and the appropriate measures taken.

In sulphuric acid, the sulphur is present with its highest oxidation number (+6) with the results that this acid and its compounds cannot be further oxidized. Still less, are they flammable. The danger associated with sulphuric acid lies primarily in its great tendency to become hydrated (its 'avidity' for water) which gives it the ability to extract water not only from substances containing water in some form but also from substances simply containing hydrogen and oxygen in forms other than water such as the structural materials and tissues of living organisms. When such tissues are 'necrotized' by sulphuric acid, almost only the carbon remains.

Carbon disulphide, in which the oxidation number of sulphur is (−2), finds itself chemically at the other end of the scale (with all other sulphur compounds between these two extremes). It has a structure which is akin to those of organic substances, it is hydrophobic, volatile, can be oxidized, and is therefore also flammable.

Although it is apolar on account of its purely covalent bonds which, in turn, arise from the small difference between the electronegativities of its two components, carbon disulphide is a liquid, albeit a volatile one, at normal temperatures on account of its high molecular mass, and its vapour is heavier than air. There is therefore an inclination for bodies to become bathed in its vapour which is also carried into the lower zones of the biosphere where life is most highly developed.

The most immediate dangers arising from the presence of carbon disulphide are:

* the formation of explosive mixtures with air over wide limits,
* its high flammability upon detonation and its ready ignition upon simply increasing the temperature,
* the complete solubilization of cuticular lipids and, in particular, the lipids of human skin which it brings about,
* the tendency it exhibits to be absorbed through the skin and the ease with which it can be inhaled,
* the deactivation of the functional lipids of the tongue and the nervous system which it brings about by preferentially dissolving them.

It is therefore concluded from a study of industrial accidents and the toxicology associated with the use of CS_2 and, above all, poor working practices during its use (while smoking, creating electrical or mechanical sparks, not ventilating the working area, and so on) that accidents occurring in the presence of CS_2 are among the most marked in the chemical industry.

It is not possible here to describe the complex accident prevention regulations associated with the treatment and the use of certain sulphur compounds.

The following are rules of a general character:

- ventilation of areas where such compounds are stored, employed, or handled in any way whatsoever;
- the setting up, if they are not already present within the framework of the general health service, of efficient first aid posts or of hospitals in the proximity of industrial units concerned with sulphur and its compounds;
- taking care to install systems for the scrubbing of sulphuriferous compounds from industrial effluent gases;
- not smoking in the presence of directly flammable materials (e.g. CS_2) or those which may give rise to flammable compounds (H_2), and ensuring that, when drums

are opened, especially when they contain acids and show signs of being swollen, this is done a long way away from any flames.

In the case of sulphuric acid in particular, it is extremely important that it should be kept a long way from water if it is concentrated, and, even more so, if one is dealing with oleum. This is to avoid contacts which give rise to sprays of fine droplets causing burns which necrotize living tissues and produce holes in textiles.

The measures to which recourse frequently has to be made can also be indicated with reference to the treatment of injuries caused by sulphuric acid.

After sulphuric acid has come into contact with the skin, the affected part must first be wiped and then thoroughly washed with running water. If the contact is in any way prolonged or it is with a delicate part of the body, it must be wiped, thoroughly washed with water, and then with soap and water or a 2% solution of $NaHCO_3$.

Sulphuric acid sprays which have come into contact with the eyes, regardless of their concentration, are treated by holding the eye of the patient open and rinsing the eye with a copious flow of water for 5 to 10 minutes. The eye must subsequently be irrigated, always keeping it open, with a tepid solution of 2% bicarbonate. After a further 3–4 minutes, 2–3 drops of Pantocain must be administered to the affected eye, and, after the same amount of time has elapsed, the rinsing of the eye with water must be resumed.

If sulphuric acid is ingested, the patient is treated by the administration of milk together with egg white or an aqueous suspension of magnesia. If nothing else is at hand, the affected party must drink copiously, but vomiting must not be induced, since it will lead to further damage to the throat and mouth. The treatments which have just been advised apply only to patients who are fully conscious.

Any spillages of concentrated sulphuric acid must not be cleared up by absorbing the liquid with sawdust or rags, as this can lead to fires. The acid should be first drowned with water and then neutralized with lime water, before it is washed down the drain.

Finally, sulphuric acid must always be diluted by slowly pouring the acid into the water with constant stirring so as not to allow any dangerous local accumulations of the large amount of heat which is evolved. **Water must never be poured into concentrated sulphuric acid.**

7

The industrial utilization of air

7.1 THE COMPOSITION AND PROPERTIES OF AIR

Air mainly consists of two gases, nitrogen and oxygen, which are practically considered to constitute $\frac{4}{5}$ and $\frac{1}{5}$ of air by volume respectively.

The most recent and precise measurements assign a value of 20.95 vol.% to the oxygen and 78.08 vol.% to the nitrogen in air.

In order of their percentage abundance in air, then follow:

argon at 0.934%;
carbon dioxide at 0.033%;
neon at 0.0018%;
helium at 0.0005%;
methane at 0.00015%;
krypton at 0.0001%;
hydrogen at 0.00005%;
xenon at 0.000009%.

As well as being given as a percentage, the composition of air with respect to the rarer components may also be expressed in ppm, that is, in parts per million or in cm^3/m^3. For example, xenon is present to an extent of 0.09 ppm.

All of the components which have been mentioned up to now are present in air in amounts which are practically constant (with the exception of CO_2 which has actually shown a decennial increase of the order of 0.0008%).

However, water vapour and traces of ozone and iodine are present in air in variable amounts. Obviously, the composition of air also depends on the altitude and how near it is to the sea.

Finally, in the neighbourhood of industry, built-up urban areas, and where there are events associated with volcanic phenomena, other gases such as CO, CO_2, H_2S, and NO_2 are also present in air.

Air is therefore essentially a mixture of gases with the following physical constants:

- melting point of the solid $-216.2°C$;
- solidification temperature $-212.9°C$;
- boiling point of the liquid at 1 atm., $-194.5°C$;
- condensation temperature of the vapour at 1 atm., $-191.5°C$.

There exists for air, as for every other gaseous substance, a critical temperature above which, no matter how high the pressure to which it is subjected, it retains the capacity to occupy the whole of the volume which it has at its disposal.

The critical temperature of air, which is free from water and carbon dioxide, is $-140.8°C$, hence, under normal conditions, air is a gas. At low pressures and far from its critical temperature the gas law $pV = nRT$ holds for air.

Technologically, the properties are exploited in that it is possible to condense it under a suitable pressure at temperatures equal to or lower than $-140.8°C$ to produce liquid air from which nitrogen, oxygen, and the noble gases (argon) may be recovered.

7.2 PRELIMINARY STATEMENTS RELEVANT TO THE STUDY OF LINDE PROCESSES WHICH LEAD TO THE LIQUEFACTION OF AIR

Under normal conditions air is a gas below its inversion temperature. It therefore exhibits a positive Joule–Thomson effect in that it is heated by compression and cooled by expansion.

It may be recalled from physical chemistry that the Joule–Thomson effect is the variation (the increase or decrease) in the temperature during the expansion of a real gas.

It is found that, for every pure gas or gaseous mixture, there exists a (maximum) temperature, the so-called inversion temperature, below which expansion always leads to cooling of the gas (a positive Joule–Thomson effect) and above which expansion always leads to heating of the gas while compression leads to its cooling (a negative Joule–Thomson effect).

The Linde methods lead to the liquefaction of air by means of repeated cycles of compression, removal of the heat of compression, and expansion. From the point of view of industrial economics these liquefaction processes require the determination of the compression yield and the calculation of the work of compression.

The compression yield is given by the change in temperature ΔT which is brought about by allowing the gas to expand under specified conditions, and can be calculated by means of the empirical Joule–Thomson formula:

$$\Delta T = \alpha \cdot \left(\frac{273}{T}\right)^2 \cdot (p_2 - p_1) \tag{7.1}$$

where T is the temperature at which the expansion begins and α is the Joule–Thomson coefficient* the mean values of which, over a pressure range from 1–200 atm. and at various temperatures, are shown in Table 7.1

*The reader is referred to physical chemistry texts for a more profound treatment of the significance of this coefficient which is put forward here as a simple statement of fact.

Table 7.1—Values of the Joule–Thomson coefficient

Temp. °*C*	α	*Temp.* °*C*	α	*Temp.* °*C*	α	*Temp.* °*C*	α
50	0.14	−25	0.214	−100	0.32	−130	0.33
0	0.18	−50	0.26	−120	0.31	−140	0.28

The work of compression, that is, the amount of energy expended in bringing a fixed amount of gas at an initial temperature T from a pressure p_1 to a pressure p_2 $(p_2 > p_1)$ is proportional to the logarithm of the ratio of the final and initial pressures of the gas; that is:

$$L = \beta \cdot \log \frac{p_2}{p_1} \tag{7.2}$$

The coefficient of proportionality β is a constant for a fixed amount of a given gas at a fixed temperature.

7.3 THE LINDE PROCEDURES FOR THE LIQUEFACTION OF AIR

The first procedures for the liquefaction of air were perfected by K. P. G. Linde, a German engineer.

The first Linde process consisted of the compression of the gas, followed by cooling with water and its re-expansion to normal pressure, while the second Linde process consisted of the compression of the gas, its refrigeration with liquid ammonia, and the re-expansion of the compressed gas to intermediate pressures of 30–50 atmospheres.

The first Linde method

Let us look in detail into how the first method perfected by Linde was developed.

By compressing purified air, from which all water and CO_2 has been removed, at a pressure of 200 atm., cooling it down to 12°C by a deluge of water, and then allowed it to expand to a pressure of 1 atm., its temperature falls by an amount which can be calculated from equation (7.1):

$$\Delta T = 0.17^* \times \left(\frac{273}{285}\right)^2 \times (200 - 1) \simeq 31°.$$

Therefore, as a result of this treatment, the temperature of the air falls from 12°C to −19°C.

The air which has been cooled by expansion is recycled around the coil through which further air which has been compressed and cooled to 12°C passes (Fig. 7.1). Under the assumption that the exchange defect is 8 degrees†, the compressed air arriving in the coil is cooled from 12°C to −11°C (−19 + 8). Therefore this air starts

*This and the following values of the coefficient are approximately estimated from the data given in Table 7.1.

†The difference between the amount by which the temperature is actually lowered (−11°C), and the reduction in the temperature which is predicted theoretically (−19°C) is referred to as the 'exchange defect'.

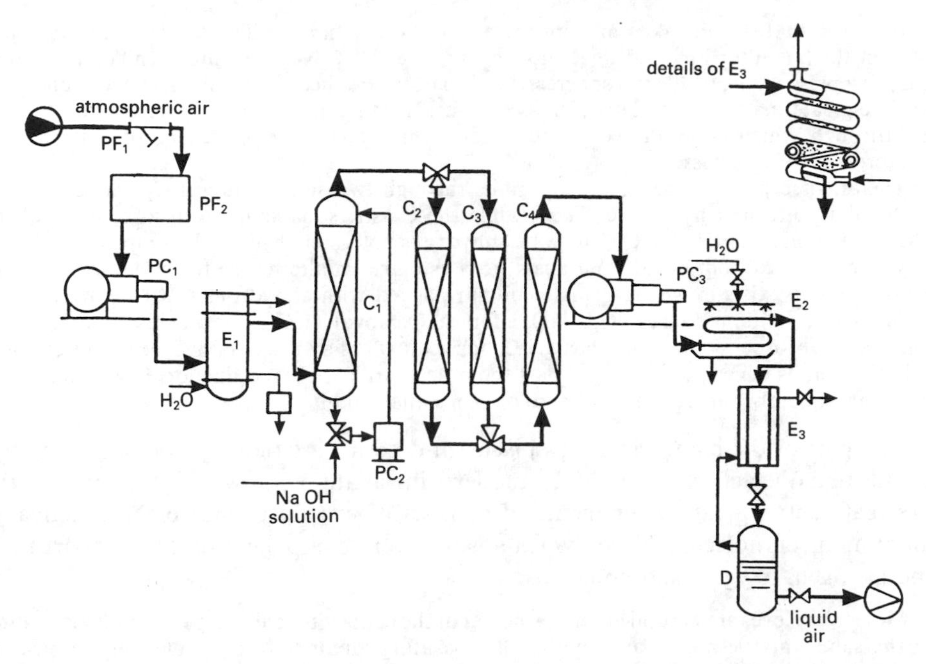

Fig. 7.1—Scheme for the first Linde method. From the air intake up to the absorber C_4 the apparatus is that used traditionally in all methods for the purification of air. The whole of the plant is described in the text, and, as can be seen from the detail shown, the heat exchanger E_3 is of the 'tube within a tube' type.

to re-expand not from 12°C but from −11°C, undergoing a reduction in its temperature of about:

$$\Delta T = 0.20 \times \left(\frac{273}{262}\right)^2 \times (200 - 1) \simeq 43°.$$

As can be seen from Fig. 7.1, the expanded air, after it has cooled the new air arriving in the coil, is exhausted into the atmosphere.

As a consequence of the expansion which lowers the temperature by a further 43°, the air will now have a temperature of −54°C [−11 + (−43)] and will now be able to cool new compressed air arriving from the water cooling stage to −46°C*.

Upon expansion, the air which arrives at −46°C (227 K) has its temperature lowered by:

$$\Delta T = 0.25 \times \left(\frac{273}{227}\right)^2 \times (200 - 1) \simeq 71°,$$

reducing it to a temperature of −117°C [−46 + (−71)].

It is evident that, by continuing in this manner, the air, having become a vapour below its critical temperature, will start to condense.

Fig. 7.1 shows the scheme of the first Linde method preceded by the classical apparatus required for intake of the air and pre-purifying it.

*Under the permanent assumption that the average exchange defect between the air being cooled and that bringing about the cooling is 8°C.

It is seen that the air taken in from the atmosphere is first purified by the removal of fine dust in the line filter PF_1 and then with the air filter PF_2. Next, having been compressed in PC_1 to a pressure of 30–40 atmospheres, the air passes into the water cooler E_1 which condenses out the greater part of the water which it contains. It is then passed into the absorption column C_1 through which concentrated caustic soda, which has been put under pressure by the compressor PC_2, is passed.

The air then passes in series or in parallel through two small columns C_2 and C_3 packed with solid potassium hydroxide. Thus, while the caustic soda solution absorbs most of the CO_2 and other acids (H_2S, NO_2, etc.), the anhydrous potassium hydroxide completely removes any remaining carbon dioxide and acids from the air. An alternative to the use of a double KOH column exists in which a single column packed with anhydrous $CaCl_2$ is arranged in series with a column packed with KOH for the removal of CO_2 and acids. Finally, there follows an absorber containing silica gel C_4 for removal of all traces of hydrocarbons and any other materials which may have got past the other purification and fixation stages and which may subsequently damage the structure or block the conduits.

The part of the plant which has just been described is that traditionally common to all air liquefaction methods. In modern installations, however, this part of the pretreatment is greatly simplified and consists of just dust removers and activated alumina or of molecular sieves which adsorb every component of the air apart from the nitrogen, oxygen, and noble gases.

Molecular sieves are natural aluminosilicates of the chabazite zeolite type or synthetic zeolites of the same varieties. They are capable of separating mixtures of molecules on the basis of their shape, dimensions, and chemical nature, and can be repeatedly regenerated.

Both in the case of alumina and molecular sieve adsorbers, the operations must be carried out at the appropriate temperatures for the adsorption phase and the regeneration phase.

Finally, when Fränkl regenerators are adopted as components in the liquid air fractionation stage, the task of removing any water and carbon dioxide from the air, as we shall see (page 249), may be left to these devices.

After the air has been purified and compressed to about 200 atm. by means of the multistage compressor PC_3, it is cooled with water in E_2 and then sent for expansion in the tank D whence it returns to cool more air, which has been purified, compressed, and cooled back to a normal temperature, in E_3. Finally, it is allowed to escape into the atmosphere. This dispersion of air is a serious drawback to the method as the work of compression is lost, and it is particularly serious until the process has attained its normal operation conditions since, at this point, the energy losses are fairly limited. The normal working conditions are reached when, after repeated stages similar to those which have just been described, liquid air is formed under normal pressure in the tank D.

The second Linde method

The second process conceived by Linde for the liquefaction of air puts greater emphasis on precooling of the air and uses liquid ammonia for this purpose rather than water. Furthermore, in the new method, the compressed air is expanded from 200 to about 40 atm. rather than from 200 to 1 atm.

In this way the compression yield, calculated from equation (7.1), turns out to be much greater: both because the value of the Joule–Thomson coefficient α is greater at the temperature at which the expansion commences (see Table 7.1) and because the temperature T appearing in equation (7.1) is reduced.

The second Linde method also turns out to be more convenient from the point of view of the work of compression because, in equation (7.2), the logarithm of the ratio of the pressures turns out to be reduced to less than one third under the new conditions. Against the substantial advantages of the new method which have just been mentioned, one must balance the major costs entailed in the running of a liquid ammonia plant and the 25% reduction in the cooling effect calculated from equation (7.1), because the $p_2 - p_1$ term is smaller in the second method than is found by applying the comparable formula to the first method.

In terms of the plant required, the second Linde method, in the first place, uses the conventional purification system which converts normal air into purified air (see Fig. 7.1). This air enters into the gas mixer MS (Fig. 7.2) where it is reunited with expanded, recycled air which is under approximately the same pressure. The mixed air then passes to the compressor PC where its pressure is raised to about 200 atm. The heat of compression is removed from it first by cooling in a deluge of water in E_1 and then by cooling with liquid ammonia which lowers its temperature to $-40°C$. It is then successively passed into the exchangers E_3 and E_4 after which it is expanded to 30–50 atm. through the valve V_1. This causes a large part of it to be condensed in the tank D_1. The uncondensed vapour is recycled to the compressor PC, after having given up its refrigerating units* in the exchanger E_3 and turning into gas

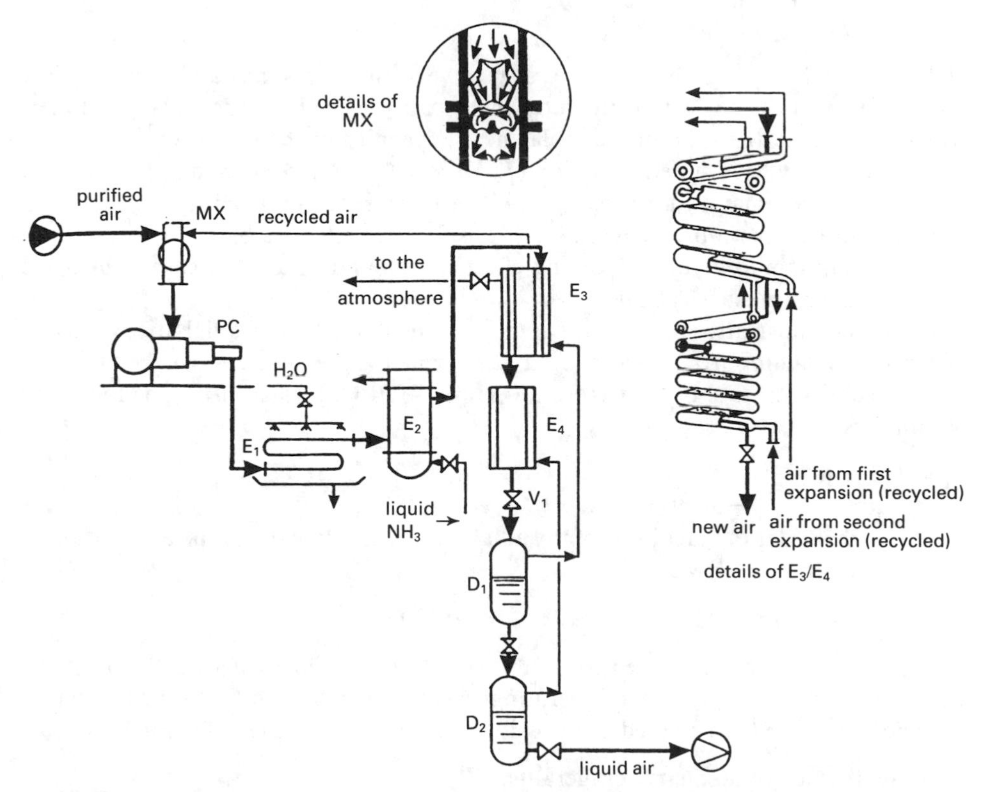

Fig. 7.2—The second Linde method for the liquefaction of air. This is described in the text. It leads to a reduction in the amount of liquid air vapour dispersed into the atmosphere and thereby to greater yields. The detail enclosed in the circle illustrates the mode of operation of the mixer MX which is characterized by the crossings of the gas streams which are brought about.

*The refrigerating unit has the same magnitude as a calorie but the opposite sign. The giving up of refrigerating units therefore leads to the donating system becoming warmer.

again, by uniting it with new air in the mixer MX. The mixing process ensures that the recycled gas is at an appropriate temperature for its subsequent treatment in PC.

The liquid air which is formed in D_1 then passes, upon expanding from a pressure of 30–50 atm., to a pressure of 1 atm., into the tank D_2 where it is under normal pressure. The vapour which rises from the liquid air acts as a refrigerating fluid in E_4 and E_3 and is then released to the atmosphere.

7.4 THEORETICAL BASIS OF THE CLAUDE METHOD FOR THE LIQUEFACTION OF AIR

If a gas is expanded in a machine under almost adiabatic conditions, that is, without the exchange of any heat with the surroundings, it does work and is greatly cooled. A lowering of the temperature of the gas therefore takes place which can be calculated (correctly only in the case of an ideal gas) by applying the formula:

$$\frac{T_2}{T_1} = \left(\frac{p_1}{p_2}\right)^{\frac{1-\gamma}{\gamma}} \tag{7.3}$$

where p_1 and p_2, T_1 and T_2, are respectively the pressures and the absolute temperatures of the gas before and after the expansion, and γ is the 'adiabatic index' which is equal to the ratio of the molar heat capacities of the gas (C_p/C_v).

As has already been stated, equation (7.3) is strictly valid only for an ideal gas, but is a good approximation for a real gas which is far removed from its condensation (liquefaction) temperature. Nevertheless, this formula also enables one to establish the order of magnitude of the cooling effect on air and other real gases which are made to expand in such a device.

Thus, as an illustrative proof of this, it may be calculated that an adiabatic expansion of air in a machine from 30 atm. to 1 atm., which necessitates work being done, lowers the temperature of the gas from 0°C to −170°C. In fact, using the following parameters:

$T_1 = 273$ K (0°C);
$p_2 = 1$ atm. and $p_1 = 30$ atm.,
$\gamma = 1.4$ (the order of magnitude of the adiabatic index for air) and the application of equation (7.3) yields:

$$T_2 = 273 \times 30^{(-0.4/1.4)} \simeq 103 \text{ K} = -170°\text{C}.$$

In practice, the gas is cooled to −130 to −150°C when treated in this manner. Therefore, by exploiting the refrigerating effect associated with the phenomenon on which equation (7.3) is based, it is possible to bring about the liquefaction of air:

- in the absence of auxiliary refrigerants,
- by using relatively low pressures and thereby expending only a small amount of work in compression,
- recovering in part the energy spent in compressing the air in the form of mechanical work.

In particular, it has been very difficult to find lubricants which are resistant to the extremely low temperatures at which the expansion motor operates, and this fact has held up the adoption of the Claude method for the liquefaction of air. A remedy has been put forward for overcoming this obstacle:

* the adoption of petroleum ether as the lubricant in the first stage;
* by subsequent recourse to expansion in fast turbines (Kapitza) which do not suffer from packing and lubrication problems.

7.5 PLANT FOR THE IMPLEMENTATION OF THE CLAUDE PROCESS FOR THE LIQUEFACTION OF AIR

The plant which implements the Claude process also consists of two quite distinct parts which are, respectively, intended for:

- the purification of atmospheric air,
- converting the air to vapour and subsequently condensing it.

In the first section, air which has been freed from dust is treated, to remove CO_2 and water, in one of the ways which have already been discussed when describing the Linde method. Here, there is definite preference toward the adoption of molecular sieve purifiers or the use of Fränkl regenerators.

In a plant such as that shown in Fig. 7.3, the air is freed from dust in PF_1 and in PF_2, it is slightly compressed in PC_1, and freed from carbon dioxide and dried in the twin gas bottles B_1 and B_2 which are packed with molecular sieves. A system of valves ensures that the two bottles can alternate between their operational and regeneration phases.

The dry, purified air is first compressed in PC_2, then cooled with water in E_1, and mixed in MX with recycled air which has already been compressed in PC_3 to the pressure at which the new air is available.

Upon leaving the mixer, the air flows through the crossed countercurrent exchanger E_2. This comprises tubes which have been formed into a helix around a core, as shown in the detailed insert to Fig. 7.3. They are contained in a cylindrical jacket. The air is then divided into two streams: one passes directly to a motor, or a turbine or a pump G, to expand almost adiabatically while doing work; the other stream goes into the exchanger E_3 which is similar in construction to E_2. From E_3, the air then passes into E_4 whence it undergoes expansion through the valve V. After expanding, the air goes into D as a liquid at atmospheric pressure. Any liquid air which vaporizes in the condensation stage is first recycled into E_4 and then passes into the exchanger E_3 after having been reunited with the air which has undergone expansion in G. The whole of this air then passes out into the exchanger E_2 where it produces a pronounced cooling of the air originating from MX which, as has already been stated, crosses in a countercurrent. The recycled gas is finally recompressed, as stated, in PC_3 and mixed with new air in MX. In this way, there is total recovery of the fluid in bulk, and the conditions which have already been attained are completely exploited in this process.

7.6 THE PROPERTIES AND USES OF LIQUID AIR

Air, which has been transformed into a liquid lighter than water (in fact, it has a specific density of $0.87\ g/cm^3$), is extremely mobile and has a transparent blue appearance. The fumes which liquid air appears to give off are of carbon dioxide and water vapour from the surrounding air which have been solidified by sublimation as a result of the strong cooling of these vapours.

When ordinary liquids come into contact with liquid air they solidify, while organic tissues, both vegetable and animal, become stiff owing to the total freezing of every one of their physiological fluids. All bodies solidified by liquid air are as fragile as glass.

Liquid air modifies the properties of metals and alloys, rendering nickel–alloy steels

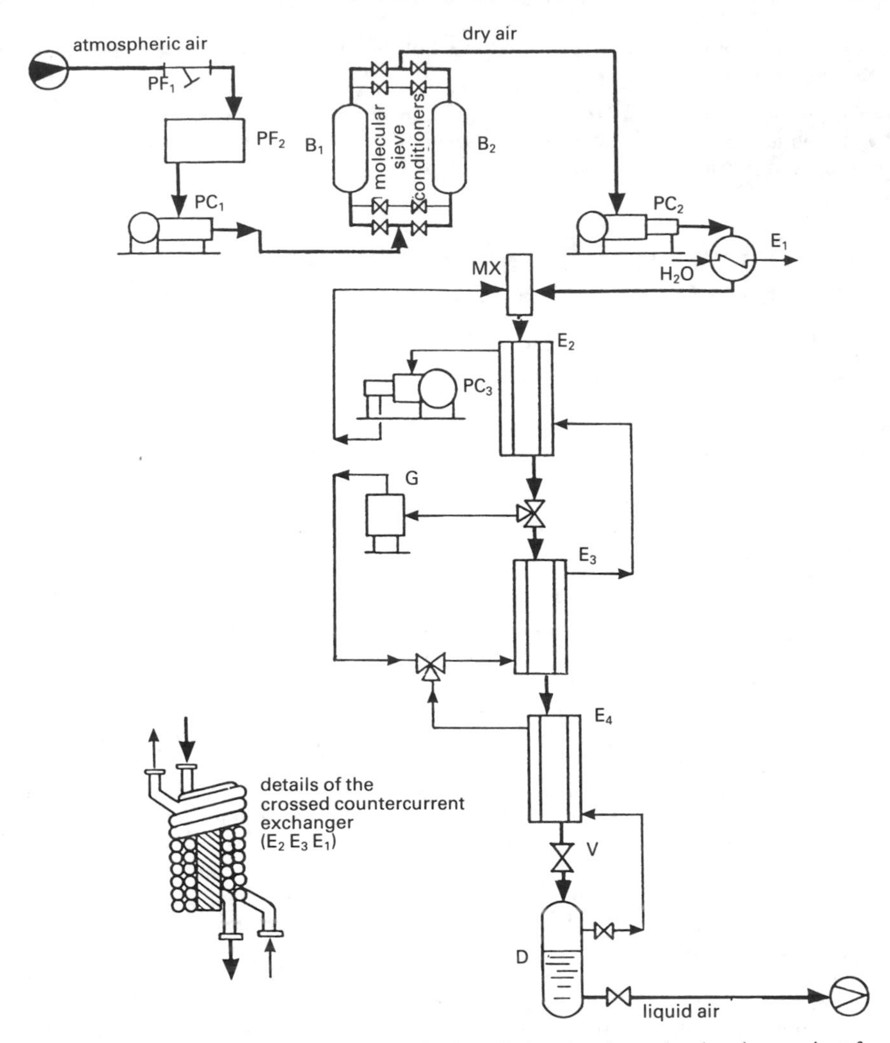

Fig. 7.3—The Claude method for the liquefaction of air using the molecular sieve variant for purification. In this method the reduction in the temperature of the air is brought about primarily by expansion with the performance of work. There is therefore no need to equip the plant with cycles that use refrigerants or to make use of the very high pressures which are employed when free expansion is used, in order to produce cooling. Instead, alternative techniques for cooling the air can be adopted such as causing liquids which can evaporate to diffuse there (it is then safe to reabsorb them when necessary).

magnetic, while both the thermal and electrical conductivities of all metallic materials increase until they become superconductors.

Liquid air is kept and transported in open vessels which are very highly insulated and are known as Dewar vessels. Owing to the effect of the heat which gets into these reservoirs, the liquid air continues to boil there while cooling itself progressively and changing its composition because nitrogen is more volatile than oxygen. In closed containers, which lack the compensation between the heat which penetrates to the interior of the vessel and the heat lost due to the evaporation of liquid air, the temperature slowly increases, leading to the development of high pressures which if they are not relieved can burst the vessels.

For a long time liquid air has been a basic medium for the attainment of low temperatures in laboratory practice and various industrial technologies such as the production of high vacua, the treatment of gases which condense only with difficulty, and, more especially, in the production of nitrogen, oxygen, and the noble gases. Liquid air is also employed in the field of explosives when absorbed on wood and carbon powders.

7.7 GENERALITIES CONCERNING THE FRACTIONATION OF LIQUID AIR

After being liquefied, the air is subjected to distillation (rectification) to separate the oxygen and nitrogen components present in it. Liquid nitrogen and liquid oxygen are miscible in all ratios and do not form azeotropic mixtures; that is, they do not arrive at a point where they form mixtures which, upon boiling, produce vapours with a composition equal to that of the liquid from which they originate.

The mode of operation of a rectification column

The operation in which a binary liquid is fractionated into its components, which is known as rectification, is carried out in a 'plate column', that is, in a rather tall cylindrical structure inside which repeated condensations and evaporations take place on shelves ('plates') which lead to a continuous change in the composition of the binary system in question throughout the length of the column until one of its pure components exists at the top of the column and the other at the bottom of the column. To understand the events to which a binary mixture is subjected in a plate column, a small section of such a column is shown schematically in Fig. 7.4. This section is formed by three plates: an intermediate plate P_n and two collateral plates P_{n-1} and P_{n+1} which are arranged below and above the plate P_n respectively.

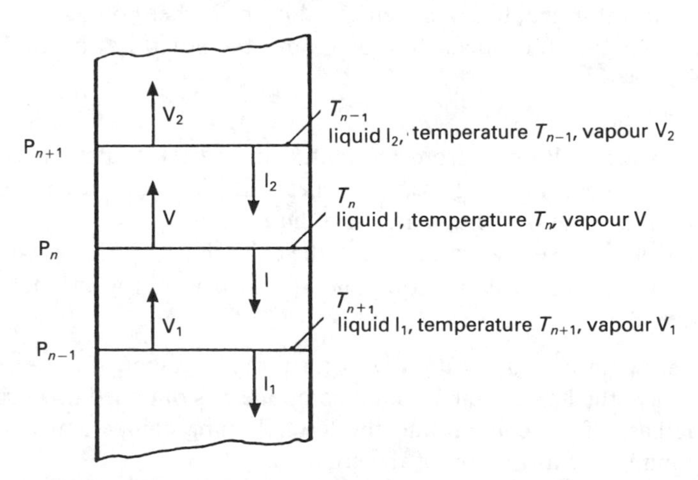

Fig. 7.4—Section of a plate rectification column showing the exchanges taking place between the plates at different temperatures as the vapours rise and the liquids descend through the column.

Let:

- V_1, V, and V_2 be the vapours which, as they pass toward the top of the column, leave the plates P_{n-1}, P_n, and P_{n+1} respectively;
- l_1, l, and l_2 be the liquids which, as they pass down the column, descend from the plates P_{n-1}, P_n, and P_{n+1} respectively;
- T_{n+1}, T_n, and T_{n-1} be the temperatures of the plates P_{n-1}, P_n, and P_{n+1}.

The existence of rising vapours and descending liquids implies that there is intercommunication between the plates.

The liquid l which descends from the plate P_n onto the plate P_{n-1} liberates a vapour V_1 which is carried onto the plate P_n at a temperature T_n. The vapour V_1 contains a higher percentage of the lower boiling point (more volatile) component than the liquid l. On the other hand, the liquid l_1, which has been formed from l together with the vapour V_1, is richer in the higher boiling point (less volatile) component and tends to descend onto another plate.

Having reached the plate P_n, the vapour V_1, which now finds itself at a temperature T_n which is lower than T_{n+1} at which it was formed, undergoes partial recondensation which deprives it of the higher boiling point component. It is thereby transformed into the vapour V which contains a greater percentage of the lower boiling point component.

Meanwhile, the liquid which has descended onto the plate P_n from the plate P_{n+1} above is transformed, by receiving the higher boiling component which has condensed out from V_1, into liquid which tends to fall again, as has already been stated, onto the plate below it.

In its turn, the vapour V rises onto the plate P_{n+1} where, upon finding itself at a temperature $T_n{-}^1 < T_n$, undergoes partial recondensation which deprives it of the higher boiling point component. As a result, it is transformed into the vapour V_2 which contains a higher percentage of the lower boiling point component and passes upward.

In the meantime, the liquid which has descended from the plate above onto the plate P_{n+1} is transformed, becoming enriched in the higher boiling point component derived from the partial condensation of V, into the liquid l_2 which falls back onto the plate P_n.

In conclusion:

- the liquids which fall down from the plates toward the heater in the base of the column (l_2, l, l_1) become progressively richer in the less volatile component (which is oxygen in the case of the distillation of liquid air);
- the vapours which rise toward the top of the column (V_1, V, V_2) gradually become enriched in the more volatile component (which is nitrogen when liquid air is fractionated).

Consequently, in a column fitted with a suitable number of plates, a position is arrived at where the higher boiling point component is obtained in a practically pure state at the base of the column and the lower boiling component is obtained in a practically pure state at the top of the column.

However, the perfect operation of a rectification column always requires that:

- the liquid to be separated into its respective components should always be introduced into the column onto a plate which supports a liquid of the same composition as that of the feedstock liquid;

- part of the distillate from the top of the column is recycled in the form of a 'reflux' with the aim of establishing conditions of repeated washing on all of the plates which refine the vapours moving towards the top of the column, according to the mechanisms which have already been described in detail.

In fact, it is clear that, as there is less of the less volatile component arriving on the plates high in the rectification column, there tends to be a shortage of liquid to ensure the refining condensations of the rising vapours until they are converted into the pure volatile component. Hence the need to create such a liquid by means of refluxing.

Why it is not easy to fractionate liquid air

It is easy (being necessary only to use water as the condensation medium) to create the reflux in rectifications carried out at temperatures above normal, but it is difficult to find a suitable medium for the refluxing of a liquid air distillate. In fact, since condensation temperatures of the order of −200°C are required for this purpose, the only possible media for the condensation of the reflux would be liquid helium or liquid hydrogen, the use of which is clearly inadmissible on both economic and operational grounds. For this reason, the separation of liquid air into nitrogen and oxygen is not as simple as the fractionation of any other binary mixture, even if the underlying principle is the same.

How it is possible to separate liquid air into its components by using a two-section fractionating tower

The problem of separating liquid air into nitrogen and oxygen has been solved in a way which is economically acceptable by designing a fractionating tower consisting of two columns which are arranged one above the other. The upper column has all the requisites of a rectification column, while the lower column functions as a simple enrichment column.

In essence, one is concerned with two columns, one being situated above the other, working at different pressures: the lower column operating at a pressure of about 6 atmospheres while the upper column operates only slightly above atmospheric pressure. There is a heat exchanger between the two columns which acts as a condenser with respect to the lower column and a boiler with respect to the upper column precisely as a result of the two different pressures which appertain in the two compartments.

More precisely: the upper column is supplied with a feedstock of a composition which is commensurate with that of the liquid situated on the plate where the feedstock is let in, and receives a suitable reflux at the top, while the lower column is fed almost normally but is not refluxed, and, instead of leading to practically pure components, it produces a liquid which is only somewhat enriched in the less volatile component (oxygen) at the bottom of the column and a liquid which is very rich in the more volatile component (nitrogen) at the top of the column.

The mode in which the two parts of the air fractioning column function is shown in detail in Fig. 7.5.

It is seen that it is the compressed air which arrives from the first section of the

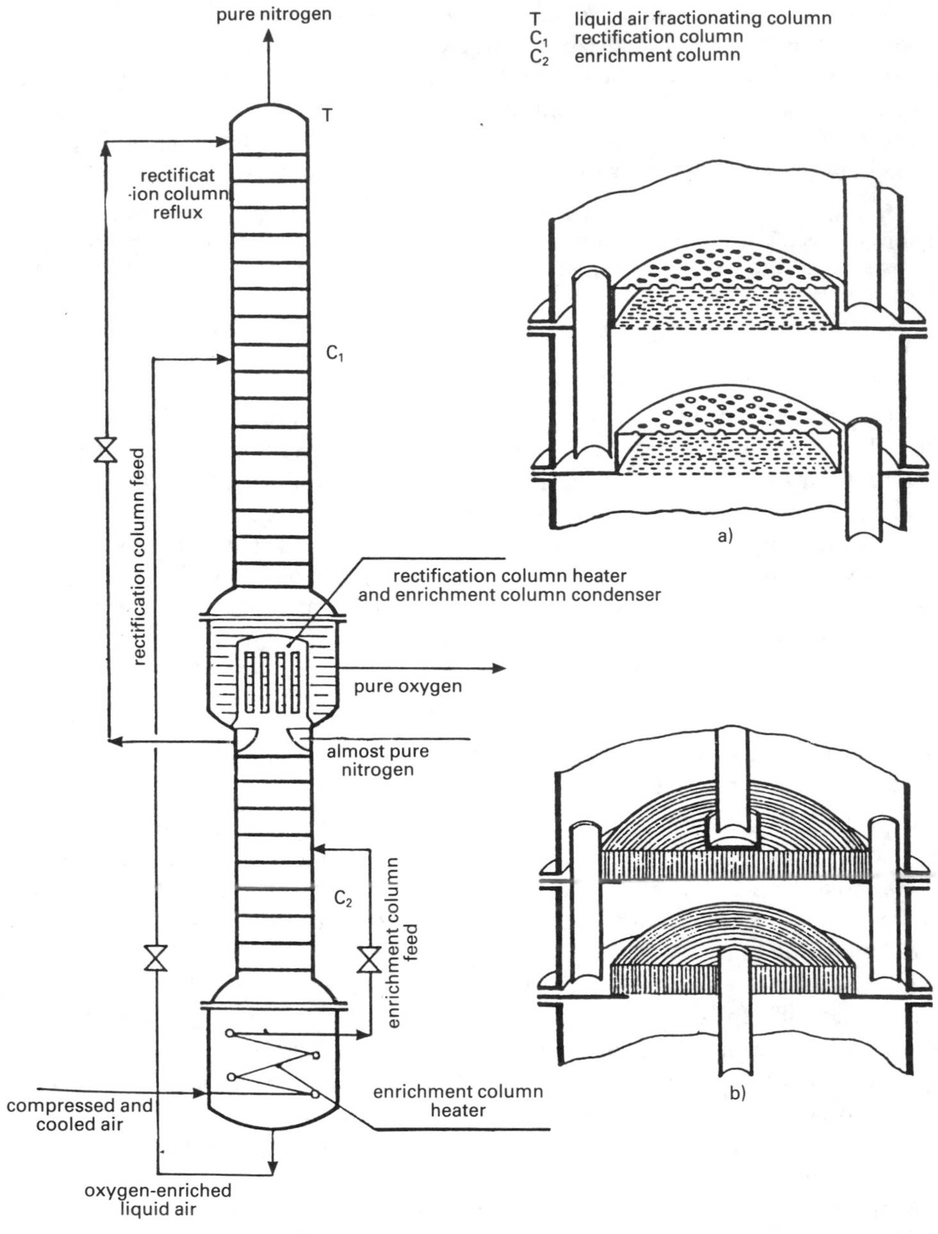

Fig. 7.5—Diagram of a liquid air fractionating tower showing the functional elements and constructional details of the heat exchange devices (plates) which are the basis for the fractionating process. The diagram on the left applies to both types of plant, whereas detail (a) refers to plants of the Linde type and detail (b) refers to Claude type plants.

plant which acts as the heating fluid in the heater at the base of the enrichment column. The same air, always contained within a tube, passes out from the lower column of the tower only to re-enter it higher up after the pressure to which it is subjected is reduced by means of a valve, resulting in the lowering of its temperature. Nitrogen with a small oxygen impurity collects at the top of the enrichment column, and, after expansion to atmospheric pressure, this nitrogen is sent to back as the reflux in the rectification column situated above. The liquid which collects in the heater at the base of the enrichment column is fed, after expansion to atmospheric pressure, onto a suitable plate of the rectification column. When it is adequately fed and refluxed, this column produces pure oxygen at the bottom and pure nitrogen at the top.

7.8 STRUCTURAL CHARACTERISTICS OF TOWERS FOR THE FRACTIONATION OF AIR AFTER IT HAS BEEN LIQUEFIED

The towers in which liquid air is fractionated therefore consist of two cylindrical sections (columns) arranged on top of one another. The two columns are separated by a heat exchanger which acts as a condenser at the top of the enrichment column and as a reboiler at the base of the rectification column.

As a bottom reboiler, the enrichment column has a boiler with a curved base. There is no condenser at the top of the rectification column, and it is closed by means of a gently curved cover with an outlet aperture. The upper column is about twice the height of the lower column, and both of them are fitted with plates spaced at intervals. The average numerical ratio of the repartitioning between the two columns is 42–25.

The shape of the plates differs according to the type of plant in which they operate. In plants of the Linde type every plate (Fig. 7.5(a)) is made up from two metal plates which are separated from one another by a certain spacing and perforated with very small apertures in the lower plate and quite large holes in the upper plate. The plates of the columns in plants using the Claude method (Fig. 7.5(b)) are strips of thin steel plate wound into a spiral with a separation of the order of a tenth of a millimetre between the spirals. On account of the capillarity due to the small apertures in the lower half of the Linde plates and the small cavities between the spirals of the Claude plates, the down-flows of the liquids are retarded, thereby favouring perfect contact between the descending liquids and the rising vapours. All the steels employed in the environment of this coming and going of liquids must be of a type which is resistant to low temperatures.

7.9 LINDE PLANTS FOR THE FRACTIONATION OF AIR

Air which has been pretreated in the ways which have already been described up to the point where it emerges from the liquid ammonia refrigerator E_3* flows through the exchangers E_4 and E_5 which are of the tube within a tube type† in a countercurrent with nitrogen and oxygen produced in a rectification column of an air-fractionating tower. Upon emerging from the exchanger E_5, the air progresses into a coil immersed in the heater liquid in the enrichment column of the tower where it acts as a heating fluid and from where it returns through a tube to the outside of the tower and, after being expanded through V, re-enters the enrichment column as feedstock for this column. The approximate values of the pressures and temperatures in the various parts of the plant are indicated in Fig. 7.6.

The feedstock introduced in this manner rises in part to attain the composition of almost

*This refrigerator here is, however, superfluous if the preceding water refrigerator is efficient and the heat exchanges with the liquid oxygen and nitrogen which is going to storage are good.

†This type of exchanger is primarily employed in apparatus where there are high pressures which correspond to relatively small volumes of gas.

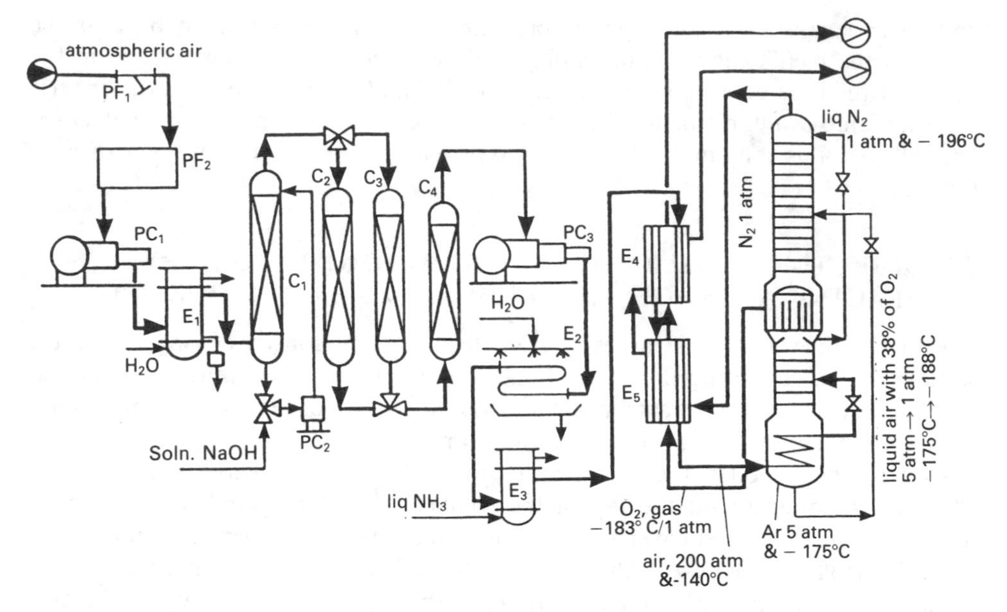

Fig. 7.6—Linde plants for the liquefaction of air and the fractionation of liquid air. The first section is the same as that employed in the second Linde method for the liquefaction of air but without the mixing and exhausting of the recycled air. The fractionating tower is the principal component of the second section.

pure nitrogen in the condenser at the top, while the other part, having been enriched in oxygen, collects in the boiler at the bottom of the column. The two products are withdrawn separately, expanded, and introduced into the upper rectification column: the liquid from the reboiler, delivered onto a plate which supports a liquid of more or less the same composition, constitutes the feedstock for the rectification column, while the condensate from the top of the enrichment column is used to reflux the rectification column. Nitrogen and oxygen are, respectively, withdrawn from the top and the bottom of this column. After having completed the refrigeration cycle in E_4 and E_5, they are stored.

7.10 CLAUDE AND OTHER PLANTS FOR THE FRACTIONATION OF AIR

Air is taken from the atmosphere and treated, as described on page 242, up to the point where it emerges from the cooler E_1. The operational cycle following this is also analogous to that pursued in the process of obtaining liquid air by the Claude method, but, as can be seen from Fig. 7.7, without the mixer MX and the compressor PC_3 (Fig. 7.3). The exchangers E_2 and E_3 are still of the crossed countercurrent type but with triple tubes in this case. There is no recycling of the air, because there is no re-evaporation which makes it necessary for the air to be recycled and also because the air which has been expanded adiabatically in the machine G is introduced as further feedstock just above the heater in the base of the enrichment column. Refrigeration of the air, compressed to a pressure of 40 atmospheres, arriving from the purification stage is ensured by the nitrogen and oxygen originating from the top of the rectification column and from the condenser at the base of the rectification column which are subsequently sent to be stored. The cycles, pressures, compositions and temperatures appertaining in the two sections of the Claude air fractionating tower are indicated in Fig. 7.7.

In recent years, after some initial resistance to their adoption associated with the costs of the materials and the installations which are required, air fractionation plants (Collins–Kellog,

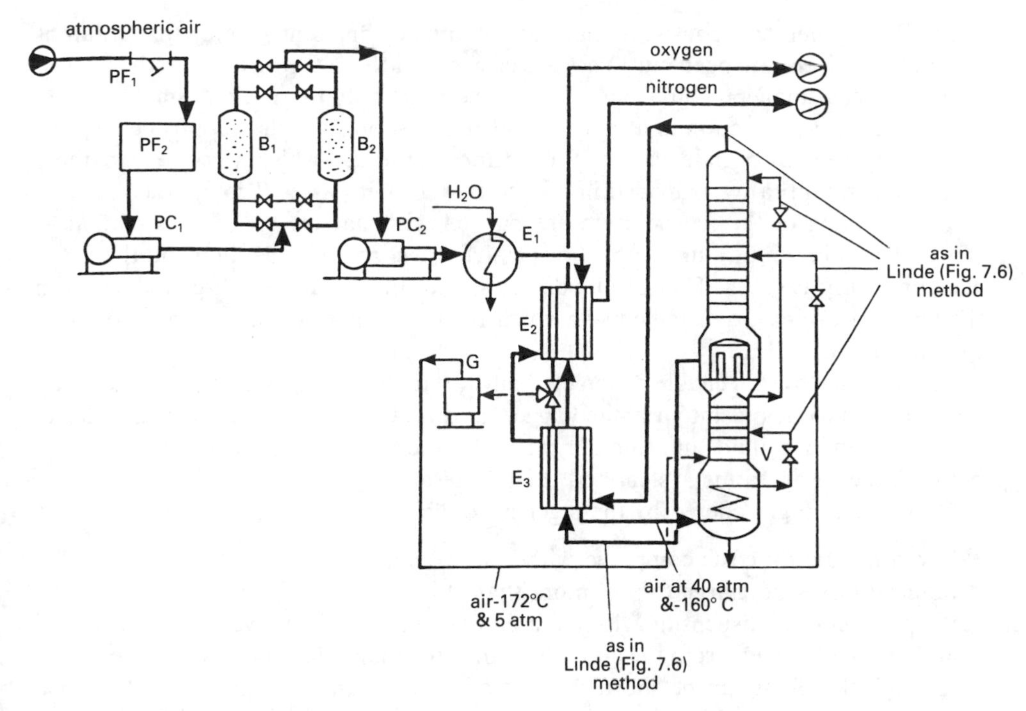

Fig. 7.7—Claude plant for the liquefaction of air and the fractionation of liquid air. The conditions under which the enrichment column is supplied with air are indicated, and it is noted that, after this stage, the conditions are identical to those employed in the Linde method.

Tonnox, etc.) which are to a greater or lesser degree different from the classical Linde and Claude plants have become more widespread. We do not, however, consider it necessary to describe them here as a knowledge of them is not of essential interest.

7.11 FRÄNKL REGENERATORS

In all plants that use liquid air technology, heat exchange is a problem of fundamental importance, primarily on account of the energy savings which can be made.

On the one hand it is noted that the values of the gas–gas transmission coefficients K (kcal/m·h·°C) are very low, as the recovery of the coldness even with the most sophisticated tube in tube and crossed countercurrent heat exchangers is at an average level of 80–85%.

Substantial improvements (with yields brought up to 98–99%) in the recovery of the coldness in the field of air fractionation have been attained with the introduction of Fränkl regenerators* which replace heat exchangers.

These devices function by storing the coldness given up by the nitrogen and oxygen originating from the air rectification column in a first phase, and then, in a second phase, giving up these stored refrigeration units to the air going to the fractionating tower. The columns are switched over from one phase to the other every 1–3 minutes.

*It would, perhaps, be more correct to speak of 'recoverers' since they permit the recovery of refrigerating units.

Fränkl regenerators consist of pairs of columns of different volume (the columns through which the nitrogen must pass are larger, and those through which the oxygen must pass are smaller), pairs consisting of one large column and one small column being alternately in action and in a state of regeneration which is controlled by the setting of the appropriate valves. Furthermore, the air which is cooled in these columns is quantitatively divided in a ratio of 4:1 as it passes through them.

The columns of the regenerators are packed with narrow spiral rolls of double strips of aluminium or aluminium alloys which are corrugated as shown in the detail accompanying Fig. 7.8. The cold fluids bring about an extremely large reduction in the temperature of the metallic mass which then rises again owing to the cooling of the 'hot' fluid.

Fränkl regenerators enable extremely high surface areas for exchange (greater than $1000\,m^2/m^3$) to be put into a limited space, and they achieve the optimal transfer of heat between one fluid and another with the thermal correction of about 1–2°C between the refrigerating system and the refrigerated system.

Other advantages offered by the regenerators are:

- low construction costs compared with heat exchangers;
- reduced losses of charge (of not more than 0.15 atm.);
- the possibility of dispensing with the use of systems for the preventive dehydration and removal of CO_2 from the air coming into the plant. This means that the routes through the plant can be shortened and adsorbers can be dispensed with (which also removes any problems with the maintenance and regeneration of these devices).

The principle of the purification of air from CO_2 and moisture within Fränkl regenerators is as follows.

When air passes into an environment which, in the preceding phase, has been cooled to a

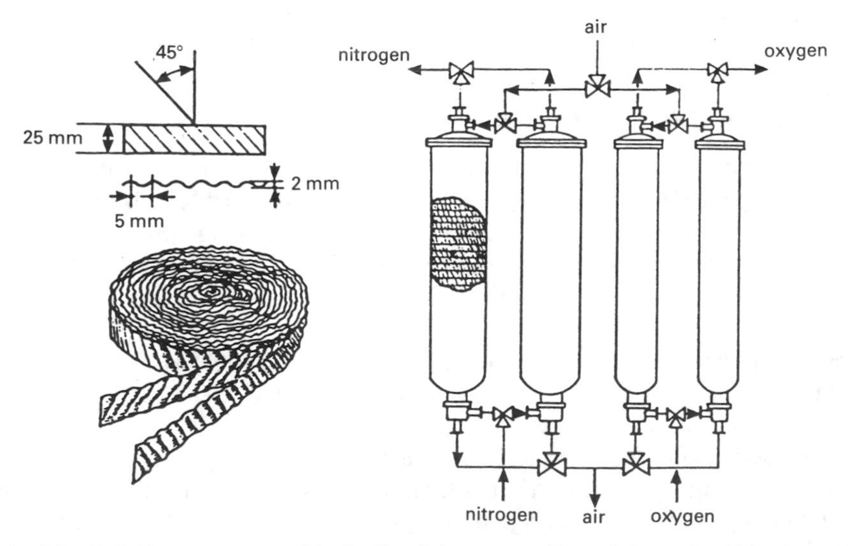

Fig. 7.8—Fränkl regenerators with details of the construction of the exchanging elements. According to the setting of the valves, one large column and one small column are operating together both in the refrigerating phase and in the regeneration phase. It goes without saying that the mechanisms by which the valves for the inversion of the flow are controlled is completely automated and perfectly synchronized.

low temperature, the impurities in this air deposit out as solids. By reversing the flow, the return gases (N_2 and O_2) finding themselves at a lower pressure than that of the incoming gas (air) remove the solid impurities.

The disadvantages of the regenerators which arise in principle, because, in practice, they can be circumvented by suitable plant modifications, are:

- acoustic disturbances as a result of the inversion of the cold gas–air flow;
- perturbations in the flow of the air feedstock to the columns (and, therefore, the rectification column);
- contamination of the nitrogen and oxygen with a part of the CO_2 and moisture.

Nevertheless, great progress has been made in the elimination of the acoustic pollution and in the regularization of the feedstock input. Moreover, if the contamination of the gases by CO_2 and H_2O is unacceptable and the requirements are not met by the usual grades of purity required in the case of nitrogen and oxygen, it is possible to adopt systems for the intermediate washing of the regenerators with about 5% losses of the returning gas.

Scheme for a liquid air fractionating plant including Fränkl regenerators

Air which has been precompressed to 5–6 atmospheres in PC_1 (Fig. 7.9) is divided into two streams: one (about 95% of the air) passes into the regenerators before being carried towards the enrichment column of the air fractionating tower. Having arrived in the proximity of this column a fraction of the chilled air (25%) is directed to the heat exchanger E_1 where it cools another fraction of the relatively compressed air to prevent it from condensing within the machine FT to where it goes to be expanded to slightly above 1 atm. before entering onto a plate of the rectification column which acts as host to a liquid of the same composition.

The modes in which the enrichment and rectification columns function are essentially the

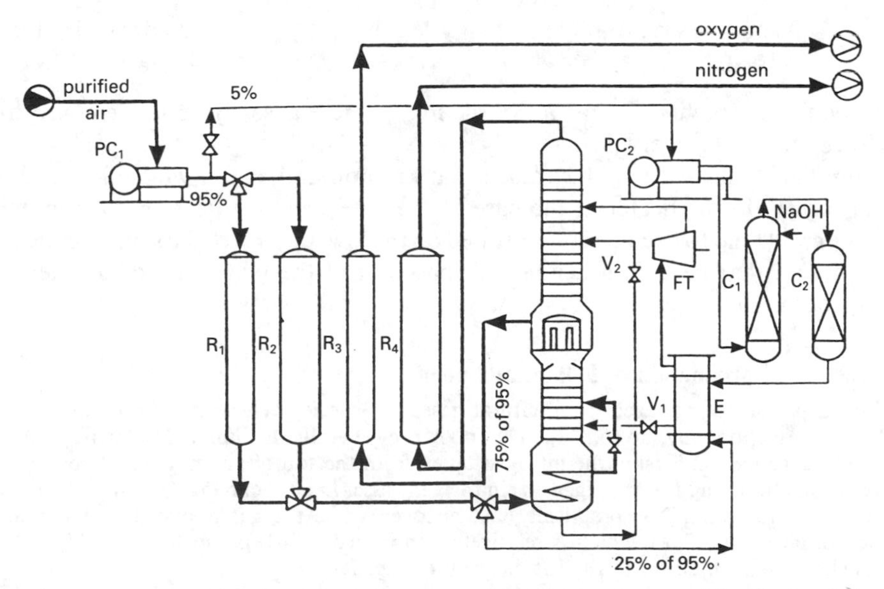

Fig. 7.9—The fractionation of air with the use of Fränkl regenerators. The details of the apparatus, and the way in which it functions, are explained in the text. For simplicity, the regenerators are 'seen' in a single stage of their operation.

usual ones even if the enrichment column has a twin air feed and the rectification column has a further supplementary feed of air which has been pre-expanded in FT. Nitrogen is withdrawn from the top of the rectification column and oxygen from the bottom.

The 5% of the air which has not been sent to the regenerators goes, after being compressed in PC_2, to a pressure of 20–25 atm., to have the CO_2 removed from it by washing with an alkaline solution in C_1 and to undergo dehydration in C_2. Next, having been precooled in E_1 by exchange with the air which has been sent to expand in the machine FT, it enters onto a suitable plate in the enrichment column after having been brought to a pressure of 5–6 atm. by expansion through the valve V_1.

The function of this 5% of the air is to compensate for the loss of refrigerating units in the plant. Meanwhile, on the one hand, it actually gives rise to nitrogen and oxygen when it is fractionated which still act as refrigerators, and, on the other hand, it does not require any contribution from the rectification products for its cooling as the lowering of its temperature is indirectly brought about by the expansion of the gas with the expenditure of work in FT.

7.12 THE PRODUCTION OF THE NOBLE GASES FROM AIR

More than twenty years ago it was discovered that the noble gases can enter into chemical combination, but this fact has not provided, up until now, the motive for any progress in the development of new technologies for the preparation of these gases from air. Hence, the noble gases are still separated from one another by fractionation making use of their different condensation temperatures and from nitrogen and oxygen by utilizing both their different boiling points and their great chemical inertness.

In decreasing order, the boiling points (in °C) of the noble gases, conveniently arranged with respect to those of the principal components of air, are as follows:

		Oxygen		Nitrogen		
		−182.97		−195.8		
Xenon	Krypton		Argon		Neon	Helium
−108.0	−153.2		−185.9		−246.1	−268.93

with the boiling points of four out of the five of these gases lying far removed from the −182 to −196 interval.

From these values it may be argued that krypton and xenon will collect with the liquid oxygen in the heater at the base of the liquid air rectification column, while the argon will tend to become concentrated on the lower plates of the same rectification column, and that neon and helium will pass out of the plant mixed together with the nitrogen.

The industrial production of helium and neon

Helium is primarily produced from natural gases in America. These gases normally contain 1–2% and, exceptionally, up to 8% of helium. More generally, helium and neon are produced from liquid air by condensing the nitrogen taken from the top of an enrichment column of a liquid air fractionating tower (Fig. 7.10) in the gas vessel D after expansion in V_1. That which condenses in the tank D is practically just nitrogen while the gas above the condensate is formed of nitrogen (49%), which is relatively rich in neon (40%), and helium (10%). All of this is shown schematically in the left diagram of Fig. 7.10.

When the nitrogen has been adsorbed from this mixture onto active carbon at −100°C, it remains to separate the two noble gases (where necessary). This is achieved by cooling with liquid hydrogen which condenses out the neon but not the helium.

The industrial production of argon

By withdrawing the liquid containing 5–15% of argon from the bottom plates of the air rectification column and feeding this into another rectification column C_3, which is refluxed with a condensate rich in argon (the diagram on the right of Fig. 7.10), a nitrogen–argon mixture contaminated by oxygen is obtained at the top of this column and a condensate consisting of almost pure oxygen is obtained at the bottom. This condensate is returned to C_2.

The ternary mixture, obtained in this manner, is passed through a condenser E where liquid oxygen which has been withdrawn from the heater of column C_2 acts, by evaporating, as a fluid for the condensation of the argon containing a small amount of nitrogen and all of the oxygen contaminating the mixture. The coolant oxygen, having become gaseous, is reunited with the main gaseous oxygen stream withdrawn from the column. The nitrogen which emerges as a gas from the top of the condenser is carried with the nitrogen which escapes from the top of the rectification column and the argon which has condensed out is partially used for the refluxing of C_3 and partially sent to be stored or, when necessary, is further purified.

For this purpose, the water formed by burning the oxygen contaminating the argon with a measured quantity of hydrogen is condensed and adsorbed, and, to obtain pure argon, the nitrogen is finally eliminated by fixing it as the nitride on a red hot mass of magnesium and calcium oxide.

The industrial production of krypton and xenon

It has already been stated that these two noble gases collect in the liquid oxygen in the air rectification column. By withdrawing this oxygen and treating it in a primary fractionating column without refluxing, a mixture is obtained at the bottom of the column which contains 0.1–0.5% of krypton and xenon.

This liquid is returned to be rectified once again without refluxing, and the two gases are concentrated in this manner to less than 5% in the heater at the base of the column.

At this point any hydrocarbons which may contaminate the gas and, especially, acetylene, are catalytically oxidized (in the presence of CuO) to prevent any subsequent explosions. The

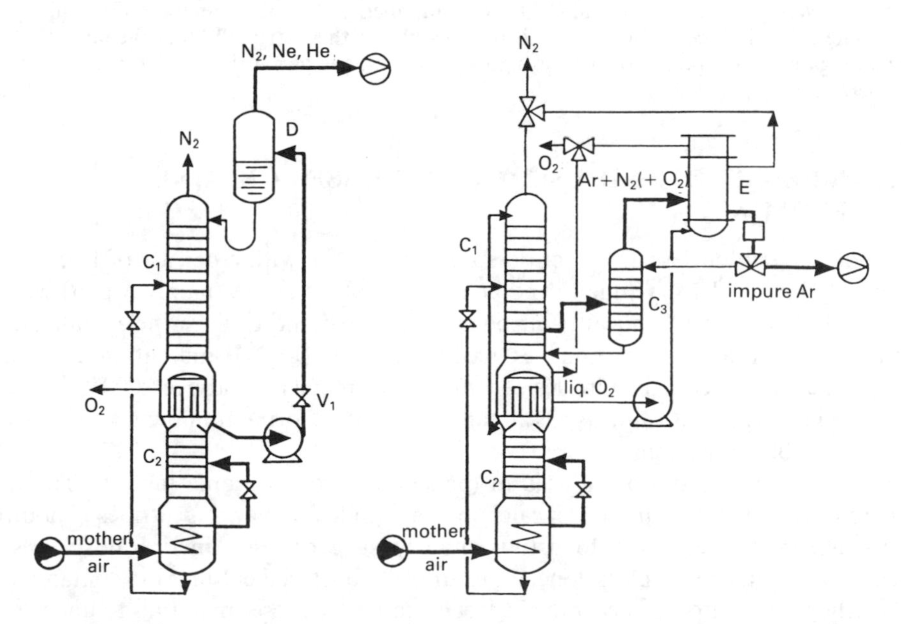

Fig. 7.10—(Left). Scheme for the production of a mixture of neon and helium in an auxiliary apparatus attached to a liquid air fractioning tower. (Right). The production of argon contaminated with nitrogen and oxygen from the oxygen–(nitrogen)–argon mixture drawn off from the bottom of the liquid are rectification column.

CO_2 and water which are formed are eliminated chemically. The oxygen containing about 5% of the two noble gases is fed, after liquefaction, into a distillation column, refluxed with oxygen (the product formed at the top of the column), whereupon a mixture containing 50–60% of krypton and xenon is obtained at the bottom of the column.

The oxygen remaining in this mixture is subsequently fixed chemically until the resulting mixture contains more than 98% of krypton and xenon which, when necessary, are separated by fractional condensation.

Separation of the noble gases using molecular sieves

It is possible, by studying the operational conditions and the nature of the sieves, to separate the individual components from mixtures of noble gases with nitrogen and oxygen and, even more so, from binary mixtures of noble gases by means of molecular sieves.

The first technique which was studied with this object in mind more than forty years ago (1947) led to the separation of oxygen from argon.

The principal factors in the separation of gaseous mixtures in this manner are:

- the different chemical affinities presented by the components in the mixtures to be separated with the sieve;
- the different molecular dimensions;
- the structural elasticity of the sieves with respect to the temperature;
- the adsorbability induced due to the gases being brought under conditions close to where there are changes of state.

It is clear how the first two of these factors can have an influence. The influence of the structural elasticity of the separators (sieves) is then proven by the fact that zeolites are known which are permeable to all the components of certain gaseous mixtures at a certain temperature but selectively permeable to certain of these components at another temperature. Finally, the phenomenon of induced adsorbability is responsible for the sharp fall-off in the normal ability of many gases and vapours to pass through a sieve in the neighbourhood of their condensation temperature.

7.13 PROPERTIES AND USES OF THE GASES OBTAINED FROM AIR

Nitrogen is a colourless and odourless gas when pure, with a density of 1.25 g/l and a slightly higher density (due to the presence of other gases, mainly argon) when it originates from air. Chemically, nitrogen is relatively inert. Nevertheless, already at ambient temperatures, it produces lithium nitride (Li_3N) and, at more elvated temperatures, it forms the nitrides of many other metals. At above 2000°C, it reacts with oxygen to give NO and, as has already been noted, forms ammonia with hydrogen under suitable conditions.

Nitrogen, apart from being used in the synthesis of ammonia (about 80% of it), is employed in the preparation of calcium cyanamide and various nitrides in addition to its uses in the freezing of foodstuffs and cryosurgery. Also, inert atmospheres are created with nitrogen. This is done, for example, if bacterial cultures are contaminated when they come into contact with air, if such contact exposes materials to undesirable oxidations and even finally to combustion, and in the execution of chemical processes and the filling and emptying of tanks for highly flammable materials.

Liquid nitrogen and, even more commonly, liquid air are also extensively used in research laboratories for the study and the utilization of the behaviour of materials at low temperatures.

Oxygen is also a colourless and odourless gas with an absolute density of 1.429 and a solubility in water of 0.0331 l/l (at 20°C). Oxygen is paramagnetic, and the amount of it present in gaseous mixtures is therefore readily recorded. Oxygen is commonly available at a purity level of only 98%. This is mainly due to the high solubility of nitrogen in liquid oxygen and the presence of noble gases which systematically tend to be occluded.

Massive amounts of gaseous oxygen are used in metallurgy and, above all, in steel making. Liquid and solid oxygen are much used in missile technology for spatial propulsion and in fuel cells. The quantities of oxygen which are consumed for feeding flames (oxyhydrogen and oxyacetylene and similar flames) and used in the many processes throughout the most widely varying branches of industrial chemistry in general, and those of petroleum chemistry and coal chemistry in particular, are very considerable.

Finally, the use of oxygen in making respiratory processes possible and raising the level of such processes, in hospitals, astronautics, diving, etc., should be mentioned.

The noble gases are characterized by their great chemical inertness and by a range of densities which enables light and various heavier inert atmospheres to be created from them. Such atmospheres are employed in aerostatic activities, in welding and brazing techniques without the formation of any compound in the molten phase formed in the process, and in other metallurgical operations. Various other chemical and parachemical industries are also interested in the exploitation of the highly inert atmospheres which are formed by the various noble gases.

Traditional general uses of the noble gases are encountered in the construction of light bulbs and fluorescent tubes with constant chromatic effects. Helium finds specific applications, and more of these are constantly foreseen, in electrotechnical and electronic technologies appertaining to the field of very low temperatures.

7.14 THE POLLUTION OF AIR

Atmospheric pollution alters the chemical composition of the air and, consequently, of its physical properties such as the partial pressures of the gases and the density of the mixture. Other physical characteristics, such as transparency and temperature, may also be changed both by the variation in the chemical composition as well as by other causes of pollution such as the output of energy into the air.

There are two aspects to the study of atmospheric pollution: that of the causes and then that of the effects.

In general, the causes of the alterations in the characteristics of the air may be natural or artificial: the first, for the most part, being due to volcanic emissions, fires due to natural phenomena, powders arising from erosion and marine aerosols; the other causes are associated with human activity: mainly in relation to fires, types of motorization and industrial plants.

Combustion plants such as power stations for the production of energy as well as for heating and for the incineration of refuse give rise to particularly high levels of pollution.

Motor transport, which, on account of its specific relevance here, is often treated quite separately from combustion processes although, fundamentally, similar processes are involved, is the cause of much pollution which arises primarily from internal combustion engines.

Industrial plants are qualitatively the most polluting in the sense that that the toxic compounds which are spilled out into the air cover a vast range of types, but, quantitatively, the pollution due to these sources is much lower than that due to the other two causes which have been mentioned.

As regards the effects of atmospheric pollutants on living organisms, it is not possible to carry out a systematic study of these here. It is only possible to sum them up by mentioning that there can be both biological effects (at the human, animal, and vegetable level) of various kinds and acuteness* as well as the tarnishing and corrosion of everything, but, especially, metallic materials. Pollutants therefore assail living organisms belonging to every biological kingdom without distinction with an aggression which may be direct (open) or subtle. In the first case, when the damage is so clearly obvious, some defence can be got ready while; in the second case, which usually occurs over a protracted period, there are no premonitory signals such as to cause alarm, and any intervention is therefore very late.

A concatenation of effects can also occur in the sense that the alteration of one kind of plant life, directly or indirectly (that is, through animals), may have an effect on man. This chaining together of the effects usually becomes larger and larger upon passing through the successive biological steps with a concomitant intensification, as a whole, of the toxic effects on the metabolic route through the food chain.

---

*It varies from irritations, undesirable disorders, and respiratory and optical constrictions to chronic allergic effects and serious pathogenic effects (cancers, bronchitis, emphysema, etc.) in the human and animal field and to lesions and phytotoxicity in the vegetable kingdom.

8

The industrial compounds of nitrogen

The principal industrial compounds of nitrogen are:

- ammonia,
- nitric acid,
- hydrocyanic acid and cyanides,
- urea,
- hydrazine.

In this chapter, as well as discussing ammonia, we shall be concerned with all of these compounds apart from urea which will be thoroughly treated in the chapter dealing with organic fertilizers.

We shall consider only the production of ammonia by synthesis here. The treatment of the recovery of ammonia from natural sources (coals, gaseous mixtures) will be dealt with in other parts of this work which can be found by referring to the indexes of the two volumes.

AMMONIA

8.1 THE STRUCTURE AND PROPERTIES OF AMMONIA

Ammonia is a colourless gas which is lighter than air, has a characteristic pungent odour, and a lacrimatory effect. The inhalation of ammonia, according to the amount, causes giddiness, suffocation, nausea, vomiting, and convulsions. The presence of 1000 ppm of ammonia in the air for only a brief period constitutes a lethal dose for animals.

The ammonia (NH_3) molecule has a large dipole moment due to its triangular pyramidal structure with the three hydrogen atoms located at three vertices and a lone pair of electrons on the nitrogen (Fig. 8.1).

The dipole moment of ammonia leads to its ready liquefaction when compressed.

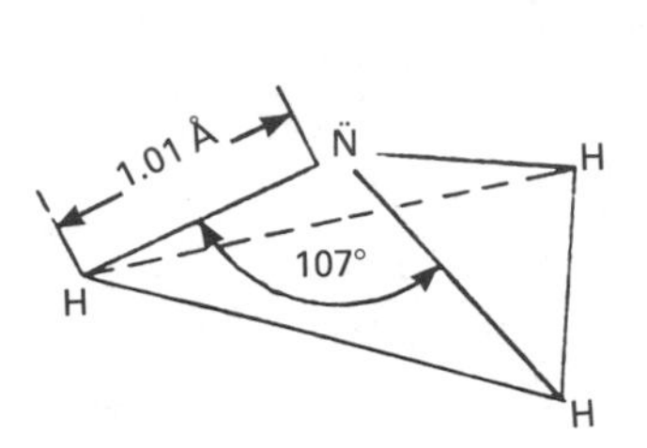

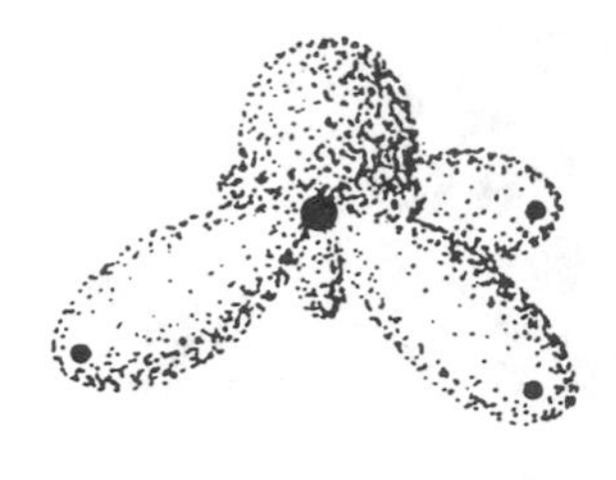

Fig. 8.1—(Left). The 'geometry' of the ammonia molecule. (Right). Its orbital structure in which the large black point indicates a nitrogen atom and the three small points indicate hydrogen atoms. The fourth sp^3 orbital contains two electrons which are located on the nitrogen atom (a 'lone pair of electrons') which are responsible for the well known basic properties of ammonia. Moreover, the large amount of electron density on the nitrogen due to the lone pair as compared to the electron density in the region of the hydrogen atoms is responsible for the 'dipolar nature' of ammonia.

Note: 1 Angstrom unit (Å) = 10^{-10} m.

Another consequence of this dipole moment, in addition to its tendency to compete with water for the preferential possession of the H^+ ion ($NH_3 + H_2O \rightarrow NH_4^+ + OH^-$), is certainly its high solubility in water. Finally, the high dipole moment is also the reason for the high heat of evaporation of ammonia which permits its use as a refrigerating fluid which it forms, after having been condensed by compression at a suitably low temperature, as a colourless liquid with a density of 0.63 g/ml.

Ammonia is quite stable at ambient temperatures, but at elevated temperatures it tends to decompose into its components, especially when there are efficient catalysts present.

Upon ignition, ammonia burns in air and in an atmosphere of oxygen with the development of a large heat of combustion on account of the high stability of the products which are formed (N_2 and H_2O). Ammonia forms explosive mixtures when present in air to the extent of 15.5–28% and in oxygen to the extent of 13.5–82%.

8.2 THE SYNTHESIS OF AMMONIA

All processes for the synthesis of ammonia include the reaction:

$$N_2 + 3H_2 \rightleftharpoons 2NH_3 + 22\ \text{kcal} \tag{8.1}$$

This is a typical equilibrium reaction which is exothermic with respect to the formation of ammonia and takes place with a contraction in volume. It is deduced from these facts that the synthetic yields are influenced both by temperature and pressure.

In so far as the kinetics of the reaction are concerned, the rate of reaction for the synthesis of ammonia is inherently very low on account of the high energy of activation involved in this process (about 55 kcal/mole NH_3).

Essentially, it is the very high bond energy of the triple bond in the nitrogen molecule which slows down the reaction kinetically as can be deduced from the fact that catalysts, which are renowned for their ability to adsorb hydrogen and to transform it to the atomic state (platinum, for example), do not catalyze the synthesis of ammonia. Instead, this process is favoured catalytically by agents which are able, above all, to break the triple bond in nitrogen.

The contrasting influence of temperature on the synthesis of ammonia

From the kinetic considerations which have just been considered it is deduced that, to increase the rate of reaction of the synthesis of ammonia, it is necessary to supply the high energy of activation to the reactants by increasing their temperature, for example. It follows from this that high temperatures increase the rate of the synthesis of ammonia.

On the other hand, however, the yields in the synthesis of ammonia are increased at low temperatures since the reaction involved in this synthesis is exothermic.

To add some quantitative content to this statement it is expedient to refer to some experimental data.

The expression for the equilibrium constant K_p of the ammonia synthesis reaction is:

$$K_p = \frac{p^2_{NH_3}}{p^2_{N_2} \cdot p^3_{H_2}}$$

This constant assumes, at different temperatures, the values shown in Table 8.1 in which the inverse role of temperature with respect to the yields from the synthesis is documented.

On the basis of these data it is clear that the synthetic yields are extremely low precisely at the temperatures which are favourable from the point of view of the kinetics of the synthesis of ammonia ($>900°C$).

The influence of pressure on the synthesis of ammonia

On the basis of Le Châtelier's principle it can be qualitatively expected that the synthesis of ammonia will be favoured by an increase in pressure on account of the fact that the reaction involves a decrease in volume.

A convincing proof of this assertion is provided by demonstrating that the mole fraction of ammonia present when the system is at equilibrium increases as the pressure is increased.

For this purpose it suffices to express the formula in terms of partial pressures, deduced from Dalton's law ($p_i = x_i \cdot P$) for the equilibrium constant K_p in terms of mole fractions:

$$K_p = \frac{x^2_{NH_3} \cdot P^2}{x_{N_2} \cdot P \cdot x^3_{H_2} \cdot P^3}$$

Table 8.1—K_p values for the synthesis of ammonia at different temperatures

T°C	K_p *atm.*$^{-2}$	*T°C*	K_p *atm.*$^{-2}$	*Observations*
200	6.60×10^{-1}	600	1.51×10^{-3}	It is seen that, at 1000°C the equilibrium constant
300	7.00×10^{-2}	700	9.50×10^{-4}	K_p on which the values of the synthetic yields
400	1.38×10^{-2}	800	3.6×10^{-3}	depend is reduced to about 2 ten thousandths
500	4.00×10^{-3}	1000	1.36×10^{-4}	($1.36 \times 10^{-4}/6.6 \times 10^{-1}$) of the value which it had at 200°C

and to recover the mole fraction of ammonia from this equation:

$$x_{NH_3} = x_{N_2}^{1/2} \cdot x_{H_2}^{3/2} \cdot P \cdot K_p^{1/2}.$$

The formulae in which partial pressures appear are strictly valid, as has been noted, for ideal gases and approximately valid for real gases at low pressures. Since the behaviour of H_2, N_2, and NH_3 at high pressures differs significantly from that of ideal gases, such formulae cannot constitute a useful basis for meaningful quantitative calculations which would allow one to find the percentages of the various gases and of ammonia in particular at equilibrium.

The pressures, temperatures, and percentages of ammonia at equilibrium

The behaviour of the components concerned in the ammonia synthesis equilibrium as real gases means that the following quantities must be measured experimentally:

- the percentage of NH_3 present at equilibrium as a function of the temperature at different fixed pressures;
- the percentage of NH_3 at equilibrium as a function of the pressure at various fixed temperatures.

Values obtained algorithmically by the application of physicochemical formulae for real gases agree with the values found in this manner.

Table 8.2 and graph(a) of Fig. 8.2 refer to determinations of the first type while graph(b) of Fig. 8.2 refers to the second kind of determinations. In both cases the percentage of NH_3 at equilibrium is that which has been established, starting out from mixtures of nitrogen and hydrogen in the stoichiometric ratio.

Results such as those shown in Table 8.2 and Fig. 8.2 can also be obtained from a calculation in which the values of the pressures which the H_2, N_2, and NH_3 would have if they were ideal gases (the fugacities of the gases) are substituted into the ideal gas formulae. The fugacities of the gases at fixed temperatures are obtained by multiplying the experimental pressures by the tabulated 'fugacity coefficients' (numbers which are more or less close to unity).

Table 8.2—Experimental values for the yield of NH_3 as a function of the pressure at fixed temperatures

Temperature	*% ammonia at equilibrium*					
°C	10 *atm.*	50 *atm.*	100 *atm.*	300 *atm.*	600 *atm.*	1000 *atm.*
200	50.66	74.38	81.54	89.94	95.37	98.83
300	14.73	39.41	52.04	70.96	84.21	92.55
400	3.85	15.27	25.12	47.00	65.20	79.82
500	1.21	5.56	10.61	26.44	42.15	57.47
600	0.49	2.26	4.52	13.77	23.10	31.43
700	0.23	1.05	2.18	7.28	12.60	12.87

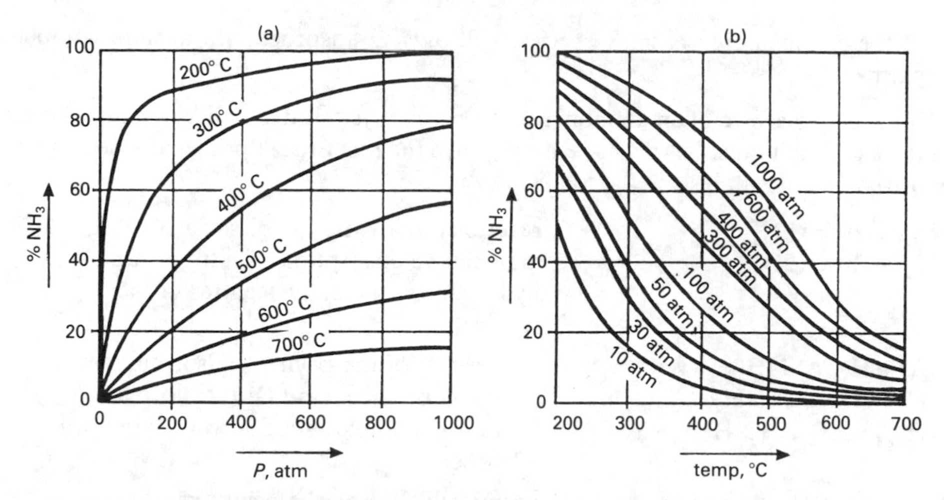

Fig. 8.2—(Left). Percentage yields of ammonia as a function of the pressure at fixed temperatures. (Right). Yields of ammonia as a function of the temperature at fixed pressures. The inverse effect exercised by the two physical variables on the yields in the gas phase synthesis of ammonia from nitrogen and hydrogen is evident.

Catalysis in the synthesis of ammonia

Premises regarding catalysis and the mechanism of the action of catalysts

A distinct advantage from the point of view of the reaction kinetics is also gained in the case of the synthesis of ammonia by exploiting the action of an efficient positive catalyst.

As has already been pointed out in connection with the ability of the catalyst to lower the energy of activation of the nitrogen molecule, the 'efficiency' in this case is primarily achieved via the dissociative adsorption of this molecule.

However, the adsorption of nitrogen on supports which permit the formation of nitrides does not, per se, mean that such supports will act as catalysts because the nitrides which are formed are extremely stable (such as TiN, for example), and a lowering of the energy of activation of the nitrogen molecule presents an obstacle to the subsequent processes which lead to ammonia.

Moreover, even the formation of labile nitrides has not been shown to lead to satisfactory catalytic effects in the synthesis of ammonia because the problem also exists as to how to lower the energy of activation of hydrogen with which the nitrogen must combine.

Today, firstly by deduction from experimental facts and subsequently by the use of sophisticated analytical techniques (such as mass spectrometry), much light has been thrown on the effective mechanisms of the catalysis appertaining to the synthesis of ammonia, and it has been concluded that, besides the principal mechanisms, collateral mechanisms must also be taken into account. The first of these mechanisms, which will be pointed out here, involves:

- reactions involving the dissociative adsorption of nitrogen and hydrogen on the active centres of the catalyst;
- the conversion to ammonia of the adsorption products by the gradual transformation

of these into more complex species, still always adsorbed, via a more complex pathway.

If the active centre of an ammonia synthesis catalyst* is denoted by *vc*, on account of the fact that it must be vacant ('empty'), then the first order ('principal') mechanism involved in this catalysis is the following:

$N_{2(g)} + 2vc \rightleftharpoons 2N_{ads}$ $H_{2(g)} + 2vc \rightleftharpoons 2H_{ads}$	reactions involving the dissociative dissociative adsorption of nitrogen and hydrogen on the active centres of the catalyst
$N_{ads} + H_{ads} \rightleftharpoons NH_{ads} + vc$ $NH_{ads} + H_{ads} \rightleftharpoons NH_{2ads} + vc$ $NH_{2ads} + H_{ads} \rightleftharpoons NH_{3ads} + vc$	conversion of the primary adsorption products by their gradual transformation into more complex species which are always adsorbed
$NH_{3ads} \rightleftharpoons NH_{3(g)} + vc$	chemically and mechanically induced desorption

The concept that a catalyst acts by subdividing a complex reaction into a number of partial reactions which are characterized by relatively low energies of activation is fully confirmed within the framework of this mechanism.

The constitution of the catalyst in the operational state and as initially prepared

Among all the metals of the first transition series which have been tested it is iron which, when suitably protected and activated, yields the best results.

The structure of the catalyst when it is in action is rather complex and involves:

- iron as a catalytic 'promoter' in so far as it is able to give some impulse to the reaction and to drive the reaction forward in accordance with the catalytic mechanisms illustrated above;
- principally†, aluminium oxide (Al_2O_3) as a 'protector' in so far, as we shall see, that it opposes any ready and rapid fall-off in the catalytic properties of the iron;
- oxides of a distinctly basic character (K_2O and CaO) as activators especially as they prevent the catalyst from becoming entangled.

To provide some idea of the increase in efficiency derived from the addition of protectors and activators, it may be stated that, under fixed operational conditions, the equilibrium percentage of ammonia was 5.48% in their absence, whereas it was 13.85% under the same conditions when protectors and activators were used.

At the start the catalysis is put into operation for the most part‡ as a product of an initial fusion and subsequent trituration and sieving, to achieve a uniform grain size, of mixtures of magnetite, protectors, and activators.

---

*More precisely, an active centre is identified here with a semi-occupied d-type orbital with a certain elasticity which permits it to interact in a 'dangling' manner with the orbitals (the π-orbitals of nitrogen) of the chemisorbed ('adsorbed') molecule.

†SiO_2, Cr_2O_3, ZrO_2 as well as other oxides are often added in small amounts as coadjuvants with Al_2O_3 in the protective action.

‡It is, in fact, also possible to prepare the catalyst in the form of a cermet (page 584) derived by sintering of the protecting and activating oxides ('ceramics') and iron (the metal).

The catalyst is then converted from its initial form into that required in the actual process, by putting the catalyst, which has been prepared as described above, into the synthesis reactor and reducing it with a current of hydrogen or, better still, with the mixture of $N_2 + 3H_2$ used in the synthesis.

The only possible reduction is that involving the reaction:

$$Fe_3O_{4(s)} + 4H_{2(g)} \rightleftharpoons 3Fe_{(s)} + 4H_2O_{(g)}$$

since the other oxides are not reduced at all under the conditions appertaining in the reactor.

The water which is formed during the course of this reduction is rapidly and completely removed in order to prevent any reoxidation of the newly reduced iron which would lead to structural modifications which are harmful as regards its catalytic activity. High flow rates of unpressurized, preheated reducing mixtures of N_2 and H_2 are employed for this purpose.

This reduction can also be performed outside the apparatus used in the ammonia synthesis (a 'prereduced' catalyst): just before putting it into the apparatus taking care to avoid all contact with air or following the reduction with stabilization. The latter consists of bringing the prereduced catalyst, at a low temperature and pressure, into contact with an atmosphere of nitrogen or nitrogen and hydrogen in the presence of traces of oxygen. The reduced iron thereby becomes superficially covered with a thin layer of iron oxide which passivates it. This superficial oxide layer is readily removed at the beginning of the synthesis in the reaction vessel.

The physical structure of the catalyst and the specific role of protectors and activators

When the catalyst used in the synthesis of ammonia is reduced, metallic iron is freed in it in the form of minute crystals ('crystallites') of α-iron. The crystallite structure which is formed in this manner involves barely no decrease in the volume of the catalyst pellets which, as the oxygen has been removed, turn out to be extremely porous. With the passage of time, and even more so at high temperatures, the structure tends to become compacted by the transformation of the crystallites into crystals of α-iron. This leads to a loss in catalytic activity.

The addition of protectors brings about, on the one hand, a reduction in the size of the crystallites which are formed as a result of the reduction of the catalyst and, on the other hand, it leads to a structure which is predominantly amorphous.

The reason for the occurrence of an 'impediment to crystallization' when Al_2O_3 (and similar other minor protectors) are present lies in the fact that, when fused, this oxide forms the compound $FeAl_2O_4$ ('hercynite') which is isomorphous with $FeFe_2O_4$ ('magnetite') and therefore forms mixed crystals with it in the form of solid solutions. It therefore does this with the principal component in the mixture from which the ammonia synthesis catalyst will be prepared. At the site of reduction it is the magnetite which is reduced, while the hercynite is not. The resulting mass is therefore completely permeated by the $FeAl_2O_4$ structure which prevents the pure iron produced from the magnetite from crystallizing.*

---

*It is known that the isomorphism (which comes into play in the case of the very well known series of spinels) exhibited by these compounds follows, from the numerical identity of the atoms with similar ionic radii in the structure.

In other terms, the most up-to-date views on the role of ammonia synthesis catalyst protectors, which have been refined within the framework of the 'theory of paracrystallinity' and confirmed by the most recent and sophisticated systems for the analysis of the solid state, suggest that the insertion of such protectors (and, especially, Al_2O_3) induces defect structures and distortions in the catalyst which, by largely preventing the formation of crystallites, hinder the transformation of those products into crystals and bring about the liberation of pure amorphous iron. The protectors maintain the structure in which they are included in a porous state and therefore a catalytically active state.

Briefly, the theory of paracrystallinity states that, when the parameters of a crystal lattice are not constant as in the case when there are alternate iron atoms and $FeAl_2O_4$ molecules, normal crystallization cannot occur. Instead, structured zones consisting of small crystals (paracrystals) are formed, and these are interspersed with highly amorphous zones.

As far as the role played by the *activators* is concerned, it should be noted that the experimental evidence all points toward the fact that they principally act in facilitating the desorption of the synthetic product from the iron. In other words, the activators preside over the elimination of the slow step in the synthesis of ammonia on a catalyst which is chemically represented as the final desorption stage in the catalytic mechanism illustrated on page 262. Mechanical desorption, on the other hand, which also prevents any hold-up in the kinetics of the synthesis due to 'obstruction of the catalyst', takes place on account of the stream of gas which is continuously drawn through the system.

The 'space velocity' factor in the synthesis of ammonia

The volume of gas, measured under standard conditions, which circulates through a unit volume of the catalyst per unit of time ($Nl/h \cdot l_{catalyst}$) is referred to as the 'space velocity'. It is obvious that this variable can affect a gas phase synthesis since it determines the extent and time of contact between the gas and the catalyst by either favouring or disfavouring the adsorption/reaction phenomena involved in the process.

Increasing the space velocity decreases the concentration of ammonia in the gases emerging from the reactor, as can be seen from the diagram on the left of Fig. 8.3.

With such an increase, however, the amount of ammonia produced per unit time increases (the right-hand diagram in Fig. 8.3) because, at low degrees of conversion, the rate of the inverse reaction (the dissociation of ammonia) is very low as the net rate of the synthetic reaction is very large.

It is therefore concluded that, if the space velocity is increased, the gas must be recycled a number of times because of the low degree of conversion which takes place during each pass, while benefit is derived from the increased production of ammonia during each unit of time, that is, the maximum degree of conversion of the gases into ammonia per unit of time.

The operating costs grow as the space velocity is increased on account of the removal of heat, greater losses of the feed stock, and the extra work required in the compression of the gases.

In industrial practice, the system is operated in such a way as to obtain the maximum degree of conversion of the gases which is economically convenient. This generally means the adoption of space velocities of 20 000–40 000 $Nl/h \cdot l_{catalyst}$ with

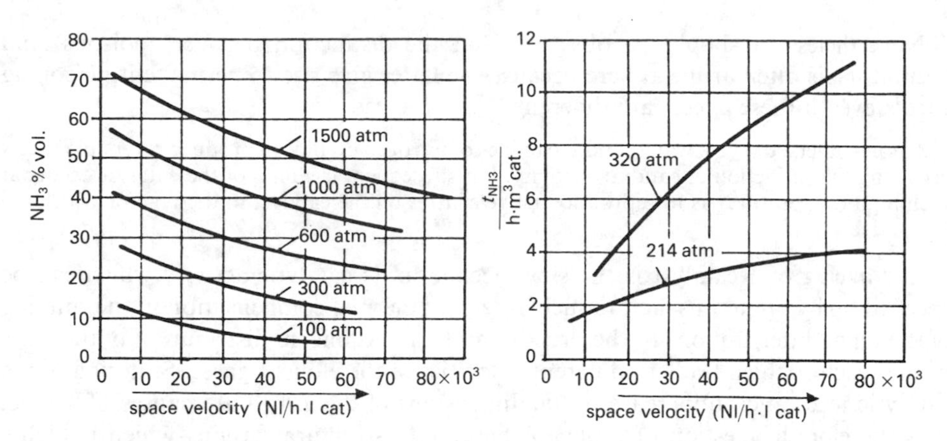

Fig. 8.3—(Left). Percentage of NH_3 in the gaseous mixture as a function of the space velocity at various pressures. (Right). Rate of production of ammonia as a function of the space velocity at different pressures.

ammonia concentrations in the gases emerging from the reactors of 10–40% as the pressure is varied from 100 to 1000 atm.

Poisons, inhibitors, and the ageing of the catalyst

The catalyst employed in the synthesis of ammonia tends to become inactivated in diverse ways for several reasons.

The principal poisons, that is, substances which, by combining with its active centres to form stable species, permanently impair the action of such a catalyst, are:

- sulphur, arsenic, phosphorus, chlorine, and their compounds (As_2O_3 and COS, for example) which can be reduced, under the conditions employed in the synthesis, to the elements or hydrides;
- hydrogen sulphide, which is extremely common and powerful, because, besides alloying with iron, it also combines with the basic oxides*;
- carbon monoxide, which tends to be transformed into methane:

$$CO + 3H_2 \rightarrow CH_4 + H_2O + 49.3 \text{ kcal/mole}$$

  with strong local heating which leads to sintering of the catalyst;
- unsaturated hydrocarbons which release heat when hydrogenated and likewise cause sintering;
  - traces of lubricants which may undergo cracking on the catalyst with the formation of carbon deposits ('poisoning by covering') and unsaturated hydrocarbons.

The most common inhibitors or 'temporary poisons' of the catalyst used in ammonia synthesis are traces of oxygen which form superficial layers of oxide which can be dissolved upon reduction with the elimination of the inhibitor. Water vapour and nitric oxide act in a similar manner to traces of oxygen.

*At 500°C, 10 ppm of H_2S is enough to rapidly poison a catalyst.

Nevertheless, it should be observed that the distinction between 'poisons' and 'inhibitors' is often artificial here because the latter give rise to permanent poisoning effects even if these effects are different.

Among others, oxygen, water, and the oxides of nitrogen do indeed produce reversible effects involving the extinction of catalytic activity, but subsequent cleaning of the catalyst occurs at such high temperatures as to reproduce active centres on the catalyst in the reactor which are less active.

The ageing of a catalyst is the slow decline in its catalytic activity which can be speeded up by numerous factors such as lack of thermal compensation in the running of the plant, interruptions to the process, and a non-uniform distribution of the mass in the reactor which leads to different velocities of the effluent gases. Sintering of the catalytic mass and an increase in the dimensions of the α-iron are causes of ageing. It is therefore a question of a phenomenon with structural origins which is highly conditioned by the manner in which the plant is managed and which progressively accelerates once it has set in as the number of crystals decreases, at constant mass, owing to a lowering of the surface tension.

Regeneration and replacement of the catalyst. Recycling to methanation operations

Catalysts which have been deactivated because of crystallization and sintering phenomena, no matter how caused (by ageing, the action of CO, etc.), can be regenerated. In principle, this regeneration consists of the re-oxidation of the iron to magnetite with water vapour:

$$3Fe + 4H_2O \xrightarrow{800_{\circ}C} Fe_3O_4 + 4H_2$$

and then re-fusing the mass in an electric furnace with the re-integration of protectors and activators.

Catalysts which have been permanently poisoned by sulphur, arsenic, etc. are not conveniently regenerated and must therefore be completely and totally replaced.

Nowdays, however, exhausted catalysts are usually sent for use in accelerating the reaction involving the conversion of methane into CO and CO_2 (methanation, as we shall subsequently call it).

Production of the $N_2 + 3H_2$ synthesis mixture

In principle, the mixture employed in the synthesis of ammonia can be produced by preparing the two gases separately and then mixing them in the correct proportions. However, this practice is not common in industry. Instead, the following methods for the preparation of this mixture are more practical:

- from coal, air, and water;
- from hydrocarbons, water, and air;
- from hydrocarbons, oxygen, and the addition of nitrogen produced by the liquefaction and subsequent fractionation of air.

In the first two cases 'raw synthesis gases', that is, the mixture $N_2 + 3H_2$ together with gases which interfere in the synthesis of ammonia (CO_2, CO, and traces of O_2), are produced in the first place, and these are converted to 'pure synthesis gases' which consist solely of nitrogen and hydrogen in stoichiometric proportions*. In contrast, pure synthesis gases are produced in the third case.

The subsequent treatment will be carried out according to the scheme:

$N_2 + 3H_2$ mixture
- from coal or from hydrocarbons with air and water
  - by preparation of the respective raw synthesis gases
  - by conversion to pure synthesis gases
- from hydrocarbons and products of the fractionation of air

The preparation of crude synthesis gas from coal

Interest in the gasification of coal for the production of gases containing the raw materials for the synthesis of ammonia progressively fell off after the Second World War, but is being sharply revived today owing to its competitiveness with the production of synthesis gas from hydrocarbons.

From a mixed gas gazogene, by mainly storing the products of cold blowing, it is possible to obtain a gas with the following percentage composition by volume:

CO_2 7%, CO 33%, H_2 37%, N_2 22%, other gases

(CH_4 and noble gases) 1%,

that is, with a H_2/N_2 ratio of 1/0.6.

If the CO contained in this mixture is 'converted' into CO_2 and hydrogen in a manner which will be explained later (page 271), by exploiting the equilibrium reaction:

$$CO + H_2O \rightleftharpoons CO_2 + H_2 \text{ 'water gas shift reaction'}$$

the mixture for the synthesis of ammonia, accompanied by other gases, is obtained.

Allowing for the fact that, since the conversion reaction is an equilibrium reaction, 2–4% of the CO is not transformed, the amount of hydrogen increases to about 67% by volume, and, since the amount of nitrogen does not change, the N_2/H_2 ratio is now approximately $22/67 = 1/3$ as a result of the conversion.

The preparation of crude synthesis gas from hydrocarbons

In this case, suitably desulphurized natural gas is generally the starting material which, for simplicity, will be identified here with methane.

'Steam reforming' or a 'partial oxidation' can be carried out. The first of these processes necessitates external heating of the reactor, while the second furnishes its own heat:

Steam reforming uses the reactions:

$$CH_4 + H_2O \rightleftharpoons CO + 3H_2 - 49.27 \text{ kcal}$$

$$CO + H_2O \rightleftharpoons CO_2 + H_2 + 9.84 \text{ kcal}$$

*The pure synthesis gases' always also contain small quantities of noble gases.

which emphasizes the fact that hydrocarbons are far better raw materials for the production of hydrogen than coal.

As has been stated, the mixture of gas which is produced becomes crude synthesis gas during a second stage when the residual methane is burnt with air which allows nitrogen into the mixture.

The main reactor consists of a tube furnace with a high heat resistance* which is packed with an alumina-supported nickel catalyst. This catalyst is readily poisoned by sulphur and sulphur compounds, and these must be completely eliminated as a preventative measure.

The process follows the path depicted schematically and interpreted symbolically in Fig. 8.4 and described in the following text.

The steam produced in CV and the methane to be reformed are preheated in E_1 to 500°C and then sent into the tubes of the furnace H_1 which are heated by the methane burner C. The fumes from the combustion are discharged from the chimney E_2 after giving up a part of their heat to the steam boiler CV and part of it to the combustion air in E_2.

The gases emerging from the reforming tubes still contain 2–3% of methane which is burnt with air in the furnace H_2. By this means, nitrogen also becomes part of the system.

After cooling to about 400°C by contact with a spray of water, the greater part of the CO in the gaseous mixture undergoes conversion to CO_2 in the converter H_3. As a result the temperature rises again to about 550°C and the gas can be used in E_1 to preheat further amounts of the mixture which are on their way to be reformed.

The final gas must have the following approximate composition:

$$H_2\ 60\%,\ N_2\ 20\%,\ CO_2\ 16\%,\ CO\ 3\%,\ CH_4 + \text{noble gases}\ 1\%$$

The plant is expensive to install, the tubes do not have a very long life, and the catalyst is extremely sensitive to poisoning which is mainly caused by sulphur.

The autothermal process is structurally more economic and less subject to wear and tear. It is based on the partial oxidation of methane:

$$CH_4 + 1/2O_2 \rightarrow CO + 2H_2 + 8.53\ \text{kcal} \tag{8.2}$$

which takes place in a muffle furnace at temperatures above 1000°C.

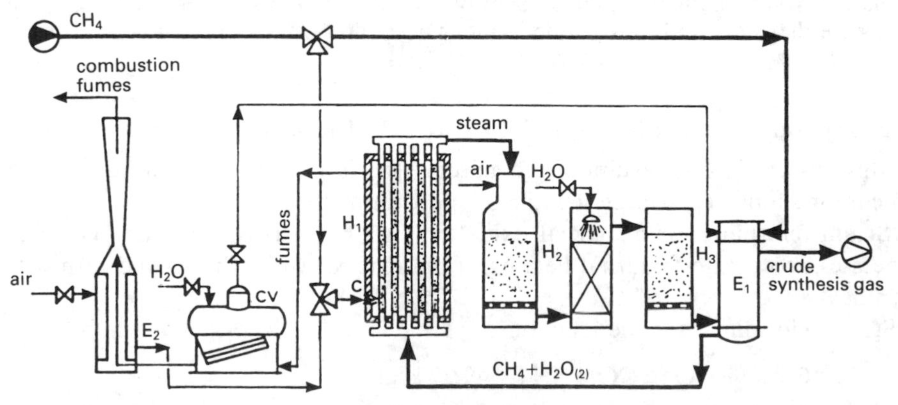

Fig. 8.4—The production of crude ammonia synthesis gas by the steam reforming of natural gas (methane).

*Tubes made of steel containing 25% Cr and 15–20% Ni are used.

To prevent the pyrolysis of methane, which would lead to the intense generation of carbon black which by blocking the catalyst (nickel supported on magnesium oxide), would rapidly reduce its activity, some water is added to the methane–air–oxygen mixture. As a result, partial oxidation takes place together with reactions characteristic of the steam reforming process.

The plant (Fig. 8.5) contains two columns C_1 and C_2 in which methane and air enriched with oxygen respectively are saturated with water vapour. The two mixtures are then preheated in the heat exchangers E_1 and E_2 before passing to the reactor/furnace H_1. The conversion reaction takes place in H_1 to such an extent that there is not more than 0.2% of methane in the emerging gases. These gases are first used to produce steam in the boiler CV, then used for the preheating of E_1 and E_2, and afterwards pass through the line filter PF where any particles of uncombusted carbon or of the catalyst support which may have been entrained by the gas are deposited. They are finally sent to the section for the conversion of CO.

The air employed in the process, which is added according to a régime, must be such that, after the conversion of the CO which produces further hydrogen, the H_2/N_2 ratio is 3/1.

The transformation of crude synthesis gases to pure synthesis gases

The gases which are finally obtained from coal or from hydrocarbons by means of reforming or by the autothermal process have some common characteristics. In fact:

- all of them contain H_2 and N_2 in the proportions of 3/1 by volume;
- they contain various amounts of the same gases which interfere with the synthesis of ammonia, that is, CO and CO_2;
- they are more or less moist.

Common treatments must therefore be carried out on such gases to recover pure synthesis gas from them, that is a gaseous mixture which solely consists of $N_2 + 3H_2$.

The gas to be converted itself reduces Fe_2O_3 to Fe_3O_4, leaving the Cr_2O_3 unaltered in an extremely highly subdivided form which is suitable for the prevention of any enlargement in the crystalline grain size of the Fe_3O_4. This type of conversion which

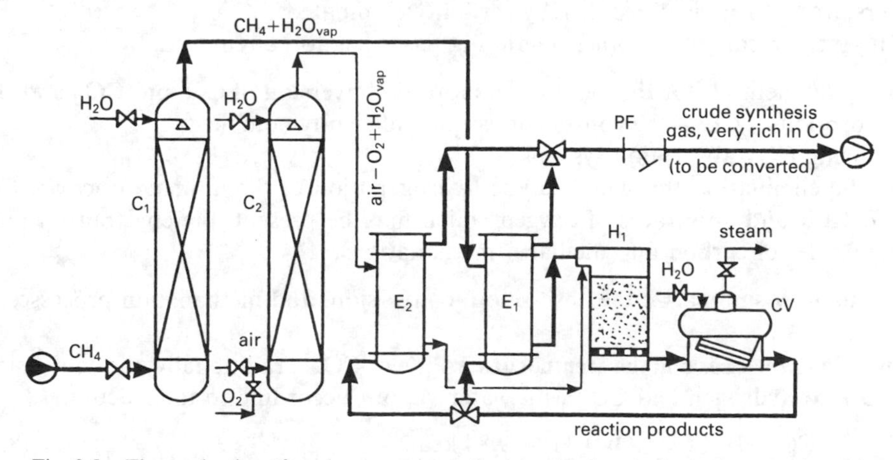

Fig. 8.5—The production of crude ammonia synthesis gas by the autothermal process involving the partial oxidation of natural gas (methane).

is carried out over a wide range of temperatures from 320 to 540°C does not lead, usually, to residual amounts of CO, in the gas treated, below 2.5%.

Consequently, the carbon monoxide has to be absorbed by a method which, while much in vogue, is used less and less nowadays owing to the new conversion techniques which are available.

The absorption process, which is carried out under pressure, uses solutions of cuprammonium compounds made up from the cuprous and cupric salts of formic acid, acetic acid, and carbonic acids complexed with ammonia.

The reactions are of the type:

$$\underset{\text{Cu(I)diamminoformate}}{NH_3 + Cu(NH_3)_2(OOCH) + CO} \underset{\text{low pressure and relatively high temperatures}}{\overset{\text{pressure and low temperatures}}{\rightleftharpoons}} \underset{\text{Cu(I)triamminocarbonyl formate}}{Cu(NH_3)_3(CO)(OOCH)}$$

Apart from the high costs of the reagents and of their regeneration, the traces of copper which are entrained by the gas in this process permanently poison the catalyst used in the synthesis of ammonia. It is therefore understandable why this method has been in continuous decline.

The new conversion methods, which try to avoid the absorption, make use of ternary $ZnO–CuO–Al_2O_3$ catalysts, activated at between 160 and 260°C which leave residual amounts of CO of the order of 0.1–0.2% in the gas after conversion.

By combining beds of traditional $Fe_3O_4 + Cr_2O_3$ catalysts, which are briefly referred to as H.T.S. (high temperature shifting) with beds containing the new class of catalysts L.T.S. (low temperature shifting) and operating under pressure which promotes an increase in catalytic activity, degrees of CO conversion of 99.95% can be attained.

Firstly, it is to be noted that, in any case, the moisture does not constitute a problem: both because by always working under pressure, there is pronounced condensation of the vapours, and also because, in all the synthetic processes, the reaction mixture is cooled with liquid ammonia before being sent into the synthesis towers, and any moisture is thereby certainly eliminated.

The greater part of the other interfering gases has to be removed:

- by using methods for the more or less forced conversion of CO into CO_2 and the absorption of CO if the conversion is not sufficiently complete;
- by using CO_2 absorption systems;
- by the elimination through catalytic hydrogenation of the ultimate traces of CO and CO_2 and any traces of oxygen which may be present, thereby transforming the oxides of carbon into methane (methanation).

Let us now summarize the absorption, conversion, and methanation processes.

Conversion processes and the eventual absorption of CO. Traditionally, the conversion of CO into hydrogen and CO_2 with water vapour according to the reaction:

$$CO + H_2O \rightleftharpoons CO_2 + H_2 + 9.84\ \text{kcal} \qquad (8.3)$$

has been carried out by using Fe_2O_3 catalysts containing some Cr_2O_3.

Carrying out the reaction under pressure, on the other hand, is not a serious economic disadvantage in as much as pressure must be applied subsequently, that is, both in the absorption of CO_2 in water as well as in the later operational phases to which the synthesis gas is subjected.

The diagrams of the plants shown in Figs 8.4 and 8.5 simply refer to the traditional single bed process for the conversion of CO.

Instead of this, modern reactors have three beds or layers of catalyst (Fig. 8.8.6). The first (which is located high in the reactor when the gas follows a path from the top to the bottom) holds the H.T.S. catalyst, while the other two beds consist of the L.T.S. catalyst.

As the reaction has to be carried out in the presence of an excess (150–200%) of steam, water is sprayed in between the three beds to regulate the temperature, especially between the first and second beds as not only must the heat released due to the exothermic nature of the reaction be removed, but·it must also be ensured that the temperature is brought down from that required for the H.T.S. to the appropriate conditions for the L.T.S. Alternatively, the plant can be appropriately fitted with efficient heat exchangers.

The overall process is known as methanation, and corresponds to the reactions:

$$CO + 3H_2 \rightarrow CH_4 + H_2O + 49.27 \text{ kcal} \quad (8.4)$$

$$CO_2 + 4H_2 \rightarrow CH_4 + 2H_2O + 37.53 \text{ kcal} \quad (8.5)$$

$$1/2O_2 + H_2 \rightarrow H_2O + 54.5 \text{ kcal}$$

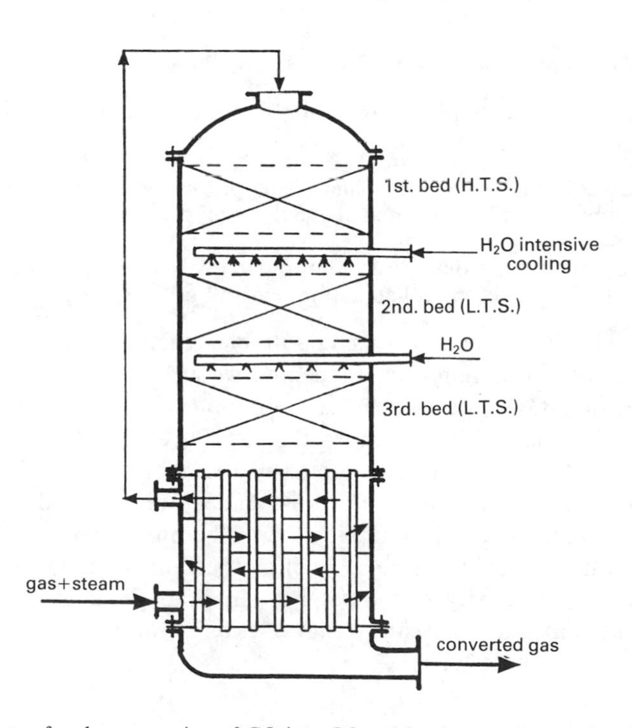

Fig. 8.6—Reactor for the conversion of CO into CO_2 with a heat exchanger for the creation of the conditions required for H.T.S.

Traces of oxygen are always present in the crude ammonia synthesis gases. In their turn, the CO_2 and CO, at the point where the methanation occurs, must be present in the gases being treated in such amounts that the quantity of methane produced does not exceed 1% since, otherwise, it has a deleterious effect on the synthesis of ammonia.

For economic reasons methanation may be carried out by using the exhausted catalysts from the ammonia synthesis plant (page 266), but the best results are obtained by using catalysts, moulded into pastilles, based on nickel oxide (7–8 parts) and aluminium oxide (2–3 parts*).

Traditionally, when spent (old or exhausted) catalysts are used for methanation in small towers which precede the tower in which the catalytic synthesis of ammonia is carried out, one speaks of the use of 'precatalyst' towers. These operate at a temperature of 350°C which is maintained by electrical resistance heating.

The absorption of CO_2. This requires a multistage process (Fig. 8.7) because it is always necessary first to absorb the greater part of the CO_2 from it by washing the gas with water under pressure, using the equilibrium process:

$$CO_2 + H_2O \rightleftharpoons H^+ + HCO_3^-$$

and then to remove the remainder of the CO_2 either by absorbing it in aqueous solutions of basic compounds, such as K_2CO_3 which involves the reaction:

$$CO_3^{2-} + CO_2 + H_2O \rightarrow 2HCO_3^-$$

or by absorption using solutions of ethanolamines which react with CO_2 according to the scheme:

$$2HO\text{–}R\text{–}NH_2 + CO + H_2O \rightarrow (HO\text{–}R\text{–}NH_3)_2CO_3$$

which leads to an extremely high degree of absorption of the CO_2 if organic solvents are also added.

For instance, the alkanolamine–sulpholane–water process, which was perfected by the Shell petrochemical company and is known as the 'Sulphinol' process, is very commonly used for this purpose.

Absorptions are also carried out by using only organic solvents: with methanol (the 'Rectisol' process) and with N-methyl-2-pyrrolidone (the 'Purisol' process), etc.

Methanation. The hydrogenation, using suitable catalysts, of the CO, CO_2, and oxygen leads to the formation of methane and water.

Since the methanation process includes exothermic reactions, the reaction system is maintained at the relatively low temperature of 300°C. For this purpose it is only necessary to send the gases into the reactor at a temperature of 230–270°C.

Finally, the need to carry out the process under pressure is dictated by the fact that the reactions take place with a reduction in the number of moles. However, the pressure values used should not be so high as to be uneconomical and should still be favourable to the activity of the catalyst. In practice, a pressure of 60 atm. is applied, but with maxima which can attain values of 230 atm. and minimal operational values of about 30 atmospheres.

---

*It should be noted, however, that the 'iron' catalysts tolerate the presence of small amounts of sulphur compounds in the gas undergoing methanation, while no such tolerance is exhibited by nickel catalysts.

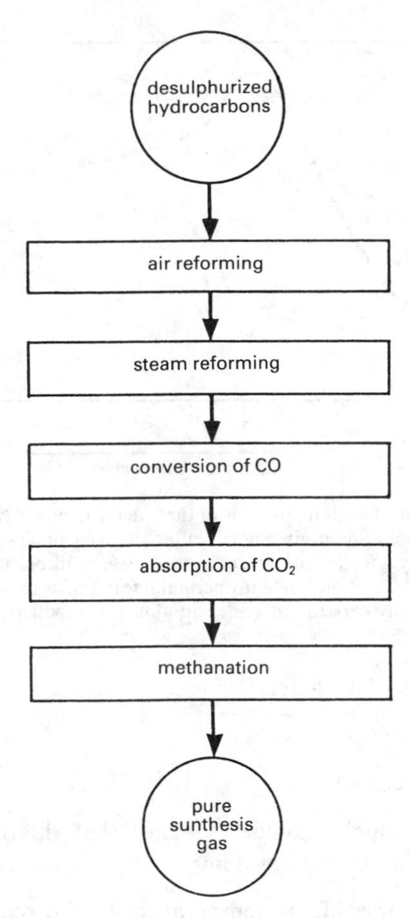

Fig. 8.7—Block diagram linking the phases in the preparation of crude synthesis gas by the autothermal process and of pure synthesis gas by means of conversions and absorptions.

Preparation of the pure $N_2 + 3H_2$ mixture from hydrocarbons and from the products of the fractionation of liquid air

When natural gases (we shall refer to methane for simplicity here) or light hydrocarbon fractions are oxidized by using the oxygen obtained from the distillation of liquid air on a catalytic bed consisting of nickel supported by a complex ternary mixture of refractory oxides, TiO_2–CaO–Al_2O_3, the following reaction occurs during the first stage:

$$CH_4 + 2O_2 \rightarrow CO_2 + 2H_2O + 210\ \text{kcal} \tag{8.6}$$

On account of the heat which is released, the temperature rises sharply to a point where the cracking of the methane tends to occur:

$$CH_4 \rightarrow C + 2H_2 \tag{8.7}$$

The following endothermic reactions then take place:

$$CH_4 + H_2O \rightleftharpoons CO + 3H_2 \tag{8.8}$$

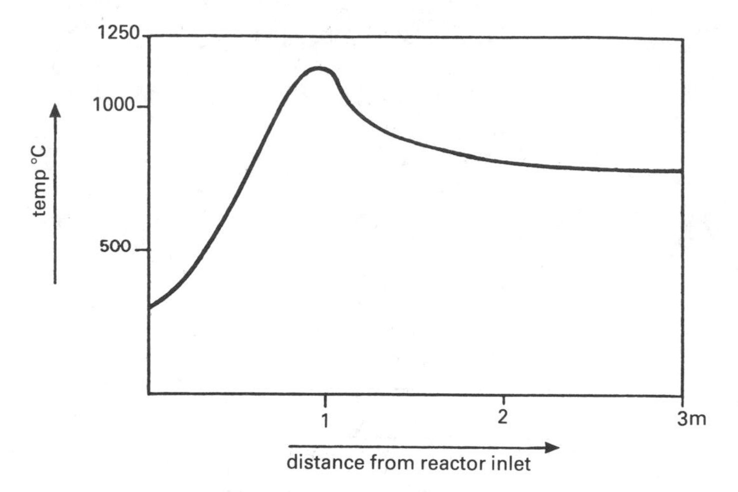

Fig. 8.8—Temperature profile along the axis of the reactor in which hydrocarbons are oxidized with oxygen obtained from liquid air. Shortly after the reagents have entered the reactor the temperature rises sharply and then decreases more slowly until becoming constant toward the outlet from the reactor. As a whole, an 'autothermal reforming' is realized in this manner, which is easily carried out under an almost adiabatic régime.

$$CH_4 + CO_2 \rightleftharpoons 2CO + 2H_2 \quad (8.9)$$

$$C + H_2O \rightleftharpoons CO + H_2 \quad (8.10)$$

$$C + CO_2 \rightleftharpoons 2CO \quad (8.11)$$

There are two facts which provide evidence that this is the mechanism of the chemical evolution of the reaction system:

- the (experimental) course of the temperature in the reactor, which is shown in Fig. 8.8
- the average composition of the gas which is obtained from the process: H_2 60.9%, CO 34.5%, CO_2 2.8%, CH_4 0.4%, $N_2 + Ar$ 1.4%.

The methane percentage of 0.4% present in the gases emerging from the reactor is relatively high, but it can be lowered by reducing the amount of carbon black which is liberated. This can be achieved by the addition of steam to the reagents which also prevents the composition of the mixture from falling within the explosive limits, which is a phenomenon favoured by the catalytic environment.

Finally, it should be observed that it is carried out under pressure, even if this reduces the yields somewhat, owing to the increases in volume resulting from the reactions. The reasons which are decisive in the selection of such conditions are: the enhanced activity of the catalysts when they are under pressure, the availability of raw materials which are already pressurized (natural gases and refinery gases), and the automatic increase in this pressure (a doubling or tripling of it) as the reaction takes place. All of these factors act successively to minimize the energy costs of the compression of the gas.

Instead of using natural gases and light hydrocarbon fractions, it is possible to obtain products which are qualitatively similar to those mentioned above by starting out from heavy liquid hydrocarbons and putting them 'to the flame' at a temperature of 1200–1400°C. From a quantitative point of view, the percentage of hydrogen in the effluent gas from this process is lower (46–52%), while the amounts of CO (42–47%) and CO_2 (4.4–4.9%) are increased. However, in this case no carbon black is formed even when the process is carried out without any water vapour.

Having obtained the abovementioned gaseous mixture, the customary treatments for the conversion of CO and the absorption of CO_2, and the simultaneous absorption of H_2S and COS if hydrocarbons originating from naphtha, heavy oils, etc.* were employed as raw materials, are carried out on it. Methanation, however, is then avoided, and it is replaced by washing with liquid nitrogen which is available together with the oxygen with which the main oxidation has been carried out.

For this purpose the gas is washed in a suitable column (built to be resistant to low temperatures) with a stream of liquid nitrogen which causes the CO, O_2, CH_4, and CO_2 to condense out, thereby leaving H_2 containing N_2 at the top of the column.

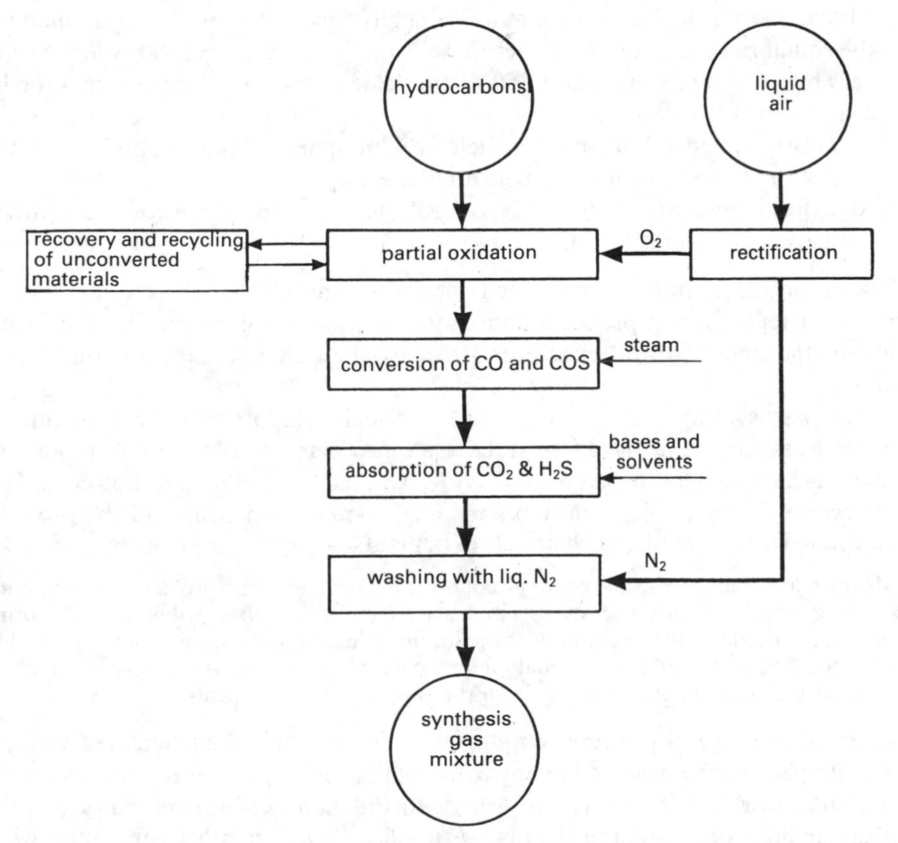

Fig. 8.9—Block diagram of the route for the production of ammonia synthesis gas mixture from hydrocarbons (for the great part, heavy hydrocarbons) and oxygen and nitrogen obtained by the fractionation of liquid air.

*With regard to these sulphur compounds, it may be noted that COS is hydrolyzed in a pronounced manner under the conditions used in the conversion of CO: $COS + H_2O \rightleftharpoons CO_2 + H_2S$, and that H_2S is absorbed together with the CO_2.

The gas washed with liquid nitrogen is so pure as to require no purging of inert substances in the synthetic cycle. In fact, the small amounts of inert substances remaining after the washing do not accumulate because they are present in an amount which approximately corresponds to that which the liquid ammonia can carry away.

In this case, the last operation which has to be carried out, before compression and the routeing of the nitrogen/hydrogen mixture to the synthetic stage, is the integration of the appropriate amount of nitrogen into the mixture in order to attain the correct ratio required for the synthesis.

The operations leading to the mixture for the synthesis of NH_3 from those which have just been discussed are diagrammatically shown in Fig. 8.9.

Plant conditions for the synthesis: determining factors

Thermodynamic and kinetic calculations regarding the equilibrium constant for the synthesis of ammonia have allowed the following facts to be ascertained:

(1) Above 550°C, the yields of ammonia determined from the value assumed by the equilibrium constant K_p under these conditions are unacceptably low, even at very high pressures, and, below 350°C, the velocity of the reaction becomes too low even when it is catalyzed.
(2) At pressures under 140 atm., the yields of ammonia are practically inadmissible even if the lowest possible temperatures are used.
(3) At a fixed pressure, the most favourable temperature range for the synthesis lies between 370 and 530 °C.

The criteria regarding the operating temperatures have always been quite uniform, while those regarding the pressures and, more generally, the constructional characteristics and the mode of operation required from the plants have undergone considerable evolution.

In the past, syntheses carried out under extremely high pressures (>700 atm) (the Claude, Mississippi Chemical Co, and Casale processes) to obtain high yields have enjoyed a high reputation. Nowadays, however, such solutions are not considered to be economically viable, mainly because the compression costs and the problems encountered in maintaining such pressures in plants of every increasing dimensions.

Meanwhile, in fact, in a socio-economic context not concerned with problems of competitive costs in general and energy costs in particular, facts based on scientific canons formed pre-eminent criteria in the selection of the solutions which were adopted. Today, investment costs, operating costs, and the potential lifetime over which the plants will operate efficiently are regarded as matters of priority by both the suppliers and the clients.

The widest range of pressures employed (140–320 atm.) is encountered today in plants supplied by the largest company constructing and supplying ammonia synthesis plants on a world scale, that is, the American Pullman–Kellogg company*.

The conditions employed in the plants supplied by all the other companies which are active in this market also lie within this range of pressure values.

In fact, however, it is not the temperatures and pressures which are employed which form the main factors as to whether a certain plant is to be preferred or not, but mainly:

*The Pullman–Kellogg company, having arrived rather late in the market for ammonia plants as compared with their European rivals, in competition with about twenty active suppliers, is close to having erected 50% of all the ammonia synthesis plants which are operational in the World today.

* its flexibility, that is, the potential capacity which the plant has to be readily adapted for various types of feedstock (refinery gas, coal, oil from bituminous schists, etc.);
* the rational utilization of the heat of reaction to maximize the autonomous character of the plant;
* the dimensioning of the synthesis columns within the framework of the potential of the productive unit which it serves and the stratagems for simplifying the maintenance (with regard to the changing of the catalyst, for example) and lowering the operating costs associated with the running of the plant. This includes minimizing the losses of the charge in the reactors when there are specific gas flows passing through them. This may take the form of an energy saving or, more frequently, the possibility of the adoption of extremely finely divided catalysts with the highest activity.

Internal and external reactor conditions for the improvement of synthetic yields

Constant temperature curves (isotherms) which depict the course of the yields from the synthesis of NH_3 at 250 atm. as a function of time, are shown in Fig. 8.10.

According to the rules governing reaction isotherms (page 54), the graph plotted in Fig. 8.10 shows that:

(1) the percentage of ammonia which is attainable is higher, the lower the operating temperature;
(2) the thermodynamic yield under certain temperature conditions is attained in every case in a suitable time;
(3) increasing the temperature shortens the times required for the attainment of the thermodynamic yields.

This clearly demonstrates the inverse kinetic–thermodynamic role played by the temperature on the course of the synthetic reactions.

It can be seen from Fig. 8.11 how a modern ammonia synthesis reactor ('synthesis tower') of the Fauser–Montedison type is constructed in accord with the abovementioned theoretical requirements.

The temperature profile, depicted on the left of the figure, to which the synthesis gases in

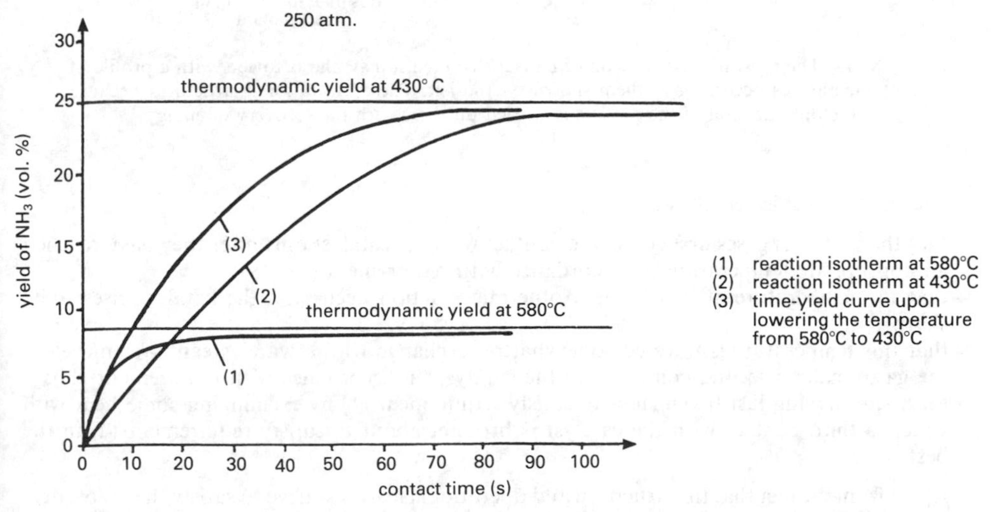

Fig. 8.10—Reaction isotherms for the synthesis of ammonia showing that the maximum yield and the time required to attain it decrease as the temperature is increased. Other points are mentioned in the text.

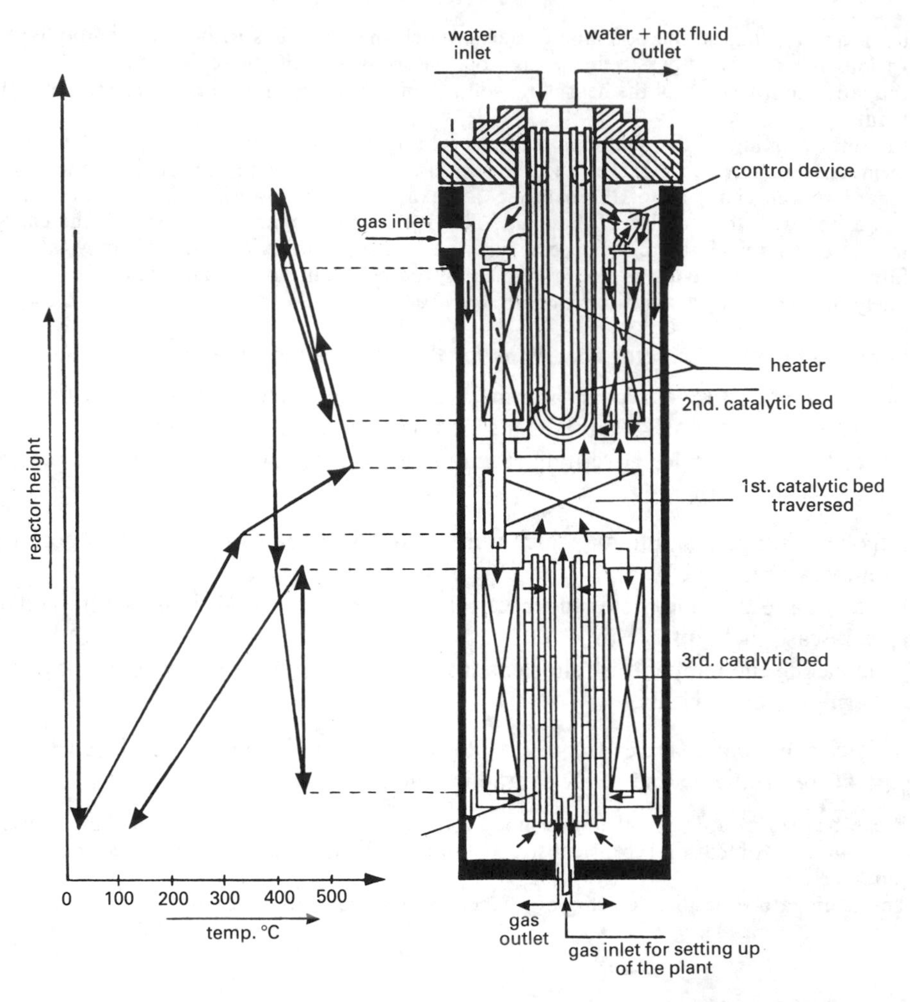

Fig. 8.11—The most modern version of Fauser–Montedison synthesis towers with a profile of the temperatures occurring in them shown on the left. These are rational structures because the optimization of the process is carried out here with the recovery of energy.

the reactor are subjected shows:

* that the synthesis gases first come into contact with the catalyst only after they have reached an appropriate temperature (in accordance with requirement (3));
* that the temperature at which the exothermic reaction occurs on the catalyst rises quite sharply;
* that this temperature is lowered somewhat by exchange with a water/steam system before the gases make a second contact with the catalyst (in accordance with requirement (1));
* that, after having just taken steps to satisfy requirement (1) by exchanging some heat with water, a third contact with the catalyst is brought about to satisfy requirement (2) in the best way.

Apart from the fact that the structural and operational features strive to satisfy the theoretical requirements specified above, these columns are built in such a way as to embody solutions which are practically convenient such as that of balancing the pressures of the cooling fluids and the synthesis fluids so as to minimize the thickness of the exchanger tubes.

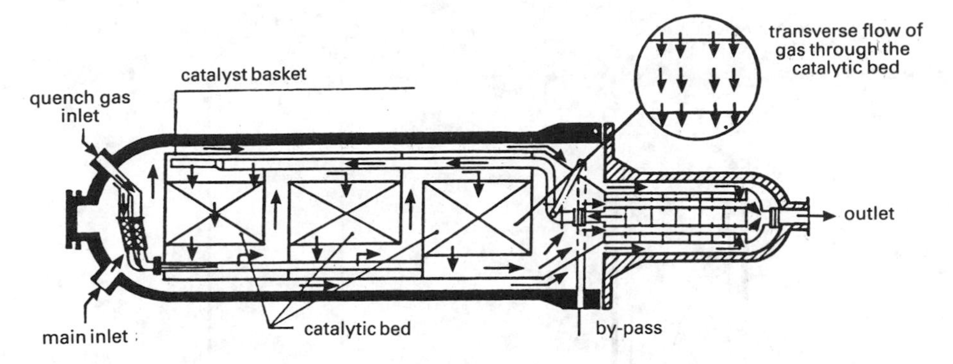

Fig. 8.12—A Pullman–Kellogg reactor with several adiabatic catalytic beds with quenching by the cold reagents between the beds. Today, this is the system which is generally preferred over a partitioning system, in plants with large productive capacities. The annular space between the pressure vessel and the compartment containing the catalyst baskets contributes to the preheating of the reagents and to the maintenance of the pressure vessel at not too high a temperature.

A diagram of one of the latest models of ammonia synthesis column supplied by Pullman–Kellogg is reproduced in Fig. 8.12.

The synthesis gases are sent into such a reactor in a double stream: the first stream is sent to be preheated in the zone at the top of the column and then returns backwards before successively passing through the various baskets holding the catalyst, while the second stream (the 'quenching' gas) is sent to be mixed in a controlled manner with the gaseous reaction mixtures which consecutively leave the first and second baskets holding the catalyst in order to cool them.

The whole of the synthesis mixture which is obtained after the second mixing is enriched, by passing it through the third basket containing the catalyst, so as to attain the maximum possible amount (within the limits of the process) of ammonia. After this, it is directed to the outlet, while, as stated earlier, preheating at the top of the synthesis gases in the main stream introduced into the reactor.

It is evident that this plant design solution, unlike that adopted in Fauser–Montedison plants which solely permits energy recovery externally to the column, also satisfies all the requirements which, as can be seen by studying the graph of the reaction isotherms, theoretically tend to favour the synthesis.

By using a normally structured catalyst, an advantage is obtained with these Kellogg columns in that the losses of the charge as a result of the passage of the gases through the catalyst itself are lowered, while, by making use of more finely divided catalysts the yields of the products of the synthesis are raised with respect to the contact time on account of the enhanced catalytic activity. This second solution is that which is generally exploited. It leads to advantages which are so pronounced as to constitute one of the principal reasons underlying the preference shown today for this type of ammonia synthesis column.

Other advantages offered by these columns are the ease with which the catalyst can be loaded and unloaded, since the baskets form a single carriage which can be removed from the body of the column together with the housing which contains them, and the ability of being adapted to provide a plant of any size simply by lengthening the pressure vessel by the addition of new sections, using a system of flanged joints with suitable gaskets.

* * *

Fig. 8.13 shows the solution based on 'adiabatic reactors' with alternating intermediate heat exchangers proposed by Braun, a company which has only quite recently entered the field of constructors of large-scale synthetic plant and, in particular, ammonia synthesis plants.

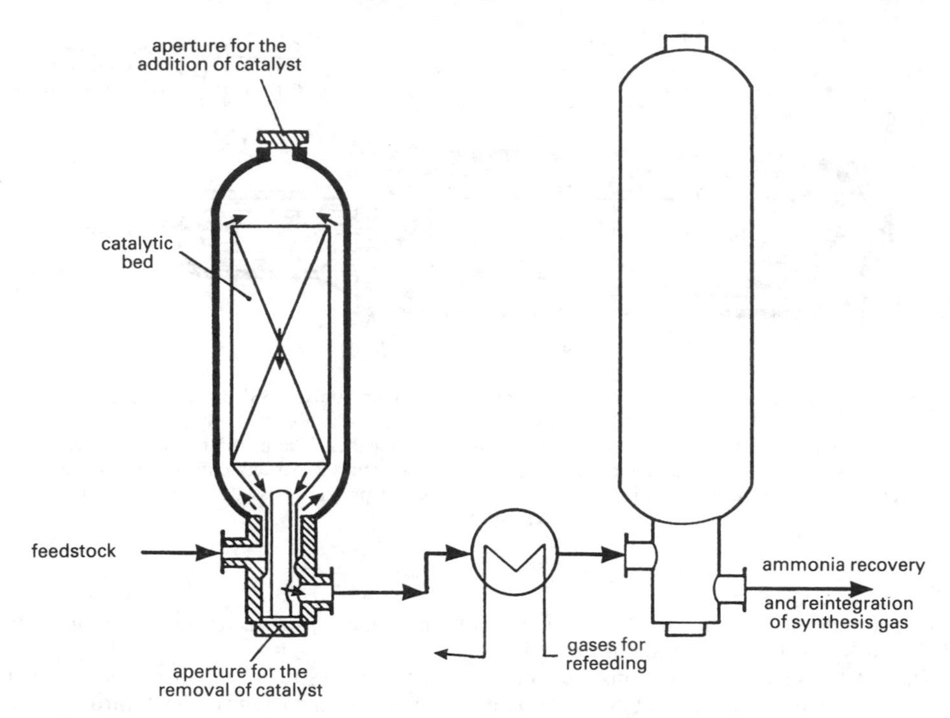

Fig. 8.13—Braun adiabatic reactors with heat exchangers between them. The temperature at the inlets into the two reactors is about 400°C; 510–540°C at the outlet from the first reactor and 460–500°C from the second reactor. The first reactor where the synthesis takes place more intensively has smaller dimensions than the second reactor, and, as a whole, they correspond to a single reactor with two adiabatic beds with an intermediate quench.

No heat exchange occurs within these Braun reactors, and they are therefore simple to construct and easily maintained. Moreover, the catalyst in them can be loaded and unloaded without any need to take out the baskets containing it.

A coil furnace for the preheating of the synthesis gas is pre-installed in the first column for the starting up of the plant; and, when the plant is in operation, preheating is ensured by the heat exchange which is always provided for between the two reactors.

Structural problems associated with synthesis circuits

By the term 'synthesis circuits' we understand the columnar reactors in which the synthetic reaction occurs and the heat exchangers in which conditions appertain which are comparable with those inside the reactors.

The severe pressure and temperature conditions under which the synthesis of ammonia is carried out necessarily restrict the choice of materials out of which the synthesis circuits are constructed. This particularly applies to those parts where the action of the aggressive reagents, that is, hydrogen and ammonia, is more vigorous.

In fact, hydrogen tends to decarburize steels with the formation of methane:

$$Fe_3C + 2H_2 \rightarrow 3Fe + CH_4$$

by, as we shall see, an aggressive attack on the cementite in these steels.

This phenomenon, which leads to a sharp fall-off in the mechanical properties of the structure, is counteracted by the use of chromium steel, a metal which when alloyed with low carbon

(0.1% max.) iron makes the structures resistant to hydrogen attack on account of the formation of stable carbides (Cr_7C_3, for example).

However, steels which are solely alloyed with chromium are relatively difficult to forge and are difficult to work in general.

Even if they are more costly, low-alloy steels containing various combinations of Mo, V, and W as well as chromium are better from every point of view*.

Ammonia attacks metals by decomposing with the liberation of nascent nitrogen which leads to the formation of nitrides and structural fragility. This phenomenon is of no great importance when it occurs on the surfaces of thick sections as it induces surface passivation processes.

The 18/8/Ti and 18/8/Mo (AISI 316) stainless steels are the materials which are the most resistant to both causes of corrosion in synthesis circuits.

The pressure vessel of a synthesis column can be made out of a single casting, or it may be an assembly of several cast sections using advanced technology, or even be formed from sheet material and subsequently welded.

To prevent the types of corrosion which have just been discussed, the interior of the pressure vessel is lined with a sheath of corrosion-resistant alloy steel of an appropriate thickness. The heat exchangers and the baskets containing the catalyst are made from the same alloys.

The synergistic action due to the two factors, temperature and pressure, which induces mechanical fatigue in the structures is obviated in modern plants by the adoption of special thermal insulation schemes apart from creating, as is always done, an annular space through which the reagent gases flow with preheating between the pressure vessel and the space occupied by the catalyst.

Cyclic reconstruction of the synthetic mixture and the separation of ammonia

In all processes for the synthesis of ammonia from its elements it follows that the ammonia has to be separated from the other gases which are subsequently recycled. For this purpose, the relative ease with which ammonia can be liquefied, in comparison with the other components in the mixture, is exploited.

From a plant point of view, this separation can be carried out in several ways, as indicated in the following table which shows the operational cycles which are carried out in the largest plants.

Separation
- *single-stage*
  - before compression for recycling
  - after compression for recycling
- *two-stage* — compression for recycling between one stage and the other

All of these solutions will now be illustrated and discussed.

Single-stage separation before compression for recycling

As is shown in Fig. 8.14, the synthesis gas is compressed in a series of centrifugal compressors (the number of these depends on the desired compression ratio). Two of these, the first and the last, are shown in the figure.

*For greater detail regarding this point, see A. Cacciatore & E. Stocchi, *Impianti Chimici Industriali*, Vol. 1 (10.1).

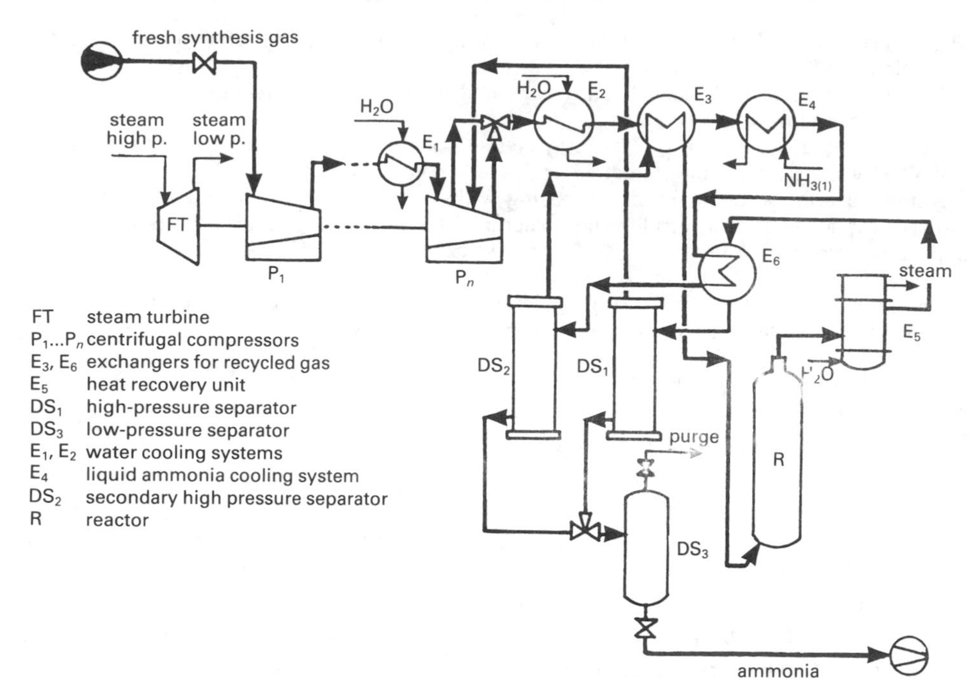

Fig. 8.14—Single-stage plant for the separation of ammonia. The separation is carried out before compression for recycling.

The fresh synthesis gas from the chain of compressors enters the final compressor. After being compressed to a pressure of the same order of magnitude as that of the output stream from the reactor R from which the heat has already been recovered in E_5, this fresh synthesis gas is mixed with the output stream to furnish a gaseous mixture which is cooled in several steps, and, in particular, with liquid ammonia. As a consequence, the ammonia formed in the recent synthetic step condenses out. The separation of the liquid from the recycled $N_2/3H_2$ mixture and the newly arrived mixture takes place in the high-pressure separator DS_1 which produces liquid at the bottom and the gas at the top. This, after having made a contribution in the cooling of the refrigeration circuit within E_3, returns to the last stage of the compressor P_n from where, having been brought up to the correct pressure, it is sent to the synthesis tower R.

It should further be noted that the liquefied ammonia, having descended into the low-pressure separator DS_2, is freed from its vapours and from the gases which are dissolved in it (H_2, N_2, CH_4, and Ar) and then sent to be stored.

Moreover, the subsequent separation, from the purge gases, of the ammonia which is exhausted into the bottom of the plant is achieved either by exploiting the large difference in their boiling points and thereby condensing it out by cooling or by its preferential adsorption given that ammonia is, by far, the most reactive component in the mixture.

Single-stage separation after compression for recycling

In this case, as is shown in Fig. 8.15, the mixing of the synthesis gas with the output stream (from the synthesis tower R), which has already been cooled in the heat recovery unit E_5, takes place in the final compressor P_n, from which the mixture emerges in a suitable state in order to be sent firstly to undergo multistage cooling in E_2, E_3, and E_4 and then for the high-pressure separation of the liquid ammonia in DS_1. It is then finally sent, after having contributed to the cooling of the refrigerating loop in E_3, into the synthesis tower R.

Here, also, the liquid ammonia is sent for storage after its pressure has been reduced in DS_2 and it has been purged. The gases arising in the purging of the liquid ammonia are subsequently

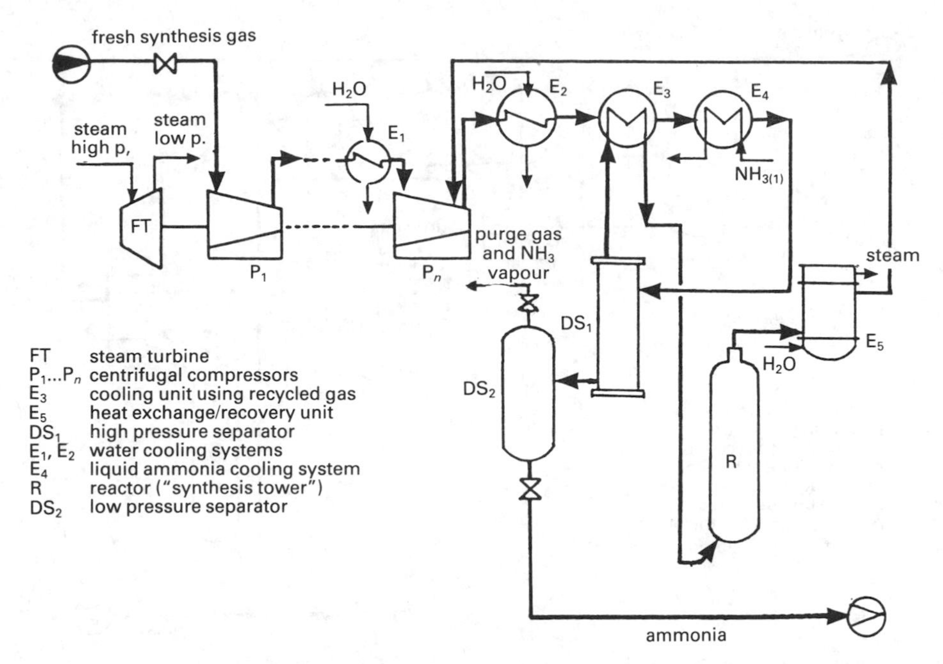

Fig. 8.15—The separation of ammonia from the reactor output stream, after having already recompressed the mixture of the output stream itself with the fresh synthesis gas, before recycling.

freed from the ammonia still remaining in them in one of the two ways which have already been discussed.

Two-stage separation with compression for recycling between the stages

The mixture of gases emerging from the reactor is first cooled in E_5 in order to recover heat and then in E_6 by indirect contact with the cold gases emerging from the liquid ammonia refrigerator E_4. The aim of this heat exchange step is to condense out the greater part of the ammonia which is separated from the unreacted gases in a primary high-pressure separation unit DS_1. Compounds remaining in the gaseous state are sent back to be recompressed in P_n with the admixture of new synthesis gas which has passed through the battery of compressors preceding P_n. The resulting mixture is cooled with the water in E_2, with the gases which are proceeding back to the synthesis tower through E_3, and with the liquid ammonia in E_4. After exchanging heat in E_6 with the mixture of gases, coming from the synthesis unit, which has already been cooled in E_5, the ammonia which has condensed out is separated in the secondary separation unit DS_2 which, like the first separation unit, operates under high pressure.

The ammonia which has been separated from the gases, having been reunited with that originating from the primary separation unit, passes, as usual, into the low-pressure separation unit DS_3 before being sent for storage.

The gases from the secondary separation unit, after the ammonia has been unloaded, go into E_3 where they contribute to the cooling of a mixture of a type of which they themselves are formed. Finally, they pass into the reactor R.

8.3 THE COMMERCE AND USES OF AMMONIA

Since thermally insulated systems are readily available nowadays, it is preferable, especially in large plants, to store liquid ammonia at atmospheric pressure and, therefore at low temperatures ($\simeq -33°C$) to obviate the need for the use of a

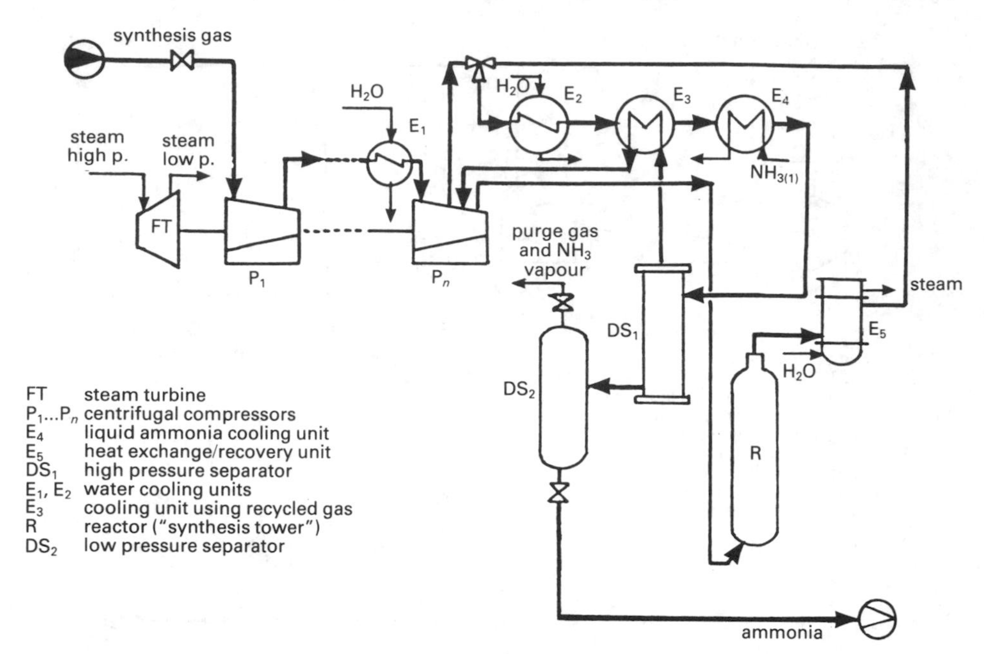

Fig. 8.16—Plant for the separation of ammonia in two stages with intermediate compression. This type of cycle is adopted when working at very high pressures because it allows one to avoid the useless compression of the major part of the ammonia produced. On the other hand, the complications of the plant design encountered here increase both the investment and running costs.

pressurized tank. Under the storage conditions appertaining in this case the ammonia is delivered by the adiabatic evaporation of part of the liquid, thereby providing cold vapours which are used in the refrigeration systems of the same plants.

On the other hand, in companies concerned with th production and use of liquid ammonia, it is stored at normal temperatures in spherical vessels ('orthospheres') at a pressure of 20 atm., while it is stored, either in dry or moist form, in gasometers, in which case, it is isolated from water with Vaseline oil.

Liquid ammonia under pressure, which is stored in tanks or in steel cylinders which can withstand pressures up to about three times that under which they effectively operate, is also always available commercially. Commercial ammonia, dissolved in water with concentrations of 22–28 Bé, is available in barrels and carboys.

Ammonia is a raw material of fundamental importance for a whole series of applications, of which the main ones are:

- the production of nitric acid;
- the preparation of fertilizers and, more generally, ammonium salts destined for use in various fields.

Very large amounts of ammonia are also used in the synthesis of a wide range of amines and amides, hydrazine, cyanides, etc., directly in the servicing of the dyeing, tanning, detergent, elastomer, and fertilizer–disinfestation industries, and are absorbed by a very wide range of technologies such as refrigeration, metallurgy, Solvay soda, and so on.

Table 8.3—The toxicological effects of ammonia

		Concentr	
Effect	*Exposure times*	*ppm*	
Tolerable	1 h	390	228
Dangerous	30–50 min	2500–3500	1900–2560
Lethal	Few seconds	5000–10 000	3800–7600

Ammonia can also be proposed as a potential fuel provided that its oxidation, using a suitable catalyst, leads to the formation of nitrogen and water.

8.4 THE DANGERS OF AMMONIA AND REMEDIES

First of all, the fact should not be overlooked that NH_3 forms, within restricted limits, explosive mixtures with air, and this is a cause of dangerous accidents.

Ammonia, whether in the liquid or gaseous state, causes burns to the eyes, swelling of the eyelids, irritation to the throat, pulmonary oedema, and, in serious cases, asphyxia.

On the skin, ammonia causes burns owing to its caustic action and the effects of cooling.

The maximum tolerable limit when there is a certain systematic repetition of the dose is 100 ppm which is equal to 70 mg/m^3. The data presented in Table 8.3 are of particular interest to workers in industrial sectors in which ammonia is produced and used.

Pulmonary oedema caused by ammonia is made worse by artificial respiration which should therefore be avoided. The inhalation of pure oxygen is the best antidote against ingestions of ammonia via the respiratory tract. Contacts with the skin are treated by a lengthy rinsing with water, followed by the application of a 2% solution of acetic acid and, finally, further rinsing.

NITRIC ACID

8.5 PHYSICAL AND CHEMICAL PROPERTIES OF NITRIC ACID

Pure nitric acid (HNO_3) melts at -41.6°C to produce a colourless liquid but then assumes more or less rapidly, depending on the temperature and the degree of exposure to light, a colour ranging from yellow to a brownish red. This is a consequence of its decomposition according to the reaction:

$$2HNO_3 \rightleftharpoons 2NO_2 + H_2O + 1/2O_2 \quad (8.12)$$

For this reason, vessels holding nitric acid are always under some measurable pressures.

Nitric acid is miscible with water in all proportions, and its aqueous solutions tend to form an azeotrope with a maximum boiling point ($bp_{760\,torr}$)* of 121.9°C which contains 68.7% HNO_3 by weight (Fig. 8.17).

Although the phenomenon is not of great practical importance, nitric acid forms two hydrates with the formulae $HNO_3 \cdot H_2O$ and $HNO_3 \cdot 3H_2O$ with water at low temperatures. The relevant phase diagram is shown in Fig. 8.18.

*1 torr = 1 mm Hg = 1.33322 mb.

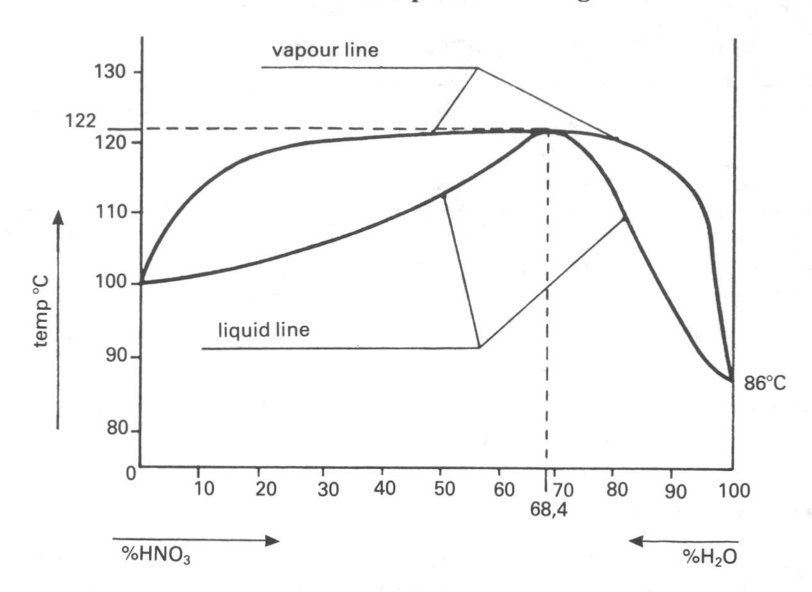

Fig. 8.17—Equilibrium liquid–vapour curve for nitric acid. It can be seen that it is impossible to prepare practically pure nitric acid by the distillation of its aqueous solutions (apart from the problems due to its decomposition according to the reaction $4HNO_3 \rightleftharpoons 4NO_2 + 2H_2O + O_2$). On account of this, as will be seen, it has to be concentrated by means of dehydration.

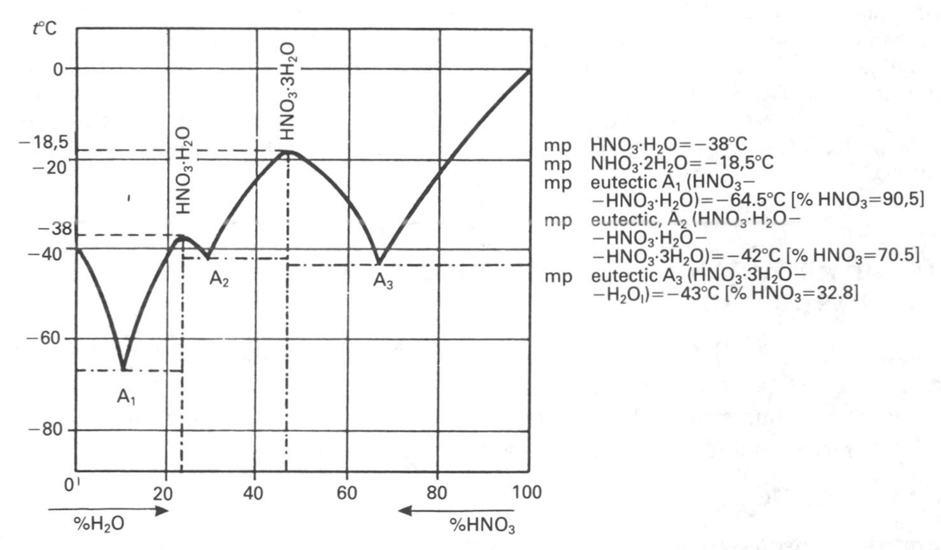

Fig. 8.18—Liquid–solid phase diagram of the HNO_3–H_2O system. The hydrates which form at certain ratios between the two components have the character of compounds which melt without decomposition.

Nitric acid dissociates in water, behaving as a very strong acid with $\alpha = 0.82$ in a 1 M solution, $\alpha = 0.84$ in a M/2 solution, $\alpha = 0.93$ in an M/10 solution, and $\alpha = 1$ in an M/100 solution. The activity coefficients of these solutions are all about 90% of the respective degrees of dissociation.

Nitric acid behaves as a strong oxidizing agent to the point that, with metals which are less noble than hydrogen, it behaves as an oxidant in addition to behaving as an acid (with the evolution of H_2).

The main half-reaction which lies at the foundation of the oxidizing power of nitric acid is:

$$NO_3^- + 4H^+ + 3e^- \rightleftharpoons NO + 2H_2O \qquad E^\ominus = +0.96\ V$$

Only the most noble metals such as Pt, Au, Rh, and Ir are resistant to the oxidizing action of nitric acid. Nitric acid behaves as a passivating agent with metals, in particular aluminium, which form adherent, compact surface layers of insoluble oxides in the surroundings where they are formed.

Finally, nitric acid is a nitrating agent for aromatic compounds and an esterifying agent for alcohols on account of the tendency which it has to furnish the 'nitronium ion' (NO_2^+):

$$HNO_3 \rightarrow HO^- + NO_2^+$$

both in 'fuming' nitric acid and also when it is mixed with concentrated sulphuric acid or oleum.

The nitronium ion has an appreciable probability for existence because it is a resonance hybrid of three canonical structures:

$$\{|\underline{O}^+—\bar{N}=\underline{\bar{O}} \leftrightarrow |O=N^+=O| \leftrightarrow \underline{\bar{O}}=\bar{N}—\underline{O}|^+\}$$

8.6 THE INDUSTRIAL PRODUCTION OF NITRIC ACID

Today, production costs restrict the preparation of nitric acid on an industrial scale to the use of ammonia as the raw material.

Historically, nitric acid was first prepared by the double decomposition of sodium nitrate, obtained from Chile saltpetre, and sulphuric acid:

$$NaNO_3 + H_2SO_4 \rightarrow NaHSO_4 + HNO_3$$

The reaction was carried out at 150–170°C in furnaces of the same type as those employed for the production of bisulphate (page 217) from which the acid emerged (since the boiling point of HNO_3 is 86°C) at a concentration of 95–96% provided that pure $NaNO_3$ and 93–94% H_2SO_4 were used.

Nitric acid was later obtained from nitric oxide prepared by the 'arc process' (the methods of Birkeland, Eyde, Wisconsin, etc.)

$$N_2 + O_2 \underset{\text{arc}}{\overset{\text{electric}}{\rightleftharpoons}} 2NO - 43\ \text{kcal}. \qquad (8.13)$$

The conversion of NO to HNO_3 was carried out by means of oxidation and hydration processes which could be applied to a large extent in achieving the same conversion of the product obtained from the oxidation of ammonia.

In the arc process, apart from the cost of the electrical energy which is consumed, there were enormous amounts of gas in circulation compared to the low concentration of NO which was formed (about 2% on the average) on account of the fact that the high temperatures also promote the reverse 'dissociation' reaction.

Actually, as has already been stated, nitric acid is now prepared industrially only from ammonia. For this purpose, it is first necessary to oxidize the raw material to nitric oxide with air:

$$4NH_3 + 5O_2 \rightleftharpoons 4NO + 6H_2O + 216.24\ \text{kcal}. \qquad (8.14)$$

A number of considerations impinge on this oxidation process, the realization of which constitutes the basis of the technology for the production of nitric acid.

Research into favourable operational conditions for the oxidation of NH_3 to NO

The reactions

$$4NH_3 + 3O_2 \rightleftharpoons 2N_2 + 6H_2O + 302.64 \text{ kcal} \quad (8.15)$$

$$4NH_3 + 6NO \rightleftharpoons 5N_2 + 6H_2O + 432.25 \text{ kcal} \quad (8.16$$

$$2NO \rightleftharpoons N_2 + O_2 + 43 \text{ kcal} \quad (8.17)$$

act as the main processes which compete with reaction (8.14).

There is a high probability that all of these reactions will take place to some extent because, since they occur with a large enthalpy decrease and (apart from the last reaction) an increase in the number of moles and thereby an increase in the entropy, large negative values of ΔG occur.

However, it is not only the competition from other reactions which hinders the formation of NO from ammonia in accordance with reaction (8.14) but also the fact that ammonia/air mixtures exhibit explosion limits. At normal temperatures and a pressure of 1 atm. the explosion limit for a gaseous mixture of ammonia and air is from 15.6% ammonia, while at temperatures above 600°C and a pressure of 1 atm., this limit is lowered to 10.5%.

To ensure that the NH_3 is converted into NO under a safe regime it is necessary:

(1) to influence the thermodynamics of the reactions competing in the process in an unfavourable manner;
(2) to speed up the kinetics of reaction (8.14) to make it take place preferentially with respect to the other reactions;
(3) to work under conditions whereby the ammonia concentration is such as to preclude any danger arising from explosions.

The thermodynamics of the competing reactions (8.15) and (8.16) are rendered unfavourable by working (since they are strongly exothermic reactions) above 500°C, while the thermodynamics of reaction (8.17) are not favoured if the process is carried out under 1200°C.

The kinetics of (8.14) are speeded up by the use of catalysts, and this is also done with the aim of preventing any reduction in the velocity of the reaction brought about by the presence of the inert gas N_2 in the reaction zone.

In fact, when the reagents are diluted with nitrogen, as happens when reactions (8.15), (8.16), and (8.17) all occur freely, the nitrogen molecules tend to reduce sharply the number of effective collisions between NH_3 and O_2 which is postulated to favour reaction (8.14).

The explosion limits are avoided by employing a quantity of air such that the amount of ammonia mixed with it is appreciably less than 10.5 vol.% of the total volume, that is, in particular with about a 40% excess of air.

The operating temperatures for the oxidation of NH_3 to NO

The most favourable temperature for the oxidation of NH_3 to NO lies between 500 and 520°C, but, to achieve these conditions, it would be necessary to work with an excess of air for kinetic reasons and at a ratio of the concentration of NO in the final gases which would be inadmissible. The most suitable operating temperatures for the oxidation of NH_3 to NO, also, with due consideration to the catalysts which are employed, have been determined experimentally. They are at 830–870°C for processes which are carried out at atmospheric pressure, and from 920–950°C for processes operating under pressure.

It should be noted that these temperatures are clearly outside both the temperature limits which are favourable to reactions (8.15) and (8.16), and the range over which the probability of the dissociation reaction (8.17) occurring is high.

To attain the abovementioned operational temperatures during the oxidation under the working conditions, regarding the composition of the reagent mixture and the flow rate of the gases into the reactors, it follows that it is necessary to preheat the air/ammonia mixture to about 200 or 250°C, respectively, according to the pressure values employed.

The catalyst for the oxidation of NH_3 to NO: structure and action

The only industrial catalyst which is used today for the oxidation of ammonia to nitric oxide is formed from a platinum/rhodium alloy containing about 10% of rhodium. The rhodium improves the catalytic properties of the platinum and, even more importantly, both the mechanical and anti-abrasive properties of the material under the operating conditions.

Structurally, the metallic alloy catalyst is fashioned into very fine threads ($d \simeq 0.05$ mm) which are woven into meshes with more than 1000 stitches per cm^2 (and even up to 3000). Two to four or even more of these meshes are placed on top of one another inside the reactors when these are put into operation.

Initially, the threads are smooth and bright and they are not catalytically very active. They then become dull and wrinkled whereupon their activity rises to a maximum, and finally they become spongy with the activity falling off to an ever greater extent until it arrives at a point when the meshes have to be replaced.

When the catalyst is in its most active state, the ammonia is quantitatively transformed, and yields of up to 98% of NO are obtained.

The mechanistic details of the action of the catalyst have recently been re-studied, and have turned out to be different from those postulated up until the sixties. Among other things, the presence of the atomic species, N, in the gas phase, even as a transient event, and the formation of a true oxide of platinum (PtO), have both been precluded. The new theoretical mechanisms can be formulated as follows:

$$O_{2(g)} + Pt \rightleftharpoons Pt_{O_2(ads)}$$

$$NH_{3(g)} + Pt \rightleftharpoons Pt_{NH_3(ads)}$$

$$Pt_{NH_3(ads)} + Pt \rightleftharpoons Pt_{NH_2(ads)} + Pt_{H(ads)}$$

$$Pt_{O_2(ads)} + Pt_{NH_2(ads)} \rightleftharpoons Pt_{NO(ads)} + Pt_{H_2O(ads)} \rightarrow 2Pt + NO_{(g)} + H_2O_{(g)}$$

More water is also formed by the reaction between $O_{2(g)}$ and the hydrogen adsorbed on the platinum, $Pt_{H(ads)}$.

Catalysts: poisons, abrasive products, and recovery

Chlorides, sulphates, H_2S, COS, As, and As_2O_2 are permanent poisons of the catalyst used in the combustion of NH_3 to NO. Hydrocarbons of the type of acetylene and ethylene are temporary poisons. In general, however, the presence of any of these poisons is very unlikely when the air is taken from the atmosphere with suitable precautions.

The catalyst is rather intolerant towards suspensions of lubricants, fats, fine dust, and abrasive powders. These are all materials which either diminish the surface activity of the catalyst or lead to unacceptable porosity in the threads of the mesh.

Suspensions of Fe_2O_3, which can originate either from rust carried by the ammonia and from rust which is present in air, are also particularly harmful.

Apart from being abrasive, these powders are responsible for the lowering of the catalytic activity because, in their turn, they catalyze the oxidation of NH_3 to N_2 in accordance with reaction (8.15).

It is therefore necessary, besides providing efficient filtration systems for the gaseous air/ammonia mixture and employing magnetic separators, to use the maximum number of components made out of aluminium and stainless steel.

In this way it is also possible to avoid the effects of the catalyzed dissociation of ammonia into nitrogen and hydrogen which is promoted by the iron in ordinary steels.

* * *

Given the high price of the catalyst, systems should be provided, especially in plants which operate under pressure, for the recovery of the platinum which is carried away by abrasion. This is estimated at 0.05 g/ton HNO_3 in normal plants and 5–6 times as much in plants operating under pressure.

The recovery units may be direct or indirect: the first type intercepting the metal which leaves the mesh while the other kinds reduce the amount of material abraded from the meshes.

The outlet tubes from the reactor, the mechanical filters upon which the reaction products impinge after leaving the Pt/Rh meshes, or, more recently, meshes made out of palladium and gold alloys, that is, metals with which platinum readily forms alloys, act as direct recovery devices. Indirect recovery devices, on the other hand, consist, for example, of systems which regulate the distribution of the gases on the whole section of the bundle of catalyst meshes to exploit the high resistance to wear and tear of the whole of the area presented by the threads of each mesh.

The flow rate of the gases and the effect of pressure on reaction (8.14)

One of the reasons for the erosion of the threads of the catalyst mesh is the high velocity of the gases in the reactor. This is of the order of 0.4–4 m/s or more, such that the contact times on the catalyst are about 5×10^{-4} s.

By operating under pressure, the amount of ammonia which is catalytically oxidized per unit time is increased owing to the greater contact between the reagents and the catalyst. It should, of course, be noted that pressure per se is unfavourable to the overall process because:

- the conversion yield drops off appreciably under pressure in so much as the reaction increases the number of moles;
- the ammonia–air explosion limit is also lowered;
- the wear and tear on the catalyst increases sharply.

Operating under pressure, which is carried out in plants which work at pressures up to 8 atm., is justified by the concomitant reduction in the volume of the apparatus and by the simpler process for the absorption of the nitrous gases which can be installed in such plants. In pressurized plants, small amounts of water vapour are mixed with the gases to avoid dangers arising from explosions, and they are equipped with efficient systems for the recovery of the catalyst powders to reduce the running costs beyond any savings because the retention of platinum in these plants is already lower.

Technology of the oxidation of NH_3 to NO

Although they now embody improvements regarding, for example, the recovery of the catalyst, the burners in use for the conversion of NH_3 into NO are still those of the classical Frank–Caro/Bomag, Parsons, and Du Pont/Uhde types.

The Frank–Caro/Bomag burner. This burner is fed from the base with air and gaseous ammonia which have been respectively preheated in the heat exchangers E_3 and E_2 into which they enter in the form of orthogonally crossed streams (Fig. 8.19) after the air has been passed through the sleeve filter PF and the ammonia has been vaporized in the evaporator E_1.

Before coming into contact with air, the stream of ammonia is dispersed by a series of small tubes to promote its mixing which is completed by means of a series of spirals suitably arranged on the base of the lower cup-shaped part of the burner. Before passing through the meshes which are supported on a stainless steel grid, the air/ammonia mixture passes through a series

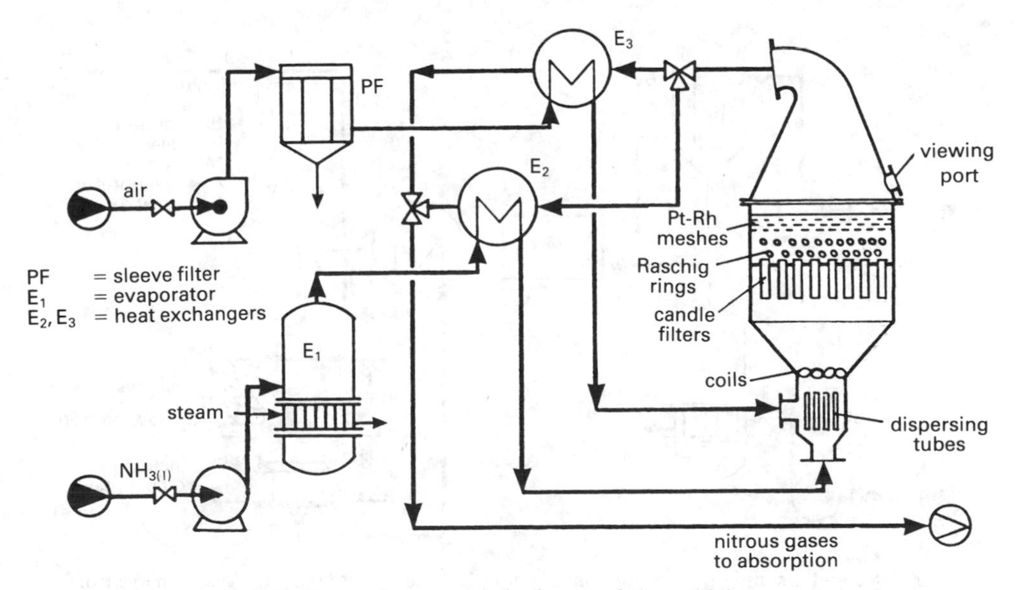

Fig. 8.19—Frank–Caro/Bomag burner and the layout of the sudsidiary apparatus for the production of NO from NH_3.

of candle filters followed by a thin layer of Raschig rings. Oxidation occurs on the layer of meshes, and the temperature is checked by using optical pyrometers through the sealed viewing apertures (made of quartz or mica).

The lower cup-shaped section of the burner is lined with aluminium, and the upper part is made out of ordinary steel which exhibits excellent resistance to the action of nitric oxide provided that no water condenses out. The internal metal structures of the burner are covered on the outside with an insulating material.

The combusted gases emerge from the top of the burner at above 800°C and are sent to feed the heat exchangers E_2 and E_3.

Other versions of Frank–Caro/Bomag burners, as shown in the first section of Fig. 8.20, are equipped in such a way as that the air and the ammonia can be mixed outside the burner, and this mixture can subsequently be preheated as a single entity.

In the more recent versions of this type of burner Au/Pd meshes are also arranged above the Pt/Rh meshes to intercept, by alloying with it, any platinum which is eroded or volatilized from the latter meshes.

Parsons burners. The mixing of the reagents which constitute the feedstock for these burners is always carried out outside the burner, which employs a single heat exchanger for the preheating of the mixture. The gaseous mixture passes from the top to the bottom, and the catalyst meshes are shaped in the form of a basket with the bottom sealed by a quartz plate to compel the gases to pass through the lateral walls of the cylinder.

Fig. 8.20 shows the classical type of these burners with the addition of the other features that have just been described.

Du Pont burners. These burners have the form of Frank–Caro/Bomag burners but use the feeding techniques of Parsons burners. The greatest difference between them is the fact that they operate under a pressure of 6– atm. (Fig. 8.21(a)) and are specifically equipped by just preheating the air.

The truncated conical shapes of the two sections out of which they are constructed has the aim of promoting a uniform distribution of the reagents on the catalyst which may consist of packs formed from up to 20–30 meshes of Pt/Rh threads. This is because, when the first meshes have been consumed, other meshes may come into operation without having to close down the burner.

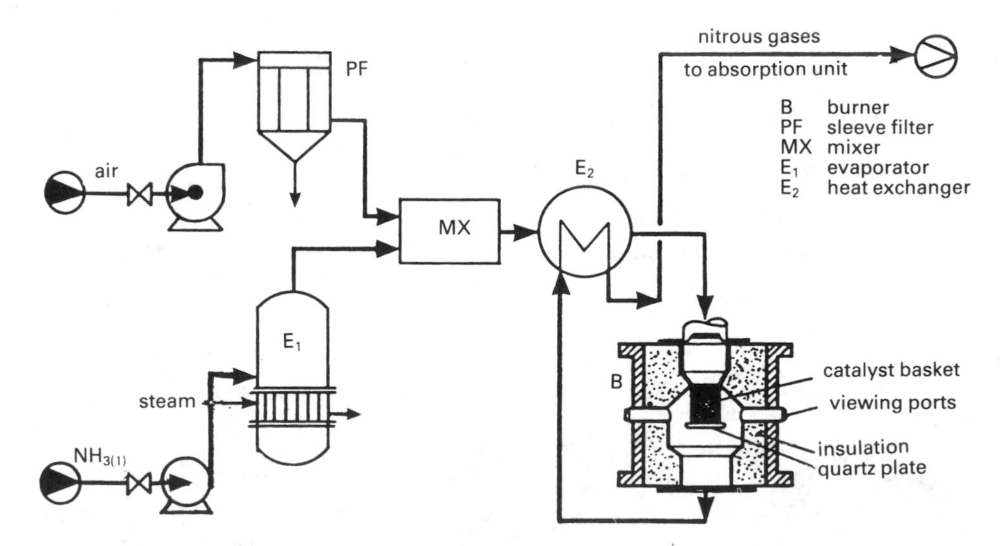

Fig. 8.20—Parsons burner for the conversion of NH_3 to NO. Making the gases emerge from the side walls of the cylindrical catalyst mesh leads to a considerably lower consumption of the platinum out of which the meshes are made.

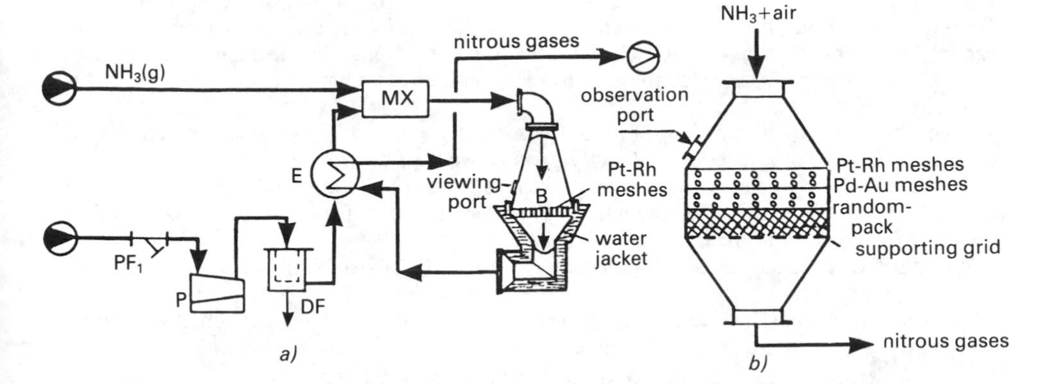

Fig. 8.21—(a) Du Pont pressurized plant for the production of NO from NH_3; (b) modern version of an ammonia burner designed to minimize the losses of platinum. In addition to the customary Pd/Au meshes, this burner also contains a highly porous layer made from a nickel/chrome alloy with a random arrangement of holes in it. This system not only intercepts the material escaping from the catalyst but also makes the passage of the gas over the whole area of the meshes homogeneous and thereby reduces the wear and tear on the catalyst meshes.

The most recent versions of these burners have a system fitted to the top inlet for the thorough purification, filtration, and intense mixing of the reagents.

Finally, there are the reactors where the greatest care is taken in the direct or indirect recovery (page 290) of the catalyst, because, as is well known, it is in these reactors that there is the greatest danger of losing it. Among other things, the modern versions of these burners systematically make use of the composite device for the maximized recovery of the catalyst from the various reaction zones as is shown schematically in Fig. 8.21b. Moreover, this device is also being adopted in all other reactors for the oxidation of NH_3 to NO.

Theory of the chemical processes used in the conversion of nitric oxide into nitric acid

Before presenting an account of how the nitric oxide contained in the gases from the combustion of ammonia is converted into nitric acid, let us write down the reactions participating in this transformation and deduce from them, on the basis of the mechanism by which they occur, the most favourable conditions for their realization.

Firstly, nitric oxide must be oxidized further to nitrogen dioxide:

$$2NO + O_2 \rightleftharpoons 2NO_2 + 27.5\ \text{kcal} \tag{8.18}$$

With water, nitrogen dioxide forms a mixture of nitrous and nitric acids:

$$2NO_2 + H_2O \rightleftharpoons HNO_2 + HNO_3 \tag{8.19}$$

The nitrous acid subsequently undergoes a further disproportionation to give back nitric oxide, which is recycled, and nitric acid which is the final product:

$$3HNO_2 \rightarrow HNO_3 + 2NO + H_2O \tag{8.20}$$

The oxidation of nitric oxide [reaction 8.18]

This is the most complex reaction of the three. The smooth course of the transformation of nitric oxide into nitric acid depends, above all, on this reaction, and the formation of nitrogen dioxide constitutes the 'rate limiting step' in the process.

This reaction, that is, (8.18), is a strange reaction, not because raising the temperature at which it is carried out drives it to the left (which is logical, as we are dealing with an exothermic reaction) but because it takes place relatively readily, even though it is formally a trimolecular reaction, and, above all, because the rate of reaction decreases as the operating temperature is raised.

It would therefore appear to contravene the second general rule of chemical kinetics which states that the rate of a reaction is directly proportional to the temperature.

To account for the reasons behind this fact it is necessary to start off again and consider the mechanism which lies at the foundation of this reaction which can be understood by taking account of the electronic structure of nitric oxide.

The NO molecule has an odd number of valence electrons so that one of these electrons has an unpaired spin.

On account of this structure the NO molecule, under favourable experimental conditions, tends to dimerize according to the scheme:

$$\underline{\overline{O}}=\overline{N}\cdot + \cdot\overline{N}=\underline{\overline{O}} \rightleftharpoons \underline{\overline{O}}=\overline{N}:\overline{N}=\underline{\overline{O}}$$

which is normally written as:

$$2NO \rightleftharpoons N_2O_2. \qquad (8.18a)$$

By assuming that NO is not oxidized but it is the dimer N_2O_2 which is oxidized according to the reaction:

$$N_2O_2 + O_2 \rightleftharpoons 2NO_2 \qquad (8.18b)$$

reaction (8.18) is interpreted as being the sum of (8.18a) and (8.18b), that is, the sum of two reactions, each of which is bimolecular and thereby kinetically facile. It should also be noted that the first of these two reactions (8.18a) is clearly favoured (by treating it as an association) by calm conditions and by a reduction in the temperature at which it occurs.

Because (8.18) turns out to be the sum of (8.18a) and (8.18b), the condition which must be satisfied is that the temperature should be low (or can be made so). In fact, it is only at quite low temperatures that there is a high probability of the nitric oxide aggregating to form dimers. The reason why the rate of (8.18a) falls off as the reaction temperature is increased is therefore clear.

The hydration of nitrogen dioxide [reaction 8.19]

Nitrogen dioxide is also a molecule in which there is an odd number of valence electrons and which therefore possesses the ability to dimerize to dinitrogen tetroxide by the coupling of the unpaired electrons on two molecules in accordance with the reaction:

$$\begin{array}{ccccccccc} |\underline{\overline{O}} & & & & \overline{O}| & & |\underline{\overline{O}} & & \overline{O}| \\ & \diagdown & & /\!/ & & & & \diagdown \quad /\!/ & \\ & & N\cdot + \cdot N & & & \rightleftharpoons & & N{:}N & \\ & /\!/ & & \diagdown & & & & /\!/ \quad \diagdown & \\ |\underline{O} & & & & \underline{\overline{O}}| & & |\underline{O} & & \underline{\overline{O}}| \end{array}$$

which, then written in the conventional manner, corresponds to the reaction:

$$2NO_2 \rightleftharpoons N_2O_4 + 13.8\ \text{kcal} \qquad (8.19a)$$

The disproportionative hydration:

$$N_2O_4 + H_2P \rightarrow HNO_2 + HNO_3 \qquad (8.19b)$$

follows this coupling when water is present, and, when this reaction is combined with (8.19a), the overall reaction is the same as that represented by (8.19).

In this case also, one is therefore dealing with a reaction, (8.19), which is bimolecular, fairly simple kinetically, and certainly favoured by the conditions which promote reactions (8.19a) and (8.19b).

To be precise: reaction (8.19a) is both kinetically (as it is a dimerization) and thermodynamically (since it is an exothermic reaction) favoured by low temperatures, while reaction (8.19b) is stimulated by an increase in the contact between the liquid absorbent phase and the absorbed gaseous phase.

The final disproportionation of nitrous acid [reaction 8.20]

This reaction is also kinetically bimolecular because it is the sum of the partial reactions:

$$2HNO_2 \rightleftharpoons H_2O + N_2O_3 \quad (8.20a)$$

$$HNO_2 + N_2O_3 \rightarrow HNO_3 + 2NO \quad (8.20b)$$

Reaction (8.20a) is driven to the right by high concentrations of nitric acid and (8.20b) is favoured by an increase in temperature (whereupon a gas is evolved). The overall reaction (8.20) formed by the sum of these two partial reactions will therefore be favoured by increasing concentration and temperature.

It is deduced from these considerations that:

- the oxidation of nitric oxide to nitrogen dioxide and the dimerization of the latter are favoured by low temperatures;
- the absorption of dinitrogen tetroxide by water is facilitated by intimate contact between the liquid phase and the vapour phase;
- in hot, concentrated solutions, nitrous acid tends to be completely converted into nitric acid.

Let us now examine how the structure of the plants and the operational conditions within them, for the conversion of nitric oxide into nitric acid, satisfy these requirements which arise from the theoretical treatment.

Plants for the conversion of nitric oxide into nitric acid

There are a number of different technologies for the production of nitric acid from NO, and the choice depends on the final destination of the acid and, thereby, on the concentrations which the acid should have. Plants are in operation for the production of 50–60% nitric acid with the possibility of concentrating it up to 96–98%, should it be requested, using either sulphuric acid or magnesium nitrate. Then, there are plants for the direct production of nitric acid at concentrations covering a range from 94 to 99%.

Here, examples of several practical solutions will be considered, but the reader's attention is drawn to the fact that the possible variants are extremely numerous.

Plant for the conversion of NO into HNO_3 under normal pressure. Concentration of the acid

First of all, let us consider the common method of production of nitric acid with the subsequent description of possible methods for its concentration.

The following description refers to Fig. 8.22.

The gases from the combustion of ammonia (NO, H_2O, N_2, and O_2) which are available at about 600°C after they have exchanged heat with the NH_3 and air being sent to the burners,

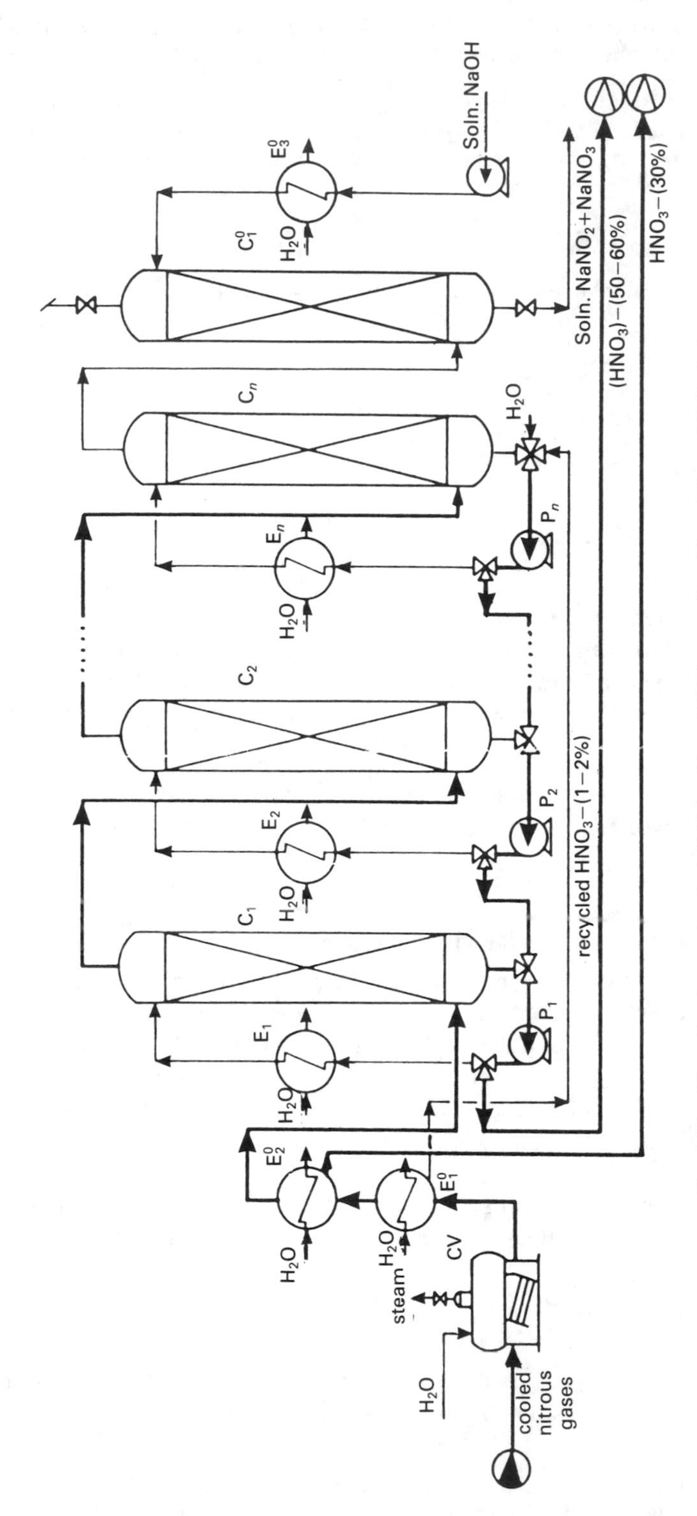

Fig. 8.22—Plant for the production of dilute (about 30%) or rather dilute (about 56%) nitric acid and a mixture of nitrites and nitrates from which only the nitrites are obtained industrially by fractional crystallization. Instead of using caustic soda (NaOH) for this production, Solvay soda (Na_2CO_3) can be used. The nitrites of other elements or nitrite esters which are of interest in preparative organic chemistry (ethyl and amyl nitrites, for example) can then be prepared from sodium nitrite by direct or indirect exchange (using silver nitrite, among other reagents).

that is, the cooled nitrous gases, undergo a further reduction in temperature by producing steam in CV. They are then brought to a temperature of 60–70°C in E_1^0 where, since a large part of the water contained in them condenses out, a very dilute solution of nitric acid is produced via reactions (8.18), (8.19), and (8.20)*. A further reduction in the temperature in E_2^0 to 30°C makes the other water containing an appreciable concentration of nitric acid (about 30%) condense out.

The gaseous mixture then enters into the proper cycle for the production of HNO_3 in stainless steel columns packed with a ceramic material, C_1, C_2, and C_n with the heat exchangers E_1, E_2, and E_3 interposed between them to systematically promote the necessary dimerizations, absorptions, and hydrations. The technique employed in this circle to make use of the NO component in the combustion products of ammonia, until the former is almost exhausted, is evident. An extremely dilute aqueous solution of nitric acid, after having collected some of the tail product from C_n, is pumped upstream in the plant by the pump P_n and subdivided into a liquid feed to the top of the same column C_n and another liquid stream which acts in an obvious way in the rest of the plant until approximately 53% nitric acid is produced after column C_1.

The column C_1^0 practically consumes the whole of the nitrogen content bound up in the residual products from the combustion of ammonia. Here, the mixture of nitrous and nitric acids, which is formed under conditions of high dilution and low temperature, is fixed by washing with a solution of NaOH cooled in E_3^0, where the heat of neutralization is removed†.

If there is no interest in the production of nitrites, which can be separated from nitrates by fractional crystallization, the nitrites can be converted to nitrates by oxidation with nitric acid:

$$3NO_2^- + 2H^+ + 2NO_3^- \rightarrow 3NO_3^- + H_2O + 2NO \qquad (8.21)$$

with recycling of the resulting nitric oxide.

* * *

When 96–98% nitric acid is required, the main product from the process which just been described can be dehydrated by using either 98% sulphuric acid or 70–72% magnesium nitrate.

The sulphuric acid process. To concentrate nitric acid by using sulphuric acid, ratios of the dehydrating agent to substance being dehydrated of 1:6 must be employed because the live steam for the reheating which is sent into the column where the two acids are in contact, in order to make the nitric acid distil over, must be absorbed.

This consumption of sulphuric acid is in itself unsuitable, but it cannot be left out because, if indirect heating were to be adopted, it would be necessary to use coils made out of an iron–silicon alloy which are mechanically very inferior, or tantalum alloys which are extremely expensive and difficult to work; that is, both of the materials are unsuitable for the construction of heat exchanger coils.

The plant in question (Fig. 8.23) consists of a packed column C_1 into the top of which 98% sulphuric acid is passed. The nitric acid to be concentrated is fed in slightly lower down the column. The column is made of iron–silicon or tantalum alloys which are resistant to the combined action of the two acids, and the packing consists of elements made of the iron–silicon

*The reason for the poor production of HNO_3 in this phase is the high temperature which hinders the dimerizations and absorptions which are prerequisites in this production process.

†It may be recalled, in fact, that dilution drives reaction (8.20a) backwards and that (8.20b) is not promoted by low temperatures. On the other hand, nitrous acid was absent downstream from E_1^0 because the relatively high temperature favoured reaction (8.20b) and, hence, reaction (8.20). It was also absent from the condensate in E_2^0 because the temperature was not very low so that reaction (8.20b) was still quite favoured and the concentration turned out to be high enough to promote the (8.20a) component of (8.20).

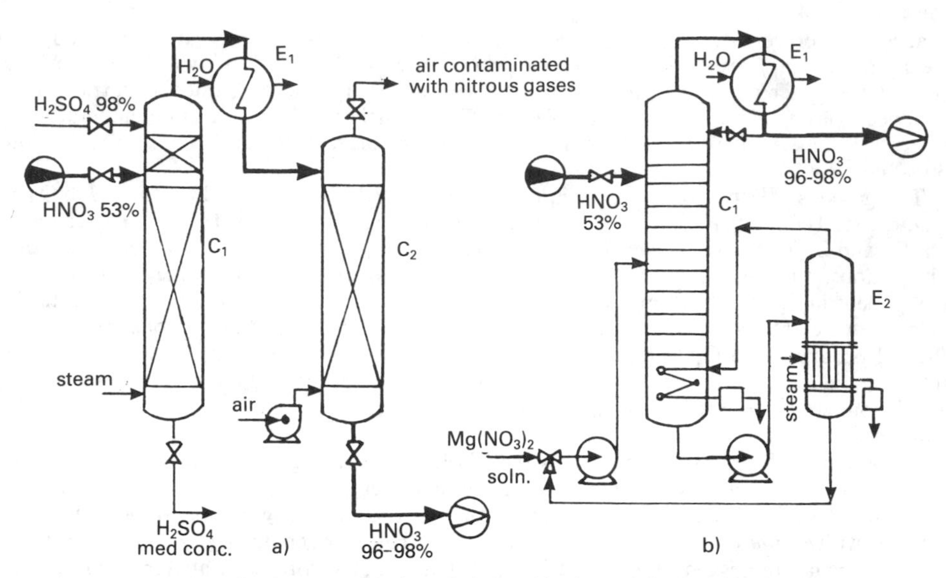

Fig. 8.23—(a) plant for the concentration of nitric acid using sulphuric acid; (b) plant for the concentration of nitric acid using magnesium nitrate. Other alkali or alkaline earth nitrates may also be used. However, magnesium nitrate is the most widely used concentrating agent, but not the only one, where one wishes to avoid the use of sulphuric acid.

alloy or quartz. As has already been stated, the reheating is carried out with live steam which is fed into the column.

The vapours of nitric acid distil and emerge from the top of the dehydration column to be condensed in E_1. Any nitrous gases present in the condensation product (96–98% nitric acid) are removed (it is 'bleached') in C_2 with a stream of air which is subsequently used for various purposes.

The sulphuric acid, which is slightly contaminated with nitrogen containing compounds and flows from the column C_1, is an ideal reagent for the production of fertilizers but may also be reconcentrated.

The magnesium nitrate process. Here, one is concerned with a process which, as a whole, is more economical than the sulphuric acid process and is particularly suited for use in high-capacity plants.

The magnesium nitrate employed in this process is prepared by treating magnesium carbonate with 53–65% nitric acid and then concentrating the resulting solution to yield a 70–72% solution of the salt in an apparatus which is analogous to that which we shall see used in the reconcentration of the dehydrating agent employed (the concentration unit E_2 of the plant scheme shown in Fig. 8.23(b)).

Magnesium nitrate has a very pronounced tendency to hydrate, and it permits the use of indirect steam for the reheating of the acid during its concentration. Consequently, the amount of water which this salt must absorb in the nitric acid concentration plant is less than that which the sulphuric acid must absorb in the analogous process.

The nitric acid to be concentrated is sent onto a plate in the upper zone of C_1 (Fig. 8.23(b)) while the solution of magnesium nitrate is fed in near the centre of the column. Nitric acid at a concentration of 96–98% distils into the top of the column and magnesium nitrate at a concentration of about 60% is taken off from the bottom. The column is heated by means of a coil through which the steam produced in the magnesium nitrate concentrator E_2 is passed.

After condensation in the water-cooled refrigerator E_1, the greater part of the nitric acid goes off to be stored and the rest is recycled as a reflux into the column C_1 from which it originated.

The magnesium nitrate solution which flows from the bottom of C_1 is reconcentrated in E_2 from where it emerges from the bottom to be returned to act as the absorbent in C_1.

Pressure plants for the transformation of NO into HNO_3

The graphs in Fig. 8.24 show that the rate of oxidation of NO is directly proportional to the pressure, hence the time required for the oxidation of NO becomes shorter as the pressure is increased.

As a consequence of operating under pressure it is possible to reduce the overall volume of the oxidation/absorption columns.

From the theoretical point of view, these results are justified by the fact that both the dimerization reactions (8.18a) and (8.19a) as well as the reactions involving the combination of the gaseous compounds (8.18b) and (8.19b) are favoured by pressure.

Since, in former times, materials which are resistant to the action of the oxygenated compounds of nitrogen were not available for the construction of the compression units, methods of the Du Pont type were developed. In these methods the non-corrosive air/ammonia mixture which is to be burnt is compressed upstream, and, in this way, further compression of the gases which are subsequently formed, oxidized, and absorbed is not required

On the other hand, after high-alloy steels with such a high chemical resistance that they could be used in the construction of units for the compression of the combustion products of ammonia had been developed, the practice of compressing these products also became very common, thereby avoiding the disadvantage arising from the lower yield when the ammonia is burnt under pressure.

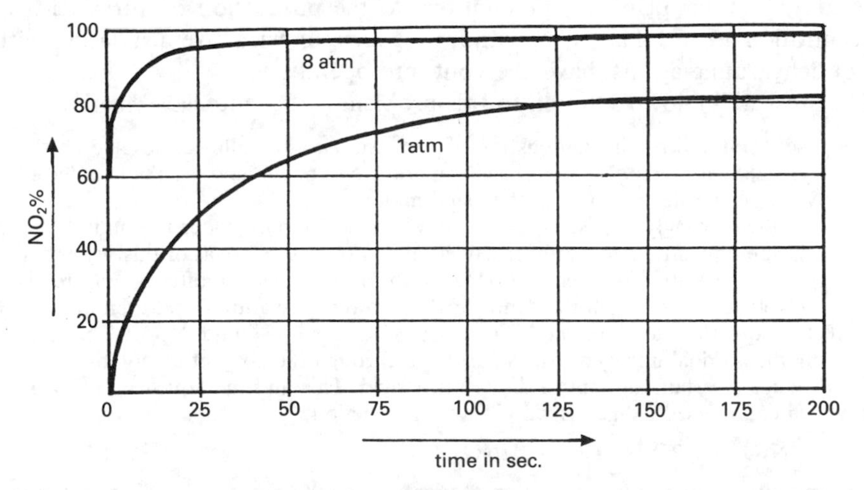

Fig. 8.24—The effect of pressure on the rate of reaction for the oxidation of NO to NO_2 under the concentration conditions which are commonly adopted in processes for the production of HNO_3. Under all concentration conditions, the importance of an increase in rate as the pressure is increased lies in the fact that it may also be possible to operate at higher temperatures under larger pressures.

The Fauser–Montecatini plants have been based on this solution, and Fig. 8.25 presents a complete panoramic view of such a plant including the ammonia combustion section where Frank–Caro/Bomag burners have been adopted with external mixing of the reagents.

The ammonia which has been vaporized in E_1 and the air which has been purified in PF_1 and PF_2 are combined in MX, preheated in E_2, and sent to be burnt in R_1–R_2.

The nitrous gases which are produced pass into E_2, produce steam in CV, and give up their heat to the exhaust gases in E_3 and to the cooling water in E_4. The condensate which is formed in E_3 and E_4 is first collected in D_1 from where the gaseous substances pass before going to the compressor P, which is driven by the turbine FT, and being pumped to one of a sequence ($A_{n/2}$) of absorption units. The gases which have been compressed to 3–5 atm. give up the heat released during their compression in E_5 and, having been enriched with oxygen, pass through the oxidation zone D_1 before passing to the first of a series of horizontal cylindrical absorbers where they encounter nitric acid with a concentration greater than 50% which originates from the higher absorbers. When brought into contact with the fresh nitrous vapours which are rich in oxygen, the concentration of the acid is raised to about 65%. It is this acid which is the product of the plant.

The battery of absorbers, cooled by the shower of water falling on to the upper parts of them, operates in a counter-current mode with a rising stream of gas and descending liquids. The first liquid consists of the water which has been pumped into the uppermost absorber A_n. Gases which are practically devoid of nitrous vapours bubble up into this water, and having been reduced to a gaseous mixture of nitrogen and oxygen, they first act as the cooling fluid in E_3 and are then allowed to expand in a turbine coupled to the compression system to recuperate some of the energy expended in their compression.

The extremely dilute acid solution which is formed in A_n descends into A_{n-1} by a simple overflow mechanism, receives the nitrous gases originating from A_{n-2} which are poor in the absorbable components, and, after bubbling through, go up to A_n. This process continues in the same way up to A_1 with the sole variant that the waters already containing nitric acid originating from D are received into the intermediate absorber $A_{n/2}$.

The direct production of concentrated nitric acid

Nowadays, various plant design solutions to the production of nitric acid with a concentration greater than 98%, without recourse to extractive distillation with the use of dehydrating agents, have been put into operation.

The most well known solution to this problem is the 'autoclave' method.

The gases produced by the combustion of ammonia are first allowed to give up their heat in E_1 (Fig. 8.26) with a reduction in their temperature to 600°C and are then cooled to about 200°C by producing steam in CV in the usual manner.

They are then strongly cooled in E_2 to allow practically pure water to condense out, part of which is subsequently used as absorbent liquid. After the removal of this water, the gases are compressed to 8 atm. in P, cooled once again in E_3, and then, after having received the oxygen-rich effluent gases from the denitrification tower, pass into T and then into T_1. Here, at −10°C, more than 80% of the NO_2 is transformed into liquid N_2O_4. The gas stream containing the residual nitrogen oxides passes on through the absorption columns C_1 and C_2 which produce a certain amount of dilute nitric acid. This and the liquefied N_2O_4 pass into R to which compressed oxygen is fed. There, the reaction:

$$N_2O_4 + 1/2O_2 + H_2O \rightarrow 2HNO_3$$

occurs and nitric acid at a concentration of greater than 96% is formed. After decompression and denitrosation in C_3, the acid is finally sent to storage.

By suitably regulating the amounts of N_2O_4 which are condensed and, therefore, the concentration of the nitric acid mixed with it in the autoclave, it is possible to

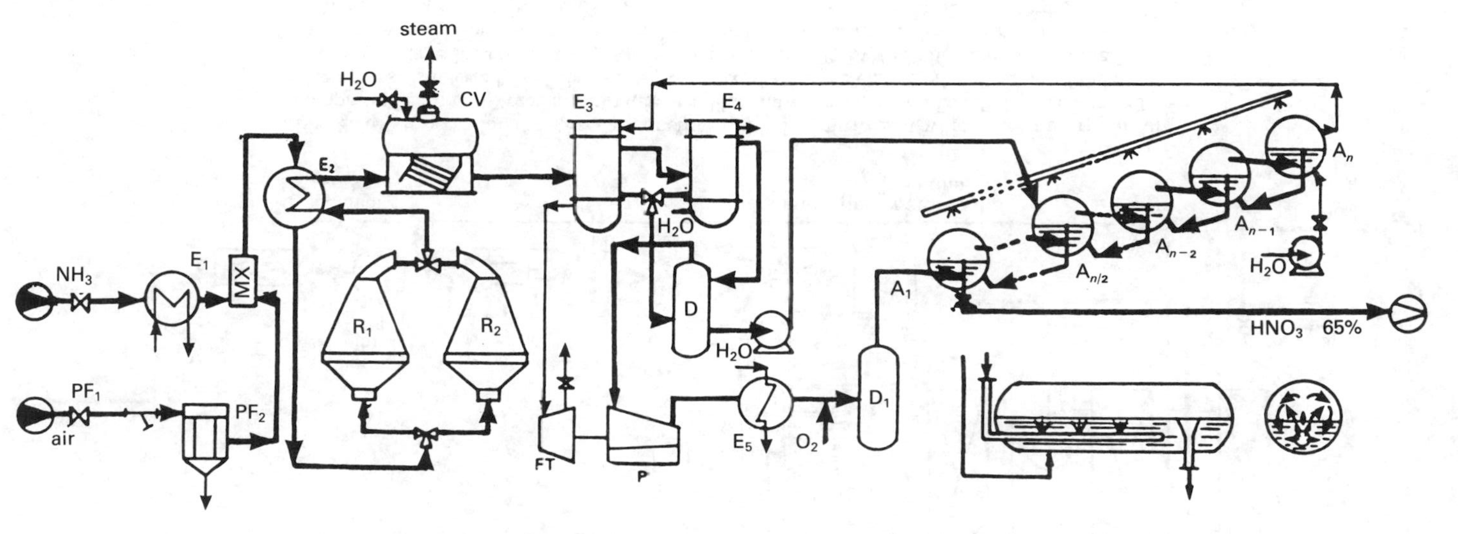

Fig. 8.25—Fauser–Montecatini (now Montedison) plant for the production of HNO_3 from NO under pressure. The procedure is described in the text. However, it should be noted that the gas to be absorbed arrives at $A_1, \ldots, A_n$ by passing through ducts with many small holes in their lateral walls. This creates turbulence which promotes absorption as is shown in the detail on the right hand side of this figure. Actually, dispersion of heat evolved during the absorption of the acid is favoured in this method because of the heat exchange between the absorber and the water that pours down their exteriors.

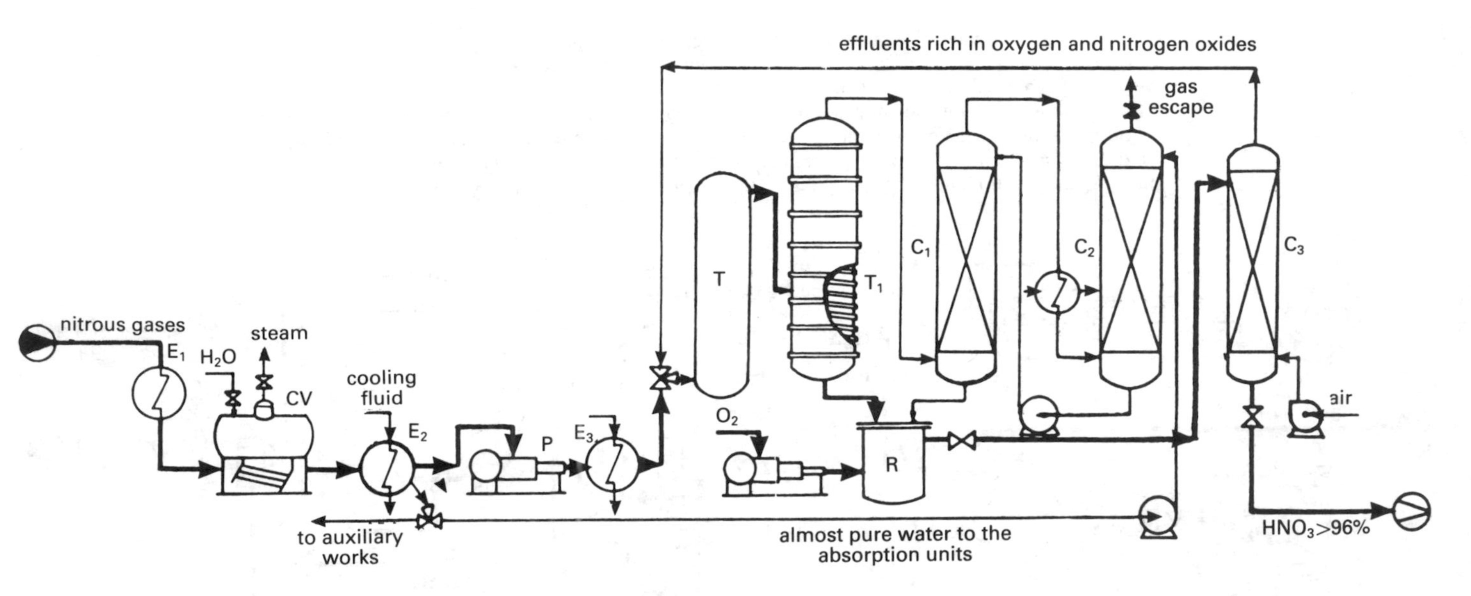

Fig. 8.26—Process for the direct production of fuming nitric acid which uses absorption of compressed gaseous oxygen in a mixture of a dilute nitric acid and liquified dinitrogen tetroxide. The process is referred to as the 'autoclave process' since the characteristic feature of the plant is the autoclave in which N_2O_4 is converted into HNO_3 by the intervention of water and oxygen. By regulating the pressure and lowering the temperature in T_1 (but not below −12°C as the N_2O_4 would solidify), it is possible to successfully condense high percentages (up to about 80%) of the N_2O_4 which is to be sent to the autoclave.

convert all of the NO obtained from the direct oxidation of ammonia into concentrated nitric acid without using recycling from C_3 to T.

8.7 COMMERCIAL FORMS, TRANSPORT, AND USES OF NITRIC ACID

Various grades of nitric acid are available commercially as can be seen from Table 8.4.

50–60% nitric acid has to be transported in vessels made out of stainless steel, stoneware, or well-protected, darkened glass.

The grades of nitric acid with concentrations of greater than 90% are stored, transported, and marketed in glass vessels protected by Kieselguhr or in passivated aluminium tanks.

Nitric acid is mainly used for the production of fertilizers and in the form of nitro-derivatives and nitrate esters (explosives). Other nitro-derivatives are used as propellants.

Nitric acid serves as a nitrating and oxidizing agent, and, indirectly, as an aminating reagent in all the chemical and parachemical industries covering the manufacture of dyes, resins and fibres, varnishes, pharmaceuticals, printing products, etc.

Technologies employed in the passivation of metals and those which make use of nitrous gases are also consumers of nitric acid.

8.8 THE HARMFUL EFFECTS OF NITRIC ACID AND THE OXIDES OF NITROGEN. ANTIDOTES

Nitric acid penetrates into the tissues with which it comes into contact, thereby causing irritation, burns, and ultimately, their destruction. The damage caused is proportional to the concentration of the acid and the time for which it is in contact with the tissue. On the skin, nitric acid immediately provokes the 'xanthoproteic' reaction which turns the tissue yellow and constitutes the beginning of the necrotization of the skin.

For an atmosphere which is inhaled for a period of 8 hours per day, the maximum tolerable limit of nitric acid is 10 ppm (25 mg/m^3).

The oxides of nitrogen ($NO + NO_2$) have an action on tissues which is similar to that of nitric acid. However, the most serious cases of poisoning by these gases are due to their inhalation. For exposures of 8 hours per day, the maximum tolerable limit for nitrous vapour is 5 ppm which is equal to 9 mg/m^3.

Table 8.4—Principal commercial grades of nitric acid and their properties

Denomination	*Characteristics*	*Density in g/cm³*	*Density in °Bē*	*Approx. %*
Commercial nitric acid	almost colourless	1.33	36	53
	yellow to reddish	1.49	47	93
Fuming nitric acid	yellow-reddish	1.50–1.508	48	96–98
Pure nitric acid	colourless	1.512	48.4	99.8–99.9

At concentrations of 100–200 ppm they cause expectoration, vomiting, and pulmonary oedema. This pathology is immediate when the concentrations are within the range from 200–700 ppm. Such exposure also rapidly leads to facial cyanosis (that is, the greenish-blue coloration of the face due to the formation of metahaemoglobin) accompanied by tachycardia and then comatose collapse which is a prelude to death.

The remedy for all skin contacts with nitric acid and nitrogen oxides is the immediate and prolonged irrigation of the affected parts, from which any clothing has been removed, with large amounts of water. The subsequent application of packs of lint saturated with sodium thiosulphate is advised.

Contacts with the eyes are also treated by thorough washing with water followed by repeated rinsing with a 2–5% solution of $NaHCO_3$.

The administration of emetics, liquid suspensions of aluminium and magnesium hydroxides, and of soapy suspensions of very dilute lime in water is used as an antidote against ingestions of these chemicals.

Among foodstuffs, milk is the one which is best swallowed as an antidote against the oxygenated compounds of nitrogen.

8.9 THE CLEANING UP OF THE GASEOUS EFFLUENTS FROM NITRIC ACID PLANTS

Polluting nitrogen oxides, N_xO_y, where, generally, $x = 1$ and $y = 1$ or 2, are emitted from all plants which produce or use nitric acid.

The legislation appertaining in different countries tends, quite justly, to lay down ever greater restrictions regarding the levels in ppm which may be present in such effluent gases on account of the ecological damage arising from the presence of nitrous gases which is detrimental to all flora and fauna.

There are two types of system used to clean up such effluent gases:

- catalytic reduction of the polluting nitrogen oxides to nitrogen by low molecular weight alkanes or ammonia;
- selective absorption in concentrated nitric acid followed by the deabsorption of these oxides.

The first method is the most widely used on account of the definitive manner in which it removes the potential pollutants. If, as usual, we take methane as an example of a low molecular weight alkane, this method is based on reactions of the type:

$$CH_4 + 4NO_2 \rightarrow 4NO + CO_2 + 2H_2O \tag{8.22}$$

$$CH_4 + 4NO \rightarrow 2N_2 + CO_2 + 2H_2O \tag{8.23}$$

or of the type:

$$4NH_3 + 6NO \rightarrow 5N_2 + 6H_2O$$

Nitrogen-containing pollutants are thereby reduced to innocuous nitrogen which can be allowed to escape into the atmosphere.

These reactions, which are catalyzed by platinum and palladium supported on acidic oxide substrates, are carried out in such a way as to recover both mechanical and thermal energy.

If oxygen is also present in the polluting gases it is inevitable that it will also react with the hydrocarbons and even preferentially over the oxides of nitrogen. On the other hand, if ammonia is used as the reagent to clean up the gases, it does not react with the oxygen, provided that the conversion temperature is kept below 500°C and the reaction is suitably catalyzed.

HYDROCYANIC ACID AND CYANIDES

8.10 PROPERTIES AND USES OF HYDROCYANIC ACID

Hydrogen cyanide, which is technically generally referred to as hydrocyanic acid (HCN), is a colourless, volatile (b.p. 25.6°C) liquid which is miscible with water and extremely toxic. As an acid, HCN is very, very weak with $K_a = 5 \times 10^{-10}$.

The cyanide ion has a mesomeric structure:

$$\{|C{\equiv}N| \leftrightarrow |C{=}\bar{N}|\}^-$$

and, in reality, the two hydrides HCN and CNH which give rise to two series of esters, the nitriles R—CN and the isonitriles (carbylamines) C=N—R, turn out to correspond to these two canonical structures.

Raman spectra, however, demonstrate that the classical hydrocyanic acid consists almost entirely of the HCN form.

It should also be noted that the claimed volatility of HCN is apparent only because its boiling point of 25.6°C is, in fact, very high when compared with that of CO (b.p. −191°C) which is an isostere* of the cyanide ion.

This, apparently, is a consequence of the hydrogen bonds which HCN can form.

Hydrocyanic acid is of great industrial importance in connection with the development of the production of the most widely varying acrylic resins, nitriles, metal cyanides, and in disinfestation processes which make use of cyanides. Traces of H_2SO_4 or of SO_2 (which extend its action to the gaseous state), are used as antipolymerization agents in HCN solutions.

8.11 THE INDUSTRIAL PRODUCTION OF HCN

The classical industrial source of hydrocyanic acid is 'coke plant gases'; the hydrocyanic acid usually being fixed as cyanide at the point where these gases were purified. However, most of this acid which is produced industrially uses methane as the source of carbon, ammonia as the source of nitrogen, and air for its production:

$$2CH_4 + 2NH_3 + 3O_2 \rightarrow 2HCN + 6H_2O \qquad (8.24)$$

The reaction is carried out at a temperature of 1100°C, using platinum meshes as the catalyst. Under these conditions, the above reaction is somewhat favoured, but it is always accompanied by the conversion of the methane by a part of the water which is formed, by dissociation of the ammonia and by various oxidations. The resulting reaction mixture, which is extremely complex, therefore makes it necessary to install a series of pieces of apparatus for the production of pure HCN.

A diagram of a plant for the production and purification of hydrocyanic acid, starting out from methane, ammonia, and air, is shown in Fig. 8.27.

After the gaseous reagents have been filtered in PF_1, PF_2, and PF_3, they are mixed in MX and transported into the reactor R. They are rapidly preheated in the upper conical part of the reactor by the reverbatory action of the refractory walls. The gases then pass directly downward toward the bottom onto platinum meshes where they undergo reaction. Some of the reaction products are withdrawn through a side outlet, and, by producing steam in CV, have their temperature lowered. By returning these products to the reactor lower down, they

*Chemical species which have the same total number of protons in the nuclei of their atoms and, consequently, an equal number of extranuclear electrons are isosteres.

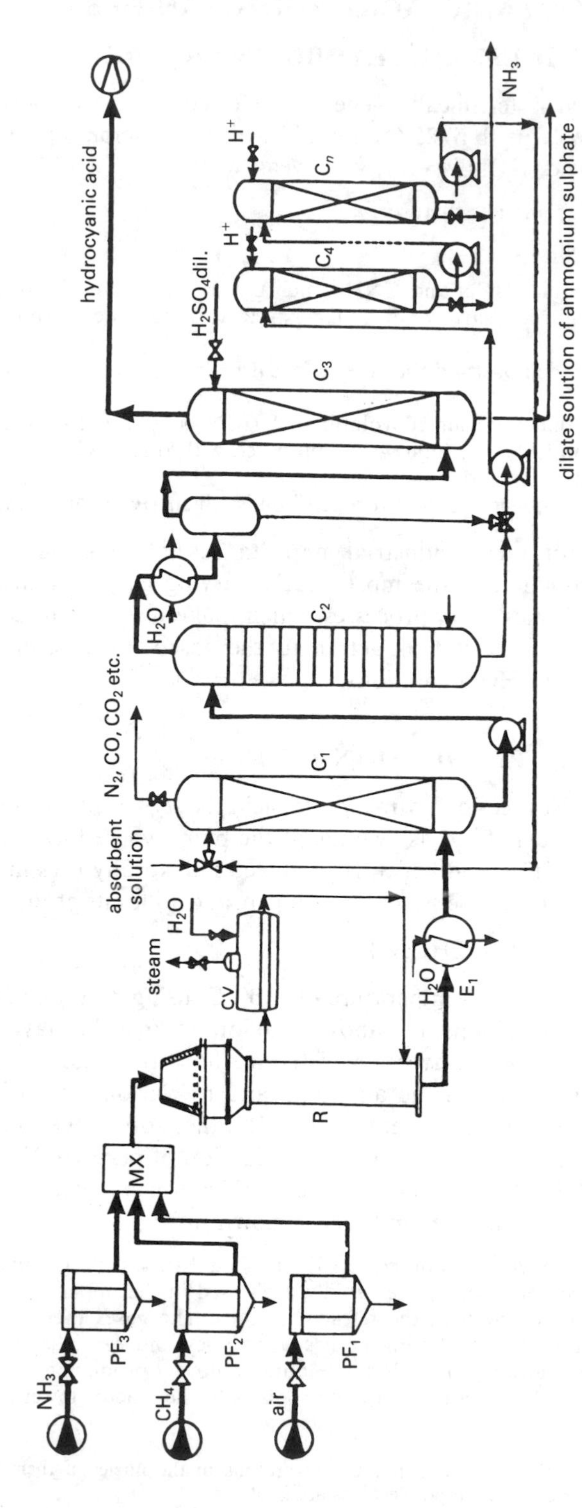

Fig. 8.27—Plant for the production of HCN by oxidative synthesis from CH_4, NH_3 and air, the separation of the reaction products, and the recovery of the unreacted ammonia. Nowadays, apart from being produced by this route (the 'Andrussov process'), which is the most commonly used, hydrocyanic acid is also produced from CH_4 and NO (Pt catalyst) and CH_4 and NH_3 (Pt catalyst, temperature $>1000°C$). Finally, HCN is obtained as a by-product in the production of nitriles such as acrylonitrile from propylene and ammonia and propylene and nitric oxide.

can be used to quench the reactivity of the reaction mixture at the point where they are about to leave the reactor. This quenching is completed in E_1 which is followed by an absorption column C_1 containing an aqueous solution of pentaerythritol (8.5%) and boric acid (2.5%). This acid chemically fixes any ammonia which has not reacted in R, and the polyalcohol–water mixture dissolves the hydrocyanic acid. Inert gases (N_2, CH_4, CO, etc.) escape from the top of this column.

The hydrocyanic acid is stripped in column C_2, while the greater part of the ammonia remains combined.

The packed column C_3, into which sulphuric acid is sprayed, subsequently removes every trace of ammonia and liberates hydrocyanic acid which is sent to be dried and stored.

The solution containing ammonium borate and pentaerythritol passes into a series of ion exchangers $C_4, \ldots, C_n$ filled with acid ion exchange resins, which remove the ammonia and regenerate the solution used for the absorption stage in column C_1.

The ammonia is then removed from the ion exchange resins, which are successively regenerated, in the form of an ammonium salt. This may be sent either to be used as a fertilizer or for displacement with alkali with the aim of recycling the ammonia through the same plant.

8.12 THE INDUSTRIAL PRODUCTION OF CYANIDES

The cyanide of greatest practical interest is sodium cyanide, followed by potassium and calcium cyanides. These are very soluble compounds which undergo pronounced hydrolysis and are marketed in the form of oval nut-shaped pieces containing 96–98% cyanide.

The cyanides, in general, and sodium cyanide in particular, are used particularly in precious metal metallurgy. Very considerable amounts of cyanides are also consumed in flotation processes and certain electrochemical processes (galvanic techniques and metal refining). Alkali metal cyanides are also used in the nitriding of steels.

Sodium cyanide is prepared industrially by synthesis in steel reactors (Fig. 8.28).

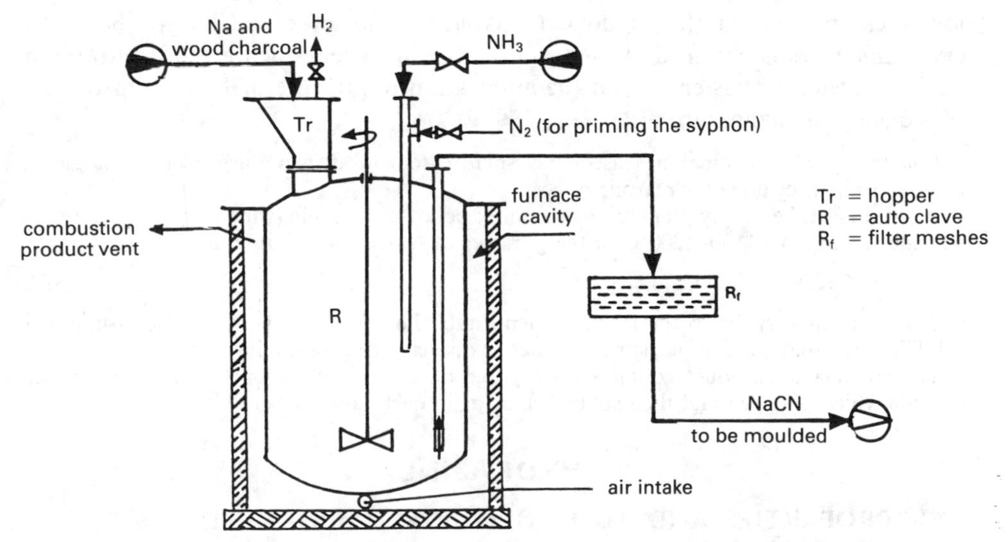

Fig. 8.28—Autoclave for the synthesis of NaCN, fitted with a mixer and devices for loading the reagents (Na, C, and NH_3), the release of H_2, and the discharge of the product NaCN. The insulated space in which the gases or petroleum fractions are burnt in a modulated manner to heat the reactor, can also be seen.

These are first loaded with metallic sodium and pre-calcined wood charcoal. Ammonia is then introduced and the apparatus is heated externally by the burning of gas or combustible oils. This leads to a rapid rise in temperature to about 500°C after which it is allowed to increase at a lower rate as the following exothermic reactions take place:

$$2Na + NH_3 \rightarrow 2NaNH_2 + H_2 \qquad (8.25)$$

$$2NaNH_2 + C \rightarrow Na_2CN_2 + 2H_2 \qquad (8.26)$$

Above 650°C, the endothermic transformation of sodium cyanamide into sodium cyanide also becomes ever more probable:

$$Na_2CN_2 + C \rightarrow 2NaCN \qquad (8.27)$$

At 700–750°C, the resulting mass consists of a suspension of carbon in a molten mixture of sodium cyanide (90%) and sodium cyanamide.

By further heating, the mass is brought up to a temperature of 800–850°C at which it is maintained for about one and a half hours. During this period the percentage of cyanide increases to 96–98%.

The contents of the autoclave are then removed by siphoning and filtered by passing through fixed metal meshes or layers of carbon. The filtrate is poured into dies where it is allowed to cystallize in the shapes which are marketed commercially.

The yields are high, and, in fact, almost 95% of both the sodium and ammonia which are reacted are converted into cyanide.

Potassium cyanide is mainly prepared by the neutralization of potassium hydroxide with hydrocyanic acid at such a concentration of the alkali as to yield about a 30% solution of the cyanide. At the end of the process there must still be a slight excess of the alkali present. Otherwise, during its later concentration, the hydrocyanic acid would tend to polymerize, turning the whole mass red.

The concentration is carried out at relatively low temperatures (up to 40°C) and low pressures. All of this is done to avoid any losses of HCN. At the end a crystalline mass is obtained in the form of a fine powder which is made into lumps agglomerated by moistening and the application of pressure or fused with the aim of reducing the danger from it.

Calcium cyanide is sometimes used to a small extent in all metallurgical technologies in place of sodium cyanide for reasons of cost.

It can, in fact, be simply prepared by heating calcium cyanamide (which, as has been noted, occludes carbon) to 1100–1200°C in the presence of NaCl. The equilibrium:

$$CaCN_2 + C \rightleftharpoons Ca(CN)_2 \qquad (8.28)$$

which is driven very far to the right, is then established. The molten $Ca(CN)_2$ produced is solidified by rapid cooling of the mass which is discharged from the furnaces.

The commercial product contains an average of 42% (a maximum of 50%) of calcium cyanide which is stored and then sold in chips or in dark grey blocks ('black cyanide').

HYDRAZINE

8.13 PROPERTIES AND TECHNOLOGICAL IMPORTANCE OF HYDRAZINE

Hydrazine (NH_2–NH_2) bears the same relation to ammonia as hydrogen peroxide does to water. Hydrazine even has the spatial structure of hydrogen peroxide.

Meanwhile, however, while hydrogen peroxide tends to lose one of its two oxygen atoms and become reduced to water, hydrazine produces only the very stable nitrogen molecule which accounts for the fact that its decomposition is accompanied by the release of a large amount of energy.

In the anhydrous state hydrazine is a viscous liquid which is metastable* at ambient temperatures but may explode with great destructive violence at 200–300°C on account of its high energy content which confers on it an explosive power which is greater than that of trinitrotoluene.

Hydrazine, which has been known for many years as a reagent for aldehydes and ketones in organic chemistry, finds important technological applications today as:

- a synthetic intermediate in organic chemistry;
- a blowing reagent for expanded resins and elastomers;
- a reducing agent in metallurgical technologies and in the treatment of boiler feed waters where oxygen must be excluded;
- a component of rocket propellant mixtures on account of the high specific impulse which it is known to confer to an ascending body.

8.14 THE INDUSTRIAL PRODUCTION OF HYDRAZINE

In spite of all the other means of preparing hydrazine which have been examined, the Raschig method still remains the most convenient on account of low cost and the high overall yield of the process.

From a chemical point of view, two reactions are involved in the Raschig process: the oxidation of ammonia to chloramine with hypochlorite:

$$NH_3 + OCl^- \rightarrow OH^- + NH_2Cl \tag{8.29}$$

and the dehydrochlorination of the chloramine–ammonia mixture:

$$NH_2Cl + NH_3 + OH^- \rightarrow Cl^- + H_2O + NH_2\text{–}NH_2 \tag{8.30}$$

The slow step in the process is represented by the second reaction which takes place in competition with the relatively fast reaction involving the dismutation of chloramine to yield nitrogen and ammonia:

$$3NH_2Cl + 3OH^- \rightarrow N_2 + NH_3 + 3H_2O + 3Cl^- \tag{8.31}$$

which is obviously promoted by a high concentration of chloramine and/or of alkali. It is therefore necessary to use a low concentration of alkali metal hydroxide. The reaction, which is parasitic with respect to the dehydrochlorination reaction leading to hydrazine, is an oxidation–reduction reaction which involves the decomposition of the hydrazine which has been formed:

$$2NH_2Cl + H_2N\text{–}NH_2 \rightarrow N_2 + 2NH_4Cl \tag{8.32}$$

Heavy metals ions (Cu^{2+}, Fe^{3+}, Ni^{2+}, etc.) catalyze this reaction. However, the effect of these ions is eliminated by the addition to the reagents of small quantities of complexing protein (casein or gelatine) mixtures.

---

*Metastable in so far as it is not thermodynamically stable even if, for kinetic reasons, it does not undergo any detectable decomposition.

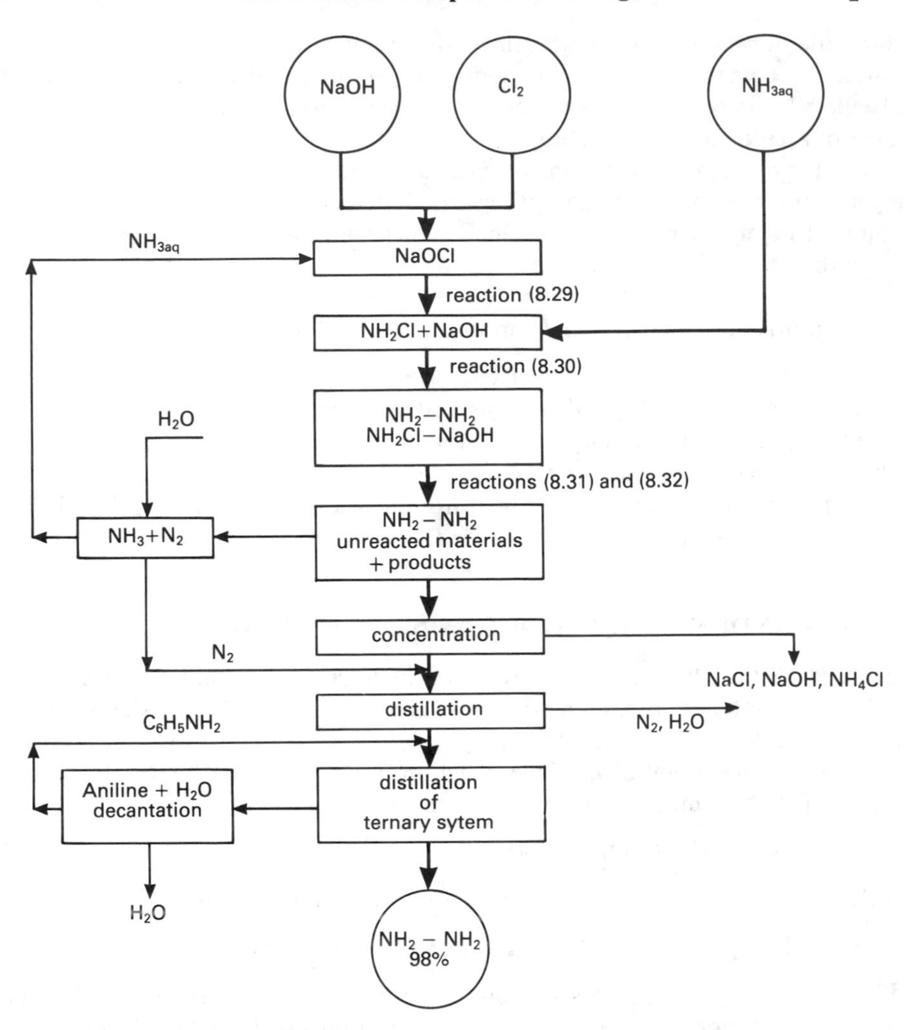

Fig. 8.29—Block diagram of the synthesis of hydrazine by the Raschig method.

The optimal preparative conditions in this case have been shown to be: a relatively high reaction temperature of 130–140°C and the use of a considerable excess, that is, a two to one ratio, of ammonia to hypochlorite.

The various steps in plants for the industrial production of hydrazine by the Raschig method are shown by the block diagram in Fig. 8.29.

Hypochlorite is formed in the first reaction which then oxidizes recycled ammonia to chloramine. After the addition of more ammonia, the alkali removes HCl from the $NH_2Cl + NH_3$ system (dehydrochlorination). Hydrazine is thereby formed.

On account of the competing and parasitic reaction which takes place, the system is composed of N_2, a large amount of ammonia (the excess ammonia and that produced in the competing reaction), and various secondary residual components and the reaction products.

The ammonia, which has been freed by the alkali present in the system, escapes with the nitrogen. The rest of the mixture is sent to be concentrated, when the electrolytes are crystallized out.

The new hydrazine–water system, which is distilled under nitrogen to reduce the decomposition of the hydrazine, forms an azeotrope with water which contains 68% of hydrazine. When recycled aniline is added, the azeotrope forms a ternary system which, when distilled, leads to almost anhydrous hydrazine. Meanwhile the aniline is removed by the water from which, after condensation, it can be separated by decantation and can thereby be recycled.

9

The halogens within the framework of industrial inorganic chemistry

The applied importance of the elements fluorine, chlorine, bromine, and iodine (the 'halogens') has become ever greater in the field of industrial organic chemistry. Nevertheless, their production and direct exploitation and the preparation and uses of their hydrides and salts are questions which are always still of concern in the field of industrial inorganic chemistry.

Among all the halogens, chlorine stands out above the others in importance in every field, followed, nowadays, by fluorine which is finding applications at an ever-increasing rate.

9.1 THE PROPERTIES OF THE HALOGENS

From a chemical and physical point of view, the halogens are among the most characteristic elements as typified by the darkening in colour, the change from gas to liquid and then solid, the decrease in the electronegativity, reduction potential, and reactivity upon passing from the lightest to the heaviest, and their other particular properties which we shall consider later.

The most pertinent physicochemical properties of the halogen subgroup are summarized in Table 9.1.

By inspecting some of the values presented there it can be seen that, at normal temperatures, fluorine (F_2) is a yellowish gas, chlorine (Cl_2) is a green vapour, bromine (Br_2) is a red liquid, and iodine (I_2) is a greyish violet solid exhibiting metallic reflection. The course of these properties and the sharp decrease in electronegativity which is encountered at the same rate in the halogens leads one to conclude that there is a sharp increase in metallic character in these elements as their atomic number increases*.

*it is well known that the progressive increase in metallic character in a subgroup is among the most characteristic of the periodic properties of the elements. Nevertheless, here this property is particularly marked.

Table 9.1—Summary of the properties of the halogens

Property \ *Halogen*	*Fluorine*	*Chlorine*	*Bromine*	*Iodine*
Atomic number	9	17	35	53
Atomic weight	19	35.46	79.91	126.9
Colour	yellow–green	green	dark red	violet vapour, grey metallic solid
Degree of dissociation (% at 1000 K)	–	0.035	0.23	2.8
Melting point (°C)	−223	−103	−7.2	+113
Boiling point (°C)	−188	+34.6	+58.8	+184.4
Critical temperature (°C)	−156	+145	+303	+552
Relative density (air at 0°C and 1 atm. = 1)	1.31	2.46	–	–
Specific weight (g/cm^3) (of solids)	1.108	1.57	3.19	4.93
Electrochemical reduction potential (volts)	+2.8	+1.36	+1.08	+0.58
Electronegativity	4	3	2.8	2.5

From the structural point of view, the series of properties which are manifested by the halogens as one passes down the group is reflected with the increase in the delocalizability of their extranuclear electron clouds with increasing distance from the atomic nucleus. It is this greater facility to give up electrons and, hence, a smaller tendency to gain electrons, which permits the development of intermolecular van der Waal's forces.

From a strictly chemical point of view the halogens are characterized by a progressively increasing affinity for oxygen upon passing from fluorine to iodine, and a concomitant decrease in their affinity for hydrogen. As a consequence of this, the electrochemical displacements which are encountered among the halogens are rather unusual. It is observed that:

- when the halogens have an oxidation number of −1, the lighter halogen (i.e. the more electronegative) displaces the heavier halogen (the less electronegative),
- when the halogens have a positive oxidation number the heavier halogen displaces the lighter halogen.

Examples of these two rules in action are provided by the reactions:

$$2KCl + F_2 \rightarrow 2KF + Cl_2$$

$$2KClO_3 + Br_2 \rightarrow 2KBrO_3 + Cl_2$$

$$KClO_3 + KI \rightarrow KIO_3 + KCl$$

In other words, a lighter halogen oxidizes a heavier halogen ion, while a heavier halogen ion reduces the oxygenated compounds of a lighter halogen.

Finally, it is noted that the extreme reluctance on the part of fluorine to give up electrons does not allow it to possess high positive oxidation numbers.

In fact, the oxidation number of fluorine in the compound F_2O (oxygen difluoride) is -1, and the only extremely unstable fluorate which is known practically ($AgFO_3$) can exist only by virtue of the fact that there are a total of three oxygen atoms in it which succeed in very slightly overcoming the resistance to fluorine to give up electrons. The compound is, however, exceedingly labile.

9.2 THE INORGANIC FLUORINE INDUSTRY

Fluorine is an element which is quite diffuse in nature and occurs in the Earth's crust with an abundance of approximately $2.7 \times 10^{-2}\%$. The fluorine minerals which are used for industrial purposes are fluorite (fluorspar) CaF_2, fluorapatite $3Ca_3(PO_4)_2 \cdot CaF_2$, and cryolite Na_3AlF_6.

In fact, from the point of view of their industrial utilization, it may be said that:

- all the fluorite is destined, firstly, to the production of HF and, secondly, to the production of fluorine,
- besides a little CaF_2 which is used for the same purposes as natural fluorite, fluorapatites indirectly furnish all of the fluorosilicic acid and sodium fluorosilicate via the industries concerned with phosphorus and its derivatives. These fluorosilicates serve for the production of the many hexafluorosilicates which are required in the most diverse chemicotechnical applications,
- the cryolite, the natural variety of which is in short supply at the present time, is all used (with a need for even more) in the metallurgy of aluminium.

Fluoride-containing by-products which are made available by the industries producing phosphorus and its derivatives (page 359), and which are not used in hexafluorosilicates or in the production of HF from CaF_2, are employed in the production of artificial cryolite.

Hydrofluoric acid

The cycle for the production of hydrofluoric acid, starting out especially from natural CaF_2 and, less often, that recovered from fluorapatite, is shown schematically in Fig. 9.1.

The reaction which takes place in the rotating furnaces mentioned in this scheme is:

$$CaF_2 + H_2SO_4 \rightarrow CaSO_4 + 2HF$$

The furnaces used for the reaction are constructed from cast iron, and the parts which come into contact with anhydrous hydrofluoric acid are of ordinary steel since iron is a metal which, by becoming passivated, offers good resistance towards anhydrous hydrofluoric acid.

Aqueous solutions of HF of various concentrations (up to 60% by weight) are also marketed. These solutions are kept in lead vessels when there is no H_2SiF_6 present, in Monel vessels in the absence of O_2 and S, or vessels made of graphite or bronze.

Pump rotors, valves, and various pieces of instrumentation are protected against the action of hydrofluoric acid by providing them with coverings of polytetrafluorethylene (PTFE).

The main uses of hydrofluoric acid are in the production of the halogen, simple and polymeric fluorinated organic compounds, inorganic fluorides, and in the nuclear industry (metallurgical treatments of uranium and the transuranic elements).

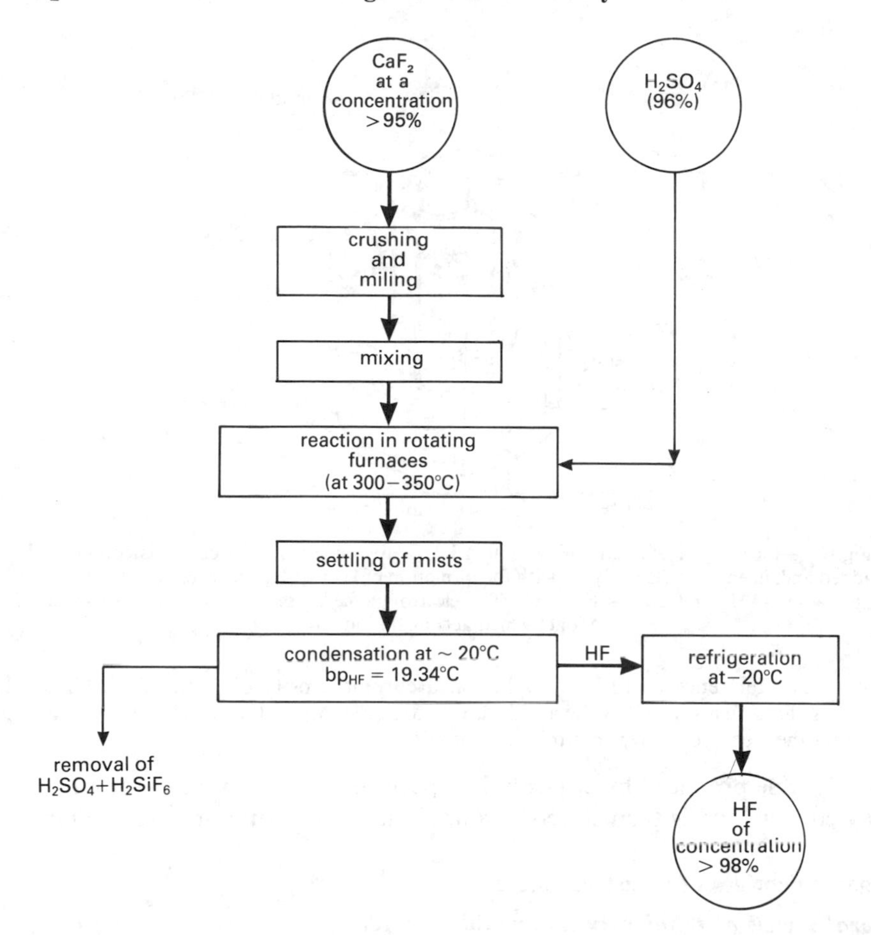

Fig. 9.1—Diagram of the production of HF in a practically anhydrous state. However, the anhydrous acid and the dilute acid are often both produced in the same plant by suitable absorption of the exhaust and tail gases.

Fluorine

Fluorine is prepared on a substantial scale by the electrolysis of HF. However, this preparation must be carried out in the complete absence of any oxygenated compounds, including water, as the anodic discharge (oxidation) of oxygen would be inevitable, were they to be present, given that fluorine has the highest redox potential of any element.

In practice, a solution of HF in KHF_2 is electrolyzed, after having prepared the potassium hydrogen fluoride by the reaction:

$$K_2CO_3 + 4HF \rightarrow 2KHF_2 + CO_2 + H_2O$$

from which the acid salt to be electrolyzed is obtained in a crystalline form.

When put in the molten state into a steel or magnesium–manganese alloy double-walled cell with the thermostatting coils in the space between the two walls (Fig. 9.2), KHF_2 remains molten above 100°C, in spite of the fact that its melting point is 239°C, if small amounts of LiF (1.2%) are added to it, and a stream of HF is continuously bubbled into it in such a way as to maintain a molar HF/KF ratio of 2:1.

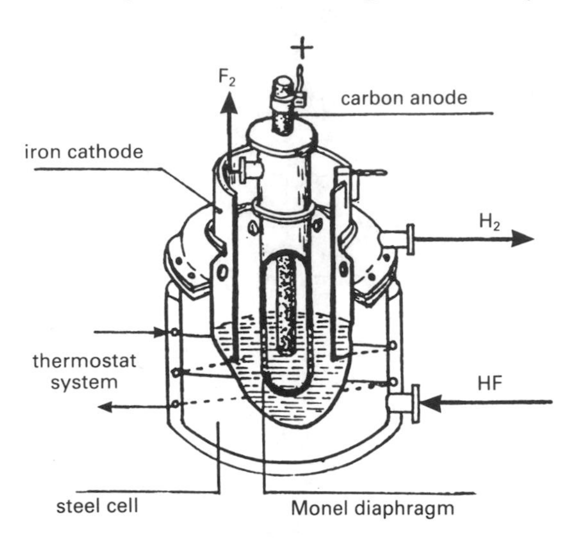

Fig. 9.2—Electrolytic cell containing molten KHF_2 to which a stream of HF is continuously added and, in effect, is electrolyzed with the formation of H_2 and F_2 according to the reactions: $2H^+ + 2e^- \rightarrow H_2$ and $2F^- \rightarrow F_2 + 2e^-$. The electrolytic cells used here can also be made of Monel which acts as the cathode.

A potential difference of 12 V is applied at a current density of 9–13 A/m^2. Under these conditions, fluorine is formed at the anode and hydrogen at the cathode. The Monel diaphragm separating the electrodes prevents mixing of the two gases.

The fluorine produced by electrolysis is purified by the removal of any HF which it may contain, and is then stored as a liquid under pressure in a steel bomb.

Inorganic fluorides of industrial interest

General review of tehnically important fluorides

Apart from the fluorides of other halogens (BrF_3, ClF_5, IF_7, etc.), there are, among the highly reactive compounds of fluorine, various metallic fluorides: CoF_2, Ag_2F, Hg_2F_2, MnF_3, and so on. Fluorination reactions are carried out with these compounds, thereby avoiding the use of fluorine which is too violent.

There is also a solid, non-corrosive substitute for HF among the fluorides which is more manageable than HF. This is ammonium bifluoride which is normally prepared from a stoichiometric mixture of the two reagents:

$$NH_3 + 2HF \rightarrow NH_4HF_2$$

Ammonium bifluoride is used in the grinding of glass and for the removal of oxides, silica, and silicates in metallurgical processes since it reacts like HF at elevated temperatures.

On the other hand, sulphur hexafluoride, SF_6, which has already been discussed (page 229), is an inert inorganic compound like nitrogen.

Other fluorides exhibit various degrees of reactivity. For instance, SiF_4 forms fluorosilicates with metal fluorides. The most important of these is Na_2SiF_6 (since other fluorosilicates can be prepared from the sodium salt by exchange). Sodium fluorosilicate is a blocking additive which is added to detergents, a disinfectant for drinking waters, a protective agent for stone structures exposed to inclement weather, a substitute for cryolite in glazes, for example, and so on.

The fluorides of boron and aluminium

BF_3 and HBF_4. Boron trifluoride is classified as one of the highly reactive compounds of fluorine and can be prepared from fluorosulphonic acid ($HF + SO_3 \rightarrow FSO_3H$) by the reaction:

$$3FSO_3H + H_3BO_3 \rightarrow BF_3 + 3H_2SO_4$$

This strong 'Lewis acid' may be used for the preparation of fluoroboric acid:

$$4BF_3 + 3H_2O \rightarrow H_3BO_3 + 3HBF_4$$

which is of considerable interest to the electrochemical industry. However, the principal use of BF_3 is as catalyst in condensation reactions, cationic polymerization, and isomerization.

AlF_3 and Na_3AlF_6. The metallurgy of aluminium requires considerable amounts of both aluminium trifluoride and trisodium aluminium hexafluoride ('cryolite'). At the present time, the white crystalline salt AlF_3 is mainly prepared by reacting alumina with hydrofluoric acid:

$$Al_2O_3 + 6HF \rightarrow 2AlF_3 + 3H_2O$$

which is carried out in special towers made of Inconel where the gas comes into contact with alumina as it rises.

The sodium salt is the most important of the 'fluoroaluminates', that is, the salts which can be prepared by reacting alkali fluorides with AlF_3. The only workable mines where Na_3AlF_6 is found are located in Greenland, and these are on the way to becoming exhausted; and Na_3AlF_6 is prepared at the present time by bubbling HF into a suspension of alumina in caustic soda:

$$6NaOH + Al_2O_3 + 12HF \rightarrow 2Na_3AlF_6 + 9H_2O$$

The gaseous effluents from the production of both phosphoric acid and of fertilizers from fluorapatites (page 358) may be used as the source of hydrofluoric acid. The cryolite which precipitates out, is filtered, washed, and calcined at a temperature of approximately 600°C.

Cryolite, which is given its name on account of the translucent crystals, similar to ice, which it forms, is used in many other branches of metallurgy in addition to its utilization in the metallurgy of aluminium. It is, furthermore, of notable importance in the ceramic industries and in the production of glazes.

9.3 CHLORINE AND ITS INDUSTRIAL INORGANIC COMPOUNDS

The halogen which is used to the greatest extent by far in industry is chlorine. It is estimated that the annual world production at the present time stands at more than 30×10^6 tons. Moreover, it is also the halogen which is the most abundant in nature where it is one of the most abundant elements in absolute terms being placed about tenth in the list with an abundance of 0.2%.

The most important of all salts, NaCl, is formed from chlorine and its importance has reached the point where it is simply known as 'salt' and should not be considered as one of the industrial compounds of chlorine in so far as the halogen itself is recovered from salt and not the other way round.

The industrial products of chlorine are hydrochloric acid* and certain of its salts: the hypochlorites, chlorates and perchlorates.

Chlorine

The halogen, chlorine, is produced industrially by means of electrochemical processes starting out from sodium chloride (salt). The overall reaction which, at first sight, appears to be easy to realize (but with the expenditure of energy) is:

$$2NaCl + 2H_2O \rightarrow Cl_2 + 2NaOH + H_2 \tag{9.1}$$

In reality, this process is complicated by the fact that the main products (Cl_2 and NaOH) remain in contact after they have been formed and thereby tend to react with one another to yield oxygenated compounds of chlorine, as we shall see.

On the other hand, the fact that chlorine can also return to react with hydrogen does not give rise to any problems because it is easy to insert a diaphragm between the respective compartments.

For this reason, special electrolytic cells have had to be developed in which reaction (9.1) is made to occur while the reverse reactions are more or less impeded.

In their historical order, the types of cells which have been developed are:

- diaphragm cells,
- mercury cathode or 'amalgam' cells,
- membrane cells.

Much attention has been paid to the development of membrane cells in recent years on account of the need to avoid both the ecological troubles caused by working with mercury cathode cells and the professional illnesses arising from the technology used in the construction of porous asbestos diaphragms.

The development of the new cells is closely linked with the perfection of their membranes.

The production and purification of brine for the production of chlorine

Quite apart from the type of cell which is employed for the production of chlorine, provision must be made in advance for the preparation of a suitable brine† to be electrolyzed. For reasons which we shall state, particular attention is paid to this operation if amalgam methods are adopted, and even more so in the case of membrane methods. As will be seen, the latter process requires final purification stages in which Ca^{2+}, Mg^{2+}, and SO_4^{2-} ions are removed by means of ion exchange resins.

The first phase of the preparation of the brine comprises the dissolution of sea salt and rock salt (page 418), with vigorous mixing, in water and diluted brine which has been recovered from the cells. This is done in dissolution and decantation vessels, from the bottom of which larger pieces of insoluble materials are discharged. The amount of material which is initially dissolved is measured out on the basis of its NaCl content in relation to the amount of solvent and the amount of NaCl contributed by the recycled brine, which, as we shall see, is obviously added so as to adjust the

*As in the case of hydrofluoric acid, the term hydrochloric acid is used in industry to indicate every form of HCl, whether pure or dissolved in water.

†The term 'brine' is the technical name of the aqueous solution of NaCl which is fed into the cells.

concentration of the brine to that which it must have when introduced into the cells. Somewhat more concentrated solutions are required for diaphragm and membrane cells than for amalgam cells.

The brine is purified (Fig. 9.3):

- by the addition of Na_2CO_3 and NaOH with the aim of precipitating Ca^{2+}, Mg^{2+}, Fe^{3+}, and Al^{3+} ions;
- by the addition of NaOCl if, for any reason, there is any ammonia present, and $BaCl_2$ to eliminate the SO_4^{2-} ions. Accurately monitored amounts of $BaCl_2$ must be added as residual Ba^{2+} ions cannot be tolerated.

The second stage in the preparation of the brine comprises its clarification, filtration, the addition of recycled brine, adjustment of the pH value, and its collection into the tanks which feed the cells. After this, provision is made for the brine feedstock to be brought to the correct temperature.

Diaphragm cells

Since the prototype was built long ago in 1892, experiments have been carried out on many types of 'diaphragm' cells. A common feature of all of these cells has always been the existence of a porous diaphragm between the electrodes and the flow of brine from the anodic(+) region to the cathodic region(−) through this diaphragm.

On the other hand, there are many differences in the ways in which the cells are constructed, the electrode separation, and the materials used for the anode and the diaphragm.

Electrodes. While the construction of the cathode from a steel mesh or a perforated steel sheet has remained unchanged, the anode has changed from a graphite composition to a structure which, for the most part, is made of titanium plates covered within thin layers of Group VIII oxides with metallic conductivity. The advantage of these replacement electrodes lies both in the reduction of the overvoltage for the development of chlorine and in the dimensional stability of such electrodes.

The graphite electrodes were not dimensionally stable owing to the partial occurrence of the reaction:

$$C + 4OH^- \rightarrow CO_2 + 2H_2O + 4e^-$$

on them as a result of which the electrodes were consumed and the plants had to be closed down while the replacement and realignment of the electrodes was carried out.

Diaphragm. As regards the diaphragm which was traditionally made out of asbestos fibres variously deposited* onto the side of the cathode plate facing the anode, the modifications consist, firstly, of the inclusion of artificial fibres and then of fluorinated polymers which are thermally sintered together with the asbestos to enhance the mechanical strength of the system and to minimize the diffusion of gaseous hydrogen through the diaphragm which would lead to the contamination of the chlorine.

Even today, however, the diaphragms must be periodically replaced because the

---

*The technique which is most commonly used for this purpose consists of the immersing of the cathodes in a suspension of fibres of suitable length and aspirating them from one side in such a way that a uniform layer, which acts as a diaphragm, is formed on the other side.

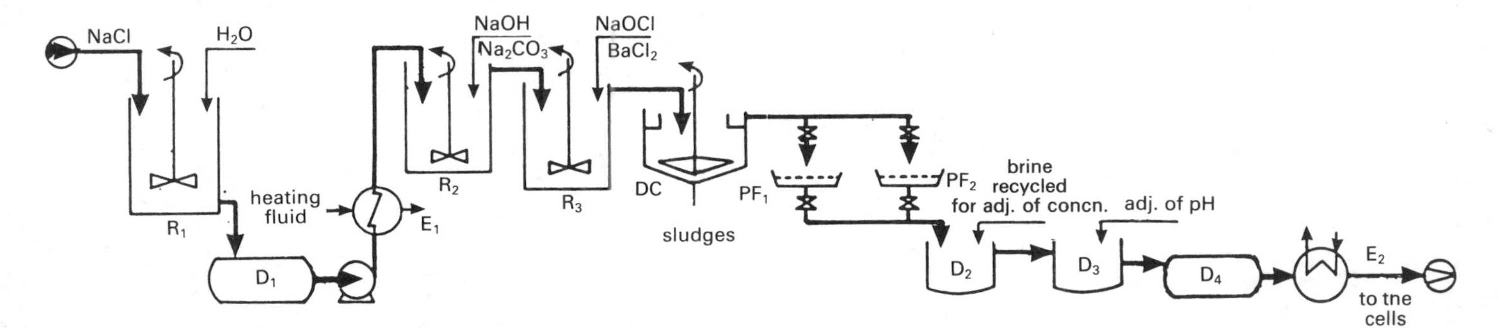

Fig. 9.3—Diagram of a plant for the preventive purification of brine destined to be used in Cl_2–NaOH plants. After being prepared in R_1 and collected in the tank D_1, the brine is preheated by passing it into the exchanger E_1 and then softened in R_2, freed from NH_3 and SO_4^{2-} in R_3, decanted and cleansed in DC, filtered in either PF_1 or PF_2, enriched with recycled brine to bring it to the correct concentration in D_2, has its pH value adjusted in D_3 and is collected in the tank D_4. It passes from this tank through the heat exchanger E_3 to the electrolytic cells.

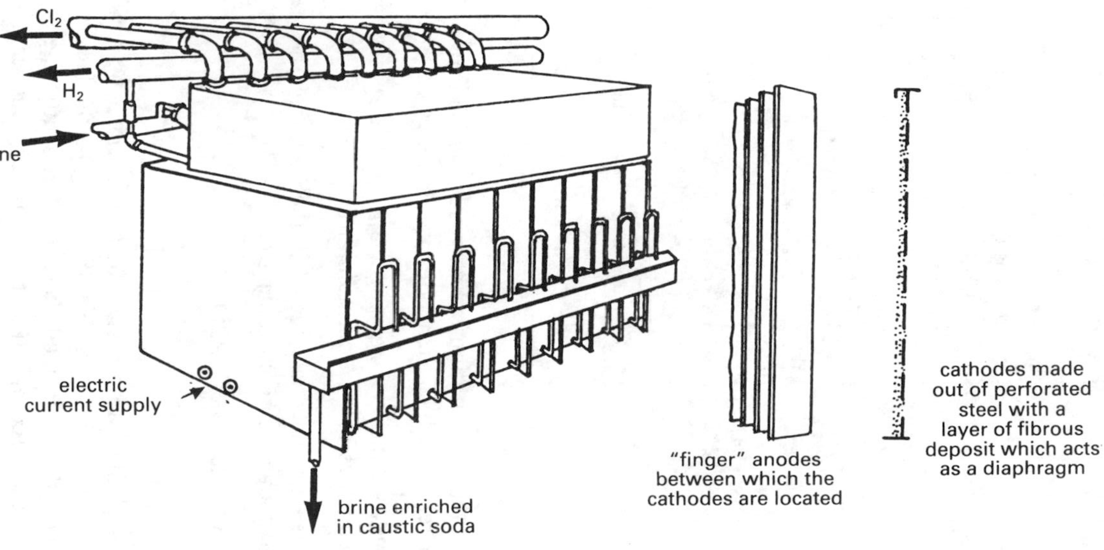

Fig. 9.4—A battery of diaphragm cells shown in perspective with the equipment required for addition of the feedstock and removing the products which are formed (Cl_2, H_2, and NaOH). The structure of the electrodes is illustrated on the right. The cathodes are made of perforated sheet into the holes in which the fibres of the diaphragms deposited onto one of their faces have been blown under vacuum during construction. These cathodes are inserted in pairs with their diaphragms turned away from each other, between the 'fingers' of the anodes.

Table 9.2—Average conditions and characteristics for diaphragm cells

Operating conditions		*Functional characteristics*		
Type	*Values*	*Type*		*Values*
Brine concentration	315–330 g/l	E.M.F.		2.95–3.8 V
Impurities: CaO	5 ppm	Current		15–150 kA
Impurities: MgO	0.8 ppm	Current density		1.18–2.9 kA/m^2
Impurities: SO_4^{2-}	0–0.3 g/l	Current efficiency		93–98%
Temperature	90–105°C	Energy consumption		2200–2900 (kWh/ton Cl_2)
pH	10.5–11	Life-time: anodes	C*	240–280 days
		Life-time: anodes	Me*	5 years
Concentration of effluent	12–14% NaOH; 14–16% NaCl	Life-time: diaphragms	A*	4–5 months
		Life-time: diaphragms	A + P*	24–36 months
Electrolysis unit	Battery of cells	(*) C = graphite, Me = metals. A = asbestos, A + P = asbestos + polymers.		

impurities in the brine and the precipitation of hydroxides which is promoted by the alkalinity in the cathodic region reduce their porosity to a greater or lesser degree.

Operating parameters. Table 9.2 shows the average values of the operating conditions and the functional characteristics of a modern diaphragm cell.

Reactions. The primary reactions, which occur as the result of the passage of a 'pulsating' current (rectified alternating current) are:

- at the cathode(−): $2H_2O + 2e^- \rightarrow H_2 + 2OH^-$ (9.2)
- at the anode(+): $2Cl^- \rightarrow Cl_2 + 2e^-$ (9.3)

The solution becomes globally impoverished in NaCl, while the contents of the cathodic compartment become enriched in NaOH owing to the fact that the Na^+ ions migrate there and that OH^- is formed there. Consequently, the pH rises to a value of more than 7 in the cathodic compartment (the environment becomes basic).

The secondary reactions which, in general, lower the current efficiency* turn out to be anodic processes. These are mainly the following:

$$Cl_2 + H_2O \rightleftharpoons H^+ + Cl^- + HOCl \rightleftharpoons H^+ + ClO^- \quad (9.4)$$

*It is recalled that the current efficiency is given by the ratio of the theoretical quantity of electricity to the practical quantity of electricity which is expended per unit of production. When one speaks of lowering it, it means that the theoretical quantity remains constant while the quantity consumed in practice increases.

$$2H_2O \rightarrow O_2 + 4H^+ + 4e^- \quad (9.5)$$

$$2OH^- + Cl_2 \rightleftharpoons ClO^- + Cl^- + H_2O \quad (9.6)$$

$$12OH^- + 12ClO^- \rightarrow 4ClO_3^- + 8Cl^- + 3O_2 + 6H_2O + 12e^- \quad (9.7)$$

the OH^- being derived from the migration of the electrolyte in the cathodic region.

As a consequence of reactions (9.4) and (9.5) the pH value will be reduced to below 7 (the environment becomes acidic), and, as a consequence of reactions (9.5) and (9.7), oxygen is evolved which contaminates the chlorine.

The current efficiency may then be lowered by these reactions because (9.4) and (9.6) directly remove chlorine while (9.7) removes chlorine indirectly and (9.5) uses current for the production of oxygen rather than chlorine.

Plants for the electrolysis and the separation of the products. The electric current originating from an a.c. generator or from a normal industrial electrical line is suitably transformed and then rectified before being supplied as energy to the cells.

Purified brine is the material which is fed into the cells as is shown diagrammatically in Fig. 9.5. The cells are of the type sketched in Fig. 9.4. In order that the primary reactions should take place in these cells while the secondary reactions which have already been described only occur to a slight extent, the moist Cl_2 (with traces of oxygen) and H_2 gases are taken off from the top of the cells while the residual brine containing 12–14% of NaOH and small amounts of the oxygenated compounds of chlorine emerge from the bottom.

After dehumidifying and cooling in E_1 and E_2 respectively, the hydrogen is sent to storage, while the chlorine is first washed with sulphuric acid in a packed column to dry it and then also sent, after it has been freed from the incondensable oxygen by liquefaction, to storage.

In the next subsection it will be seen how the caustic soda produced in the process is purified and concentrated, which uses design technology of a general applied character.

Mercury cathode cells

The great disadvantage in the use of diaphragm cells for the production of chlorine is that of the relative dilution of the caustic soda which is produced by this method

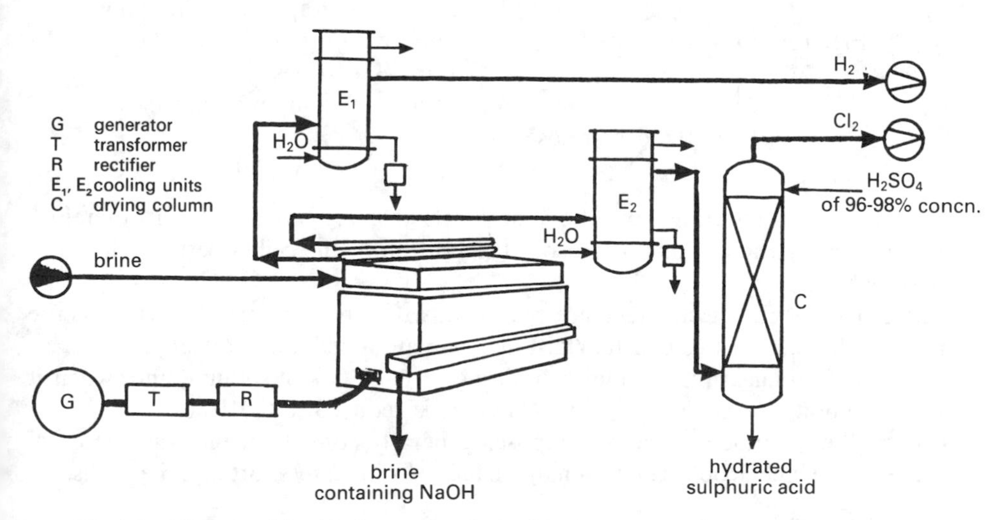

Fig. 9.5—Diagram of the chlorine–sodium hydroxide process using diaphragm cells. The plant consists of structures for feeding in the materials and the energy, the battery of cells, and the conditioners of the gases which are sent to the storage area.

at the present time. This compels one to carry out onerous concentration operations on chemically aggressive caustic solutions.

Moreover, for many industrial uses further purification of the caustic soda is required which, although now concentrated, still contains 2–3% of NaCl.

These are the reasons why, starting from the first years of this century, the use of mercury cathode cells started to compete with the use of diaphragm cells and then succeeded in becoming more widely employed.

However, ecological knowledge, based on reasons which we shall expound later, has recently greatly changed the situation, and it is to be expected that the tendency to proscribe the use of mercury will continue into the future.

The essential parts of the plant. The production of chlorine and caustic soda by the 'amalgam' method uses two cells: the 'electrolyzer' and the 'decomposer' cells.

The electrolyzer is an electrolytic cell which consists of a large steel container in the form of a rectangular parallelepiped with lined walls and under a covering of flexible and anti-explosive rubber*.

Chlorine is evolved from the electrolytic solution of Na^+ and Cl^- (purified brine) in the electrolyzer, while the sodium forms an amalgam with the mercury.

The decomposer is virtually a short-circuited galvanic cell and generally consists of a small cylindrical steel tower which is divided into two parts. The amalgam is semi-decomposed in the upper section, and the decomposition is completed in the lower part. In both compartments the mercury/sodium amalgam is therefore decomposed† by water with the formation of caustic soda, the reformation of mercury, and the development of hydrogen.

The nature and arrangement of the electrodes. The electrode systems employed in the two cells used in the amalgam method may be schematically shown as follows:

Electrodes
- in the electrolyzer
  - cathode(−) of mercury
  - anode(+) of graphite or titanium sheet covered with a thin film of Group VIII oxides
- in the decomposer
  - anode(−) of sodium amalgam
  - cathode(+) of graphite

The mercury constituting the cathode of the electrolyzer is thin and must therefore be uniform everywhere. For this reason, the bottom of the cell is perfectly flat and undeformable.

The anodes of the electrolyzer are placed at a few millimetres (2.5–3) from the mercury cathode, and this constant distance is maintained automatically.

The sodium amalgam is a liquid containing 0.16–0.20% of sodium, and, when it acts as the anode in the decomposer, this value is reduced to 0.01%.

The graphite cathode of the decomposer generally consists of piled-up pieces of graphite onto which droplets of the amalgam fall to form many short-circuited cells.

---

*Under exceptional circumstances deflagrations can occur in the primary cell owing to a build-up of hydrogen in the chlorine.

†It is for this reason that this cell is referred to as the decomposer.

Reactions. The primary reactions occurring in the electrolyzer are:

- at the anode(+): $2Cl^- \rightarrow Cl_2 + 2e^-$ (9.3)
- at the cathode(−): $Na^+ + nHg + e^- \rightarrow NaHg_n$ (n = 60–70) (9.8)

Chlorine is therefore formed at the anode while sodium amalgam is formed at the cathode.

The fact that the amalgam is formed rather than hydrogen being evolved is explained as follows. Given the low reduction potential of sodium ($E^{\ominus}_{Na^+/Na} = -2.71$ V), there is no possibility of any appreciable reduction of sodium at a cathode in aqueous solution ($E^{\ominus}_{H_2O/OH} = -0.83$ V), especially if it is acid ($E^{\ominus}_{H^+/H_2} = 0.00$ V). However, in the presence of mercury:

- given the high overpotential exhibited by this metal in the reductive process in (9.2) and by the other reductive process which is possible in the brine of the electrolyzer of mercury cathode cells:

$$2H^+ + 2e^- \rightarrow H_2 \quad (9.9)$$

- the strong depolarization effect of mercury on the electrochemical discharge of sodium (9.8) due to the very great tendency for an amalgam to be formed,
- the discharge with the formation of an amalgam becomes more likely than either (9.2) or (9.9).

The secondary reactions which are possible in the electrolyzer are, primarily, the equilibria (9.4) and (9.6) at the anode and the reduction (9.9) at the cathode.

It is especially important to suppress the parasitic cathodic reduction (9.9), that is, the process is carried out with the highest possible pH (without arriving at the point, however, when the alkalinity would favour the parasitic anodic reactions (9.6) and (9.7)), and working at a relatively high temperature.

The reactions occurring in the decomposer which, as has already been said, acts as a short-circuited battery are:

- at the anode(−): $2NaHg_n \rightarrow 2Na^+ + 2nHg + 2e^-$ (9.10)
- at the cathode(+): $2H_2O + 2e^- \rightarrow H_2 + 2OH^-$ (9.2)

When a system is said to act as a short-circuited battery, this means that the galvanic system in question furnishes thermal (heat) rather than electrical energy.

It is what happens when, instead of forming a conventional Daniell cell, zinc is put into direct contact with a copper sulphate solution. Then, in fact, the electrochemical process directly reduces Cu^{2+} ions by a part of the zinc solely with the production of heat. The role of the amalgam in a decomposer is played by zinc in this example, while the role of the water in the decomposer corresponds to that of the copper ions in the example given.

The reduction of water by the amalgam in the decomposer corresponds to the global reaction:

$$2NaHg_n + 2H_2O \rightarrow 2NaOH + H_2\uparrow + nHg$$

The heat which is developed by the short-circuited battery favours the rapid decomposition of the amalgam even in a solution of NaOH which is becoming ever more concentrated.

Electrolytic-galvanic plants and the recycling and storage of the products. The primary cell is slightly inclined to allow a downward flow-off of the brine reagent and the mercury which is transformed into an amalgam which are fed in further upstream (Fig. 9.6). When reactions (9.3) and (9.8) take place, the gaseous chlorine is taken off from the top of this cell while the

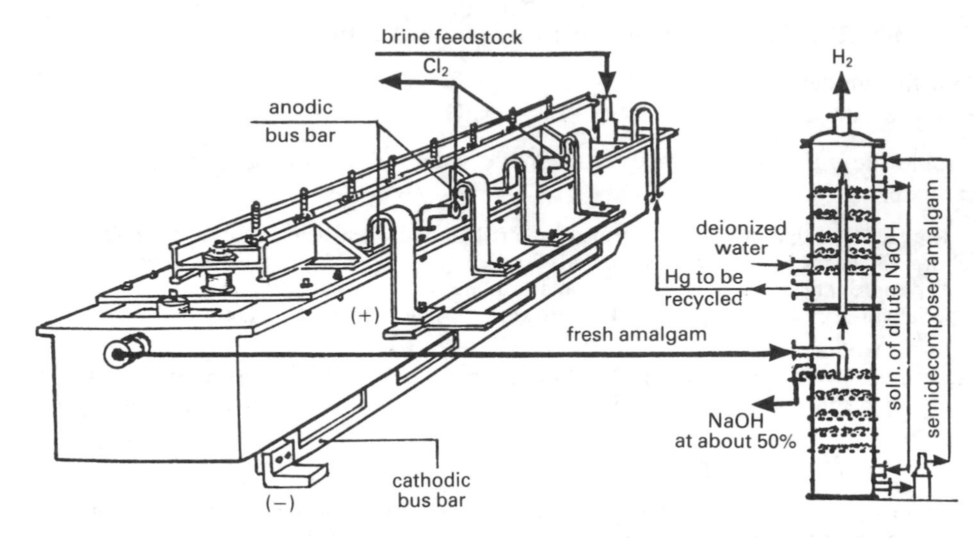

Fig. 9.6—Electrolytic cell (left) and galvanic cell (decomposer) which together produce chlorine and caustic soda by the amalgam method on an industrial scale. A special characteristic of the plant is the decomposer which takes the form of a vertical packed tower. The packing consists of pieces of graphite onto which the amalgam, which has been subdivided into droplets, falls to form many short-circuited batteries when they come into contact with the graphite.

amalgam is pumped out along a conduit as soon as it has been formed. The chlorine is sent to be stored, after it has been dehydrated by cooling, dried by treating it with concentrated H_2SO_4, and liquefied to free it from traces of incondensable H_2. The amalgam is conveyed to the decomposer.

The amalgam is decomposed in two stages: it is partially decomposed in the first section of the decomposer and completely decomposed in the other section (above, where it comes into contact with pure water). The presence of the graphite on the shelves of the two semicolumns into which the decomposer is subdivided is depicted diagrammatically in Fig. 9.6. As has been stated, this graphite acts as the cathode in the decomposition process, in which the amalgam functions as the anode.

The pure mercury obtained at the end of the stripping process is recycled back to the primary cell. The hydrogen is sent to be dried and then on to be stored, while the pure and extremely concentrated solution of caustic soda is either sent to where it will be used or subjected to further concentration.

Operating parameters. The average operating conditions and functional parameters for an amalgam cell are presented in Table 9.3.

Mercury cells and environmental pollution. Mercury is an element which is ecologically extremely harmful because, as a result of biological methylation processes which occur in an anaerobic environment (in practice, on the sea bed), the metal which is dispersed from chemical process effluents enters into the food chain, via plankton, and accumulates in the adipose tissue of fish. Consequently, fish products act as a vehicle in conveying mercury compounds with neurotoxic properties to man. The numerous cases of poisoning, including the ingestion of lethal doses, which have taken place for this reason in Japan during the years from 1950 to 1965 are a demonstration of this.

Although the production of chlorine and caustic soda by using amalgam is not the only process responsible for the dispersion of mercury in waters (processes using catalysts which are promoted by mercury compounds also contribute to an appreciable measure here), there is no doubt that the operation of mercury cathode electrolytic cells has been the principal cause of such pollution during the last few decades.

Table 9.3—Conditions and characteristics of De Nora amalgam type cells (El = Electrolyzer, Dec = Decomposer)

Operating conditions		*Functional characteristics*	
Type	*Values*	*Type*	*Values*
Brine concentration	300–320 g/l	Voltage (El)	4–4.5 V
Impurities feedstock	CaO < 5 ppm MgO < 3 ppm	Current (El)	380–420 kA
	SO_4^{2-} < 2 g/l	Current density (El)	12–12.5 kA/m^2
pH (El)	3–5	Current efficiency	95–97%
Temperature (El)	75–85°C		
Data concerning the mercury	% Na at inlet-0.01%	Energy consumption	3300–3450 kWh/*T*
	% Na at outlet-0.16–0.20% amount 4.7 t/cell	Anodes: type number lifetime	Cl_2 DSA 40–45 3–6 years
Effluent concentration (El)	NaCl 270 g/l Cl_2 0.5 g/l		
Effluent concentration (Dec)	NaOH 48–54% NaCl < 50 ppm	Production at full load	Cl_2 13–14 t/day NaOH 15.6 t/day H_2 0.4 t/day

DSA = dimensionally stable anodes (titanium anodes coated with Group VIII oxides).

Until about the middle sixties, about 300 g of mercury became dispersed in waters for every ton of chlorine produced. Subsequent drastic interventions have reduced these dispersions to minimum values of 5–10 g/ton. It is very difficult to bring about a further improvement in this situation. The only true remedy is therefore to stop using this process.

Membrane cells

The inability to prevent losses of mercury, leading to ecological damage, and the high costs which already have to be met in carrying out the necessary measures to establish the actual levels of the dispersion of the metal, have persuaded companies to consider the progressive abandonment of the amalgam process for the production of chlorine and caustic soda.

However, the asbestos used in the fabrication of the porous diaphragms which are employed in the other electrochemical method for the preparation of these products has also been indicated as a certain cause of neoplasms and silicosis.

In particular, to counteract these negative aspects of the processes which have traditionally been used in this industrial sector, a system has been implemented which combines the advantages of the low consumption of electrical energy which already characterizes the diaphragm process and the production of caustic soda which is essentially free from chlorides as in the amalgam process.

However, to pursue these advantages it is necessary that ion-selective membranes of low electrical resistance, high efficiency at high NaOH concentrations, with mechanical and chemical stability, reasonable production costs, and suitable operational lives, should be available. Furthermore, it is necessary to ensure the absence of Ca^{2+}, Mg^{2+}, and SO_4^{2-} ions which are antiselective and harmful in the long term to the efficiency of the membrane.

There have been a number of studies on membranes for electrodialysis and fuel cells which have oriented chemical technology toward looking for those membranes which are suitable for electrolytic cells intended for the production of chlorine and caustic soda.

This research has been piloted in the USA and in Europe by the Du Pont company through the preparation of fluorinated polymers ('Nafion') with cross-links consisting of ether bridges and terminal sulphonic acid groups. The structure of these polymrs may be briefly represented as:

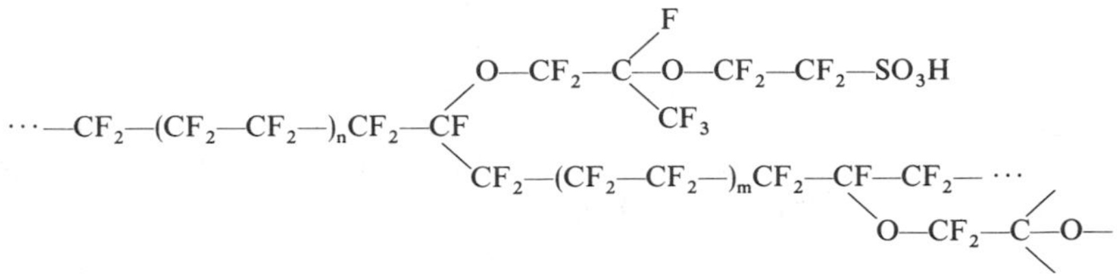

```
                                              F
                                             /
                                 O—CF2—C—O—CF2—CF2—SO3H
                                /            \
···—CF2—(CF2—CF2—)nCF2—CF                     CF3
                                \
                                 CF2—(CF2—CF2—)mCF2—CF—CF2—···
                                                      \       /
                                                       O—CF2—C—O—
                                                              \
```

In Japan, ion-selective membranes made of fluorinated polymers have been prepared by the Asahi Chemical Company. They contain the weakly anionic groups (–COOH) rather than strong (–SO_3H) groups. The plants which use these membranes are responsible for about 25% of the chlorine and caustic soda produced in Japan, and there is considerable potential for further large development.

Other concerns (such as Solid Polymer Electrolyte) have prepared membranes which, by means of the special materials which have been electrodeposited onto these in the form of bands of opposite polarity, function as a bipolar electrode, on the external faces of which chlorine and hydrogen are developed.

In an electrolytic membrane cell the Na^+ ions in the brine, which is fed into the anode compartment where the chlorine is discharged:

$$2Cl^- \rightarrow Cl_2\uparrow + 2e^- \tag{9.3}$$

pass through the ion-selective membrane into the cathodic region where hydrogen is developed by the discharge of the deionized water with which this cell compartment is filled:

$$2H_2O + 2e^- \rightarrow 2OH^- + H_2\uparrow \tag{9.2}$$

A solution containing more than 20% of NaOH is formed in this region which emerges from the cell at a temperature os 75–90°C and permits, when added to the contribution from the dilute brine which leaves the anode compartment, enough heat to be recovered for the concentration to be brought up to more than 35% without any further expenditure of energy. Further concentration is then carried out if required, as will be described in Chapter 12.

A diagram of a plant for the production of chlorine and caustic soda, using a non-bipolar* membrane, is shown in Fig. 9.7.

*Apart from the different configuration of the electrodes in the cell, this plant scheme is also valid where bipolar electrodes are used.

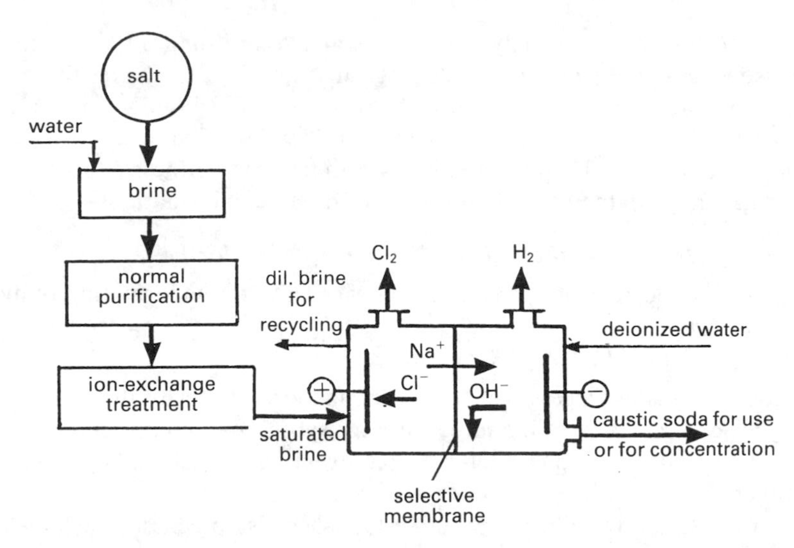

Fig. 9.7—The essential structure of a plant for the production of chlorine and caustic soda, using the ion-selective membrane process. It is seen that Na^+ ions pass through the membrane while OH^- ions do not.

The brine is made up and purified in the normal way, but, before entering the electrolytic cell, it must be thoroughly deionized by passing it through ion exchange resins. It then enters the anode compartment of the membrane electrolytic cells where it is diluted by the discharge of the Cl^- ions and the passage of the Na^+ ions across the membrane. The dilute brine passes out of the cell and is recycled.

Deionized water is sent into the cathode compartment where it undergoes reduction to H_2 and OH^-. The hydrogen produced rises and is sent to be dried and stored. The OH^- ions form NaOH with the Na^+ ions originating from the anodic region, and the alkali produced in this way is sent to its destination.

The commerce and uses of chlorine

Chlorine is generally marketed in an anhydrous and liquefied state in steel bombs with a useful volume of approximately 36 litres and a weight of 45 kg. Steel tanker wagons are required for its large-scale transport.

Numerous inorganic compounds, acids, and salts, both oxygenated and non-oxygenated, are firstly prepared with chlorine; but the greatest amounts are employed in the production of both simple chlorinated organic solvents (e.g. trichloroethylene and CCl_4), pesticides (e.g. γ-hexane), high performance lubricants, etc. and derivatives of such chlorides in the fields of plastics (PVC) and synthetic fibres (e.g. Movil). Direct use is made of chlorine in the paper industries, water treatment, bleaching, and so on.

Hydrochloric acid

Industrially, hydrochloric acid is met with either in the form of the colourless gas (hydrogen chloride) which fumes in air as it becomes hydrated or as a more or less concentrated aqueous solution of this gas. In solution, HCl is almost completely dissociated, and it behaves as a strong acid (hydrochloric acid).

Hydrogen chloride is extremely soluble in water (more than 50 000 Nl in 100 l of water) dissolving with the evolution of a large amount of heat. The resulting solution of hydrochloric acid forms an azeotrope with a maximum boiling point (110°C) at a concentration of 20.17 wt.% (about 1 mole in 8 moles of water).

At the present time HCl is produced by two routes: it is synthesized from hydrogen and chlorine, and obtained as a by-product of organic chlorinations.

Production of HCl by synthesis from hydrogen and chlorine

The direction combustion of hydrogen in chlorine leads to the formation of hydrogen chloride:

$$H_2 + Cl_2 \rightarrow 2HCl + 44 \text{ kcal} \qquad (9.11)$$

Reaction (9.11) is a process which is much favoured both by the large evolution of energy which accompanies it and because the synthesized compound escapes from the reaction mixture as a gas, thereby circumventing the fact that an equilibrium would otherwise be attained.

The very fact that an equilibrium is not established also precludes the large increase in temperature from having a negative effect on the yields of this reaction, which is highly exothermic.

The exothermic nature of the phenomenon of the direct combination of hydrogen and chlorine is such as to raise the temperature of the reagents, and the reaction products to a point where they are incandescent. The device in which this reaction, accompanied by a greenish flame at 2400°C, is carried out in practice is therefore referred to as a 'burner'. This device is contained in a cylindrical steel apparatus if the initial reagents are dry H_2 and Cl_2, or a similar apparatus made out of sintered steel or tantalum if there is any moisture present.

The combustion chamber is then cooled externally with flowing water, and a gas-tight lid is fitted at the top of the reactor which suddenly opens to allow the gases to escape in an emergency (Fig. 9.8).

On account of the existence of a large energy barrier to the reaction, a mixture of molecular hydrogen and molecular chlorine is stable at ambient temperatures and in the absence of radiation of suitable wavelengths.

Photons with frequencies which are high enough to bring about this chemical reaction by furnishing the 'activation energy' can be produced by creating an electrical spark in a mixture of molecular hydrogen and chlorine or by first burning a mixture of hydrogen with oxygen and then gradually replacing the oxygen with chlorine.

In every case, the process is initiated by the reaction:

$$Cl_2 + h\nu \rightarrow 2Cl^{\cdot} \qquad (9.12)$$

that is, by the homolytic fission of a chlorine molecule with the formation of free radicals. Once these have been produced, the chain reaction:

$$Cl^{\cdot} + H_2 \rightarrow HCl + H^{+} \qquad (9.12a)$$

$$H^{\cdot} + Cl_2 \rightarrow HCl + Cl^{\cdot} \qquad (9.22b)$$

and so on, is established.

The chain can be terminated by the occurrence of the following reactions:

$$\begin{aligned} &Cl^{\cdot} + Cl^{\cdot} \rightarrow Cl_2 + \text{heat} \\ &H^{\cdot} + H^{\cdot} \rightarrow H_2 + \text{heat} \qquad (9.13) \\ &H^{\cdot} + Cl^{\cdot} \rightarrow HCl + \text{heat} \end{aligned}$$

which implies the destruction of the chain propagators and the active chain intermediates.

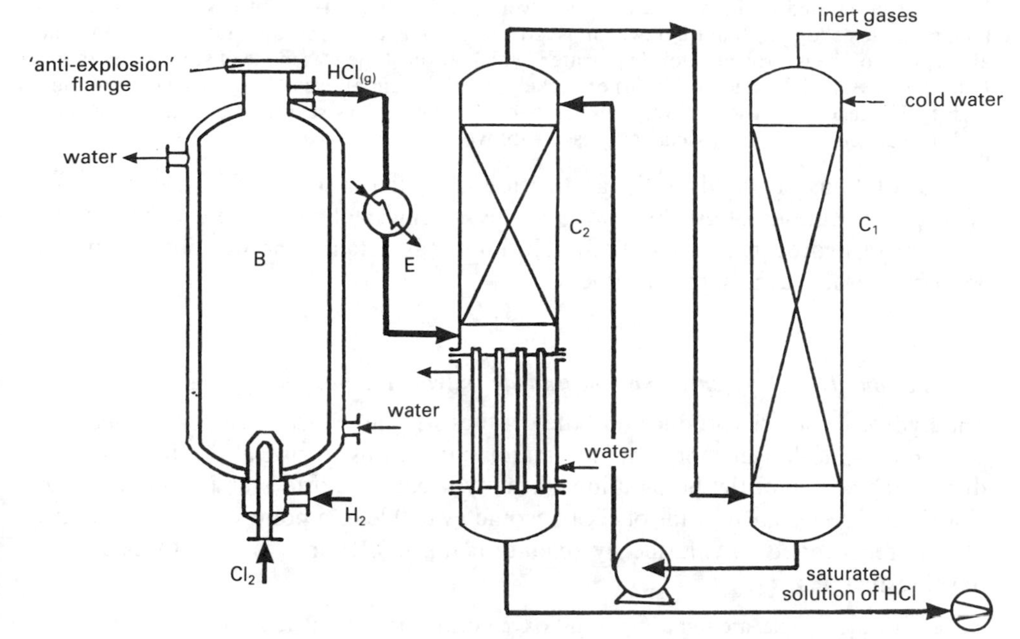

Fig. 9.8—Diagram of a plant for the production of a saturated aqueous solution of hydrochloric acid under normal pressures. The hydrogen chloride is synthesized in the burner B which is enclosed in a water jacket to ensure that it is efficiently cooled. After further cooling in E, the HCl is absorbed in the columns C_2 and C_1 by a dilute solution of hydrochloric acid and water respectively. The concentration of the aqueous solution prepared in this manner therefore tends toward the maximum possible concentration

Actually, because a large amount of heat is developed both from the chain propagation reactions and from the chain terminating processes, the continued renewal of the chain propagators by the thermal route is ensured over the long term. In brief, reaction (9.11) is a chain reaction with a 'high quantum yield'*.

To ensure that all the chlorine reacts, excesses of about 10% of hydrogen chlorine are used. This, in itself, is not enough for this purpose, and it is also necessary that the combustion chamber and the length of ducting which leads the gas to the absorbers should be sufficiently spacious, otherwise the HCl finally produced will contain free chlorine.

From a physicochemical point of view, this is interpreted as being due to a 'wall effect' which occurs if the walls of a reactor in which a chain reaction takes place are such (by their very nature, development, shape, and orientation) as to affect the chain carriers. When this is the case, the walls tend to interrupt the process by promoting the chain-breaking reaction (9.13). Physicochemically, it is customary in this and in all analogous cases to say that the 'chain terminators' act as a third body in a system which already consists of the reactant bodies.

A plant for the synthesis of hydrochloric acid from the elements, and the absorption in water of the hydrogen chloride produced, an operation which is carried out when the product is not to be used for the preparation of anhydrous hydrogen chloride, is shown diagrammatically in Fig. 9.8.

*The quantum yield of a photochemical process is expressed as the ratio between the number of molecules which are formed in the process and the number of quanta which are absorbed. In the case of hydrogen chloride, the quantum yield $\eta_q = \text{molecules}_{HCl}/\text{no. quanta} = 10^5$–$10^7$.

The reactors used in this synthesis, which were previously made of Vitreosil, are nowadays made of Hastelloy or tantalum which is strongly cooled. In the apparatus used for the absorption of the hydrogen chloride in water, which at one time was made of stoneware, today Durichlor (81.65% Fe and 14.5% Si) or nickel–molybdenum alloys ('Chlorimet') or tantalum and the Hastelloys are used. At very low temperatures ($<50°C$) structures made out of plastics and, in particular, polypropylene, are also employed.

The solutions prepared by the absorption of hydrogen chloride in water usually have concentrations of 20–30 wt.%, but, by strong cooling to reduce the vapour pressure of hydrogen chloride, such solutions may attain concentrations of up to 35–37% during the absorption process.

Production of HCl by chlorination and dehydrochlorination

The hydrochloric acid produced by the synthesis route, which is valued on account of its purity, no longer represents the major part of this industrial product. In fact, during the course of the preparation of many organic chlorides and monomers for chlorinated resins great yields of HCl become available to a point where a large part of it is reconverted to chlorine by oxidation (page 79), or it is used to carry out oxychlorinations.

Here, as an example, are some important oxychlorinations which are carried out industrially by a catalytic pathway with appropriate ratios of the reactants:

$$CH_2{=}CH_2 + 4HCl + 2O_2 \rightarrow CCl_2{=}CCl_2 + 4H_2O \quad \text{perhalogenation}$$

$$CH_2{=}CH_2 + HCl + \tfrac{1}{2}O_2 \rightarrow CH_2{=}CHCl + H_2O \quad \text{vinylation}$$

$$CH_2{=}CH_2 + 3HCl + \tfrac{3}{2}O_2 \rightarrow CHCl{=}CCl_2 + 3H_2O \quad \text{various halogenations}$$

All of these are hydrochlorination reactions which are followed by dehydrogenating oxidation reactions, and, as has been said, they consume HCl.

Among the organic reactions which, as they are carried out on a vast scale, lead to the production of enormous amounts of HCl, we would mention:

$$CH_4 + Cl_2 \rightarrow \underset{\text{methyl chloride}}{CH_3Cl} + HCl \quad \text{halogenation of alkanes}$$

$$CH_2Cl{-}CH_2Cl \rightarrow \underset{\text{vinyl chloride}}{CH_2{=}CHCl} + HCl \quad \text{dehydrochlorination of 1,2-dichloroethane to vinyl chloride}$$

$$CHCl_2 + CHCl_2 \rightarrow \underset{\text{trichloroethylene}}{CHCl{=}CCl_2} + HCl \quad \text{dehydrochlorination of } \textit{sym}\text{-tetrachloroethane to trichloroethylene}$$

The hydrogen chloride is absorbed by a counter current of water in towers lined with polypropylene, provided that the temperature does not exceed 100°C. Otherwise, it is carried out in towers constructed from special steels, and to an ever decreasing extent in graphite towers owing to the delicate work required in their upkeep.

By carrying out the absorption process under adiabatic conditions, the chlorinated compounds accompanying the HCl which are insoluble in water do not condense, and can thereby be taken from the top of the absorption column.

Anhydrous hydrogen chloride

By exploiting the combustion of hydrogen in chlorine it is possible to obtain the commercial anhydrous form of HCl by passing the somewhat cooler, but still hot, gases originating from the burner over anhydrous $CaCl_2$, or washing them with 98% sulphuric acid and then compressing the cold gas to a pressure of 60 atm. in steel cylinders. The purity of this product is about 99.9%.

On the other hand, there are other routes to the preparation of anhydrous HCl. Here, it is first necessary to absorb the gas in water and to distil the resulting solutions to produce a concentration of about 36% if one is to obtain 97% HCl at the top of the column. The 36% hydrochloric acid is then cooled to −12°C, and an aqueous liquid containing 50% HCl is left to condense, while the residual gases, when they have been denebulized and compressed to 60 atm., are of a purity exceeding 99.5%.

The commerce and uses of hydrochloric acid (hydrogen chloride)

Anhydrous gaseous hydrogen chloride is marketed in steel vessels, while aqueous solutions of hydrochloric acid at 22–28 Bé (32–37 wt.%) are transported in demijohns with plastic bungs or in steel tanks lined with PVC or polypropylene. Exposure to sunlight and heat should always be avoided, and care should be taken not to overfill the tanks.

The major uses of hydrochloric acid are met with in the production of chlorides (NH_4Cl, $ZnCl_2$, etc.) in extractive metallurgy (metals are recovered electrochemically from the chlorides which have been formed), in pickling systems, in the dye, tanning, and food industries, and in mechanical technologies (as 'muriatic' acid).

Some chlorides of particular interest

Some of the most important industrial salts are numbered among the chlorides. Thus, NaCl and KCl, especially the former, find extensive and varied uses to a point where it is appropriate to dedicate special sections to their treatment. Other chlorides, such as NH_4Cl and $CaCl_2$, are the by-products of numerous and very large-scale industrial technologies, and these will therefore be discussed in conjunction with the particular technologies which lead to their production. Finally, there are chlorides which are of considerable importance as a whole, even if they are not specifically prepared or used in an industrial environment. It is this last group of chlorides which it is expedient to consider here.

Aluminium chloride

The use of anhydrous aluminium chloride as a catalyst in laboratory and industrial syntheses necessitates the production and marketing of appreciable quantities of this compound.

$AlCl_3$ is a crystalline mass which volatilizes at normal temperatures and sublimes at 183°C. Being exceedingly hygroscopic, aluminium chloride is extremely soluble in water from which the hexahydrate ($AlCl_3 \cdot 6H_2O$) crystallizes when the aqueous solution is evaporated in the presence of an excess of hydrochloric acid.

Aluminium chloride fumes in air on account of its reaction with moisture which hydrolyzes it to $Al(OH)_3$ and HCl after the formation of intermediate hydroxy chlorides.

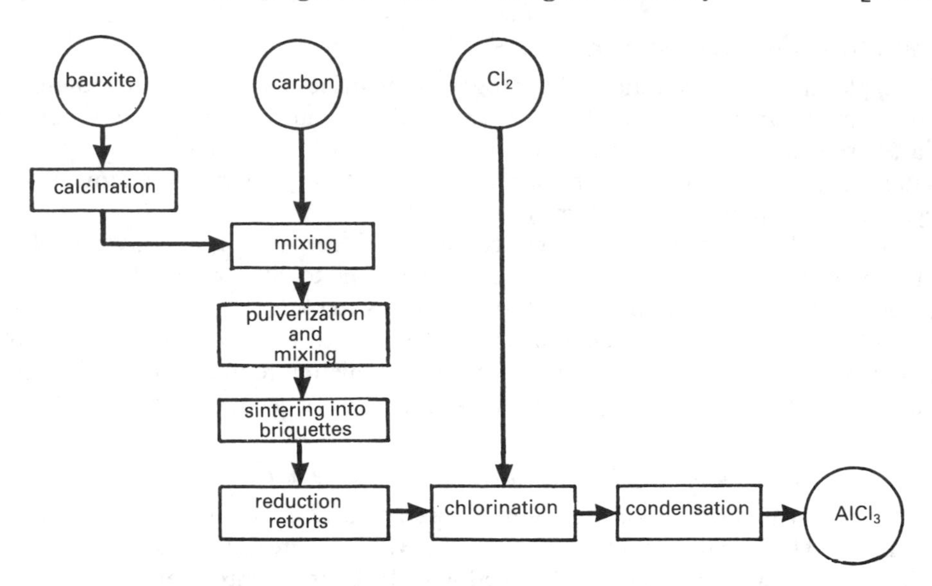

Fig. 9.9—Block diagram of the essential steps in the production of $AlCl_3$.

Fig. 9.9 is a flow diagram of the industrial preparation of aluminium chloride from bauxite, chlorine, and carbon, according to the overall reaction:

$$Al_2O_3 + 3C + 3Cl_2 \rightarrow 2AlCl_3 + 3CO$$

Actually, the preparation may start off either from pure calcined alumina or from bauxite which is practically free from iron and has been precalcined in rotating furnaces at 1000°C. These materials are then added to powdered carbon ($3Al_2O_3$:1C) and intimately mixed with it. The resulting mixture is sintered into small briquettes, and the development of an active metal mass in them is promoted by heating them in retorts at 1000–1200°C. After this, the resulting mass which is piled up in the chlorinating furnace is permeated by a descending stream of chlorine.

The aluminium chloride which is formed sublimes and is condensed out in the condensation column, while the residual gases are purified by using absorption processes in columns where they are washed with lime water.

Iron chlorides

The dyeing, printing and pharmaceutical industries, the technology for the clari-flocculation of waters, and the production of pigments and certain catalytic processes, absorb appreciable amounts of iron chlorides, especially $FeCl_3$, both in the anhydrous form and as various hydrates.

Anhydrous *ferrous chloride* is prepared as a white sublimate by the treatment of iron turnings with dry HCl, while the same salt crystallizes to form various hydrates from solutions prepared by dissolving iron in aqueous hydrochloric acid. This must be done in the complete absence of air both during the attack and evaporation stages.

Ferric chloride is prepared by the chlorination of iron turnings, until they are red, in furnaces lined with a ceramic material:

$$2Fe + 3Cl_2 \rightarrow 2FeCl_3 + 192 \text{ kcal}$$

The salt which is formed sublimes and is subsequently condensed in steel cylinders which are cooled with water externally and vibrated to prevent the formation of crusts. The residual vapours are 'deactivated' with lime water. Hydrated iron (III) chlorides can be prepared by bubbling chlorine into solutions of ferrous chloride and separating them by crystallization.

The chlorides of Group II metals

Magnesium chloride will be discussed within the framework of the metallurgy of this metal, while the production of *calcium chloride* is described in connection with the preparation of Solvay soda.

Barium chloride, which is of considerable importance in the photographic, tanning, and textile industries, is prepared by heating barium sulphate with carbon powder and $CaCl_2$ in rotating furnaces at 900°C when the reaction:

$$BaSO_4 + 4C + CaCl_2 \rightarrow BaCl_2 \rightarrow BaCl_2 + CaS + 4CO$$

takes place. The mass which is removed from the furnace is crushed, pulverized, and leached. When the resulting solution is concentrated in vacuo, the salt separates out in the form of tabular crystals with the formula $BaCl_2 \cdot 2H_2O$ which can be dehydrated by heating up to 200°C.

Zinc chloride, a salt which can be prepared in a pure state by heating zinc clippings in a stream of chlorine, is used as a condensing agent and a catalyst in synthetic reactions, as a mordant, as a wood preservative, and as an ingredient in rapidly setting cements. When it is prepared by dissolving the metal in hydrochloric acid followed by evaporation of the solution, an impure zinc hydroxychloride product is first obtained which formerly served most of the technologies in which $ZnCl_2$ was required.

Zinc chloride is a white salt which is hygroscopic and extremely soluble in water, whereas zinc hydroxychloride does not dissolve.

Industrially, the *chlorides of mercury* are prepared, for the most part, sequentially as shown in Fig. 9.10.

The synthesis of *mercuric chloride* (corrosive subliminate) is carried out in boilers of glazed iron maintained at 20–90°C by immersion in water so that the product does not sublime. *Mercurous chloride* (calomel) is synthesized by distilling, in similar boilers, 3 parts of Hg with 4 parts of $HgCl_2$ and then condensing the vapours.

Calomel has pharmaceutical, veterinary, ceramic, and pyrotechnic uses, while corrosive sublimate (mercuric chloride) is a starting point for the preparation of other mercury compounds, serves as a wood preservative and for anatomical preparations, and is used in textile technologies (the printing of materials) and, especially, in medicine and surgery.

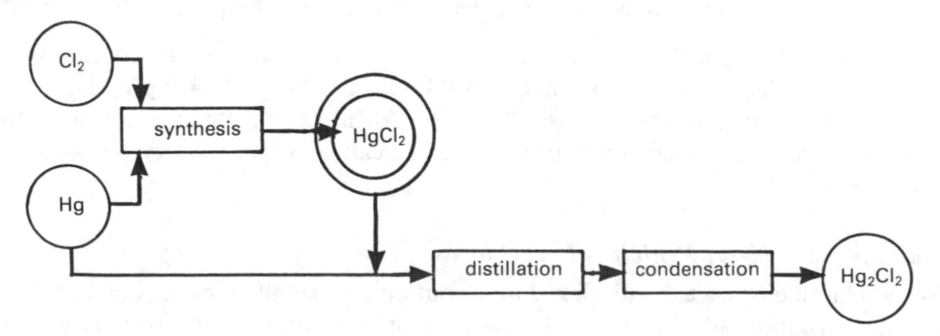

Fig. 9.10—Coupled synthesis of $HgCl_2$ and Hg_2Cl_2. The double circle around $HgCl_2$ denotes that part of this product is sent to be stored while the other part is employed in the production of Hg_2Cl_2.

Industrial products based on oxygenated compounds of chlorine

Under various chemical and physical conditions (reagents, pH, temperature, and voltage), various compounds of chlorine are produced on a more or less vast industrial scale according to the degree of commercial interest in them.

Solutions containing 'active chlorine' and alkali hypochlorites

Electrolytic processes in solutions of drinking water, to which pure NaOCl has been added, and sea water are carried out on a large scale for the sterilization of the water in swimming pools and of water for refrigeration processes.

In this way, by working without a diaphragm between the electrodes at a temperature of 20°C which is maintained by the circulation of water through coils, hypochlorite solutions containing a low level of active chlorine are formed via reactions (9.2), (9.3), and (9.6).

Both here and in other cases where this concept is applied, the term 'active chlorine' refers to all of the chlorine involved in the production process, that is, that which possibly remains free and that which is combined either as chloride or as hypochlorite.

This is because the latter exerts an oxidizing action which is equal to that of all the chlorine which, by disproportionating, has produced the hypochlorite together with the chloride via (9.6). In fact, taking the oxidation to 1 mole of iodine as the example, both the species ($Cl_2 + 2KI \rightarrow 2KCl + I_2$ and $2KI + NaOCl + H_2O \rightarrow NaCl + 2KOH + I_2$) operate independently of the fact that, by forming NaOCl, half of the chlorine reagent molecule may have become chloride.

Still greater amounts of solutions of disproportionated chlorine in alkalis are produced industrially by the direct chemical route:

$$2NaOH + Cl_2 \rightarrow NaOCl + NaCl + H_2O$$

for the whitening and industrial bleaching of papers, fibres, yarns, etc., and for the supply of domestic bleaching agents. About 20% solutions of NaOH on the one hand, and, on the other hand, most of the chlorinated streams of air formed in the purging processes in the production of chlorine and soda are used for this purpose.

The operation is generally carried out in towers made of carbon steel which have been lined with chemically resistant plastic materials, given that the temperature is kept under 30°C by cooling. The towers are irrigated from above with fresh or recycled solutions of caustic soda so long as the percentage of NaOH does not fall below 0.5%, and come into contact with the chlorine-containing streams of gas rising from the bottom. However, the pH must remain basic*.

Commercial hypochlorite solutions have the following average composition: NaOCl 150–170 g/l; NaCl 115–140 g/l; NaOH 2– g/l; Na_2CO_3 4–7 g/l.

The amounts of iron and other similar metals (Mn, Cr, Ni, etc.) present must not exceed 10 ppm. to prevent them from acting as catalysts for the decomposition of the hypochlorite.

Calcium chloride and calcium hypochlorite

When chlorine is reacted with slaked lime containing a small excess of water (3–4%) either the mixed salt $CaOCl_2$ (Cl–Ca–OCl) or a mixture of the two basic salts

*Either the pH or the temperature of the solutions being treated is controlled in such a way as to prevent the disproportionation of hypochlorite into chlorate and chloride.

$Ca(OCl)_2 \cdot 2Ca(OH)_2$ and $CaCl_2 \cdot Ca(OH)_2 \cdot H_2O$, containing disproportionated chlorine, is formed.

The resulting system, after the addition of CaO to reduce the water content is known as chloride of lime and contains about 35% of active chlorine. This product is commercially available in powder form which can be readily transported and stored provided that it is kept away from heat and light.

Chlorine of lime is widely used as a decolorant, disinfectant, and bleaching agent: in ponds, pools, hospitals, toilets, and the agricultural/food sector dealing with bones, fibres, the slaughter of animals, etc.

Chlorination under pressure increases the percentage of calcium hypochlorite which is present in the product of the reaction between slaked lime and chlorine up to a point where the active chlorine attains a maximum value of 99.5%. In general, however, there is up to about 65% of active chlorine in commercial calcium hypochlorite.

Since it is possible to quantitatively liberate the chlorine used in its production from calcium hypochlorite:

$$Ca(OCl)_2 + 4HCl \rightarrow CaCl_2 + 2H_2O + 2Cl_2$$

this product is a more convenient and safe way of transporting chlorine over large distances than in the form of the liquefied gas. Moreover, various grades of calcium hypochlorite are used in domestic detergents and also as an alternative to chloride of lime.

Compounds in which chlorine has a high oxidation state

Electrochemical technologies lie at the foundation of the production of compounds in which chlorine has the oxidation states of +5 and +7. Compounds in which the chlorine is in an oxidation state of +3 or +4 are then prepared from compounds in which the oxidation state is +5.

Chlorates. These are prepared by the electrolysis of alkali chlorides under slightly acidic conditions (pH = 5.8–6.4) in diaphragm-less electrolytic cells.

In recent years, the strong demand for these products (2.2×10^6 tons in 1983) has led to the perfection of a method for their production with an increase in the current efficiency and a reduction in the amount of energy consumed.

These improvements consist of a more precise study and control of the pH conditions, how to prevent the chlorates which have been formed from being reduced again at the cathode, and, especially, the adoption of DSA anodes (page 328) which suppress the evolution of oxygen in accordance with (9.7), whereas, in the old plants, this reaction was induced by the graphite anodes even when working at a relatively low temperature.

The cathodic reaction, if one starts out from NaCl, is (9.2), and the primary reaction occurring at the anode is (9.3). Reaction (9.4) also takes place at the anode with its second equilibrium displaced (to the left) toward formation of HClO: that is, the weak electrolyte which is the keystone in the oxidation reaction:

$$ClO^- + 2HClO \rightarrow ClO_3^- + 2H^+ + 2Cl^- \qquad (9.14)$$

and is now favoured by running it at the highest possible temperature as a result of the adoption of the new anodes.

The acidity which arises on account of reaction (9.14) tends to neutralize the cathodic alkalinity which diffuses into the cell. Since, however, the acidity is also consumed in the reduction of the chromate which is added to the cell to suppress reduction of the chlorates by means of the formation of a cathodic layer of $Cr(OH)_3$, it is necessary to restore the appropriate level of acidity by the addition of hydrochloric acid until the solution has the weak acidity which is characteristic of this process (a pH value of about 6).

Some of the operating conditions are shown schematically in Table 9.4 so that they may be contrasted with those used in the production of perchlorates, while a diagram of the whole of a standard plant for the preparation of chlorates is reproduced in Fig. 9.11.

Nowadays, the greatest amounts of sodium chlorite which are consumed in industry are employed in the production of perchlorates, chlorine dioxide, and chlorites. However, appreciable quantities of it are also required as a general weedkiller and for the production of dyestuffs. Potassium chlorite is also employed in the fabrication of matches, pyrotechnic and explosive powders, and galenical mixtures.

Chlorine dioxide and chlorites. The sterilization of drinking waters, the bleaching of fibres, of flour, and other products from the farming/food sector, with an action which is less degrading as regards the structure of the materials and which changes their organoleptic properties to a lesser extent as compared with chlorine and hypochlorites, has led to an enormous increase in the production of *chlorine dioxide*. This compound can be prepared by the reaction of chlorates with SO_2, oxalates, methanol, etc. For example, the reaction:

$$6ClO_3^- + CH_3OH + 6H^+ \rightarrow 6ClO_2 + CO_2 + 5H_2O$$

can be used.

Sodium chlorite is prepared industrially from chlorine dioxide:

$$2ClO_2 + Na_2CO_3 + H_2O_2 \rightarrow 2NaClO_2 + CO_2 + H_2O + O_2$$

It is a salt with a powerful oxidizing action which can be transported in the solid state and has a low tendency to degrade cellulose fibres and saccharide molecules in

Table 9.4—Structural differences and process variables for electrolytic cells for the production of chlorates and perchlorates

Conditions	*Chlorates*	*Perchlorates*
Anodes	DSA	Platinum
Cathodes	Ordinary steel ('iron') rendered unsuitable for chlorate reduction by the addition of chromates which are converted to Cr(III)	The same steel as for chlorate cathodes and likewise treated to prevent the reduction of chlorates and perchlorates
Voltage	3.0–3.8 V	6 V
Current density	3000 A/m^2	4000 A/m^2
Current efficiency	95%	85%
Temperature	80°C	40–50°C
Energy consumption	5000–5400 kWh/t	2800 kWh/t

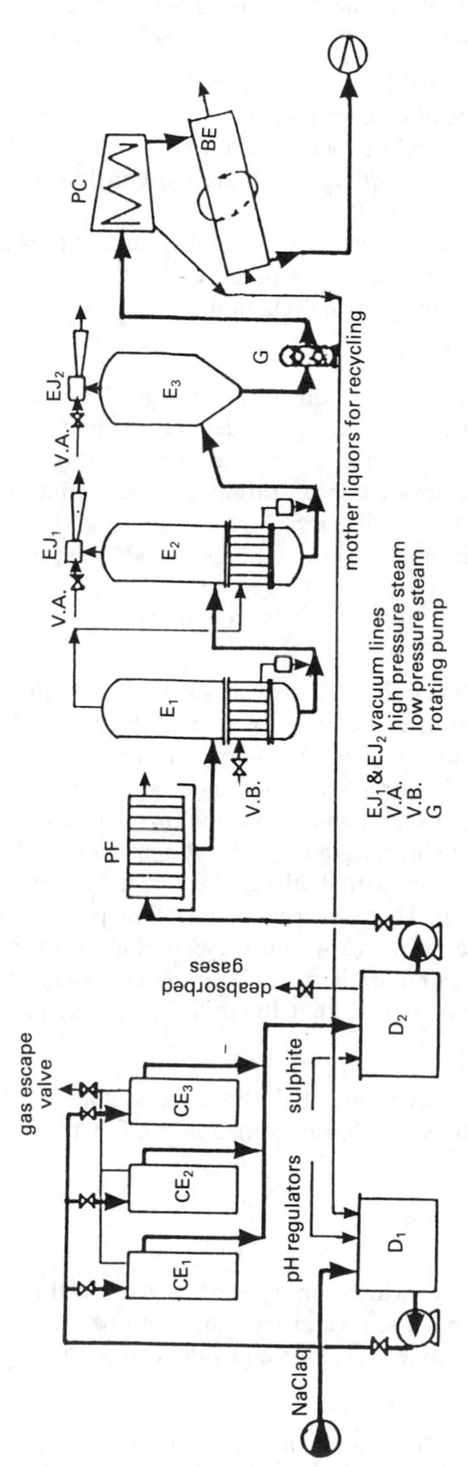

Fig. 9.11—Diagram of a plant for the production of chlorates. The electrolytic solution, the make-up of which can be seen by reference to D_1, is pumped into the electrolytic cells CE_1, CE_2, and CE_3. After electrolysis, the solution is reductively freed from hypochlorites in D_2, filtered in PF, evaporated under the double effect of E_1 and E_2, and crystallized in vacuo in E_3. The crystals are separated in the centrifuge PC and dried in the drum-type desiccator before being sent to storage. Owing to the discriminating action of the pH value, only the hypochlorites are selectively reduced by the sulphite which is added. The chlorates remain unaffected.

general. Above all, sodium chlorite is the basis for the advertised bleaching agent which does not damage the materials.

Perchlorates. Some of the major fields in which perchlorates are currently used are those of mining explosives (cheddites), propellants for airborne jets and missiles, pyrotechnics, the refining of oils, and the depolarization of batteries. These products are in the form of white crystalline powders which are more or less deliquescent and soluble in water.

Sodium perchlorate from which the less soluble potassium and ammonium salts can be prepared by double decomposition, is prepared by the anodic oxidation of the corresponding chlorate according to the reaction:

$$ClO_3^- + H_2O \rightarrow ClO_4^- + 2H^+ + 2e^- \qquad (9.15)$$

The electrolytic bath is fed with a solution of 380–430 g/l of sodium chlorate and a similar amount of perchlorate which acts as a depolarizer and allows a solution to be discharged which has more than twice the concentration of the original perchlorate, and in which the chlorate concentration has been reduced to about one tenth of the amount added.

Table 9.4 contrasts the operating conditions for the production of chlorates with those applying to perchlorates.

9.4 THE MINOR HALOGENS

Bromine is considered as an element from the hydrosphere, and *iodine* is considered as an element from the lithosphere, because bromine, which is 1000 times more abundant ($6.5 \times 10^{-3}\%$) than iodine ($5 \times 10^{-6}\%$) in sea water, is mainly recovered from such waters, while iodine is extracted from deposits on terra firma.

The properties of both halogens are similar, but the properties of bromine bear a greater resemblance to those of chlorine than to the properties of iodine.

For instance, bromine combines with hydrogen when irradiated with light, or heated, or in a catalyzed reaction, whereas iodine requires the presence of an efficient catalyst in order to react. The extent to which they combine with metals and the tendency to undergo disproportionation in water becomes ever smaller upon passing from bromine to iodine to such a degree that the iodine–water system:

$$I_2 + H_2O \rightarrow HI + HIO$$

is practically neutral, as the reaction indicated above barely takes place at all.

The stability of the compounds which are formed falls off at the same rate as the tendency to form them.

Bromine

Production and uses

In slightly more than the 16 litres which, on average, remain in the mother liquors from the salt works for every m^3 of sea water treated, it can be calculated that there is approximately 61.38 g of bromine* which is equivalent to 3.8 kg of Br_2 per m^3 of the mother water.

*These values may vary over quite wide limits because, for example, the waters of the Tyrrhenian sea are somewhat less brominated than those of the Ionian sea, while the waters of the Dead Sea actually contain around 5 kg/m^3 of the mother water.

The bromine exists in these waters as NaBr and $MgBr_2$ from which the halogen is recovered industrially by displacement with chlorine in accordance with the reaction:

$$2Br^- + Cl_2 \rightarrow Br_2 + 2Cl^- \tag{9.16}$$

A plant for the complete recovery of bromine from the mother liquors from salt works* is shown in Fig. 9.12.

The mother liquors, originating from the collection vessel D, pass through the heat exchanger E_1 where they are preheated at the expense of the exhausted waters emerging from the stripping column and which are subsequently discarded.

After preheating, the mother liquors are fed into the packed column C_1. Gaseous chlorine is fed into the bottom of this column together with a recycled mixture consisting of chlorine, bromine, materials which cannot be condensed, and, without fail, steam.

Reaction (9.16) takes place in C_1, and while the liquid bromine and a little of the chlorine remain dissolved in the descending mother liquor, the excess of chlorine and the materials which cannot be condensed emerge into the top of the column from where they are sent for recovery of the chlorine and the traces of bromine. The liquid in the bottom of column C_1 (the mother liquors charged mainly with liquid bromine) falls like rain onto the packing of column C_2 into which live steam is fed from the bottom. This strips out the bromine and chlorine, leaving the liquid exhausted. The mixture of gases and vapours at the top of column C_2 is largely condensed in E_2 and is then made to pass into the separator SD.

The materials which cannot be condensed, bromine vapours, and, especially, the chlorine, are returned to the chlorination column C_1. The aqueous phase in SD which floats on the moist and chlorine-contaminated layer of liquid bromine which has collected at the bottom of the separator is sent into the stripping column C_2, while the lower layer of crude bromine is sent as feedstock to the next stripping column C_3 which is equipped with a bottom heater. This column discharges from its base pure bromine, which is sent to storage, while chlorine containing traces of bromine and the usual non-condensable materials is taken off at the top as a mixture of gases and vapours which is returned to the chlorination column C_1.

At the present time, the main use of bromine is in the production, on account of its ready addition to a double ethylenic bond, of 1,2-dibromoethane ($CH_2Br{-}CH_2Br$) which is a component of the petrol antiknock additive based on tetraethyl-lead. Since the $PbBr_2$ which is formed is volatile, it prevents the build-up of lead deposits in the cylinders of internal combustion engines.

Bromine finds other applications in the preparation of pharmaceuticals, inorganic sedatives, and organic anaesthetics. Relatively large amounts of bromine are used in the preparation of silver bromide (AgBr) which is prepared, for the photographic industry, by the metathetical reaction between $AgNO_3$ and KBr. Finally, appreciable quantities of bromine are used in the preparation of the bromo-derivatives of organic dyestuffs, insecticides, lacrimatory agents, and for other purposes.

Production of inorganic compounds of bromine

PBr_3 is readily prepared by synthesis from the elements, as is its analogue PCl_3. These are both used in the halogenation of organic compounds.

The following scheme shows how hydrobromic acid† is prepared and the many

---

*The process used for the extraction of bromine from minerals in terrestrial deposits is similar after the material has been crushed, enriched if necessary, and the solids have been dissolved in a little water.

†A term used in technological jargon to indicate both gaseous HBr (which, correctly, is hydrogen bromide) as well as HBr in aqueous solution.

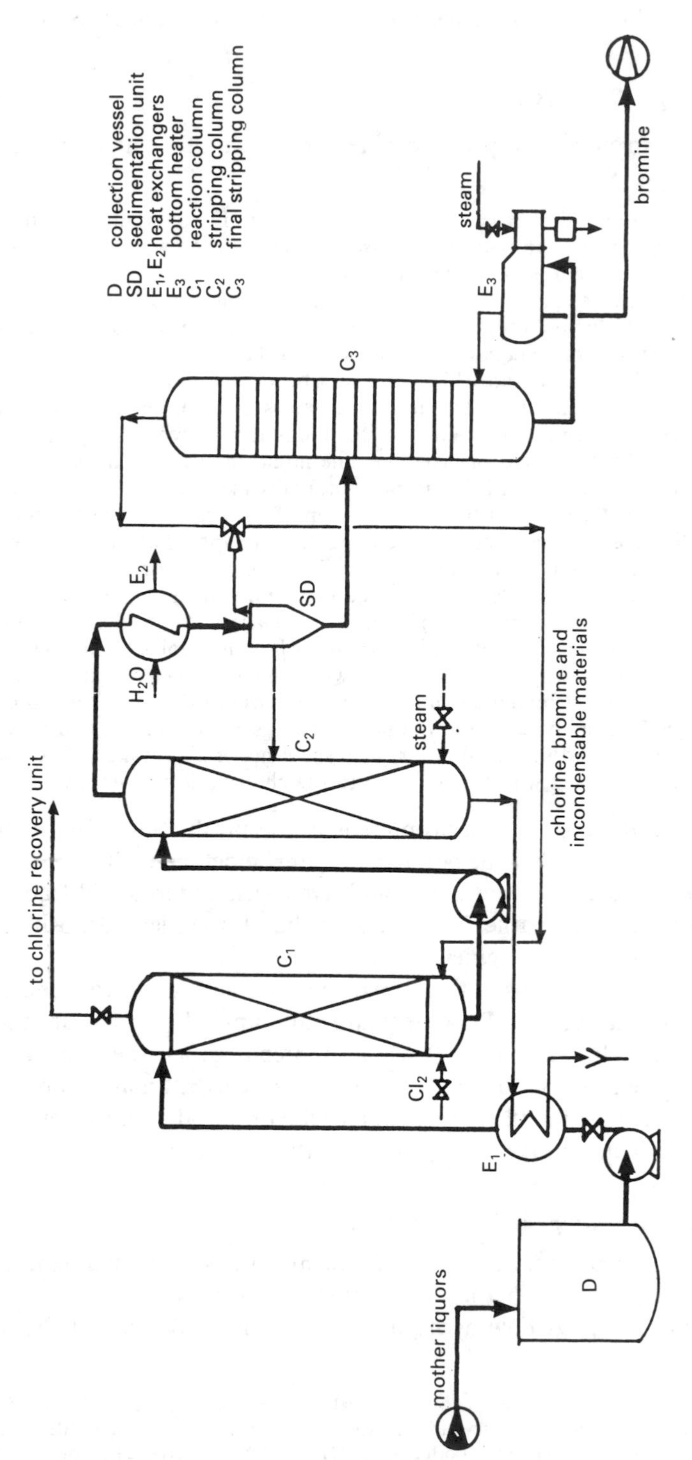

Fig. 9.12—Plant for the production of bromine from the mother liquors from salt works and for the working of terrestrial deposits of salts originally precipitated from the ocean and based on a chloride or sulphate matrix. In Europe, these are found in Alsace and, especially, in the Stassfurt district of Saxony. Finally, it is also possible to prepare bromine by starting off from sea water itself, in which case, owing to the costs, the bromine, after chlorination, is carried away in a stream of air rather than steam.

ways in which it is used in the preparation of alkali metal bromides which are intermediates in the preparation of other compounds of bromine.

Bromine	$\xrightarrow[\text{light}/200°\text{C}]{+H_2}$	HBr	$\xrightarrow[-(CO_2+H_2O)]{+CO_3^{2-}}$	Br^-	alkali bromides
	$\xrightarrow[\text{cold}]{+OH^-}$	$Br^- + BrO^-$	$\xrightarrow[-(SO_2+H_2O)]{+H_2S}$	Br^-	alkali bromides
	$\xrightarrow[\text{hot}]{+OH^-}$	$Br^- + BrO_3^-$	$\xrightarrow[-(CO_2+H_2O)]{+HCOOH}$	Br^-	alkali bromides
	$\xrightarrow[\simeq 100°\text{C}]{\text{Fe}}$	Fe_3Br_8 $(2FeBr_3.FeBr_2)$	$\xrightarrow[-\text{basic iron carbonates}]{+CO_3^{2-}}$	Br^-	alkali bromides

The bromates are prepared in a pure state by the metathetical reaction between the chlorine of chlorates and bromine (the basis of this reaction has been discussed on page 313):

$$2ClO_3^- + Br_2 \rightarrow 2BrO_3^- + Cl_2 \tag{9.17}$$

working in an autoclave where the liquid bromine and the powdered chlorates are heated. The bromates find only specialized uses, including those in the field of analytical chemistry.

Iodine

The natural sources of iodine are quite numerous, but they contain only a low percentage of the halogen and are in an extremely dispersed state. Iodine is mainly found as sodium iodate (but also as the periodate) in Chile saltpetre which contains from 0.02 to 0.2% of iodine. In addition, the 'varek' algae of Normandy and Scottish and Norwegian 'kelp' are considered as exploitable sources of iodine. Finally, this halogen is extracted from the alkaline salts of certain thermal waters such as those of Salsomaggiore and Monticelli (Parma) in Italy.

The methods of extraction vary, depending on the raw material. This is shown by the three parallel schemes depicted in Fig. 9.13, where references are given to the reactions taking place in those processes which can be properly regarded as being chemical.

Let us now pass to a brief review of the uses which are made of iodine and indicate the preparation and uses of its principal compounds.

Iodine, dissolved in alcohol (tincture of iodine) or solubilized in water by complexation:

$$I_2 + I^- \rightarrow I_3^-$$

is used as an antiseptic pharmaceutical. Furthermore, agents for use in radiology, diagnostic inorganic and organic compounds (after the halogen has been rendered radioactive in the channels of nuclear reactors), colouring agents, catalysts, and silver iodide for the photographic industries, are produced from iodine.

More specifically, among the organic iodo-derivatives, the alkyl iodides are very commonly

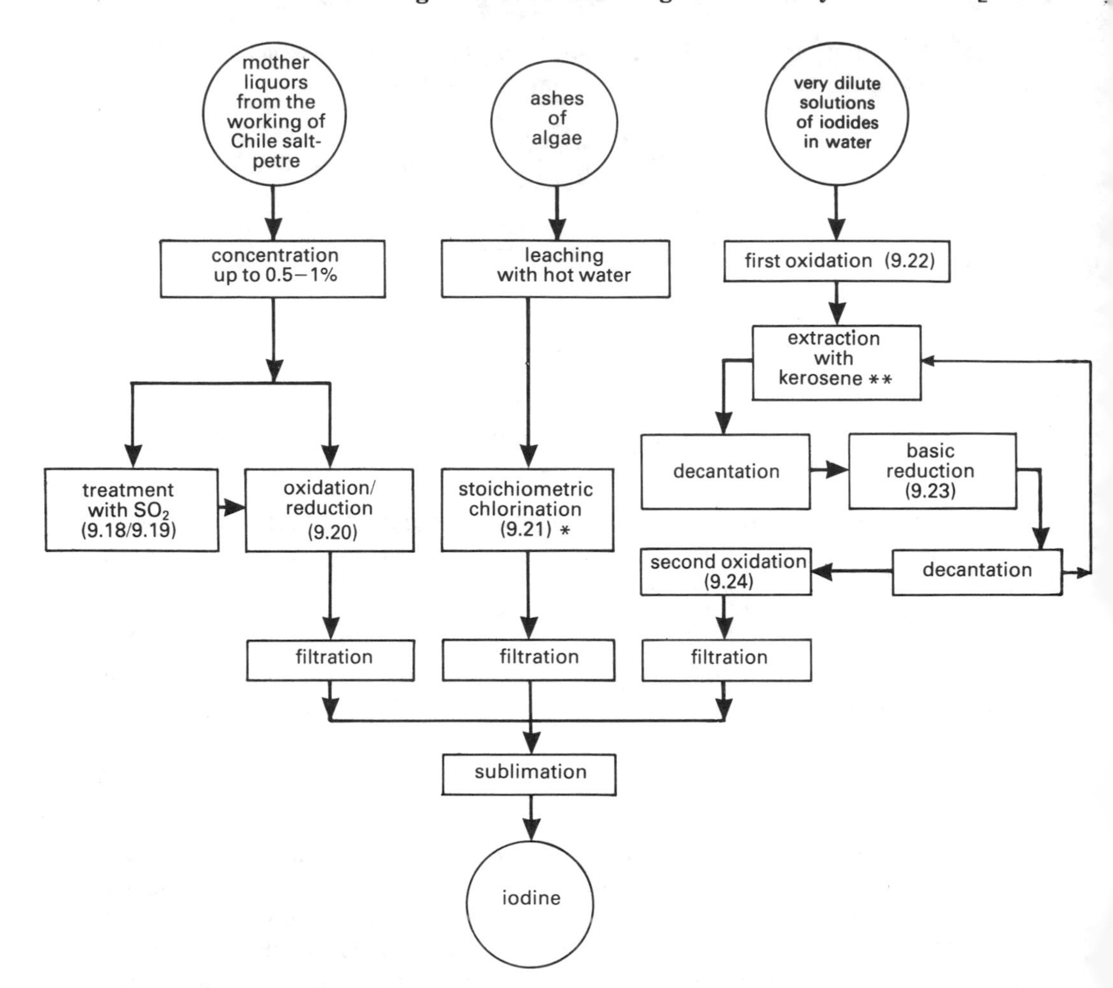

Fig. 9.13—Scheme for the production of iodine from its principal natural sources. The reference numbers refer to the following chemical reactions:

$$IO_3^- + 3SO_2 + 3H_2O \rightarrow 3SO_4^{2-} + I^- + 6H^+ \quad (9.18)$$

$$IO_4^- + 4SO_2 + 4H_2O \rightarrow 4SO_4^{2-} + I^- + 8H^+ \quad (9.19)$$

$$5I^- + IO_3^- + 6H^+ \rightarrow 3I_2 + 3H_2O \quad (9.20)$$

$$2I^- + Cl_2 \rightarrow 2Cl^- + I_2 \quad (9.21)$$

$$2I^- + 2NO_2^- + 4H^+ \rightarrow I_2 + 2NO + 2H_2O \quad (9.22)$$

$$I_2 + SO_3^{2-} + 2OH^- \rightarrow 2I^- + SO_4^{2-} + H_2O \quad (9.23)$$

$$6I^- + 2CrO_3 + 12H^+ \rightarrow 2Cr^{3+} + 3I_2 + 6H_2O \quad (9.24)$$

*As is well known, chlorine first displaces the iodine and then, if used in excess, the bromine. Finally, it is also capable of oxidizing iodine to iodate.

**The distribution coefficient for the partitioning of iodine between kerosene and water is 400; that is, iodine is 400 times more soluble in kerosene than in water.

employed in various syntheses and as alkylating agents on account of their special reactivity. Other high molecular weight organic iodo-derivatives containing a large amount of iodine find use in X-ray diagnostic techniques such as cholecystography, cholangiography, and pyelography as contrasting agents.

The iodides which are used mainly in analytical chemistry are prepared in a similar manner to the bromides, while *hydriodic acid*, which is strongly reducing, is prepared in a better way in the pure state by the reduction of silver iodide to silver with hydrogen.

The reducing power of all these compounds is attributable to the fact that the iodide ion (I^-) can be readily oxidized to elemental iodine which can be detected when present in very small quantities by means of the intense blue coloration which it gives with starch.

9.5 THE ACTION AND PREVENTION OF THE EFFECTS OF THE PRINCIPAL TOXIC INDUSTRIAL COMPOUNDS OF THE HALOGENS

The major inorganic toxic halogen compounds are chlorine and hydrochloric acid, hydrofluoric acid, and the fluorides and bromine.

While iodine is also, in fact, poisonous (a maximum tolerable concentration in the surrounding atmosphere of 0.1 ppm, that is, 1 mg/m^3) and fluorine is very poisonous (a maximum limit of 0.1 ppm, that is, 0.2 mg/m^3), the probability of having to take precautions against the action of these two halogens is very low. The same may be said of hydrobromic and hydriodic acids which have maximum tolerable limits of 3 ppm (10 mg/m^3) and 4 ppm ($\simeq$20 mg/m^3) respectively.

Chlorine

Since the specific weight of chlorine is greater than that of air, clouds of chlorine can be formed at ground level which are difficult to disperse. Chlorine may be emitted from all plants where chlorinations are carried out (factories producing chlorinated solvents, pesticides, allyl chloride, etc.) as well as from plants concerned with the manufacture of soda by electrochemical processes.

Chlorine dispersed in water or in air not only possesses an intrinsic toxicity, but also assumes the aggressiveness and toxicity of many of its derivatives since it is capable of yielding hydrochloric acid, hypochlorous acid, the chlorinated derivatives of organic compounds contained in 'photochemical smog', and so on. The specific physiological effects of many of the substances derived from the dispersion of chlorine into the environment have been made evident, and, in particular, the various allergy-provoking and carcinogenic properties of these compounds have been established.

The tolerable limit for chlorine in a populated area is 1 ppm, that is, 3 mg/m^3. At this limit, chlorine can be readily smelt.

Table 9.5 summarizes the gravity of the toxic effects of chlorine as a function of the exposure time.

In the gaseous state chlorine induces acute poisoning by acting both as an oxidant and as a chlorinating agent with respect to tissues.

Chlorine first causes irritation of the respiratory tracts and a feeling of suffocation. The acute poisoning causes lachrimatory effects, conjunctivitis, skin lesions, collapse, and cyanosis.

The therapy administered in the case of chlorine poisoning consists of maintaining the patient in a state of complete repose and warmth. Pure oxygen must be inhaled and 20 ml of calcium gluconate must be injected intravenously. Artificial respiration is contra-indicated.

Hydrochloric acid

Hydrochloric acid, which, fortunately, can be removed relatively easily, is emitted from all plants where chlorinations and dehydrochlorinations are carried out and from sites where

Table 9.5—The seriousness of chlorine poisoning as a function of exposure time and dose

		Doses dispersed in air	
Effects	*Exposure time*	*ppm*	*mg/m³*
Without consequences	60 min	4	12
Serious or fatal consequences	30–60 min	40–60	120–180
Certain death	a few minutes	1000	3000

chlorine comes into contact with hydrogen, water, etc. Very large amounts of hydrochloric acid may also originate from plants where refuse is incinerated.

When very concentrated or gaseous HCl comes into contact with the skin it produces acute burns and lesions. The action of this acid on the cornea, where it exerts an action similar to injury by burning, is particularly distressing as is its action on the glottis which it causes to swell (oedema of the glottis).

The tolerable limit in an environment occupied by people is 5 ppm, which is equal to 7 mg/m^3.

Hydrochloric acid can be smelt at a concentration of 1 ppm, and its toxic effects usually depend on the exposure time and the concentration, as is shown in Table 9.6.

The remedies are:

- when the skin, mouth, etc. has been in contact with the acid, the affected area must be washed with copious amounts of drinking water or, better still, with distilled water;
- if appreciable amounts have been ingested, magnesia, milk, and egg white should be administered. Carbonates and bicarbonates are contra-indicated.

On the other hand, bicarbonate is a good remedy against small ingestions of hydrochloric acid which is known to be already in the stomach.

Hydrogen fluoride and fluorides

The main sources of contamination by F_2, HF, and fluorides are the phosphate fertilizer industries and related industries (e.g. those producing cryolite). However, many companies dealing with ceramics and cement also produce greater or smaller amounts of these substances from the fluorinated calcareous rocks which they use, and the toxicity of these substances is such that it cannot be ignored.

Table 9.6—The seriousness of HCl poisoning as a function of exposure time and dose

		Doses dispersed in air	
Effects	*Exposure time*	*ppm*	*mg/m³*
Tolerable	several hours	10–50	14–70
Tolerable	60 min	50–100	70–140
Dangerous	30–60 min	1000–1300	1400–1820
Lethal	a few minutes	1300–2000	1820–2400

Hydrofluoric acid stimulates effects of the same type as those caused by hydrochloric acid, but they are far more pronounced.

The toxic action of fluoride ions which get into the blood stream is due, above all, to the powerful complexing action which these ions exhibit when they encounter the oligo elements which are present in the blood and which are essential for the maintenance of the function of many protein-enzyme systems. Fluorides only formally have a tolerance limit (2.5 mg/m^3) below that of hydrofluoric acid (3 ppm which is equal to 2.7 mg/m^3) because, actually, when this limit is referred to the weight of fluorine present, it is higher by virtue of their higher molecular weight. This fact is explained by the complete dissociation of fluorides; and vice versa, the hydrogen-bonded molecular association which characterizes hydrofluoric acid.

The remedies against the physiological damage caused by hydrofluoric acid are practically analogous with those used against the action of hydrochloric acid. As a remedy for ingested fluorides, one can also employ a stomach pump, make use of emetics, and introduce calcium phosphate preparations either orally or intravenously.

Bromine

The odour of bromine is extremely irritating and unpleasant, and its action causes inflammation of the respiratory passages (the nose, the back of the mouth, and the lungs) in a way which is far more painful and lasting than that due to chlorine.

Nevertheless, bromine can be smelt only at a concentration which is much higher than that which can be tolerated in the atmosphere over a long period. It can, in fact, be smelt only when it is present in amounts greater than 1 ppm, whereas the tolerability limit is 0.1 ppm, which is equal to 0.7 mg/m^3.

The best prophylaxis both against bromine and other halogens in the gaseous or vapour state and against their toxic and/or aggressive compounds consists of the use of gas masks, ventilation of the area, and the wearing of protective garments.

If bromine has been ingested, the use of emetics and a stomach pump is advised, together with the administration of a dilute solution of sodium thiosulphate mixed with bicarbonate. When bromine reacts with this mixture, innocuous sulphates and bromides are formed:

$$S_2O_3^{2-} + 4Br_2 + 10HCO_3^- \rightarrow 2SO_4^{2-} + 8Br^- + 10CO_2 + 5H_2O$$

This mixture of thiosulphate and bicarbonate is also a good antidote against ingestions of fluorine and chlorine, for the same reason.

On the other hand, ingestions of iodine are treated with a dilute solution of thiosulphate on its own, as iodine participates in the well known reaction:

$$2S_2O_3^{2-} + I_2 \rightarrow 2I^- + S_4O_6^{2-}$$

in which the thiosulphate is oxidized to tetrathionite, a relatively innocuous compound.

10

The phosphorus and boron industries

The ever increasing development of the use of phosphorus compounds during the last two decades, especially in the industrial fields of detergents, pesticides, foodstuffs, and plastics, as well as in the technologies employed in the treatment of waters, the dressing of textiles, and industrial emulsifiers, has led us, unlike in traditional schemes, to treat phosphorus and its derivatives separately from phosphate fertilizers, and, what is more, from fertilizers in general.

The treatment of boron is coupled with that of phosphorus. The former is a non-metallic solid of considerable importance in industry, and which forms compounds which are destined for use in the same sectors (such as glasses and detergents, for example) as phosphorus derivatives.

Part 1 PHOSPHORUS AND ITS DERIVATIVES (EXCLUDING FERTILIZERS)

10.1 PHOSPHATE MINERALS

All rocks which, by virtue of their calcium phosphate [$Ca_3(PO_4)_2$] content, may act as raw materials for phosphorus-based industrial compounds are referred to as phosphate minerals in technical jargon. Even bones may be considered as phosphate minerals, and, because they also contain nitrogen, they are particularly suited for use as fertilizers. In order of their importance, the principal basic chemical components ('apatites') of phosphate minerals are:

$3Ca_3(PO_4)_2 \cdot CaF_2$ fluoroapatites
$3Ca_3(PO_4)_2 \cdot Ca(OH)_2$ hydroxyapatites
$3Ca_3(PO_4)_2 \cdot CaCO_3$ carbonatoapatites
$3Ca_3(PO_4)_2 \cdot CaCl_2$ chloroapatites
$3Ca_3(PO_4)_2 \cdot CaSiO_3$ silicatoapatites

Very often, and even quite generally, however, these pure compounds occur in natural mixtures with one another and also combined with one another as definite chemical species such as 'dahlite':

$$2Ca_3(PO_4)_2 \cdot Ca_3(CO_3OH)_2 \cdot Ca(OH)_2$$

There exist on earth mineral phosphate deposits in the form of eruptive (igneous) rocks which are said to be of primary origin and are particularly rich in fluoroapatites.

Of course, the majority of the phosphate minerals are of the sedimentary type which have undergone metamorphosis and originate from various types of redeposition of phosphorus compounds which have earlier been solubilized by thermal waters, marine waters, or attack on rocks by living organisms and the metabolic products of these organisms. This redeposition may have been skeletal, by recrystallization, by reprecipitation brought about by calcium salts, or, more frequently, mixed, as is revealed by the deposits found in the coastal marine zones of California and Mexico where deposits of animal origin are combined with precipitates which are clearly of marine type.

The consequent metamorphosis of the early deposits is mainly due to the phenomena of recrystallization under conditions of high pressure and over extremely long periods.

More than 80% of the production of phosphate minerals is concentrated in the USA, the USSR, and Morocco.

10.2 THE PREPARATION AND MARKETING OF PHOSPHORUS

Details and course of the preparation

While, in the preparation of phosphorus, preference was given in the past to the use of rich phosphate minerals from specific deposits (especially American 'land pebble'), minerals obtained from any source whatsoever are used for this purpose today, and these, when necessary, are enriched by flotation.

However, in addition to phosphate minerals, quartzites and coke are the raw materials for the production of elementary phosphorus. The basic reaction for the preparation of elementary phosphorus is:

$$2Ca_3(PO_4)_2 + 10C + 6SiO_2 \rightarrow 6CaSiO_3 + P_4 + 10CO \qquad (10.1)$$

If this preparation is to be successful, it is important to ensure that the three components of the raw material are thoroughly and homogeneously mixed with one another. To do this, especially if they are obtained from flotation processes, the phosphate minerals must be suitably agglomerated.

Nowadays, for this purpose, the phosphate mineral is, for the most part*, converted into nodules after the addition of a small amount of quartzite in rotating furnaces which are heated by utilizing the combustion of the carbon monoxide, which is formed in reaction (10.1), in conjunction with that of fuel oils.

Once prepared, the raw materials with a homogeneous speckled appearance are loaded into electric arc-resistance furnaces in mineral/quartzite/coke weight proportions of 100:35:16.

The arc-resistance furnaces are responsible for providing the energy, by the conversion of electricity into heat, required for reaction (10.1). This reaction is highly endothermic and requires 5894 kcal/kg.

A diagram of a complete plant for the production of phosphorus is shown in Fig. 10.1.

*Other methods of agglomeration are also possible, such as the compression of the moist powdered mineral.

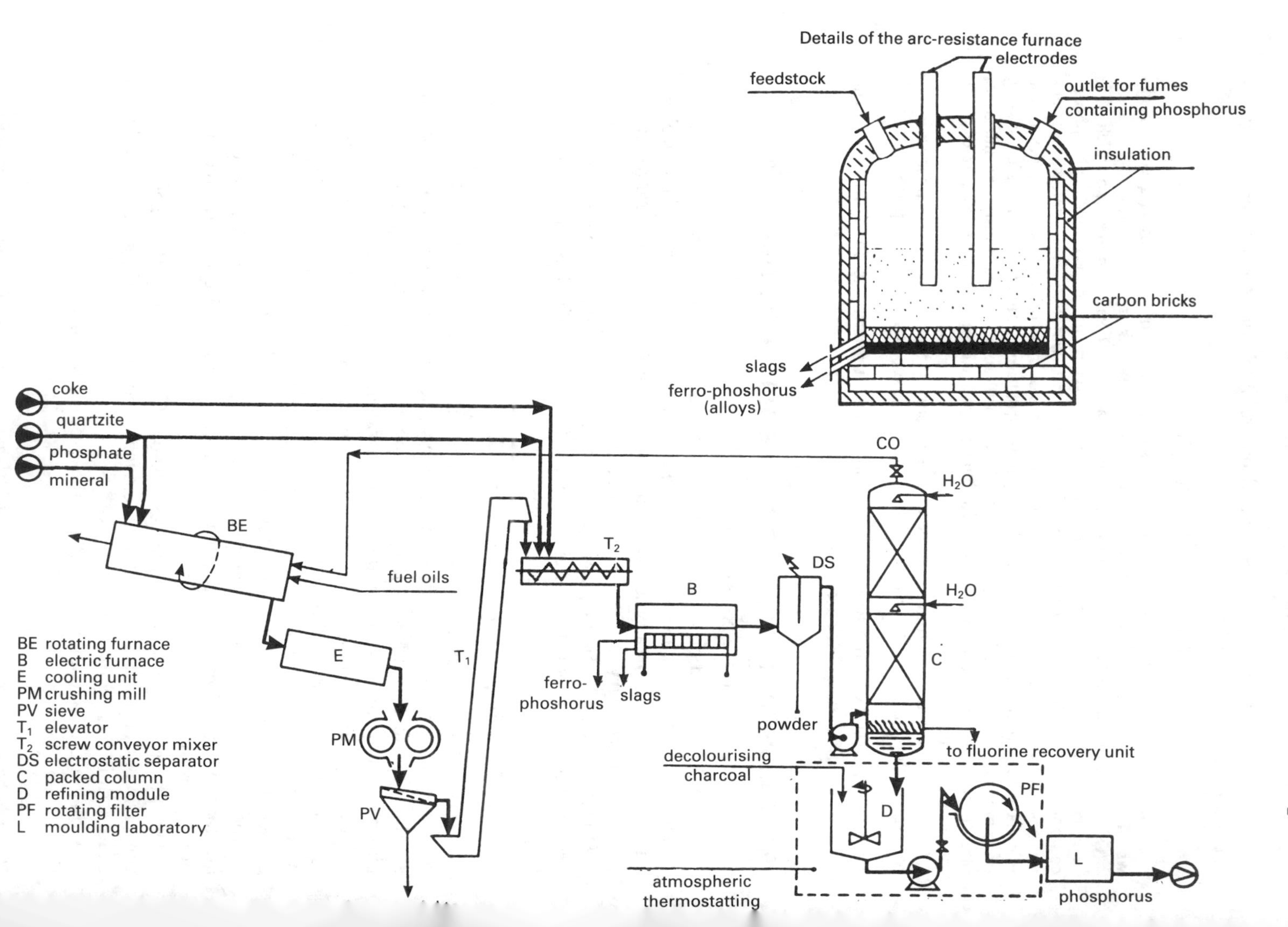
Details of the arc-resistance furnace
electrodes
outlet for fumes
containing phosphorus
insulation
carbon bricks
feedstock
slags
ferro-phoshorus
(alloys)
CO
H_2O
H_2O
C
DS
to fluorine recovery unit
PFl
L
phosphorus
D
powder
decolourising
charcoal
atmospheric
thermostatting
B
slags
ferro-
phoshorus
T_2
T_1
fuel oils
E
PM
PV
BE
coke
quartzite
phosphate
mineral
BE rotating furnace
B electric furnace
E cooling unit
PM crushing mill
PV sieve
T_1 elevator
T_2 screw conveyor mixer
DS electrostatic separator
C packed column
D refining module
PF rotating filter
L moulding laboratory

Recovery of the product and by-products

The fumes which emerge from the furnace are first freed from any mineral and other fine reagents which may have been carried up at such a temperature (>300°C) that the phosphorus itself is not condensed out. The remaining gaseous products, which consist of phosphorus vapours and the gases CO and SiF_4, are sent to the bottom of a tower in which they are sprayed with water at two different heights while taking care that the temperature does not drop below 60°C. The phosphorus thereupon condenses out but does not solidify, and is collected under water which reacts with the SiF_4 gas, converting it into a solution of metasilicic and fluorosilicic acids.

$$3SiF_4 + 3H_2O \rightarrow H_2SiO_3 + 2H_2SiF_6$$

The *fluorinated component* is subsequently recovered from this solution, as we shall see (page 359).

The *carbon monoxide*, which has been completely freed from phosphorus and fluorine compounds, is then dried by cooling and subsequently used as a fuel.

The *slags*, essentially consisting of $CaSiO_3$, which are produced in the furnace and subsequently discharged from its base, are good additives for cements and bituminous antiskid conglomerates for road and airport runway construction.

The iron alloy, *ferro-phosphorus*, which is used in metallurgical processes, is also discharged from the base of the furnace.

The liquid phosphorus, after decoloration with active carbon, is pumped to undergo filtration and is subsequently solidifed in the 'white' form and moulded under water or under an atmosphere of CO_2 into the commercially supplied shapes (commonly in the form of sticks). The white phosphorus, which is always stored under water, is then used in this form.

Like the vapour, both liquid and solid phosphorus contain P_4 molecules.

The commerical forms and uses of phosphorus

The phosphorus obtained from the preparation furnaces is therefore available in the form of white phosphorus. In this form it is utilized for the greater part in the establishments where it is produced while, in commerce, it is usually marketed as red phosphorus, which is practically stable in air, and is sealed in steel drums or cans.

White phosphorus is converted into red phosphorus by the prolonged heating of the element in boilers which have been pressurized up to 2 atmospheres, using nitrogen or carbon dioxide.

For this purpose it is first heated at 230°C for 2.5 hours and then for as long again at 260°C. It is finally brought up to 320°C and kept at this temperature for almost 8 hours. The times and temperatures necessary for this transformation are, however, reduced by working in the presence of traces of iodine.

Fig. 10.1—Diagram of a plant for the production of phosphorus. The phosphate mineral is made into nodules with a little quartzite in BE, cooled in E, reduced to pieces of a suitable size in PM, sieved in PV, and charged by means of T_1 and T_2 where it is homogenized together with the other raw materials which have a similar size distribution. The phosphorus is produced in B. The fumes from which any dust particles have been removed are separated into CO which emerges from the top of the separation column, impure phosphorus, and a solution of H_2SiO_3 and H_2SiF_6 which emerges from the base of this column. The phosphorus separates out from the solution under gravity, and the supernatant liquid is sent on for the recovery of the fluorine. The purified phosphorus, which has been filtered to remove the adsorbents and impurities, is suitably moulded into pieces and stored.

Table 10.1—Salient features of the commercial forms of phosphorus

Type	*Effective colours*	*mp* (°C)	*bp* (°C)	d_4^{20}	*Special properties*
white	white–yellow	44.1	282	1.84	soluble in CS_2, poisonous and autocombustible
red	red tending to violet	590 (43 atm)	–	2.25	insoluble in CS_2, practically stable in air

The red phosphorus is finally washed with CS_2 and NaOH to dissolve any residues remaining in it and to carry away, in the form of salts, any traces of oxidized compounds which may have been formed.

Some properties of the two forms of phosphorus are shown in Table 10.1.

Phosphorus is mainly used for the production of phosphoric acid and then for the preparation of the halides in phosphorus (in particular, the chlorides) and the sulphides of phosphorus. The whole range of the various chemical derivatives of these compounds: phosphate, polyphosphates, superphosphates, esters of phosphoric acid, etc., are widely used in industry. Finally, both directly and through its compounds, phosphorus is used in the fabrication of matches, incendiary bombs, and pyrotechnic materials, as well as in the production of gases for the control of infestations (PH_3) and for the deoxidation of melts in many metallurgical processes.

10.3 THE PHOSPHORIC ACID INDUSTRY

There are two industrial methods for the preparation of phosphoric acid:

- the thermal process which comprises the burning of phosphorus and the hydration of the resulting tetra-phosphorus decaoxide;
- the wet process which uses the reaction of phosphoric acid with phosphate minerals.

At one time the thermal process was carried out by starting directly from the phosphorus vapours emerging from the electric furnaces or by burning the white phosphorus after it had been purified and condensed. Nowadays, the second of these variants of the thermal method has been almost exclusively adopted: because the demand for pure phosphoric acid has arisen, or because it is necessary to prevent any pollution of the environment, or, finally, because one is forced to recover the fluorine which accompanies the phosphorus emerging from the electric furnaces in the form of SiF_4.

The production of H_3PO_4 by the thermal route

The liquid phosphorus is mixed with air as it is injected and is burnt in the tower type combustion chamber. The reaction is:

$$P_4 + 5O_2 \rightarrow P_4O_{10} + 2535\ \text{kcal/kg}_{P_4O_{10}} \qquad (10.2)$$

This combustion chamber is lined internally with silicaceous or trachitic refractories or is constructed from special steel which is internally passivated by the film of the concentrated H_3PO_4 solution which continuously descends over it but is not shown in Fig. 10.2.

On account of the very highly exothermic nature of this reaction, the ambient temperature is lowered by spraying water or a dilute solution of phosphoric acid into the chamber. Further water is sprayed onto the device, following the combustion chamber in the flow diagram, which acts as a water circulation heat exchanger. Finally, as much of the P_4O_{10} as possible is hydrated in a succeeding column by irrigation with water sprayed into it from different heights.

The residual clouds of P_4O_{10} are swept down in precipitators or in special scrubbers which furnish the dilute solutions of the acid which are sprayed into the combustion chamber.

The most significant impurities in the acid produced in this manner consist of arsenic compounds. If the acid is destined for use in food industries, these arsenic impurities are precipitated out by bubbling H_2S through the solution, which is subsequently filtered.

The production of H_3PO_4 by the wet route

Generalities

In the wet procedure (Fig. 10.3) mineral phosphates are attacked with recycled phosphoric acid and sulphuric acid.

The main reactions taking place are:

$$Ca_3(PO_4)_3 + 4H_3PO_4 \rightarrow 3Ca(H_2PO_4)_2 \quad (10.3)$$

$$Ca(H_2PO_4)_2 + H_2SO_4 + 2H_2O \rightarrow CaSO_4 \cdot 2H_2O + 2H_3PO_4 \quad (10.4)$$

$$Ca_3(PO_4)_2 + 3H_2SO_4 + 6H_2O \rightarrow 3CaSO_4 \cdot 2H_2O + 2H_3PO_4 \quad (10.5)$$

Actually, calcium hydrogen phosphate ($CaHPO_4$) can also be formed in (10.3), and then, depending on the temperature at which the reaction is carried out, it is also possible to produce $CaSO_4 \cdot \frac{1}{2}H_2O$ and even anhydrous $CaSO_4$ at above 125°C by the modification of (10.4) and (10.5).

In every case, in addition to the reactions including the attack of the phosphate component of the mineral, reactions also occur in which the other components of the mineral phosphates (CaF_2 and $CaCO_3$) participate as well as the oxides in the gangues (Fe_2O_3, Al_2O_3, etc.).

Reactions which lead to fluorinated products are also of interest:

$$CaF_2 + H_2SO_4 + 2H_2O \rightarrow CaSO_4 \cdot 2H_2O + 2HF \quad (10.6)$$

$$4HF + SiO_2 \rightarrow SiF_4 + 2H_2O \quad (10.7)$$

$$3SiF_4 + 3H_2O \rightarrow 2H_2SiF_6 + H_2SiO_3 \quad (10.8)$$

On the other hand, reactions which lead to the loss of sulphuric acid by yielding compounds which are of little commercial value or which contaminate the solutions with extraneous ions, are considered as being parasitic reactions:

$$CaCO_3 + H_2SO_4 + H_2O \rightarrow CaSO_4 \cdot 2H_2O + CO_2 \quad (10.9)$$

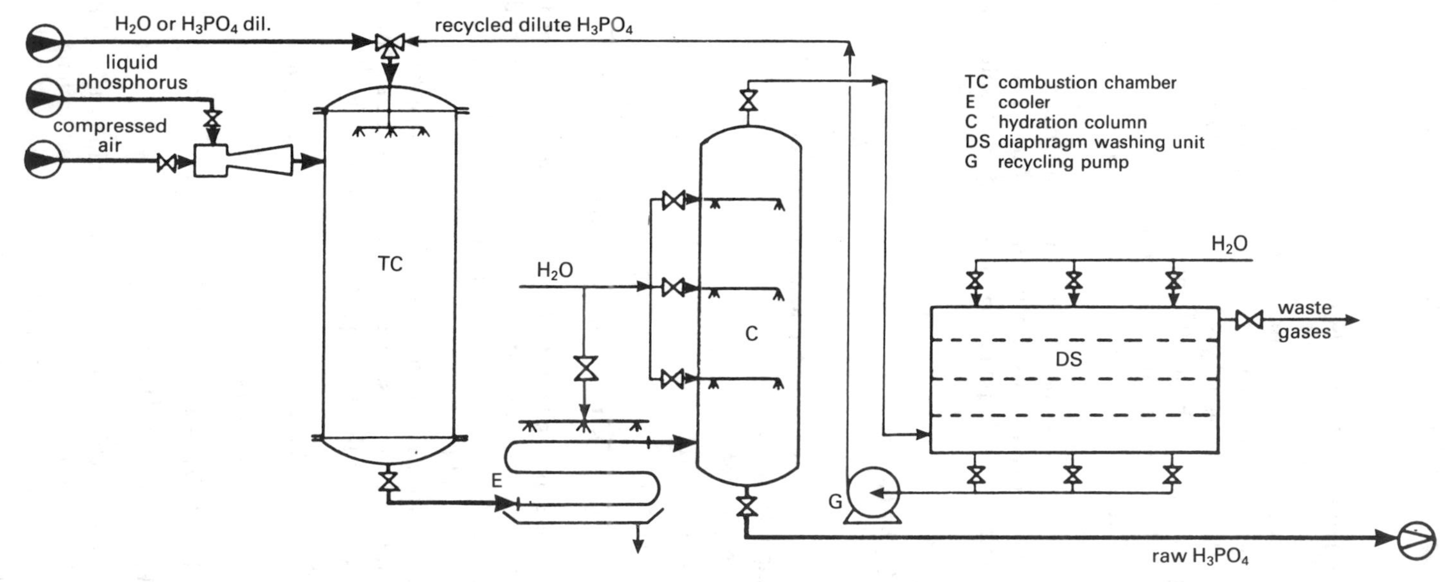

Fig. 10.2—The production of phosphoric acid by the thermal route. The amount of air employed results in there being an excess with respect to the quantity which is theoretically necessary. This, like the water which is sprayed in, contributes to the cooling of the tetraphosphorus decaoxide (P_4O_{10}) which is produced. The cooler DS is constructed from graphite and scrubs out the clouds of P_4O_{10}. Finally, the phosphoric acid, which is still quite hot (70–80°C) is collected in stainless steel tanks which may also be of the open type on account of the intrinsic stability of the product.

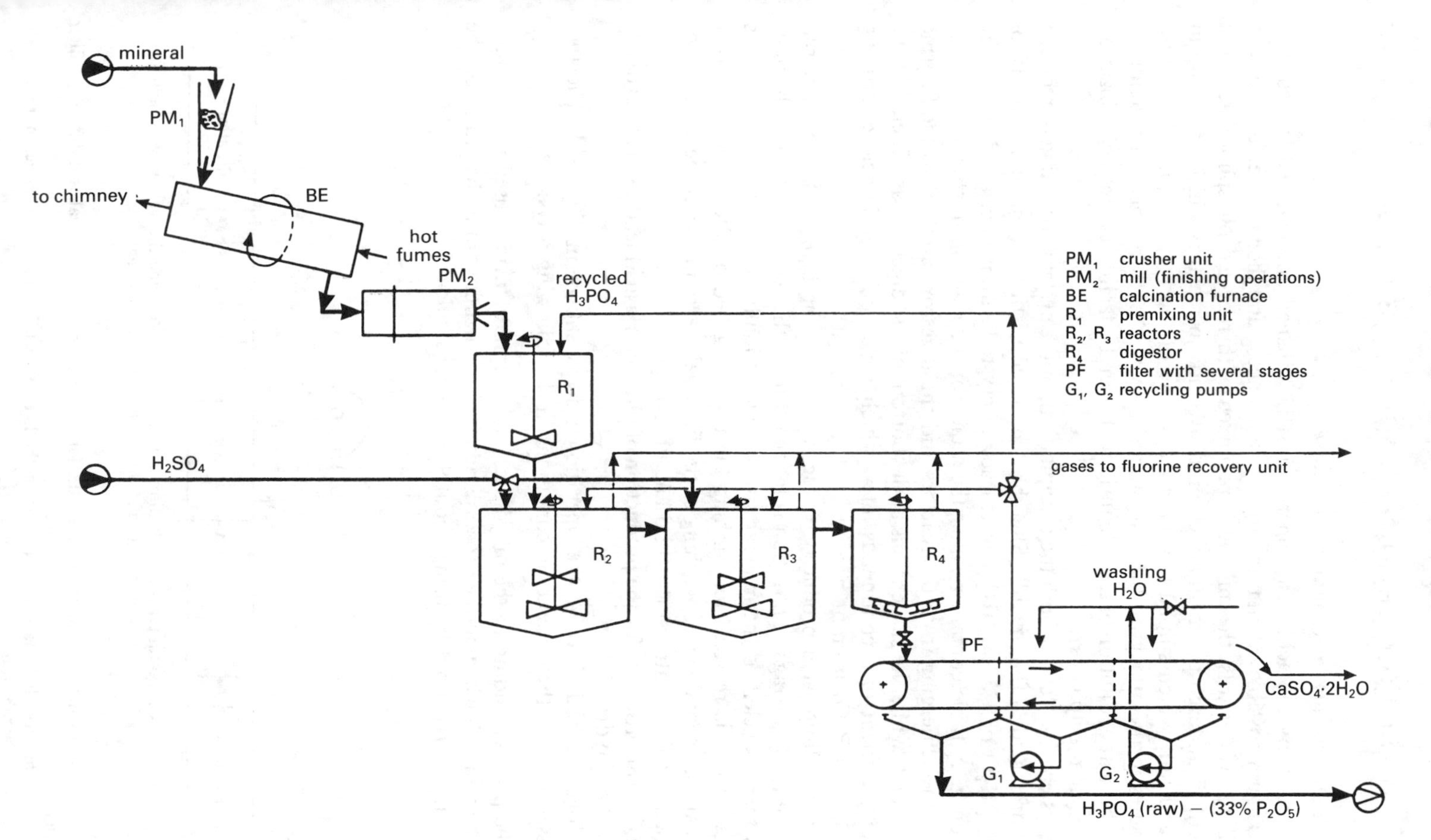

Fig. 10.3—Production of phosphoric acid by the wet route with attack by means of sulphuric acid and crystallization of the salt dihydrate $CaSO_4 \cdot 2H_2O$.

$$Fe_2O_3(Al_2O_3) + 3H_2SO_4 \rightarrow 2Fe^{3+}(2Al^{3+}) + 3SO_4^{2-} + 3H_2O \qquad (10.10)$$

Technologies of current production processes

Before attack by the acids, the mineral must be crushed and calcined to free it to the maximum possible extent from organic substances* and then reduced to a powder by milling. Afterwards, the mineral is premixed with recycled phosphoric acid and then sent to undergo the main attack on it by acids in reactors which are set up in series and into which sulphuric acid is fed. Some of the recycled phosphoric acid contributes to the attack on the mineral and disperses both the heat of reaction as well as the heat of dilution of the sulphuric acid, thereby facilitating the crystallization of the calcium sulphates†.

These steps also favour both the precipitation, in the system consisting of recycled phosphoric acid, of a readily filterable chalk consisting of minute crystals of $CaSO_4 \cdot 2H_2O$ or $CaSO_4 \cdot \frac{1}{2}H_2O$ which act as crystallization nuclei, as well as the digestion in suitable containers, after the attack of the reaction mixture.

The fumes emerging from the reactors and the digestors are sent to the fluorine recovery unit, while the suspensions at the bottom of the digestors are filtered in the first compartment of a continuous filter, which yields phosphoric acid containing about 33% of P_2O_5 from its base.

The more dilute acid produced by the washing of the layer of filtered material formed in the first stage of filtration is taken off from the second section of the filter and is then recycled to the mineral attack stage. Finally, some very dilute acid is obtained which is recycled for washing the travelling bed of filtered material as it arrives at the second stage. Washing of the filter cake arriving from the latter point is carried out with water in the third and final stage.

The current processes for the production of phosphoric acid lead to the 'dihydrate' and the hemihydrate'.

The two methods involve the operational steps shown in Fig. 10.4. The two processes differ in their operating conditions in the following ways:

- in the process leading to the dihydrate, 78% sulphuric acid is used in the attack on the mineral, while, in the process leading to the hemihydrate, the attack on the mineral is carried out by using about 95% sulphuric acid;

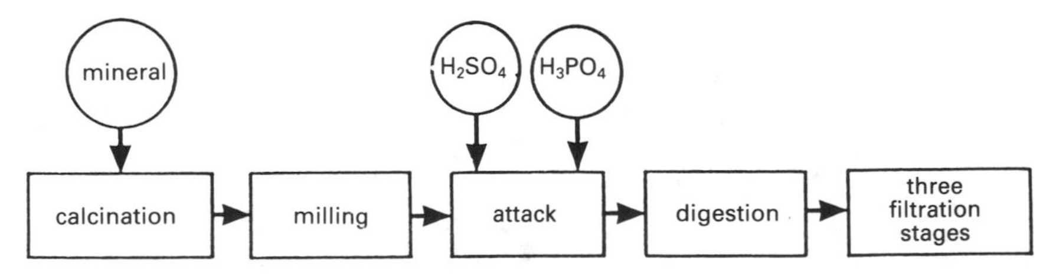

Fig. 10.4—Flow diagram for the production of phosphoric acid, using the wet process leading to the hemihydrate and dihydrate of calcium sulphate.

*The presence of organic substances promotes the formation of foams which makes it difficult to filter off the calcium sulphate.

†The plural indicates that, depending on the method, it is possible to obtain either normal chalk $CaSO_4 \cdot 2H_2O$ or hemihydrated chalk $CaSO_4 \cdot \frac{1}{2}H_2O$.

- in the dihydrate process the operating temperature does not exceed 80°C, while, in the hemihydrate process, it is conducted at a temperature of 100°C;
- in the dihydrate process there are fewer problems concerning the resistance of the materials (in the case of reactors, digestors, filter gauzes, etc.);
- the reaction and digestion times are shorter in the hemihydrate process.

Obviously, the results are also different:

- phosphoric acid containing an average of 33% P_2O_5 is obtained from the dihydrate process, while phosphoric acid containing an average of 38% P_2O_5 results from the other process;
- the removal of phosphates from the mineral is higher by about 2% in the case of the hemihydrate process which achieves an average global yield of 97–98%;
- the quantities of calcium sulphate produced are smaller in the hemihydrate process, resulting in lower transport costs.

We shall consider how the large amounts of calcium sulphate obtained in these processes are utilized within the framework of the modern chemistry and industrial technology of chalk (page 612).

The purification of phosphoric acid produced via the wet route

Phosphoric acid which is intended for use in fertilizers does not require any purification, while that which is destined for chemical use and in food products must be purified.

Qualitatively, there are many purification processes, and they vary in their effectiveness. One process is illustrated in Fig. 10.5.

Traditionally, the product is concentrated to obtain phosphoric acid which is chemically acceptable (to the detergent and plasticizer industries, etc.). This must be done by using a flash or spraying technique or by the bubbling of hot gases (especially air) to avoid the incrustation of the heat exchangers by precipitates. During concentration, the major part of the impurities, occluded by the abundant calcium sulphate which precipitates, is separated as phosphates or sulphates.

Whenever concentration is not required, methods of chemical purification using precipitants (such as H_2S, for example, with the aim of removing arsenic compounds), reducing agents (such as iron when vanadium(IV) compounds are to be removed), and, above all, pH modifiers such as $Ca(OH)_2$, are commonly employed.

Under the basic ambient conditions which are produced in this way, precipitates are formed which occlude Fe^{3+}, Al^{3+}, V^{3+}, Mn^{2+}, and Mn^{3+}. After the removal of these precipitates, the excess alkali is neutralized with sulphuric acid which leads to the precipitation of various calcium sulphates. The phosphoric acid is subsequently suitably concentrated. Further precipitates are formed during this process which have to be filtered off, while, for the most part, the residual fluorides pass into the vapour state.

Solvent extraction methods (using organic solvents such as mixed esters, ethers, and alcohols), which preferentially extract the phosphoric acid, are also employed. These solvents are recovered by stripping them both from the acid which has been treated and from the layer containing the impurities, and they are then recycled. The

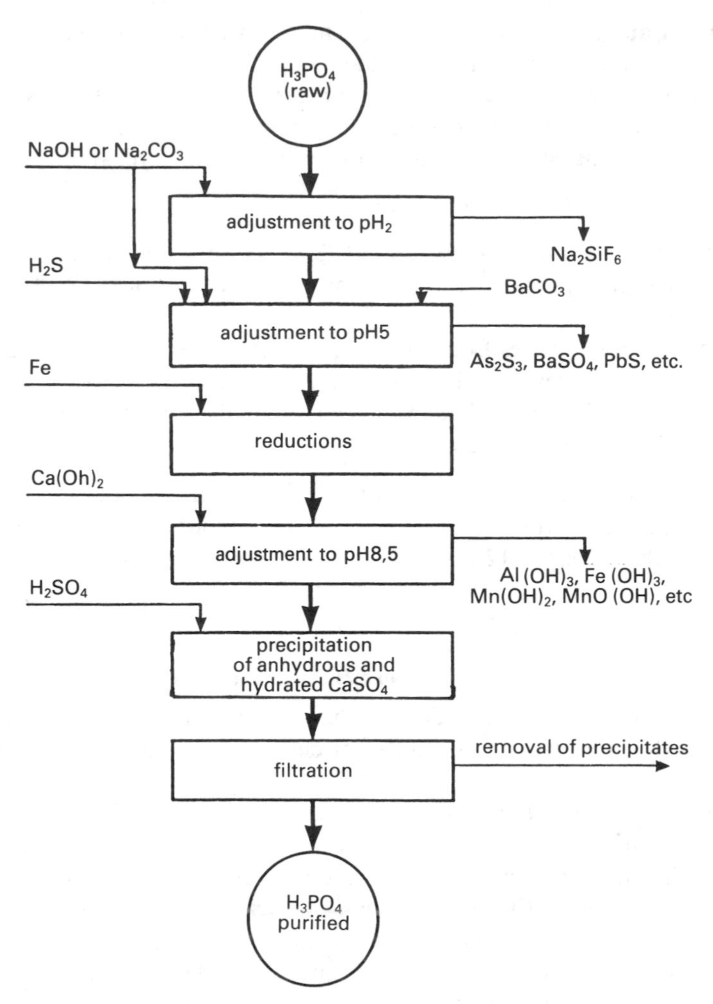

Fig. 10.5—Block diagram of a possible process for the purification of phosphoric acid.

acid purified in this manner has a composition which is similar to that produced by the thermal route starting out from very pure phosphorus.

The part of the acid which is not extracted and contains the impurities is used in the production of fertilizers.

The recovery of fluorinated compounds from the products of reaction with fluoroapatites

Forms in which fluorine is liberated

Fluorine is an industrially important element which, being a constituent of fluoroapatites, is present in mineral phosphates. The fluorine is removed from these minerals when phosphorus is produced electrothermally and when they are subjected to acid treatments; that is, during the production of phosphate fertilizers and, above all, when phosphoric acid is produced by the wet route.

The fluorinated products formed in such treatments of phosphate minerals are shown in Table 10.2, and the fluorine compounds contained in such products are indicated.

As can be seen, of all the fluorine compounds produced, hydrofluoric acid (HF) and gaseous silicon tetrafluoride (SiF_4), which is extremely reactive and capable of producing skin sores which are very slow to heal, predominate.

Methods of recovering fluorinated compounds where necessary

Not all the products of the working of phosphate minerals in which the fluorine finishes up lend themselves at present to the economic and convenient recovery of the halogen. From this point of view, the following sources are not exploited:

- the fluorinated drosses which are discharged from the furnaces used in the production of phosphorus;
- the insoluble fluorides occluded in the fertilizer mixtures and the chalks formed in the technologies used for the production of phosphoric acid.

As has already been stated (page 353), a solution of H_2SiF_6 and H_2SiO_3 is formed from the fumes, from which the dust has been removed, emerging from the electric furnaces in which phosphorus is produced. Na_2SiF_6 precipitates out when this solution is neutralized with NaOH. After filtration, this important fluorosilicate is sent to the industry dealing with fluorine compounds (page 316).

Na_2SiF_6 is also obtained from the gaseous effluents from the phosphate fertilizer industry and the wet method for the production of phosphoric acid upon reaction

Table 10.2—The working of phosphate minerals which produce fluorine compounds by different routes

Processes	*Fluorinated compounds, possible distribution of the fluorine and fields where they are produced*
Production of phosphorus	• SiF_4 in the fumes • fluorosilicates in the dross at the bottom of the furnaces
Production of phosphoric acid by the wet method	• approx. 17% of the fluorine in the initial mineral passes into the effluent gases as SiF_4 and HF • approx. 54% of the initial fluorine is carried as H_2SiF_6 and, much less, as unattackable fluorosilicates under the the reaction conditions in the raw phosphoric acid • approx. 29% of the total fluorine finishes up in the form of H_2SiF_6 in the water imbibed by the calcium sulphate, while the rest remains occluded in this solid as CaF_2 or as fluorosilicates which have not been attacked owing to the incompleteness of the reaction
Preparation of of phosphate fertilizers	• SiF_4 and HF in gaseous effluents • fluorosilicates and fluorides of calcium in chalks and in the fertilizers produced

with caustic soda:

$$SiF_4 + 2HF + 2NaOH \rightarrow Na_2SiF_6 + 2H_2O$$

This product is also sent to the producers of fluorine compounds, as previously stated.

As an alternative, cryolite (page 317) and other technically important fluorides may be prepared from the abovementioned gaseous effluents.

If, on the other hand, the fluorine is available in the form of H_2SiF_6 as in the waters included in the chalks precipitated during the preparation of phosphoric acid and can be recovered by thorough washing of the masses in which it is contained, a dilute solution of it can be sent directly to the producers of fluorinated compounds.

Finally, fluorine is recovered during the chemical purification of raw phosphoric acid:

- as Na_2SiF_6 at the point where the pH is first modified with NaOH or with Na_2CO_3 (Fig. 10.5):

$$Na_2CO_3 + H_2SiF_6 \rightarrow Na_2SiF_6 + CO_2 + H_2O$$

- as SiF_4 and HF from the vapours produced in the later concentration of the phosphoric acid.

From what has been seen it is clear that the maximum recovery of the fluorine contained in the industrially exploited fluoroapatites is carried out at the site where phosphoric acid is produced by the wet method.

The block diagram in Fig. 10.6 shows the percentages of the fluorine initially present in the mineral being treated (100%) which are obtained during the different stages of this kind of production.

By comparing the percentages shown in this scheme with those presented in Table 10.2, it is deduced that practically all the fluorine is recovered from the gaseous effluents, that approximately 3% of the halogen is dispersed in the phosphoric acid solution, and that the losses in the chalks are about 12%.

Composition, behaviour, and the uses of phosphoric acid

The composition of industrial phosphoric acid is variable both with respect to the

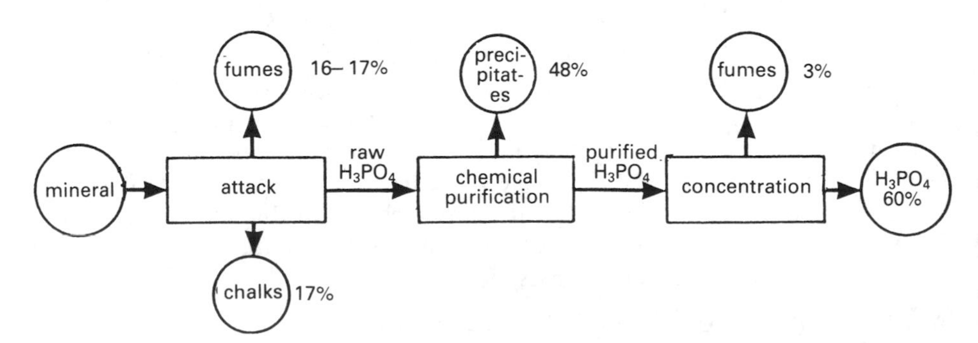

Fig. 10.6—Simplified block diagram showing how H_3PO_4 is produced, together with the gaseous and solid phases from which the fluorine contained in the starting phosphate minerals is recovered, as a certain percentage of an original content corresponding to 100%.

amount of phosphoric acid present and the extraneous components which are present in it.

As regards its constitution, the forms with the following densities are alternative ways of expressing the reference percentages, are sold commercially,

d_4^{20}:	1.024	1.052	1.113	**1.18**	**1.133**	**1.573**	1.685	1.874	2.045
$\%H_3PO_4$:	5	10	20	**30**	**50**	**75**	85	100	115
$\%P_2O_5$:	3.62	7.25	14.50	**21.74**	**36.21**	**54.34**	61.57	72.42	83.30

The forms shown by the emphasized numbers are preferred commercially, and the impurities which can be tolerated vary according to the uses to which the acid is to be put. While the impurities are of little concern when the solutions of the acid are to be used in the production of fertilizers, it is necessary that certain limits should be imposed on the impurities in those stocks of acid which are destined for chemical uses, while stocks of acid to be employed in the food industry must be practically free of impurities.

More precisely, in the case of phosphoric acid intended for the food industries, phosphite ions must be absent, arsenic, lead, vanadium, and fluorides must be almost absent, the amount of iron and aluminium must be very low, and the concentration of sulphate ions must be reduced.

It is the industries producing the acid which generally undertake to supply users with phosphoric acid having the special compositional characteristics required for their activities.

The behaviour of solutions of phosphoric acid is such that, at normal temperatures (0–30°C), all phosphoric acid solutions with H_3PO_4 concentrations of up to 75% are completely liquid. It is for this reason that 75% is usually the maximum concentration of industrial solutions of this acid.

On the other hand, solutions containing various amounts of a sediment of $H_3PO_4 \cdot \frac{1}{2}H_2O$ (the acid hemihydrate) and even anhydrous H_3PO_4 exist over the same temperature range when higher concentrations of H_3PO_4 are present.

At temperatures above 30°C (or, more precisely, 29.3°C) and up to 42.2°C (the melting point of H_3PO_4), solutions of phosphoric acid only contain anhydrous H_3PO_4 as a sediment at H_3PO_4 concentrations above 95% (more precisely, 94.8%).

For the exact study of the behaviour of industrial batches of phosphoric acid with concentrations from 0 to 100%, account must be taken of the following data:

* 100% phosphoric acid solidifies at +42.4°C;
* the chemical $H_3PO \cdot \frac{1}{2}H_2O$, which corresponds to a composition of 91.6% of the acid (and 8.4% of water), solidifies out at a temperature of 29.3°C;
* the eutectic $H_3PO_4 \cdot \frac{1}{2}H_2O + H_3PO_4$, which contains 61.82% of the hemihydrate and corresponds to a composition of 94.8% of the acid and 5.2% of water, solidifies out at about 21°C;
* the eutectic ice + $H_3PO_4 \cdot \frac{1}{2}H_2O$, which corresponds to 62.5% of the acid and 37.5% of water, solidifies out −85°C.

Likewise, by taking account of the fact that ice melts at 0°C, the temperature ranges over which solutions of phosphoric acid of various concentrations exist can be precisely delineated on the basis of the data given above.

To understand the composition of industrial batches of solid phosphoric acid it must be kept in mind that, when the pure acid is heated, dehydration reactions take place which can be represented as:

$$n\ \begin{matrix}HO & & OH\\ & P & \\ O\!\!/\!\!/ & & OH\end{matrix} \xrightarrow{-(n-1)H_2O} \begin{matrix}HO & & O\\ & P & \\ O\!\!/\!\!/ & & OH\end{matrix}\left[\begin{matrix} & & O\\ & P & \\ O\!\!/\!\!/ & & OH\end{matrix}\right]_m \begin{matrix} & & OH\\ & P & \\ O\!\!/\!\!/ & & OH\end{matrix}$$

polyphosphoric acid ("pyrophosphoric")

with $n = 2$, 3 and 4 and $m = 0$, 1 and 2.

Both the fact that commercial phosphoric acid can have a percentage concentration which is potentially greater than 100% and a corresponding content of P_2O_5* which is higher than 72.42% (which is equivalent to the concentration of this acid anhydride in pure phosphoric acid) can be understood on the basis of this behaviour.

The uses to which the phosphoric acid produced by the chemical industry are put can be expressed in terms of the following percentages:

- 67% for fertilizers;
- 18% for alkali phosphates in general and for polyphosphates in particular;
- 15% for various other purposes.

Apart from agrochemicals, the industries which make use of the most derivatives of phosphoric acid are, the detergent, food, textile, pharmaceutical, and parapharmaceutical (e.g. dental hygiene) industries. Among the industries which consume compounds derived from phosphoric acid, the electrochemical (e.g. pickling and etching), water treatment (especially in the thermotechnical field), and flotation process industries stand out. Phosphoric acid also finds direct uses in the corrosion protection of iron alloys and as a catalyst in many sectors of the petrochemical industry.

10.4 COMPOUNDS DERIVED FROM ELEMENTARY PHOSPHORUS

Two types of compound are derived from the two forms of elementary phosphorus which are of importance in various branches of the chemical industry, especially as synthetic intermediates and reagents. Table 10.3 includes the most important of these compounds.

Production of phosphorus chlorides and their derivatives

Phosphorus trichloride and analogous halides

The main trihalide of phosphorus is the trichloride PCl_3 which is prepared by reacting measured quantities of white phosphorus and chlorine in a reflux reactor. The reaction is carried out with the temperature thermostatically controlled at the boiling point of the product (76.1°C). The compound is distilled over phosphorus to remove any traces of the pentachloride. A diagram of a complete plant for the synthesis of PCl_3 is shown in Fig. 10.8.

Besides being used, as will be mentioned later, as an intermediate in the production

*Here and in the diagram (Fig. 10.7) phosphoric anhydride is indicated in the traditional way as P_2O_5 rather than, more correctly, as P_4O_{10}.

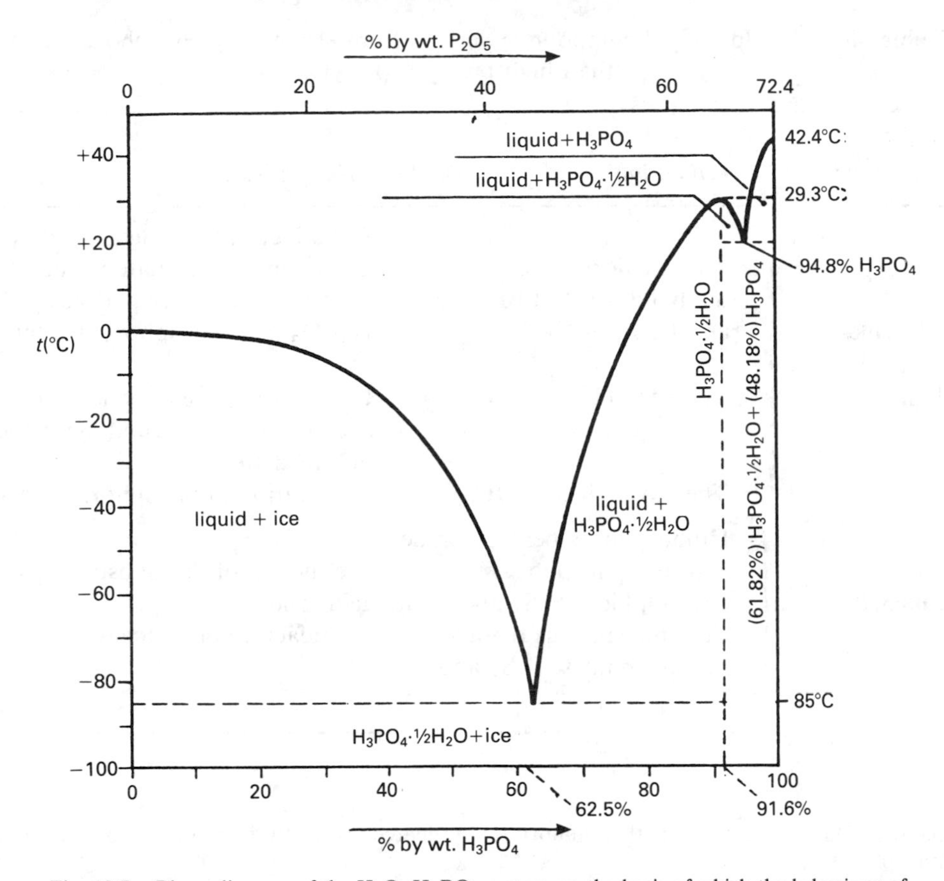

Fig. 10.7—Phase diagram of the H_2O–H_3PO_4 system on the basis of which the behaviour of commercially marketed batches of phosphoric acid can be interpreted. In particular, it is found that, above its melting point of 42.4°C, phosphoric acid commences to dehydrate to form polyphosphoric acids which are used in the production of mixtures of polyphosphates which have various uses.

of other chlorides, phosphorus trichloride finds use in:

- the preparation of phosphorous acid by its hydrolysis:

$$PCl_3 + 3H_2O \rightarrow H_3PO_3 + 3HCl\uparrow$$

- the preparation of the esters of phosphorous acid by alcoholysis:

$$PCl_3 + 3ROH \rightarrow R_3PO_3 + 3HCl\uparrow$$

which contain three radicals with phosphorus–oxygen bonds [R–O–P(OR)$_2$].

Phosphorous acid, a dibasic acid, is a strong reducing agent, and the phosphorous esters which are formally derived from the tribasic structure of the acid are both polymerization inhibitors, because they are antioxidants, as well as polymer stabilizers.

PI_3 and PBr_3 are prepared by the action of iodine and bromine respectively on phosphorus dissolved in carbon disulphide. The solvent is subsequently evaporated off to obtain the solid

Table 10.3—The principal compounds derived from elementary phosphorus (other than high purity H_3PO_4)

Type of compound	*Specification*	*Principal uses*
Chlorides and other halides	The trichloride PCl_3 and the halides PI_3, PBr_3 and and PF_3	All halogenating reagents. PCl_3 is also an intermediate in the production of other halides, H_3PO_3, and phosphorous esters
	The pentachloride PCl_5	Chlorinating agent, condensation catalyst, and important synthetic intermediate
	The oxychloride $POCl_3$	Preparation of phosphoric esters
Sulphides	Diphosphorus pentasulphide and tetraphosphorus decasulphide: P_2S_5 and P_4S_{10}	Production of thiophosphoric acids and esters
	Tetraphosphorus tri- and heptasulphides: P_4S_3 and P_4S_7	Manufacture of matches

halides. On the other hand, the trifluoride is prepared indirectly from fluorides and metal phosphides:

$$3PbF_2 + Cu_3P_2 \rightarrow 3Pb + 3Cu + 2PF_3$$

These compounds, especially PF_3, are powerful halogen donors in preparative and synthetic chemistry.

Phosphorus oxychloride and phosphorus pentachloride

The oxychloride of phosphorus (phosphoryl chloride) is prepared by the oxidation of PCl_3 in a reactor:

$$PCl_3 + \tfrac{1}{2}O_2 \rightarrow POCl_3 \ (bp_{POCl_3} = +105.8°C)$$

The reaction is stopped before the PCl_3 is exhausted and the excess of the reagent is distilled off ($bp_{PCl_3} = +76.1°C$).

On the other hand, *phosphorus pentachloride* is prepared by the chlorination of PCl_3:

$$PCl_3 + Cl_2 \rightarrow PCl_5 \ (mp_{PCl_5} = +149°C)$$

This preparation is generally operated continuously, since the product which tends to settle out in the crystalline state at the bottom of the liquid trichloride can be systematically extracted from the bottom of the reactor, which is fed from the top with PCl_3, by means of a screw conveyor. The thick turbid extract is then filtered under an inert atmosphere, the PCl_5 is sent for storage, and the PCl_3 is recycled.

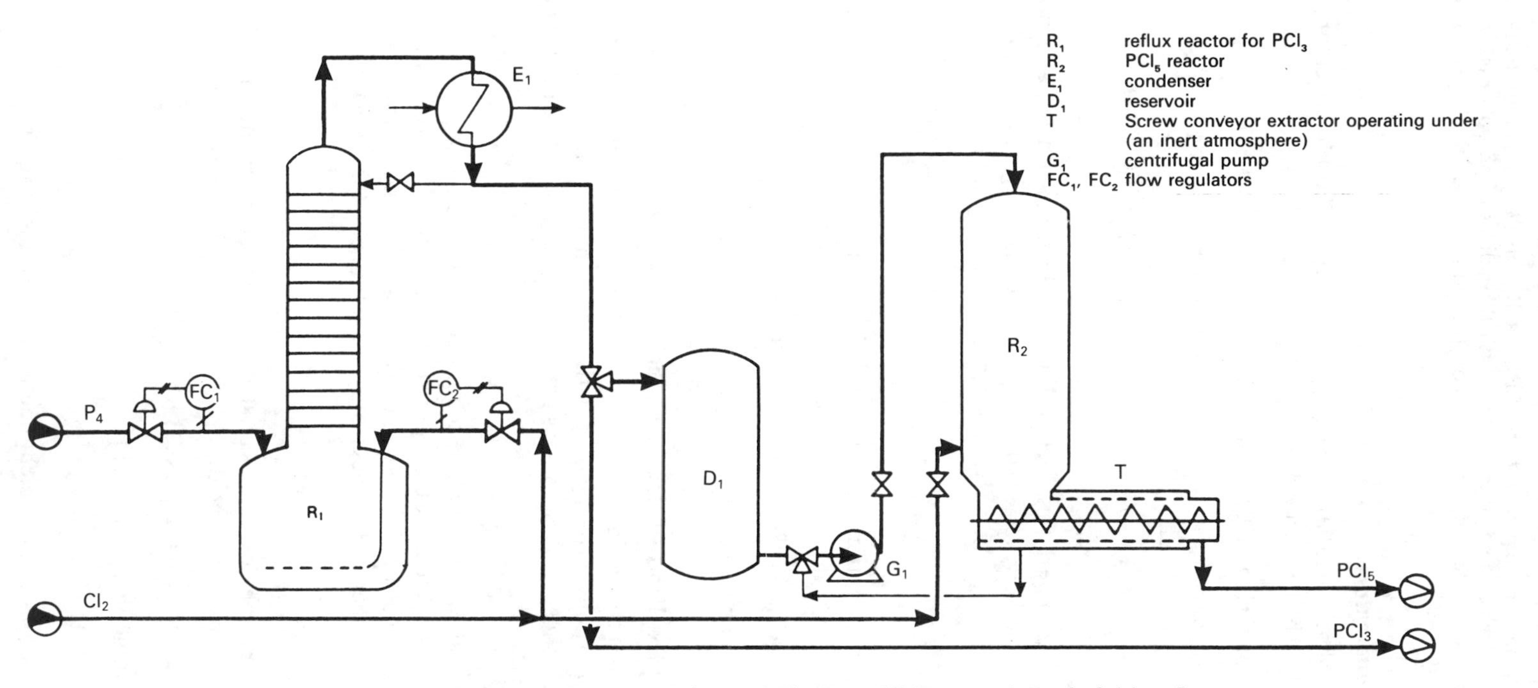

Fig. 10.8—Stoichiometric amounts of P_4 and Cl_2 ($P_4 + 6Cl_2$) are continuously fed into R_1. The PCl_3 can be stored as such or sent into D_1 from where it is pumped into the top of R_2 from the base of which Cl_2 rises. The PCl_5 thereby formed is filtered under an inert atmosphere in T, before storage. Any unreacted PCl_3 is recycled.

Fig. 10.8 shows a plant for the preparation of PCl_3 and then PCl_5. It should be noted that all sections of the plant which come into contact with the halides of phosphorus, whether in the synthetic steps or as storage tanks, must be constructed from special alloy steels or from carbon steel lined with lead.

Phosphate esters and chlorsulphonic acid

The esters of phosphoric acid are usually prepared from phosphoryl chloride and alcohols or phenols. The pertinent reaction is:

$$POCl_3 + 3ROH \rightarrow 3HCl + OP(OR)_3$$

In particular, *p*-tricresylphosphate is prepared by the reaction:

$$O{=}PCl_3 + 3HO{-}C_6H_4{-}CH_3 \rightarrow O{=}P(O{-}C_6H_4{-}CH_3)_3 + 3HCl$$

Phosphate esters are interesting fuel additives, valued plasticizers, and powerful pesticides. Their use is, however, restricted owing to their high toxicity, the action of which causes paralysis of the nervous system in animals and man.

* * *

One of the many preparations in which PCl_5 is used is that of chlorsulphonic acid from sulphuric acid:

$$H_2SO_4 + PCl_5 \rightarrow ClSO_3H + HCl + POCl_3$$

Chlorsulphonic acid has been discussed earlier (page 229).

The production of phosphorus sulphides and their derivatives

When phosphorus, usually in the red form, and sulphur are reacted in the stoichiometric proportions corresponding to the compound which is to be synthesized, mixtures of various phosphorus sulphides are always obtained with a preponderance of the desired species.

The commonest of these sulphides are P_4S_3, P_4S_7, and P_2S_5. These are yellow or pale yellow compounds which can all be separated from one another, when necessary, by fractional distillation under reduced pressure. Very often, however, as when these compounds are destined for the manufacture of matches, no attempt is made to separate the individual chemical species from the mixture of sulphides which is formed.

The preparation of thiophosphate esters

Diphosphorus pentasulphide (P_2S_5) can be used for the preparation of both the neutral esters of thiophosphoric acid:

$$P_2S_5 + 6ROH \rightarrow 3H_2S + 2S{=}P(OR)_3$$

and the esters of dithiophosphoric acid:

$$P_2S_5 + 4ROH \rightarrow H_2S + 2S{=}P(OR)_2SH$$

by suitable modifications of the stoichiometric ratio of the reagents.

The mechanisms of these two complex reactions are shown in Fig. 10.9. The esters produced in this manner are important intermediates in the production of insecticides, lubricant additives, and flotation activators.

The manufacture of matches

Even after the advent of various types of automatic lighters, matches are industrial products which are still consumed in very large numbers and may still contain phosphorus as a component, but, more frequently, phosphorus compounds.

Matches can be classified either on the basis of their stems or on the basis of the mode in which they ignite and the dangers arising from their combustion products.

With respect to their stems, matches may be classified as:

- wax matches, if they are composed of a cotton wick saturated with molten paraffin or stearine (wax substitutes);
- safety (Swedish) and sulphur matches, if the stem is a small stick of poplar with one end impregnated with paraffin in the former and with sulphur in the latter.

On the basis of the mode of ignition and the nature of the combustion products, they may be classified as:

- safety matches, the heads of which are prepared by dipping either of the two ends, if they are wax matches, or the end impregnated with paraffin or sulphur, if they are safety matches or sulphur matches, into a viscous suspension (usually containing glue which acts as an adhesive) of oxidizing agents (such as potassium chlorate and dichromate and manganese dioxide) and a reducing agent flammable sulphur);
- strike anywhere matches, which have heads consisting of red phosphorus, phosphorus sulphides (P_4S_3, P_4S_7), and antimony trisulphide (Sb_2S_3), besides the mixture which acts as the oxidant and comprises MnO_2, $KClO_3$, $K_2Cr_2O_7$, etc.;
- hygienic matches, in which the use of red phosphorus is completely avoided both as a constituent in the head or as a component of the surface on which the match is struck. The heads of these matches contain potassium chlorate, manganese dioxide, and potassium nitrate.

Safety matches can be lit only if they are rubbed on a rough surface covered with a mixture of red phosphorus and antimony trisulphide fixed to a support by means of gum arabic. The roughness is produced by coarse glass powder.

Strike anywhere matches can be struck by rubbing them against any unpolished dry surface.

Finally, hygienic matches can be struck on any surface on account of the phosphorus sulphides and sulphur phosphides (e.g. Cu_5PS) in their heads.

For a full account of matches and their manufacture see Finch & Ramachandran, *Match making: science and technology*, Ellis Horwood, 1983.

10.5 SALTS DERIVED FROM PHOSPHORIC ACID

A number of salts are prepared from orthophosphoric acid. These are mainly sodium salts but also potassium, calcium, ammonium, and magnesium salts which find a very extended range of industrial applications. Examples of these are the mono-, di-, and tri-metallic salts of phosphoric acid, while others are phosphates prepared from

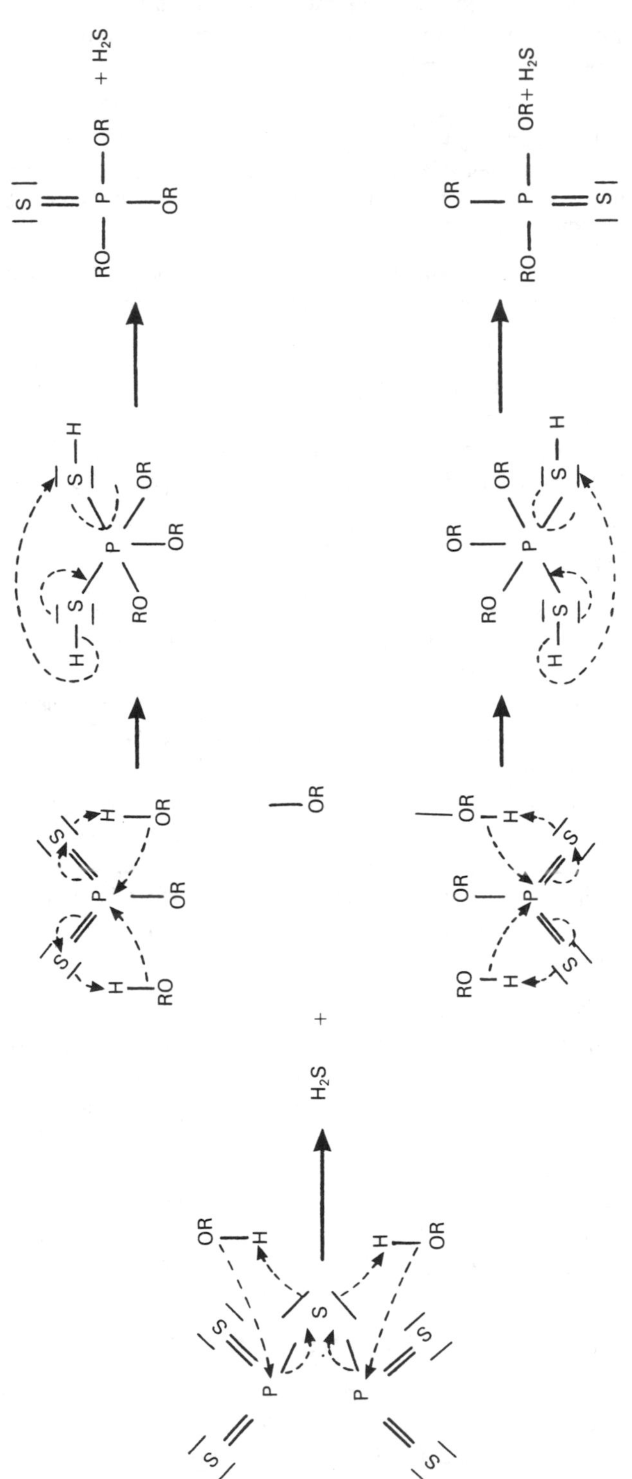

Fig. 10.9—Production of the esters of thiophosphoric acid from P_2S_5. It can be seen how the phosphorus atoms and the atoms of the other elements always correctly use their valence electrons. The solid lines represent electron pairs (either bonding or lone pairs) while broken and more or less curved lines show the electron displacements.

the polymeric chemical species produced by the condensation of several molecules of phosphoric acid. The first are referred to as phosphates, while the latter are known generically as polyphosphates.

In general, it must be known, before the production of the salt, whether there is any need to purify the phosphoric acid. This purification must be carried out in a manner consistent with the generic or specific degree of purity required.

Industrial types, preparation, and uses of phosphates

An extremely wide range of chemical and parachemical industries (excluding the fertilizer industry, to which the whole of Chapter 11 will be devoted on account of the multiplicity of economic interests, the provision of raw materials, etc.) require various quantities of:

- the whole range of sodium, potassium, and calcium phosphates;
- monoammonium and diammonium phosphates;
- trimagnesium phosphate.

Besides phosphoric acid, sodium, potassium, and calcium hydroxides, ammonia and magnesium oxide are required in the production of phosphates. The manufacture of these salts always includes some degree of 'neutralization'* specified by the Me_xO_y/P_2O_5 or NH_3/P_2O_5 ratio corresponding to the composition of the salt which is to be prepared.

In general, phosphates exist in many hydrated forms with differing stabilities varying from the anhydrous salt to the fully hydrated salt which are the two most stable species. For this reason, there is a preference, industrially, to produce either the fully hydrated salt or the anhydrous salt.

For instance, in the case of Na_2HPO_4, the forms $Na_2HPO_4 \cdot 7H_2O$ and $Na_2HPO_4 \cdot H_2O$ are known in addition to the anhydrous salt and the decahydrate of which the latter two are of the greatest industrial importance.

Besides describing their degree of hydration, phosphates are also classified commercially in terms of their purity, and, generally, 'technical' and 'for use in foods' grade materials are available.

Some purely 'technical' grade phosphates are not marketed, and only food grade calcium monophosphate $Ca(H_2PO_4)_2$ exists. Also, there is only the 'technical medicinal' grade of trimagnesium phosphate $Mg_3(PO_4)_2$ on account of the specific uses which are made of this compound.

Nowadays, the automated production of phosphates requires the attainment of certain operational pH values which are predetermined in accordance with the different stages of the neutralization of phosphoric acid, as shown in Fig. 10.10.

The programmed product is obtained by adjusting the pH to the correct value by the addition of alkali or ammonia and then crystallizing out the salt which is formed at a suitable temperature with prior concentration of the solution, if this is necessary.

For instance, disodium and dipotassium hydrogen phosphates are prepared by the addition of alkali to phosphoric acid until a pH value of 8.5 is attained, concentration of the resulting

---

*The term 'neutralization' means salt formation here, as it simply expresses the fact that a certain number of the hydrogen ions in the acid are replaced by metal ions.

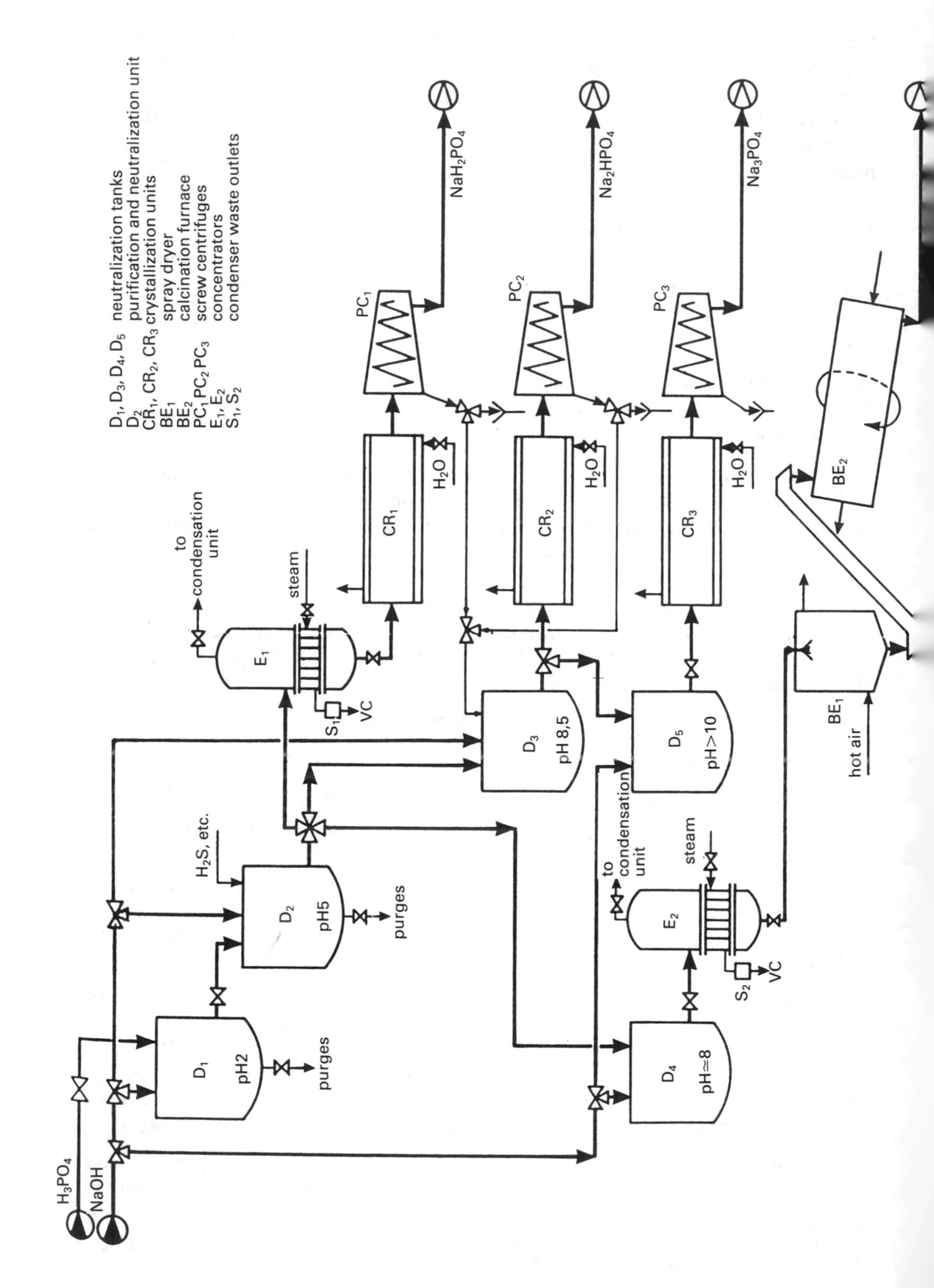

D1, D3, D4, D5 neutralization tanks
D2 purification and neutralization unit
CR1, CR2, CR3 crystallization units
BE1 spray dryer
BE2 calcination furnace
PC1 PC2 PC3 screw centrifuges
E1, E2 concentrators
S1, S2 condenser waste outlets
H3PO4
NaOH
D1
pH2
purges
H2S, etc.
D2
pH5
purges
to condensation unit
steam
E1
S1
VC
CR1
H2O
PC1
NaH2PO4
D3
pH 8,5
CR2
H2O
PC2
Na2HPO4
D5
pH>10
CR3
H2O
PC3
Na3PO4
D4
pH≈8
to condensation unit
steam
E2
S2
VC
BE1
hot air
BE2

solution if the anhydrous form is required, crystallization at a suitable temperature, and separation of the crystals by centrifugation with partial purging of the mother liquors and recycling of the rest.

For reasons of cost, sodium phosphates are much more widely used than potassium phosphates unless there are specific reasons for using the latter (generally the greater biological compatibility at the bacterial level of potassium ions compared with sodium ions).

Phosphates are mainly used for the stabilization of pH values on account of the buffering action which they exhibit, the treatment of waters and sugar solutions, the pickling and etching of metals, textile technology, the emulsion and protection of colloidal suspensions in the production of foodstuffs, and in polymerization processes. Monoammonium dihydrogen phosphate and diammonium hydrogen phosphate find specific uses as flame retardants in paper, timber, and textiles (on account of the non-flammable gases which they can evolve when hot) as well as acting as nutrients for cultures of autotrophic microorganisms intended for various purposes.

Industrial types, production, and uses of polyphosphates

On account of their ability to form stronger complexes with the ions responsible for water hardness, condensed phosphates (polyphosphates) are distinctly preferred to the phosphates in the treatment of waters and as components in detergent formulations.

Polyphosphates can be prepared either by the neutralization of polyphosphoric acids which, in their turn, are prepared by a suitable thermal dehydration of orthophosphoric acid or, more commonly, by the heating of the salts of this acid, either singly or mixed. This second method of preparation is shown diagrammatically in Fig. 10.10.

The polyphosphates which are produced industrially are, especially:

- the tetra-alkali metal ion dimers, *sodium pyrophosphate* ($Na_4P_2O_7$) and *potassium pyrophosphate* ($K_4P_2O_7$):

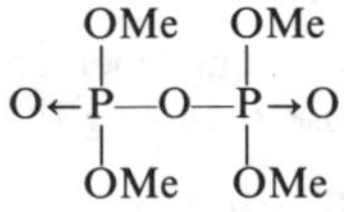

$$\begin{array}{ccccc} & \mathrm{OMe} & & \mathrm{OMe} & \\ & | & & | & \\ \mathrm{O}\leftarrow & \mathrm{P} & -\mathrm{O}- & \mathrm{P} & \rightarrow\mathrm{O} \\ & | & & | & \\ & \mathrm{OMe} & & \mathrm{OMe} & \end{array}$$

  which are used for the pickling and etching of metals and in the formulation of liquid detergents, as additives to prepared meatstuffs, and as emulsion stabilizers;
- the dimeric disodium salt, *disodium dihydrogen pyrophosphate* ($Na_2H_2P_2O_7$), which finds various uses (as artificial swelling agents in the formation of foodstuff emulsions and of sludges in mining techniques);
- the *ultraphosphates* of the general formula $(MePO_3)_n$, which are therefore of a composition which is not well defined as far as their molecular weight is concerned. Of these *sodium hexametaphosphate* ($n = 6$) is of particular importance. They are used in certain detergents, in the sweetening of waters, and as dispersing agents;

Fig. 10.10—Intuitive schematic diagram of the process for the purification of phosphoric acid and the chemical plant conditions for the subsequent preparation of various phosphates.

● the trimeric *sodium tripolyphosphate*, $Na_5P_3O_{10}$ which has the structure:

```
   ONa    O     ONa
    |     ↑      |
O←P—O—P—O—P→O
    |     |      |
   ONa   ONa    ONa
```

A diagram of a plant for the preparation of this compound is shown in Fig. 10.10.

$Na_5P_3O_{10}$, which is already widely employed as a water sweetener, is the polyphosphate which is becoming every more common in the formulation of detergent materials where it acts as a support for all the other components: alkylbenzene sulphonates, silicates, carboxymethylcellulose, bleaching agents. etc.

Sodium tripolyphosphate is prepared using the plant depicted in Fig. 10.10 and as shown schematically in Fig. 10.11.

The calcination reaction referred to in these figures is:

$$2Na_2HPO_4 + NaH_2PO_4 \rightarrow Na_5P_3O_{10} + 2H_2O$$

This takes place in this manner because the pH value of the solution from which the polyphosphate is to be produced is adjusted to be very close to the pH value leading to the formation of Na_2HPO_4 as shown in Fig. 10.10.

10.6 THE PROBLEM OF PHOSPHORUS IN THE BIOSPHERE

Among all the elements which are active in ensuring the operation of the biochemistry of the biosphere, phosphorus is the element which has presented, and still presents, the greatest problems.

Until the middle of the nineteenth century the amounts of biochemically active phosphorus compounds in soils were generally very low, and insufficient to allow enhanced yields of farming products. This was attributable both the relatively low abundance of phosphorus in the Earth's crust (about 0.1% as compared, for example, with 2.6% of potassium) and to the insolubility of the natural compounds of phosphorus in aqueous and weakly acid solutions.

Owing to a scarcity of an energy producing* element, many living organisms had developed alternative metabolic pathways to those requiring the intervention of phosphorus.

Neither manuring nor the application of animal and plant residues succeeded in the convenient reintegration of the phosphorus to farmland which had been removed in the crops, since, of the three fertilizing elements (nitrogen, potassium, and phosphorus), phosphorus is the one which is given back to the soil in the smallest amounts.

It is for these reasons that such phosphates are added to the soil as basic chemical fertilizers.

However, with the extensive and intensive use of such phosphate fertilizers and the leaching out of these materials from soils by the drainage of rainwaters, more or less stagnant waters have started to become enriched in phosphate anions and insoluble phosphates which are in such a state of subdivision as to be readily assimilated by living organisms.

---

*It is noted that phosphorus mainly forms part of the make-up of the energy reserve structure, ATP.

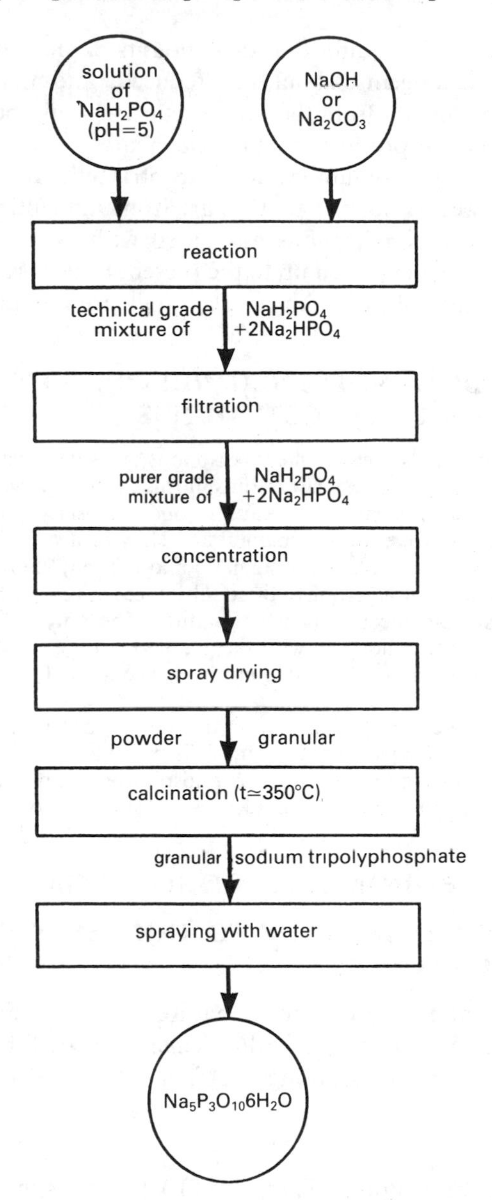

Fig. 10.11—Block diagram of the steps in the production of sodium tripolyphosphate hexahydrate.

More recently, the situation has become much worse, owing to a large increase in the use of 'builders' in the more efficient washing products, and, in particular, sodium tripolyphosphate (STPP) with the result that, today, there is an excess of phosphorus in the form of phosphates which can be assimilated as a nutrient by living organisms in too many parts of the biosphere.

The consequences are particularly serious with respect to the fertilization of surface, continental, and marine waters which consequently suffer from eutrophication phenomena (page 110). To avoid such grave damage to the environment, legislation has been introduced which requires that the amounts of phosphates in effluents, and further upstream in detergents, should be reduced, and that, together with the elimination of the phosphates from the waters, efforts should be directed toward the preparation of alternative products to the phosphates which are presently in use. Possible alternatives are trisodium citrate, nitrilotriacetic acid (NTA), and certain zeolites. It must, however, be said that, apart from difficulties of another nature, namely, the toxicological implications associated with NTA, the action of all the builders which have been developed up to the present time as replacements for STPP has been worse than that of the product which they are intended to replace.

10.7 TOXICOLOGICAL NOTES REGARDING PHOSPHORUS AND CERTAIN OF ITS COMPOUNDS

Many phosphorus compounds are extremely poisonous, as is demonstrated by the use which is made of quite a lot of these as pesticides. Phosphorus itself, especially in the white form, is both toxic and flammable. In particular, it causes bone necrosis, especially of the jaw bone*, and produces burns which are extremely painful and slow to heal.

It is not possible here to indicate the systematic and efficient steps taken to treat patients suffering from the inhalation or ingestion of these substances, given the high degree of poisoning and the lethal or paralysing effects which they induce. The only valid advice which can be given is to get the patient to a hospital with the greatest of urgency.

Table 10.4, nevertheless, shows the maximum tolerance limits for phosphorus and some of its typical compounds.

The high toxicity of the compounds shown in this table is deduced from the low maximum tolerance limits, and, on the basis of these data, it is possible to estimate approximate MTL values for other analogous compounds such as phosphorous acid (which is more toxic than phosphoric acid), various phosphorus sulphides, esters of phosphorus, etc.

Part 2 BORON AND BORON COMPOUNDS

10.8 FROM NATURAL DEPOSITS TO THE USES OF BORON AND ITS COMPOUNDS

Boron is a relatively abundant element in nature, and, in terms of its abundance of $1.4 \times 10^{-3}\%$, lies in 35th place among the elements in the lithosphere.

Table 10.5 contains data regarding the foundations of the boron and boron

Table 10.4—Maximum tolerance limits (MTL) for phosphorus and phosphorus compounds

Substance	*MTL*	*Substance*	*MTL*	*Substance*	*MTL*
Phosphorus	0.1 mg/m^3	Phosphorus pentasulphide	1 mg/m^3	Phosphorus trichloride	0.5 ppm (3 mg/m^3)
Phosphine	0.05 ppm (0.07 mg/m^3)	Phosphoric acid	1 mg/m^3	Parathion	0.1 mg/m^3

*Commonly known in the nineteenth century as 'phossy jaw'.

Table 10.5—The origin and uses of boron

The minerals			*The element*		
Formula	*Name*	*Location*	*Preparation*	*Properties*	*Uses*
$Na_2B_4O_7 \cdot 10H_2O$	Borax Tincal	China Tibet California	Hot reduction of B_2O_3 with with Mg or Na, or by treating BCl_3 with Na or H_2	Very hard, brittle, crystalline material with d_4^{20} 1.73, mp 2300°C, bp 2550°C	Deoxidizing and degassing of metals and alloys. Hardener for steels. Neutron capture material in nuclear reactor rods
$Na_2B_4O_7 \cdot 4H_2O$	Kernite (Rasorite)	California Colorado Turkey			
$NaCaB_5O_9 \cdot 8H_2O$	Ulexite	Chile Peru and Bolivia			
H_3BO_3	Sassolite	Tuscany			

compound industry whose variety of chemical formulations is a consequence of the multiplicity of acids from which they can be derived by the dehydration of differing numbers of H_3BO_3 molecules.

The technical sectors which have the greatest interest in the use of boron compounds are:

- ceramics and metallurgy in which borax and the other oxides, in general, are used;
- the pharmaceutical and food industries which make use of boric acid in particular;
- the preparative organic chemical industry, the petrochemical industry, and the so-called 'hi-tech' aerospatial and metallurgical industries which employ boranes, borohydrides, borides and boron halides;
- the field of high polymers with very high thermal stability.

However, the compounds of boron and the uses to which they are put will be described in greater detail in the following pages.

10.9 BORIFEROUS FUMAROLES ('SOFFIONI')

Generalities

In Tuscany, in the provinces of Pisa, Siena, and Grosseto, superheated water is emitted from boreholes and clefts in the ground which, nowadays, are for the most part made by drilling. The temperatures of the emerging fluids vary over quite wide limits from 150°C to 260°C with evident incremental (but non-proportional) changes in the temperature as the depth from which the fluids are taken increases.

The maximum yield from every single well can exceed 300 ton/hour, but it normally varies from 30–70 ton/hour at a pressure of 3–9 atm.

A particular characteristic of the fluids obtained from the subsoil in this region is the relatively high content of boric acid. This phenomenon is the only one of its kind in the World in so far as there are no other 'postvolcanic' manifestations (geysers, fumaroles, various geothermal vapours) which have such a high content of boric acid as is found in the Tuscan fumaroles.

The genesis of the geothermal manifestations in the areas around Larderello and Monte Amiata are considered to be associated with phenomena involving the thermalization of the meteoric waters circulating in permeable rocks which are heated by a magnetic plutonic intrusion situated at a considerable depth (5000 to 6000 metres) which has not yet been reached by any drilling operation or natural cleavage.

The plutonic intrusion would consist of material of a radioactive granitic-tourmaline-bearing nature, as similar rocks which have solidified as outcrops in the archipegalo off the Tuscan coast (in Elba, in particular) and have been subjected to the action of superheated water vapour, also yield a radioactive boron-containing liquid.

The composition of a particular boriferous fumarole (in the Larderello region) expressed in terms of kilograms of components per kilogram of fluid is as follows:

water vapour 0.945	boric acid 5.2×10^{-4}	ammonia 1×10^{-4}
carbon dioxide 0.052	hydrogen 4.2×10^{-4}	nitrogen 5×10^{-5}
hydrogen sulphide 7.6×10^{-4}	methane 3.4×10^{-4}	oxygen 1×10^{-5}

There are also traces of noble gases, and, in particular, radon which makes the system quite radioactive.

The exploitation of fumaroles

A plant scheme showing how a boriferous fumarole can be integrally exploited is presented in Fig. 10.12.

The natural vapour emerging from the wells is sent to heat exchangers of the type of E_1 where pure steam is produced by the constant release of energy from the degassed solutions which are derived from the vapour itself.

The condensate obtained from the fumarole vapours is first passed into the general degassing column C_2, from the bottom of which water emerges which, practically, contains only ammonia that is subsequently stripped out in the plate column C_1. The products from the top of the degassing column C_2 pass into the packed column C_3 which is irrigated from above with water to hold back the CO_2 and H_2S, while the hydrogen, methane, and noble gases present are sent to storage. The two acid gases CO_2 and H_2S are liberated in C_4 which discharges the degassed water into a drain while sending the products from the top of the column to be separated. The CO_2, after condensation in the cooling loop E_2, is partially stored while the remainder of it is used for the production of NH_4HCO_3. The hydrogen sulphide passes on through the plant to be burnt in B which is the first step in a regular process for the production of sulphuric acid which has already been described (page 210). A part of the ammonia which has been stripped in C_1 is used to produce ammonium bicarbonate with the CO_2 and water in R_1, while the remainder goes to be combined with sulphuric acid in R_2 in the manufacture of ammonia sulphate which, having been centrifuged in PC_2 with recycling of the mother liquors, is then finished and stored.

The water produced with the vapour from the fumarole which is derived in a degassed state from the tail of the column C_1 is sent into the exchanger E_1 where the greatest part of it is vaporized, thereby giving rise, by bringing about the condensation of the natural vapour, to solutions which are appreciably concentrated in boric acid. These are passed on to be crystallized in CR and for centrifugation in PC_1. As a result, crystalline boric acid is obtained which is collected and stored.

The steam obtained in E_1 from the degassed water coming from the bottom of C_1 passes into the steam turbine FT which drives the alternator AT to produce electrical energy. The low-pressure steam emerging from the turbine FT is condensed in a mixing condenser, which uses water which has previously had its temperature lowered in the cooling tower TF, and served by a suspended drop separator and a vacuum pump.

To summarize, the integral exploitation of a fumarole leads to the following

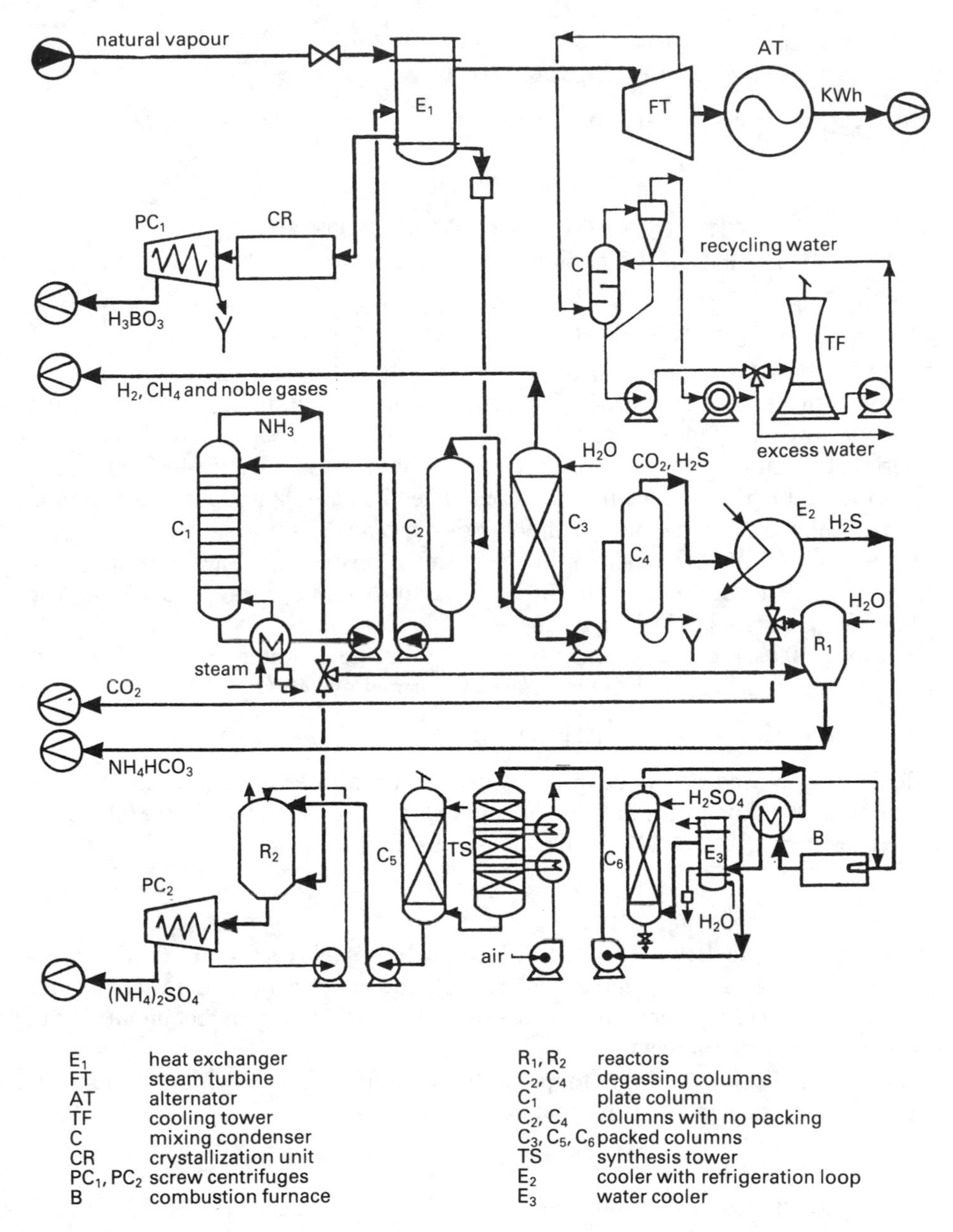

Fig. 10.12—Scheme of a plant for the complete exploitation of the chemical and energy content of the natural vapour emerging from a boriferous fumarole.

industrial products:

- **electrical energy**,
- *thermal and mechanical energy* which are utilized in operations at the site of the plant,
- **boric acid**,
- *ammonium bicarbonate* and *ammonium sulphate*,

- *carbon dioxide*,
- *hydrogen*, *methane*, and *noble gases*.

At present, the energy produced is, by far, the most valuable component.

10.10 BORIC ACID AND BORON COMPOUNDS OF INDUSTRIAL INTEREST

The 'crude' boric acid which crystallizes from the discharge waters of the evaporators serving the electrical plant, which utilizes the energy from the fumaroles, contains about 95% of the main component. This is refined by treating the saturated boiling solution with 1.5% of decolorizing carbon, filtering the solution, and then allowing it to cool slowly in lead-lined vessels or vessels which have been lined with a plastic material. The mother liquors are recycled and used to dissolve further amounts of raw boric acid which is to be recrystallized. The product obtained in this manner is a refined solid containing more than 99% of boric acid.

Boric acid, H_3BO_3, crystallizes in the form of glistening white lamellae with a solubility of 49 g/l at 20°C (about 4.5%). It is an extremely weak acid which yields salts which are strongly hydrolyzed.

At above 100°C boric acid tends to lose a molecule of water to form HBO_2 and has a solubility of 392 g/l. Tetraboric acid is formed at 140°C:

$$4H_3BO_3 \rightarrow H_2B_4O_7 + 5H_2O$$

Boric acid is used as a food preservative, as an antiseptic in pharmacy, in the ceramic industry, and in glazes. A further important use of this acid is in the fabrication of heat-resistant glasses and in woven glass fibres.

Boraxes

Boraxes, such as $Na_2B_4O_7 \cdot 10H_2O$ (the 'decahydrate'), $Na_2B_4O_7 \cdot 5H_2O$ (the 'pentahydrate'), and $Na_2B_4O_7$ ('anhydrous' borax), rather than boric acid, are used in glasses, glazes, and pigments for porcelains, and, particularly, in techniques for the welding and brazing of metals.

The hydrated compounds are prepared by treating, at first slowly, boric acid with solution of caustic soda or Solvay soda:

$$4H_3BO_3 + Na_2CO_3 \rightarrow Na_2B_4O_7 + 6H_2O + CO_2$$

and then leaving it to crystallize after the solutions, which have been cooled by various means, have attained a density of about 1.15. The pentahydrate is formed if the crystallization is carried out at above 60°C, and the decahydrate if the crystallization is carried out below 50°C. Anhydrous borax is prepared by heating the hydrates up to about 200°C.

Perborate and perborax

Sodium 'perborate' is widely used in the preparation of detergent powders and in bleaching and laundering processes. This salt is only slightly soluble but decomposes readily, especially in a weakly acid environment, with the liberation of nascent oxygen.

The structure of this perborate is not interpreted as being that of a peracid (HBO_3) but rather as containing one molecule of oxidized water (hydrogen peroxide) among the four molecules of water with which it crystallizes. It is prepared by the electrolysis of a solution of borax and soda in the presence of small quantities of chromate as well as by treating an aqueous solution of boric acid and sodium hydroxide with hydrogen peroxide:

$$H_3BO_3 + NaOH + H_2O + H_2O_2 \rightarrow \underset{\text{sodium perborate}}{NaBO_2 \cdot H_2O_2 \cdot 3H_2O}$$

As an alternative to the perborate, perborax, which is also used as a bleaching agent mainly in the formulation of detergents, may be prepared by the route:

$$4H_3BO_3 + Na_2O_2 + 4H_2O \rightarrow \underset{\text{sodium perborax}}{Na_2B_4O_7 \cdot H_2O_2 \cdot 9HO}$$

The commercial value of both the perborate and perborax depends on their content of active oxygen, that is, the volume of oxygen which is evolved in a gaseous form under normal conditions per unit of weight of the compound decomposed.

Binary compounds of boron

Boron forms various binary compounds both with metals and non-metals. Here, for the most part, one is concerned with 'interstitial' compounds in which the atoms of one of the components are located in the 'interstices' or voids in the crystal lattice. They are therefore more subject to constraints imposed by the crystal structure rather than the rules of chemical composition (stoichiometry). Among the binary compounds of boron which merit specific mention, there are:

- *boron carbide*, a compound almost as hard as diamond, which is prepared by reacting B_2O_3 with carbon in an electric furnace at 2500°C;
- *boron nitride* which is isosteric* with carbon and is therefore also dimorphous, being able to take up either the hexagonal structure of graphite or the cubic structure of diamond.

Metal borides have also been reported which are prepared by heating their components, in fine powder form, up to temperatures of about 2000°C or by reduction of the corresponding oxides with carbon.

Boron carbide is used as an abrasive and in nuclear reactor plant technology on account of the high neutron capture cross-section of boron-10.

Boron nitride, which is produced in the graphitic form by the reaction of boric oxide or boron halides with ammonia or amines and in the diamond form ('borazon') by subjecting the graphitic form to high pressures ($>60\,000$ atm), is used in the first form as a special solid lubricant and refractory and, in the borazon form, as a substitute for industrial diamonds.

Metal borides are chemically and thermally stable, hard, and conductors of both heat and electricity. They are refractories which can be used under extremely rigorous conditions (in the aerospace and nuclear industries) and occur in the formation of metal alloys.

---

*Isosteres are substances which, even if they are formed from different elements, have the same number of protons and electrons. The most famous case of isostery is that of N_2 and CO. The physicochemical properties of isosteres are similar.

Table 10.6—Principal boranes and their properties.

Name	*Formula*	*b.p. (°C)*	*Physical state*
Diborane	B_2H_6	−93	gas
Tetraborane	B_4H_{10}	+18	liquid–vapour
Pentaborane	B_5H_{11}	+65	liquid
Decaborane	$B_{10}H_{14}$	+213	solid (m.p. +99.7°C)

Boranes

The hydrides of boton are known as 'boranes'. In general, they have structures which are not in accord with the normal 'valency' rules. The best known boranes are listed in Table 10.6.

The boranes, which can be prepared from metallic borides by reacting them with hydrochloric acid* or by a catalytic reaction from diborane, have a negative heat of formation. Also, on account of the extremely high heat of combustion of boron, these compounds have the highest energy content per unit weight of any known compounds. Nevertheless, the practical use which has been made of them as rocket propellants has encountered serious obstacles in their application owing to the combined difficulties associated with their preparation and the great danger in handling them. They are autoigniting, explosive, and toxic.

Nowadays, the boranes and various compounds which are classified with them owing to the manner in which they are prepared (e.g. $2LiH + B_2H_6 \rightarrow 2LiBH_4$) are also used as synthetic intermediates of an industrial character. Among other things, they are used in the preparation of primary alcohols by the hydration of alkenes and for the direct reduction of organic acids which would be impossible without them.

Among the hydrides of boron and other elements (the 'borohydrides') the most important, other than that of lithium, are sodium borohydride ($NaBH_4$) and aluminium borohydride $Al(BH_4)_3$. All of them are soluble in various organic solvents.

Boron halides and borazoles

Among the inorganic compounds of boron, the halides and, above all, the trifluoride and the trichloride, stand out in importance.

On the other hand, the organic compounds of boron are of mainly theoretical interest. As an example, one may recall the alkylboranes (e.g. trimethylborane, $B(CH_3)_3$ and triethylborane $B(C_2H_5)_3$) are the arylboranes (e.g. triphenylborane, $B(C_6H_5)_3$). These compounds are gaseous, liquid, or solid depending on their molecular weight, and are highly flammable. There are also some highly volatile alkyl esters (borates).

Boron trifluoride is prepared by the action of fluorosulphonic acid, produced by the synthetic reaction ($HF + SO_3 \rightarrow HFSO_3$), on boric acid:

$$3HFSO_3 + H_3BO_3 \rightarrow BF_3 + 3H_2SO_4$$

Boron trichloride is prepared by synthesis from the elements or by the action of a current of chlorine on a red-hot mixture of B_2O_3 and carbon:

$$B_2O_3 + 3C + 3Cl_2 \rightarrow 3CO + 2BCl_3$$

*This reaction leads to diborane mostly being formed.

They behave as strong 'Lewis' acids and are very widely used as catalysts in industrial chemistry.

The (extremely strong) acid, *fluoroboric acid*, HBF_4 exists only in solution and is prepared by the reaction between HF and BF_3. This compound is also of considerable practical interest especially in the field of organic synthesis.

Borazole is a cyclic compound with a benzenoid type structure and can be prepared according to the reaction*:

$$3B_2H_6 + 6NH_3 \xrightarrow{200^\circ C} 2\ (\text{cyclic } B_3N_3H_6:\ BH,\ NH,\ NH,\ BH,\ BH,\ NH) + 12H_2$$

Borazole is a liquid which boils at 55°C. It is chemically very stable and its derivatives have aromatic properties. Heat-stable polymers, and refractories towards high-energy radiations, have been successfully produced from it. Upon thermal pyrolysis, fibres of such polymers are transformed into exceedingly strong fibres of boron nitride.

---

*In effect, this starts off from a borane–ammonia adduct $H_3B \leftarrow NH_3$, the so-called *borine*.

11

The chemical fertilizer industry

11.1 THE ROLE AND ASPECTS OF MINERAL FERTILIZATION

A large part of the world economy and the very solution of the most serious problems facing humanity such as those connected with hunger in the world are based on the use of chemical manures or chemical fertilizers*.

The progressive growth in the yields of farm products from each hectare is not due to the use of mineral fertilizers alone. The introduction of new breeds of plants and modern cultivation techniques have also contributed to the improved results in this field. But the use of mineral fertilizers has always been a prerequisite for the effectiveness of such innovations.

Starting from some actual data concerning the amounts of fertilizing elements required by crop-bearing land and the natural replacement of these elements, it is possible to convince oneself of this.

On the one hand it is calculated that 20–25 kg of nitrogen, 5 kg of phosphorus, and 6 kg of potassium are removed from a farm when it sells one ton of cereals, and that 2 kg of nitrogen, 0.5 kg of phosphorus, and 4 kg of potassium are taken from the land when a similar weight of sugar beet is sold.

On the other hand, the most reliable estimates lead one to believe that only about 30% of the nutritive requirements of the plants can be provided by the addition of animal excrement and by ploughing in carried out at the farm, because such means of provision are limited to a recycling of that part of the nutrients which animals discard or which is contained in the plant matter which is ploughed in.

Other data from this source lead to Table 11.1 which shows the incremental increases in fruit crops, harvests, and the use of chemical manures which have taken place during the last one hundred and fifty years when referred to a standard unit of cultivated land (ha = hectare) in the more advanced nations.

It is confirmed, among other things, that the harvests sharply increased in parallel with the growing use of manuring with mineral chemical fertilizers, that is, in relation to the enhancement

*Even if the word 'chemical' is omitted in the following discussion, it is implied. While natural manures also continue to play an essential role in agriculture, they will not be considered here.

Table 11.1—Cause and effect in the progressive use of chemical fertilizers

Year	*No. of people fed per ha of cultivated farm land*	*Harvest in quintals of cereals per ha*	*Quintals of chemical fertilizers (nitrates, phosphates, and potassium salts required per ha*
1830	0.8	7.3	–
1875	1.3	12.0	0.31
1900	1.6	18.4	1.56
1925	2.1	22.8	4.39
1950	3.3	29.8	10.19
1975	4.6	44.3	23.35
1982	4.5†	46.6	25.58

†This decrease in the number of people fed is attributable to an increase in the standard of living.

of the 'fertility' of the soil which is the amount of the crop harvested per hectare of cultivated land.

It should, however, be added that the other parameters associated with the fruitfulness of the land, such as the amount of humus present, the depth of the topsoil of the cultivated land, and the microbiological activity of the cultivated soils are also sharply improved by the use of mineral fertilizers owing to the fact that this not only leads to the production of larger amounts of foodstuffs but also to a greater mass of roots, straw, grass, and leaves which persist in the soil or which are returned to the soil in the form of litter. This increased addition of organic material to the land favours the formation of humus and the insurgence and multiplication of microorganisms with the task of converting the nutrients making up many chemical fertilizers into compounds which can be assimilated by plants.

11.2 THE COMPONENTS AND ACTION OF FERTILIZERS

Major elements and trace elements of soil fertility

Chemical fertilizers are compounds which are chemically produced which bring the chemical elements required in greater or smaller amounts (and therefore in far greater amounts than are available in the soil)* to cultivated land to ensure the germination, growth, and fruitfulness of the plants.

The chemical elements which play the role of fertility factors when they are available in conspicuous amounts in agricultural land are phosphorus, nitrogen, potassium, and, sometimes, calcium.

Plants require not only the elements mentioned above in appreciable amounts: they also require considerable amounts of hydrogen, carbon, and oxygen. However, plants cannot succeed in getting the required amounts of the abovementioned elements from their natural environment, that is, the soil in which they grow, water, and air.

*Here, it is a question of 'availability' rather than presence. Phosphorus and potassium are often present in appreciable quantities in agricultural land, but they are not available because they are combined in insoluble compounds. Nitrogen, which is likewise present, is not directly available if it is ammoniacal nitrogen.

Given that areas of land require the application of relatively large quantities of the abovementioned elements, these elements are known collectively as the major elements of soil fertility.

The elements which it is sometimes necessary to apply to land, always in small amounts, so that plants should grow quickly and heathily, are referred to as the trace elements of soil fertility. Copper, cobalt, zinc, boron, and manganese are counted among the trace elements.

Industrial chemistry is specifically concerned with the production and marketing of chemical fertilizers based on the major elements of soil fertility. For this reason, reference is always, and only vaguely, made to compounds of the major elements when discussing chemical fertilizers.

Functions and efficiency in the use of the fertilizing elements

The role of the major elements of soil fertility is either plastic, that is, it is intended to shape the plant structures, or physiological, or physicochemical.

Nitrogen and phosphorus exert a plastic action, while potassium, the ions of which promote photosynthesis, exhibits an action which is predominantly physiological. On the other hand, calcium, which acts in regulating the acidity of agricultural land and is a factor in preventing the leaching out of nitrates and potassium ions, mainly acts physicochemically.

For the most part, the trace elements enter to form parts of enzymatic systems, and their role is therefore explained at the level of biological catalysis. This explains why, in practice, they are necessary in only very small amounts.

Although the functions of the different soil fertility elements tend to be explained in a differentiated and specific manner in relation to various types of plant, it has been shown that their action is correlated in the sense that a dearth of one of them tends to have a negative effect on the utilizability of the others.

For this reason, as we shall see, the tendency in recent times has been to resort to the use of compound or complex fertilizers.

The main factors determining the efficiency of the use of fertilizers are essentially the chemical composition of the land and seasonal and climatic factors. Table 11.2, besides showing that the yields of grain are increased when fertilizers are used, also demonstrates that the environmental conditions play an important role in determining the size of the harvest.

Finally, it has been proven that the qualitative and quantitative choice of the fertilizers to be used is related to the type of crop which is to be developed when the other factors determining the efficiency of the fertilizers which are used are equal.

11.3 CLASSIFICATION, COMPOSITION, AND COMMERCIAL FORMS OF FERTILIZERS

The fertilizers produced industrially can be classified as nitrogen fertilizers, phosphorus fertilizers, and potassium fertilizers and further subdivided into:

- simple fertilizers, if they contain a single fertilizing element;
- mixed fertilizers, which are said to be binary or ternary if they contain two or three fertilizing elements respectively;

Table 11.2—Consumption of standard fertilizers and yields of grain per ha

Country	*kg of fertilizer*	*Harvest in quintals*	*Country*	*kg of fertilizer*	*Harvest in quintals*
Argentina	3	13	Japan	250	35
Australia	10	18	Netherlands	450	32
Denmark	140	35	Pakistan	2	12
France	70	22	Spain	25	11
Great Britain	150	24	Turkey	5	12
India	2	8	USA	50	22
Italy	50	20	West Germany	250	27

In their turn, mixed fertilizers are classified as:

- compound fertilizers, if they contain the fertilizing elements in the form of a single chemical compound, such as $Ca(NH_4)PO_4$;
- complex fertilizers, if they are mixtures of simple fertilizers.

Nowadays, most mixed fertilizers are of the complex type.

The grade of a fertilizer is expressed in terms of the amount of nitrogen per hundred parts in weight of the fertilizer (N%), P_2O_5 per hundred parts in weight (P_2O_5%) and K_2O per hundred parts in weight (K_2O%).

It is therefore readily understood why, for example, potassium chloride for agricultural use is sold with grades of 17%, 18.9% and, especially, 60% of K_2O.

The composition of a fertilizer, no matter how many active components it contains, is customarily stated on the packaging as three successive numbers showing the 'fertilizer units' which are conventionally written in the order:

$$\%N \qquad \%P_2O_5 \qquad \%K_2O$$

So, the following abbreviations are to be used:

$Ca(CN)_2$	22	0	0	calcium cyanamide containing 22% N
$Ca(H_2PO_4)_2$	0	18	0	superphosphate containing 18% P_2O_5
KCl	0	0	60	potassium chloride containing 60% K_2O
KH_2PO_4	0	52	34	monopotassium phosphate containing 52% P_2O_5 and 34% K_2O

and these are to be interpreted as:

- with 22 units of nitrogen fertilizer,
- with 18 units of phosphate fertilizer,
- with 60 units of potassium fertilizer,
- with 52 units of phosphate fertilizer and 34 units of potassium fertilizer.

Commercial fertilizers are most commonly marketed in a powder or granular form, but crystals, suspensions, solutions, and even gaseous fertilizers are also used.

The powdered form is preferred for products with a low solubility and where immediate action is required. On the other hand, the granular form offers the advantages of not being dispersed as it is applied and of slowly dissolving, which somewhat reduces the losses of the material caused by rain.

11.4 NITROGEN FERTILIZERS

Simple chemical fertilizers which supply nitrogen to the land are subdivided into:

- ammonia fertilizers, if they bring nitrogen to the land in the form of ammonia or the ammonium ion, regardless of the initial manner in which the ammonia or the ammonium ion may be combined;
- nitrate fertilizers, if they supply nitrogen in the form of the nitrate ion.

That important 'modulated' action nitrogen fertilizer, ammonium nitrate, occupies a unique position here. In this text it will be considered together with the nitrate fertilizers.

The action and the way in which nitrogen fertilizers behave in the soil is rather complex, and the development of these aspects of them lies outside the field of industrial chemistry.

Nevertheless, it is interesting to note:

- that a fraction of the nitrate fertilizers applied to the land is lost by either the phenomena of their retrotransformation into atmospheric nitrogen carried out by bacteria, or leaching out processes;
- that nitrates are immediately assimilated by plants, while the uptake of ammonia fertilizers, since the ammonia has first to be converted into nitrate (nitrified) by bacterial enzymes, is delayed.

Ammonia fertilizers

In the final analysis ammonia is the active component of ammonia fertilizers. Besides ammonia (which is actually used directly), the most important representatives of this class of fertilizers are ammonium chloride, calcium cyanamide, and, above all, ammonium sulphate and urea.

Ammonia

In countries with advanced agricultural economies the use of anhydrous ammonia as a fertilizer is taking hold to an ever increasing extent. This contains the maximum percentage of nitrogen (82.2%) and may be applied either in the gaseous or liquid state. Aqueous solutions of ammonia are also used, and special techniques for the transport, distribution, and protection of the operatives are also implemented, given the high toxicity of NH_3 and the danger of its dispersion and, even, of explosions.

Ammonium chloride

The following facts indicate why the adoption of ammonium chloride as a large-scale fertilizer has encountered difficulties:

- the incompatibility of Cl^- ions with the physiology of many plants;

- the corrosive action which it exhibits owing to the high degree of hydrolysis that it undergoes;
- the difficulty in storing it owing to its tendency to cake.

The use which is made of NH_4Cl is determined by:

- its low cost, as it is often directly available as a by-product from important industries such as the Solvay soda industry (page 422) and the potassium sulphate industry (page 404);
- the fact that it combats certain plant diseases and prevents others.

The pronounced acidic behaviour of ammonium chloride can be countered by mixing it with $Ca(OH)_2$ and calcium cyanamide.

Its production starts from the hydrochloric acid solutions containing about 30% HCl which are obtained by scrubbing of the hydrogen chloride, produced in synthetic organic reactions, with water. The resulting solutions are then reacted with ammonia:

$$NH_{3(g)} + HCl_{(aq)} \rightarrow NH_4Cl_{(aq)}$$

The ammonium chloride is then crystallized and separated. The crystals are dried in a similar manner to that used in the preparation of ammonium sulphate (page 218).
Here, however, it is necessary to ensure that the reactors have an acid-resistant lining, and they must not be operated above certain temperatures during the drying phase as NH_4Cl tends to dissociate. In practice, the salt is dried by circulating air or under low pressure.

Ammonium chloride is marketed commercially in a white crystalline or granular form with a nitrogen content of about 25%.

Ammonium sulphate

Ammonium sulphate is extremely widely used as a fertilizer both because, as a direct or indirect by-product of many chemical processes, it has a relatively low commercial price and also because the sulphate ion is, in general, quite compatible with the physiology of many plants.

Among the few contra-indications regarding the use of ammonium sulphate as a fertilizer, there is the case of rice fields in which a particular fungus can bring about the reduction of sulphates to sulphides with the development of toxic substances which have a deleterious effect on the plants.

Even when ammonium sulphate is purposely manufactured it is still looked upon as an industrial by-product as it is always a question of recovering and utilizing materials which have been discarded from the processes such as the sulphuric acid effluent from electrochemical industries, from processes used in the manufacture of pigments, and from the activities of refineries and the chemical gypsum produced by firms concerned with the manufacture of phosphoric acid (page 353).

The main industrial methods for the preparation of ammonium sulphate have already been stated and generally described and illustrated (section 6.8).

Ammonium sulphate is marketed in plastic and jute sacks as a white crystalline powder (granules with a diameter of 0.5 to 1 mm) which does not cake during storage or transport provided that it has been treated with anticaking agents such as diatomaceous earth and certain alkali metal salts derived from naphthalene sulphonates.

The nitrogen content in $(NH_4)_2SO_4$ almost attains a value of 21%, while the free sulphuric acid in it, which is slightly less than the amount of moisture, is up to 0.6%.

Urea

Carbonic acid is formally dibasic and can therefore form a double amide, urea:

$$\underset{\text{carbonic acid}}{HO\text{—}\overset{\overset{\displaystyle O}{\|}}{C}\text{—}OH} \qquad \underset{\text{urea}}{H_2N\text{—}\overset{\overset{\displaystyle O}{\|}}{C}\text{—}NH_2}$$

Nowadays, urea is produced industrially in large plants which make use of a reaction leading to the formation of ammonium carbamate, followed by the dehydration of the latter to yield urea:

$$CO_2 + 2NH_3 \rightleftharpoons H_2NCOONH_4 + 38.08 \text{ kcal/mole} \tag{11.1a}$$

$$H_2NCOONH_4 \rightleftharpoons H_2NCONH_2 + H_2O - 6.26 \text{ kcal/mole} \tag{11.1b}$$

$$2NH_3 + CO_2 - H_2O \rightleftharpoons H_2NCONH_2 + 31.82 \text{ kcal/mole} \tag{11.1c}$$

The process is therefore globally exothermic, and the heat which is developed can be utilized in all the other plant requirements.

For obvious kinetic reasons and from the thermochemical data presented above, it can be seen that an increase in the temperature only favours the rate of formation of ammonium carbamate, while it speeds up the kinetics and increases the yields of urea from the dehydration of the ammonium carbamate which is produced.

Increasing pressure, as can be seen from the global reaction where the states of the reactants and products are shown, distinctly favours the formation of urea.

Regarding the concentrations, it may be noted that, on the basis of the reaction coefficients, the use of an excess of ammonia favours the synthesis more than does an excess of CO_2. Furthermore, as ammonia tends to combine with water, the use of an excess of ammonia favours the production of urea, because the equilibrium involved in the dehydration of carbamate to urea is displaced in such a direction by the removal of water.

A further appreciable benefit in the synthetic yields follows when an excess of ammonia is used, owing to the fact that water*, which is seen to be an essential reagent for the occurrence of the following parasitic reactions in the synthesis of urea, is made less available:

$$NH_3 + CO_2 + H_2O \rightleftharpoons NH_4HCO_3$$

$$NH_4HCO_3 + NH_3 \rightleftharpoons NH_4HCO_3$$

$$2NH_3 + CO_2 + H_2O \rightleftharpoons (NH_4)_2CO_3 + 56.25 \text{ kcal/mole} \tag{11.2}$$

Finally, the mechanism of the conversion of ammonium carbamate into urea also

*It has, in fact, been checked that, in the absence of water, only ammonium carbamate is formed from NH_3 and CO_2, while ammonium carbonate tends to be the predominant product, especially at a low temperature, if water is available.

demonstrates that an excess of ammonia is favourable to the production of urea. Actually, in the case of the mechanism which is shown here:

$$H_2N{-}COONH_4 \rightleftharpoons H_2N{-}COOH + NH_3$$

$$H_2N{-}COOH \rightleftharpoons NH{=}C{=}O + H_2O$$

$$HN{=}C{=}O + NH_3 \rightleftharpoons H_2N{-}CO{-}NH_2$$

$$H_2N{-}COONH_4 \rightleftharpoons H_2N{-}CO{-}NH_2 + H_2O$$

there is a danger that, if the excess of ammonia is sharply reduced, the isocyanic acid ($HN{=}C{=}O$) which is formed will bind to a molecule of urea with the formation of biuret:

$$H_2N{-}CO{-}NH_2 + HN{=}C{=}O \rightarrow \underset{\text{biuret}}{H_2N{-}CO{-}NH{-}CO{-}NH_2}$$

an amide which causes damage to plants to such an extent that a maximum of 1% of it is tolerable in those urea stocks which are destined for agricultural purposes.

Everything that has been said up to now regarding the operating conditions which lead to a speeding up of the kinetics and an increase in the synthetic yields of urea, should be kept in mind in the planning and operating of industrial plants for the manufacture of urea.

Actually, there is a tendency to work at a temperature which becomes progressvely higher (given the exothermic nature of reaction (11.1a), this act would be automatic) on account of the particular benefits which are to be extracted from reaction (11.1b) in this way. The pressures are maintained at constant high values during the synthesis and the $NH_3 \cdot CO_2$ ratio is also kept high (3.6/1, for example). On the other hand, a low H_2O/CO_2 ratio is adopted (e.g. 0.6/1).

* * *

The industrial processes for the synthesis of urea which are at present in use are of three types:

- processes without any recycling in order to minimize running costs;
- processes with partial recycling to enhance the yields;
- processes with integral recycling in increase the yields still further and to reduce the dimensions of the apparatus.

Processes without recycling are particularly carried out where the residual gases ($CO_2 + NH_3$) can be usefully exploited in other preparations, either separately or as a mixture. The production of ammonium salts may be cited as an example of their use as a mixture, while carboxylations and aminations are examples of their use separately.

Processes with partial recycling involve the absorption of CO_2 with selective solvents which are indifferent towards ammonia since the latter can be recycled. The resulting absorption product is subsequently desorbed with the production of purge gas (CO_2) which can be used in various ways, and the solvent used for the absorption can be re-used.

Processes with total recycling encounter serious difficulties associated with the recompression of the recycled gases ($NH_3 + CO_2$) which leads to the formation of a solid ($H_2N–COONH_4$) which obstructs the apparatus and also gives rise to corrosion problems.

Block diagrams of the first two processes are shown in Fig. 11.1, while a scheme for a complete plant using the third process and which avoids the recompression of the recycled gases is shown in Fig. 11.2.

Liquid ammonia is supplied to the plant under pressure, vaporized in E_1, and dispatched to the reactor R into which carbon dioxide is also fed under pressure. The synthesis of ammonium carbamate and its partial transformation into urea and water, that is, reaction (11.1), then takes place in R at slightly below 200°C and at a pressure of about 200 atm.

The solution emerging from the reactor is sent into E_2 where, upon heating and decompression, almost all of the carbamate which has been carried over is decomposed into CO_2 and NH_3, and the gaseous phase is separated from the liquid phase in SV_1. The latter, which contains urea and a little residual carbamate, passes to the decomposer E_3 and the separator SV_2 which operates at about 12 atm. as compared with a pressure of 70 atm. which appertains in SV_1. The gaseous components which are separated in SV_1 pass to two condensers

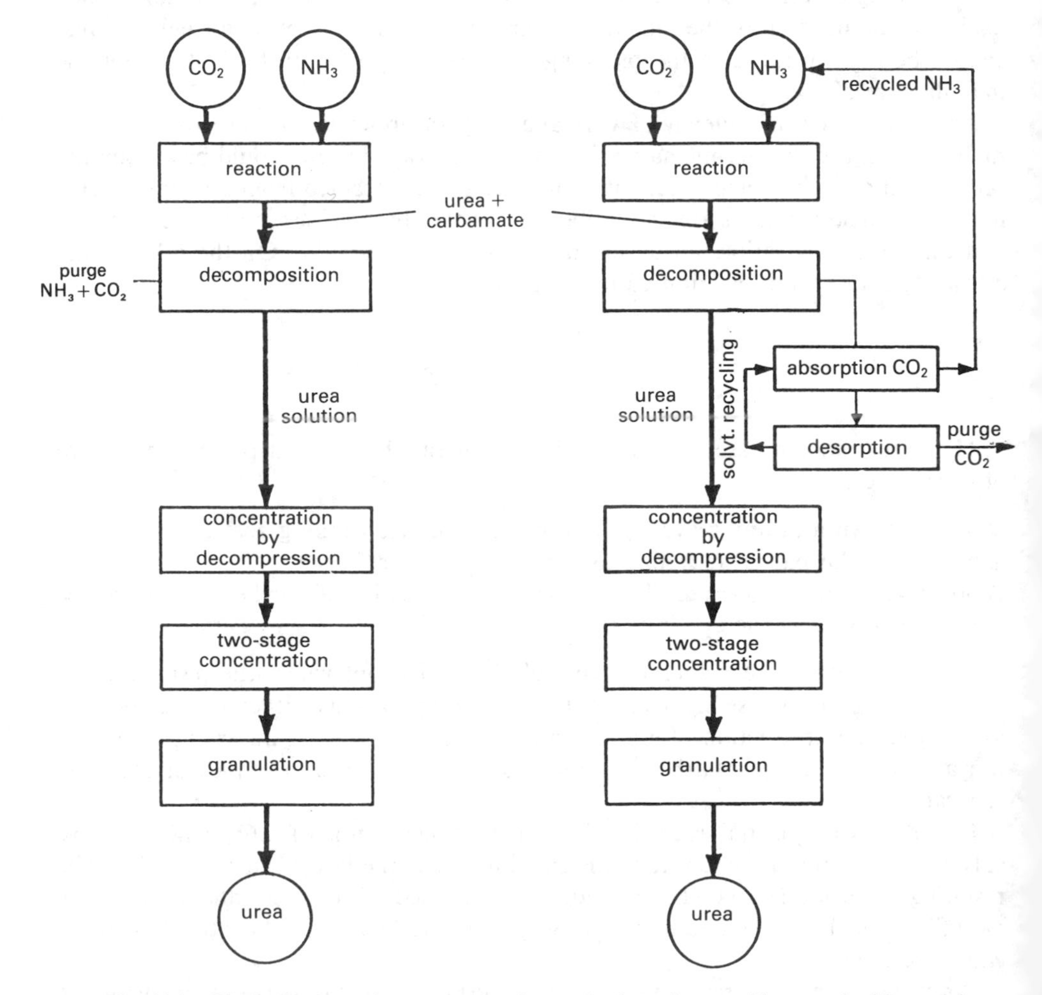

Fig. 11.1—Block diagram of processes for the production of urea without and with partial recycling.

E_4 and E_5, arranged in series, which operate at the same pressure but at temperatures of 140°C and 100°C respectively. A large part of the condensates gradually comes into these condensers, and an aqueous solution of ammonia and carbamate is thereby obtained.

On the other hand, the gases and vapours which are evolved in SV_2 are condensed in E_6 and the resulting liquid is used for washing the residual gases and vapours in C_1 which have been condensed out in E_5, while the washing of the gases and vapours which have not been condensed out in E_6 is carried out in C_2. All of the solutions emerging from E_4, E_5, C_1, and C_2 are recycled into the reactor R.

The solution of urea discharged from SV_2 is partially evaporated in E_7 and the gases and vapours emerging are washed in C_3 with the product obtained from the top of the column C_4.

A 70% solution of urea which is discharged from the evaporator E_7 is subsequently concentrated to above 99% in the two concentrators E_8 and E_9, which are arranged in series, and both operate under an increasing degree of vacuum. The vapours which are separated in E_8 and E_9 are condensed in E_{11}, and a rectification is carried out in C_4 whereupon an aqueous solution of ammonia is recovered from the top of the distillation column, while water, which is practically free from ammonia, is obtained at the bottom as it emerges from the reheater E_{10}.

The fused urea which comes out of the concentrator E_9 is sent, before being stored, to undergo granulation in the prilling tower T.

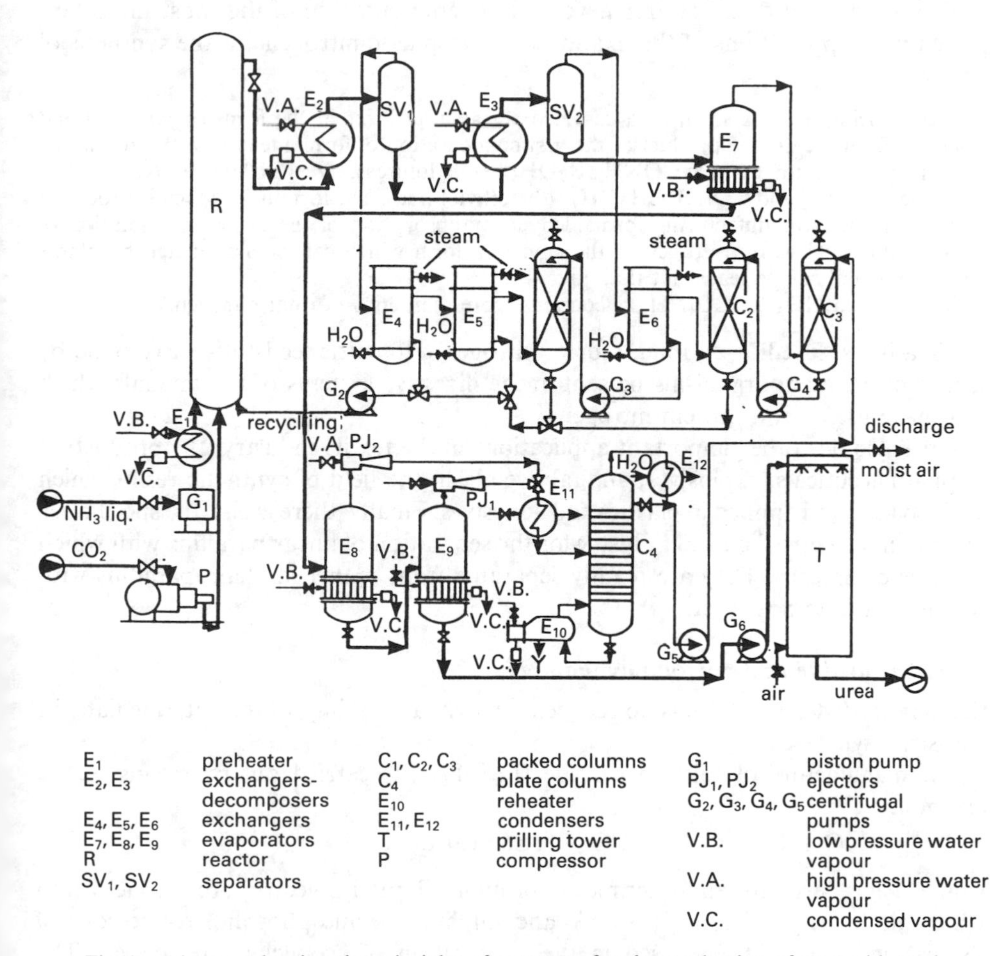

Fig. 2—Diagram showing the principles of a process for the production of urea with total recycling without recompression of the gases.

After an aqueous solution of urea has been obtained, all three processes complete the preparation in a similar manner by carrying out a preliminary concentration of this solution by decompression after reheating and two successive stages of intensive concentration until it is practically pure (greater than 99%). In this state the urea passes to a prilling tower where, upon encountering a hot ascending current of air, it is transformed into quite uniform granules which are discharged onto a conveyor belt and sent for storage.

* * *

Urea contains a high percentage of nitrogen (46.3–46.6%). It is readily hydrolyzed to form ammonia and carbon dioxide by the action of urease (an enzyme) in bacteria which occur widely in both air and soils:

$$CO(NH_2)_2 + H_2O \rightarrow 2NH_3 + CO_2$$

This also prevents the build-up of urea in soils.

The use of urea as a fertilizer nevertheless represents one of the most important agricultural applications of the fixation of atmospheric nitrogen via the synthesis of ammonia.

Besides being used as such, urea can also act as a fertilizer in the form of *formurea* and *calcurea*. Formurea is the product of the first condensation of formaldehyde with urea which results in a mixture of $NH_2CONHCH_2OH$ (methylolurea), $NH_2CONHCH_2NHCONH_2$ (methylenediurea), and $CO(NHCH_2OH)_2$ (dimethylolurea). The formurea, which is produced directly in plants manufacturing formaldehyde by absorption of the latter in a solution of urea, gradually gives up nitrogen to the land in a form which can be assimilated by plants and is more difficult to leach out than urea.

Calcurea, $Ca(NO_3)_2 \cdot 4CO(NH_2)_2$, contains more than 36% of total nitrogen.

Urea is also used as an integrator of nitrogen into cattle feedstuffs. It is made by means of the microorganisms present in the digestive systems of the animals which can make use of non-protein nitrogen.

Urea also has other important applications in the textile industry, the production of pharmaceuticals, explosives, and, above all, in the field of synthetic resins which are also used and applied in plastics and adhesives. Finally, there is also the specific use of urea in the petrochemical industry for the separation of linear paraffins with which it forms complexes. They are readily separated from branched chain paraffins with which urea does not react.

Calcium carbide and calcium cyanamide

Calcium carbide, CaC_2, has long been known as an important intermediate in industrial syntheses.

The manufacture of calcium carbide (Fig. 11.3) is carried out by means of the reaction:

$$CaO + 3C \rightarrow CaC_2 + CO - 110.5 \text{ kcal/mole}$$

which, being strongly endothermic, is brought about in electric arc furnaces into which high-purity quicklime (>95%) and sulphur-free and phosphorus-free coke (if the carbide is later to be used in the preparation of acetylene) are loaded. The lime/coke ratio in the charge is 1:0.62.

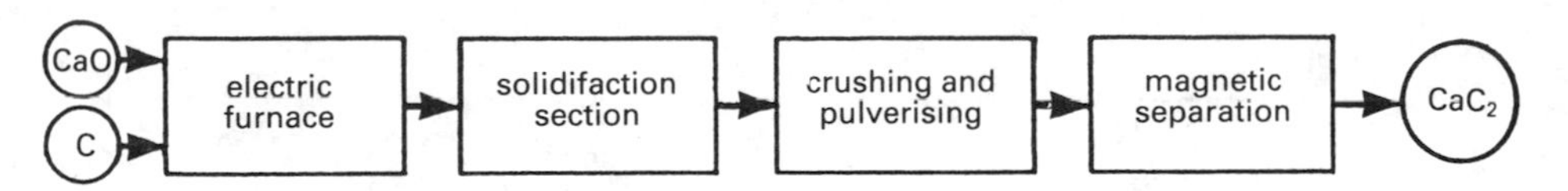

Fig. 11.3—Block diagram of the production of CaC_2.

The reaction proceeds by a fusion of the reagents with the operating temperature slightly above 2000°C even though CaC_2 on its own melts at 2300°C. This is because a eutectic is formed between CaC_2 and CaO. For every ton of carbide produced, 25 kg of graphite electrodes and almost 3000 kW of electrical energy are consumed. The production of this compound is therefore economic only in countries which have access to supplies of low cost electrical power.

The carbide is unloaded through an aperture which is made in the sealed furnace by means of an auxiliary electric arc. This opening is closed each time, after the contents of the furnace have been poured out, by the sintering of appropriately placed carbide powder. After being discharged from the furnace, the carbide is poured into steel moulds where it solidifies with a contraction in volume. It is then broken up into blocks which are subsequently crushed, granulated, or pulverized according to the form in which the material is required for commercial purposes.

The ferrosilicon alloy included in the mass originating from the iron-bearing gangue (Fe_2O_3) and the silica (SiO_2) in the limestone which stands above it, is separated magnetically.

Calcium cyanamide, $CaCN_2$ is a classical antiparasitic compound which has a corrective action on acidic soils.

In fact, the following reaction occurs in the ground:

$$2Ca{=}N{-}C{\equiv}N + 2H_2O \rightarrow Ca(OH)_2 + Ca(NH{-}CN)_2$$

Here, one is dealing with a fertilizer of the urea/ammonia type because its subsequent decomposition leads to urea:

$$Ca(NH{-}CN)_2 + CO_2 + 3H_2O \rightarrow CaCO_3 + 2NH_2{-}CO{-}NH_2$$

and then to ammonia:

$$2NH_2{-}CO{-}NH_2 \xrightarrow[\text{hydrolysis}]{\text{enzymatic}} 4NH_3 + 2CO_2$$

The industrial preparation of calcium cyanamide (Fig. 11.4) uses calcium carbide and nitrogen as the starting materials.

The pertinent reaction is:

$$CaC_2 + N_2 \xrightarrow{1100^\circ C} CaCN_2 + C + 97.8 \text{ kcal/mole}$$

Since the reaction is exothermic, it needs only to be primed by means of electrical resistances, made out of carborundum, which are encapsulated in the walls of the furnaces.

Crushed carbide is used, and the reaction is carried out continuously in rotating cylindrical furnaces which are fed from the top with CaC_2 while a current of nitrogen is supplied from the bottom. About 95% of this nitrogen is fixed by the carbide. There is no agglomeration of the material which is produced, and therefore no need for any crushing of it at the end of the operation. The calcium cyanamide is cooled under an atmosphere of nitrogen and the powdered product is agglomerated and coated with oil to reduce its dispersivity and its caustic and hygroscopic properties. Mineral oils in amounts of about 2% are used to coat the material.

The product is stored in concrete structures under a controlled atmosphere to prevent explosions due to the formation of acetylene.

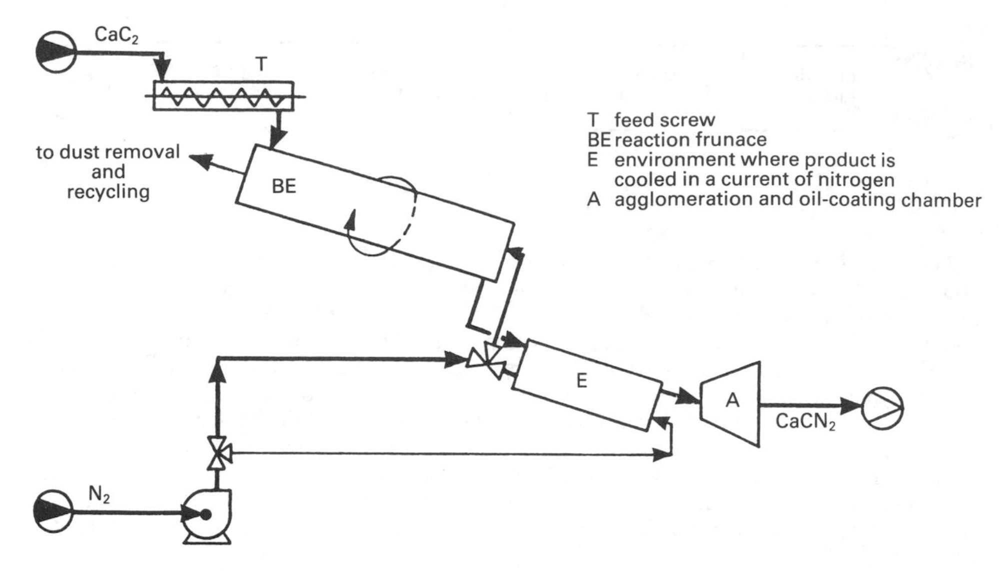

Fig. 11.4—Outline diagram of the production of $CaCN_2$.

Commercial calcium cyanamide is blackish because it contains carbon, while the pure compound is white.

White calcium cyanamide may be produced by the process:

$$CaO + 2NH_3 + 2CO \xrightarrow{750^\circ C} CaCN_2 + H_2O + 2H_2 + CO_2$$

which also leads to the evolution of hydrogen cyanide which is both extremely dangerous and difficult to separate. It is especially for this reason that the process is only rarely used.

There is 20–24% of nitrogen in commercial calcium cyanamide as compared with the theoretical value of 35%. It is also used as an agent for the correction of soil acidity and as antiparasitic compound.

Nitrate fertilizers

In agricultural soils some of the nitrosobacteria have the task of oxidizing any ammoniacal nitrogen which is present to nitrous acid and nitrites, while other bacteria have the task of oxidizing these products to nitrates whereupon they are in a form which can be assimilated by plants. If, on the other hand, the nitrogen is applied to the land directly in the form of a nitrate, this constitutes a nutrient which can be directly used by the plant.
The most important nitrate fertilizers which contain only nitrogen as a fertilizing element are the nitrates of ammonia, calcium, and sodium.

Ammonium nitrate

The ammonium nitrate industry arose on account of its use in explosives, but, nowadays, the greater part by far of the vast World production of this salt is destined for agrarian purposes. In fact, there is great agricultural interest in this fertilizer which brings to the land both nitrogen which can be rapidly absorbed and nitrogen which is slowly absorbed by plants.

Ammonium nitrate is usually produced by the direct reaction between ammonia and nitric acid. The ammonia is used in the gaseous state in conjunction with a 50% aqueous solution of nitric acid, the reaction being:

$$NH_3 + HNO_{3(aq)} \rightarrow NH_4NO_3 + 26\ \text{kcal/mole}$$

The heat of reaction increases as the concentration of the nitric acid is increased.

The plant shown in Fig. 11.5 is used for the production of ammonium nitrate. The reaction is carried out under pressure in the reactor R to which 50% nitric acid and ammonia, which has been vapourized in the exchanger E_1, are added. The heat of reaction is such that, upon emerging from R, the nitrate concentration is greater than 60%. The solution passes from R to the tank D which is heated with steam, which subsequently condenses, and is then pumped into the concentrator E_2, maintained under a vacuum by the ejector PJ, which aspirates the incondensable materials which are not deposited in C.

The final granulated product can be obtained by various methods among which the prilling tower usually turns out to be the best. After it has been concentrated up to about 95%, the solution enters this tower where it encounters an ascending current of hot air and is thereby transformed into a uniform granular solid known as prilled ammonium nitrate.

The product which is discharged is gathered by a conveyor belt T, which transports it to the fluid bed cooler E_3 from where it is passed to the vibrating sieve PV. This separates out the small pieces of scrap material from it, and then it passes to TR where it is coated with anticompacting agents (dolomite, diatomite, etc.). Finally, it is stored.

When ammonium nitrate is struck violently or heated, it is transformed explosively into exclusively gaseous products (N_2, O_2, H_2O). It is therefore used on its own as

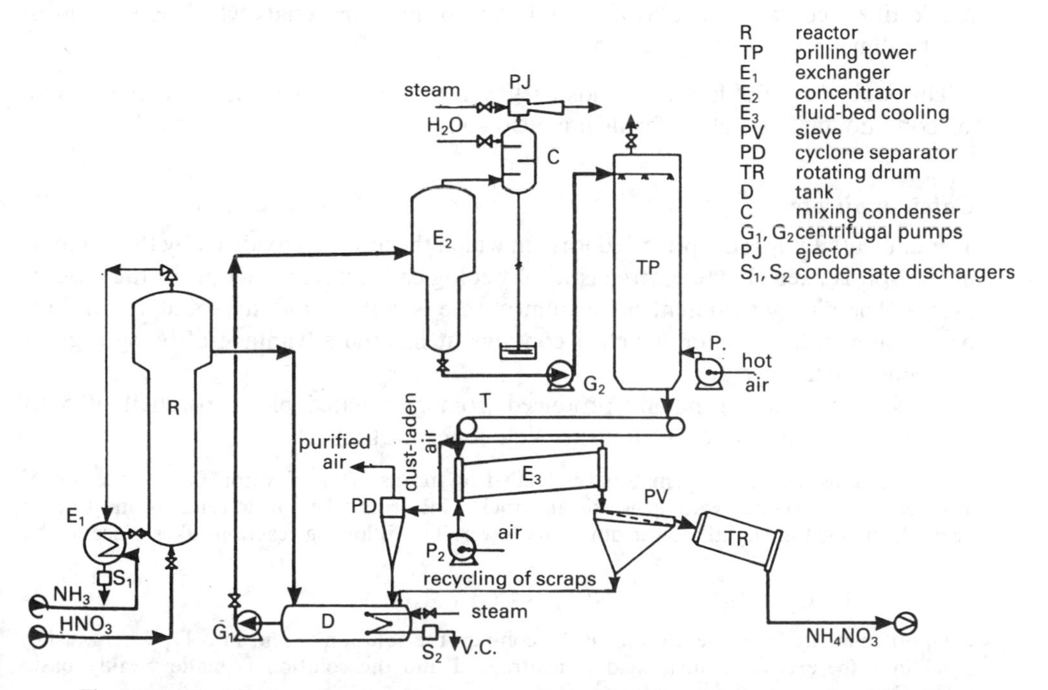

Fig. 11.5—Plant for the production of ammonium nitrate. The different methods are distinguished by how they make use of the heat of neutralization in the concentration of the salt that is produced. Most of them operate under a reduced pressure, while others bring about evaporation by the transmission of heat between the concentrically arranged reaction chamber and the part of the plant where the evaporation takes place.

an explosive. More often, however, on account of the ready manner in which it decomposes upon coming into contact with reducing agents and its high oxidizing power, it is used (always in the pure and anhydrous form) in compound explosives:

- 'safety' explosives with TNT (trinitrotoluene) and dinitronaphthalene;
- 'demolition' explosives with aluminium powder and carbon.

The reaction between ammonium nitrate and the aluminium powder which is primed by detonators is:

$$2Al + 3NH_4NO_3 \rightarrow Al_2O_3 + 3N_2 + 6H_2O$$

Given the great stability of the products formed, it is easy to imagine the amount of energy released in this reaction.

Explosive grade ammonium nitrate is not marketed for agricultural purposes since it forms lumps during storage, and it is dangerous to attempt to break these.

To adapt NH_4NO_3 to a form which is suitable for storage, inert materials are strewn over it after it has been granulated, and, to make it more stable and economical, quite large amounts of a chalk base diluent ($CaCO_3$ powder) are added to it before granulation.

Consequently, two different types of NH_4NO_3 based fertilizers are available commercially:

- a fertilizer containing 33% of N and an average of 3% of inert materials, which is known as ammonium nitrate;
- a fertilizer containing 21% of N and 40% of inert materials which is known as nitro-lime.

This second type of fertilizer lends itself well to use in somewhat wet climates and for conditioning, as well as fertilizing, acid soils.

Calcium nitrate

The salt $Ca(NO_3)_2$ is the preferred form in which the nitrogen oxidized by the bacteria in the soil is fixed. It therefore has a 'physiological' fertilizing action in the case of plants. The nitrogen content of calcium nitrate is rather modest (about 15%), but, on account of the calcium which it contains, it has the advantage of being a good soil-conditioning agent.

Calcium nitrate is generally produced from granulated pieces of chalk (0.5 to 1.5 cm) and nitric acid of medium to high concentration.

The chalk used as the raw material is loaded by means of the elevator T_1 into the top of a tower (Fig. 11.6) lined with acid-resistant bricks, TR, from where it descends to meet nitric acid which is fed in slightly lower down the tower. The following reaction takes place in the tower:

$$2HNO_3 + CaCO_3 \rightarrow Ca(NO_3)_2 + CO_2 + H_2O$$

Upon emerging from the bottom of the column, the solution is sent into D_1 where, using quicklime, the excess of nitric acid is neutralized and the solution is made weakly basic (pH = 7–9).

The solution is concentrated by sending very hot fumes into a series of rotating discs which are partly immersed in the solution. After being produced in the furnace B and passing into the discs in BE_1, these fumes encounter the solution to be concentrated in D_2 before being discharged into the atmosphere.

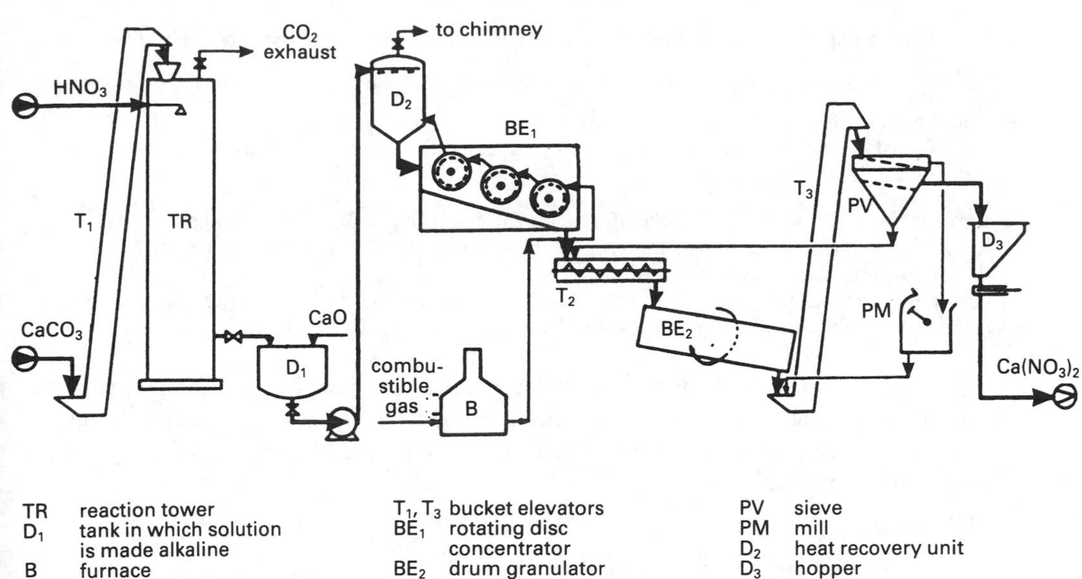

Fig. 11.6—Plant for the production of calcium nitrate. The salt is concentrated by direct and indirect heat transfer from gases.

The product discharges from the concentrator as a paste and is passed to the crystallization unit T_2 and then to the drum-type mixer–granulator BE_2. From here, the product is sent by means of the bucket elevator T_3 into the vibrating sieve PV which provides a means of recycling the finest product particles to the crystallization unit, the particles which are too large to the mill PM, and the particles with the correct grain sizes to the hopper D_3. The crushed material also finishes up in the hopper D_3 because the mill discharges its product into the screw conveyor T_2.

Nowadays, calcium nitrate is stored in plastic sacks because it is hygroscopic and even deliquescent. This also makes it necessary to work in an air-conditioned environment at the production site where the granulated material comes into contact with the atmosphere.

Various improvements have been made to certain of the production units in the case of the process illustrated above. More precisely, these are:

- the solution emerging from the tank where it is rendered alkaline is concentrated up to 85–90% by evaporation under vacuum and subsequently granulated in a prilling tower;
- before granulating the product, 3–5% of ammonium nitrate is added to the product to facilitate the operation and to obtain a product which is richer in nitrogen.

Further on (page 413), it will be seen that a method exists in which calcium nitrate is produced at the same time as 'nitrophosphate'.

Sodium nitrate

On account of its ready availability as it exists (or rather existed) in natural deposits, sodium nitrate is the oldest large-scale nitrate fertilizer. These natural deposits are (were) found especially in Chile but also in Mexico, the United States, Egypt, and Australia.

The main process for the working of natural nitrates, which contain up to 70% of $NaNO_3$ and NaCl and Na_2SO_4 as the principal gangues, is of the fractional crystallization type. It comprises the following phases:

* the crushing of the rock which is classified according to its $NaNO_3$ content and particle size;
* leaching out of the granulated material with water at 40°C which leads especially to the dissolving out of the sodium nitrate;
* cooling, decantation, filtration, and concentration until a solution with a density of 1.55 is obtained which, when crystallised, yields $NaNO_3$ with a purity of about 98.5%.

The industrial process which is now most used for the production of sodium nitrate is that which uses the reaction between 53% nitric acid and caustic soda, followed by separation of the product until it is crystalline. A diagram of the plant for this operation is shown in Fig. 11.7.

The reaction:

$$HNO_3 + NaOH \rightarrow NaNO_3 + H_2O$$

is carried out in the reactor R, and the resulting solution, having received the recycled mother liquors, is preheated in E_1 and undergoes evaporation, first in the concentrator E_2 and then E_3. This second concentrator operates under a reduced pressure which is maintained by the vacuum pump G_2. The concentrated solution deposits crystals upon cooling in CR. These crystals are separated in the screw conveyor centrifuge PC from which the recycled mother liquors are obtained. Finally, the sodium nitrate which has been given the finishing treatment in the drier BE is sent to be stored.

The use of sodium nitrate as a fertilizer is retarded both by the poor compatibility of sodium ions with plant physiology and by its relatively high cost.

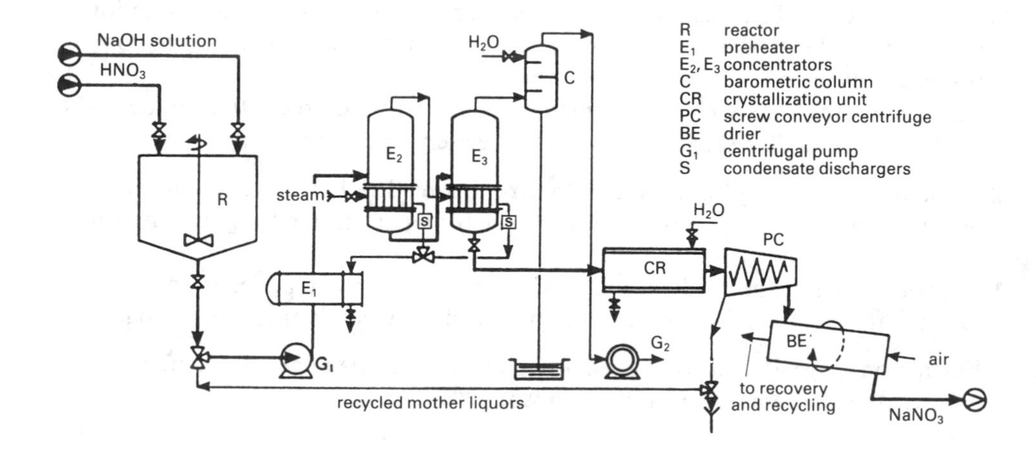

Fig. 11.7—Plant for the production of sodium nitrate. One of the possible solutions is shown here. Other solutions to this synthetic problem start off by using Na_2CO_3 or NaCl on the one hand and the oxides of nitrogen (apart from HNO_3) on the other hand.

11.5 PHOSPHATE FERTILIZERS

Phosphorus is an essential element from a biological point of view because a large number of metabolic reactions involved in the synthesis of carbohydrates in plants and the production of biological energy occur via the intercession of phosphate esters.

Moreover, phosphorus acts as a part of important lipid systems (lecithins), of enzymatic protein systems, of nucleic acids, and of the supporting tissues.

Nevertheless, neither the apatites nor bones as such can be made use of as phosphate fertilizers with a rapid effect on soils as they are not soluble either in water or in weak acids.

Nowadays, however, phosphate powders of the finest structure are also in fact used as fertilizers by exploiting their fairly rapid solubilization which is brought about by the acid–bacterial agents in the soil.

These phosphate powders are prepared by the crushing of phosphate minerals or bone, grinding them, and regrinding those fractions which are not very fine after having enriched them, when necessary, by flotation. The product is finally dried and stored in silos.

Industry is responsible for the conversion of natural phosphates into soluble chemical compounds for agricultural use, taking account of the fact that the phosphates of calcium can be classified according to their solubilities:

- *calcium dihydrogen phosphate* $Ca(H_2PO_4)_2$ is completely soluble in water and can therefore be rapidly assimilated by plants;
- *calcium hydrogen phosphate* $CaHPO_4$ is soluble in weak acids* and can therefore be only slowly assimilated by plants;
- *tricalcium phosphate* $Ca_3(PO_4)_2$ is insoluble in water and in weak acids and can be assimilated only extremely slowly by plants.

Simple superphosphate

The most used phosphate fertilizer consists of a mixture of calcium dihydrogen phosphate monohydrate and calcium sulphate dihydrate. This mixture is referred to as simple superphosphate or 'perphosphate'.

The cycle appertaining to the production of superphosphate includes the crushing and grinding of the raw material in ball mills and its subsequent treatment with about 65% sulphuric acid ('lead chamber acid'). By operating in this way, the products of the reaction, which may be schematically depicted as:

$$Ca_3(PO_4)_2 + 2H_2SO_4 + 5H_2O \rightarrow Ca(H_2PO_4)_2 \cdot H_2O + 2CaSO_4 \cdot 2H_2O \quad (11.3)$$

can be hydrated. Actually, the theoretical content of water-soluble P_2O_5 in simple superphosphate is 22%. However, this figure may drop to 13%, but is commonly around 16–18%.

There are various causes for this reduction in the quality of the product such as the occurrence

*The solubility of this compound is referred in the standard way to ammonium citrate which has been prepared by dissolving 400 g of crystallized citric acid in water, saturating the cold solution with $d_4^{20} = 0.92$ ammonia until it gives a neutral reaction, allowing it to cool down again, and, finally, making the resulting solution up to a volume of 1 litre of with water.

of the following reaction:

$$Ca_3(PO_4)_2 + H_2SO_4 + 2H_2O \rightarrow 2CaHPO_4 + CaSO_4 \cdot 2H_2O \qquad (11.4)$$

and the phenomenon known as the 'retrogradation of perphosphates' caused by the presence of Al_2O_3 and Fe_2O_3 in the raw materials. These oxides first react with H_2SO_4 to yield $Al_2(SO_4)_3$ and $Fe_2(SO_4)_3$ and then (indicating Al or Fe by Me) participate in the reaction:

$$Me_2(SO_4)_3 + Ca(H_2PO_4)_2 \rightarrow CaSO_4 + 2H_2SO_4 + 2MePO_4$$

thereby producing insoluble phosphates.

This retrogradation takes place especially during the storage of the superphosphates and even in the fertilized land. Nevertheless, on account of the high degree of subdivision of the insoluble phosphates which are formed, the P_2O_5 content of the latter can be slowly assimilated by plants in an analogous manner to the assimilation of the calcium hydrogen phosphate, which is soluble in weak acids, produced in reaction (11.4).

The superphosphates are generally manufactured in continuously operating plants, a standard example of which is shown in Fig. 11.8.

The first section of this plant consists of a crushing unit PM_1, a mill PM_2, a sieve PV, and two devices for adding measured amounts of the raw materials. The central part of the plant is made up from a cylindrical mixer–reactor RM connected to the cyclone unit DC in which any solids transported by the gases are removed, the gases themselves being recovered. This especially applies to HF and SiF_4 (page 359). There is also a series of chambers fitted with screw conveyor units and arranged in a cascade which are responsible for slowly removing the mass obtained from RM until, some hours later, it is discharged into the top of the last unit in the series T_n. In these 'prematuration' chambers the temperature rises up to 120°C and the reaction involving the attack of the sulphuric acid on the raw material is thereby completed, and the phosphoric acid formed in part in the reactor RM is recombined according to the reaction:

$$4H_3PO_4 + Ca_3(PO_4)_2 + 3H_2O \rightarrow 3Ca(H_2PO_4)_2 \cdot H_2O \qquad (11.5)$$

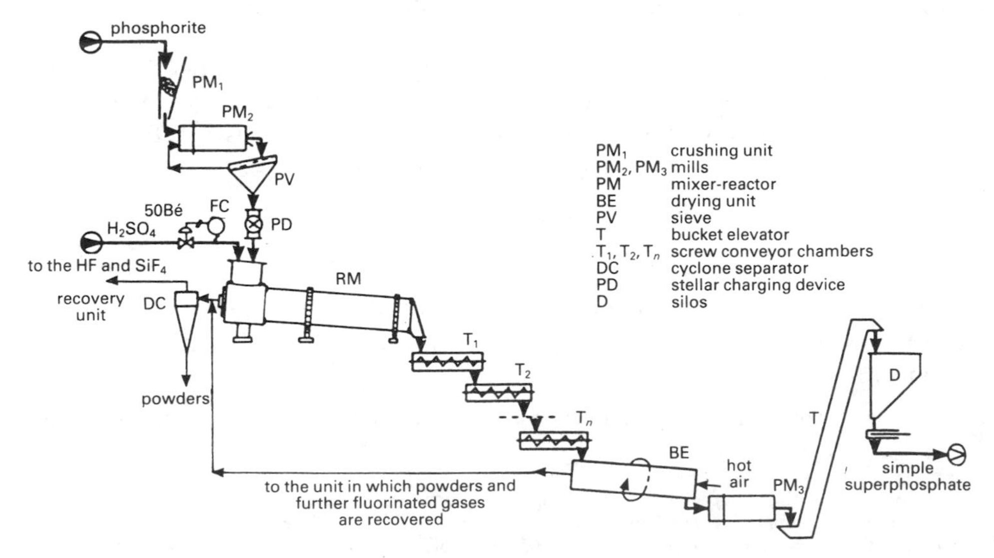

Fig. 11.8—Diagram of a plant for the production of simple superphosphates (perphosphates).

Finally, the product is dried in BE, ground up in PM_3, and stored for further maturation in a compartment of the silos D. The perphosphate is subsequently taken out of the silos, milled, granulated, and dried to a point which prevents it from caking after having been put into sacks.

The granular product is therefore the commonest commercial form of superphosphate, but it is also marketed in other small pieces which are more or less porous. Furthermore, it is also supplied and used in the form of an extremely fine powder to ensure its assimilation which is almost free from the phenomenon of retrogradation.

Double and triple superphosphates

By proceeding as in the manufacture of normal ('simple') superphosphates but also adding H_3PO_4, together with sulphuric acid, during the attack stage in an amount which is inversely proportional to the P_2O_5 content in the starting phosphate raw materials, superphosphate products are obtained which contain 32–35% of P_2O_5 and are soluble in water via reactions (11.3) and (11.5). These phosphate fertilizers are known as 'double superphosphates'.

By using suitable manufacturing plants and attacking phosphate-rich raw materials simply with H_3PO_4, a type of superphosphate which contains up to 48% of P_2O_5 and is soluble in water is obtained via reaction (11.5). This is commercially known as 'triple superphosphate'.

In practice, two types of process can be used for the manufacture of triple phosphates:

- plants in which the phosphate mineral is attacked with dilute phosphoric acid and the water is eliminated in the maturation and drying phases;
- plants which use concentrated phosphoric acid to avoid the need for subsequent drying.

The first method is the most widely used as it is cheaper overall owing to the fact that the H_3PO_4, which is impure as well as dilute, can be used, and this can be obtained by the attack of sulphuric acid on mineral phosphates rather than having to produce the phosphoric acid thermally, starting out from phosphorus.

When the dilute acid method is used it is necessary to prime the reaction which is carried out in multiple reactors by the injection of live steam and then blowing in air which contributes to the remixing of the reaction mass and the control of the heat of reaction while also favouring the liberation of the reaction gases. A diagram of the plant is shown in Fig. 11.9

It is seen that the mass is removed from the final reactor by means of a screw conveyor system (the use of which is made possible by the fact that the mass is thickened by the addition of powders and 'fines' discharged from the cyclone separator and the base of the vibrating sieve respectively). The reaction product is then treated in a hot-air drying unit BE which produces non-agglomerated triple superphosphate which is sent for storage.

Commercial double and triple phosphates are most commonly marketed in a granular form, but they are also sometimes produced in the form of small pieces with a porous structure. They are all normally sold in plastic sacks.

11.6 POTASSIUM FERTILIZERS

The by-products from the treatment of many plants constitute some of the traditional

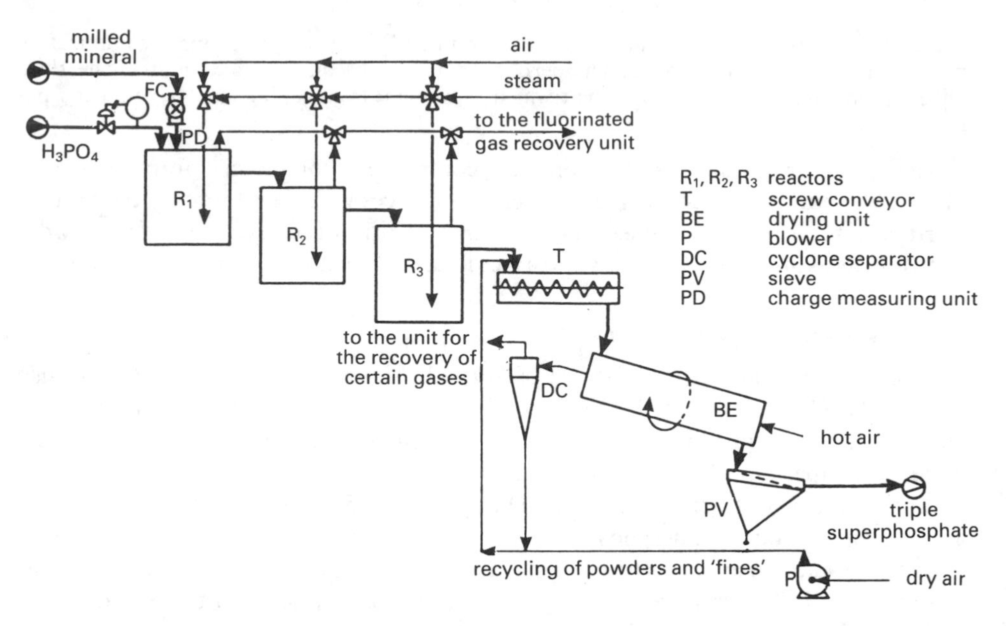

Fig. 11.9—Diagram of a plant for the production of triple superphosphate by the dilute phosphoric acid process.

mixtures of potassium fertilizers. Among these are:

- the potassium salts originating either from the alcoholic fermentation of molasses or during the removal of sugar by dialysis from molasses itself;
- the tartars obtained from the dregs of pressed grapes and from the crust and sludge in wine barrels;
- the ashes obtained from the burning of plant matter which contain an average of 16% of potassium salts, mainly in the form of K_2CO_3 and K_2SO_4.

A mixed potassium fertilizer is also the product of the very fine milling of the richest leuchitic rock (leuchite $KAl(SiO_3)_2$) from central northern Italy. By crushing and selection or hydraulic enrichment of various types, materials are produced which contain 90–95% of leucite (about 18% of K_2O).

Nowadays, however, there is a tendency to use the salts KCl and K_2SO_4 or concentrates of them as potassium fertilizers. For the most part, these are obtained from natural potassium minerals which are principally:

- *carnallite*: $KCl \cdot MgCl_2 \cdot 6H_2O$;
- *sylvinite*: a mixture of rock salt (NaCl) and sylvite (KCl) in a ratio of about 7:3;
- *hard salt* (Hartsalz): a mixture of carnallite, sylvinite and kieserite ($MgSO_4 \cdot H_2O$);
- *kainite*: $KCl \cdot MgSO_4 \cdot 3H_2O$.

Other sources of potassium salts which are used in industry in general and as fertilizers in particular, especially in highly technologically developed countries (e.g. Israel) are the mother liquors of the salt marshes and natural salt water basins such as the Dead Sea and the Great Salt Lake.

Potassium chloride

Highly concentrated potassium chloride is obtained especially from sylvinite and then from Hartsalz and carnallite.

Nowadays, the sylvite (KCl) concentrates for agricultural use tend to be ever-increasingly prepared from sylvinite or from Hartsalz by flotation in multiple cells using special additives which specifically make the KCl crystals hydrophobic so that they are carried upwards to float occluded in the froth. A substance containing 92–96% of potassium chloride can be obtained in this way which is also a good product for use in the potassium salt industry (page 419).

A second method for their preparation corresponds to the scheme, with reference to sylvinite, shown in the flow scheme presented in Fig. 11.10.

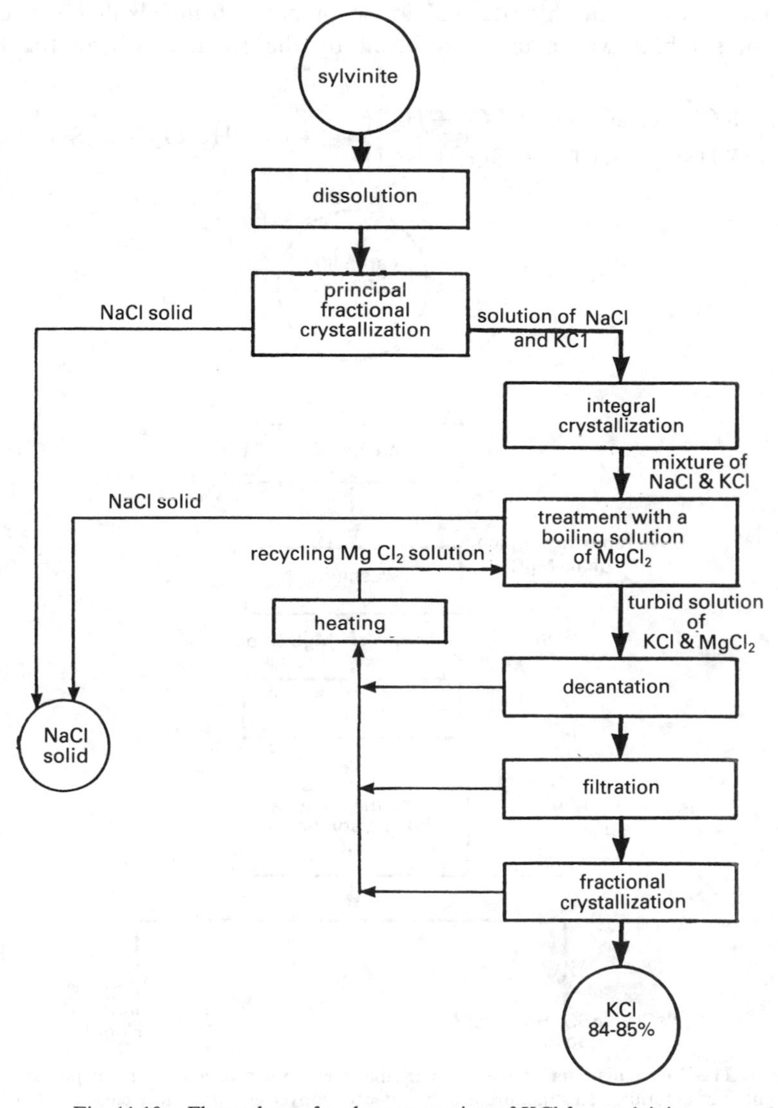

Fig. 11.10—Flow scheme for the preparation of KCl from sylvinite.

The block diagram of Fig. 11.11 shows the earlier stages of a plant for the recovery of potassium chloride from carnallite which is always accompanied by rock salt, kieserite, and anhydrite ($CaSO_4$).

This diagram stops at the stage where a turbid solution of KCl and $MgCl_2$ is produced because, from this point on, the remainder of the process is the same as that shown in Fig. 11.10.

Finally, potassium chloride fertilizer can also be obtained from Hartsalz by using a treatment which is parallel to that shown for carnallite.

Potassium sulphate

Potassium sulphate is the best available potassium fertilizer as it is more compatible with plant physiology than the corresponding chloride.

So long as hydrogen chloride had to be prepared from alkali metal chlorides, potassium sulphate was mainly produced by the route involving the following reactions:

$$\left.\begin{array}{l} KCl + H_2SO_4 \rightarrow KHSO_4 + HCl \\ KHSO_4 + KCl \rightarrow K_2SO_4 + HCl \end{array}\right\} \rightarrow 2KCl + H_2SO_4 \rightarrow K_2SO_4 + 2HCl$$

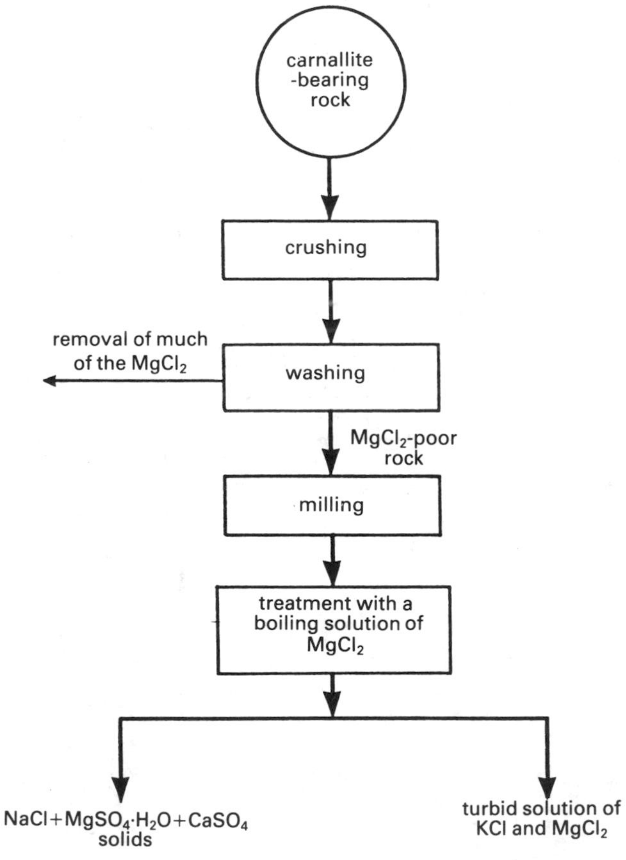

Fig. 11.11—Block diagram of the steps in the preparation of a mixture of potassium and magnesium chlorides. This mixture is subsequently converted into highly concentrated KCl in the manner which has already been shown in Fig. 11.10.

The plant is operated in the same way as that used in the analogous preparation of Na_2SO_4 which has already been described (page 215).

Nowadays, it is preferably prepared by:

- a double decomposition reaction with the calcium salt:

$$2KCl + CaSO_4 \rightarrow K_2SO_4 + CaCl_2$$

  at a low temperature ($\simeq 0°C$) and in the presence of ammonia which facilitates the crystallization of the $CaSO_4$;
- roasting a mixture of KCl and $(NH_4)_2SO_4$:

$$2KCl + (NH_4)_2SO_4 \rightarrow K_2SO_4 + 2NH_3 + 2HCl$$

  which leads to ammonium chloride as the waste product due to the combination of the vapours which are liberated during the roasting;
- a special process for the exploitation of kainite or, better still, kainite rock.

The method developed in the sixties by the ex-Montecatini to exploit the product from the mines at S. Cataldo (Sicily) in the plant at Campofranco (AG) uses kainitic rock as the starting material.

The mineral is first floated (Fig. 11.2) to separate the greater part of the NaCl which it contains whereupon 'commercial kainite' is obtained.

A first leaching is therefore undertaken at 15°C, using the waters from the subsequent leaching stage. Here, the following double exchange reactions take place during this leaching

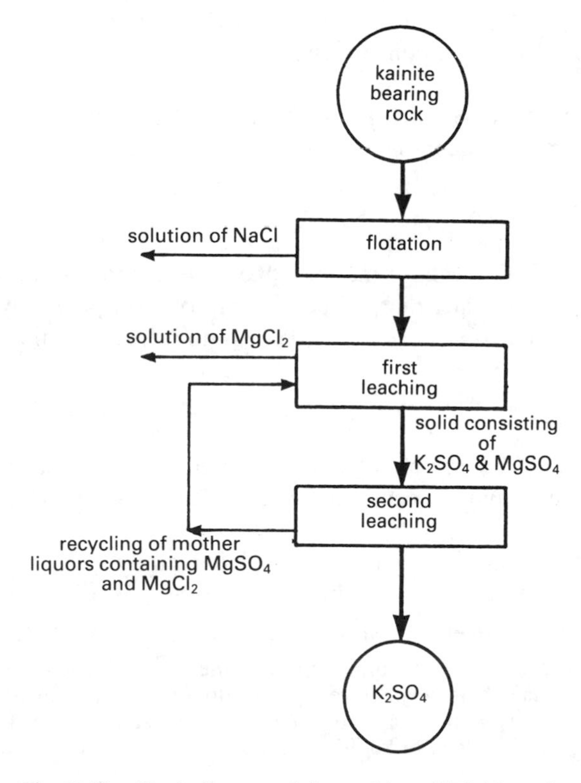

Fig. 11.12—Block diagram of the working of kainitic rock.

(the solubilities of the products in g/l are indicated below the formulae):

$$2KCl \cdot MgSO_4 \cdot 3H_2O \rightarrow \underset{111.5}{K_2SO_4} + \underset{542.5}{MgCl_2} + \underset{356}{MgSO_4} + 6H_2O$$

On the basis of these solubility values it can be understood how the relatively high solubility of magnesium salts as compared to the solubility of K_2SO_4 is exploited in the process.

A second leaching with pure water is carried out at 50°C on the residue from the first leaching which is a solid, low in $MgCl_2$. In this way a residue of high-purity K_2SO_4 is obtained which is filtered and dried.

On account of the purity of the rock found at S. Cataldo and, especially, its low NaCl content, three different grades of potassium sulphate are produced at Campofranco: 50–52% K_2O (about 96% K_2SO_4), 48–50% K_2O, and 38–40% K_2O.

11.7 MIXED FERTILIZERS

The convenience of using fertilizers which may bring several elements to the soil which promote the development of plants was intuitively recognized on the basis of the composition and effects of natural animal excrement manures, and was confirmed after it had been demonstrated that the abundance in the soil of a single fertilizing element leads to a waste of this element rather than a major assimilation of it by the plants. In other words, a dearth of a certain fertilizing element also hinders the assimilation of the others. Besides making savings in packaging materials, transport, and storage, the use of mixed fertilizers has the advantage that they can be applied in the form of a single solution, and, by measuring out and mixing the optimal amounts of the individual components, the mixed fertilizer promotes synergy between the components themselves.

As mentioned on page 385, the discussion is divided into a section on compound fertilizers and a section on complex fertilizers.

Compound fertilizers

The compound fertilizers are those endowed at a molecular level with several fertilizing elements allowing for the fact that, while phosphorus can play only an anionic role and potassium a cationic role, nitrogen can play both roles in compound fertilizers.

Potassium nitrate

Potassium nitrate, KNO_3, which is traditionally known as saltpetre or nitre, is quite widespread in nature, and it occurs in Bengal, Ceylon, China, Iran, Bolivia, Peru, and elsewhere.

The amount of KNO_3 in the natural deposits rarely exceeds 5%, the average content being 2.5%. An exception to this, however, is the Indian deposits which have a high KNO_3 content (of up to 30%).

To make use of the various natural deposits, the crushed rock is lixiviated, thereby obtaining solutions which, when crystallized, yield saltpetre contaminated with other nitrates and especially extraneous sodium salts (the chloride and sulphate). Subsequent fractional crystallization processes can lead to 90–95% KNO_3.

The practical method used to supply the agricultural and other industries with this salt for more than a century has been 'double decomposition' between $NaNO_3$ and KCl. This method was developed at the time when much use was made of Chile saltpetre, and, today, the production of potassium nitrate by exchange has been part replaced either by the method using the neutralization of nitric acid with potashes (K_2CO_3 and KOH), which is analogous to the method employed for the preparation of $NaNO_3$, or by other special processes such as that which makes use of the reaction:

$$KCl + HNO_3 \rightarrow KNO_3 + HCl$$

This is carried out in pentanol which, being a low polarity organic solvent, does not dissolve the ionic compound KNO_3 but does dissolve the polar covalent hydrogen chloride. The HCl is extracted with water as hydrochloric acid from the pentanol after the nitre has been filtered off.

Israel Mining Industries (IMI), who retain the patent on this process, subsequently use the hydrochloric acid, which is recovered, for attacking phosphate minerals in the production of phosphoric acid.

The diagram of the plant for the production of KNO_3 shown in Fig. 11.13 by the classical double decomposition process is essentially based on the reaction:

$$NaNO_3 + KCl \rightarrow KNO_3 + NaCl$$

To understand the mechanism and the consequent implementation of this process, it is necessary to consider the solubilities in gram per litre (g/l), when hot and in the

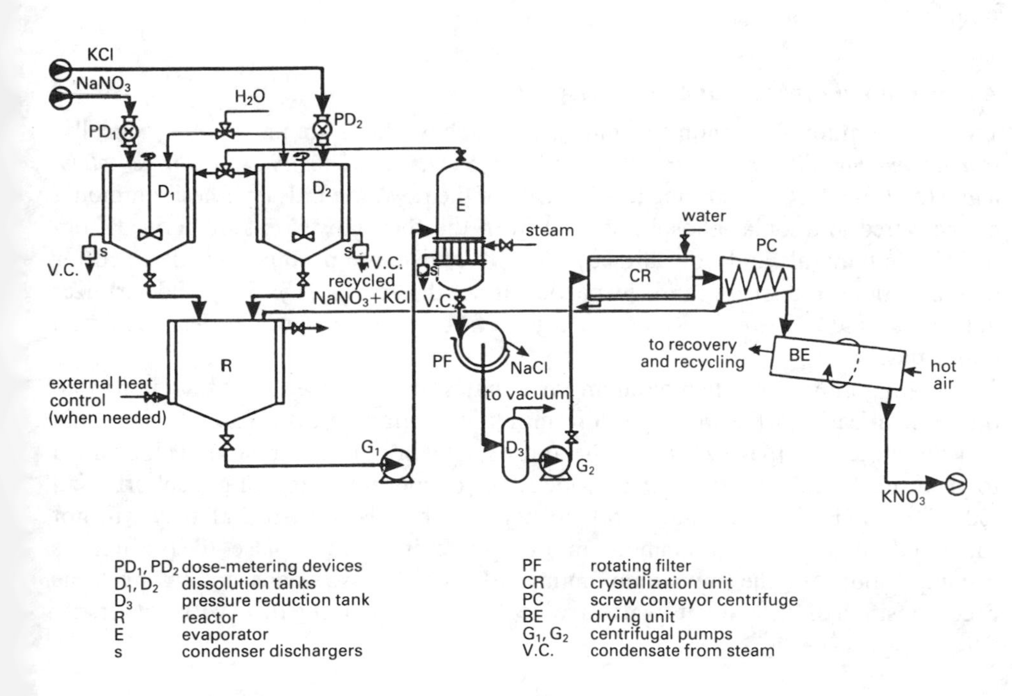

Fig. 11.13—Diagram of a plant for the production of KNO_3 by double decomposition.

cold, of the four salts in the reaction:

	$NaNO_3$	KCl	KNO_3	$NaCl$
at 100°C	1760	365	2460	396
at 20°C	800	340	320	360

It can be seen from these values that, in so far as the reaction products are concerned, KNO_3 has a high solubility when hot and that NaCl has approximately the same solubility when it is hot as when it is cold. NaCl therefore precipitates from hot solutions from which it is practically impossible to separate KNO_3.

The production plant consists of two units D_1 and D_2 with steam jackets heated by the steam coming from the evaporator E. These are used to dissolve up the reagents. A slight excess of $NaNO_3$ is used to exploit the common ion effect in the precipitation of NaCl.

The reaction is carried out in R from where the mixture is pumped into the evaporator E where, after the solution has become concentrated, NaCl precipitates out from the hot solution until the amount of it remaining dissolved is practically equal to the amount of it which would be soluble in the cold. The mixture is filtered and the solution is then sent to the crystallization unit CR which operates at between 5 and 25°C in a manner such as to precipitate the saltpetre which has now been made less soluble. The precipitate is centrifuged in PC from where the mother liquors emerge to be recycled. Meanwhile, the KNO_3 crystals which have been dried in BE are sent away to be stored.

Saltpetre is a relatively expensive fertilizer which is used on selected crops such as tobacco, sugar beet, and flowers. There are, nevertheless, many industries which require considerable quantities of it. These cover a range from the food industry to pharmaceuticals and from metallurgy to industrial organic chemistry. Furthermore, potassium nitrate is a raw material used in fireworks, pyrotechnic powders, and other explosives.

Ammonium phosphates and polyphosphates

Two salts containing ammoniacal nitrogen and phosphorus are produced practically: *monoammonium dihydrogen phosphate*, $NH_4H_2PO_4$, and *diammonium hydrogen phosphate* $(NH_4)_2HPO_4$. According to the purity of the raw materials (synthesis ammonia or recovered ammonia, phosphoric acid from the thermal process or derived from phosphate minerals and not purified), the two phosphate products are destined for use in various industries, and, in particular to the food (page 361) and fertilizer industries where there is a wide margin of tolerance regarding their common impurities.

The ammonium and diammonium phosphates are produced in a similar way to that which leads to the formation of ammonium sulphate from gaseous ammonia and sulphuric acid (page 219) with the appropriate amount of ammonia being added to yield the desired salt and using a wide range of concentrations of phosphoric acid (30–75%). It is then necessary both to dry the crystals well so that they will not cake and to store the diammonium phosphate in a cool place to prevent its decomposition to the monoammonium salt which also leads to caking. The decomposition of the two salts is also avoided by carefully drying them with hot air.

* * *

Other ammonium phosphate fertilizers are obtained by reacting ammonia with polyphosphoric acid. The general preparative reaction is:

$$(n+2)NH_3 + H_{n+2}P_nO_{3n+1} \rightarrow (NH_4)_{n+2}P_nO_{3n+1}$$

For example, if $n = 3$, $(NH_4)_5P_3O_{10}$, ammonium tripolyphosphate, is obtained. It is seen that the nitrogen/phosphorus (N/P) ratio of the ammonium polyphosphates is high.

A plant for the production of these fertilizers is shown in Fig. 11.14. It is seen that metered amounts of more or less concentrated polyphosphoric acid and gaseous ammonia which is bubbled into the solution are fed into the reactor. Polyphosphates of varying concentration, whether they are to be crystallized or not, are discharged from the bottom of the reactor which is thermostatted at 25°C and equipped with a mixing device. When the polyphosphates are to be crystallized the mother liquors are recycled.

The solutions of polyphosphates can be stored and subsequently used by scattering them over the land. Aircraft may also be employed for this spraying operation. Alternatively, the crystals which have been thoroughly dried with hot air, thereby eliminating every trace of moisture to prevent caking, are stored.

Potassium phosphates

The high solubility of KCl and KNO_3, which means that there is a danger of their being leached out from soils, presents a considerable obstacle to the use of these fertilizers. Apart from bringing a second fertilizing element to the land and not exhibiting the contra-indications of chlorides toward certain plants, potassium phosphates partly remedy this situation because they are less soluble than the abovementioned salts.

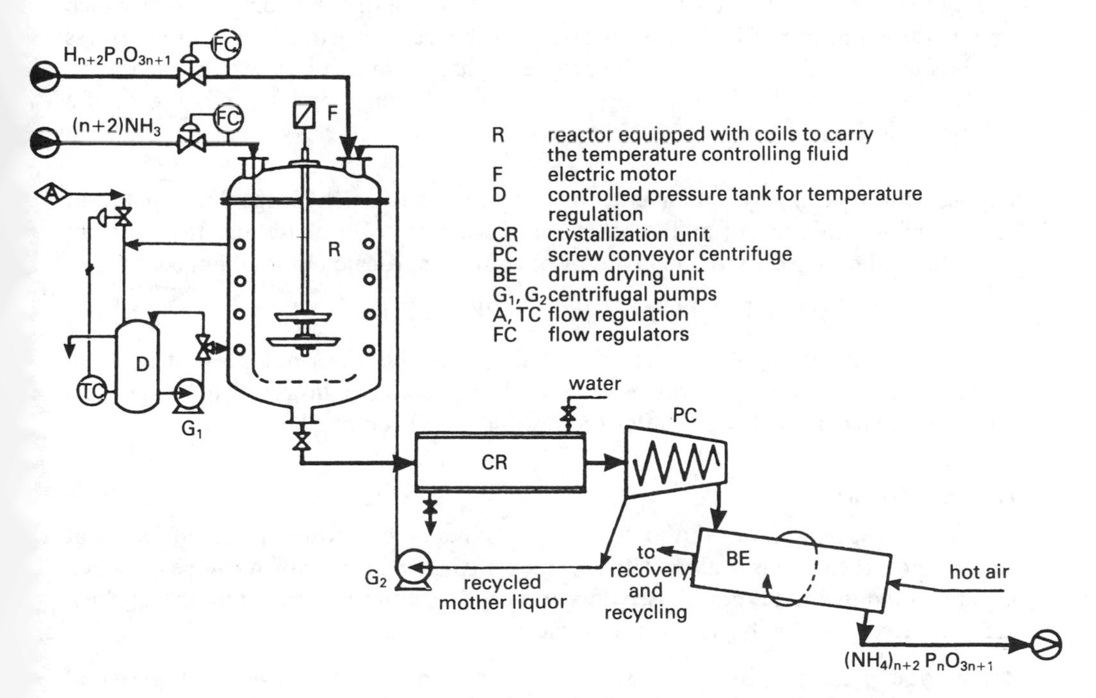

Fig. 11.14—Diagram of a plant for the production of ammonium polyphosphate.

The phosphates of potassium which are used in practice as fertilizers are: *potassium metaphosphate* KPO_3, *monopotassium dihydrogen phosphate* KH_2PO_4 and *dipotassium calcium pyrophosphate* $K_2CaP_2O_7$. These fertilizers are normally marketed in the crystalline state.

Potassium metaphosphate. To produce this compound, the reactions which must be employed are either:

$$KCl + H_3PO_4 \rightarrow KH_2PO_4 + HCl$$

or:

$$2KOH + H_2O + P_2O_5 \rightarrow 2KH_2PO_4$$

$$2KCl + 3H_2O + P_2O_5 \rightarrow 2KH_2PO_4 + 2HCl$$

When concentrating the solutions obtained in this way until they have the consistency of a paste, the dehydration of the monopotassium hydrogen phosphate starts:

$$KH_2PO_4 \rightarrow H_2O + KPO_3$$

and is subsequently completed by the treatment of the fine mass in a rotating furnace at 500°C. In this way the monopotassium dihydrogen phosphate (is first completely dried and then subsequently undergoes dehydrating calcination until it is all converted to KPO_3.

Monopotassium phosphate. This compound, as has just been seen, is the product of the first phase of the preparation of potassium metaphosphate. However, given that, during the whole process, there is no chance of attaining the high temperatures which favour the evolution of HCl, it is necessary in this case to work first with an excess of phosphoric acid or P_2O_5 and then to neutralize them suitably with potash.

Monopotassium dihydrogen phosphate is also sold commercially in the form of a solution which is generally very concentrated.

Dipotassium calcium pyrophosphate. By calcining calcium hydrogen phosphate at 1200°C with a mixture of monopotassium dihydrogen phosphate and tripotassium phosphate, the mixed pyrophosphate of potassium and calcium is obtained:

$$2CaHPO_4 + KH_2PO_4 + K_3PO_4 \rightarrow 2K_2CaP_2O_7 + 2H_2O$$

During calcination the mass partly vitrifies, but this does not prevent its total transformation into the salt which is required. It is possible, however, to extract any unreacted material with water after the product has been crushed.

Complex fertilizers

It will be recalled that, if the fertilizing elements constitute parts of different compounds, then one is dealing with complex mixed fertilizers which can be classified into two fundamental types: those which are produced mechanically and those which are produced chemically. These are, in fact:

- the type produced by the mixing of components which have been prepared separately;

- the type produced by the simultaneous interaction of suitable reagents at the preparation site.

All the finished products of both types are described by a name which sometimes reflects their chemical composition, but, mostly, it is conventional and quite general simply to state the grade, which is expressed as mention earlier (page 385).

An extremely wide range of these fertilizers is available nowadays, and, even if they are often not very different, the ways of obtaining them within the framework of each preparative technology are extremely varied. Consequently, it is not possible to present substantially summarized details concerning each type.

Mixture of simple fertilizers

Complex NPK fertilizers are prepared by the dry mixing of compounds containing the three fertilizing elements.

A complete account of three formulations of one of the best known and widely used types ('10-10-10') of these fertilizers is shown in Table 11.3.

Other triple (NPK) complex fertilizers are obtained by the dry mixing of ammonium phosphates and potassium salts.

* * *

Table 11.3—Possible alternative formulations for a '10-10-10' complex NPK fertilizer

Formulation	*Components*	*Composition by weight (kg/100 kg)*	*% of main standard fertilizers*	*Calculation of the kg/100 kg of principal fertilizers*	*Type of fertilizer*
1	ammonium nitrate	29.83	33.5(N)	29.83 × 0.335	N 10
	normal superphosphate	53.5	18.69(P_2O_5)	53.5 × 0.1869	P 10
	potassium salt	16.67	60(K_2O)	16.67 × 0.60	K 10
2	ammonium sulphate	39.02	20.5(N) }	39.02 × 0.205 +	N 10
	urea	4.46	45(N) }	4.46 × 0.45	
	triple superphosphate	9.7	45(P_2O_5) }	9.7 × 0.45 +	P 10
	normal superphosphate	30.15	18.69(P_2O_5) }	30.15 × 0.1869	
	potassium salt	16.67	60(K_2O)	16.67 × 0.60	K 10
3	ammonium sulphate	13.973	20.5(B) }	13.973 × 0.205 +	N 10
	urea	15.857	45(N) }	15.875 × 0.45	
	normal superphosphate	53.5	18.69(P_2O_5)	53.5 × 0.1869	P 10
	potassium salt	16.67	60(K_2O)	16.67 × 0.60	K 10

Table 11.4—Binary mixed fertilizers produced by the largest Italian companies in this field

Name	*Components*	*% of main standard fertilizers*
Fosfazoto M	ammonium sulphate and triple superphosphate	15 N and 32 P_2O_5
Fosfazoto A	ammonium sulphate, urea and triple superphosphate	10–20 N and 40 P_2O_5
Fosfazoto C	ammonium salts and normal superphosphate	7 N and 14 P_2O_5

Some examples of mixed binary fertilizers are shown in Table 11.4 where the name refers, as is often done, to the company which has produced them (M = Montedison, A = ANIC, C = Caffaro).

It is noted that there is a tendency not to produce complex mixed binary fertilizers containing potassium as a component, since, in most cases, this element is efficacious only in the presence of the other two fertilizing elements, and because, in specific cases, the fertilizer compounds, potassium nitrate and potassium phosphate, are convenient to prepare and lend themselves well to such use.

Products of crossed reactions

Among the complex mixed fertilizers which are obtained, containing two or three components, by a crossed reaction involving various reagents (NH_3, H_3PO_4, $Ca(H_2PO_4)_2$, K_2SO_4, etc.), we shall mention only those products arising from the 'ammoniation' of superphosphates and the formation of homogeneous granular mixtures of ammonium and potassium salts and nitrophosphates.

In its simplest form ammoniation (and there are, in fact, many operational variants of this procedure which also involve the intervention of other components) consists of the addition of ammonia to superphosphate:

$$Ca(H_2PO_4)_2 + NH_3 \rightarrow CaHPO_4 + NH_4H_2PO_4$$

As a consequence of this reaction, monocalcium dihydrogen phosphate is converted into calcium hydrogen phosphate which is soluble in ammonium citrate and, practically, in a weakly acidic medium.

The use of excess ammonia is to be avoided so as not to promote reactions which can lead to a loss of fertilizing power owing to the formation of tricalcium phosphate as, for example, occurs in the following reaction:

$$3CaHPO_4 + NH_3 \rightarrow Ca_3(PO_4)_2 + NH_4H_2PO_4$$

This leads to the conversion of calcium hydrogen phosphate which is soluble in citric acid to the completely insoluble tricalcium phosphate.

Ammoniation can be carried out in rotating cylinders (Fig. 11.15) in which solutions of ammonia or ammonium salts are sprayed onto granular superphosphate. Suitable

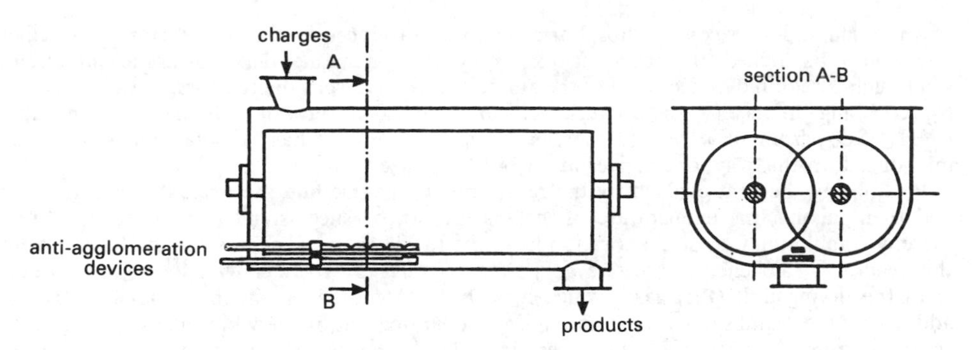

Fig. 11.15—Plan and section views of a plant used in the ammoniation process.

devices prevent the granules from agglomerating, while other devices break up any lumps which may have been formed.

The type of fertilizer produced by an ammoniation process is declared by stating the 'degree of ammoniation', that is, by indicating the amount in weight of NH_3 which has been absorbed by unit weight of P_2O_5 present in the superphosphate which has been treated.

In the case of a normal superphosphate, containing 20% P_2O_5, the degree of ammoniation is 0.2 or 0.3 which corresponds to 4 or 6 kg of NH_3 for every quintal (100 kg) of superphosphate.

* * *

The homogeneous granular mixtures of ammonium phosphates (and sulphates) and potassium salts are obtained by using plants for the type shown in Fig. 11.16.

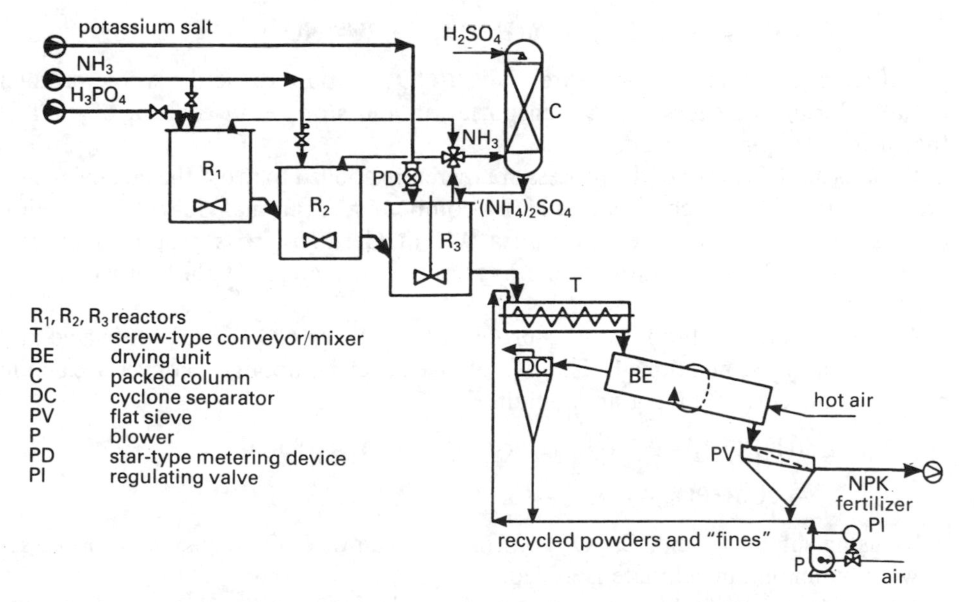

Fig. 11.16—Preparation of homogeneous granular mixtures of ammonium phosphates and potassium salts.

Ammonia and an excess of phosphoric acid are introduced into a first reactor R_1. Further ammonia is dispatched into the second reactor R_2 in such a manner that it forms the ammonium salt which is required. Since some of the ammonia escapes from the reactors, this is recovered by scrubbing the gases with sulphuric acid in a packed tower which discharges its product ($(NH_4)_2SO_4$) from its base into the third reactor R_3 into which a potassium salt is also introduced. In this way, one arrives at an NPK fertilizer.

By utilizing the heat evolved by the reactions and controlling the reagent concentrations and their mixing, the temperature of the system can be successfully maintained at 100°C, thereby stimulating intense evaporation from the third reactor. The mixture discharged from this reactor is thickened in the screw type mixer/conveyor T either with the solid products from the drying unit (BE) gases which have been separated out in the cyclone or by the addition of the crumbs of material and tailings separated out by the vibrating sieve. The paste emerging from the screw conveyor is sent into the drying unit BE where it is drained, dried and granulated. This granulated material is graded in the vibrating sieve PV and the 'fines' are discharged from the bottom while the material which does not pass through the sieve is of the correct size distribution and is sent for storage.

For example, a 12-12-12 granular fertilizer with NH_3, H_3PO_4, H_2SO_4, and KCl is prepared in this manner. The KCl, in particular, is added into the third reactor in a ratio of 20 kg of a product containing 60% K_2O to every hundred kilograms of the final fertilizer which the plant is programmed to produce.

* * *

The products obtained when nitric acid intervenes in the attack on phosphate minerals during the production of fertilizers are known industrially as 'nitrophosphates'. Monocalcium dihydrogen phosphate and calcium nitrate are formed at the beginning of the preparation of nitrophosphates:

$$Ca_3(PO_4)_2 + 4HNO_3 \rightarrow Ca(H_2PO_4)_2 + 2Ca(NO_3)_2$$

If it were to be left in the mixture, the $Ca(NO_3)_2$ would cause the whole product to cake during the course of its transportation and storage on account of the fact that it is very hygroscopic.

Various modifications of the process are therefore applied to avoid the inconvenience arising from the hygroscopic nature of this compound. A rational, but costly, solution to this problem is that used by Lonza AG (a German–Swiss group). This uses conversion of $Ca(NO_3)_2$ into its less hygroscopic trihydrate at the site where it is crystallized.

Among the other solutions to this problem those using CO_2 or sulphuric acid and ammonia may be mentioned. The use of the latter compounds converts calcium nitrate to the corresponding ammonium salt:

$$Ca(H_2PO_4)_2 + 2Ca(NO_3)_2 + 4NH_3 + CO_2 + H_2O$$
$$\rightarrow 2CaHPO_4 + 4NH_4 + CaCO_3$$

By using sulphuric acid, $CaSO_4$ is formed instead of $CaCO_3$, just as in processes in which ammonium sulphate is added.

All of the nitrophosphate fertilizers prepared in the manner which has just been described are of the binary NP type.

Ternary NPK fertilizers are obtained by allowing potassium sulphate to intervene in the calcium nitrate conversion reaction and ammonium sulphate to control the potassium content in the fertilizer.

The extremely soluble nitrophosphates, which contain both phosphorus and nitrogen in the form of nitrate, are some of the energetic* supporters of the excessive disordered growth and multiplication of aquatic plants, that is, they promote the phenomenon of eutrophication.

As a result of a partial leaching out of soils by rain waters nitrophosphates arrive in semistagnant water bodies (marshes, lakes, and ponds) where they accumulate in excessive concentrations with respect to the normal metabolic equilibrium of the plants.

As a consequence of this, the aqueous environment progressively contains less and less dissolved oxygen on account of cyclic reproduction interspersed with the putrefaction of plant matter until every possibility of aerobic life is extinguished.

*Energetic, because the two most important fertilizing elements are carried to the water body in a state where they are completely soluble and combined in a manner whereby they can be immediately assimilated.

12

Sodium, potassium, and related compounds

The reason for this specific treatment of the industrial role of sodium, potassium, and related compounds is the remarkable chemical and technological importance of these metals, and, even more so, of certain of their compounds. Moreover, they stand out sharply, in every field, regarding the uses to which other common metals are put and those to which their eventual products of combination are put.

12.1 THE MAJOR NATURAL COMPOUNDS OF SODIUM AND POTASSIUM

A rational study of this subject requires a preliminary treatment of the natural compounds from which the two alkali metals, sodium and potassium, and all of their derivatives are obtained either directly or indirectly. These two natural compounds are *sodium chloride* and *potassium chloride*.

Sodium chloride

The compound NaCl is indubitably the most important of all the salts on account of the uses which are made of it in preservation techniques, in practical zootechnics, and in foodstuffs intended for human consumption, apart from the fact that it is the basis of the sodium and chlorine industries and industries based on all of their compounds.

On the other hand it is also the salt which is the most abundant in nature, constituting about 80% of the salt content of seawaters (Table 12.1).

The compositions of sea salt and rock salt are compared in Table 12.2. It is mainly the content of magnesium salts, apart from the lower percentage of NaCl, which reduces the commercial demand for sea salt as compared with rock salt. Nevertheless, far, far greater amounts of sea salt are used than of rock salt on account of the greater availability of the former. In particular, sea salt is predominantly used in Italy.

Table 12.1—The mean composition of sea water

Components	*Percentage*	*Components*	*Percentage*
$NaCl$	78.14	$MgSO_4$	6.58
KCl	1.34	$CaSO_4$	3.60
$MgCl_2$	8.54	$CaCO_3$	0.30
$NaBr$	1.47	Fe_2O_3	0.008

Table 2—Mean composition of the commercially available forms of sodium chloride

Sea salt		*Rock salt*	
Compound	%	*Compound*	%
$NaCl$	96.2	$NaCl$	98
$CaSO_4$	0.25	$CaSO_4$	0.6
$MgCl_2$	0.2	$CaCl_2$	0.2
$MgSO_4$	0.15	$MgCl_2$	0.1
Other salts	0.2	Other salts	0.05
Insolubles	0.8	Insolubles	0.2
Water	2.2	Water	0.95

Sea salt is obtained from sea waters in plants which are known by the collective name of *salt pits*, located along coasts at sites where the rainfall is low.

In the following description we shall consider salt pits in countries with a hot climate, simply noting that, in countries with cold climates, the salt pits consist of reservoirs in which the sea water freezes, thereby increasing the concentration of the remaining saline solutions. These concentrated solutions are subsequently converted into saturated solutions by evaporating off the water, using heat.

The 'evaporation lagoons' are an essential part of a salt pit. These are the vessels with a slight incline and which are made of fine clay which has been impacted to make it impermeable*. Into them flows water which has been pumped up from the sea. By means of natural evaporation due to local climatic conditions (the atmospheric temperature and winds) and the large surface area exposed, the water is progressively concentrated, thereby causing the salts which it contains to be deposited in the order in which their solubilities become exceeded. A diagram of a salt pit is shown in Fig. 12.1.

The salt which is deposited in a crystalline form on the bottom of the 'salting

*The vessels are made perfectly impermeable by cultivating an alga (*Micrococcus corvium*) in them which, apart from making the substrate impermeable, makes it possible to collect the non-polluting salts from the soil.

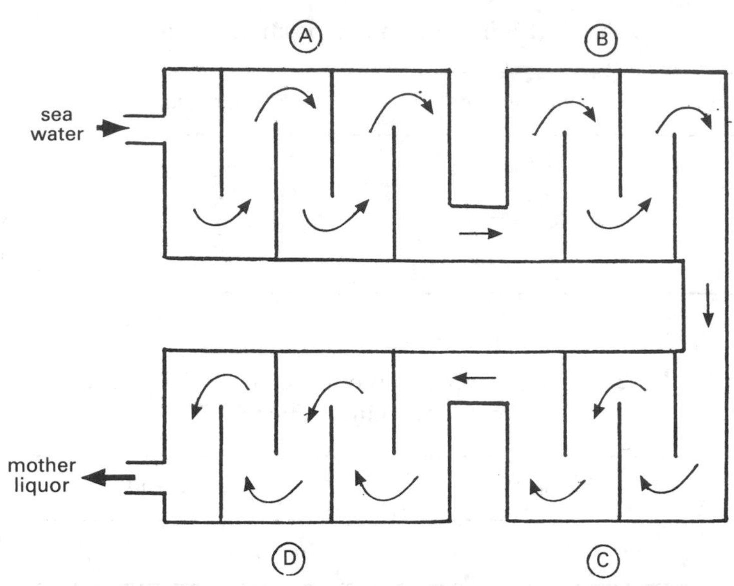

Fig. 12.1—The course of the desalination of sea water in a salt pit. The basins C and B constitute the 'salting compartments'. The mother liquors mainly contain NaCl, KCl, $MgCl_2$, KBr, $MgBr_2$, and KI.

A. A preconcentration basin and first concentration basins where iron salts are precipitated:

$$2Fe(HCO_3)_2 + \tfrac{1}{2} + \tfrac{1}{2}O_2 \rightarrow Fe_2O_3\downarrow + 4CO_2 + 2H_2O$$

B. Second concentration basins where calcium salts are precipitated:

$$Ca(HCO_3)_2 \rightarrow CO_2 + H_2O + CaCO_3\downarrow \quad \text{and} \quad CaSO_4 \cdot 2H_2O\downarrow$$

C. Third concentration basins which yield NaCl, more $CaSO_4 \cdot 2H_2O\downarrow$, and $MgSO_4 \cdot 7H_2O\downarrow$.

D. Fourth concentration basins which yield more NaCl and some magnesium salts ($MgSO_4 \cdot 7H_2O$ and $MgCl_2 \cdot 6H_2O$).

compartments' is collected at intervals and is piled into heaps to allow the mother liquors, which contain a large part of the contaminating magnesium salts, to drain off.

Commercially, sea salt is classified as:

- *cooking salt*:
  - common ($96.2\% \leqslant NaCl < 99\%$): large; milled
  - refined ($99\% < NaCl$): washing with a very pure solution of NaCl to dissolve up the minimum amount of sodium chloride during the removal of the magnesium salts
- *industrial salt*: 'zootechnics', 'for freezing brine', 'vitrifying' (for the ceramic industries) and so on. Here, one is concerned with materials which are discarded from salt pits or which are produced by the denaturing of cooking salt*.
- *very purest salt*: containing more than 99.99% NaCl and destined for chemical and medicinal uses. It is obtained by precipitating saturated solutions of cooking

*Denaturation is carried out where the salt, as in Italy, is a state monopoly. For this purpose the salt is made dirty with earth and unpleasant tasting by the addition of other salts (of iron, calcium, etc.).

salt, saturating them with a current of gaseous hydrogen chloride, and subsequently washing the precipitate, which is formed by the common ion effect, with solutions of pure NaCl in water.

Potassium chloride

The preparation of potassium chloride, KCl, from various natural raw materials (sylvinite, carnallite, and Hartsalz) in the state requested by the fertilizer industry was discussed in Chapter 11. The grades of purity attained by the various production technologies vary from a common material containing 82–84% of KCl, up to a KCl content of 92–96% if the flotation process is used.

By washing the resulting products with cold water, the KCl content of these materials can be raised to 98–99% and above 99% respectively. The waters used in this washing process are obviously recycled in the treatments of the natural raw materials.

12.2 METALLIC SODIUM AND POTASSIUM

Properties and preparations

The bright silvery and soft alkali metals, sodium and potassium, which are very readily oxidized when they come into contact with air, are mainly prepared from their chlorides.

Sodium is produced by the electrolysis of fused NaCl in Downs cells which are built of steel and internally lined with refractories. These cells have a central graphite anode and two semicircular carbon steel cathodes.

A metal mesh diaphragm separates the electrodes and impedes the bubbles of gas which are evolved at the anode from coming into contact with the cathode on which the sodium is discharged. The electrochemical processes are:

upon fusion: $NaCl \rightarrow Na^+ + Cl^-$

at the anode: $Cl^- \rightarrow \frac{1}{2}Cl_2 + e^-$; at the cathode- $Na^+ + e^- \rightarrow Na$.

The fused metal ($mp_{Na} = 97.7°C$) floats ($d_{Na} = 0.97\ kg/m^3$) on the fused salt and is collected in a side tank D which it reaches by ascending into it. The chlorine is conveyed away to the outside of the cell by means of a hood which is attached to the mesh diaphragm separating the electrode compartments (Fig. 12.2).

The molten salt bath is at a temperature of about 620°C because the addition of 0.2–0.3% of $CaCl_2$ to the sodium chloride feedstock reduces its melting point below the theoretical value of 801°C.

Since, at the operating temperature, the equivalent enthalpy of formation of $CaCl_2$ is appreciably lower than that for NaCl, the sodium chloride is preferentially electrolyzed on energetic grounds in addition to mass action. The sodium produced is therefore pure.

Potassium can be prepared by the electrolysis of its chloride in an analogous manner to the preparation of sodium, but, nowadays, it is preferred to start by using the following displacement reaction:

$$\underset{\text{b.p. } 883°C}{Na} + \underset{\text{b.p. } 1407°C}{KCl} \rightleftharpoons \underset{\text{b.p. } 1465}{NaCl} + \underset{\text{b.p. } 776°C}{K}$$

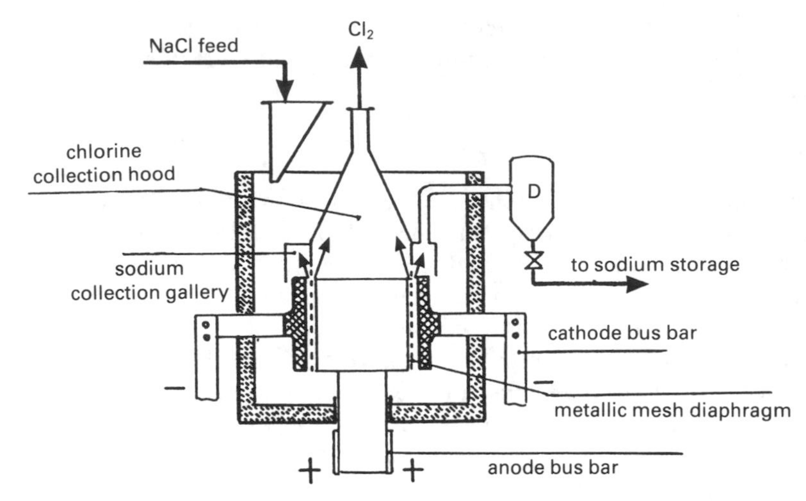

Fig. 12.2—Downs cell for the production of sodium by the electrolysis of fused chloride (NaCl). The sodium produced is aspirated as a liquid into the tank D from which it is sent to be stored.

It can be seen from the values of the boiling points that it is possible to operate in such a way that there is a mixture of sodium and potassium vapours over a fused mixture of NaCl and KCl. By distilling these vapours in a fractionating column, pure potassium can be obtained from the top of the column.

Uses of sodium and potassium

Broadly speaking, the present day uses of sodium can be subdivided as follows: 70% for the production of the antiknock agent, tetraethyllead, 15% as a refining agent for alloys and in special metallurgical processes (of titanium, in particular), 8% for the production of sodamide, sodium cyanide, and sodium peroxide; the rest is used as a reducing agent, a condensing agent (directly and especially via alkoxides), and drying agent in organic chemistry and in primary heat exchangers in nuclear reactor technology. The use of sodium–sulphur cells for the propulsion of electric vehicles is also in an advanced state of experimentation.

Potassium, apart from its use in the production of potassium compounds (the cyanide, peroxide, etc.) finds application as an alternative to sodium. It suffers, however, from the handicap of higher cost.

12.3 SODIUM AND POTASSIUM PEROXIDES

Sodium peroxide and, on a far smaller scale, potassium peroxide are prepared industrially for use as powerful oxidants, in the bleaching of fibres, waxes, and resins, and because, when treated with weak acids, they form hydrogen peroxide.

These peroxides also find other uses in breathing apparatus, in submarines, and, more recently, in the living quarters of astronauts on account of the fact that they purify exhaled air by absorbing CO_2 and releasing oxygen:

$$2Na_2O_2 + 2CO_2 \rightarrow 2Na_2CO_3 + O_2$$

The faintly yellow sodium peroxide is prepared by blowing air into a rotating double-walled metal drum containing molten sodium. The reactions:

$$Na \xrightarrow{\frac{1}{2}O_2} Na_2O \xrightarrow{\frac{1}{2}O_2} Na_2O_2$$

are exothermic; therefore, after priming, the system is cooled to maintain it at the operating temperature of 300–400°C.

The priming and subsequent thermostatting of the system are achieved by the circulation of hot and cold air respectively in the space between the two walls of the drum-shaped vessel.

Potassium peroxide, which is orange in colour, is obtained by heating potassium oxide at 400°C. The oxide disproportionates under these conditions:

$$2K_2 \rightarrow K_2O_2 + 2K$$

so that half of the potassium volatilizes.

In its turn, potassium oxide is best prepared by barely ('carefully') melting potassium nitrate with the metal:

$$2KNO_3 + 10K \rightarrow 6K_2O + N_2$$

Potassium oxide exhibits a pronounced tendency to form potassium superoxide ($K_2O_2 + O_2 \rightarrow 2KO_2$), an extremely dangerous compound which is also formed when preparing potassium peroxide by the direct oxidation of potassium, using the same technique as is employed in the preparation of Na_2O_2. Consequently, this route to the preparation of K_2O_2 is not practical.

12.4 SODIUM AND POTASSIUM CARBONATES

Sodium carbonate ('soda')

After the chloride, the carbonate is by far the most important sodium salt. It is marketed commercially in an anhydrous state in the form of white microcrystals which are exceedingly soluble in water and quite hygroscopic. An aqueous solution of sodium carbonate is strongly alkaline on account of the hydrolysis of the carbonate ion.

On the other hand, the decahydrate of soda ($Na_2CO_3 \cdot 10H_2O$), which effloresces slightly owing to its tendency to lose water, assumes the form of clear prismatic crystals.

The preparation of soda

More than one third of the present-day production of soda makes use of the exploitation of the following natural minerals:

- Na_2CO_3 *natron*
- $Na_2CO_3 \cdot H_2O$ *thermonatrite*
- $Na_2CO_3 \cdot NaHCO_3 \cdot 2H_2O$ *trona*

which occur in the desert regions of the USA, lower Egypt, Somalia, Arabia, and elsewhere.

The methods used in the working of these minerals to obtain pure soda include the dissolution/suspension of the minerals and gangues in water followed by decantation, clarification and filtration processes, and, finally, the recovery of the required solid salt by means of suitable evaporation stages.

However, these Na_2CO_3 based natural minerals can also be used directly for many purposes.

Nevertheless, the greater part of the sodium carbonate is still traditionally produced by the Solvay soda industry which is so-named because it employs the 'Solvay' process. This process has been used for many years as it meets all the criteria appertaining to the convenience of an industrial chemical process. These are:

- the exploitation of naturally occurring raw materials which are widely available;
- the efficient recovery of the intermediates formed in the process;
- the ability to operate the process in very-large-scale plants.

The Solvay process consists of three principal stages:

- the saturation of an aqueous solution of NaCl with gaseous ammonia (the preparation of ammoniacal brine) and the subsequent treatment of this with carbon dioxide, which corresponds to the overall reaction

$$2NaCl + 2H_2O + 2NH_3 + 2CO_2 \rightarrow 2NaHCO_3 + 2NH_4Cl \quad (12.1)$$

- the filtration of the slightly soluble sodium bicarbonate which is precipitated out under these conditions, followed by its thermal decomposition according to the reaction

$$2NaHCO_3 \rightarrow Na_2CO_3 + CO_2 + H_2O \quad (12.2)$$

- recovery of the ammonia by the reaction between ammonium chloride and calcium hydroxide

$$2NH_4Cl + Ca(OH)_2 \rightarrow 2NH_3 + CaCl_2 + 2H_2O \quad (12.3)$$

and of two auxiliary stages:

- the calcination of chalk

$$CaCO_3 \rightarrow CaO + CO_2 \quad (12.4)$$

- the preparation of lime water

$$CaO + H_2O \rightarrow Ca(OH)_2 \quad (12.5)$$

The overall reaction is obtained by summing equations (12.1), (12.2), (12.3), (12.4), and (12.5):

$$2NaCl + CaCO_3 \rightarrow Na_2CO_3 + CaCl_2 \quad (12.6)$$

which, even if it only represents the stoichiometry of the Solvay process, clearly shows that NaCl (sea salt) and $CaCO_3$ (chalk) are the only raw materials which are continuously resupplied in the process, and that the products of the reaction are Na_2CO_3 (soda) and $CaCl_2$ (the 'waste').

Reaction (12.1) is the most interesting, not only because it shows the role played by NH_3 and CO_2 in the process, but also because, as it constitutes the only equilibrium step in the process, it determines the yields of the final product. For this reason, the conditions which are favourable to reaction (12.1) are precisely defined. For this purpose the reaction is divided into two steps:

$$2NH_3 + 2CO_2 + 2H_2O \rightleftharpoons 2NH_4HCO_3 \tag{12.1a}$$

$$2NaCl + 2NH_4HCO_3 \rightleftharpoons 2NaHCO_3 + 2NH_4Cl \tag{12.1b}$$

It is readily understood that (12.1a), a reaction which is undoubtedly favoured by low temperatures because it requires the dissolution of a gas in water, is displaced to the right by virtue of the fact that (12.1b), which utilizes the product by substracting it from (12.1a), is displaced in the same direction. Consequently, it is the precipitation of sodium bicarbonate according to (12.1b) which is the driving force behind the entire preparation of soda by the Solvay method.

The data presented in Table 12.3 indicate that this precipitation fortunately tends to take place preferentially and with satisfactory yields.

On the basis of these data, and recalling the common ion effect on the precipitation of salts, it may be concluded that the physicochemical conditions most suitable for the displacement of (12.1b) to the right, thereby causing the $NaHCO_3$ salt product in it to precipitate, are:

(1) the lowest possible temperature in order to lower the solubility of the sodium bicarbonate (a condition which, as has already been stated, is also favourable to the displacement of (12.1a) to the right so as to increase the amount of the ion which NH_4HCO_3 has in common with $NaHCO_3$);
(2) the greatest possible concentration of one or both of the salts appearing on the left-hand side of (12.1b) with the aim of lowering still further the solubility of the $NaHCO_3$.

These conditions are nevertheless discerningly applied because they serve to bring about appreciable increases in the yields of sodium bicarbonate and permit the most effective use of the most costly reagent (NH_4HCO_3) in reaction (12.1b).

Among other things, attention is paid to the fact that, if the precipitation temperature is always kept low, the sodium bicarbonate separates in a microcrystalline form which can be filtered only with difficulty, which is very soluble during subsequent washing on the filter, and, which, furthermore, requires the use of an excess of NaCl for cost reasons which have already been mentioned.

Experimental practice has shown that the conditions which most effectively reconcile the physicochemical aspect of the phenomenon of the precipitation of $NaHCO_3$ with the economic side of the process are as follows:

- the use of about 284 g/l ($\simeq$4.9 mole/litre) of NaCl and reacting this with 76 g/l ($\simeq$4.5 mole/litre) of NH_3, instead of an equimolecular solution of the two reagents;

Table 12.3—Solubilities of the four salts appearing in (12.1b) at various temperatures

Temperature	*NaCl*	NH_4HCO_3	NH_4Cl	$NaHCO_3$
0°C	357 g/l	120 g/l	298 g/l	69 g/l
20°C	358.5 g/l	217 g/l	374 g/l	95.4 g/l
30°C	359 g/l	269 g/l	467 g/l	109 g/l

- using a relatively high temperature (60–65°C) at the start so as to allow the formation of well-developed sodium bicarbonate crystallization seeds and then increasing the volume of these same seeds to decrease the solubility of the salt, with a gradual cooling.

Let us now move on firstly to describe the individual pieces of apparatus which are typical of the Solvay process and then consider the whole plant which is used in this process.

Saturators. The aqueous solution of sodium chloride (brine) is mixed with gaseous ammonia in twin-bodied units known as saturators (Fig. 12.3a).

The upper part of a saturator consists of the plate column C_1 into which the brine enters from the top while ammonia comes in from the bottom. At the centre of the bottom of the column an overflow device allows the brine to pause and become saturated with the ammoniacal gases (70% NH_3, 15% CO_2 and 15% water vapour) which have been reformed from the lost ammonia and bubbled into the brine via a tube in the form of a perforated ring. The outside of this section of the column is cooled by the circulation of water.

At the bottom, the saturator consists of a cylindrical vessel which carries a layer of solid NaCl on a grill through which the ammoniacal brine, which has surmounted the abovementioned overflow device, percolates. The brine becomes saturated with salt in this way.

Units for the carbonation of ammoniacal brine. The apparatus for the carbonation of the ammoniacal brine which leads to the formation of bicarbonate are the Solvay towers. These

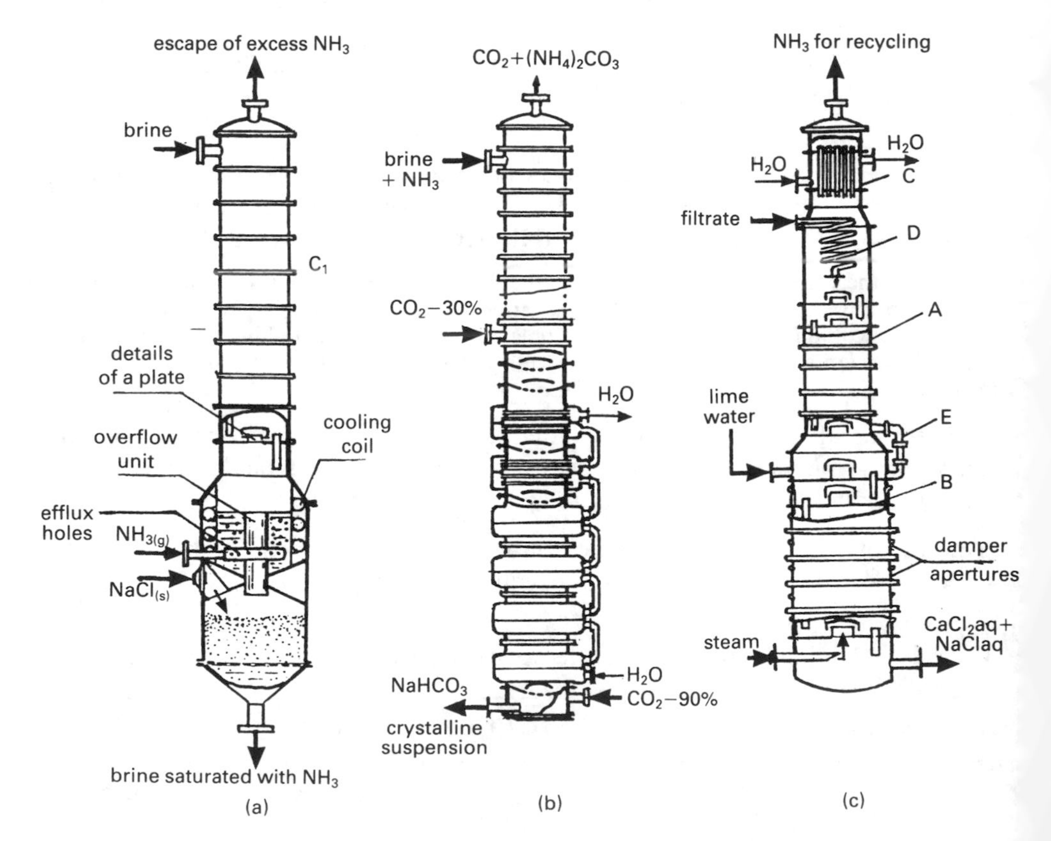

Fig. 12.3—Constructional details of the three principal pieces of plant employed in the Solvay process: (a) the saturator, (b) the carbonation unit, and (c) the causticization unit.

are columns with a height of 15–20 metres and an average width of 1.8 metres and are of the perforated plate type with the plates separated from one another by about 1 metre.

As can be seen from Fig. 12.3b, each plate is slightly concave; is covered with a convex cap; and has a series of apertures around its centre for the purpose of splitting up the gas stream (which rises toward the top of the column) and facilitating the descent of the solid sodium bicarbonate which is formed.

The gas, containing about 30–40% of CO_2, originating from a lime furnace, is fed into this column about halfway up its height after the temperature of this gas has been reduced and it has been purified to remove any other fumes present. Meanwhile, the cooling water rises from the bottom of the column and 90% CO_2 supplied by the sodium bicarbonate decomposition furnaces is blown into the solution.

The ammoniacal brine coming from the saturator, after having deposited any insoluble compounds which it may contain (calcium salts, ferric hydroxide, etc.) is sent from the top into the carbonation tower from which it descends in a counter-current with the CO_2 which is coming up from below.

The temperature rises up to 60°C within the zone in which the first crystals of $NaHCO_3$ are formed (slightly higher than halfway up the column), and is then gradually reduced to 25°C in the zone where a high concentration of CO_2 is bubbled in, by means of the water which rises through coils from the bottom.

The gases which escape from the top of this column and the stream of ammonia which has escaped from the saturators are scrubbed in a small auxiliary plate column (see Fig. 12.4) with a solution of sulphuric acid. The end product from this unit is sent to the causticization units for the recovery of NH_3, and the CO_2 from the top of the column is reunited with that originating from the decomposition furnaces.

Causticizers. The composite column structures in which ammonia is recovered from the mother liquors originating from the filters, which separate the solid sodium bicarbonate which has been precipitated, are known as causticizers.

These causticizers consist of three parts: a distillation column A, a series of causiticization plates B, and a cooling unit C (Fig. 12.3c).

The filtrate to be treated enters from the top of A. The liquid is preheated in the coil D at the expense of the heat content of the hot vapours which are rising up the column. On the various plates of section A, any uncombined ammonia dissolved in the solution is stripped from this solution by the steam which is passing upwards, from which it is subsequently separated for the greater part within the water cooling unit C where the steam tends to condense out.

The descending brine, having reached the lowest plate of column A, enters a side tube E which carries it to meet with a solution of lime water (30 Bé) on the first plate of the lower part.

A vigorous reaction then occurs between the ammonium chloride and the aqueous lime solution, leading to the strong development of ammonia which rises, entrained by the steam, towards the top. This causticization reaction also continues on the lower plates, although somewhat less intensely, until at the bottom, if the operation has been carried out correctly, an aqueous solution of the $CaCl_2$ which has been formed and the NaCl which was used in excess in the process emerges from the tower.

The steam used to strip the ammonia from the solutions is introduced into the bottom of the causticizer. This steam continuously compensates for its own heat in section B of the tower at the expense of the heat developed in the causticization reaction.

Since links between the plates of section B tend to become clogged up by the formation of calcareous incrustations, it is necessary to clean the system from time to time. This is done via the lockable inspection ports which are fitted at the level of each plate.

Decomposition furnaces. These may be of various types, but rotating cylindrical furnaces are the most common. The operating temperature in these furnaces may vary over quite a wide range (180–280°C) according to the amount of moisture in the sodium bicarbonate which is being treated. With the aim of working at the lowest temperature, the bicarbonate to be treated is dehydrated, also using centrifugation, to the greatest possible extent.

These furnaces are heated with the combustion gases which circulate in the heat jackets surrounding them.

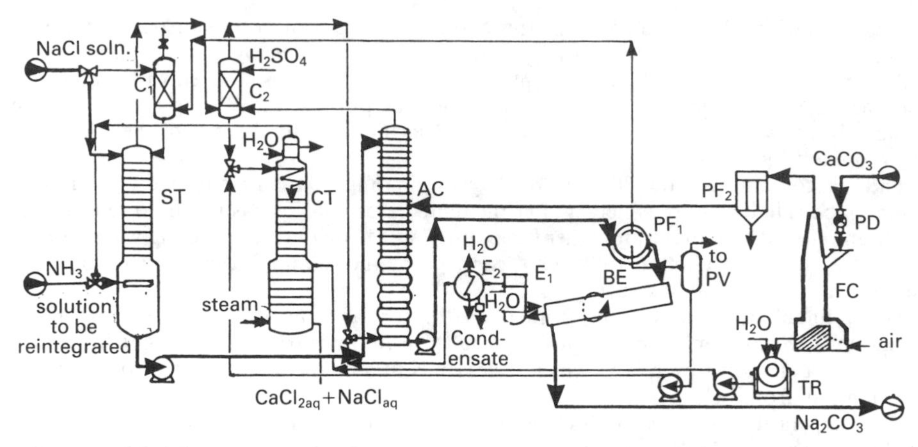

Fig. 12.4—Diagram of a Solvay process plant. A single unit of each type is shown whereas, in practice, there are a number of them. This applies, for example, to the carbonation units AC.

Overall plant technology of the Solvay process. The aqueous solution of sodium chloride is fed into the saturator ST (Fig. 12.4) either directly or via the small absorption column C_1 where the ammoniacal vapours originating from the Oliver filter PF_1 are recovered. The ammonia which escapes from the top of the saturator is recovered in the small column containing sulphuric acid C_2, and the ammonium sulphate which is formed in this column is sent into the causticizer CT. Gaseous ammonia is supplied from the top of this unit which goes to feed the saturator ST.

The aqueous solution of sodium chloride and ammonia is introduced from the top into the absorption and carbonation tower AC where it first encounters the gaseous stream of carbon dioxide coming from the lime furnace FC and then the stream of very concentrated carbon dioxide coming from the decomposition furnace BE (the coolers/water condensers E_1 and E_2 are fitted to free it from water). Further, highly concentrated, carbon dioxide also comes from the absorption column C_2 where it constitutes the volatile component in the reaction between sulphuric acid and the products which are developed by the tower AC. The fumes emerging from FC are purified in the powder-separating unit PF_2.

The precipitated suspension of sodium bicarbonate which has been removed from the bottom of the tower AC is sent to the filter PF_1, from where the precipitate which is separated passes to the furnace BE, the filtrate to the causticizer CT, and the vapours to the small column C_1 for the recovery of ammonia.

Sodium carbonate is obtained at the outlet from the furnace BE.

Finally, it should be mentioned that the quicklime obtained from the furnace FC is converted into limewater in a rotating drum TR from where the solution is then sent on to the highest plate of the lower section of the cauticizer CT.

When the process is carried out rationally, the production of 1 ton of Na_2CO_3 requires the use of 1.48 ton of NaCl, 1.2 ton of $CaCO_3$ (calculated on the basis of the limestone used), and 0.11 ton of coke for the roasting of the limestone; and there is a loss of about 1 kg of ammonia. The yield, calculated on the basis of NaCl, is slightly less than 75%, the remainder of this salt finishing up in the discharge waters.

Statistics of the production and uses of soda

More than 21 million tons of soda, intended for various industrial destinations, are produced worldwide. Almost 50% of this is required by glass industries and ceramics in general, while 20% is used in the sodium compound industry, the pigment industry, and metallurgical operations. More than 10% is used in the paper industry, and as much again is absorbed in the treatment of waters and in soapworks. The rest mainly serves tanning plants and the textile, dyeing, and pharmaceutical industries.

The soda which is obtained from the decomposition furnaces of the Solvay process in the form of a light powder consisting of very small grains is not always adaptable for all purposes. For example, the glass industry requires 'heavy soda' which does not powder, on account of the conditions under which the materials are put into the furnaces.

Suitable treatments enable the various commercial forms of soda to be made available. For instance, the 'heavy' type is obtained by converting the product of the Solvay process into the monohydrate of the carbonate and then calcining the resulting salt.

Potassium carbonate (potash)

The Solvay process cannot be adapted to the production of potassium carbonate (potash or pearl ash) on account of the high solubility of the key compound $KHCO_3$.

Traditionally, potassium carbonate is mainly prepared by the carbonation of potassium hydroxide ($2KOH + CO_2 \rightarrow K_2CO_3 + H_2O$) which has been produced electrolytically in the same manner as NaOH. This method entails the expenditure of large amounts of energy in the electrochemical process for the production of KOH, in the dechlorination (from KCl) of the brines discharged from the cells, and, finally, to concentrate the carbonated solutions.

On account of this, the Engel method for the preparation of K_2CO_3 has grown in importance. This has two stages:

(1) the suspension of natural or artificial* 'hydromagnesite' in a solution of potassium chloride in the presence of CO_2 to precipitate the double salt $KHCO_3 \cdot MgCO_3 \cdot 4H_2O$:

$$3MgCO_3 \cdot Mg(OH)_2 \cdot 3H_2O + 2KCl + 2CO_2 + 5H_2O$$
$$\rightarrow 2[KHCO_3 \cdot MgCO_3 \cdot 4H_2O] + MgCl_2 + MgCO_3$$

(2) decomposition of the double salt which has been separated by filtration:

$$4[KHCO_3 \cdot MgCO_3 \cdot 4H_2O] \xrightarrow[\text{aqueous suspension}]{\text{heating in}}$$
$$2K_2CO_3 + 3MgCO_3 \cdot Mg(OH)_2 \cdot 3H_2O + 3CO_2 + 14H_2O$$

After filtration, the basic magnesium carbonate is recycled and the residual potassium carbonate solution is suitably concentrated.

Potassium carbonate is a white powder which is deliquescent in air. It is marketed as a hydrate (80–85%) or calcined (95–98%). It is extremely soluble in water and yields a strongly alkaline solution owing to the hydrolysis of the carbonate ion.

Potassium carbonate is used especially in the manufacture of crystal glasses, in the production of colorants, in the textile industry in general, and the wool industry in particular. A further characteristic use of potassium carbonate is in the preparation of soft soaps, lubricants, thickeners, and anti-adhesives for moulds.

---

*Artificial hydromagnesite is prepared by precipitating a soluble magnesium salt with sodium carbonate:

$$4Mg^{2+} + 3CO_3^{2-} + 2OH^- + 3H_2O \rightarrow 3MgCO_3 \cdot Mg(OH)_2 \cdot 3H_2O$$

12.5 THE HYDROXIDES OF SODIUM AND POTASSIUM

Sodium hydroxide (caustic soda)

Another fundamental sodium compound which is produced in quantities equal to about 1.5 times the amount of sodium carbonate which is produced is caustic soda, NaOH.

Moreover, the present-day development of the industrial compounds of chlorine has led to the concomitant production of caustic soda in sufficient quantity to meet commercial demands. As a consequence, the method using the causticization of sodium carbonate with lime water has become of less importance as an industrial source of NaOH, and is simply reduced to a (not infrequently carried out) process for the regeneration of this product by converting it into Na_2CO_3 in the various industrial processes which include the fixation of CO_2 with caustic soda (in the course of the metallurgy of aluminium, for example).

In the following text we shall sequentially develop the theory and practice of the causticization of sodium carbonate for the production of caustic soda, and shall treat some of the processes for the concentration of caustic soda fundamentally.

The causticization of sodium carbonate

Concentrations of reagents favourable to the process. The following chemical equilibrium lies at the foundation of the causticization method for the production of NaOH:

$$Ca(OH)_2 + Na_2CO_3 \rightleftharpoons CaCO_3 + 2NaOH \tag{12.7}$$

the equilibrium constant of which is:

$$K_c = \frac{[CaCO_3][NaOH]^2}{[Na_2CO_3][Ca(OH)_2]} \tag{12.8}$$

Since calcium carbonate and calcium hydroxide are only slightly soluble, their solutions are always saturated, and the concentrations of the two components in the solution are therefore constant. Equation (12.8) can therefore be written as:

$$K_c' = \frac{[NaOH]^2}{[Na_2CO_3]} \tag{12.8a}$$

The yield in sodium hydroxide is given by the ratio between the number of moles of the hydroxide which are effectively present at equilibrium and those which would be there if all of the carbonate were to be converted into hydroxide.

Hence,

$$\eta_{NaOH} = \frac{[NaOH]}{[NaOH] + 2[Na_2CO_3]} \tag{12.8b}$$

Upon dividing the terms in this fraction by the concentration of the hydroxide, it is found that:

$$\eta_{NaOH} = \frac{1}{1 + 2\dfrac{[Na_2CO_3]}{[NaOH]}} \tag{12.8c}$$

The ratio which appears in the denominator of (12.8c), when use is made of (12.8a), is equal to the other ratio $[NaOH]/K_c'$. On the basis of this, (12.8c) becomes:

$$\eta_{NaOH} = \frac{1}{1 + 2\dfrac{[NaOH]}{K_c'}},$$

that is:

$$\eta_{\text{NaOH}} = \frac{1}{1 + K_c''[\text{NaOH}]} \tag{12.8d}$$

It is seen from (12.8d) that the yield of sodium hydroxide is high when the concentration of the same hydroxide at equilibrium is low; that is, when the starting concentration of sodium carbonate is small. The experimental data presented in Table 12.4 confirm this conclusion.

In practice, it is necessary to work with starting solutions which are not too dilute in order to avoid the excessive cost of concentrating the caustic soda solutions which are produced. Generally, solutions containing 12–14% of Na_2CO_3 are used.

The required operating temperature. As always, the temperature here affects both the equilibrium yields and the rate of reaction. In this case, it influences the rate of decantation of the $CaCO_3$ formed for the greater part. The reason for this latter effect is that the temperature acts both on the solubility (and, hence, the dissociation) of $Ca(OH)_2$ and on the activity of the NaOH, thereby modifying the ambient pH value; and since the type of product precipitated varies with pH, the ease of sedimentation of $CaCO_3$ is precisely a function of pH.

It has been found experimentally that the optimal temperature at which it is necessary to operate is about 100°C, that is, at a temperature which is almost at the boiling point of the solution.

Reaction time. The maximum possible yields under the concentration and temperature conditions adopted are obtained when the reagents are contacted for about one and a quarter hours. However, the reaction time to be adopted is also a function of the quality of the calcium hydroxide used.

Mixing of the suspension. Another factor which is of concern, if the causticization process is to be successful, is the controlled agitation of the mixture because this favours the formation of a calcareous precipitate which is readily filterable and washable on the filter.

The causticization plant. The soda is brought into solution in the dissolution unit D_1 (Fig. 12.5) which uses, as far as possible, the waters employed in the washing of the calcareous sludges, and is then mixed with lime which has been slaked in a rotating drum TR.

The mixture then passes very slowly into the reaction kettles R_1, R_2, and R_3. These are arranged in cascade and are fitted with stirrers and heating systems. The muddy solution which emerges from the last kettle passes into the decantation unit D_2 from which a clear solution continuously overflows. This solution is sent to be concentrated.

The sludge which is obtained at the bottom of the decantation unit is removed by means of a pump and dispatched into a mashing kettle D_3 where it is brought into suspension in water. It passes from this unit to the Dorr classifier D_4 which returns any larger pieces of insoluble material to the mashing unit while sending the fine material which succeeds in remaining in suspension to be filtered in the Oliver filter PF.

After they have been washed and air dried, the cakes removed from the filter are sent into the rotating furnace BE where the quicklime, CaO, is regenerated for subsequent recycling.

Table 12.4—Yields of caustic soda as a function of the initial percentage of Na_2CO_3

$\%Na_2CO_3$	$\eta_{\text{NaOH}}.100$	$\%Na_2CO_3$	$\eta_{\text{NaOH}}.100$
2	99.4	14	94.5
5	99	16	93.7
10	97.2	20	90.7
12	96.8	25	84.2

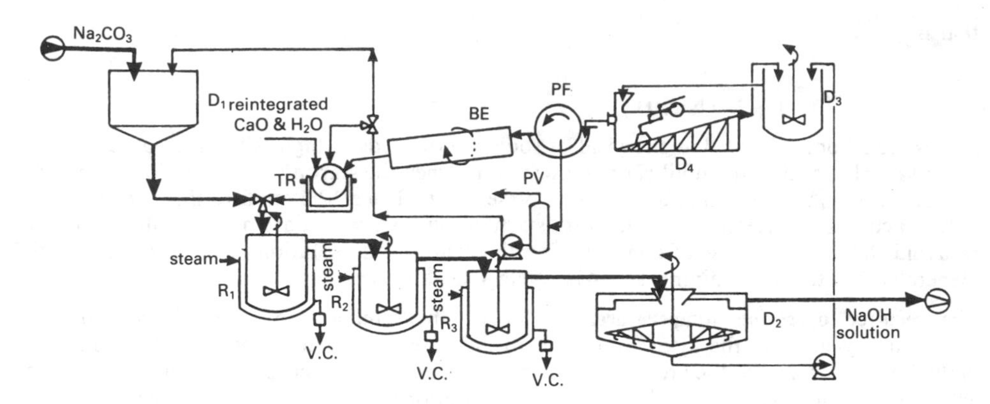

Fig. 12.5—Plant for the causticization of sodium carbonate.

On the other hand, the filtrate and the washing waters which the filter unit furnishes are used for the reformation of the initial solution of lime water and sodium carbonate. It is therefore possible to carry out very precise washings in view of the fact that the effluents can be completely reutilized.

The concentration of caustic soda

The caustic soda solutions obtained either by causticization or electrochemically frequently need to be concentrated. This is also necessary because they are desalinated in this manner.

The problems in relation to the number and the complexity of the operational phases of the concentration of caustic soda solutions differ, depending on the amount of it which is present in the starting solution, which, in turn, depends on the production process and the NaOH content which is required in the final product. It will therefore be assumed that one starts off from solutions containing the lowest concentration which is practically available (9–10%). The whole course of the process in which such solutions are converted into solid crystalline caustic soda will be described. The phases of the above description which are pertinent to the solutions of NaOH obtained by the various industrial processes will then be shown schematically.

Since one is dealing with solutions containing slightly less than 10% of caustic soda (caustic soda from causticization units), the following operational phases must be carried out in order, first of all, to bring them up to the highest concentrations which are usually required of solutions ($\simeq 50\%$):

(1) concentration up to 14–16% in 'Kestner evaporators'. On account of their large evaporation surface, these evaporators are well suited for the removal of large amounts of water from solutions, and the lengths of their tubes are adapted for the treatment of solutions (which tend to foam) such as NaOH solutions;
(2) evaporation in multistage units or (if the cost of electrical energy permits it) in thermal compressors up to concentrations of the order of 30% at which contaminating salts such as NaCl and Na_2CO_3 are completely precipitated;
(3) further concentration up to approximately 50% in vacuum evaporators and/or

heating by the circulation of superheated steam because such solutions boil at about 140°C.

Much less water is eliminated in these final evaporators than in the multistage evaporators, although the increase in concentration which occurs may be distinctly larger.

It may be calculated that a solution containing 106 kg of NaOH per m^3 of water is a 9.6% solution of caustic soda, and correspondingly a 16% solution contains 106 kg of NaOH in 550 kg of water, a 30% solution contains 249 kg of water, and a 50% solution contains 106 kg of water. Hence, to concentrate the solution from 16% to 30%, it is necessary to eliminate $(550-249)=301$ kg of water, while to concentrate the solution from 30 to 50%, $(249-106)=143$ kg of water have to be removed from it.

Concentrations higher than 50% are not conveniently attainable by heating with superheated steam since the high pressures which are required necessitate the use of thick-walled conduits which are always less capable of ensuring good heat exchange characteristics.

Two other phases must therefore be carried out in order to bring about the further evaporation of the solutions until solid crystallized caustic soda is attained.

(4) fuming the solutions in open vessels constructed of nickel-base alloys ('converters–finishers') until the solutions are concentrated up to approximately 60%;
(5) dispatching the new solutions to 'basins', that is, to hemispherical cast iron* boilers which are heated directly with a flame where the caustic soda gradually melts. It is kept for several hours in the fused state before the emission of dangerous spurts of steam ceases. In the meantime the temperature rises to about 700°C, the mass becomes every more fluid, and leaves a deposit of the iron oxide formed from the corrosion of the vessels on the bottom of the basins. After this the temperature stabilizes and the fused caustic soda is poured into moulds made of autogenically welded sheet steel.

The Fe_2O_3 deposits lead to a reduction in the quality of the caustic soda which is produced. To avoid these deposits entirely, the caustic soda is fused in apparatus made of nickel or silver which resist (and this applies for silver) attack by fused alkalis. The plant design shown in Fig. 12.6 is capable of concentrating sodium hydroxide solutions containing less than 10% of NaOH up to their maximum concentration.

The phases out of those which have just been described which are required to concentrate the various types of industrially available caustic soda up to the maximum concentration are shown in Table 12.5.

Uses of caustic soda

The principal destinations for caustic soda, which has an annual production level exceeding 30 million tons, are:

- approximately 40% in the production of sodium compounds (including Na_2CO_3 by carbonation) and, especially, in metallurgical processes;

*A typical cast iron for basins of this type has the composition: 3.4% C; 0.04% Mn; 0.7% Si; 0.62% Cr; 1.8% Ni. The lifetime of these vessels can be greatly prolonged by passivating their surfaces by the addition of approximately 0.1% of chlorates to all the solutions which are treated in them.

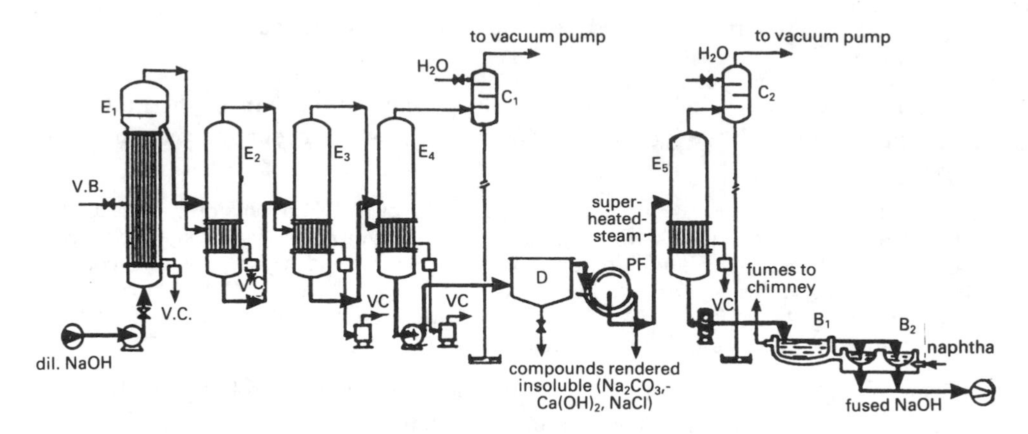

Fig. 12.6—Diagram of a plant for the concentration of caustic soda from 9–10% to the state of fused caustic soda to be solidified in various shapes.

E_1	Kestner evaporator	D	crystallization–sedimentation unit
E_2, E_3, E_4	multistage evaporators	B_1	concentrator–finisher
E_5	vacuum evaporator	B_2	basin
C_1, C_2	mixing condensers	VB	low-pressure steam
PF	rotating Oliver filter	VC	steam condensate

Table 12.5—Concentration states carried out on industrial caustic sodas

Caustic sodas		*Average concentration upstream*	*Admissible steps*†	*Observations*
From causticization		9–10%	(1), (2), (3), (4), (5)	The first two steps are necessary to remove Na_2CO_3 from it
Electrolytic	diaphragm cells	12–15%	(2), (3), (4), (5)	The first stage (2) is necessary to eliminate NaCl
	membrane cells	20%	(2), (3), (4), (5)	Stage (2) is necessary as heat is recovered
	amalgam cells	50	(4), (5)	

†The numbers refer to the numbered steps in the preceding text.

- approximately 20% in the preparation and finishing of artificial fibres (rayon) and film (Cellophane), for the mercerization of cotton and in the production of dye intermediates and plastic materials;
- about 20% in the paper industry;
- more than 8% in detergents, bleaching, and whitening products and soaps.

The rest goes to an extremely wide range of destinations including alkali pickling, the recycled rubber industry, and the refining of petroleum fractions and vegetable oils.

***Potassium hydroxide* (*'caustic potash'*)**

Caustic potash (KOH) is prepared in solutions of various concentrations by the electrolysis of potassium chloride, in a similar manner to the production of sodium hydroxide. The methods employed use diaphragm and membrane cells.

After concentration, which is also carried out in a similar manner to the concentration of sodium hydroxide, the solid potassium hydroxide has a purity of approximately 90% owing to the greater difficulty encountered in dehydrating this compound as compared with NaOH, which may be up to 98% pure.

The two hydroxides may be prepared in a pure state by dissolving the technical products selectively with respect to the salts which contaminate them in alcohol, filtering the resulting solutions, and then evaporating off the alcohol and water before fusing the salts.

Nickel or silver crucibles must be used for the fusion of KOH as this compound is highly corrosive towards every type of cast iron.

Caustic potash (KOH), a very strong base, absorbs water vapour and carbon dioxide from the atmosphere to produce concentrated solutions of K_2CO_3. Under the same conditions, NaOH produces solid Na_2CO_3.

When fused, KOH absorbs oxygen, thereby stimulating oxidative processes which are exploited, for example, in the production of dye intermediates. Besides oxidative alkaline fusions, fused caustic potash is also used as a dehydrating agent. Potassium compounds such as the permanganate, sulphides and polysulphides, nitrate, and many other salts, are also prepared from KOH. The preparation of soft soaps and the electrolytic solutions in fuel cells and cells for the electrochemical production of hydrogen and oxygen also require the use of conspicuous amounts of KOH.

12.6 THE MINOR INDUSTRIAL COMPOUNDS OF SODIUM

Apart from NaCl, Na_2CO_3, and NaOH which have been discussed in this chapter and the various sodium salts which have been considered earlier in the book (the sulphate, nitrate, borate, hypochlorite, etc.), sodium silicate and sodium bicarbonate are of noteworthy importance.

Sodium silicate. When SiO_2 and Na_2CO_3 in proportions of 3:1 are fused at 1300–1500°C, usually in a cylindrical rotating furnace, the following reaction occurs:

$$3SiO_2 + Na_2CO_3 \rightarrow Na_2SiO_3 \cdot 2SiO_2 + CO_2$$

which leads to the production of sodium silicate ('water glass').

When cold, the resulting mass assumes the form of a transparent glass with a yellowish-green coloration which is soluble in water when treated with steam at a pressure of 3 atmospheres. The commercial forms have various water contents according to the degree of concentration brought about by evaporation.

The pasting of paper, the preparation of mastics, cloth finishes, glazes, soaps, fire-retarding agents for the treatment of woods, and the surface treatment of constructional stones, are but some of the many technologies in which sodium silicate is employed.

Sodium bicarbonate. Large overall amounts of pure $NaHCO_3$ are used in the manufacture of effervescent powders, artificial leavening agents, and pharmaceutical products. This bicarbonate is prepared from crude bicarbonate from the Solvay process by dissolving the latter in hot water (65°C), filtering the solution, and then saturating the filtrate with CO_2. The crystalline precipitate formed in this manner is filtered and then dried at only 40°C in a stream of carbon dioxide.

12.7 OTHER INDUSTRIAL COMPOUNDS OF POTASSIUM

Besides the potashes K_2CO_3 and KOH and the salts used in explosives, fertilizers, etc., potassium permanganate and potassium peroxydisulphate ('persulphate') are of industrial interest.

Potassium permanganate. This compound, which is used in bleaching processes, in pharmaceutical preparations, and in analytical chemistry, is prepared in two stages:

- the oxidation by alkaline fusion of pyrolusite in the presence of chlorates and perchlorates, with the injection of oxygen until the development of a fused green mass:

$$2MnO_2 + 4KOH + O_2 \rightarrow 2K_2MnO_4 + 2H_2O$$

- electrolysis, in cells with nickel anodes and iron cathodes, of the solution obtained when the cooled green mass of manganate is leached with water and filtered through glass wool.

Electrolysis leads to the formation of hydrogen at the cathode ($2H^+ + 2e^- \rightarrow H_2$) while the manganate ions are oxidized to permanganate at the anode ($MnO_4^{2-} \rightarrow MnO_4^- + e^-$). Some dissociated KOH therefore remains in solution which, by lowering the solubility of the permanganate, brings about its precipitation on the bottom of the cell. After it has been collected, filtered on glass wool, and carefully dried, it is packaged appropriately for the uses to which it is to be put.

Potassium persulphate. When a collectively extremely concentrated solution of ammonium sulphate acidified with sulphuric acid (250–280 g/l H_2SO_4 and 200–220 g/l of $(NH_4)_2SO_4$) is electrolyzed in cells with lead cathodes and smooth platinum anodes (a high oxygen overpotential) at a low temperature (20–30°C) and a high current density, a single reaction is favoured at the anode: $2H_2O \rightarrow 4H^+ + O_2 + 4e^-$, and the occurrence of the following reaction is stimulated:

$$2HSO_4^- \rightarrow S_2O_8^{2-} + 2H^+ + 2e^-$$

Suitable diaphragms separate the anode compartments and cathode compartments (where hydrogen is evolved), and cooling water is circulated through coils of lead pipes since one is dealing with a low-efficiency electrochemical process in which a large amount of heat is evolved.

Since the NH_4^+ is not discharged, the solution emerging from the anodic compartments consists almost entirely of $(NH_4)_2S_2O_8$.

To obtain the corresponding potassium salt, which is much less soluble, the resulting solution is treated with potassium bisulphate:

$$(NH_4)_2S_2O_8 + 2KHSO_4 \rightarrow (NH_4)_2SO_4 + H_2SO_4 + K_2S_2O_8$$

The solution to be recycled to the electrolytic cells is regenerated in this way. Potassium peroxydisulphate (potassium persulphate) is only slightly soluble and therefore precipitates. It is collected, filtered, dried, and stored.

12.8 THE HYDROGEN PEROXIDE INDUSTRY

Potassium persulphate, which was formerly used both in the immediate preparation of hydrogen peroxide and its preparation at some later time at sites remote from where the potassium compound was prepared, has become of considerably lesser importance for this purpose.

The reaction at the basis of the production of hydrogen peroxide from persulphate at 70–80°C is the following*:

$$K_2S_2O_8 + 2H_2O \rightarrow 2KHSO_4 + H_2O_2$$

This must be carried out under vacuum in order to distil over an aqueous solution containing 30–35% of H_2O_2.

Instead of proceeding via potassium persulphate, hydrogen peroxide can also be produced by the direct hydrolysis of the ammonium persulphate emerging from the anodic compartments of the electrolytic cells in which it is prepared. The reaction is analogous to its production from $K_2S_2O_8$.

Nowadays, in fact, hydrogen peroxide is mainly obtained as the product of the oxidative phase of processes using the oxidation and reduction of secondary alcohols and of particular phenols.

Most commonly, the process starts out from 2-alkylanthraquinols which, when dissolved in benzene and higher alcohols, can be oxidized to quinonoid diketones by air enriched with oxygen:

OH, R, OH $+O_2 \longrightarrow$ O, R, O $+H_2O_2$

By extracting the hydrogen peroxide produced from the reaction mixture by means of a countercurrent of water, solutions containing approximately 24% of peroxide can be obtained.

The anthraquinol is regenerated by reduction under mild conditions (hydrogen at 1.5 atm. on Raney nickel) at room temperature:

O, R, O $+H_2 \xrightarrow[\text{catalyst}]{}$ OH, R, OH

The anthraquinol is then recycled. A diagram of the plant for this process is shown schematically in Fig. 12.7. Pure hydrogen peroxide, which is a transparent liquid (m.p. −0.9°C, b.p. +151.5°C, d_4^{20} 1.46), is an unstable chemical species which is

*Actually, the reaction would proceed in accordance with the following steps:

$$K_2S_2O_8 + H_2SO_4 \rightarrow KHSO_4 + KHS_2O_8 \xrightarrow{H_2O} KHSO_4 + H_2SO_5 \xrightarrow{H_2O} H_2O_2 + H_2SO_4$$

which can be established by analyzing the intermediate compounds.

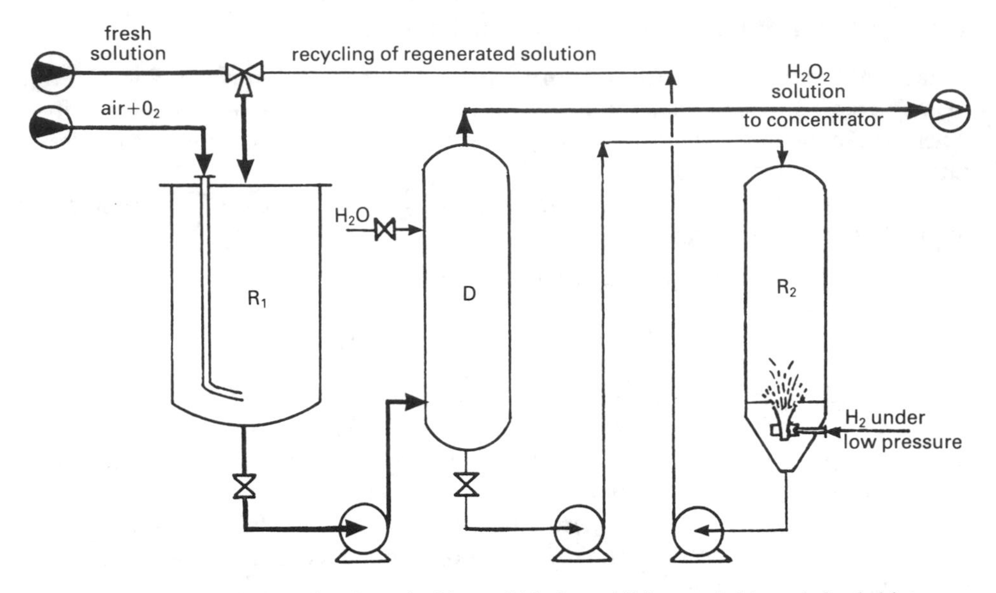

Fig. 12.7—The solution of anthraquinol into which the oxidizing gas is blown is loaded into the reactor R_1. The hydrogen peroxide formed and the benzene/alcohol mixture containing the diketone which is produced pass into the extraction unit D in which H_2O_2 is dissolved in the water which constitutes the light phase of the system. The heavy phase is pumped into the reactor R_2 where the diphenol is regenerated by mild hydrogenation and then recycled into R_1.

rapidly decomposable into water and oxygen both when hot and in the presence of certain ions or MnO_2. At high concentrations of hydrogen peroxide (>85%), the decomposition is violent, and the stabilizers which are added to it (H_3PO_4, pyrophosphates, aromatic acids, etc.) vary according to the uses to which it is to be put.

Hydrogen peroxide, which is completely miscible with water at all concentrations, is commercially available mainly at 3%, 35%, and 80% solutions. The different concentrations are obtained by the distillation of very pure solutions under a reduced pressure.

A 3% solution of hydrogen peroxide has a content of gaseous oxygen (which is evolved if it is decomposed) of 12 litres at S.T.P. per litre, while a 30% solution corresponds to 100 litres at S.T.P. per litre, and a 90% solution corresponds to 412 litres at S.T.P. per litre. Such contents are known as the 'volumes' of hydrogen peroxide, and these are commonly referred to in commerce and industry.

Hydrogen peroxide finds different uses depending on its concentration. For instance, 412 volume hydrogen peroxide is employed as a propellant, 100 volume H_2O_2 is used in whitening and bleaching processes, while 12 volume hydrogen peroxide serves as a disinfectant. Hydrogen peroxide is utilized in many chemical production processes and in food, textile, tanning, metallurgical technologies, and so on.

Hydrogen peroxide is manufactured in vessels whose construction depends on the concentration being produced. At 30–36% it is produced in special glass containers (to avoid catalysis which leads to its decomposition). At 50–90% it is produced in non-hermetically sealed aluminium vessels, and so on.

12.9 THE DANGERS ARISING FROM ALKALIS AND REMEDIES FOR THEIR TREATMENT

It is pertinent to recall that the alkalis are substances which act as bases and that they are therefore the antagonists of acids whose action they neutralize. In current usage, the term 'alkali' is reserved for the most part to the hydroxides of sodium and potassium. But it also refers to the carbonates of these metals, and the hydroxides of the alkaline earth metals (Ca, Sr, and Ba) are frequently classified as alkalis. This classification, of course, is to be propounded or accepted respectively when it is a question of the dangers presented by alkalis to the physiological integrity of living organisms.

From this point of view, all of the oxides of the alkali and alkaline earth metals should be classified as alkalis in so far as they are capable of stimulating the same effects. In the latter case, however, these effects are even more exacerbated by the removal of the water of hydration as they tend to aggregate during their transformation into the corresponding hydroxides.

The working of alkalis, and with alkalis, exposes operatives to the dangers of their frequent contact with the skin. This is dangerous because these compounds tend to saponify the fats which lubricate the skin and solubilize the fatty acids produced owing to salt formation, and to dehydrate and denature, to various extents, the skin and tissues exposed by the cracking and desquamation of the skin.

Contact with alkalis is first characterized by a sensation of greasiness to the touch. When this warning symptom is recognized, the affected part, after the removal of any clothing when necessary, must be immediately and thoroughly washed with a copious stream of water. The post-contact drying off of the skin, which is possibly accompanied by smarting or by considerable irritation, is soothed by the application of creams or fat, or silicone or pharmaceutical grade glycerine-based ointments.

Prolonged contact with articles of clothing which have come into contact with alkalis, even if they are dilute, must be avoided. Where necessary, leggings, boots, shoes, and gloves made out of alkali-resistant elastomers must be worn. In many cases, safety spectacles are also of use, especially during pouring, the inspection of reactors, and the starting-up of stirrers submerged in alkaline solutions.

So far as the possible ingestion of alkalis is concerned, it must be stated that the lethal dose varies with the concentration of the solution which is ingested, bearing in mind that alkalis in the stomach and then in the intestine first neutralize the gastric juices, then decompose the tissue (by dehydration, deamination, desulphurization of proteins, the saponification of fats, etc.) and finally perforate the internal organs.

The symptomatology of poisoning by the ingestion of alkalis varies with the concentration and the amount ingested. The commonest manifestations of alkali poisoning are the swelling of the lips and the oral mucosa, blood-stained expectorations, and the passing of blood-containing diarrhoea.

The principal remedies against ingestions of alkalis are the swallowing of dilute acetic acid, lemon juice, raw egg, or milk. The use of a stomach pump is *only admissible immediately after ingestion*. Otherwise, this remedy increases the danger of perforations.

* * *

To complete this panoramic discussion of the dangers arising from the chemical products treated in this chapter, it should be noted that hydrogen peroxide and, what is worse, other peroxides (owing to the alkalis which are also formed) are dangerous irritants, sensitizers, allergy inducers, and, frequently, skin causticizers. Moreover, their chronic irritative action leads to hyperpigmentation, hypercheratosis, rhagades, and even skin tumors.

Finally, one must not disregard either the dangers of violent decomposition (explosion) at the point of production, concentration, and transport of hydrogen peroxide, nor the fact that, in the presence of this chemical product at a concentration higher than 80%, only low mean concentration limits are tolerable in a working environment. In the case of 90% hydrogen peroxide the tolerable limit is 1 ppm which corresponds to 1.4 mg/m^3.

13

Chemical aspects of industrial metallurgy

13.1 CHEMISTRY IS NOT INVOLVED IN EVERY ASPECT OF METALLURGY

Metallurgy comprises the techniques which lead:

- from minerals to metals and primary alloys (primary metallic materials),
- from primary metallic materials to those of a definite composition,
- from metallic materials of defined composition to metallic materials with their properties adapted to satisfy various mechanical requirements.

Correspondingly, metallurgy is subdivided into:

- extractive metallurgy,
- the metallurgy of alloys,
- physicomechanical metallurgy.

While extractive metallurgy uses mechanical technologies which serve chemical processes, and the metallurgy of alloys uses chemical phenomena in a context of mechanical technologies, physicomechanical metallurgy lies almost completely outside of the interests of chemistry.

It follows from this that limits may be conveniently set in the treatment of metallurgy in a book on industrial chemistry where it is necessary to provide some preliminary treatment of the structure of metallic materials in addition to the treatment of physicochemical and chemicometallurgical processes in order to enable the reader to appreciate the consistency and the consequences of the heat and mechanical treatments, the interest in which is, nevertheless, technological.

In this chapter we shall omit the mechanical–technological aspects of metallurgy and of the treatments of metals and alloys, from tempering to all the processes which are complementary and surrogate to this, from metallographic process controls to microanalytical techniques, and to those connected with electron microscopy. As a result, only specific uses of metallic materials in any branch of the metallurgical and mechanical industries will be emphasized.

Part 1 THE STRUCTURE OF METALS AND THE CONSTITUTION OF ALLOYS

13.2 THE METALLIC STATE

The physical properties of metals

With the exception of mercury, the metals, which are solid chemical elements at ordinary temperatures, are physically characterized by the following properties:

- of being lustrous and shiny when fractured (lustre),
- of being opaque even when in the form of a thin layer (opacity),
- of having high thermal and electrical conductivities (conductivity),
- of being capable of being worked into thin sheets or drawn into threads (malleability and ductility),
- of undergoing limited deformations under the action of a force and returning to their original shape as soon as the force is removed (elasticity),
- of resisting more than every other substance multidirectional mechanical forces, that is, compression, traction, and flexure (toughness),
- of consisting of giant* molecules in the solid state and of monatomic molecules in the gaseous state,
- of acting as cations in electrolytic solutions, thereby demonstrating a distinct tendency to lose electrons.

Of these physical properties the one which is the most characteristic of metals is electrical conductivity. This is not only because it is higher than that of any other type of element with the exception of graphitic carbon, but also because the resistivity of metals increases without exception when the temperature is raised. This does not occur in any other substance, including graphite.

The metallic bond

The physical properties of metals as a whole cannot be explained within the framework of the theories of normal covalent or ionic bonds, or, for that matter, in terms of heteropolar bonds. They can be accounted for only by assuming that a specific type of bond, the 'metallic bond', lies at the foundation of the structure of the metallic state.

This type of bond does not involve the sharing of electrons between pairs of atoms or between discrete groups of atoms making up the metallic body, but among all the atoms forming the entire body.

In fact, while the atomic nuclei and the non-valence electrons are confined to certain positions in the crystal lattice, the valence electrons are delocalized over all the atoms, making up the metallic body in such a way as to constitute an 'electron gas' which diffuses throughout the metal (Fig. 13.1).

It is recalled that all the electrons in the outermost electron shell of an atom, and those in the penultimate shell of atoms of elements located in the Periodic System within the zone where this shell is incomplete (transition elements), are valence electrons.

A metal is therefore made up from a multitude of positive ions with charges equal to the number of electrons which each of them confers to the electron gas. The

*It is recalled that substances which are formed from an enormous number of atoms which are bound into a single entity have the structures of giant molecules if they are of the inorganic type, and macromolecules if one is dealing with organic substances.

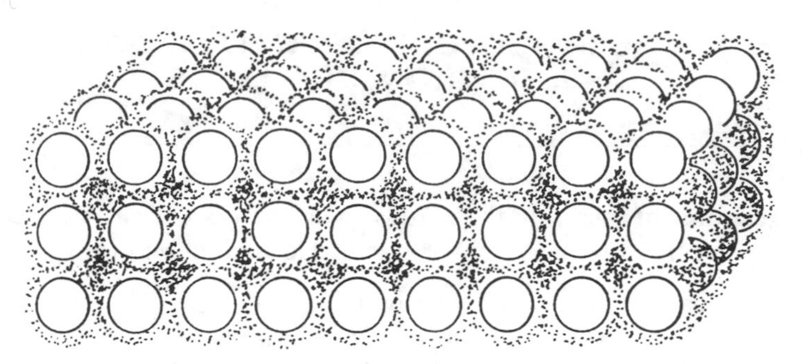

Fig. 13.1—The diffuse delocalization of electrons over the entire metallic structure may therefore be conceived as a lattice of positive ions immersed in an electron gas which statistically neutralizes the positive ions; that is to say, as a whole rather than as single ions. The electron gas which permeates them therefore binds the positive ions of the metallic crystal lattice, and this therefore results in a neutral and compact system as a consequence of a general electronic envelopment.

positive ions are bound into a single framework by all the free electrons which have been released by them. Such a system has no limits: it develops entirely as the enormous collection of atoms making up the body.

What is more, metallic cohesion, which is an index of the force which binds the atoms within the framework, increases as the number of electrons delocalized by the various atoms increases (Fig. 13.2). As the cohesion increases, so does the density which passes from minimum values in the case of the alkali metals (e.g. $d_{Na} = 0.97\ g/cm^3$) to low and medium values (e.g. $d_{Al} = 2.7$ and $d_{Fe} = 7.86\ g/cm^3$) and then on to the highest values (e.g. $d_{Os} = 22.48\ g/cm^3$). The highest densities are encountered in the transition elements with high atomic numbers. This is because, in these metals, there is both a large number of delocalized electrons and a large degree of delocalization as the delocalized electrons find themselves at large distances from the nuclei of their respective atoms, which greatly diminishes the attractive force.

The crystal structures of metals

In metals, the atoms or, rather, the positive ions formed by the general pooling of the electrons, are not arranged in a completely random manner but occupy definite positions which correspond to the lattice points of various types of crystal lattices.

Several ordered structural configurations found in metals are shown in Fig. 13.3..

It is seen that metals most commonly crystallize in the body-centred cubic, face-centred cubic, and hexagonal close-packed structures.

These and the other crystalline forms of high symmetry are realized by the repetition in the space within the crystal of dense assemblages of nodal points located in reticular layers which are arranged one above the other.

Fig. 13.4 shows such layers of atoms in the case of the easily representable body-centred cubic lattice.

The delocalized electrons responsible for the metallic bond travel between these atomic planes, thereby almost forming a layer of 'lubricant' which allows the planes to glide over one another without ever giving rise to a repulsion between the positive ions constituting the planes which would be capable of causing them to move apart such that the crystal would fracture.

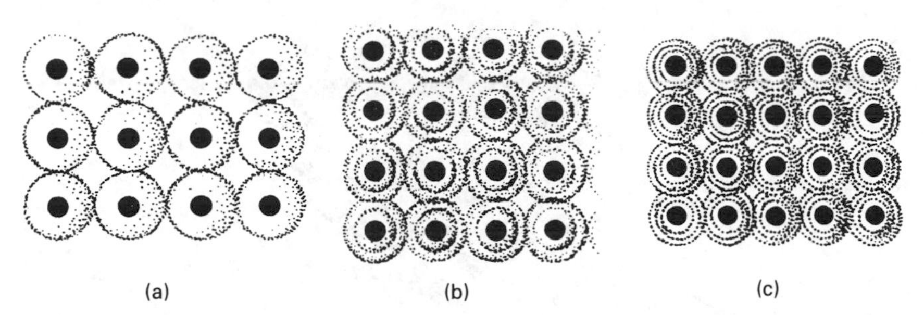

Fig. 13.2—The valence electrons of a metallic atom are numerically equal to those which are delocalized in the metal owing to the formation of bonds: (1) (as in a) for metals of the sodium type, (2) (as in b) for metals of the magnesium type, (3) (as in c) for metals of the aluminium type. It is clear that metallic cohesion increases along this series. However, this property is also a function of the coordination of the metal in the crystal. In the transition elements, the number of delocalized electrons increases to a maximum and then, as their melting points show, falls off as the number of spin-paired electrons in the d-subshell of the penultimate shell increases. This is illustrated, for example, by the following melting point data: W 3380°C, Re 3170°C, Os 2700°C, Ir 2443°C, Pt 1769°C, and Au 1063°C.

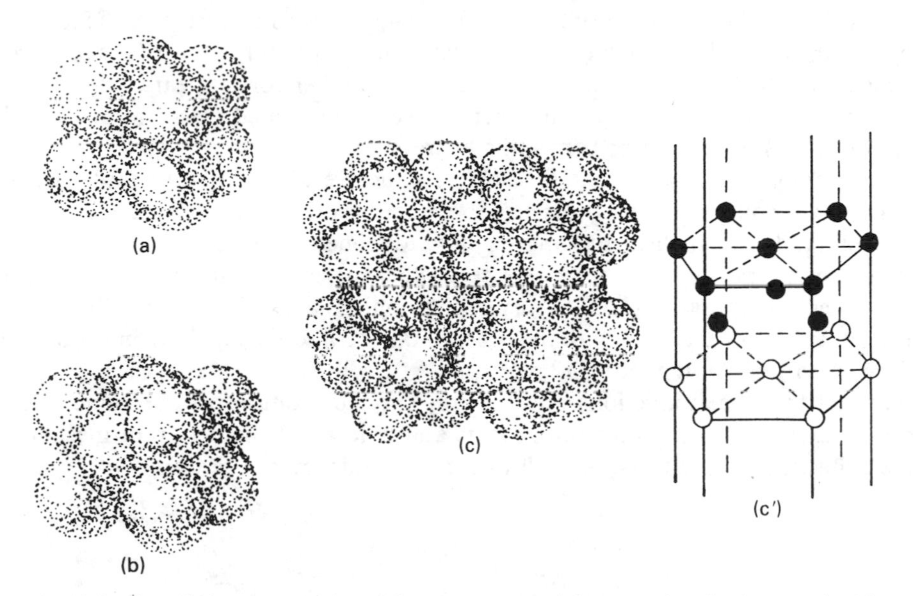

Fig. 13.3—A realistic representation of the arrangement of the atoms in a 'body-centred cubic lattice' (8 atoms at the vertices + 1 atom at the centre of a cube) is shown in (a). A 'face-centred cubic lattice' (8 atoms at the vertices and six atoms at the centres of the faces) is shown in (b), and the configuration of a 'hexagonal close-packed metallic lattice' is shown in (c). For clarity, a schematic diagram of (c) is presented in (c′).

Justification of the physical properties of metals

Certain properties of metals can be immediately understood on the basis of the nature of the metallic bond interpreted in terms of the state of the bonding electrons.

For instance, the electrical conductivity follows from the fact that the application of an electric field leads to a movement of the delocalized electrons in the direction of the field (an electric current), while the thermal conductivity is a consequence of

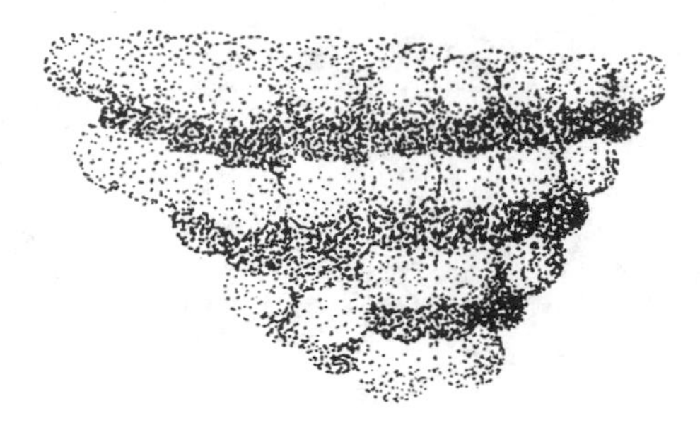

Fig. 13.4—Fragment of a body-centred cubic metallic lattice formed from close-packed assemblages of nodal points which are characteristic of this type of crystal structure and form a series of layers arranged one upon the other.

the facile transmission of disorganized energy (heat) received at one point to all the atoms over the whole system through the agency of the electron gas.

In their turn, metallic opacity and lustre are also interpreted on the basis of the theory of delocalized electrons. Actually, light which is incident on a metallic body finds electrons which can be excited to higher energy levels by virtue of the fact that they are delocalized. It is owing to this fact that the photons do not travel across the metal on which the light is incident, and the metal therefore appears to be opaque. Then, by returning to their normal energy levels, the excited electrons re-emit the light which has been absorbed (metallic lustre).

Other properties require that the atomic structure of the metal should also be taken into account so that they may be understood. This is so in the case of malleability, ductility, elasticity, and metallic toughness in general.

Malleability and ductility, that is, the coefficients of metallic plasticity, are the consequence of the ability of the crystallographic planes, formed by the regular arrangement of metal atoms, with the concomitant creation of an electronic lubricant, to glide over one another (Fig. 13.5).

The fact that the atomic ions must occupy specific positions in the framework of a crystal system is responsible both for its ability to resist the action of mechanical forces, that is, its toughness, as well as for the tendency of these ions to return to

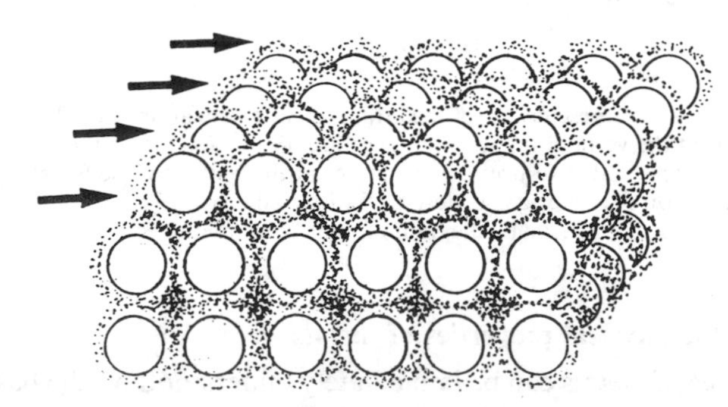

Fig. 13.5—Mechanical forces applied to a metallic body bring about a regular gliding of the planes because every component finds itself once again in its old environment in each new position. These glides and deformations, which are allowed by the malleability and ductility in metals, are facile and do not entail any breakdown of cohesion. On the other hand, if rigid bonds exist such as those in diamond and silica, similar actions lead to the fracture of the bodies.

their previous sites (elasticity) when a force, which does not exceed certain limits, displaces them from such positions.

Finally, it is the concomitant effect of both the state of the binding electrons and the vibrational capacity of the atoms which is responsible for the increase in the electrical resistance of metals as the temperature is raised, since, when heat is taken up as mechanical energy which causes the atoms to vibrate with ever-increasing amplitudes in a disordered manner, the vibrating metal ions at the nodal points of the crystal lattice oppose the motion of the flux of delocalized electrons in the direction of the electric field (the electric current) in such a way that the motion becomes more difficult as the amount of heat received (the temperature) increases.

13.3 CHEMICAL PROPERTIES OF METALS

From a chemical point of view, metals have three fundamental properties:

- of dissolving when they come into contact with certain reagents (corrosibility);
- a predisposition to become chemically inert, that is, for an autoprotection mechanism to become activated which stops any further corrosion (an ability to become passivated);
- a tendency to form alloys (alloy-forming ability).

More generally, the tendency of metals to dissolve is a measure of the ability of metals to lose electrons.

The tendency of metallic materials to corrode

The mobile electrons which are present in the metallic elements can be lost when such metals come into contact with substances which exhibit a pronounced tendency to acquire electrons. On account of this, metals are oxidized, that is, they corrode. Corrosion is therefore the result of a contest for the possession of electrons where it is the metallic material which loses them.

As can be seen from the series of reduction potentials presented in Appendix A2 to this volume, among the common agents present in the environment, there are:

- oxygen in the presence of moisture, which implies the reduction half-reaction: $2H_2O + O_2 + 4e^- \rightarrow 4OH^-$, with $E^\ominus = +0.4$ V;
- acidity which allows the fundamental reduction half-reaction: $2H^+ + 2e^- \rightarrow H_2$, with $E^\ominus = 0.000$ V to take place;
- oxygen in the presence of acidity which involves the reduction half-reaction: $4H^+ + O_2 + 4e^- \rightarrow 2H_2O$, with $E^\ominus = +1.23$ V.

The third reaction is the strongest acceptor of electrons, followed by the first and then the second.

More precisely, it is established that:

- all the metals (with the exception of gold) lose electrons when they are in an aerated acidic environment;
- only the metals which are quite noble (and these, therefore, do not include copper) do not lose electrons in a humid, oxygenated environment;
- all the noble metals (lying below hydrogen in the electrochemical series) do not lose electrons in an environment which is simply acidic.

The ability of metals to become passivated

It is found in practice that some metals which have a low reduction potential (chromium, $E^{\ominus} = -0.74$ V) or a very low reduction potential (aluminium, $E^{\ominus} = -1.67$ V) are actually only slightly oxidizable, that is, they are only slightly corroded.

In fact, it is well known that metallic components which have been well chromium plated are extremely resistant to environmental agents (acidity, oxygen, water) and that aluminium surfaces are and remain a bright silver colour which barely changes over long periods. This behaviour of chromium and aluminium is the consequence of a specific (in the sense that it is not the same for all metals) capacity of metals to protect themselves by coating themselves with an extremely thin layer of their compounds. These compounds are generally oxides, but, in the case of some metals, it may be layers of their salts. The state of autoprotection attained by a metal is referred to as 'passivation'.

Passivation can be brought about in metals either by using chemical reagents or by making them the anode of electrolytic cells where oxygen is developed (anodic passivation). If the passivation is to be good, the protective layer of oxide or salt must be extremely compact and adhere very strongly to its support. In fact, it is only when there is perfect symbiosis between the two media that the support finishes up by exhibiting the typical behaviour of its coating with respect to agents present in the external environment.

The ability of metals to form alloys

As is well known, an alloy is defined as a mixture of various (two or more) metals and small quantities of non-metals, provided that the mixture, as a whole, exhibits metallic characteristics*.

We shall subsequently return to the constitution of alloys and to the reasons why it is necessary to produce them, but, for the moment, it is desirable to explain how these important systems can be formed.

The structure with mobile electrons which is characteristic of the metallic state readily permits different metals atoms to form a metallic structure with a common electron cloud. It is precisely this fact which lies at the foundation of the formation of alloys.

In the case when the alloy includes a non-metal, the latter element, in general, enters into the structure in such a way that it occupies the interstices between the atoms of the alloying elements, and always involves an element of relatively small atomic dimensions such as carbon, nitrogen, or boron.

13.4 METALLIC ALLOYS

Solid metals therefore have the structures of crystals with the binding between the components being ensured by the 'metallic bond' (Fig. 13.6).

However, various atomic substitutions may occur in metallic crystals, leading to inclusions of smaller atoms and the formation of bonds which differ from the metallic bond or are hybrids with the metallic bond. The opportunity for the occurrence of

*Individual metals and alloys constitute metallic materials as a whole, and metallurgy deals with these collectively.

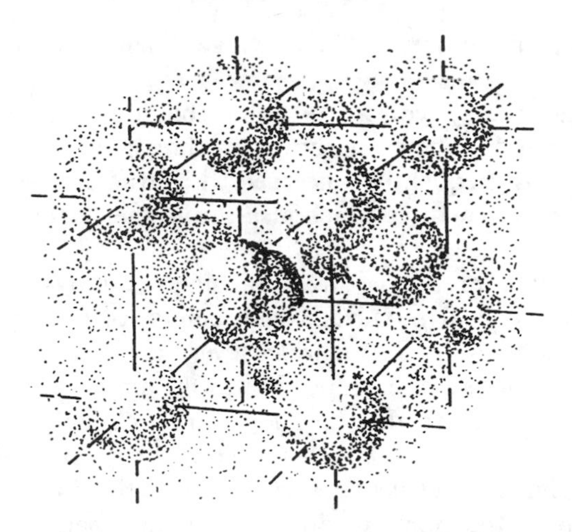

Fig. 13.6—The two characteristics of the metallic state: crystallinity, that is, the ordered arrangement of the atoms, and the electron cloud which is delocalized over the whole of the metallic body. The distance between the particles at the lattice points is exaggerated for the sake of clarity.

alloys exists in all of these possible 'deviations' from the structure of the normal metallic state.

The composition and basic properties of alloys

In an alloy, some metallic component, referred to as the base metal, is always present to the largest extent. This is accompanied by smaller amounts of other components known as alloying elements. The base metal determines the type of alloy, and one therefore speaks of iron alloys, copper alloys, lead alloys, etc.

The alloying elements in alloys are not casual impurities but additions of other components to the base metal which have been carefully examined both qualitatively and quantitatively and are capable of bringing about profound changes in the properties of the base metal such as the electrical conductivity, the chemical resistance, and the hardness.

In particular, the action of atoms of non-metals which are occluded in the crystal lattice of the base metals of alloys is such as to hinder any attack on the system on the part of chemical reagents, to harden their matrices, modify their malleability, and, more generally, their toughness.

For instance, the addition of small amounts of carbon is sufficient to transform iron into steel. These modifications, while not leading to a loss of metallic character as a whole, are profound from a chemical and a mechanical point of view.

The modifications induced in the base metal by the alloying elements depend on the nature and the amounts of the latter. Alloys therefore also have the advantage over pure metals that they can be prepared in a vast range of types which are adapted to a great variety of uses.

The structure of alloys

By 'structure', one does not mean the elementary composition of an alloy but rather the manner in which the atoms of the alloying elements are located with respect to the base metals.

Structurally, alloys may consist of solid solutions, mechanical mixtures, or intermetallic compounds. These are, however, three limiting structural states, and, in practice, one is most frequently concerned with structures which are intermediate between them, that is, structures which result from their combination in pairs or of all three.

The simplest of these three basic structural states of alloys to comprehend and to study is that of mechanical mixtures which consist of heterogeneous mixtures of components of various grain size in which every constituent tends to manifest its own individual characteristics in proportion to the amount of it which is present.

For instance, it is observed that an alloy which is simply a mechanical mixture of the crystals of the two metals has an electrical conductivity (the product of conductivity and volume) which lies between the conductivities of the two components.

There is therefore little point in dwelling on the treatment of alloys with the structures of mechanical mixtures, as the study of the two other types of alloy structure is indubitably of greater importance.

Solid solutions

The fusion of several elements which are generally metallic but also metals and non-metals, transfers their components in various ways into the solids which are thus formed. One of these ways involves the formation of solutions also in the solid state. Such solutions may either be of the 'substitutional' or 'interstitial' types.

Substitutional solid solutions. When two metals with comparable atomic dimensions, the same coordination capabilities, and polarizabilities are crystallographically similar, they tend to yield the same crystals. It is therefore very likely that, if the atoms of such metals find themselves in the same fusion mixture, they will enter, when solidification occurs, as parts of the very same crystal, subject to the criterion that the metal which is present in the greater amount will occlude that which is present in a lesser amount.

The structure of the basic metallic matrix is not changed by this mixture of chemically different species, and the melt therefore crystallizes with a habit which is the same as that of the pure occluder.

Since solid mixtures of crystallographically similar metals change their composition in a continuous manner as a function of the composition of the melt from which they are derived, one is dealing with true solid state solutions, bearing in mind that a solution is every system which is a mixture of substances of a composition which is gradually variable, even if within more or less wide limits, with changes as small as may be desired.

For example, the fusion of silver and gold leads to the formation of solid substitutional solutions. In the solidification phase, it may happen that the silver atoms in a crystal of this metal are replaced by gold atoms without, however, any change in the actual crystal structure of the silver, when the melt contains less gold than silver. On the other hand, if silver is added to molten gold, a certain number of atoms of the host metal may be replaced by an equal number of guest atoms in the crystals which are formed at the site of solidification.

The case of silver and gold is one of the most favourable for the formation of solid solutions because the affinities which the two metals have for one another is such as to make them completely miscible in all proportions.

Normally, however, the solubility in the solid state also cannot go beyond certain limits in accordance with the general definition of a solution which has just been stated. For example, silver and copper are only partially miscible.

* * *

When the atoms occluded in solid solutions randomly occupy the nodal points of the crystal lattice of the occluding metal, one is dealing with simple substitution solid solutions, while, if the distribution of the guest atoms is perfectly regular and ordered, that is, absolutely uniform, the resulting crystal system is defined as a phase with a superstructure (Fig. 13.7).

The formation of simple substitution solid solutions rather than phases with superstructures depends on the degree of likeness between the atoms, by their numerical ratio in solution, the temperature at which solidification occurs, and, to the greatest degree, on the time over which the solidification process takes place. The likelihood of the formation of this type of solid solution is less than that for the formation of simple substitution solid solutions on account of the need for all of

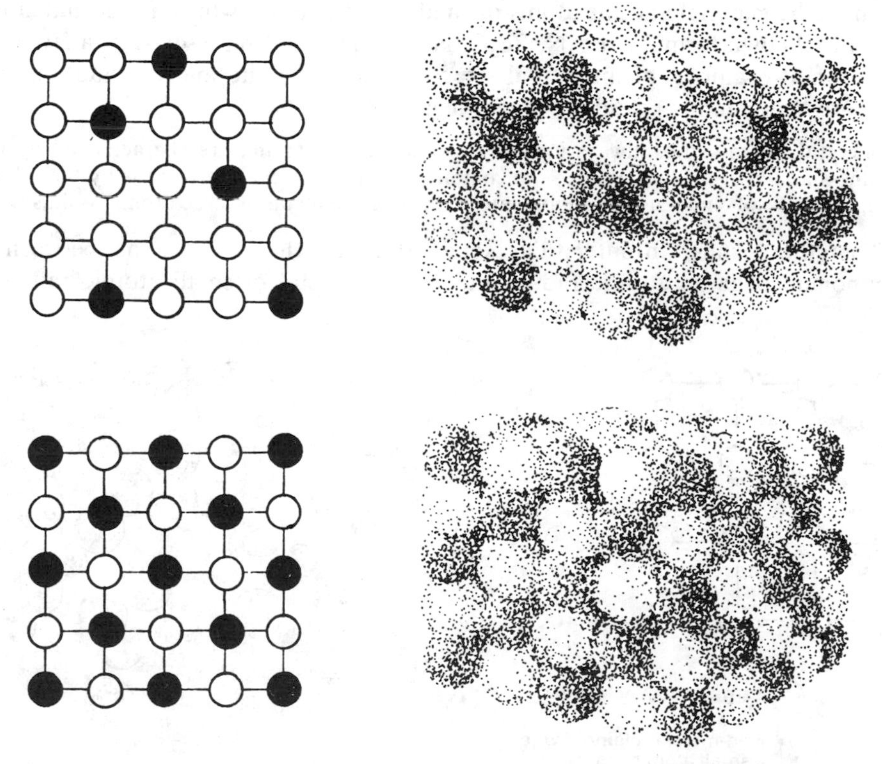

Fig. 13.7—Above. Schematic diagram (on the left) and a more realistic diagram (on the right) of the arrangement in a metallic body consisting of a simple solid solution. Below. The same metallic body when a phase with a superstructure has been formed.

these factors to correspond to particular values if phases with superstructures are to be produced.

Interstitial solid solutions. If the extraneous atoms included in a melt include elements of small atomic radius (C, H, B, and N) then these atoms may be inserted into the cavities in the basic crystal lattice during solidification. Interstitial solutions formed in this manner are depicted in Fig. 13.8.

As a result of the discontinuities induced in the electronic clouds of a metal by non-metallic atoms which are inserted interstitially into the cavities of a metallic crystal lattice and as a consequence of the mechanical stresses which such insertions give rise to in the metal matrix, as well as the reduced mutual mobilities of the metal atoms which result, the chemico-physico-mechanical properties of materials involving interstitial solid solutions are very different from those of the pure metal.

The behaviour of austenite, an interstitital solid solution of carbon in γ-iron, provides a typical example of the degree to which the properties of these structures differ from those of the base metal.

Sometimes interstitial solid solutions also possess the particular structures which are representative of 'interstitital compounds' in an analogous manner to that in which substitutional solid solutions may give rise to phases with superstructures. In practice, it is found that, either during the solidification of melts or during the annealing of existing interstitital solid solutions, structures are formed with definite and constant atomic ratios between the (metallic) element constituting the matrix of the metallic material and the atoms of small atomic radius which are occluded in it. The well known formulae Fe_2N, Fe_3C, NiH_2, and Fe_2B represent but a few of the commonest examples of interstitial structures with definite and constant (because they are 'compounds') ratios between their components.

Interstitial compounds always include transition elements and are characterized by great hardness and high melting points. The commonest way of preparing them is by the diffusion of the element of small atomic radius into a mass of the metal of large atomic radius.

Therefore, while, in interstitial solid solutions the proportions between the component of large atomic radius and the non-metal of small atomic radius are

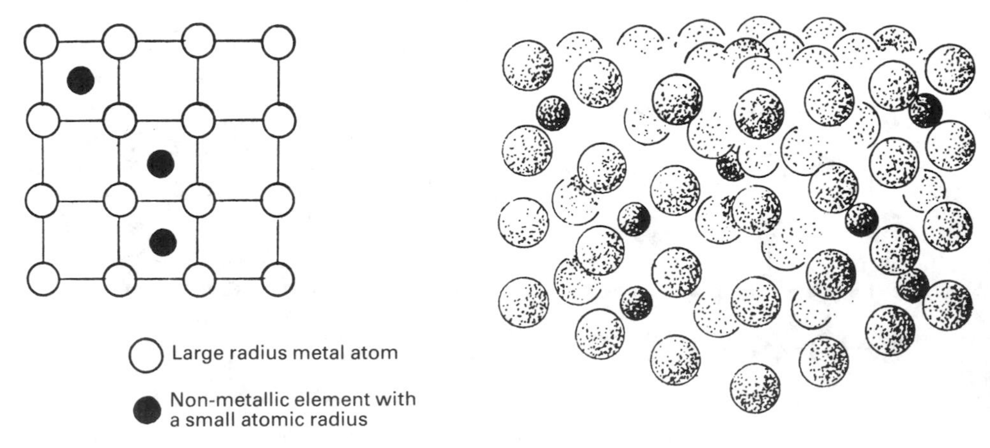

Fig. 13.8—Diagram (on the left) and a more realistic representation (on the right) of the structure of a metallic body formed by interstitial solid solutions.

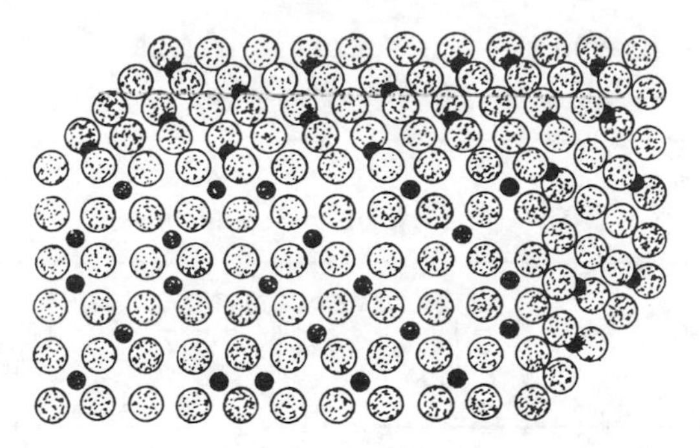

Fig. 13.9—Interstitial solid solutions sometimes have the structures of interstitial compounds, that is, of systems in which the (non-metallic) atoms of small atomic radius which are occluded between the metal atoms form assembles with those which are characterized by definite and constant atomic ratios (hence the term 'compounds'). The structure of an Me_3X type compound is depicted in this figure, where Me is a metal (e.g. Fe) and X is a non-metal (e.g. C). In the representation shown here, a random distribution of the X atoms among the Me atoms has been adopted.

variable, a fixed ratio between the matrix atoms and the occluded atoms exists in interstitial compounds (Fig. 13.9).

Intermetallic compounds

Frequently, when melts containing only metals are allowed to solidify, chemical species separate out which are characterized by definite and constant compositions even if they do not correspond to any traditionally known valency rule.

These structures are referred to as 'intermetallic compounds' regardless of the fact no ordinary valence rules are observed in them, and that, when there is parity between the components, Dalton's law of stoichiometry is found not to be obeyed in them.

An example of this is the intermetallic compounds formed between copper and tin:

Cu_5Sn $Cu_{31}Sn_8$ Cu_3Sn

On the basis of Dalton's law of multiple proportions, one should find simple ratios between the amounts of one of the components which forms compounds with a fixed mass of the other. Actually, however, ratios (which are anything but simple) between the numbers of copper atoms forming compounds with 8 atoms of tin of 40:31:24 are found.

Intermetallic compounds therefore do not obey Dalton's law, and they are known as 'non-Daltonide compounds'.

Further, intermetallic compounds are not simply random combinations, but they are subject to a statistical law (the Hume–Rothery rule(s)) according to which the ratio between the sum of the valence electrons* of their components and the sum of the number of atoms from which they are formed assumes characteristic constant values to each of which there corresponds a 'phase'.

The commonest Hume–Rothery ratios are 3/2 (the β-phase) 21/13 (the γ-phase) and 7/4 (the ε-phase), and these, for example, apply in the case of the compounds shown in Table 13.1.

*Among the valence electrons which must be taken into account here are those in the outermost shell, while the Group VIII elements are, in addition, considered as being zerovalent.

Table 13.1—Examples of formulae of intermetallic compounds belonging to different 'phases'

Hume–Rothery phases	*Representative compounds*
β-phase	CuBe, CuZn, Cu_3Al, Cu_5Sn, FeAl
γ-phase	Cu_9Al_4, Cu_5Zn_8, Ni_5Zn_{21}, $Cu_{31}Sn_8$, Au_5Cd_8
ε-phase	$CuZn_3$, Ag_5Al_3, $FeZn_7$, $AuZn_3$, Cu_3Sn

It is easy to check that the intermetallic compounds shown in Table 13.1 obey the Hume–Rothery rule. By randomly selecting three of these compounds (one from each phase), FeAl, $Cu_{31}Sn_8$, and Cu_3Sn, and bearing in mind that iron is considered as being zerovalent because it belongs to Group VIII of the Periodic Table, it is found that:

$$\frac{\text{Sum of the valence electrons}}{\text{Sum of the constituent atoms}} = (0+3)/2,\ (31+32)/39,\ (3+4)/4$$

$$= 3/2,\ 21/13,\ 7/4 \text{ respectively.}$$

It should also be noted that:

- on account of their structural correspondence to a calculation involving the number of electrons, intermetallic compounds are also known as 'electron compounds';
- all the intermetallic compounds of the β-phase crystallize as face-centred cubic crystals, while those belonging to the γ-phase crystallize in a complex cubic form with basic groupings of 52 atoms, and those belonging to the ε-phase crystallize in a hexagonal form with a lattice of somewhat unusual dimensions.

Justification for the natural existence of intermetallic compounds has been given by means of quantum mechanical calculations which have led to it being demonstrated that potential energy minima are associated with just such structures, and it would therefore be expected that these compounds would have the greatest likelihood of existence with respect to other intermetallic atomic distributions which are theoretically conceivable.

Imperfections and defects of crystal lattices

It is of interest to have some knowledge of the nature of the lattice abnormalities which can arise in the crystals of metallic materials, because they determine various aspects of the behaviour of the crystals and compel one to carry out operations to improve the crystals which tend to eliminate or diminish the number of such lattice abnormalities.

These abnormalities may be either imperfections or defects.

A crystal lattice is imperfect if it contains zones, of various sizes, which are irregularly structured, with atoms displaced from their normal positions. This arises from distortions due to innate malformations or is caused by the action of external agents (traction, pressure, etc.) on crystals which have already been formed.

On the other hand, a crystal lattice has defects when there are atoms randomly distributed in interstitial positions or when some atoms occupy lattice points which, for reasons of symmetry, should be occupied by atoms of a different type. The absence

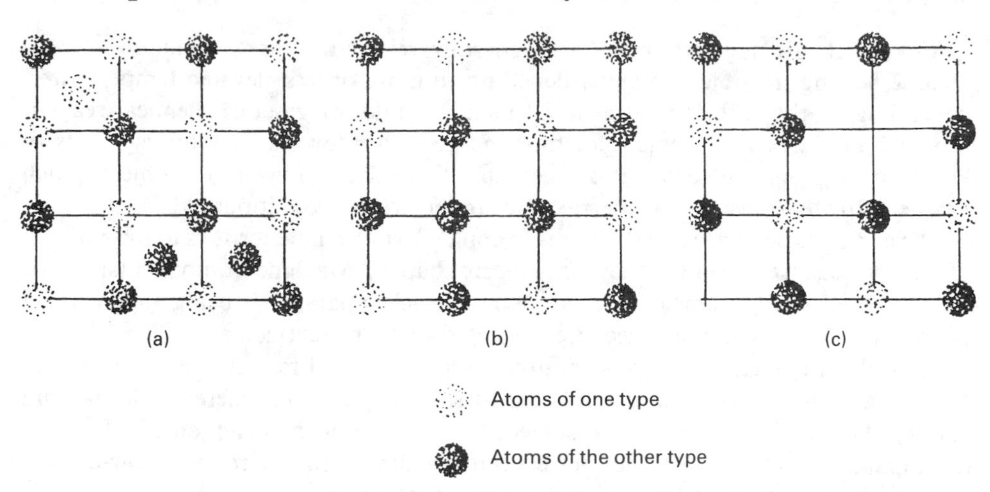

Fig. 13.10—In (a) atoms of pure metals (in which case they are all the same) and of alloys are distributed irregularly in interstitial positions: in (b) atoms which are necessarily different (and so this applies only to alloys) mutually 'usurp' each other's positions: in (c) certain lattice points of the metals or alloys are empty. These are some 'lattice defects'.

of atoms from certain lattice points also constitutes a lattice defect. Some types of lattice defect are shown in Fig. 13.10.

Lattice defects of types (a) and (c) can also be realized in crystals of pure metals, whereas type (b) defects are possibly only in alloys.

Given the great tendency of metals to crystallize, lattice imperfections due to congenital crystal malformations as well as by permanent distortions brought about by the action of mechanical forces have little effect upon the properties of the alloys. The consequences of lattice defects caused, especially, by some of the atoms being displaced into interstitial positions during the formation of the mixed crystals, and atoms of various kinds exchanging their positions within the lattice, are both conspicuous and frequent.

In contrast to what happens in mixed salt crystals where the need to maintain electrical neutrality strongly opposes any exchange in the positions of the ions, in metallic crystals it is only the equal probabilities for the distribution of the atoms which reduces this phenomenon. These equal probabilities are readily rendered unequal by the random forces produced by thermal states, mechanical actions, stray electrical currents, and chemical actions which modify the ordering of the components in the lattice. Defects of this kind are induced, for example, by the presence of hydrogen, H_2S, O_2, and other gases in boiler waters acting under pressure, and by the preliminary treatments when metallic catalysts are put into operation. In the latter case, such treatments are intentionally carried out because the catalytic activity itself, or, more often, the increase in the catalytic activity of the material, may be connected with the existence of lattice defects.

The formation of alloys by the action of reagents on metallic surfaces

It is of interest to consider how the phenomenon of alloying due to the formation of intermetallic compounds and interstitial solid solutions on the surfaces of metals and alloys can be brought about, with particular reference to ferrous alloys on account of their special technological importance.

Superficial alloying treatments are generally referred to as 'cementations', and include heating the object to be hardened up to more or less elevated temperatures, depending on the type of treatment, in a solid, liquid, or gaseous chemical reagent in which the body must remain for the time necessary for one or more elements to be absorbed from the cementation medium. The final cooling of the object which has been treated must also be carried out under precise conditions.

The aim of these surface treatments is to confer certain characteristics to the surfaces of the objects treated after they have been shaped. Machine components, gears, change-speed shafts, camshafts, pump axles, which, apart from being tough, must possess excellent wear resistance, are subjected to such treatments.

Inter alia, the teeth of many gears need to have internal resilience combined with superficial hardness when in use. That is to say, they require different internal and surface states if they are not to be subject to fracture under the action of a force or an impact. Therefore, once they have been machined from a resilient low-carbon steel, their surfaces must be hardened by cementation.

Carburizing. This is a process for the hardening of ferrous alloys by the formation of carbides and, especially, cementite, Fe_3C carried out with solid, liquid, or gaseous 'iron-working cement'.

Solid iron-working cement is a mixture of wood charcoal granules and activating constituents consisting of alkali and alkaline-earth carbonates. These prime the reaction by liberating CO_2 by dissociation which, when it reacts with the carbon, is reduced to CO:

$$CO_2 + C \rightarrow 2CO$$

Upon coming into contact with red-hot iron (the process is always carried out at a temperature exceeding 930°C), CO liberates active carbon by disproportionating:

$$2CO \rightarrow CO_2 + C^*$$

The CO_2 passes back into the cycle, while the active carbon forms cementite:

$$C^* + 3Fe \rightarrow Fe_3C$$

which diffuses into the iron, sharply modifying its characteristics.

Carburizing with solids is carried out in steel casks arranged and heated in furnaces. These casks are closed and sealed with clay so that air will not enter, and the objects to be treated are kept separated by means of structures made of concrete.

Liquid iron-working cement is usually a fused mixture of alkali and alkaline earth carbonates with cyanides. The temperature at which the process is carried out is higher than 1000°C, which stimulates the endothermic reaction:

$$Me(CN)_2 + 3Fe \rightarrow Fe_3C + MeCN_2$$

as a result of which cementite is formed which hardens the metal by diffusing into the structure which is being treated.

The pieces to be hardened are immersed in the molten iron-working cements which are protected from contaminants and air. The process proceeds at a much faster pace than when it is carried out with solids.

Gaseous iron-working cement is made, for the most part, from a mixture of methane and carbon monoxide. The CO also functions here, as previously, by disproportionation, while methane carburizes the metal directly:

$$CH_4 + 3Fe \rightarrow Fe_3C + 2H_2$$

The treatment is carried out in steel retorts arranged in furnaces or heated electrically. The gaseous cementation agents circulate in these retorts. They act less rapidly than liquid cementation agents which are more controllable, as the composition and charge can be readily changed.

In every case, the amount of carbon absorbed in a properly hardened component must be greatest in the surface and then gradually and uniformly decrease toward the interior of the component to ensure the best possible adhesion between the hardened border of the component and its inner core.

Direct tempering (more or less rapid cooling of the treated pieces) or indirect tempering (cooling the pieces slowly at first, then returning them to be heated up again, followed by rapid cooling) always follows the cementation operation if the hardened components are not to be subjected to any further mechanical working.

Nitriding. When alloy steel components are heated up to 500°C in an atmosphere of ammonia, the latter is dissociated with the release of 'nascent' nitrogen which combines with the iron and with the alloying elements to give nitrides of the Fe_2N, Fe_4N, AlN, CrN, VN, and Mo_2N which harden the steels which are treated, and make them both corrosion and heat resistant.

Nitriding is a hardening operation which is carried out by using gaseous reactants which is extremely slow in comparison with the corresponding carburizing process, and it is also necessary in this case to pay greater attention to the physicochemical and mechanical state required of the components being treated. These, in fact, apart from containing various metals which form nitrides, must present surfaces that are receptive to the reagent, and all mechanical working must have already been completed in so far as, while there is little deformation at the relatively low temperatures used, the nitrided compounds are intolerant towards mechanical forces after the treatment.

Carbonitriding. When operating in gaseous mixtures of air and methane at a temperature of 700–880°C, a gaseous mixture containing CO and nitrogen is obtained which is capable of carburizing and nitriding alloys. The operating temperature has been arrived at in such a way as to favour both processes: nitriding being dominant at the lower temperatures and carburizing at the higher temperatures at which the process is conducted.

Apart from the lower working temperature, the carbonitriding process conditions are those for carburizing with gaseous reagents, and the process is specific for the hardening of components of small thickness which could be deformed at the higher temperatures at which carburizing alone is carried out.

Part 2 GENERAL METALLURGY

13.5 NATURAL METAL RESOURCES. PERSPECTIVES

Almost three quarters of the naturally occurring elements are metals, but the large number of such elements contrasts markedly with their low quantitative abundance. There is presumably less than 1% of the Universe and 20–25% of the Earth's crust consisting of metals.

Table 13.2—Natural resources of technologically important metals

Metals	%	*Metals*	%	*Metals*	%	*Metals*	%
Aluminium	7.52	Vanadium	0.01	Niobium	0.002	Bismuth	0.0002
Iron	4.8	Zinc	0.01	Lead	0.002	Mercury	0.00005
Magnesium	1.95	Nickel	0.008	Molybdenum	0.001	Silver	0.00001
Titanium	0.48	Copper	0.007	Beryllium	0.0007	Cadmium	0.0001
Manganese	0.095	Tungsten	0.007	Germanium	0.0007	Palladium	0.000001
Chromium	0.02	Tin	0.004	Uranium	0.0004	Gold	0.0000005
Zirconium	0.02	Cobalt	0.002	Tantalum	0.0003	Platinum	0.0000005

The natural abundances in the Earth's crust of the metals, which are predominantly of technological rather than chemical interest and which will be discussed in this section, are shown in Table 13.2.

On the basis of these tabulated values it is seen that, apart from those of aluminium and iron which are, however, for the greater part (especially the former) present in deposits which are non-exploitable* for the purpose of obtaining metals from them, the natural abundances of the metallic elements being considered are either low or very low. Fortunately, both the few metals which occur in the native state (essentially gold, platinum, mercury, silver, and palladium) on the Earth and the combined metals are concentrated in local accumulations which constitute mineral deposits containing amounts of metals which are considerably higher than the average amounts of these metals present in the Earth's crust.

However, the richest mineral deposits are disappearing because of the growth in their exploitation, which has increased at an exceptionally fast pace during the course of this century.

From a production of the order of less than ten million tons of metals at the beginning of this century, it had passed to 100 million tons in 1938 and to an order of a thousand million tons at the present time. Metals which were entirely disregarded such as titanium, cobalt, vanadium, zirconium and niobium, have today leapt into the forefront to a point where the overall annual productivity of these metals is now of the order of tens of millions of tons.

The intensive exploitation of mineral deposits has brought into perspective a serious problem which will face humanity in the not too distant future, namely, a shortage of metal-bearing materials from which essential raw materials can be extracted.

Forecasts† based on the assumption that the demand for metallic materials will continue to grow in proportion to demographic growth and to the increase in the per capita demand as the standard of living, especially in the developing countries, is raised, are that the mineral deposits of all the technologically important metals will be exhausted by 2020 with the exception of chromium which is forecast to run out in 2050 and iron which will presumably last until the decade from 2060–2070.

*A mineral source is 'exploitable' when the costs of working it, transporting the ore, and extracting the metal from it are compatible with the commercial price of the product. For instance, although the price of gold is very high, the cost of extracting it from sea waters which contain the greater part of it is still prohibitive. That is, the sea is a non-exploitable gold deposit.

†The possibilities of exploiting the mineral deposits on ocean floors are largely still not known, and it cannot be foreseen to what extent progress made in the techniques for the recovery of metals from scrap materials, etc. could effectively improve the position regarding the demand for metallic materials produced by a metallurgical route.

13.6 THE ENRICHMENT OF METALLIFEROUS DEPOSITS

The basic components of a mineral deposit are the mineral which is to be exploited (the 'mineral') and the 'gangue', that is, the entirety of all the other non-utilizable materials accompanying the mineral.

Nowadays, as it is becoming ever more necessary to process deposits containing smaller and smaller amounts of the mineral, the gangue has to be separated before proceeding with the actual metallurgical processes. Differences in the physical properties of the granulated particles, or even more so, of the powders of the two basic components of the deposit, are usually exploited for this purpose.

The enrichment processes which are of the greatest interest are:

- separation by gravity* which exploits differences in the specific weights of the components;
- magnetic separation, which exploits the differences in the magnetic properties of the components;
- the 'sink and float' treatment, which utilizes the separating action of a liquid system with a density which is intermediate between that of the mineral and that of the gangue;
- the flotation process, which exploits differences in the natural and induced aerophilic and hydrophilic properties of the two components.

For the theory and practice of the hydraulic and flotation classification processes the reader is referred to treatises on industrial chemical plant. Here, we shall dwell on the two other processes for the enrichment of metalliferous minerals which have been mentioned.

Magnetic separation. The mixture, in a granular state or, better still, in a state of fine subdivision such that there are almost pure fragments of the components, is carried on conveyor belts in front of the poles of an electromagnet. The different extents to which the components are attracted to the poles of the electromagnet, which depends on their magnetic susceptibilities, causes them to be separated from one another and therefore to fall into different compartments of the collecting hoppers (Fig. 13.11).

Sink and float. Granulated material with a suitable particle size distribution is introduced into a bath containing a liquid of suitable density. Generally, the mineral sinks while the gangue floats (Fig. 13.11, right).

A 'sink and float' bath is not a chemically homogeneous medium but an aqueous suspension of extremely finely milled particulate materials among which ferrosilicon (85% Fe and 15% Si) and galena (PbS) stand out in importance. All the agents used to increase the density of sink and float baths must be of a high stability in this environment, that is, with a tendency to remain in intimate contact with the liquid which envelops them.

The material extracted from the mineral, after it has been crushed and granulated, is introduced in a continuous mode into the separators which are in the form of an

*A rudimentary process in which materials are classified by using gravity is that of 'levigation' which consists of the treatment, with jets of water, of the precrushed (or disaggregated by the same jets of water) material in such a way as to convey the gangue some distance, while the mineral of interest remains deposited at the site.

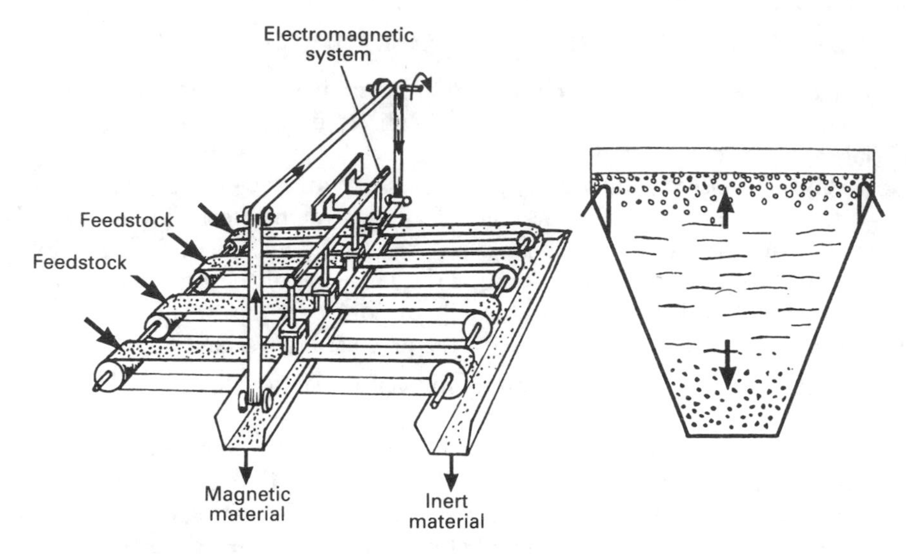

Fig. 13.11—Left. Diagram of a magnetic separation plant. Right. The gangue particles of a practically homogeneous aqueous suspension of accumulating powders tend to float, while particles of the metal-bearing mineral sink.

inverted cone with a capacity of about 10 m^3. Little by little, as the separation takes place, the light material collects in the upper part of the vessel from where it is removed, while the heavier material, which collects at the bottom, is recovered by means of a bucket conveyor which carries it away from the solution.

A panoramic overview of the operations which are intended to remove the gangue from mineral products is shown in Fig. 13.12 with the final carrying over of the concentrate which is to be formed into pellets in a truncated cone apparatus (page 458) of a suitable size for use in the subsequent metallurgical treatments.

13.7 PREMETALLURGICAL TREATMENTS

Having obtained concentrated minerals by enrichment, these are customarily subjected to treatments which, by chemically transforming the useful mineral, get it into a more suitable form for submission to the subsequent actual metallurgical treatments.

The main premetallurgical treatments are leaching, roasting, and agglomeration, the last-named sometimes being carried out at the same time as the roasting.

Leaching

The mined product, after possibly undergoing some suitable preparation, is thoroughly washed with water or treated with chemical reagents of an acidic nature. More or less concentrated saline solutions are obtained in this manner.

Leaching can be carried out by:

- irrigating the rocks containing soluble minerals with a solubilizing reagent;
- placing beds of minerals in the form of suitably sized pieces in concrete vessels

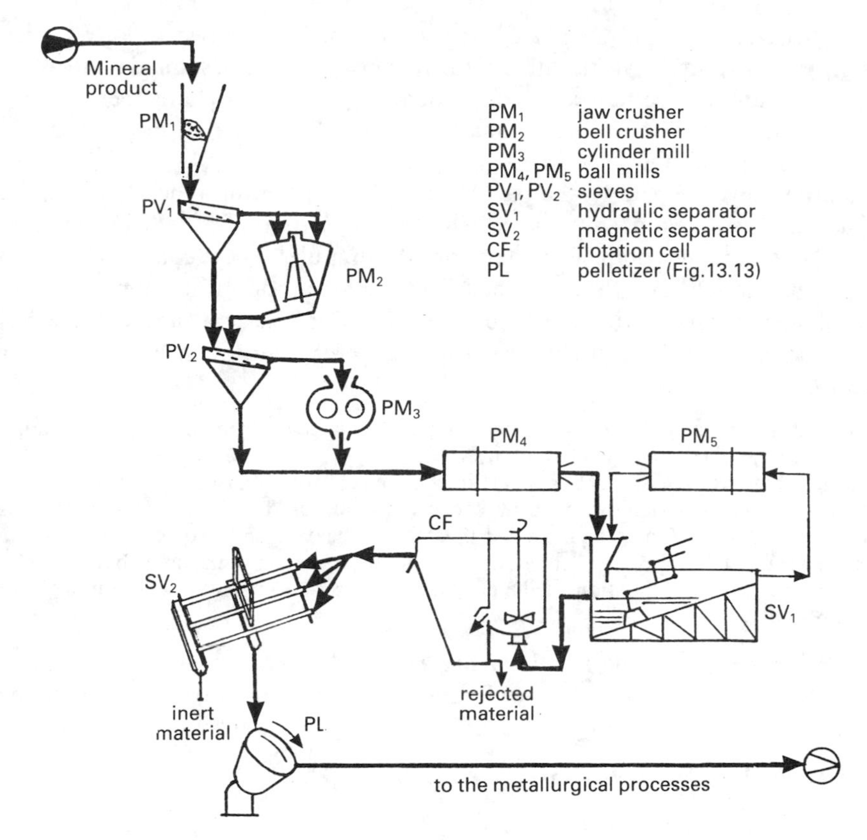

Fig. 13.12—Process scheme for the wet enrichment and coarse pelleting of mined minerals. The process uses the magnetically levitated fraction.

(percolators) with perforated bottoms, and allowing leaching agents to pass over these beds;

- keeping the crushed and milled minerals in suspension in the leaching agents by means of agitation with compressed air or mechanical mixing.

The effectiveness of leaching operations is very dependent on the fineness of the material being treated, the contact time, and the temperature at which it is carried out.

Roasting processes

In principle, roasting is a metallurgical operation in which an enriched mineral is heated in some way and to a certain degree such that, generally, it is made to react with air which is blown in.

The simplest roasting process is calcination. This is simply heating the system to a suitable temperature, and is generally carried out with the aim of dehydrating hydrated oxides (such as those of bauxites) or to dissociate carbonates or even sulphates.

Calcination is carried out in cylindrical rotating furnaces or in kettle type furnaces as used in the production of lime.

Desulphurization employs oxidative roasting which is carried out by using air, possibly enriched with oxygen, with the aim of converting metal sulphides into oxides.

In this operation, certain sulphides, such as pyrites FeS_2, are thermally autosufficient, while others must be continually resupplied with heat by means of naphtha, methane, or carbon powder burners.

Attention must be paid to the particle size distribution of minerals which are to be roasted so as to avoid, so far as possible, the formation of powders which are too fine, and to ensure the complete oxidation of the granules. These constitute conflicting requirements which have to be reconciled in the most opportune manner.

Desulphurization roasting is carried out in multiple hearth furnaces, Herreshoff type furnaces (page 192), in chamber (injection) furnaces, or in fluid bed furnaces (page 193).

In chamber furnaces, the finely ground material is injected from the top by means of a current of compressed air (to 1–1.15 atm.).

A large amount of air (secondary air) is introduced from the bottom in the countercurrent together with the material to be burnt because the air used to transport the material is insufficient to burn it. The secondary air must be preheated, and provision must be made, where necessary, to supply supplementary heat to the system by means of burners. Heat is recovered by recycling a fraction of the combustion gases to increase the amount of SO_2 in them (up to about 9%).

Another type of roasting is *chloridizing roasting* which is carried out in order to recover the valuable by-products of large-scale operations. For example, copper is recovered in this way from pyrites ashes according to the reactions:

$$CuO + 2NaCl + S + 3/2O_2 \rightarrow Na_2SO_4 + CuCl_2$$

$$CuS + 2NaCl + 2O_2 \rightarrow Na_2SO_4 + CuCl_2$$

This operation enables the volatility and/or solubility in water of the chlorides which are formed to be exploited in the recovery of the metal of interest.

Agglomeration processes

The basic aim of processes employing the agglomeration of metallurgical products is to transform materials with a fine or extremely fine particle size into products with dimensions which are suitable for treatment in metallurgical furnaces. Nevertheless, this process can also be used for the purpose of oxidizing sulphide minerals.

The two versions are commonly known as pelletizing and agglomerating roasting precisely according to the aims of the process. However, in both types of process, when powders to be agglomerated are heated at a temperature below their melting point, it is a matter of 'sintering' methods.

Pelletizing

The water content of the material to be sintered is adjusted to a certain level, and binders (for the most part, bentonite clays, lime, or gypsum) are added to it. It is then put into drum-type, plate-type, or truncated cone-type inclined units (Fig. 13.13) which rotate at a constant velocity (60–100 rpm). Nuclei are formed in this manner by the accumulation of particles which, by rolling on the walls of the containers

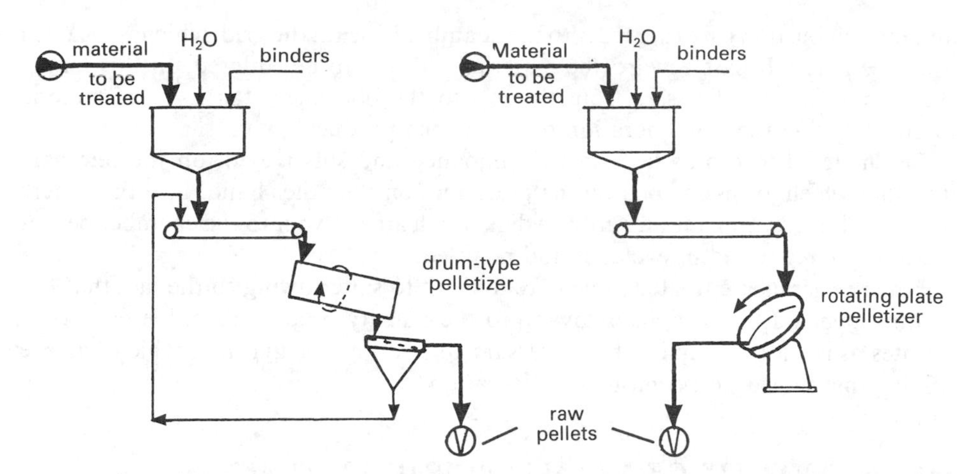

Fig. 13.13—Drum-type and rotating plate pelletization systems.

while held in contact with them by centrifugal force, increase in size until they become small spheres with diameters of 8 to 10 mm which are referred to as 'green raw pellets'.

After emerging from the bottom of these units, the small spheres are sent to a vibrating sieve which separates the 'fines' and allows them to be recycled.

Finally, the fired pellets with the elevated mechanical properties necessary for their subsequent treatment in metallurgical furnaces are obtained by heating the green pellets to the sintering temperature in other rotating furnaces or in tunnel furnaces, with the green pellets laid out on conveyor belts.

Pelletizing processes lend themselves particularly well to the sintering of the fine materials originating from the enrichment treatments used in mining.

Agglomerating roasting

A bed consisting of large pieces (1.5–2.5 cm) of material (grit) is spread onto grids operating as a slow conveyor belt (Fig. 13.14) onto which an automatic distributor deposits a 10–20 cm layer of the extremely fine mineral to be sintered.

Since sulphides (of Cu, Pb, Zn, etc.) are treated in this way, it is possible to prime their combustion at 1000–1200°C with a naphtha or coal burner. Air (the oxidant)

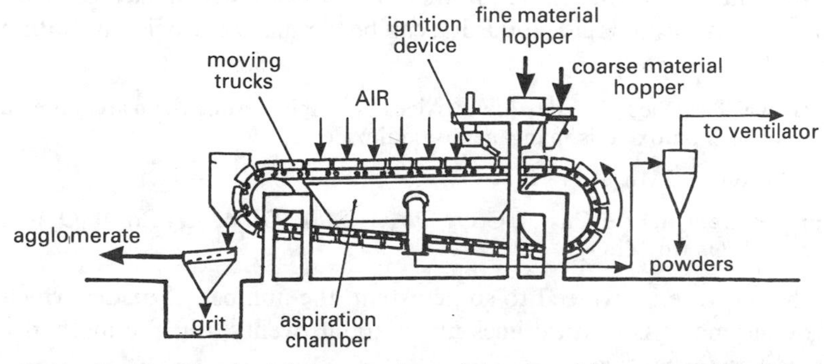

Fig. 13.14—Dwight–Lloyd unit for the roasting and sintering of the most widely varied sulphide minerals. Agglomeration by firing on a grid is particularly used in iron and steel making plants.

and the combustion gases are fed into the chamber beneath the grid, which is located in a depression, and pass toward the material which is to be sintered. In this manner the combustion front passes from the top to the bottom of the layer of material, thereby causing the agglomeration of the refined product.

The material from this last stage is unloaded and subdivided, on the one hand, into that which forms the bed on the grid; and, on the other hand, into the sintered metallurgical grade material. Any powders which are blown into the chamber beneath the grid are recovered in cyclones and recycled.

When minerals are treated which are difficult to sinter owing to the fact that their melting points are too high, or owing to the paucity of a combustible component (pyrites ashes, for example), it is necessary to add lime or gypsum, which has been suitably moistened, to promote agglomeration.

13.8 THE NATURE OF METALLURGICAL TREATMENTS

After enrichment and suitable premetallurgical treatments, the materials derived from mineral deposits pass on to the actual processes of extractive metallurgy (metallurgical treatments). These essentially consist of three classes of chemical or physicochemical processes which are optionally carried out by using reagents or, at least dissolving agents:

- reductive and slag-forming processes;
- galvanic processes;
- electrolytic processes,

which are complemented by various refinement techniques.

Reduction processes

The theory of reduction and reducing reagents

In principle, it is possible to convert metal oxides to the metal by means of dissociation reactions of the type:

$$Me_xO_y \rightarrow xMe + y/2O_2 - A\ \text{kcal} \tag{13.1}$$

which are endothermic to a greater or lesser degree and are characterized by the evolution of gas. They are therefore favoured by high temperatures and low pressures.

Thermodynamic calculations show that only oxides which have a low heat of formation, such as Ag_2O, HgO, and PdO, can be thermally dissociated at atmospheric pressure.

The oxides which can be reduced to lower oxides by simply heating them are more numerous. An example of such an oxide is manganese sesquioxide:

$$3Mn_2O_3 \rightarrow 2Mn_3O_4 + \tfrac{1}{2}O_2$$

Other examples are: CuO (to Cu_2O), CrO_3, Pb_3O_4 (to PbO), As_2O_5 (to As_2O_3), MnO_2 (to Mn_2O_3), and MoO_3 (to MoO_2).

When the pressure is reduced to some extent, the number of oxides which can be reduced by thermal dissociation goes up to about twenty, but the method is rather difficult and very expensive.

The rational resolution of the problem of the reduction of oxides is still that which has been used since the time when chemistry was a simple collection of empirical ideas. It uses reducing agents which therefore have an affinity for oxygen, and which, by constantly removing a reaction product, enable the equilibrium (13.1) to be displaced to the right.

Metals and non-metals, the oxides of which are more stable than the oxide of the metal which is to be reduced, are more or less good reducing agents.

For instance, the heats of formation of the oxides of aluminium, magnesium, and calcium, which increase in this order, are greater than those of the commonest technologically important metals and semi-metals, and may therefore be used for the reduction of Bi_2O_3, CoO, Fe_2O_3, FeO, ZnO, Cr_2O_3, SiO_2, SnO_2, and so on.

A large amount of energy in the form of heat is made available in all of these reduction processes because the pertinent reactions, as has been stated, convert less stable compounds (the oxide to be reduced) into more stable compounds (the oxide which is formed). For this reason one is dealing with energetically autosufficient processes which need only to be primed to produce the products.

In its own right, hydrogen is the best reducing agent both on account of the fact that its oxide has a large heat of formation and also because the formal equilibrium position is never attained in the reduction reaction, owing to the constant removal of water from the reaction mixture in the form of a gas. However, the cost of hydrogen and the dangers inherent in its use have restricted its use for practical purposes to a few special metallurgical processes.

In practice, carbon has also been, and remains, the reducing agent of the greatest importance by far, both directly in the form of carbon, and especially in the form of CO, since carbon monoxide assumes the role of the fundamental reducing agent in metallurgy.

There are various reasons for this preference accorded to carbon and carbon monoxide as reducing agents:

- ready availability and relatively low cost;
- the great stability of the oxide which is formed (CO_2);
- the volatility of the reduction product which leads to all the oxide being reduced until it is completely exhausted.

Given the great importance of carbon and carbon monoxide as reducing agents, the ensuing discussion will preferentially refer to these.

The preference which has just been stated is due also to the fact that, while many tried and tested structures are available for carrying out reductions with C and CO, no specific pieces of apparatus exist for carrying out reductions that use metals and non-metals. In fact, for this purpose one either works in some type of smelting furnace (in particular, electrical furnaces and, especially, induction furnaces (page 481) to avoid the formation of the carbides of the metals which are produced* which may occur in arc furnaces) or in the same furnaces as those employed in the desulphurization of minerals by the addition of deoxidants and, possibly, slag forming agents (page 463) to the mass when it is still molten. At the end, before the metal produced is extracted or poured from the furnace, the oxides formed from the deoxidants and any slags which float together on the molten mass are removed.

*It is noted that recourse has often to be made to the action of deoxidants other than carbon on account of the fact that carbon forms undesirable carbides with the metals to be prepared.

Apparatus for the reduction of metals

Oxides originating from mineral deposits and which have been simply enriched, or oxides which have been formed after enrichment at the site where the premetallurgical treatments are carried out, are reduced in furnaces which will first be briefly described and schematically classified:

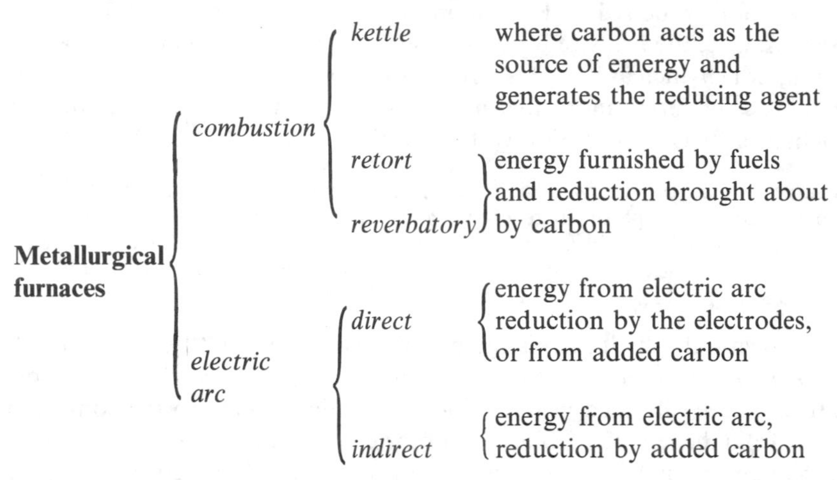

There are various versions (the blast-furnace, water-jacket, horizontal retort, vertical retort furnaces, etc.) of all these furnaces, and, in particular, of kettle combustion furnaces, the latter being by far the most important. Here, however, only a cursory descriptive mention will be made of such furnaces; the presentation of more detailed information concerning them will be postponed until they are considered with regard to specifical metallurgical processes.

Kettle combustion furnaces. In operation, these are characterized by the charges which descend into their interior and rising air which, by burning the carbon, provides the heat for the system and for the formation of the CO which acts as the reductant. The gaseous products pass out from the top, while the liquid (molten) products are drawn off from the bottom. Particular attention is paid to this type of furnace in connection with the metallurgy of lead and, especially, iron.

Retort combustion furnaces. An intimate mixture of carbon and the oxide mineral is loaded into retorts, which are externally heated. The metal liberated during the reduction process falls in drops outside of the retorts, and the slag is then removed, in a discontinuous manner if they are horizontal retorts, and in a continuous manner if they are vertical retorts. Details of these furnaces will be furnished in connection with the metallurgy of zinc.

Reverbatory furnaces. These are equipped with a pit into which the mixture of the mineral to be reduced and carbon are loaded. They are mainly heated by the radiation (reverberation) produced by their refractory canopy which is heated by the combustion fumes produced in a lateral hearth. A diagram of these furnaces is presented in Fig. 13.15.

Direct arc electric furnaces. These are usually fitted with three electrodes which penetrate into the furnace crucible (Fig. 13.16a) and strike short circuited arcs between their lower extremities and the molten charge with the production of a great deal of thermal energy. Contact between the electrodes and the molten mass brings about reduction reactions which are promoted by the electrode carbon.

Indirect arc electric furnaces. Here, the arc is struck between the two carbon electrodes shown (Fig. 13.16b) and the mass formed from the oxide mineral and the carbon reacts upon being heated by the radiation from the arc. The wear and tear on the refractory walls in this type

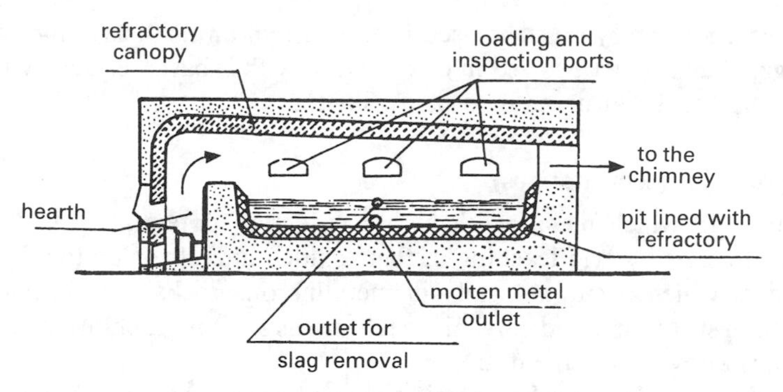

Fig. 13.15—Section of a reverbatory furnace in which the charge is heated by convection and irradiation. The charges vary, depending on the operation which is to be carried out in the furnace.

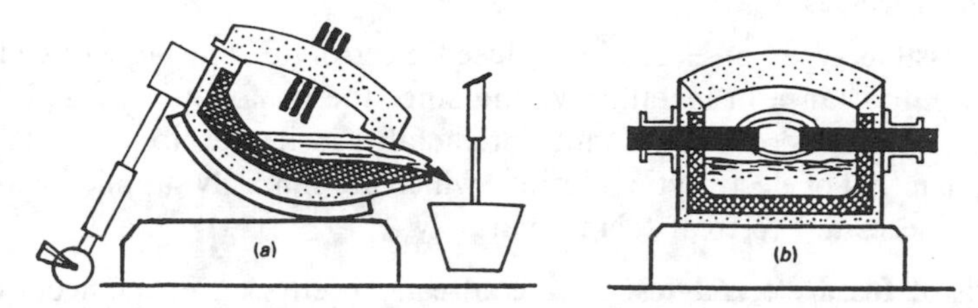

Fig. 13.16—Left: (a) A crucible electric furnace with three (three-phase) electric arcs. The carcass is made of sheet steel and lined internally with a refractory. The furnace is shown in the pouring position when the slag is separated from the alloy in the version of the furnace shown here. Operation of the hydraulic ram attached to the rear of the furnace puts it into the pouring phase by rotating it on the circular base on which it rests. Right: (b) refining phase in an indirect arc electric furnace. A section is shown through the furnace at a point where the electrodes, between which the arc is struck, are located.

of furnace is much higher than that in direct arc furnaces. It is for this reason that they are much less used.

Slag formation

The gangue which accompanies oxide minerals, even if they are enriched, is taken off from the metallic material which is to be produced. For this purpose it is made to combine with suitable reagents (especially, SiO_2 and CaO which are added to the mineral furnaces when they are operating. These are referred to as slag-forming reagents. Slags are light products which float on the top of the fused metals and are therefore removed from the furnaces by skimming or by passing out of the furnaces through laterally disposed conduits. Since the melting point of a gangue is lowered when it reacts to form a slag, slag-forming reagents are also known as fluxes.

The nature of the slagging agent is of importance because it may determine the outcome of the metallurgical process. In fact, the use of a basic slagging agent, such as lime and magnesium oxide, promotes good desulphurization and dephosphorization of the molten metal, while an acid slagging agent favours deoxidization. The type of refractory with which the furnaces are lined also depends on the nature of the slagging agent. An acid refractory must be used in the case of an acid slagging agent; and a basic slagging agent must be used in conjunction with a basic refractory if attack on the wall linings of the furnaces is to be avoided.

Galvanic processes (cementations)

The operation in which metals with higher reduction potentials, which have been converted into salts by acid leaching of their oxide minerals or by the dissolution of their powders which occur naturally in metalliferous rocks, are displaced by less noble metals in short-circuited galvanic reactions, as seen in the chlorine–soda process (page 325), is known as 'cementation'.

In a metallurgical plant, the 'cements' obtained by this route are, after an immediate purification process when required, smelted and poured into ingots and then sold for direct use or, more frequently, to be refined.

Electrolytic processes

The two possible types of electrolytes, fused electrolytes and aqueous electrolytes, are both of importance in metallurgy, and since the behaviour of fused electrolytes which are subjected to an emf is more predictable in its development and linear in its course than that of electrolytic solutions which are similarly supplied with electrical energy, it is necessary to treat them separately.

Electrolysis of fused electrolytes. As has already been seen in connection with the production of metals of predominantly chemical interest (page 419), from a fused electrolyte containing positive metal ions it is possible to bring about the cathodic discharge of such ions, with the production of the corresponding metal. Electrolysis in fused electrolytes is also of great interest in the production of technologically important metals, especially aluminium and magnesium.

Electrolysis in aqueous solution. In an aqueous solution it can only be guaranteed* that electrolysis of salts of the noble metals, that is, those which follow hydrogen in the reduction potential series, will lead to the discharge of the cations at the cathode. Meanwhile, electrolyses using metal cations which are less noble than hydrogen are problematic with respect to the efficiency of the electrode phenomena.

In general, it may be stated that, in aqueous solution, cathodic reduction leading to the metal to be produced is possibly only if the combined effect of the high concentration of the cations of these metals (which is positive) and the 'hydrogen overpotential' (which impedes the evolution of hydrogen) is such as to bring about an inversion in the normal order in which the competing chemical species are discharged.

It is therefore necessary, from case to case, to modify the operating conditions (that is, the type of electrodes, the concentrations of the solutions, especially, with respect to their pH values, the temperature, etc.) in an appropriate manner so as, in practice, to bring about an inversion in their positions in the reduction potential series in order to favour the cation which it is desired should be discharged in preference to every other chemical species present in the solution.

*Here, it is a matter of a 'guarantee' which has to be constantly maintained by taking care to keep the concentration of the electrolyte component which is to be discharged up to its correct value.

The refining of metals

The metallic masses obtained from the metallurgical processes, which have been described, frequently need to be purified in order to remove elements which are considered to be undesirable. The refining processes employed for this purpose can be classified as follows:

Refining
- *Pyrolytic*
  - using partitioning processes
  - using oxidation processes
  - using differences in physical properties
- *Electrolytic* using differential electrolytic processes (electrolytic refining)

There are therefore various technological processes in metallurgy by means of which one or more substances are partially or completely eliminated from an alloy in the molten state.

Refining by partitioning. This is based on the fact that a component I in a molten metal is capable of partitioning itself, in a manner similar to that occurring in solvent extraction, between two immiscible phases A and B. At equilibrium, the ratio between the concentrations in the two phases is a constant k which is known as the partition coefficient.

Therefore if the concentration of I in phase A is indicated by C_A and the concentration of I in phase B by C_B, the following expression is valid:

$$\frac{C_A}{C_B} = k \tag{13.2}$$

This type of refining therefore exploits the Nernst partition law applied to molten mixtures.

Refining by partitioning is more successful as a process, the greater the amount of the component to be removed from the metal being purified which is 'captured' by the added phase in comparison with the amount which remains in the molten metal. It is obvious that this 'capture' corresponds to the greater solubility which the component to be extracted must have in the extracting solvent during the solvent extraction process. Here, the term capture has been used because it may be a matter of an affinity to react chemically as the sulphur, contaminating metallic materials, reacts with sodium carbonate, for example.

The satisfactory outcome to this process is subject to kinetic factors which permit a complete and rapid attainment of the partition equilibrium described by (13.2).

The partition refining process is, per se, only partial, and its efficiency can be improved by repeating the operation several times, just as in the case of solvent extraction.

Refining by oxidation. By selectively oxidizing the elements to be removed from a metallic material in a controlled manner, they can be converted into solid, liquid, or gaseous phases which are insoluble in the molten metal from which they can therefore be separated by various methods (skimming, blowing inert gases into the mass to carry it out with the gas stream, etc.).

Elements such as silicon, manganese, chromium, aluminium, iron, and phosphorus, and especially carbon, can be eliminated by this route during the refining of metal and alloys.

The oxides formed are intrinsically solids (e.g. SiO_2, Al_2O_3, MgO) or gases (e.g. CO, CO_2, CO_2, P_2O_5) for the most part under the conditions of their formation within the molten metal which is being treated; but it is actually a liquid phase which must be removed, as the oxides tend to react among themselves to form liquid slags containing ferrites, aluminates, and, above all, silicates.

It is not possible to refine metallic materials from elements which are less oxidizable than those which must remain in the metallic phase to be purified. For example, nickel and copper cannot be eliminated from ferrous alloys by selective oxidation.

Refining by using differences in physical properties. By utilizing their different boiling points, cadmium impurities (b.p. 764°C) are separated from those of zinc (b.p. 906°C); some amalgamated metals are liberated from mercury on account of the fact that they sublime at different temperatures; and lead and tin, for example, are purified from many metals or vice versa by using the differences in their melting points. Finally, because of its higher density, copper sinks in aluminium.

Electrolytic refining. There may be a large number of metallic impurities in an ingot, a plate, or a melt or unrefined metal (copper, nickel, silver, aluminium, etc.). In every case, these impurities are always divided into two categories: those elements which are electrochemically more noble than the metal present in an enormous excess in the system to be refined, and those of them which are electrochemically less noble.

By making the melt, ingot, or plate of unrefined metal the soluble anode of an electrogalvanic system, it is possible to oxidize (firstly) all the less noble metals (of lower reduction potential) in the metal to be refined, and (successively) also the metal itself. On the other hand, on account of the large amount of the latter present, there is no way in which the more noble contaminants (with higher reduction potentials) of the material being refined can be oxidized, with the result that their particles will gradually become detached from the cathodic mass as it is consumed and will gather on the bottom of the cell in the form of a slime (anodic slime).

At the same time, since they have the largest reduction potential, the cations of the metal to be refined will be discharged at the cathode in preference to every other form of cation which has been formed at the anode. The cations of the less noble contaminants will therefore remain in solution. Care must be taken, however, to ensure that the amounts of these present do not increase excessively in the long term, as a high concentration is also a factor which favours cathodic reduction.

It is therefore concluded that electrolytic refining essentially consists of the anodic dissolution of the metal to be refined and the cathodic discharge and deposition of the ions of the same metal which is rendered extremely pure in this process, with the mechanical deposition of the more noble contaminants at the anode and permanent solubilization of the less noble.

13.9 THE COORDINATION AND CHEMISTRY OF METALLURGICAL PROCESSES

In the preceding section, metallurgical treatments have been studied in 'sealed compartments' in so far as the oxidation of sulphides, the reduction of oxides, and metallurgical electrolytic processes have been treated, as if they were ends in their own right. Moreover, the problem of separating two metals which are contained in the same mineral, as arises, for example when the iron has to be separated from the chromium in chromite ($FeCr_2O_4$), has not been considered directly.

In reality, however, metallurgical processes, on the one hand, are often coordinated and sequential, while, on the other hand, it often comes to having to work mineral compounds using the same criterion with which gangues mixed with minerals are treated: that is, by making use of the fact that metals which are combined in a single chemical species are noble to different extents and therefore separable because the oxide of the less noble metal passes preferentially into the slag.

The basic metallurgical processes are described in Table 13.3, and examples of the relevant chemical reactions are shown.

A classic example employing both coordinated and sequential metallurgical processes (oxidation–reduction–refining) as well as the exploitation of the differences in the ability of their metals to pass into slags or dross, is offered by the need to treat the minerals bornite ($Cu_2S \cdot CuS \cdot FeS$) and chalcopyrite ($CuS \cdot FeS$) to recover the copper from them.

Part 3 IRON AND STEEL MAKING

That part of metallurgy which is concerned with the production and the working of ferrous alloys and ferroalloys is called 'siderurgy'.

The particular and specific name given to this branch of metallurgy is justified by the enormous technological importance of ferrous alloys, and, indirectly, in so far as they serve to prepare such alloys, of ferroalloys.

13.10 THE RAW MATERIALS OF IRON AND STEEL MAKING

The production of the iron alloy which is known generically (and as we shall see there are many varieties of this) as pig iron (cast iron) lies at the foundation of iron and steel working processes. Pig iron is traditionally produced by using a blast furnace.

The raw materials required for the operation of a blast furnace are: the mineral, coke, air, and fluxes. If the air is excluded from them, they represent the only costs of supplying the furnace.

The minerals used to charge blast furnaces

The minerals used to charge blast furnaces always consist of oxidized compounds of iron which, for the greater part, originate from the natural compounds listed in Table 13.4.

The compounds listed in this table form part of the composition of 'iron minerals', a term which usually indicates both such compounds and their gangues, that is, mainly, SiO_2, CaO, MgO, and Al_2O_3. On the basis of their iron content, iron minerals are technically subdivided into rich ($Fe > 52\%$), medium rich ($28\% < Fe < 52\%$), and poor ($Fe < 28\%$) types, while, on the basis of their gangue, they are subdivided into basic gangue minerals (mainly in the case of CaO and MgO), acidic gangue minerals (especially in the case of SiO_2 and Al_2O_3), and autofluxing minerals when the acid/base ratio is more or less equal.

The value of iron minerals depends on the possible contents of certain elements and the possession of certain physicomechanical properties.

Table 13.3—Examples of general methods used in metallurgy

Types of minerals	*Descriptive note regarding procedure*	*Chemical reactions*	*Alternative to process and electrolytic refining*
1. Native 'noble' metals	Oxidative complexing with cyanide (a), then displaced electrochemically	(a) $4Au + 8CN^- + O_2 + 2H_2O \rightarrow 4Au(CN)_2^- + 4OH^-$ (b) $4Au(CN)_2^- + 2Zn \rightarrow 4\mathbf{Au} + 2Zn(CN)_4^{2-}$	They can also be amalgamated with mercury which is subsequently removed by distillation.
2. Simple oxides	Reduction with carbon (c) or with CO (d), Al (g), H_2(e), etc.	(c) $SnO_2 + 2C \rightarrow \mathbf{Sn} + 2CO$ (d) $Fe_2O_3 + 3CO \rightarrow 2\mathbf{Fe} + 3CO_2$ (e) $GeO_2 + 2H_2 \rightarrow \mathbf{Ge} + 2H_2O$	The oxides can also be converted into salts: $ZnO + H_2SO_4 \rightarrow ZnSO_4 + H_2O$ which, when electrolyzed yields the metal at the cathode: $Zn^{2+} + 2e^- \rightarrow \mathbf{Zn}$
3. Mixed oxides	The less noble component is converted into slag (f); the process is then as in 2 (g)	(f) $FeCr_2O_4 + SiO_2 \rightarrow \underset{slag}{FeSiO_3} + Cr_2O_3$ (g) $Cr_2O_3 + 2Al \rightarrow Al_2O_3 + 2\mathbf{Cr}$ *'aluminothermic process'*	* * *
4. Simple sulphides	Are oxidized (h) and then then reduced as in the case of 2, as exemplified by (i)	(h) $ZnS + \frac{3}{2}O_2 \rightarrow ZnO + SO_2$ (i) $ZnO + C \rightarrow CO + \mathbf{Zn}$	In **electrolytic refining** the metal of interest is dissolved from the *anode* of an electrolytic cell (e.g. $Cu \rightarrow Cu^{2+} + 2e^-$). All the less noble metals (e.g. Fe, Ni, Zn) are thereby oxidized, while the more noble metals (e.g. Ag, Au) are deposited on bottom of the cells ('anodic slime'). Only the ions of the metal of interest are discharged at the *cathode* because it is the most noble and abundant of all those oxidized at the anode. In the case of copper: $Cu^{2+} + 2e^- \rightarrow \mathbf{Cu}$ with the formation of a deposit of very pure copper.
5. Mixed sulphides	The enriched minerals are oxidized (j) and reactions of type (k) are made to occur in analogy with case 3.	(j) $CuFeS_2 + 3O_2 \rightarrow CuO + Fe + 2SO_4$ (k) $\begin{cases} FeO + SiO_2 \rightarrow \underset{slag}{FeSiO_3} \\ CuO + Mg \rightarrow \mathbf{Cu} + MgO \end{cases}$	
6. Halides	These are first purified by crystallization, then fused (l) and electrolyzed (m)	(l) $MgCl_2 \xrightarrow{fusion} Mg^{2+} + 2Cl^-$ (m) $\begin{cases} \textit{cathode}: Mg^{2+} + 2e^- \rightarrow \mathbf{Mg} \\ \textit{anode}: 2Cl^- \rightarrow Cl_2 + 2e^- \end{cases}$	

Table 13.4—Characteristics and deposits of the principal natural compounds of iron

Name	*Formula*	*%Fe in pure compound*	*Appearance*	*Main producing countries*
Magnetite	Fe_3O_4	72.5	black, opaque	Brazil, Liberia, Mauritania, Sweden, Norway
Haematitte	Fe_2O_3	70	steel grey (blood red powder)	Angola, Australia, Spain, Chile, South Africa, Venezuela, and magnetite producers
Gothite	$Fe_2O_3 \cdot H_2O$	63	yellow	Peru, Sierra Leone, USSR, USA, and all haematite producers
Limonite	$2Fe_2O_3 \cdot 3H_2O$	60	yellow-brown	Algeria, Mexico, New Zealand, France, West Germany, China, and all goethite producers
Pyrites	FeS_2	46.6	yellow-gold	N. Korea, Jugoslavia, Romania, Italy, and all limonite producers
Siderite	$FeCO_3$	48.2	yellowish white	widespread, a little everywhere, but retained and not readily exploitable

There are a few extraneous elements which increase the value of these minerals (mainly manganese and small amounts of phosphorus of the order of 0.7 0.8%), while there are many extraneous elements which depress their value (S, As, Ti, Cu, P (above 0.8%), and even chromium and nickel).

Among the physical properties of iron minerals, porosity is valued provided that it is of such a nature as to enhance the reducibility of the mineral while not impairing the compressive strength and anti-abrasion qualities of the mineral.

Structurally, the best iron minerals for use in blast furnaces are therefore those which do not become friable, have good mechanical strength, and do not swell when heated

The mineral charge to be introduced into a blast furnace must have:

- a high iron content which is exclusively in the form of oxides; and
- be in pieces with sizes of the order of 3–8 cm depending on its compactness.

To ensure that these requirements are met, recourse has been made in recent decades to the use of enrichment operations, especially magnetic and flotation processes on account of the necessity to work mineral deposits which are becoming even poorer. In addition, the traditional treatments have been improved upon and intensified. These consist of an aerial oxidation followed by a calcining roasting in the case of siderites, of integral desulphurization burning in the case of pyrites, of hot sintering in the case of fine minerals, and of pelletization of pyrites ashes and of mining concentrates.

It is also important to ensure that the mineral charge is chemically homogeneous, since appreciable inhomogeneities may change the course of processes in an uncontrolled manner.

Iron and steel making coke

Blast furnace coke must be free from powders and crumbs, it must consist of uniform pieces (6–10 cm), and have a high and sufficient mechanical strength (120–170 kg/cm^2). Moreover, it should be porous and infusible, have a low content of ash (7–12%), of volatile substances (2% max.), of sulphur (1–1.2%), and an extremely low phosphorus content (0.005%). Finally, such a coke has a specific weight of 0.4–0.5 ton/m^3 and a calorific content of the order of 7500 kcal/kg.

The combustion air

The air to be introduced into the bottom of a blast furnace through the tuyères is first precompressed in centrifugal compressors (blowers) driven by gas turbines and then preheated to 1100–1200°C in heat recovery devices (Cowper stoves).

Cowper stoves are cylindrical towers with a height of 15–30 m and a diameter of 6–12 m which are constructed from steel plates lined with refractories, and divided into a free section and a section occupied by refractories which are piled up in a chequerwork pattern. The amount of coke consumed in a blast furnace is inversely related to the efficiency of these Cowper stoves.

With two or four units functioning, the stoves act alternately in pairs as heat accumulators and as preheaters. In the accumulation phase, some of the gas produced in the blast furnace is burnt at the bottom of the empty section so as to develop a flame from which fumes are derived, which, by passing through the subsequent stacks of chequerwork brick transfer their heat to it. In the subsequent phase, the air to be preheated passes backward along the pathway of the fumes, and the flame and is thereby brought up to the temperature required for it to act as an oxidant in the blast furnace.

An auxiliary Cowper stove is always available to substitute any other such stove which may become damaged.

Fluxes (slag-forming agents) for a blast furnace

To increase the fusibility of the gangue which accompanies the iron minerals, it is always necessary to some extent to introduce fluxes into a blast furnace which cause the gangue to form low-melting-point slags.

The addition of such slag-forming agents also has the purpose of promoting processes such as desulphurization and dephosphorization which may be of primary interest in the process.

Fluxes may be subdivided into basic and acidic fluxes. The first consist, for the greater part, of limestones and, less often, of dolomites, while the others consist of quartzites.

To calculate the amounts of these corrective agents or of their derivatives (CaO, MgO, sands) which must be added, it is necessary to refer to the chemical analysis of the materials which are introduced with regard to the different components from the oxides of iron in the minerals and the carbon of the coke.

13.11 THE PRODUCTION OF HOT METAL IN A BLAST FURNACE

The structure and functioning of a blast furnace

In essence, a blast furance is a shaft furnace which is continuously loaded from the top with coke and which has a fusion bed consisting of a mixture of iron minerals and the flux.

With an overall height of 25–30 m which is supported by iron pillars and lattice work, a blast furnace has a maximum diameter of 7–8 m according to the ratio between the shaft and the cylinder ('belly'). The hearth below the zone into which the air enters is restricted to a diameter of 5–7 m. The interior of the furnace is lined with silicoaluminate refractories except in the zone which attains the highest temperature at the level of the tuyères, where it is lined with a graphite–clay refractory. Externally, it is wrapped in thick sheet steel which is cooled in the lower parts in order to extend the lifetime of the refractory. The tuyères are made of phosphor-bronze and cooled by the forced circulation of air.

The charge in a blast furnace slowly descends, to pass through the whole structure in 6 to 12 hours, while becoming ever hotter owing to the action of the heating and reducing gases which pass it as a countercurrent. These gases (N_2 and CO) originate either from the incomplete combustion of the coke of the lower part of the blast furnace by the preheated air which is blown into the furnace or (CO_2) from the progressive reduction of the iron oxides by the CO.

The air enters into the blast furnace from nozzle injectors (tuyères) which are arranged in a circle around an annular tube. As has been stated, beneath the zone into which the air is blown, there is a hearth in which the fused products of the blast furnace collect. The slags can be removed from the upper part of the hearth via the slag notch, while the hot metal is intermittently withdrawn from the bottom through the metal notch.

The thermal and chemical conditions in a blast furnace

A diagram of a blast furnace is shown in Fig. 13.17 flanked by the average temperatures which appertain in it at various heights. This enables one to interpret the chemical phenomena which are occurring at the different stages.

It is of particular interest to note that the reduction of iron by CO proceeds gradually from the higher oxidation states to the lowest oxidation state and then to metallic iron, and that this reduction takes place below 800°C because, above this temperature, CO shows a tendency to remain stable at the expense of CO_2. In fact, at above 800°C there is also a more or less rapid transformation of all the CO_2 into CO, by the decomposition of the gangue carbonates or the flux, or from the immediate impact at the tuyères between the carbon and the combustion air which enters into the blast furnace. It is also to be noted that the less reducible oxides (MnO, SiO_2, and P_2O_5) are reduced at high temperatures by thc direct action of C.

Among the processes concerncd in the chemical settlement of the hot metal in the hearth at the surface of contact between the metal and the slag which are mentioned in the reaction scheme attached to Fig. 13.17, the desulphurization of iron due to the action of bases such as CaO stands out in importance:

$$FeS + CaO \rightarrow CaS + FeO$$

which is followed by the direct reduction of FeO by C:

$$FeO + C \rightarrow Fe + CO,$$

a highly endothermic reaction which contributes to the lowering of the temperature with respect to the very hot zone above it.

Blast furnace products and auxiliary apparatus

Crude hot metal, slags, and fumes are discharged from blast furnaces. On account of this, externally and upstream of the blast furnace, there are units for manipulating the hot metal and slags, and, laterally, plants for cleaning the fumes. The entirety of these structures, in addition to the Cowper stoves, the silos where the raw materials are collected, and the apparatus for loading the raw materials into them, constitute the accessories of a blast furnace. We shall proceed to describe the products in detail.

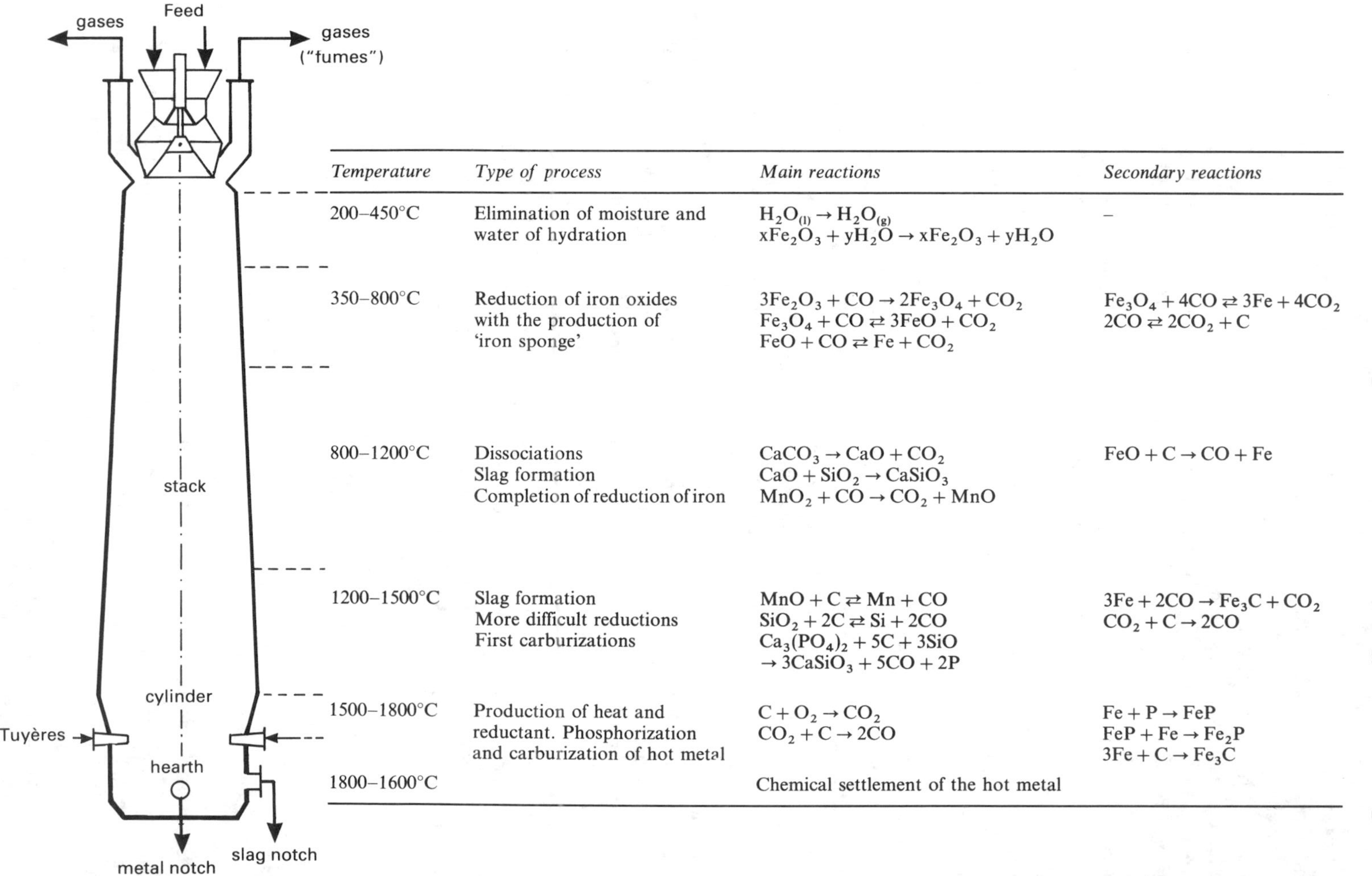

Temperature	Type of process	Main reactions	Secondary reactions
200–450°C	Elimination of moisture and water of hydration	$H_2O_{(l)} \rightarrow H_2O_{(g)}$ $xFe_2O_3 + yH_2O \rightarrow xFe_2O_3 + yH_2O$	–
350–800°C	Reduction of iron oxides with the production of 'iron sponge'	$3Fe_2O_3 + CO \rightarrow 2Fe_3O_4 + CO_2$ $Fe_3O_4 + CO \rightleftarrows 3FeO + CO_2$ $FeO + CO \rightleftarrows Fe + CO_2$	$Fe_3O_4 + 4CO \rightleftarrows 3Fe + 4CO_2$ $2CO \rightleftarrows 2CO_2 + C$
800–1200°C	Dissociations Slag formation Completion of reduction of iron	$CaCO_3 \rightarrow CaO + CO_2$ $CaO + SiO_2 \rightarrow CaSiO_3$ $MnO_2 + CO \rightarrow CO_2 + MnO$	$FeO + C \rightarrow CO + Fe$
1200–1500°C	Slag formation More difficult reductions First carburizations	$MnO + C \rightleftarrows Mn + CO$ $SiO_2 + 2C \rightleftarrows Si + 2CO$ $Ca_3(PO_4)_2 + 5C + 3SiO$ $\rightarrow 3CaSiO_3 + 5CO + 2P$	$3Fe + 2CO \rightarrow Fe_3C + CO_2$ $CO_2 + C \rightarrow 2CO$
1500–1800°C	Production of heat and reductant. Phosphorization and carburization of hot metal	$C + O_2 \rightarrow CO_2$ $CO_2 + C \rightarrow 2CO$	$Fe + P \rightarrow FeP$ $FeP + Fe \rightarrow Fe_2P$ $3Fe + C \rightarrow Fe_3C$
1800–1600°C		Chemical settlement of the hot metal	

Fig. 13.17—The blast furnace showing the temperature ranges in the various sections and the principal and secondary reactions which take place over each of these temperature ranges. These temperature bands in the furnace are only approximate to those present in the zones of the furnace to which they [illegible] the chemical reactions referring to them occur. The chemical settlement of

The design structure of a blast furnace and of its auxiliary apparatus is shown in Fig. 13.18.

The hot metal. With blast furnaces with a capacity of about 1000 m^3, it is possible to produce more than 900 tons of hot metal per day, while, in the larger, modern blast furnaces with more than 2000 m^3 of useful volume, the production of hot metal increases more than proportionally to attain a value of about 4000 ton/day.

The liquid hot metal is tapped from the blast furnace at regular intervals of about ten hours and at a temperature of about 1500°C. This is generally done by breaking off the refractory plug which closes the metal notch of the hearth with a bar. However, there are blast furnaces in operation which allow a continuous (slow) flow of the hot metal. Having been run off into channels, the hot metal is collected in refractory-lined containers which may be large buckets ('ladles') or of a special elongated form ('torpedo-shaped travelling metal mixers') if it is intended for steel making. It is distributed in various other ways to make it into ingots or pigs if it is to go to other destinations.

The hot metal has a very variable composition with regard to elements other than iron. The limits for the main elements are:

C 3–4%, Si 0.5–3.5%, Mn 0.5–1%, P 0.05–2%, S 0.05–0.15%

Slags. Slags are a by-product of the production of the hot metal in a blast furnace. They are usually cast in large water-filled vessels where they granulate in the form of sand.

The analysis of these slags is of great technological interest with regard to the operation of a blast furnace because the course of the fundamental reactions such as the reduction of manganese and the desulphurization of the hot metal is dependent on the type of slag.

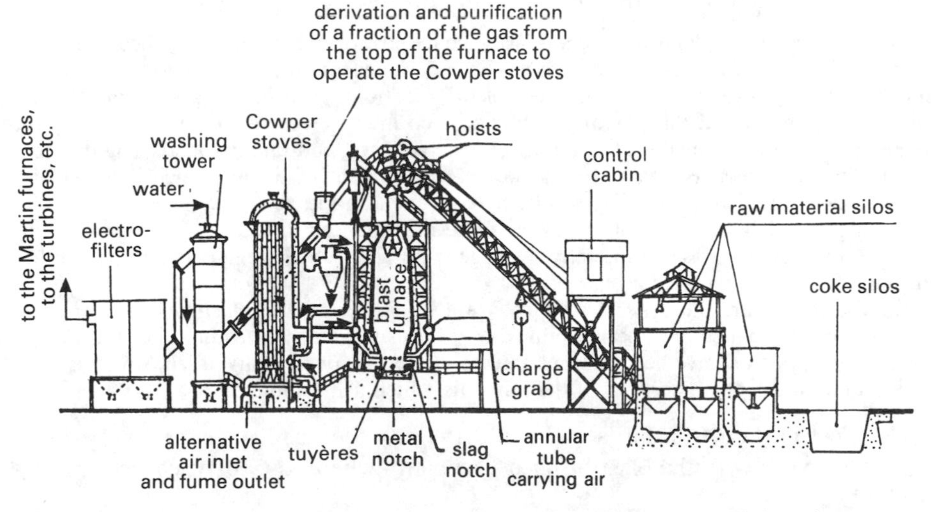

Fig. 13.18—Panoramic diagram of a blast furnace and its auxiliary units. Of these the most important is the Cowper stove which is half filled with refractories and is alternately traversed (from right to left) by the flames of the burning gases from the blast furnace and (from left to right) by the air to be preheated before its enters into the blast furnace.

Given their low specific weight, the volume of the slags from a blast furnace is very great, attaining a value of 0.3–0.6 ton per ton of hot metal.

Slags are of noteworthy technological importance. They serve, according to their nature, for the preparation of cement clinker, as road paving materials, as thermo-acoustic insulation after blowing, and for the production of low-quality glass. Their main destination, however, is as blast furnace cements.

Fumes. The dust-laden gases which escape from the top of blast furnaces have mean powder contents of 15–30 g/m^3. After the larger material has been removed in cyclones or in velocity reducers, they are sent to a scrubber or centrifugal washers, and finally, if required, they are passed through electrofilters. The powders which are recovered, which consist mainly of iron oxides and carbon, are sintered or turned into pellets.

The purified gas is combustible because it contains CO, but it has a low calorific content (800–1000 $kcal/m^3$), an approximate composition of 15% CO_2, 30% CO, and 55% N_2, and a volume of up to 2500 m^3(STP)/t of hot metal.

About 20% of the blast furnace gas finds use in driving the blowers which force air into the blast furnace, about 30% of it is used in the preheating of the air by the Cowper stoves, and the remainder serves various other technologies within the ambit of steel and iron making activities.

13.12 IMPROVEMENTS MADE TO THE BLAST FURNACE, AND EXPERIMENTAL MODIFICATIONS TO BLAST FURNACES

Various attempts have been made to reduce the costs of producing hot metal with a blast furnace in order to maintain its advantages at the first stages of iron and steel-making activities over other processes. The aims of these attempts include economization of the cost of the fuel-reducing agent, shortening the time of the operation, and improving the quality of the product.

The improvements planned for the blast furnace are of a management, mechanical, and chemical character. Considerable benefit has been derived from the introduction of automation and the computer-controlled management of the functioning of the furnace. It has also been positively demonstrated that the application of a counterpressure at the fume outlet reduces the powder content in the fumes by reducing their velocity, and that, by increasing its contact with the mineral, increases its reducing power. Enrichment of the air which is blown in by adding oxygen has also made it possible, by concentrating it, to increase the activity of the reducing gas and reduce the volume of the fumes. Finally, an at least partial replacement of the coke by reducing gases (especially hydrogen) has improved the quality of the products, although there are increased costs.

For the most part the processes which are alternatives to the blast furnace process use rotating furnaces which have been modified in various ways as well as fluid bed furnaces. Up to now such systems have been demonstrated to function locally only, a fact which is mostly attributable to the characteristics of the minerals which are exploited.

The electric reduction furnace (Fig. 13.19) has been shown to be competitive in various respects with the blast furnace, especially where low-cost electrical energy is available.

By exploiting the thermal energy developed by electric arcs established between carbon electrodes and a charge consisting of a mixture of the mineral and coke, very high temperatures can be attained in these furnaces. The direct reduction of iron

V_1 hydraulically controlled reservoir
C washing column
P ventilator
DS electrofilter

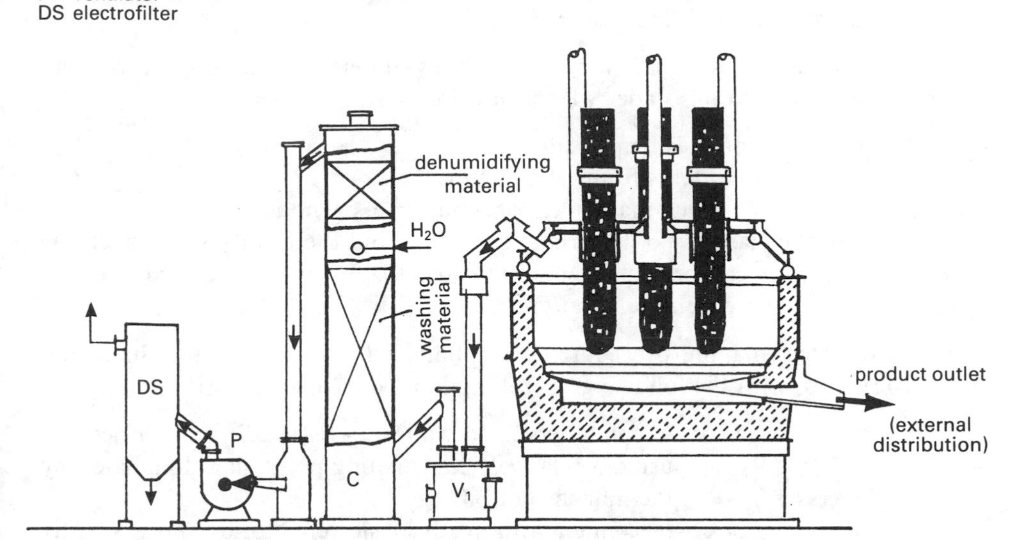

Fig. 13.19—Electric arc furnace of the type which replaces a blast furnace (electric reduction furnace). Using a three-phase electric current, three electrodes are employed. In these furnaces, where temperature control is simple (although costly, because electrical energy is used) non-agglomerated minerals and those with a low iron content can also be treated. As can be seen from the auxiliary apparatus, care is especially taken to recover the almost pure CO which comes out of the furnace. This is a rich fuel which indirectly permits a reduction to be made in the energy costs.

oxide with carbon:

$$FeO + C \rightarrow Fe + CO$$

takes place there with the consequent evolution of considerable quantities of CO which, when suitably made use of, partly compensates for the high cost of the electrical energy consumed in the process.

13.13 THE PHYSICOCHEMICAL PHENOMENOLOGIES OF THE CONVERSION OF CAST IRON INTO STEEL

Steel is essentially a refined alloy with respect to the elements (especially carbon, sulphur, and phosphorus) which accompany the iron in the products from the blast furnace. It is often alloyed with particular metals.

The product from the blast furnace (hot metal) is therefore converted into steel essentially by refining it and possibly by adding alloying elements.

Similarities and differences in the procedures

All steel production processes have the following operational stages in common:

- a preliminary smelting of the charge;
- oxidation of the elements to be removed and the elimination of any eventual excess of oxygen from the smelts which are treated;

- desulphurization by partitioning between the slag and the metal;
- the possible addition of alloying elements;
- the final manipulation of the steel produced.

However, these processes may vary considerably in detail owing to the variability of the particular conditions under which they are carried out, such as:

- the different raw materials employed;
- the use of different oxidants from time to time;
- the energy required, which may be endogenous or exogenous;
- variations in the elasticity of the whole operation and the facility with which the necessary controls can be implemented in order to permit the production of the many types of steel which are required.

In general, steel making procedures are subdivided into acidic procedures and basic procedures, with the following operational and expedient criteria:

Processes
- acid: in containers lined with low-cost acidic refractories and using a relatively low-melting-point slag dominated by the influence of SiO_2
- basic: in contact with costly basic refractories and a slightly fluid slag, but with the possibility of eliminating sulphur and phosphorus.

On the whole, the basic processes are preferred by far, and particular attention will therefore be paid to their treatment in the following discussion.

Possible chemical processes

The conversion of hot metal into steel employs a series of chemical reactions, especially oxidation reactions, for which the environment in which they occur must be suitable.

Highly pure oxygen (once upon a time it was air) or an oxidizing slag (scraps of ferrous materials) may be the oxidizing agents. When the gas is used, the diffusion of oxygen into the mass in which the oxidations must occur is extremly rapid, while the use of oxidizing slags entails the rather slow process of the diffusion of the oxygen through the metal–slag interface.

The principal phenomena taking place via chemical reactions, during the conversion of hot metal into steel are: the removal of silicon, decarburization, demanganization, dephosphorization, and desulphurization.

Removal of silicon

On account of its high affinity for oxygen, silicon tends to be oxidized first in the processes for the refinement of hot metal:

$$Si + O_2 \rightarrow SiO_2,$$

and the silica formed comes to the surface and collects in the slags on account of its relatively low specific weight.

Further downstream in the refining process, the silicon content which is required in the steels which are produced can be restored by the addition of ferro-silicon alloy.

Decarburization

Given the temperature at which it is carried out, the sole oxidative decarburizing process which is possible under these conditions is:

$$2C + O_2 \rightarrow 2CO$$

In fact, this reaction also corresponds to the sum of the processes:

$$C + O_2 \rightarrow CO_2; \qquad CO_2 + C \rightarrow 2CO$$

which tends to take place when the oxidant and the fuel first come into contact.

The gas phase in this process acts as a thermal equalizer of the mass, accelerates the removal of inert slags from the metallic phase, and improves the contact between the metal and the active slag, thereby favouring partitioning processes between the two phases.

Demanganization

The oxidizability of manganese during the processes for the refining of hot metal increases as the decarburization proceeds, that is, as the amount of carbon in the melt is lowered.

When the amount of oxygen available in the molten bath exceeds the value required for the noticeable establishment of the equilibrium:

$$Mn + \tfrac{1}{2}O_2 \rightleftharpoons MnO$$

there is a greater chance of manganese oxide being formed, which subsequently collects in the slag.

Dephosphorization

The relatively large amount of oxygen required for the oxidation of phosphorus according to the reaction:

$$4P + 5O_2 \rightleftharpoons P_4O_{10} + 720\ \text{kcal} \qquad (13.3)$$

means that dephosphorization during the refining of hot metal also requires, above all, the ready availability of oxygen in its environment, and this is possible only when the residual carbon contents are low.

Since reaction (13.3) is highly exothermic, it takes place more readily at relatively low temperatures.

Finally, to drive the equilibrium (13.3) to the right, it is necessary to make use of a substance which 'fixes' the 'phosphoric anhydride' which is formed. A strongly basic slag (arising from CaO) can be used for this purpose:

$$P_4O_{10} + 8CaO \rightarrow \underset{\text{tetracalcium phosphate}}{2Ca_3(PO_4)_2 \cdot CaO}$$

This is also a precaution which removes the product and prevents any return of the phosphorus from the slag to the metal during the subsequent deoxidation process carried out on the refined material.

Desulphurization

The elimination of sulphur arising from various metal sources during the refining of hot metal to produce steel is a consequence of the establishment of the following equilibrium:

$$S + CaO \rightleftharpoons CaS + O \qquad (13.4)$$

Hence, the prerequisites for satisfactory desulphurization are:

- a high CaO content in the slag;
- a low level of ambient oxidation,

as both of these factors contribute to drive the equilibrium (13.4) to the right with the formation of an insoluble CaS phase in the iron which therefore passes into the slag.

While all the processes used in the conversion of hot metal into steel assist desulphurization owing to the basicity factor, the reductive environments found in electric arc furnaces or furnaces in which metal powders and/or silicon-based reducing agents can be usefully introduced are particularly favourable to this process.

Converters

In order of merit regarding their importance as methods for the conversion of hot metal into steel, energetically endogenous oxygen converters have a lead in the World today, with a figure of 70%, followed by energetically exogenous electric furnaces (25–28%), and then almost exclusively by the also energetically exogenous Martin–Siemens furnaces.

Oxygen converters

By using high-purity oxygen (>99%) for the conversion of hot metal into steel, the occurrence of nitrogen in the resulting steel is avoided as is, indeed, the inconvenience resulting from the use of air* when one can treat only hot metal with a high phosphorus content which, by oxidizing, can supply the considerable amount of heat to the nitrogen which is required to bring it up to the operating temperature. In oxygen converters, the charge may consist of ferrous scrap, apart from liquid cast irons which also have low contents of the thermogenic elements silicon and phosphorus. Furthermore, these converters also permit one to stop the decarburization process at any level of the carbon content.

A general characteristic of oxygen converters is that of blowing the gas which emerges at a high pressure from the nozzles at the head of water-cooled lances onto the fused hot metal to which scrap iron has been added. These converters are metallic containers, lined with a dolomite refractory containing a tar-based binder. When they are in action, two fundamental versions of oxygen converters exist: vertical profile converters and inclined profile converters.

The Linz–Donawitz (L.D.) converter. This is the prototype, and the best known converter which was put into operation in 1949–1950 in the steelworks of Linz and Donawitz in Austria.

*Air was used in the older Bessemer (acid) and Thomas (basic) converters where, among other things, it was not possible to re-melt scrap owing to the non-availability of sufficient heat.

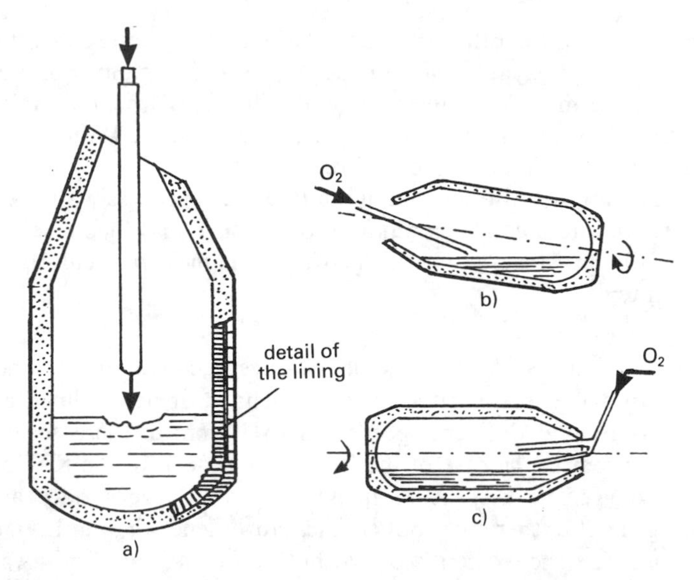

Fig. 13.20—(a) An L.D. converter in the vertical position, charged and with oxygen being blown in. A cowl (not shown) which picks up the combustible CO, is fitted above the converter. The traditional refractory consists of dolomite bricks and MgO bonded with a tar-based binder. Nowadays, however, a magnesia/chromite refractory is also used. (b) the Kaldo converter and (c) the Rotor rotating converter which has less capacity than the L.D. converters with the same refractories in so far as the one is inclined and the other almost flat.

As can be seen from Fig. 13.20a, this operates in a static vertical position after having been charged in an inclined position.

The impact of the high-pressure jet of oxygen on the molten metal creates a large amount of turbulence which causes the temperature to rise slowly to very high values to bring about the oxidation of the iron itself. The FeO which circulates in the metallic mass then reacts according to the equation:

$$FeO + C \rightarrow Fe + CO$$

Decarburization therefore proceeds rapidly, the turbulence becomes ever greater owing to the development of the gas, and all the oxidizable elements (Si, Mn, and P) which are either directly reached by the oxygen or by the FeO are also oxidized.

Basic fluid slags are made to form by putting in slagging agents (CaO) and fluxes (CaF_2) via runways which pass around the mouth of the converter. Delay in the addition of the correcting agents which have have been mentioned is reasonable only if irons with low phosphorus contents are being treated.

The data obtained from rapid analytical checks indicate any possible modifications which are required to bring the metal to their final composition, after which the mass is poured into the casting ladle while adding an appropriate amount of deoxidants to it.

Oxygen lime powder (O.L.P) and L.D.-adding calcareous converters (L.D.-A.C.). These are pieces of apparatus for the conversion of hot metal into steel which are similar to one another and derived with minor modifications from the L.D. converter with the aim of treating hot metals with a high phosphorus content. In the first of these (the Oxygen Lime Powder converter) the oxygen jet directly picks up a flux of powdered lime, while, in the other (the L.D.-Adding Calcareous converter), there is a lance coupled with the oxygen lance for the addition of calcium compounds (CaO, $CaCO_3$, CaF_2) to the melt. With these methods, the silica which is rapidly formed, even as the first blowing starts, passes into the slag, and the course of the dephosphorization is also more rapid and regular.

A two-stage blowing can also be used in the L.D.-A.C. process with A.C. restricted to the first stage, an intermediate elimination of the slag, and then a second stage in which only one L.D. process is used which corrects the melt, particularly with regard to its carbon content.

Kaldo and Rotor converters. A second series of oxygen converters includes pieces of apparatus with an axis which is either slightly inclined or not inclined at all to the horizontal but rotating at different speeds (up to 30 rpm in the case of the Kaldo converter and 1–2 rpm in the case of the Rotor converter). In the Kaldo converter, the pressurized oxygen strikes the bath, which has a large surface area, almost tangentially (Fig. 13.20b) thereby producing less turbulence than in L.D. converters. In the Rotor converter the oxygen is blown both onto the bath and away from the bath (Fig. 13.20c).

There are appreciable differences between these converters and L.D. converters. Firstly, these inclined converters provide a greater degree of contact between the metal and the slag, thereby improving the dephosphorization and desulphurization. In these converters there is not only the turbulence created by the blasting in of oxygen and the CO which is evolved within the mass which remixes the bath, but also the rotation of the apparatus which constantly renews the surface exposed to the action of the oxygen. Finally, in the Kaldo converter, and even more so, in the Rotor converter, the CO which is evolved from the melt burns to yield CO_2 so that the thermal balance is improved, which allows one to push up the composition of the charge to a point where the scrap/cast iron ratio is slightly lower than 1/1.

However, the operation is slower than in L.D. converters, and the consumption of the lining refractory is much higher.

Electric furnaces

Arc and induction electric furnaces are used for the conversion of cast iron into steel and in the production of high-alloy steels by melting suitable amounts of the components together. Arc furnaces are readily adapted for conversion processes, while induction furnaces are better suited for melting operations.

Electric furnaces, which are available in a vast range of sizes which allow amounts from a few hundred kilograms to many tens of tons of materials to be handled, are valued both on account of their particularly pure atmosphere (as compared with atmospheres consisting of combustion effluents and air residues) which is also reducing (with respect to those in oxygen converters) and for their great thermal flexibility which is made possible by the use of readily controllable electrical energy as the source of heat.

The constraints imposed on the use of these furnaces in steelworks is solely of an economic nature because the use of electrical energy for heating purposes is relatively expensive.

Arc furnaces. The use of electric arc furnaces in metallurgy has already been mentioned, both with regard to the apparatus employed for the reduction of metal oxides (page 463), and in the use of alternative devices to the blast furnace. The mode of operation of these furnaces has been described diagrammatically in Fig. 13.16.

In steel manufacturing processes using these furnaces, the charges almost always consist entirely of solid scrap iron, rolling mill scale, iron pigs, and, possibly, coke and anthracites if there is not enough cast iron present. On the other hand, if material is to be decarburized, desilicized, etc., $>99\%$ oxygen is prevalently used. In this way, the conversion operations are speeded up.

The lining of the hearth of arc furnaces is the same as that used in oxygen converters: a basic lining of dolomite or magnesia bricks with a tar-based binder. The canopy, however, can also be lined with silica ('Dinas') bricks if scrap iron is mainly handled in the furnace, that is, if the use of an energetic basic slag is not required.

Induction furnaces. Furnished with an acid lining, these are used, as almost always happens, for melting and alloying. Otherwise, a basic lining is employed and the charge is similar to that used in arc furnaces.

They consist of a vertical basin around which an induction coil is wound which is fed by a low- or high-frequency current according to the type of furnace which it serves. The charge in these furnaces forms the short-circuited secondary winding of a transformer, that is, it converts the electrical energy, which would be withdrawn from the secondary winding in a normal electrical transformer, into heat.

The temperatures attainable in these furnaces are of 1500–2500°C, and the consumption of electrical energy falls off as the power of the furnaces increases.

Martin–Siemens (M.S.) furnaces

Martin–Siemens (M.S.) furnaces are flame reverbatory furnaces using liquid or gaseous fuels which, by recovering heat for the purpose of heating the combustion air and even (if necessary for increasing its performance) the fuel gas, attain temperatures of 1600–1700°C which are

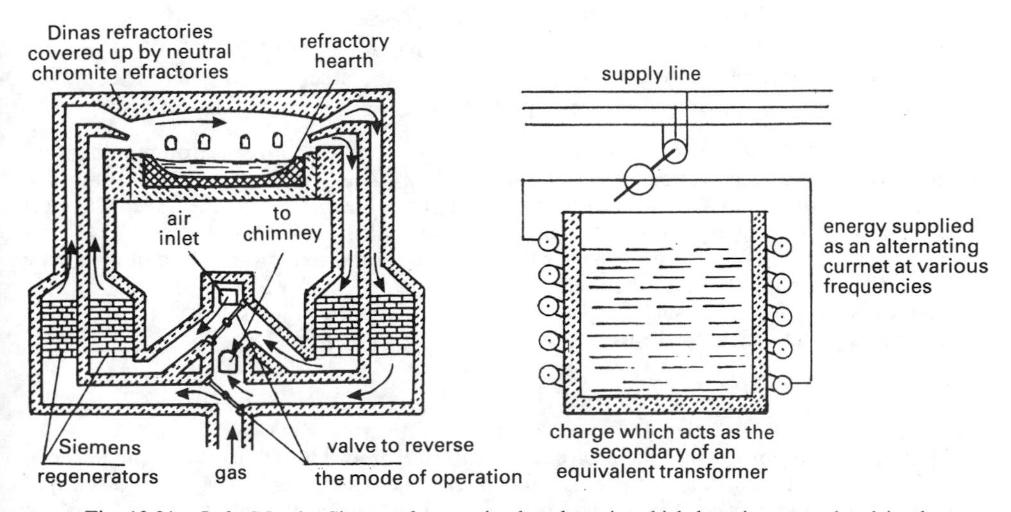

Fig. 13.21—Left: Martin–Siemens furnace in the phase in which heat is accumulated in the heaps of refractories on the right and lost from those on the left. Right: Crucible induction furnace.

necessary to ensure the melting and refining of cast iron and ferrous scrap so as to produce liquid steel.

The heat is recovered in two stages: by heating one stack of refractory material up to 1100–1350°C, using the appreciable amount of heat in the fumes which pass out from the furnace, and then causing this same refractory to be traversed by the combustion air and, possibly, the poor fuel gas moving in the opposite direction to the fumes coming out of the furnace. There are two piles of silica–alumina refractories so that, by means of a valve, it is always possible to keep one of them with the hot fumes passing over it while the other is preheating the air, and vice versa.

Martin–Siemens furnaces have a hearth made of dolomite which has been prebaked, pulverized, and then formed into a single block by pressing—which the steel itself sinters during the first few melts carried out in the furnace. The canopy, which runs the risk of being considerably damaged (apart from by the thermal stress to which it is subjected) by the lime which is used as a slag-forming agent and sublimes, is lined with chrome–magnesite refractories, and the walls are paved with magnesite bricks.

The furnace operates in a continuous manner. It is charged at calculated intervals, and the molten steel produced is almost completely discharged at the end of each production cycle, which lasts from five to seven hours. The furnace charge, more rarely, consists of cast iron and oxide minerals of iron (the 'mineral process'), but it is more frequently cast iron and ferrous scrap (the 'scrap process'). The limestone which is added constitutes about 5–6% of the charge.

The reactions occurring are varied and complex. The most interesting are:

$$Fe_2O_3 + Fe \rightarrow 3FeO \qquad FeO + C \rightarrow Fe + CO \quad \text{oxide reduction}$$

$$\left.\begin{array}{l} 7FeO + 2P + Si \rightarrow 7Fe + P_2O_5 + SiO_2 \\ P_2O_5 + SiO_2 + 5CaO \rightarrow Ca_3(PO_4)_2 \cdot CaO + CaSiO_3 \end{array}\right\} \quad \text{slag formation}$$

Since the length of the operation permits it, samples of the molten material can be withdrawn to be analyzed in various ways in order to make corrective modifications to the process.

The installation of M.S. furnaces is costly and their rate of operation is slow. Nevertheless, they do permit steels with constant and controlled characteristics to be produced on a large scale.

The simple casting of ingots and deoxidation of the steel

After the steel has been produced it must be transformed into blocks of various shapes ('ingots') before passing on to be subjected to other operations within foundries, rolling mills, forges, etc., that is, the iron and steel working activities which form the subject of the study of mechanical technology.

The casting of the steel into ingots is carried out by pouring it into cast iron moulds from large casting ladles, using various techniques (Fig. 13.22). The ladles are fashioned from thick steel plate and are internally lined with silica–alumina refractories, and they keep the steel liquid for more or less long periods after they have been filled. Combustion of part of the gas from a blast furnace can furnish the heat which is required to keep the steel in a molten state in a ladle before pouring it into the ingot moulds.

The resulting ingots, after solidification and removal from the moulds, have various shapes and dimensions according to the uses to which they are to be put.

* * *

Apart from the normal technique of casting the steel into ingots, there are some special techniques. Outstanding among these, is vacuum casting, which has undergone a very rapid development in recent years on account of the possibility which it offers of deoxidizing the steel produced with oxygen furnaces. In this technique the environment in which the steel is poured from the furnace into a casting ladle or from a ladle into an ingot mould is kept under

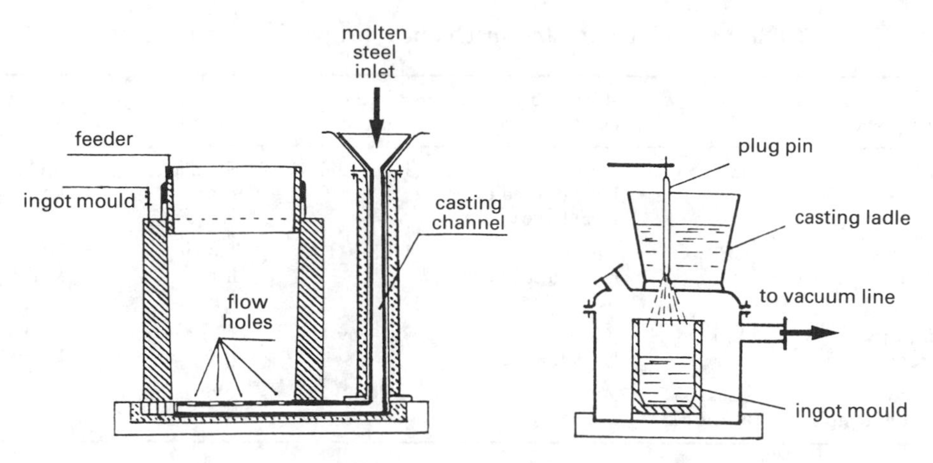

Fig. 13.22—Techniques for the casting of steel into ingots. Left: Cast in a rising mode, that is, by bottom (or uphill) casting. Parallel to this technique is that of direct (top) casting. Right: the casting of steel under a vacuum to degas it.

a vacuum by means of various devices (mechanical pumps, ejectors, etc.) with the aim of thoroughly degassing it* (Fig. 13.22).

Since a stream of molten steel is dispersed into minute drops when entering an environment where there is essentially a vacuum, the gases can readily escape from the whole mass of the steel. Moreover, as CO is removed from the steel under the reduced pressure, the latter is deoxidized both by the direct subtraction of oxygen and by the displacement of the equilibrium:

$$C + \tfrac{1}{2}O_2 \rightleftharpoons CO$$

to the right in such a way that oxygen is consumed.

13.14 IRON AND THE PRINCIPAL STRUCTURAL CONSTITUENTS OF ITS ALLOYS

The physicochemical and mechanical properties of iron

In its own right, iron is a metal with rather poor chemical and mechanical characteristics.

Chemically, it is a very reactive metal which displaces hydrogen from acids, rusts in moist air to become covered with a layer of basic ferric carbonate which is crumbly and therefore indefinitely renewable by progressive corrosion, and which can be passivated to give a glossy finish only by immersion in concentrated nitric acid or chromic acid.

Just one favourable chemical property stands out, that is, its ability to form alloys, particularly with carbon which also sharply improves its mechanical behaviour.

*Account may be taken of the fact that, in principle, the solubility of a gas in molten steel is subject to Sievert's law which is expressed by the formula

$$C_g = k\sqrt{p_g}$$

where C_g is the percentage concentration of the gas in the steel, p_g is the pressure of the gas in the steel, and k is a proportionality constant which depends on the type of gas and the type of steel.

Table 13.5—Physical and mechanical properties of pure iron

Properties	*Values*	*Allotropic forms and their structure*	*Range of existence*	*Properties*	*Values*
Melting point	1536°C	δ-iron body-centred cubic (bcc) lattice	1536–1392°C	Specific heat (0–100°C)	0.111 kcal/kg
Boiling point	≃2800°C	γ-iron face-centred cubic (fcc) lattice	1392–911°C	Latent heat	70 kcal/kg
Density (room temp.)	7.87 kg/dm^3	β-iron bcc lattice	911–769°C	Ultimate stress (room temp.)	27 kg/mm^2
Hardness (room temp.)	Mohs:4.5 Brinell:67	α-iron bcc lattice	under 911°C	Elongation (room temp.)	40%

On account of the very large variety of types and their range of properties, iron–carbon alloys are by far the most important industrial metallic materials. Data regarding the physical and mechanical properties of pure iron are presented in Table 13.5.

Among all the properties tabulated, the most important are those regarding the allotropic forms. Of these, α-iron is essentially ferromagnetic, δ-iron and γ-iron are paramagnetic, and α-iron is partially paramagnetic. In connection with this latter statement, it is observed that, between 911 and 769°C, α-iron likewise continues to be paramagnetic. It was therefore thought that another allotropic modification of iron (β-iron) might exist between 769°C, the Curie point at which iron loses its ferromagnetism (iron is 'demagnetized') and 911°C. Since, however, allotropic transformations entail some modification of the crystal lattice (Fig. 13.23), while the transition which iron undergoes at 769°C consists solely of a change in the distribution in the various energy levels, 'β-iron' is solely considered as non-magnetic α-iron.

These hypotheses are supported by the fact that, while allotropic modifications manifest themselves as discontinuities from a thermal point of view (that is, as a halt on a heating curve upon passing from one modification to the other), the transformations of magnetic α-iron into non-magnetic α-iron takes place over a certain temperature range. That is to say, the value of 769°C is the approximate temperature at which the change occurs and not a 'critical point'.

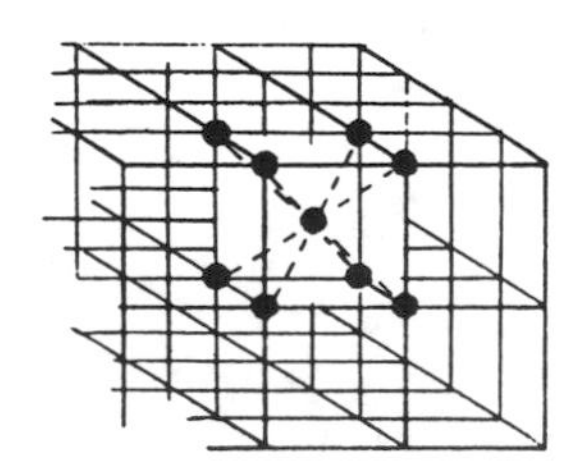

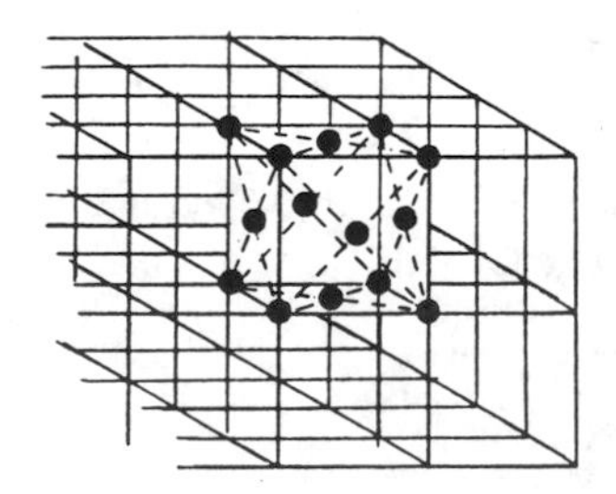

Fig. 13.23—Left: Crystalline frameworks of δ-iron and α-iron with their body-centred cubic lattice picked out. Right: The same framework with the face-centred cubic lattice of γ-iron picked out.

The principal components of iron–carbon alloys

Interstitial compounds, mechanical mixtures, and solid solutions take part in the make-up of iron–carbon alloys. More precisely, the most important components are the following:

Cementite. This is the interstitial compound, Fe_3C, which contains 6.67% of carbon. It is said to primary cementite if it separates as the first solid from melts, but secondary cementite if it is formed by the decomposition of solids which have been previously formed.

Austenite. This is an interstitial solid solution of carbon in γ-iron. This solid solution, which is more or less rich in carbon, attains the maximum degree of saturation at 2.06% of carbon at a temperature of 1147°C.

The Fe atom/C atom ratio in austenite at saturation is found by solving the equation:

$$\frac{12}{12 + 56 \cdot n} \cdot 100 = 2.06$$

where 12 is the relative atomic mass of carbon and 56 is the relative atomic mass of iron. The value of n turns out to be 10.2 since, within austenite, which is saturated to the maximum degree with C, there is, on average, slightly less than one atom of carbon for every ten atoms of iron. Consequently, the number of carbon atoms for every total of 100 atoms in the alloy (the carbon content in atomic.%) is obtained from the proportion:

$$1:(10.2 + 1) = x:100 \Rightarrow x \simeq 8.9\%$$

Ledeburite. This component is a eutectic mixture (and, for that reason, a mechanical mixture) of cementite and austenite saturated with carbon. The total carbon content in ledeburite is about 4.30%.

Ferrite. This is the name given to iron with a body-centred cubic lattice which is soft, ductile, and malleable. An α-ferrite and a δ-ferrite therefore exist.

In α-ferritic iron, the solubility of carbon is at a maximum of 0.02% at 769°C. Its solubility at room temperature is practically zero.

Pearlite. This is a euctectoid mixture of ferrite and cementite. In practice, a eutectoid mixture has the same properties as a eutectic mixture. However, the eutectic is formed from cooling of a melt, while a eutectoid is formed by the cooling of a solid. Under a microscope, perlite has the iridescent appearance of mother of pearl, from which it derives its name.

Martensite. This is a mass of thin needles which are densely interwoven with each other in an intricate manner. It is very hard and mechanically resistant. It consists of an interstitial solid solution of carbon in α-iron.

However, since, under equilibrium conditions, carbon has only an extremely low solubility in this type of iron, martensite is a metastable supersaturated phase.

Troosite and sorbite. These two structural components of iron–carbon alloys consist, respectively, of ferrite supersaturated with carbon and of the finest, submicroscopic grains of cementite which starts to separate out (troostite), and a mass characterized by a fine uniform dispersion of cementite in ferrite (sorbite).

Martensite, troostite, and sorbite are not shown in the corresponding phase diagram (see below), although they are possible structural components of iron–carbon alloys. However, they are formed (as well as other components such as bainite and osmondite) during the heat treatment of iron–carbon alloys of which they are metastable structural phases.

Since each of these structures has its own characteristic properties, their individual existence in such alloys or, what is more, the numerous combinations of them which can occur, enable one to understand the vast range of properties which are exhibited by these alloys in practice.

13.15 THE IRON–CARBON PHASE DIAGRAM

The graphical representations of the temperature and concentration fields in which the structural components of iron–carbon alloys have a stable existence is known as the iron–carbon phase diagram.

The upper region of the Fe–C phase diagram

If the treatment is to be complete, it is necessary to describe this part of the diagram, although the corresponding phases may be of little practical interest. This zone is shown in greater detail in Fig. 13.34 from which it may be deduced that, starting off from a melt with less than 0.11% C, only crystals of 'δ-ferrite', that is, solid solutions of δ-iron and carbon, separate out.

When the carbon content lies between 0.11 and 0.23%, solid solutions of δ-ferrite first separate out when melts (type T) are cooled and then, as the system cools and this separation continues, the melt reaches point B at which crystals of austenite with a composition corresponding to point J also appear. The line HJB is the 'peritectic line': such that the liquid which reaches point B reacts with the δ-ferrite thereby partially transforming it into austenite. When all the liquid has solidified, the ratio between the masses of the two types of solid (austenite and δ-ferrite) which have been formed with respect to an initial liquid T is found by applying the 'lever rule': δ-ferrite/austenite = T″J/T″H*.

Starting off from a composition R, δ-ferrite separates out at R′. As a result, the composition of the liquid is displaced towards B where the ferrite which has crystallized out reacts with the remaining liquid in such a way that the resulting solid is all austenite. Furthermore, all the liquid remaining after this will also crystallize entirely as austenite so that, at the end,

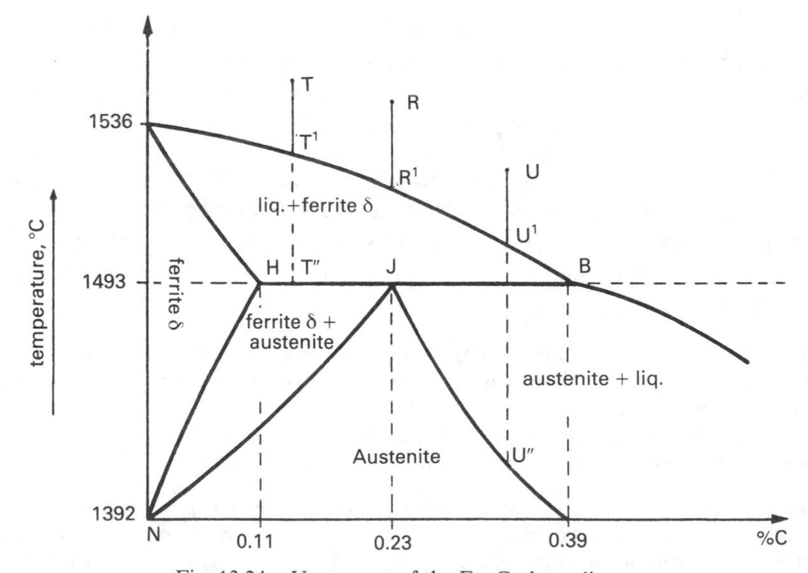

Fig. 13.24—Upper part of the Fe–C phase diagram.

*It may be noted that, according to the lever rule, the relative amounts of the solids which are formed are inversely proportional to their distances from T″, T″J being the austenite distance and T″H the δ-ferrite distance.

austenite is the only solid phase which is obtained. Starting out from a composition lying between J and B (such as U), the δ-ferrite which first separates out reacts with the liquid as soon as the composition of the latter reaches B to give austenite until the δ-ferrite is completely exhausted. After all the δ-ferrite has disappeared, the liquid recommences cooling with the separation of further austenite.

When either the austenite which is first formed or the new austenite is cooled extremely slowly, they gradually modify their composition when they are in contact with the liquid until, at the end of the transformation, it corresponds to that of austenite with the composition U″.

At below 1493°C, δ-ferrite has an ever-decreasing carbon content in accordance with the curve HN, while producing austenite with a composition which is found from the curve JN (the phenomenon of 'segregation').

The lower part of the Fe–C phase diagram

Melts containing from 0.39 to 2.06% C solidify to yield only austenite because they do not reach the eutectic point C (Fig. 13.25) in so far as that, provided they evolve under equilibrium conditions, they finish solidifying before reaching C*. If, on the other hand, the carbon content lies between 2.06 and 4.30%, austenite first solidifies out and then, when the liquid has reached the eutectic point at C, ledeburite crystallizes. Finally, melts containing between 4.30 and 6.67% of carbon precipitate primary cementite when equilibrium conditions appertain, that is, in the absence of any actions which promote the separation of graphite along the outline C′D′ and then, upon reaching C, they yield ledeburite.

The behaviour of iron–carbon alloys which has just been described enables one to subdivide them into two large groups: those which contain less than 2.06% of carbon which, after complete solidification, consist only of a solid solution (austenite) and those containing more than 2.06% of carbon which, when completely solidified, are formed of two components: austenite and eutectic ledeburite or from free ('primary') cementite and ledeburite. The steels belong to the first group and the cast irons to the second group.

The types of alloys of just iron and carbon which can exist in a thermodynamically stable state at ordinary temperatures are shown in Table 13.6.

In principle, steels are therefore alloys of iron and carbon with less than 2.06% C which inherently contain only austenite, and cast irons are alloys of iron and carbon with more than 2.06% of carbon and which inherently contain either austenite or ledeburite (cast irons used in practice) or primary cementite and ledeburite (cast irons not used in practice).

Austenite is unstable below 1147°C. If its carbon content lies approximately between 0% and 0.8%, non-magnetic ferrite and then magnetic ferrite separate in the solid state, while, if it contains from 0.8 to 2.06% of carbon some secondary cementite separates. When the composition of the austenite reaches S, it forms a mechanical mixture of α-ferrite and secondary cementite (pearlite eutectoid).

However, pearlite is also unstable, and, below 723°C, more cementite separates from the ferrite which it forms along PZ.

Below 1147°C, alloys containing between 2.06 and 4.30% of carbon which, as has already been stated, initially solidifies with the formation of austenite and terminates

---

*It is known from the physical chemistry of phase diagrams that this statement can be demonstrated by applying the lever rule to calculate the ratio of the liquid component to the solid component(s) in the system at the point where solids are formed which corresponds to the attainment of the composition of the original melt.

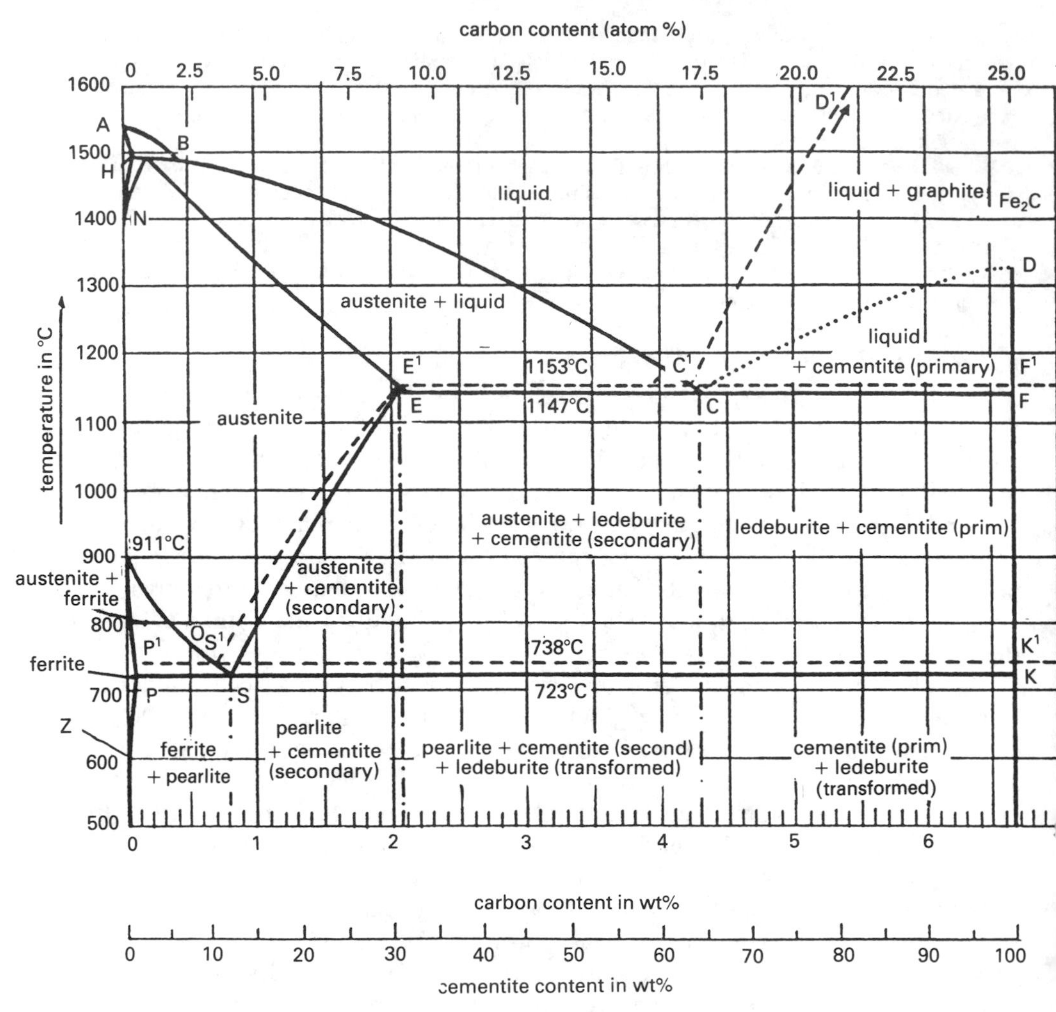

Fig. 13.25—Part (lower) of the Fe–C phase diagram which allows one to define and classify the alloys belonging to this binary system.

with the production of ledeburite, form secondary cementite along the curve ES until, at a temperature of 723°C, they form pearlite. Consequently, at temperatures below 723°C, these alloys are formed of cementite, pearlite, and transformed ledeburite.

Alloys containing more than 4.30% of carbon which first solidify with the separation of primary cementite and then ledeburite, also undergo some modifications at a temperature below 1147°C, and, in fact, the austenite component of the initial ledeburite appears to be transformed, at ambient temperatures, into secondary cementite and pearlite.

The relative limits and absolute importance of the Fe–C phase diagram

Cementite is not a stable compound and, on account of this, graphite may crystallize from melts of iron and carbon instead of cementite, while graphite may separate from austenite instead of secondary cementite. The formation and separation of graphite follow the respective

Table 13.6—Nomenclature and stable physico-chemical constitution of iron–carbon alloys at low (< 723°C) temperatures

Alloys	*Nomenclature*	*% of carbon*	*Constitution*
Steels	hypoeutecoid	< 0.80	ferrite + pearlite
	eutectoid	0.80	pearlite
	hypereutectoid	0.8 < C < 2.06	pearlite + secondary cementite
Cast irons	hypereutectic	2.06 < C < 4.30	pearlite + secondary cementite + transformed ledeburite
	eutectic	4.30	ledeburite
	hypereutectic	4.30 < C < 6.67	primary cementite + transformed ledeburite

broken lines in the phase diagram which run at slightly higher temperatures than the corresponding lines for cementite. It follows from this that an iron–graphite phase diagram exists which is superimposed on the iron–cementite diagram (Fig. 13.25).

In every case, the appearance of graphite is a consequence of the decomposition of cementite into carbon (graphite) and iron.

This decomposition is catalytically retarded when mangangese is present, and is facilitated by other elements such as silicon, titanium, and aluminium. Owing to these catalytic effects, the Fe–C phase diagram is a representation of merely theoretical values since, on account of the practical complexity of the constitution of cast irons and steels, it is inevitable that such catalytic effects will occur.

Another factor which is capable of modifying the theoretical forecasts is the rate at which the melts are cooled and the solids are formed since, by varying this, the equilibrium thermodynamic states described by the Fe–C phase diagram as a consequence of the evolution of a system containing specific amounts of iron and the purest carbon under conditions of extremely slow cooling may only partially be attained or not attained at all.

In spite of its limited practical validity, if taken literally, the iron–carbon phase diagram provides a sound foundation of knowledge regarding the possible behaviour of metallic materials consisting of the base metal, iron, and in which the alloying element or, at least, the most typical alloying element, is carbon.

For example, it may be intuitively perceived that, simply because the transformations which this phase diagram describes are the result of extremely slow cooling, it should be possible, by lowering the temperature rapidly, to 'freeze' the equilibrium involving the transformation of stable structures outside of the field where they normally exist, and, as a result, be able to exploit their physicochemical and mechanical properties in some technical or practical area.

Similar deductions are possible with regard to the catalytic effects induced by the introduction of particular elements into iron–carbon alloys. Apart from acting as stabilizers, they may also act to enlarge or diminish the fields where the phases described by the Fe–C phase diagram exist.

In treating iron alloys, as we shall do in the next sections, we shall have occasion to recall specific examples of structural situations which demonstrate the importance of knowing which of the basic phases, described and correlated by the Fe–C diagram, are present in such alloys.

Moreover, as well as the fact that it explains their structures and illustrates their interdependence, the reason for the absolute validity of the Fe–C phase diagram lies in the fact that it implicitly demonstrates that the coefficients of a structure are the chemical composition and the thermal modifications, hence these are the coefficients of the physical, mechanical, and chemical properties of Fe–C alloys.

13.16 CAST IRONS

According to the European Iron and Steel Community standards, cast irons of practical use are alloys of iron and carbon containing the alloying element (that is, the carbon) in amounts varying from 1.9 to about 4%.

In general, it may be said that it is precisely the carbon content which is the fundamental parameter which determines the characteristics of cast irons. This carbon may be in the form of free carbon, lamellae, globules of graphite, in combination with iron in free cementite, in a eutectoid pearlite mixture, or in a eutectic lederburite mixture.

Various criteria are employed in the classification of cast irons as can be seen from Table 13.7.

Fig. 13.26 provides an explanation of how the microstructures of cast irons vary in appearance and mechanical properties.

The occurrence of carbon in various forms in cast irons (as has been stated above, confirmed in Table 13.7, and exemplified in Fig. 13.26) is a consequence of the type of cooling to which it is subjected and/or catalytic effects such as those due to cerium and magnesium in spheroidal cast irons, for example.

13.17 STEELS

According to the European Iron and Steel Community standards, steels are alloys of iron and carbon containing an amount of the alloying element (C) between 1.9 and 0.02%. However, the steels which are used in practice are restricted to a carbon content between 1 and 0.1%.

Commercial classification of steels

Commercially, steels are primarily named in such a way as to indicate their cost. This is seen from the following scheme.

Steels

- *carbon* with the (inevitable) random presence of elements other than Fe and C
  - *common* (>0.1% [P + S])
  - *quality* (P < 0.06%, S < 0.06% and [P + S] < 0.1%)
  - *special* (P < 0.035%, S < 0.035% and [P + S] < 0.06%)
- *alloy* with the deliberate addition of alloying elements other than Fe and C
  - *low alloy* (if no element except Fe is present in them in an amount greater than 5%)
  - *alloy* (if they contain at least one element apart from iron in an amount greater than 5%)

Table 13.7—Classification of cast irons and their properties

Criterion	*Types*	*Origin, distinctive characteristics, and principal applications*
Preparation	First melt	Products of poorly controlled composition from blast furnaces, possibly used in the manufacture of low-value castings.
	Second melt	Products of the remelting and refining with adjustment of the composition of blast furnace cast iron in cupola or electric furnaces.
Appearance	White	Bright clear fracture. Prepared by rapid cooling and possibly contains other elements such as silicon, aluminium, nickel, and phosphorus.
	Grey	Dark grey fracture. Prepared by slowly cooling melts and possibly contains elements such as silicon, aluminium, nickel, and phosphorus.
Microstructure	Cementite	White cast iron as the result of the formation of an iron–cementite structure.
	Graphitic	Grey cast iron as the result of the formation of an iron–graphite structure.
	Mottled	A white matrix containing primary cementite with grey speckles of graphite.
	Spheroidal	Clear matrix with secondary cementite, some pearlite, and graphitic spherules.
	Meehanitic	Very fine graphite lamellae uniformly dispersed in a pearlite matrix.
Mechanical properties	White	Intrinsically, hard and brittle. Properties can be improved by adding Mn, Ni, Cr, V, etc.
	Grey	Relatively soft, malleable, and workable. Good compressive strength but low tensile strength.
	Mottled	Intermediate between white and grey cast irons.
	Malleable	Properties intermediate between grey cast irons and steels ($R_{tensile(av)} = 38$ kg/mm^2).
	Spheroidal	Tough ($R_{tensile} \simeq 84$ kg/mm^2) and ductile (max. elongation, 15%), similar to steels.
Uses	White	For refining (production of steels) and remelting (production of malleable cast irons). Not used much directly if not at least partially transformed (e.g. mottled cast irons).
	Grey	Radiators, boilers, gears, taps and fittings, heating devices, stoves, etc.
	Malleable	Motors, automobiles, tractors, textile machine components, connectors, and valves.
	Spheroidal	As malleable cast irons but in thick-walled castings, hydraulic cylinders, compressor components, drainage pipes, large gears, wheels, Diesel engines, and machine tools.
	Alloyed	If grey (allowed with Mo, Cr, Ni, and Cu), they have enhanced wear, high temperature and corrosion resistance. If they are remelted white cast irons (alloyed with Ni, Cr and Mn), they are hard and tough. If spheroidal (alloyed with Mg, Ni, and Cu), they have high tensile and impact strengths.

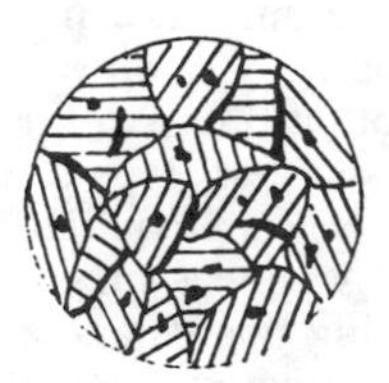

Fig. 13.26—Examples of the microstructures of grey cast irons. The parts in white are occupied by ferrite, the parts in black are occupied by (laminar or globular) graphite. The hatched areas represent pearlite, while ledeburite is to be found in the small white circles in the interior (in the diagram on the extreme right).

The common steels are those which are on average by far the most used, and are subdivided as follows (R = tensile strength):

● very low carbon	$C < 0.15\%$	$R \leqslant 38\ \text{kg/mm}^2$
● low carbon	$0.15\% < C < 0.30\%$	$R \simeq 45\ \text{kg/mm}^2$
● medium hard	$0.30\% < C < 0.45\%$	$R \simeq 65\ \text{kg/mm}^2$
● hard	$0.45\% < C < 0.65\%$	$R \simeq 75\ \text{kg/mm}^2$
● extra hard	$0.65\% < C < 1.9\%$	$R \geqslant 80\ \text{kg/mm}^2$

The hardness and tensile strength of common steels therefore become greater as the percentage of carbon increases. On the other hand, it should also be noted that the workability and impact strength of these steels fall in parallel with the increase in the former properties.

Official classification of steels

The classification and coding of steels according to UNSIDER, the official Italian bureau for iron- and steel-making standards, which is affiliated to EURONORM (the European Standards Institute), are shown in Table 13.8.

Greater detail regarding this matter should be sought either in up-to-date books on metallurgy or metallography or, better still, from the point of view of authoritativeness, in source material (by consulting the most recent periodical publications of the specific tables published by the above institutes).

Technological uses of steels

Technologically, steels are usually classified on the basis of their characteristic uses. Here, we shall summarize the most important technological types of steel.

Automatic steels. 'Automatic' steels are, for the most part, low-carbon steels containing sulphur and lead which are adapted for work involving the removal of swarf in operations such as turning, milling, chasing, by automatic machine tools.

Steels for machine construction. These are so named because they are typically employed in the construction of machine components such as levers, driving mechanisms, connecting rods, and pinions.

They must be tough and have a martensitic structure which is obtained by means of special heat treatment processes, the description of which lies outside of the compass of this book. They may be carbon steels or high-alloy steels depending on the severity of the forces to which they will be subjected.

Steels for building construction. In the building industry, the carbon steels are by far the most used.

The steel rods for reinforcing concrete are made of very-low-carbon or low-carbon steels, while structural sections, laminated components and tubes are made from low carbon or medium hard steels. Finally, the wires used in the construction of modern 'prestressed concretes' are made out of hard or extra hard steels.

However, high alloy, 'stainless' type steels have also started to be used by the building industry when there are special requirements which, nowadays, are becoming ever more numerous and perceived. Architectural coverings and linings, window and door frames and fittings, frameworks, skirting boards and even entire external structures are the fields where high alloy steels are in the greatest demand in the building industry.

Steels for use in chemical plants. These steels are produced on a vast scale as they are required for the chemical, petroleum, food, and pharmaceutical industries, etc. and require treatment in accordance with the specific plant environment in which they are to be used. There are

Table 13.8—Distribution and official.representation of steels

Groups	*Subgroups*	*Classes*	*Nomenclature*
I contains steels designed on the basis of their physical characteristics. *They are all indicated by the symbol Fe*	I.1 contains steels designated on the basis of their mechanical properties.	I.1.1 on the basis of the tensile strength	Symbol Fe followed by the value (kg/mm^2) of the tensile strength (e.g. Fe 42)
		I.1.2 on the basis of the yield point	Symbol Fe followed by E and the minimum value of the yield point (kg/mm^2) (e.g. Fe E 36)
		I.1.3 on the basis of tensile strength and a suitably added alloying element	Symbol Fe followed by the tensile strength value (kg/mm^2) and symbol of alloying element (e.g. Fe 40 Ni)
		I.1.3 on the basis of the yield point and a suitably added alloying element	Symbol Fe followed by E by the value (kg/mm^2) of the minimum yield point and the symbol of the alloying element which characterizes the steel (e.g. Fe E 42 Mn)
	I.2 contains steels designated on the basis of particular characteristics		Symbol Fe followed by letters and numbers which indicate the special properties characterizing the steel (magnetic losses, conductivity, etc.)
II contains steels designated on on the their composition. *No general symbolism*	II.1 contains non-alloyed steels. (Fe and C only)		Symbol C with C% × 100 (e.g. C 35 is a non-alloyed steel containing 0.35% C
	II.2 contains alloy steels	II.2.1 low-alloy steels (every alloying element $<5\%$)	No particular symbol. %C × 100 and % of alloying elements multiplied by 4 or 10 (*).
		I.2.2 alloy steels (content of at least one alloying element $>5\%$)	General symbol X. % of C multiplied by 100 and and actual percentages of alloying elements (**).

(*) For instance, a low-alloy steel containing 0.12% C and 1% Ni can be indicated as 12 Ni 4 or as 12 Ni 10.
(**) e.g. an alloy steel containing 0.1% C and 13% Cr is indicated as X 10 Cr 13.

many technological categories of steels which are of great interest in this field, and they will be listed here. These, in particular, include stainless steels and high- and low-temperature steels.

Spring steels. These are characterized by a high elastic limit which is achieved by means of various heat and mechanical treatments. Spring steels may belong either to the hard and extra hard carbon steels or to both categories of alloy steels. It is essential that they should be highly alloyed, especially with Cr, W, and Mo, if they are destined to become springs which operate at high temperatures (valve springs in internal combustion engines, for example).

Steels for carburizing and nitriding. Those used for carburization are low-carbon steels ($\leqslant$0.20%) of the 'common' or alloyed types, while those to be nitrided contain 0.3–0.4% of carbon and Al, Cr, and Mo, that is, elements which readily form nitrides.

Tool steels. These must always be particularly hard when cold and often up to a red heat, as well as being abrasion resistant, indeformable, and tough. They are used in the manufacture of cutting tools, wire drawing dies, boring tools, saws, etc. The most important of these steels are the high-speed steels which permit machining operations to be carried out at high speeds and/are, to a greater or lesser degree, highly alloyed with W, Mo, and Co.

Typical formulations of steels used in this field are shown in Table 13.9.

Stainless steels. This is a category of steels without any specific destination but which are particularly resistant to corrosion.

Metallographically, stainless steels may be of many different types (austenitic, martensitic, etc.) but the one characteristic which they will have is that of being highly alloyed especially with chromium and then nickel.

Stainless steels are included, in particular, amongst the steels employed in the construction of chemical plant. For a deeper knowledge of these steels the reader is referred to texts on this subject.

High- and low-temperature steels. Steels which do not lose their mechanical properties (and, at high temperatures, their resistance towards chemicals) to any pronounced extent above and below normal temperatures respectively are included in these categories.

In general, there many types of steel which are adapted to exhibit good resistance qualities at high temperatures. A common characteristic of these steels is their strong resistance to 'viscous creep'*.

Even though quality carbon steels are resistant at operating temperatures below 450°C, under conditions of modest mechanical stresses and a non-aggressive chemical environment, it is necessary to change to low alloy molybdenum steels, even at the same temperatures, if the mechanical stresses are greater or when the operating temperatures may go as high as 520°C.

As the conditions become more severe, the following steels are successively required:

- low-alloy chrome-molybdenum steels with a greater percentage of the first alloying element for operations up to 650°C in an environment which is not particularly aggressive;
- high-alloy chrome steels from 500 to 700°C under oxidizing and chemically aggressive conditions;
- nickel–chromium stainless steels of a particular composition intended to make them proportionally resistant both to mechanical stresses and an aggressive chemical environment when called upon to operate at temperatures up to 900°C;
- superalloys where iron is no longer the major element and in which there are ever greater amounts of elements of the type of nickel and cobalt.

Low-alloy nickel ferritic steels (3–4.5%) or alloy (9%) nickel ferritic steels are very suitable for operating at low temperatures (in the manufacture of tanks for the transportation of liquefied gases, for example).

Electromagnetic steels. Apart from the purest annealed iron (C $<$0.05%), annealed steels containing various percentages of silicon (1.4 and 4.5%), high-alloy cobalt (36%), and molybdenum (8.5%) steels, and also steels containing Cr and W, are employed in various technologies concerned with the construction of electromagnetic apparatus. Silicon steels are primarily used in the manufacture of the laminations for alternators and transformers, while cobalt steels are employed in the construction of permanent magnets.

Non-magnetic steels, which are therefore not attracted by a magnet, are also included among the steels for electromagnetic applications. They obviously have an austenitic structure and are highly alloyed with Cr–Ni or Mn or only Ni

---

* *Viscous creep* is the phenomenon of deformation under 'viscous' conditions to which stressed materials are subjected at a certain temperature. They finally fracture after a specific limiting degree of viscous creep has taken place.

Table 13.9—The composition of high-speed tool steels and substitutes for them which have been developed to economize on the amounts of tungsten and cobalt used

Original	*C%*	*Cr%*	*W%*	*Mo%*	*V%*	*Co%*
Medium–high speed	0.6–0.8	4.0	10.0	small amounts	small amounts	–
High speed	0.6–0.8	4.0	18.0	0.8–0.1	small amounts	–
Ultra–speed	0.6–0.8	4.0	18.0	0.5–1.0	0.5–2	4–9
Substitutes	*C%*	*Cr%*	*W%*	*V%*	*Mn%*	*Si%*
Commonly used as semi-high-speed and high-speed steels	0.9	4.5	8.0	2.2	–	–
Profiling and planing tools under high-speed and ultra-high-speed conditions	1.30	4.5	11.0	4.3	0.35	0.35
Ultra-high-speed advance	0.80	4.5	14.0	2.2	–	–

Structural steels. Also known as constructional steels, these metallic materials are used in the constuction of plates, structured sections, small girders, tubes, and reinforcing rods. These steels are generally carbon steels. Less commonly, quality steels are employed and, even more rarely, low-alloy steels.

13.18 FERROALLOYS

Metallic materials which contain iron, carbon, and one or more other elements in very high proportions are known as ferroalloys.

Table 13.10 presents examples of the complete composition of ferroalloys which are used to correct the composition of cast irons and steels, to carry out operations in non-ferrous metallurgy, and for the manufacture of alloy steels.

Ferroalloys are generally prepared by reduction of the oxide of the metal which predominates in them with coke in the presence of an iron mineral and a slag-forming agent or ferrous scrap. This is, for the most part, carried out in electric arc furnaces, utilizing the high temperatures which they can attain.

Usually, the elements of ferroalloys exhibit a pronounced tendency to form carbides which are converted to the metal by prolonged high-temperature contact with the oxides of the same metal. For instance, manganese carbide loses carbon when it is oxidized with manganese dioxide:

$$2Mn_3C + MnO_2 \rightarrow 7Mn + 2CO$$

However, ferroalloys are also prepared by the aluminothermic and silicothermic

Table 13.10—Composition of the most important iron alloys (iron completes to 100)

Symbolic name	*Mn*	*Si*	*C*	*S*	*P*	*Cr*	*Ca*	*Al*	*W*	*Mo*	*V*	*Ti*
Fe–Si	–	90–85	<0.20	<0.10	<0.10	–	–	<2.00	–	–	–	–
Fe–Si	–	75–80	<0.30	<0.10	<0.10	–	–	<2.0	–	–	–	–
Fe–Si	–	45–50	<0.40	<0.15	<0.10	–	–	<1.50	–	–	–	–
Fe–Si	–	20–25	<1.50	<0.15	<0.10	–	–	–	–	–	–	–
Fe–Si	–	10–12	<2.00	<0.15	<0.10	–	–	–	–	–	–	–
Fe–Mn	75–80	<2.00	<8.00	<0.40	<0.40	–	–	–	–	–	–	–
Fe–Cr	–	<3.00	6–8	<0.15	<0.10	60–70	–	–	–	–	–	–
Fe–Cr	–	<3.00	4–6	<0.15	<0.10	60–70	–	–	–	–	–	–
Fe–P	<0.20	1.50–2.00	<0.25	–	22–25	–	–	–	–	–	–	–
Fe–Si–Al	–	45–50	–	–	–	–	–	15–20	–	–	–	–
Fe–Si–Cr	–	40–45	<0.10	–	–	35–40	–	–	–	–	–	–
Fe–Si–Mn	65–70	20–25	<2.00	<0.05	<0.20	–	–	–	–	–	–	–
Fe–Si–Mn	45–50	20–25	<2.00	<0.05	<0.20	–	–	–	–	–	–	–
Fe–Si–Mn	4–6	14–16	<2.50	<0.05	<0.20	–	–	–	–	–	–	–
Fe–Si–Mn–Al	19–21	<20	–	–	–	–	–	9–11	–	–	–	–
Si–Ca	–	60–70	<1.00	<0.10	<0.10	–	20–25	<3.00	–	–	–	–
Si–Ca	–	60–65	<1.00	<0.10	<0.10	–	30–35	<3.00	–	–	–	–
Fe–Ti	–	<2.00	–	–	–	–	–	4–6	–	–	–	22–25
Fe–Ti	–	<2.00	–	–	–	–	–	7–8	–	–	–	37–40
Fe–Ti	–	<3.00	–	–	–	–	–	10–11	–	–	–	45–50
Fe–Mo	<0.75	<1.00	–	<0.20	<0.10	–	–	–	–	60–65	–	–
Fe–V	–	<1.00	–	<0.10	<0.10	–	–	<2.00	–	–	50–60	–
Fe–W	<1.00	<1.00	–	<0.06	<0.06	–	–	<0.50	75–85	–	–	–
Fe–Mn	80–85	<2.00	<1.00	<0.04	<0.40	–	–	–	–	–	–	–
Fe–Mn	80–85	<1.00	<0.50	<0.04	<0.40	–	–	–	–	–	–	–
Fe–Cr	–	<2.00	<2.00	–	–	60–70	–	–	–	–	–	–
Fe–Cr	–	<2.00	<1.00	–	–	60–70	–	–	–	–	–	–
Fe–Cr	–	<2.00	<0.50	–	–	60–70	–	–	–	–	–	–
Fe–Cr	–	<2.00	<0.10	–	–	60–70	–	–	–	–	–	–

processes. An example of the first of these processes, which leads to the preparation of ferrotungsten, is:

$$3FeWO_4 + 8Al \rightarrow 3Fe + 3W + 4Al_2O_3$$

Electric furnaces are also the most suitable means of carrying out this type of chemicometallurgical operation.

Part 4 THE NON-FERROUS METALS

We shall consider the non-ferrous metals that are discussed in this book in three groups:

- metals of large-scale technology such as copper and nickel;
- metals employed in specialist technologies, such as zirconium and uranium;
- the noble metals such as gold and silver.

13.19 THE METALS OF LARGE-SCALE TECHNOLOGY

In terms of the amounts of the metal produced or used (chromium and manganese), the wide variety and interests of the uses (tin), for example, or for both reasons (especially aluminium and copper), the most important metals from a technological point of view are aluminium, copper, nickel, zinc, lead, chromium, manganese, and tin.

Aluminium

Properties and natural state

Aluminium is the most abundant metal in nature (page 454) and, after iron, it is the most used metal. This eminent position has been attained only during the present century, and it is due largely to the availability of electrical energy for its production.

Aluminium has a melting point of 659.7°C, a boiling point of 2057°C, and adensity, $d = 2.7\ g/cm^3$.

It is among the best electrical conductors and, in spite of its low reduction potential, has a high resistance towards chemicals (perfect resistance towards common atmospheric reagents) because it is passivated by covering itself spontaneously with a very thin, adherent, and compact layer of Al_2O_3 which protects it from further attack. This can also be achieved in a faster and more secure manner by anodic oxidation in a suitable electrolytic bath (anodic passivation).

Although it is exceedingly abundant in rocks (aluminosilicates, clays, micas, etc.), there are few aluminium minerals which can be usefully exploited for production in the metallic state. It is extracted on a global scale from a rock called bauxite with the reference formula $Al_2O_3 \cdot nH_2O$, but is always accompanied by a predominant silicaceous gangue ('white bauxite') or iron-bearing gangue ('red bauxite') as well as by TiO_2 and other minor minerals.

An aluminium compound which occurs naturally in a very pure state is cryolite (Na_3AlF_6). This is never used as a source of aluminium but, as we shall see, it is essential in the production of aluminium to a point when, having exhausted its natural deposits, it is now prepared artificially (page 317).

In various countries attempts have been made to utilize concentrates of local aluminium minerals for extracting the metal. For instance, there is interest in nepheline ($NaAlSiO_4$) in the

USSR, andalusite (Al_2SiO_5) in Switzerland, and leucite $KAl(SiO_3)_2$ in Italy. It is, however, difficult to foresee any successes in the exploitation of alternative sources while the market still offers good quality bauxites.

The metallurgy of aluminium

The metallurgy of aluminium is a two-stage process: high purity (99.5–98.8%) Al_2O_3 (alumina) is first recovered from the bauxites, and then the 'first electrolysis' metal is obtained by suitable electrolysis of this oxide. From this product up to 99.99999% aluminium can be subsequently prepared by means of electrolytic refining of the fused metal.

The preparation of aluminium. The operating technology which is most commonly used for this purpose is the 'Bayer process' which employs the selective dissolution of the amphoteric component of bauxite (Al_2O_3) in concentrated caustic soda and the precipitation, of $Al_2O_3 \cdot 3H_2O$ under controlled pH conditions and with seeding of the crystallization followed by the final dehydration and calcination of the filtered precipitate.

A block diagram of the Bayer process is shown in Fig. 13.27. This process leads to the preparation of very pure alumina which is free from iron and silicon which tend to codeposit at the cathode with aluminium.

Bauxite, which is poor in SiO_2, is precalcined at 450°C to destroy any organic substances which it may contain, reduced to 'bauxite flour' by intensive milling, and treated in autoclaves with a hot (150–200°C) concentrated (280 g/l Na_2O) solution of caustic soda, which operate under pressure (5–20 atmospheres) for a time which is a function of the operating conditions. The following reaction occurs:

$$Al_2O_3 \cdot 3H_2O \cdot RS + 2NaOH \rightarrow 2NaAl(OH)_4 + RS$$

where RS indicates the 'red sludges' which are insoluble in the alkali and which consist of SiO_2, Fe_2O_3, and TiO_2.

After cooling, the insoluble and the unsolubilized* material is filtered off, the precipitate is washed, and then disposed of.

The filtrate is diluted† with washing waters (from the filtered sludges and the hydrated aluminium oxide which has been successively precipitated and then washed) in order to lower the pH value somewhat, without reaching the isoelectric point, to a pH value of about 6 which should cause the residual dissolved silica to precipitate. Effects which promote the phenomenon such as seeding with crystallization nuclei, allowing the solution to stand for a long time (about 100 hours), and slow stirring, more then compensate for the effect of the pH which is not very favourable to the precipitation of Al_2O_3.

The hydrolytic precipitation process uses the reaction:

$$2NaAl(OH)_4 + aq \rightarrow 2NaOH_{(aq)} + Al_2O_3 \cdot 3H_2O$$

The precipitate is filtered off and washed with the production of weakly alkaline waters which are recycled for the dilution of the hydroxyaluminate, as mentioned above.

If necessary, the filtrate has caustic soda added, and, without fail, it is concentrated in three stages before being suitably reintegrated to make up for the losses of NaOH and then recycled

---

*Actually, when it is hot and under pressure, SiO_2 has an appreciable solubility. However, by decompressing the system and cooling it, a large part of the dissolved silica is reprecipitated as sodium aluminosilicate, while the rest of it, if the pH value does not fall below 10, remains dissolved in the form of sodium silicate.

†So as not to have to dilute it so much, use is also made of the practice of blowing CO_2 into the caustic soda solution. This, however, has been found to be generally inconvenient owing to the subsequent costs of recausticization.

to the attack on the bauxite flour. The precipitate (Al_2O_3, $3H_2O$) is washed, dehydrated, and then calcined at 1200°C in rotating furnaces which discharge very pure alumina.

Production of the 'first electrolysis' metal. Pure alumina, Al_2O_3, is added to a molten mixture of fluorides (in which, apart from the large amount of cryolite, Na_3AlF_6, there is also present AlF_3 and a little CaF_2) in rectangular electrolytic cells, shaped from sheet steel covered with a refractory which, in its turn, is lined with monolithic cathodic lining material, an electrical conductor made from an anthracite-coke and tar paste. The alumina forms a eutectic with the cryolite and is thereby able to melt and remain molten at only 960°C (in spite of its theoretical melting point of 2046°C). Heat is supplied to this system both by the combustion of the anodes and, mainly, by the Joule effect*.

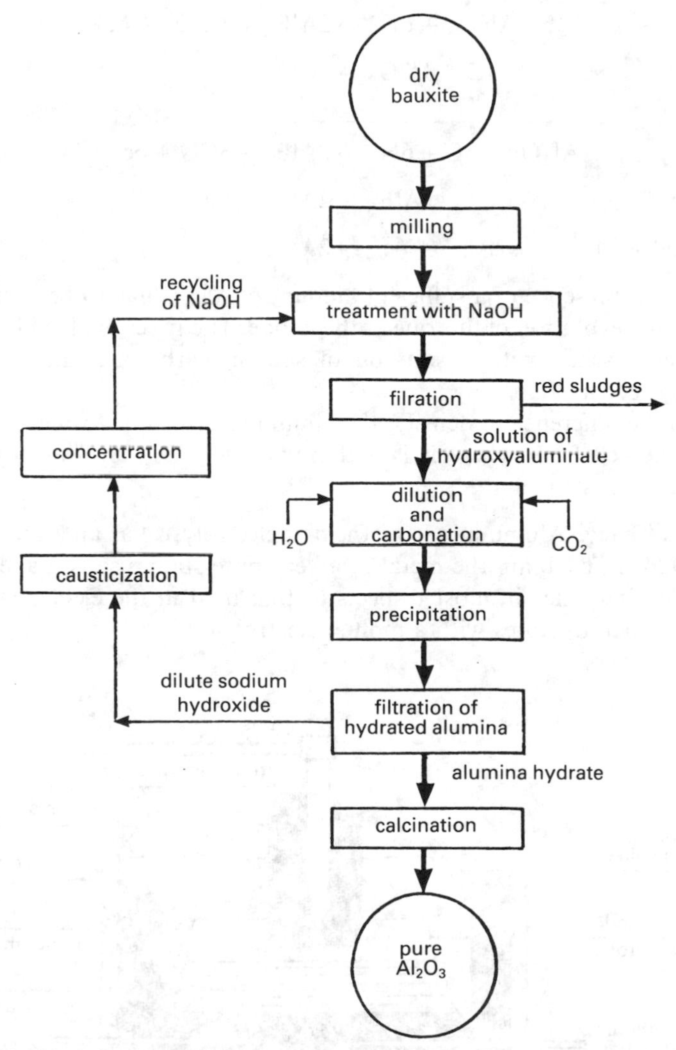

Fig. 13.27—Block diagram of the preparation of pure alumina (Bayer alumina) from bauxite.

*The production of heat by the passage of a current through an electrical resistance is controlled by varying the distance between the cathode and the anode.

In theory, the potential to be applied is 1.7–1.8 V, but, in practice, a potential of 6–7 V must be used on account of the numerous phenomena which lead to a drop in the potential.

Single or multiple anodes may be used, but, as the second type lends itself to greater production levels, it is much preferred nowadays.

The single electrode (Fig. 13.28), called a 'Södeberg electrode' is cast *in situ* by the baking of a mixture of pregraphitized carbon powder from petrochemical production (especially, cracking operations) and tar, using the heat irradiated from the cell itself. When loaded into a steel jacket, this paste becomes a single block during its slow descent towards the bottom, while receiving current from metal rods. The electrochemical processes are:

at the anode(+): $$Al_2O_3 + 6F^- \rightarrow 2AlF_3 + 3/2O_2 + 6e^-$$

$$3C + 3/2O_2 \rightarrow 3CO$$

$$Al_2O_3 + 3C + 6F^- \rightarrow 2AlF_3 + 3CO + 6e^-$$

in the melt: $$2AlF_3 \rightarrow 2Al^{3+} + 6F^-$$

at the cathode(−): $$2Al^{3+} + 6e^- \rightarrow 2\mathbf{Al}$$

The constant presence of a sufficient amount of Al_2O_3 has to be ensured in order to minimize the evolution of fluorine at the anode. The gases evolved from the anode are, however, washed with a solution of sodium carbonate, and the fluorine is subsequently recovered.

Owing to the difference in density, the aluminium which is formed collects on the bottom of the cell from where it is withdrawn every 24–28 hours by means of a vacuum suction system.

Electrolytic refining. Aluminium from the first electrolysis has an aluminium content of about 99.5%. To obtain the metal in a very pure state, use is made of a refining process which is similar in most respects to that used in the electrolytic refining of copper, but which operates with a molten electrolyte.

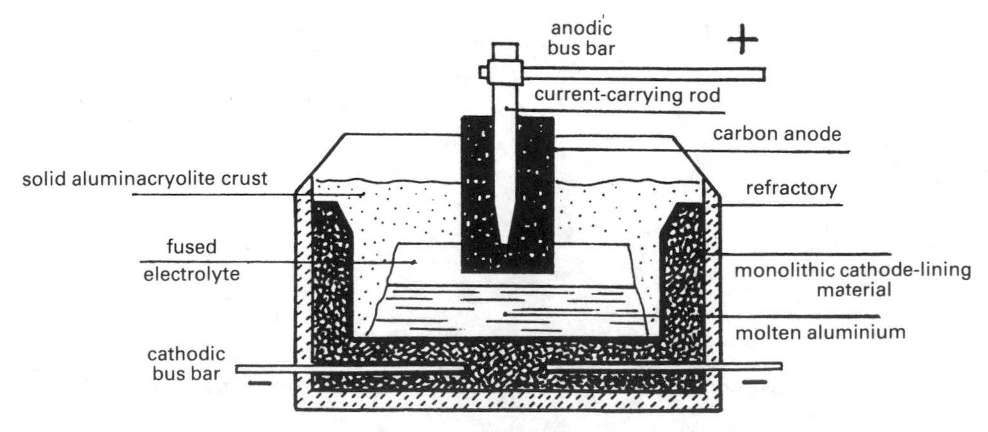

Fig. 13.28—Section of an electrolytic cell for the production of aluminium from fused alumina in cryolite. A single (Södeberg) anode system is shown.

The anode for this process is made by alloying copper and aluminium into a relatively dense metallic material which is carried on to the bottom of the cell. The cathode is made of the purest aluminium and floats on the molten fraction of the complex mixture (of cryolite, BaF_2, $CaCl_2$, and alumina) which acts as an electrolytic bath of such a density that the cathode floats while the anode sinks, as is required.

Once the appropriate potential difference of about 6.5 V has been established, all the metals constituting the less noble impurities (Na, Mg, Ca, etc.) dissolve at the anode together with the aluminium:

$$Al \rightarrow Al^{3+} + 3e^-$$

while Fe, Zn, Cu, etc. remain undissolved.

Only aluminium is subsequently deposited at the cathode:

$$Al^{3+} + 3e^- \rightarrow Al$$

because it is the most noble metal of all those present as cations in the electrolyte. The floating metal separates and is protected from oxidation reactions due to contact with air by the solid crust which covers the surface of the electrolyte.

Uses of aluminium

The properties of aluminium which are exploited in practice are, above all, the ease with which it can be passivated, its non-toxic nature, electrical conductivity, lightness, and combustibility at high temperatures.

Its readiness to become passivated lies at the foundation of the uses which are made of it in chemical and parachemical plant engineering (the fermentation, pharmaceutical, and food industries), in boat construction, tanker wagons, and railway carriages. This property and its lightness also favour its use in building construction and, when taken together with its non-toxic nature, have promoted its imposing overall consumption in the domestic utensil industry, the dairy and cheese-making industry, and the pharmaceutical–food–cosmetics industry: in bottles, aerosols, flexible and rigid tubes, in bottle caps, as foils for the wrapping of soft foodstuffs, in ring-pull cans for drinks, and packaging for conserved foodstuffs, and so on.

Electrical technology exploits the conductivity of aluminium in various ways, while its combustibility is exploited in different ways in metallurgy (the aluminothermic reaction), the pyrotechnics industry (coloured flames), and certain technologies associated with the arms industry (napalm). Finally, the graphic and varnish industries make use of aluminium mainly in the form of foils and powders respectively.

However, most of the aluminium which is produced (more than 60% of the entire production) is used in the form of alloys in the fields of motor car construction, the construction of other kinds of road transport and rail transport, and the shipbuilding, aeronautical, metallurgical, and mechanical engineering industries (the manufacture of machines, of structural metalwork for buildings and electric transmission lines).

The many aluminium alloys can be subdivided into three types, on the basis of their composition and properties:

- electrical alloys of the Aldrey type (*c.* 0.5% Si, 0.5% Mg, and 0.2% Fe), which has an electrical conductivity which is only slightly lower than that of aluminium but has considerably improved mechanical properties. On account of this, high-voltage electrical transmission lines are made out of these alloys;

- casting (foundry) alloys, which are suited for pouring into complex moulds and also where shapes with thin sections are to be cast;
- alloys for plastic working (for construction), which are for the manufacture of laminates, structured sections, and extrusions.

The abbreviated names of foundry alloys start with GAl, to which is added the symbol of the principal alloying element and its percentage, as well as the symbols of any other elements which are present. For instance, GAlSi12.72FeCu indicates a foundry alloy which contains 12.72% of Si as well as some iron and copper.

For plastically worked alloys the prefix is PAl, followed by the symbol of the principal alloying element and by its percentage, as well as by the symbols of the other alloying elements. For example, the alloy duralumin, which contains 4.5% Cu, 1% Mg, 0.8% S, and 0.75% Mn, is indicated as PAlCu4.5MgSiMn.

Finally, aluminium alloys are also technologically subdivided into light and ultra-light alloys. Since magnesium predominates in the latter, they are more correctly considered as alloys of this metal.

The light alloys are subdivided into:

- alloys which can be hot-aged, that is, by heating for one day at about 130°C;
- alloys which can be cold-aged, that is, by heating them at a temperature not exceeding 60°C for ten days.

In every case, ageing has the main aim of hardening these alloys by bringing about their recrystallization with an increase in the tensile strength.

Alloys for constructions, such as ergal and anticorodal predominantly belong to the alloys which can be hot-aged, while the casting alloys such as silumin and the alloys for plastic working such as duralumin and its numerous modifications (avional, lautal, and so on) belong to the class of alloys which can be cold-aged.

Copper

Properties and natural state

Copper, with its salmon-pink colour, medium hardness, and great ductility and malleability is a metal which is much used in the pure state on account of its very special thermal and electrical conductivity characteristics, its relatively high resistance to corrosion, and its outstanding workability both when hot and cold.

Copper becomes hard when cold-worked, almost doubling its mechanical strength and minimizing its percentage elongation.

While it is not attacked by non-oxidizing acids, a layer of the basic carbonate (verdigris) forms on it in moist air.

The physical properties of copper are as follows: melting point 1083°C, boiling point 2595°C, and density $d = 8.92\ g/cm^3$. Copper is also found (but it is extremely rare nowadays) in the free state in nature, and its minerals, which are qualitatively extremely numerous, are usually classified as oxides and sulphides: the latter constituting more than 80% of copper minerals.

Among the oxides, only cuprite Cu_2O is of any importance in the production of the metal, while, among the sulphides, there is great interest in chalcocite Cu_2S and, even more so, in chalcopyrite, $CuFeS_2$.

Copper minerals such as malachite $CuCO_3 \cdot Cu(OH)_2$, azurite $2CuCO_3 \cdot Cu(OH)_2$, and brochantite $CuSO_4 \cdot 3Cu(OH)_2$ constitute the outcrops of copper compound deposits, and minerals such as bornite Cu_5FeS_4 and enargite, Cu_3AsS_4, are accompanied by other sulphides.

The four largest copper mining districts in the world are in the USA, Chile, Central Africa, and the USSR.

The metallurgy of copper

After an enrichment of the levigation and/or sink and float type in the case of oxides, and using flotation in the case of the sulphides, copper cementation (the displacement of copper from its solutions by iron) is carried out on the oxide minerals, while a metallurgical treatment using persistent oxidation and a final refining reduction is most frequently applied to sulphide minerals.

Fig. 13.29 shows the basic stages in the metallurgical processes which are carried out on copper minerals.

While the chemistry underlying the wet processes has already been adequately explained on the basis of the material presented on page 468, the dry process requires a more detailed explanation than that which has already been mentioned under (5) in Table 13.3. In particular, the problem of the oxidation and removal of the gangue as a slag should be treated, since this problem barely exists when chemical attack and electrochemical displacement is used.

The production of 'matte'. After the agglomeration of the floated mineral (if the operations are carried out in a 'blown' cast iron shaft furnace—internally lined with a silicoaluminate refractory), or without agglomeration of the available powders (if the work is carried out in a reverbatory furnace), the mineral, sand, and coke are loaded into the appropriate furnace. The amount of coke added varies, depending on a number of factors: the composition of the mineral on which the sulphur content depends, the amount of air used, and in the limit, whether oxygen is used instead of air. In the latter case it suffices to add just that amount of coke which is necessary (about 1% of the mineral treated) to prime the process.

If the mineral being worked contains chalcopyrites, the main reaction is:

$$2CuFeS_2 + 5/2O_2 \rightarrow Cu_2S + FeS + FeO + 2SO_2 \quad (13.5)$$

followed by the removal of the FeO in the slag:

$$FeO + SiO_2 \rightarrow FeSiO_3 \quad (13.6)$$

Both of these reactions are exothermic, and so, with the remainder of the heat which is required being supplied by the heat of combustion of the coke, the mass melts at 1000–1100°C and separates into layers with the slag (with a density of 2.5–3 $g \cdot cm^{-3}$) at the top and an approximately equimolar mixture of Cu_2S/FeS (with a density of 4.5–5.5 $g \cdot cm^{-3}$), which constitutes the 'matte', at the bottom.

From the 'matte' to 'black copper'. After the slag has been removed, the matte is loaded with a silicaceous slag-forming agent either into furnaces which are similar to modern pear-shaped oxygen converters of the Bessemer type or, preferentially, into cylindrical furnaces called 'Peirce–Smith converters'. The two possible plant solutions to this problem are shown in Fig. 13.30.

In both types of reactor the air is blown in over the fused mass and the reactions occurring are:

preferentially: $FeS + 3/2O_2 \rightarrow FeO + SO_2$

partially: $Cu_2S + 3/2O_2 \rightarrow Cu_2O + SO_2$

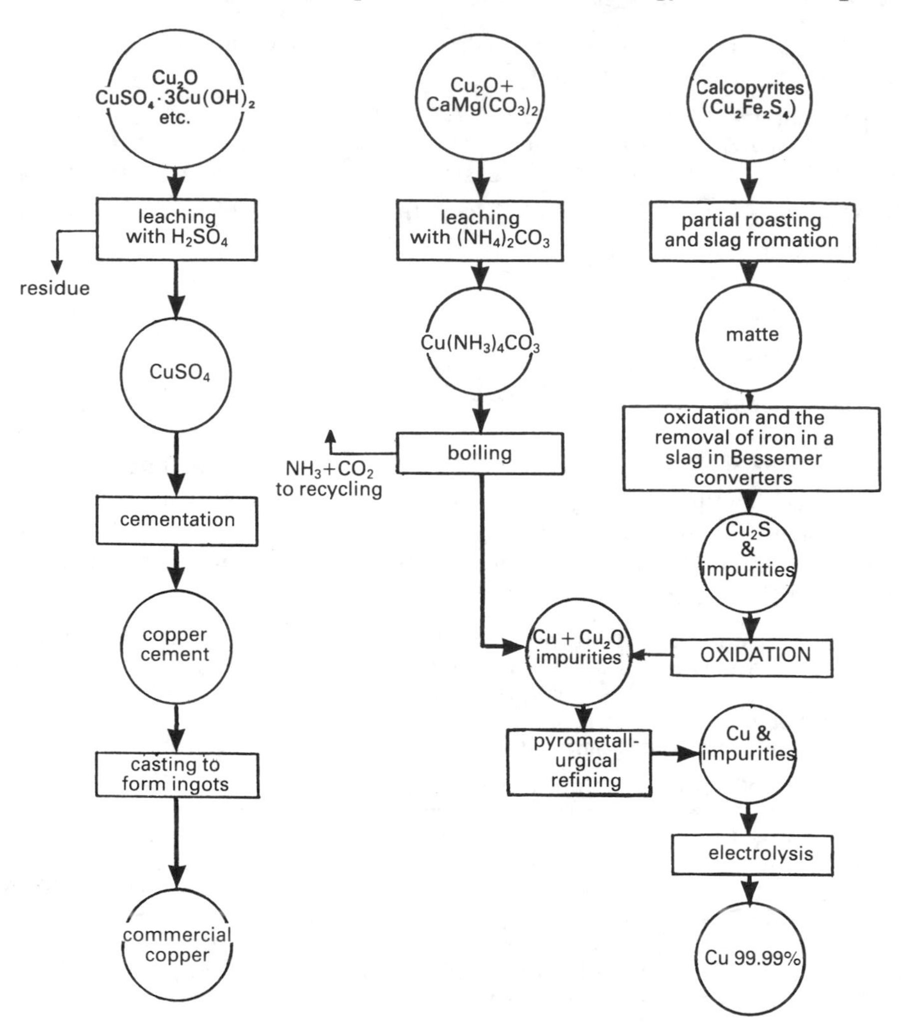

Fig. 13.29—Block diagram of the 'wet', 'wet and dry', and 'completely dry' metallurgical processes which are carried out, according to their nature, on copper minerals.

and, subsequently:

$$FeS + Cu_2O \rightarrow Cu_2S + FeO \tag{13.7}$$

followed by the formation of a $FeSiO_3$ slag according to (13.6).

Reaction (13.7) is a consequence of the fact that copper has a much greater affinity for sulphur than iron, so that, while FeS is present, Cu_2S is not stable converted to the oxide, and the Cu_2S therefore finishes up by becoming practically free of FeS.

At this point the slag* is poured off and, while the apparatus is inclined in this position, air is blown into the molten mass. The following reactions then occur:

$$2Cu_2S + 3O_2 \rightarrow 2Cu_2O + 2SO_2$$

$$Cu_2S + 2Cu_2O \rightarrow 6Cu + SO_2 \tag{13.8}$$

At the end of the process it is inevitable that there is some unreduced Cu_2O close to the copper, and traces of Cu_2S, FeS, and other sulphides (PbS, ZnS, etc.) and noble metals (e.g. Au).

*This slag is further worked because it contains 2–5% Cu on average.

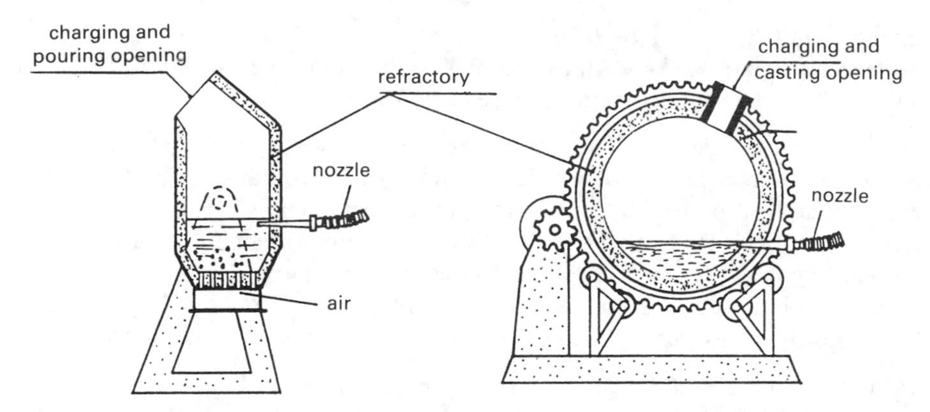

Fig. 13.30—Left. Bessemer converter which can be rotated about an axis. Right. Peirce–Smith converter which can be rotated by means of a geared rack system. Both types are equipped with an acidic refractory lining and a circular band of nozzles for the injection of air into them close to the surface of the charge when the furnaces are in the positions shown in the figure. However, when the furnaces are suitably rotated, the air bubbles into the bulk of the fused mass being worked.

The mass, which is called 'black copper',* contains more than 94% of Cu. As the type of mineral which is initially used, and the mode in which the operations are carried out, can have a very large effect on this percentage, it is also possible to reach a value of 97–98%.

The refining of copper. The copper, which is now in the form of black copper, is refined by using a two-stage (oxidation and then reduction) process which is carried out in reverbatory furnaces. This refining process can also be used for scrap copper.

The first phase of the process is carried out in an excess of air or oxygen over the amount which is theoretically necessary to produce the flames which lead to the reverberation from the canopy of the furnace. The Cu_2O, which is already present in the melt and that which is subsequently reformed there, eliminates all the sulphur via reactions of the type of (13.7) and (13.8) and thereby transforms all the metals which are less noble than copper (Zn, Ni, etc.) into their oxides. These oxides are subsequently converted into easily removable silicates by the addition of sand.

The second (reduction) phase was prevalently carried out by dipping massive trunks of green pine trees (the method of 'poling') into the hot mass, when hot CH_3OH, among other compounds, distils from these trunks, and methyl alcohol is very readily oxidized to formaldehyde in the presence of Cu_2O:

$$Cu_2O + CH_3OH \rightarrow 2Cu + CH_2O + H_2O$$

A phosphorus–copper alloy was once solely used in the manufacture of copper for bronze which had to be deoxidized to the greatest possible extent. Nowadays, silicon–manganese alloys, carbon powder, as well as copper–phosphorus alloys are generally used as deoxidants. The latter alloys react in the following manner:

$$Cu_3P_2 + 5Cu_2O \rightarrow 13Cu + P_2O_5$$

After refining, the copper has the following average contents of extraneous chemical

*The term 'black copper' is reserved for the less pure products containing from 94 to 98% of Cu, while the products containing from 97–99% Cu are known as 'blister copper' (because it contains an appreciable number of blowholes).

elements: O_2 0.3–0.35%, Fe 0.05–0.07%, Pb 0.01–0.03, Ni + Co 0.1–0.15%, As 0.04–0.06%, Sb 0.06%, Ag + Au 0.0045–0.006%, other elements (Zn, Bi, Pt, Ir, etc.) 0.007–0.009%. The copper content exceeds 99.3%.

The electrolytic refining of copper. With the development of electrical energy, there has been an ever-increasing demand for copper which contains at least 99.99% of the metal, and, in particular, is free from impurities such as Fe, P, and As which lower its electrical conductivity very greatly. Electrolytic refining is carried out to obtain such copper, following the criteria described on page 468.

The electrolytic cells into which ingots of various shapes, obtained by the casting of pyrolytically refined copper, are put in order to act as anodes, include the following:

- cathodes made of thin sheets of extremely pure copper at the start;
- an electrolyte which contains an aqueous solution of 13–20% H_2SO_4 and 10–17% of $CuSO_4 \cdot 5H_2O$, a salt which is cyclically regenerated during the process;
- an applied voltage of 0.5 V and a current density of 160–250 A/m^2;
- a temperature of about 50°C.

A diagram of the structure of a cell for copper refining is shown in Fig. 13.31. The copper anode contains:

metals less noble than copper (Fe, Ni, Pb, etc.)	**Cu**	elements more noble than copper and compounds (Sb, Ag, Cu_3As_2, Cu_2Se, etc.)

Owing to the effect of the electrolysis, the less noble metals pass into solution first of all, and at the same time as the copper:

$$\text{Fe, Ni, \ldots, } \mathbf{Cu} \rightarrow Fe^{2+} + Ni^{2+} + \cdots + Cu^{2+} \quad \text{anodic dissolution}$$

The elements which are more noble than copper and the compounds are anodically insoluble and precipitate on the bottom of the cells beneath the anodes as 'anodic

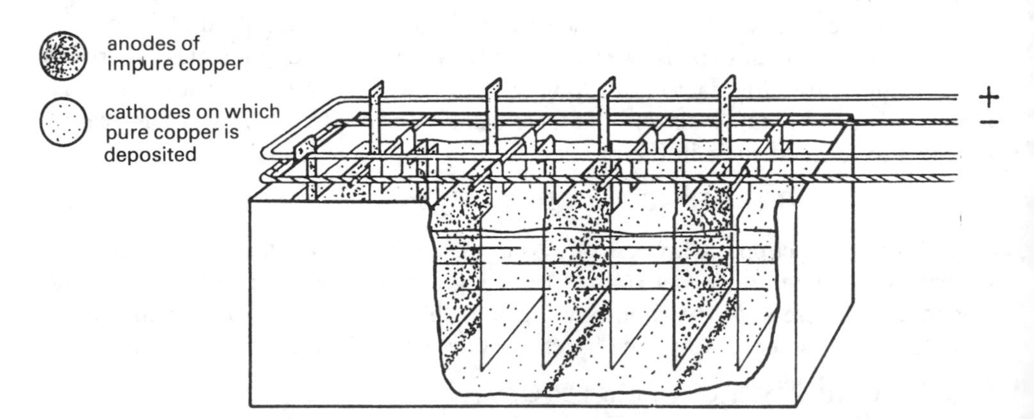

Fig. 13.31—Diagram showing the structure of the cells in which the electrolytic refining of copper is carried out. The anodes and cathodes are connected to the + and − current lines respectively. The 'anodic sludges' are shaded in, and some idea of the greater purity of the copper cathode as compared with that of the anode is shown by the smaller density of spots on the cathodes. It should be added that, to reduce the effects of the ohmic drop, the anodes and cathodes are closer together than indicated in the figure.

slimes' together with the PbO_2 produced by anodic oxidation and the products from the hydrolysis of antimony and bismuth salts.

At the same time, only Cu^{2+} is practically discharged at the cathode:

$$Cu^{2+} + 2e^- \rightarrow Cu \quad \text{cathodic deposition,}$$

provided that the precaution is taken of purging the baths to prevent high concentrations of cations which are less noble than Cu^{2+} building up, which would lead to them competing in the cathodic deposition reaction.

After electrolysis, the cathodes are remelted under controlled reducing atmospheres and then cast into ingots which are subsequently coverted into rods or drawn into wire.

The uses of copper

Almost 60% of the copper which is produced is used as an electrical conductor and in heat exchange technology, while about 10% is used in the preparation of $CuSO_4 \cdot 5H_2O$ which is capable of conversion into other copper compounds. The remainder of the metal is used in alloys, given the fact that copper can be very readily converted into alloys. In fact, the following alloys are produced:

- alloys for plastic working (brasses). These are both hot-worked and cold-worked, and contain from 57 to 97% of Cu. The remainder is zinc and small quantities of other elements (Al, Sn, Ni, Mn, and Pb) which are added to modify the mechanical and/or chemical and/or workability characteristics;
- alloys which, for the most part, contain 4–25% of tin. These bronzes are hard and mechanically strong but cannot readily be worked. If the tin contents is below 8%, they can be plastically deformed (antifriction and coinage bronzes), while those containing from 10 to 20% of Sn* are used for bearings, gears, and taps and fittings;
- cupronickels, which contain nickel which readily forms alloys with copper (to an extent that 'naval bronzes' with 4–6% of Sn and 4% of Ni have been manufactured for many years). Alloys are produced containing from 10 to 30% of nickel and are used in chemical plant engineering, telephone systems, and for the striking of coins;
- silicon copper (1.5–3% Si), aluminium copper (5–10% Al and small amounts of Fe, Ni and Si), chrome copper (0.1–1.0% Cr and a smaller amount of silicon), and beryllium copper (1.5–2.0% of Be and also Cr), are alloys which have outstanding combinations of pairs of mechanical/chemical/electrochemical properties such as workability and hardness, weldability and low conductivity, corrosion resistance and wear resistance, etc.

Finally, the German (nickel) silvers are widely used copper alloys containing a high percentage of both Zn (18–28%) and nickel (12–20%).

Where copper finds direct uses, its workability and mechanical properties can be enhanced by the addition of traces of alloying elements (Tl, Cd, Pb, etc.).

---

*Nowadays, bronzes are no longer used in these areas, having been replaced by steels and aluminium alloys. They are also no longer used for other purposes (such as bells, where the percentage of Sn, for the most part, exceeds 20%). Overall, the traditional importance of bronzes has greatly declined.

Nickel

Properties and natural state

Nickel is a bright, tough, and conspicuously ferromagnetic silvery white metal. Electrically, it is a poor conductor (with about $\frac{1}{7}$ of the electrical conductivity of copper). It has greater wear resistance than copper but is less ductile and malleable.

Nickel melts at 1455°C, boils at 2900°C, and has a density, $d = 8.9$ g/cm^3.

It is slowly dissolved by hydrochloric and sulphuric acids with the evolution of hydrogen, but is readily oxidized and dissolved by nitric acid to form the corresponding salt. It is also resistant even to very concentrated hot alkalis, and, as it is readily passivated, it is stable in air also at quite elevated temperatures.

In general, nickel forms alloys very readily, and the tendency of nickel powders to absorb hydrogen is quite remarkable.

The most important minerals of nickel are:

- the pyrrhotites (Fe_nS_{n+1}, with $n = 5$–16) with 2–4% of nickel because they contain pentlandite (Ni, Fe)S and some copper due to the presence of chalcopyrite, $CuFeS_2$;
- the garnierites (Ni, Mg)$SiO_3 \cdot nH_2O$. These are rocky minerals containing 4–7% of Ni because it is accompanied by iron oxides and silica;
- the rare niccolite, NiAs, which is always impure owing to the presence of other arsenides (rammelsbergite, $NiAs_2$, and gersdorffite NiAsS).

The largest nickel mining area in the world is around Sudbury in Canada (nickel- and copper-bearing pyrrhotites), followed by New Caledonia and Brazil (garnierites), and the USSR with its mixed deposits of sulphides and arsenides.

The metallurgy of nickel

The extraction of nickel from its minerals is quite different, depending on whether one starts out from nickel-bearing and copper-bearing pyrrhotites or from garnierites, that is, from sulphidic minerals or oxidized minerals.

Metallurgy relating to nickel–copper-bearing pyrrhotites. Of the three basic components of these minerals, only the pyrrhotite is appreciably magnetic, while the chalcopyrite-bearing component is much more readily floated than the pentlandite-bearing component. Thus:

* after crushing and grinding the mineral, the pyrrhotite is separated magnetically;
* a large part of the chalcopyrite-bearing component is separated out by selective flotation;
* the pentlandite concentrate is roasted in such a way as to burn a large part of the sulphur, which is combined with iron, to produce SO_2 which is suitable for the production of H_2SO_4;
* melting the desulphurized mineral is performed in reverbatory furnaces together with silica so that a large part of the iron is removed in the slag;
* oxidation of the remaining FeS is performed in Bessemer or in Peirce–Smith converters and the formation of another slag by the addition of silica to yield 'Bessemer matte';
* the matte is melted and cast into ingots which are allowed to cool very slowly so as to form well-shaped crystals of Cu_2S and Ni_2S, and, finally, with the solidification of the impurities which consist, for the greater part, of magnetic FeS, which occludes the precious metals;
* the resulting material is ground so that the impurities may be separated magnetically. A powdered mixture of Ni_2S_3 and Cu_2S is thereby obtained from which almost all of the copper-containing component can be separated by selective flotation;
* the Ni_2S_3 concentrate is thoroughly oxidized on moving Dwight–Lloyd grids so as to obtain a mixture of the oxides NiO and CuO. When this mixture is leached with cold sulphuric acid, the CuO dissolves preferentially to form $CuSO_4$. At this point, the subsequent metallurgical process may follow one of two routes as shown in Fig. 13.32.

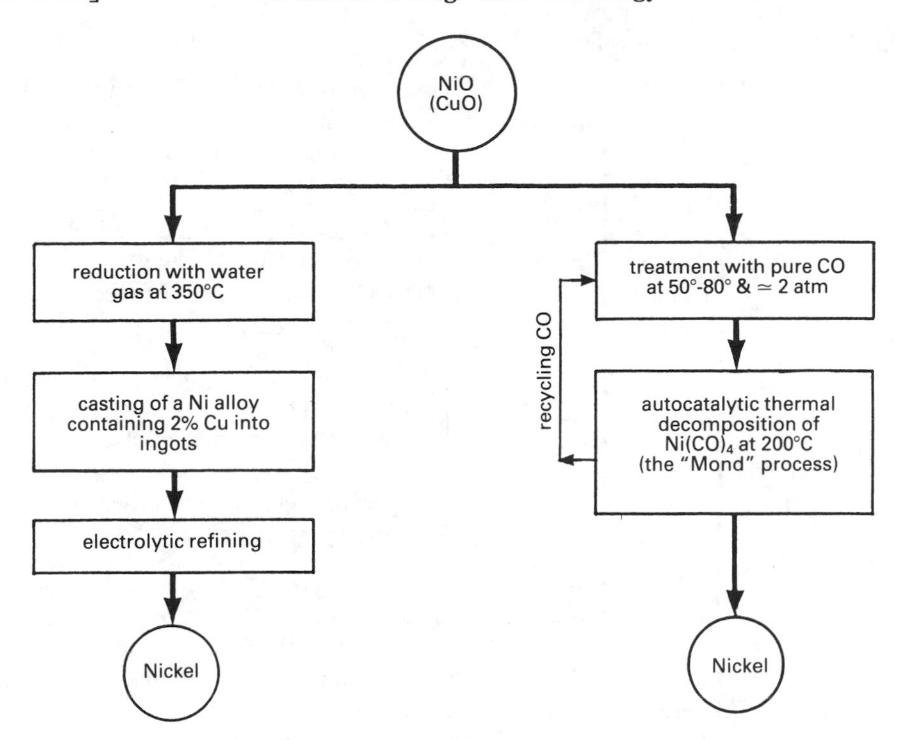

Fig. 13.32—The two routes followed to prepare nickel from NiO containing CuO as an impurity. One pathway uses reduction of the oxide with a mixture of H_2 and CO followed by electrolytic refining, while the other uses the Mond process.

There are a number of variants of the classical process for the treatment of the Bessemer matte which has just been described. The most important of these included its complete oxidation followed by the reduction of the oxides (after the composition has been suitably adjusted) with carbon so as to yield the important alloy containing 63–68% Ni and 28–30% Cu known as Monel.

The metallurgy of garnierite-containing silicates. The treatment which leads from garnieritic rocks to nickel or to an iron alloy of nickel may include dry processes, wet processes, or mixed processes which make use of the steps shown diagrammatically in Fig. 13.33.

The reaction which, in the dry process, leads from rocky garnierite to a mixture of the sulphides of nickel and iron by means of its treatment with calcium sulphate and carbon, includes the formation of a calcium–magnesium silicate slag with the intervention of both the silicate silica and the silica gangue.

The uses of nickel

Nickel is used in the pure or almost pure state both as a passivated metal for the lining of, specifically, carbon steels onto which it is either deposited electrolytically or as cladding, as well as in vessels for the chemical industry, in electrotechnic and electronic apparatus, and in the fabrication of electrodes for fuel cells.

Nevertheless, the major uses of nickel are found in alloying metallurgy where thousands of metallic materials containing more than 1% of nickel have been noted.

Nickel alloys* may be classified as follows:

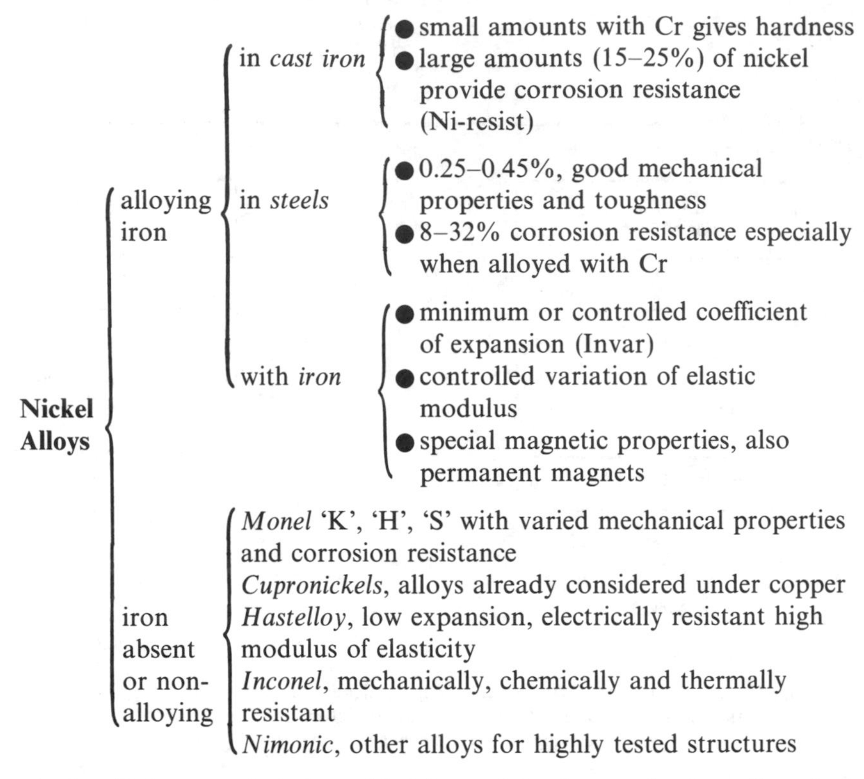

Nickel Alloys
alloying iron
in *cast iron*
● small amounts with Cr gives hardness
● large amounts (15–25%) of nickel provide corrosion resistance (Ni-resist)
in *steels*
● 0.25–0.45%, good mechanical properties and toughness
● 8–32% corrosion resistance especially when alloyed with Cr
with *iron*
● minimum or controlled coefficient of expansion (Invar)
● controlled variation of elastic modulus
● special magnetic properties, also permanent magnets
iron absent or non-alloying
Monel 'K', 'H', 'S' with varied mechanical properties and corrosion resistance
Cupronickels, alloys already considered under copper
Hastelloy, low expansion, electrically resistant high modulus of elasticity
Inconel, mechanically, chemically and thermally resistant
Nimonic, other alloys for highly tested structures

Finally, it should be recorded that nickel is used in the production of a whole range of well-known catalysts and in the construction of alkaline iron–nickel and cadmium–nickel cells.

Zinc

Zinc is a metal with a rather low melting point (420°C), a low boiling point (906°C), and with a low electrochemical reduction potential ($E^{\ominus} = -0.76$ V). It has a low to medium density, $d = 7.14$ g/cm^3.

Zinc, which has a bright bluish-white colour when freshly cut, rapidly turns grey upon exposure to air. It is ductile, malleable, and can be passivated in the pure state, but is hardened by even small amounts of certain metals (Fe and Cd, for example) and rendered fragile by other inclusions (Sn). It is corroded by more noble metals (Cu). The natural minerals of zinc from which it can be usefully extracted are: blende (sphalerite) ZnS, smithsonite $ZnCO_3$, and calamine $(ZnOH)_2SiO_3$.

The metallurgy of zinc

Zinc minerals must be enriched up to 40–60% by flotation or sedimentation before

*It is necessary to use the plural in the indication of these alloys. In fact, while each corresponds to a certain range of composition, there are numerous varieties which are sometimes obtained with small modifications of the basic structure.

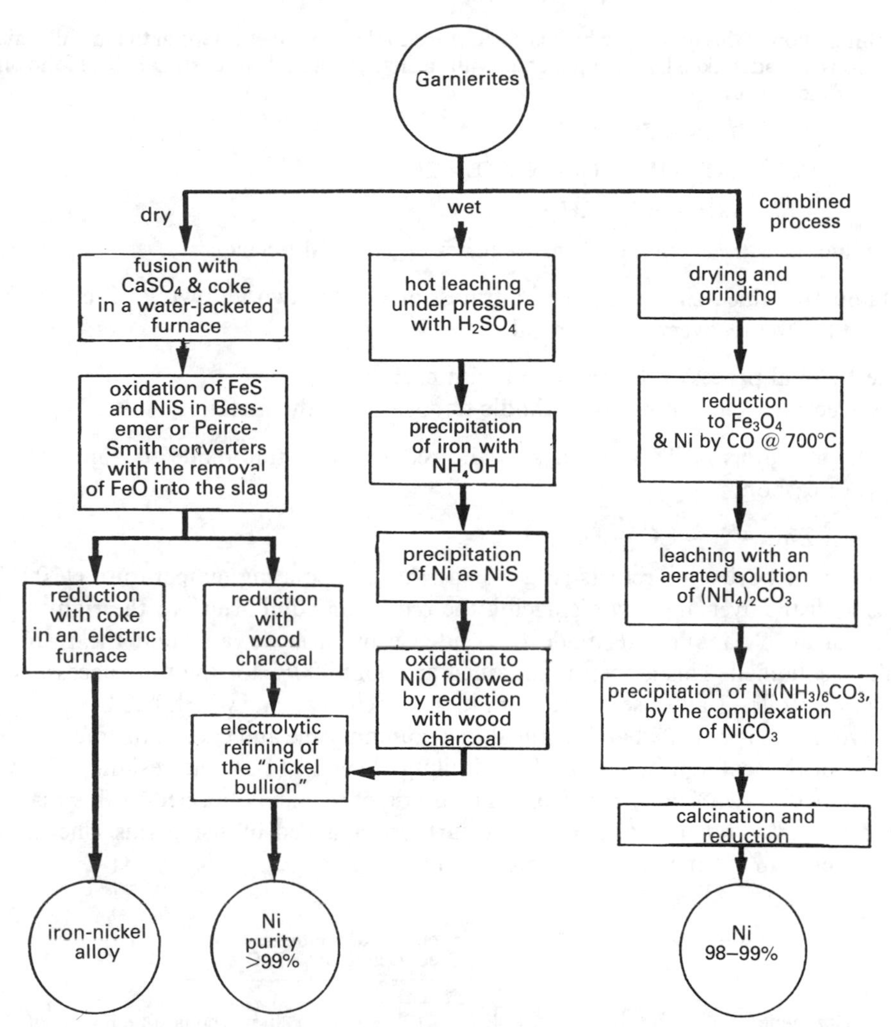

Fig. 13.33—The three possible processes for the preparation of nickel from garnierites. They are readily explained in terms of the steps, which are more correctly chemical, on the basis of (1) what has been said regarding the sulphides (taking account of the fact that, in the dry process, $CaSO_4$ in the presence of coke reduces iron and nickel to the sulphides and transforms the silica into a slag), or on the basis (2) of simple analytical chemical concepts (the precipitation of $Fe(OH)_3$, the precipitation of NiS from $NiSO_4$ with H_2S, and the complexation of Ni with $(NH_4)_2CO_3$ in the presence of oxygen).

commencing extractive metallurgical operations on them. If they consist of blendes, they are successively preroasted in multiple hearth (Herreshoff) furnaces and then roasted and agglomerated on Dwight–Lloyd machines or thoroughly roasted, with agglomeration of the charge, in fluid bed furnaces.

If, on the other hand, smithsonite or calamine are being treated, which are always associated, they are calcined in rotating shaft furnaces with the elimination of CO_2 and H_2O. In this manner, both free ZnO and ZnO which is combined with silica are obtained.

Often, and this applies especially in the case of calamines, they are associated with CaF_2.

The elimination of this fluorine, which is deleterious to the subsequent steps in the metallurgical operations, is undertaken by mixing blende with the mineral and then roasting it. The following reactions then occur:

$$ZnS + 3/2O_2 \rightarrow ZnO + SO_2$$

$$CaF_2 + SO_2 + H_2 + \tfrac{1}{2}O_2 \rightarrow CaSO_4 + 2HF$$

$$4HF + SiO_2 \rightarrow SiF_4 + 2H_2O$$

The fluorine is then recovered from the fumes by the usual techniques.

Having obtained zinc oxide concentrates, one of the two following processes can be used for the recovery of the metal:

- the thermal process with reduction with carbon;
- the electrolytic process with cathodic deposition of the metal.

The thermal process. This consists of the reduction of zinc oxide to zinc with an excess of carbon:

$$ZnO + C \rightarrow CO + Zn$$

in vertical or horizontal retorts (Fig. 13.34). At the operating temperature (1200°C), the zinc distils over, thereby displacing the reduction equilibrium to the right.

The distilled vapours are made to condense in various ways, and the resulting liquid is solidified. This is then maintained at about 400°C for the time necessary to allow any molten lead to separate from it. The technical zinc which has been largely freed from lead is remelted and distilled in a column. Zinc and cadmium emerge from the top of this column, while the high boiling elements (Pb—the residue), Fe, Cu, etc.) are withdrawn from the bottom. The grade of the zinc obtained in this manner is about 99.5%, but this purity can be further enhanced by using this zinc as the raw material in the metallurgy of cadmium.

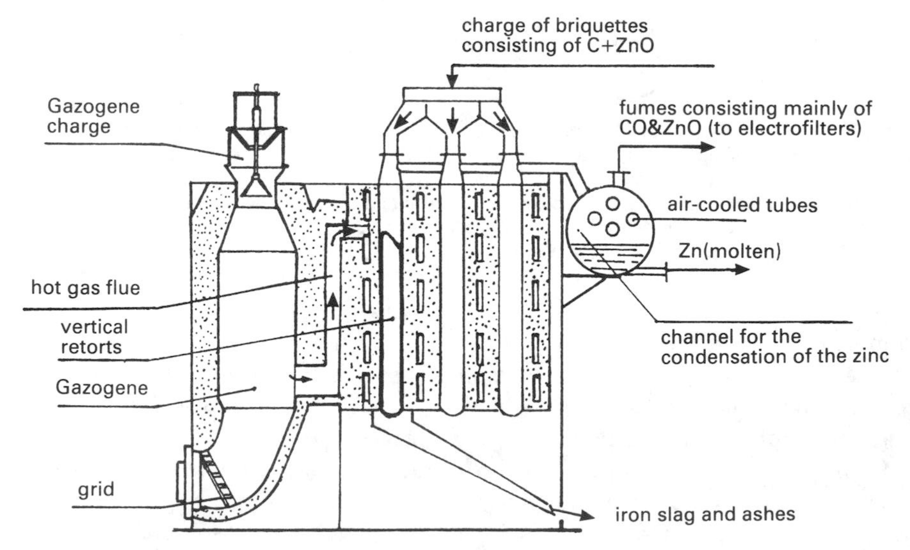

Fig. 13.34—Vertical retorts for the reduction of zinc oxide with carbon. The walls of the retorts are externally heated with Gazogene mixed with purified gas originating from the distillation, and are then made to burn to fire the briquettes to be introduced into the furnace.

The electrolytic process. In addition to electrolysis, this also includes dissolution and purification stages, as can be seen from the block diagram in Fig. 13.35.

The electrolysis is carried out in cells made of resistant cements or of lead-lined wood with lead anodes and aluminium cathodes from which a part of the zinc which has been deposited is periodically removed. To lower the hydrogen overpotential to the greatest possible extent, the temperature is maintained at about 35°C by circulating water through lead tubes. Furthermore, the pH value is controlled in such a way as to disfavour the development of hydrogen (pH $\simeq$4.5).

The voltage used is 3.5 V, and the energy consumption is some 3–4 kWh.

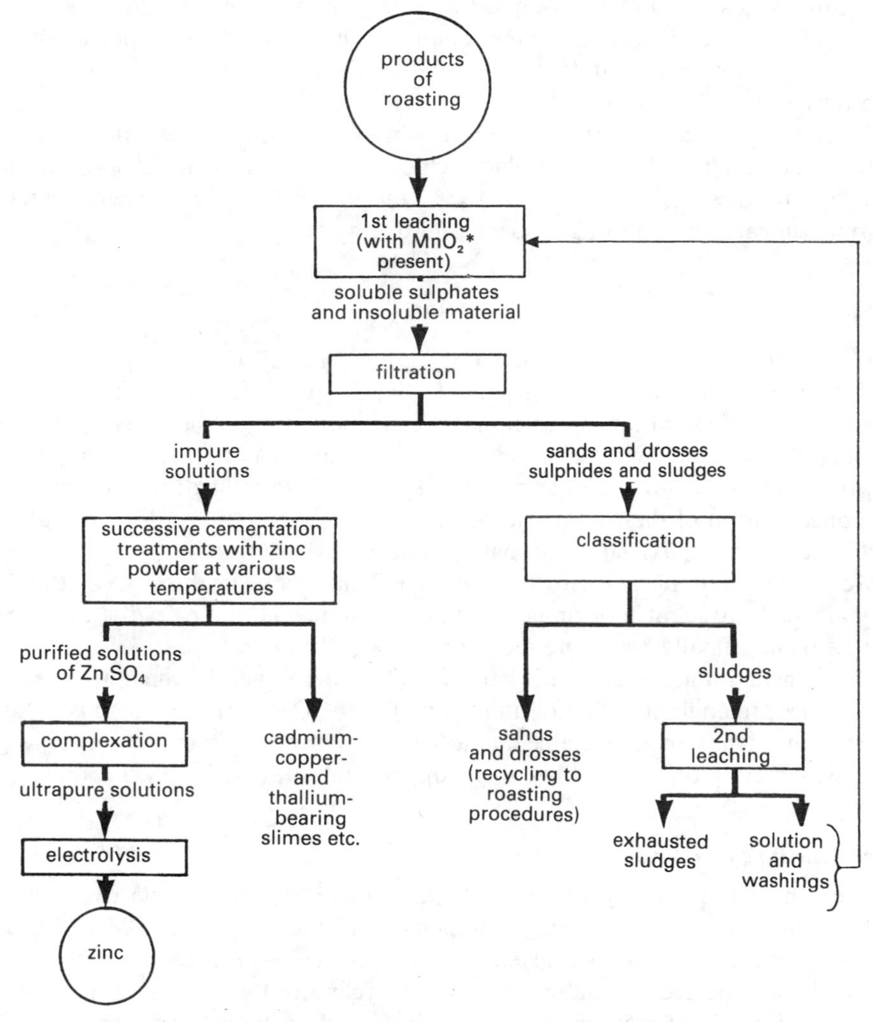

Fig. 13.35—Block diagram showing the operations in the pretreatment of the enriched and roasted zinc minerals followed by the electrolytic deposition of the metal.

*This is added with the aim of producing $Fe(OH)_3$:

$$2FeSO_4 + MnO_2 + 3ZnO + H_2O + 2H_2SO_4 \rightarrow MnSO_4 + 3ZnSO_4 + 2Fe(OH)_3$$

which occludes arsenides and other dispersions thereby facilitating filtration and removing substances (Sb, Ge, etc.) which would possibly interfere with the electrolysis.

The uses of zinc

The zinc, which is deposited in various ways onto carbon steel wires and sheets (by immersion in molten zinc, electrolytically, by sherardization), acts as a soluble protective coating for them. Much use is made of zinc alloyed with very small quantities of mercury (less than 0.5%) in the protection of buried tubes, naval structures, plants subjected to stray currents, and other cathodic structures. Here, it is a matter of 'sacrificial anodes'.

The use of zinc in alloys, mainly with copper and aluminium, follows the use of pure zinc. More precisely, the alloys with copper are brasses, while those with aluminium and small amounts of magnesium are 'Zam' alloys. There are also many special alloys: with copper and aluminium in almost equal quantities (3–4%) for casting, with copper (about 1.2%), and with titanium (0.15–0.30%) which have a high tensile strength, and so on.

Considerable quantities of zinc are also employed in the production of chemicals such as the sulphates, oxide, sulphide, chlorides, and chromate. There are many fields where these compounds are used: in pigments, in vulcanizing catalysts, pharmaceuticals, rapid setting binders, and dyeing aids.

Lead

Lead is a soft, relatively dense ($d = 11.35$ g/cm^3) metal with a low melting point (327.5°C) but with a high boiling point (1740°C). When freshly cut it has a bluish mother of pearl coloration which rapidly turns grey in air. When finely subdivided, it is readily attacked by mineral acids, especially nitric acid. In the bulk state it is slowly attacked by nitric acid and especially hydrochloric acid; but, provided that the concentration of the acid does not exceed 60%, it is not corroded by sulphuric acid as it becomes passivated by coating itself with $PbSO_4$.

Moist air containing CO_2 slowly transforms it into white lead, $2PbCO_3 \cdot Pb(OH)_2$. Organic acids attack it to form complexes. When heated in air, lead is transformed into its oxides, finally becoming red lead, $Pb_3O_4(2PbO \cdot PbO_2)$.

The principal mineral of lead is galena PbS. Other minerals which are of notable importance are anglesite, $PbSO_4$, and cerussite, $PbCO_3$. Moreover, lead is obtained, apart from galenas, in appreciable quantities as a byproduct from other metallurgical processes, mainly those involving zinc, copper, and antimony.

The metallurgy of lead

After crushing and grinding, the lead sulphide mineral is generally selectively enriched by flotation, using activators and depressors, until it contains 60–70% of PbS.

The galena concentrate is roasted and sintered in Dwight–Lloyd machines. The agglomerate is passed through sieves, with recycling of the fines, and the prevalent larger sized pieces of material are loaded into water-jacketed furnaces (Fig. 13.36) together with coke, re-used slags, and silica or lime to correct for the composition of the gangue.

Four different operating zones can be distinguished in the furnace:

- the drying and dehydration zone (150–400°C);
- the indirect reduction zone (350–700°C): $PbO + CO \rightarrow Pb + CO_2$;

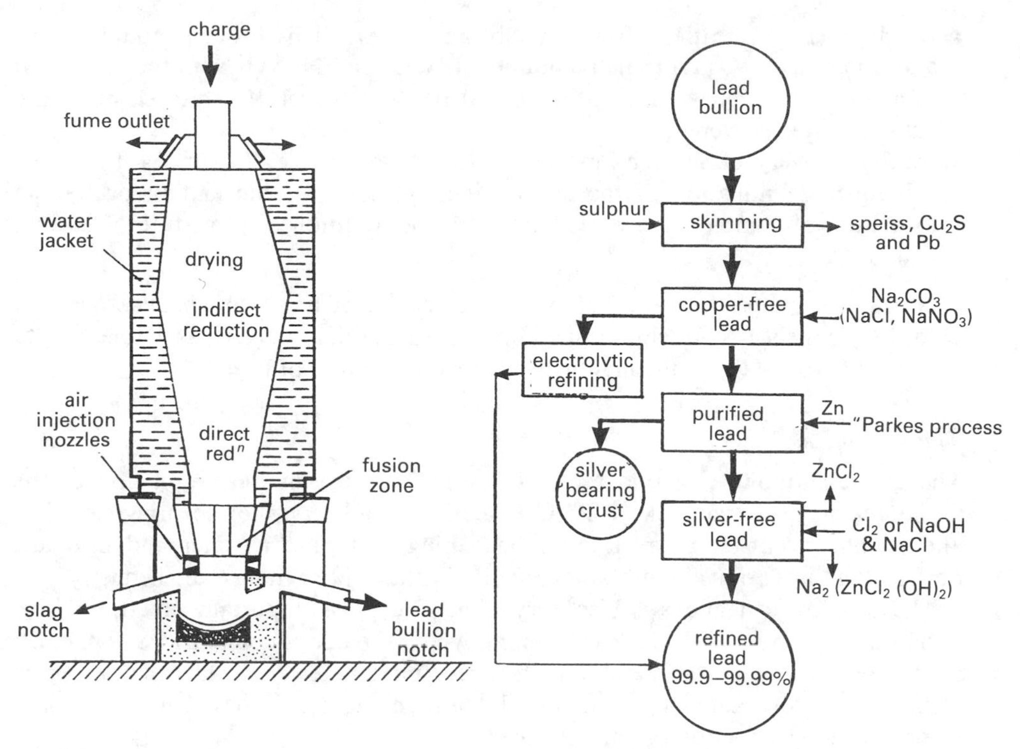

Fig. 13.36—Left. Cross section of a water-jacketed furnace. Right. Block diagram of the working of lead bullion in order to obtain the technically 'pure' metal.

- the direct reduction-with-carbon zone (600–1000°C), and decomposition of carbonates and sulphates*;
- the fusion zone (900–1300°C) where the reduction of PbO, of the sulphates to sulphides and of zinc oxide to zinc is completed.

The following are discharged from the water-jacketed furnace:

- lead bullion containing various elements which are either free (Sn, Ag), combined as oxides and sulphides (Cu, Fe), or as 'Speiss' (the sulpho-arsenides and sulpho-antimonides of copper, iron and lead);
- fumes containing zinc sublimate which, partially reoxidized, is trapped in sack filters and separately worked;
- iron and calcium silicate slags which are rich in so far as they occlude lead, Cu_2O, Cu_2S, and some speisses. These 'rich slags' are therefore separated in reverbatory furnaces into copper matte, speiss, lead silicates, lead bullion, and inert slags. The latter are removed, while the other components are either recycled or worked separately.

The lead bullion is refined:

- by prolonged melting with stirring at 350°C and the addition of a little sulphur to produce a dross of speiss, containing metallic lead and the sulphides of copper, on the surface;

*It is from this point that the metallurgy of $PbCO_3$ and $PbSO_4$ (cerussite and anglesite) also start off.

- by chemical purification* from Sn, Sb, and As which is brought about by the addition of Na_2CO_3 and small amounts of NaCl and $NaNO_3$ with the formation of a stannate, arsenate, and antimonate dross from which the three elements are subsequently recovered;
- by the recovery of silver by means of the 'Parkes process' (page 544) and the subsequent elimination of the excess zinc by selective chlorination or by the formation of complex hydroxychlorides by the addition of a mixture of NaOH and NaCl.

Nowadays, copper-free lead bullion is also refined by an electrochemical method according to the usual principle, using sulphamic acid (NH_2SO_2OH) as the electrolyte and taking advantage of the high hydrogen overpotential of lead.

The uses of lead

The greatest amounts of lead are still destined for use in the production of the antiknock agent, tetraethyllead, $Pb(C_2H_5)_4$. Pure lead serves for covering electrical and telephone cables, as a vibration-deadening material, in water and drainage outfalls, and in various linings and conduits in the field of industrial chemistry.

The field of lead alloys is also very rich. The most important are those with antimony (which hardens the lead), among which the alloys containing 8–9% of Sb used in lead accumulators stand out when account is taken of the fact that, in terms of the amount of lead which is employed, the manufacture of these batteries is only second to the production of tetraethyllead.

Other lead alloys also have characteristic uses:

- the alloys for shot gun pellets containing 0.5% As, 0.1% Sn, and 0.1% Bi;
- the alloys for printing type (Pb 67–87%) which vary in composition according to the kind of printing in which they are used (hand, Linotype, Monotype, etc.). The use of these has substantially fallen off owing to the great expansion of lithographic technologies;
- lead antifriction alloys which are usually composed of lead (65–85%), Sn, and Sb, but also with small amounts of copper (0.5–1.5%) and arsenic (0.1–1.0%).

Lead, in varying percentages, is also a constituent in the structures of many other alloys such as bronzes, brasses, complex fusible alloys, and soldering alloys.

Finally, an appreciable amount of lead is used in the preparation of lead compounds which are employed in various technologies: glasses and glazes (PbO, litharge, paints (Pb_3O_4, red lead), pigments and industrial oxidation processes (PbO_2, lead dioxide). On the other hand, the use of white lead as a pigment is limited on account of its high toxicity and the equally good covering power which is offered by other pigments (TiO_2, $BaSO_4 + ZnS$, etc.).

Chromium

Chromium is a bright white metal which is readily passivated to a point where, even if it is heated in air, it shows no superficial changes. This is a consequence of the fact

*This is a modern process for the purification of copper-free lead which was traditionally carried out by means of an oxidizing treatment of the mass in a reverbatory furnace, to first produce tin antimonite and then lead arsenite and lead antimonite as a dross.

that, upon becoming passivated, its reduction potential passes from a value which is lower than iron ($E^\ominus = -0.51$ V) to a value which is similar to that of gold ($E^\ominus = +1.3$ V).

It is dissolved by hydrochloric and sulphuric acids, but not by nitric acid since it is strongly passivated by the latter acid.

When heated or melted, it combines with halogens, sulphur, nitrogen, silicon, and boron, which enables one to explain their state and behaviour in an alloy.

The phase transitions of chromium occur at relatively high temperatures (m.p. = 1890°C and b.p. = 2200°C), and it has a medium/low density ($d = 7.2$ g/cm^3).

The metallurgy of chromium

The only mineral of importance as a source of chromium is chromite, $FeCr_2O_4$ (chrome spinel). There is therefore a tendency to prepare ferroalloys* rather than the pure metal. These ferroalloys are supplied to the cast iron manufacturing sector, and to those areas in steel-making where chromium is required as an alloying element.

To produce *ferrochrome*, chromite is treated in electric arc furnaces with silica and coke in suitable proportions. Part of the iron passes into the slag, while the remainder forms part of the intended product which predominantly consists of chromium. This operation can also be carried out in two stages: by first preparing a ferroalloy containing a large amount of carbon, and then partially decarburizing it.

Metallic chromium and *chromium compounds* are produced (Fig. 13.37) by first preparing sodium chromate from chromite which has been roasted in air in the presence of lime and soda:

$$4FeCr_2O_4 + 8CaO + 8Na_2CO_3 + 7O_2$$
$$\rightarrow 2Fe_2O_3 + 8\mathbf{Na_2CrO_4} + 8CaCO_3$$

Sodium chromate is an important reagent in chrome plating technology and an industrial chemical intermediate in the preparation of chromium compounds.

Inter alia, chromium oxide is obtained by lixiviating sodium chromate with H_2 and SO_2:

$$2CrO_4^{2-} + 3SO_2 + H_2O \rightarrow \mathbf{Cr_2O_3} + 3SO_4^{2-} + 2H^+$$

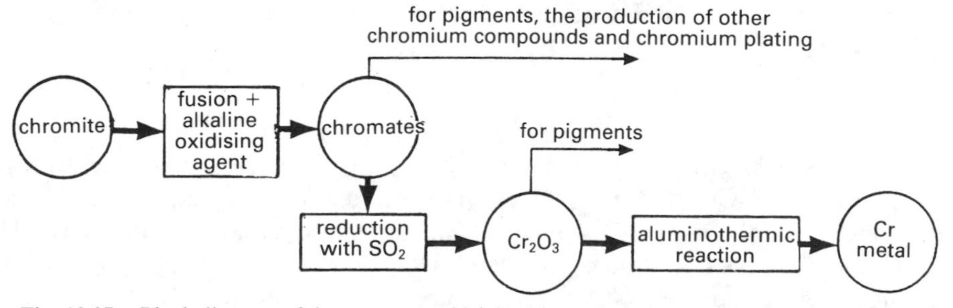

Fig. 13.37—Block diagram of the processes which lead from chromite to chromium compounds and the metal.

*There are two types of ferrochrome: low carbon and high carbon. The first contain not more than 2% of carbon, while the others contain not less than 4%.

which is the raw material for the aluminothermic preparation of 99% chromium:

$$2Al + Cr_2O_3 \rightarrow Al_2O_3 + 2\mathbf{Cr}$$

Metal produced in this manner can be purified to yield metal of the highest purity by chlorination, followed by the reduction of the $CrCl_3$ formed with magnesium:

$$2CrCl_3 + 3Mg \rightarrow 2\mathbf{Cr} + 3MgCl_2$$

The uses of chromium

Large amounts of chromium are used in the manufacture of:

- chrome cast irons which are resistant to attack in an oxidizing atmosphere at temperatures below 900°C ($Cr < 12\%$) and at more than 900°C (Cr 12–30%);
- ferritic and martensitic steels (Cr 13–25%), austenitic stainless steels (Cr 18–25%), and austenitic-ferritic stainless steels (Cr 20–30%).

Chromium is a constituent in many other alloys: chrome–molybdenum steels, nickel and cobalt superalloys, steels with special magnetic properties, and so on.

On account of the extremely favourable effects arising from its very pronounced tendency to become passivated, pure chromium is employed in very large amounts globally for deposition, in very thin layers, (2–4 μm) on the most widely varied metallic materials, but particularly on common steels which have previously been copper plated and nickel plated in order to ensure good adherence to the substrate. This is because copper alloys well with iron and nickel, and nickel bonds tenaciously to chromium.

An electrolytic chromium plating bath contains sodium dichromate in a solution which is acidified by means of sulphuric acid. In this way, a constant concentration of Cr^{3+} ions can be guaranteed, these ions being continually regenerated by the reduction of the dichromate. Using this technique, deposits are obtained which are uniform, of minimal cost, and of guaranteed corrosion resistance.

The amounts of chromium used as a component in chemical products of industrial interest are also noteworthy. Among these products, mention may be made of: Cr_2O_3 and various chromates which are used as pigments, Cr_2O_3 itself, and chromic anhydride which are components in catalysts, and, also chromic anhydride and dichromates which are made use of in oxidation processes.

Finally, chromium compounds are used to quite a considerable extent in the refractory industry.

Manganese

Manganese is a bright, silvery metal which is hard and brittle and readily forms alloys (with Fe, Al, Cu, etc.). Its surfaces are oxidized in air and take on an iridescent appearance since the oxide film which is formed is not of a uniform thickness. This produces a labile passive state of the metal which allows it to dissolve in all acids.

It has a melting point (1260°C) which is appreciably lower than that of iron, and a boiling point (1890°C) which is very much lower, while the density of manganese is the same as that of chromium ($d = 7.2$ g/cm^3).

The bulk metal burns in air at elevated temperatures, and it is pyrophoric in a finely subdivided state.

The metallurgy of manganese

The principal mineral of manganese, which is the object of large commercial exchanges between the various producer countries (USSR and South Africa are at the top of the league) and the consumer countries, is pyrolusite, MnO_2. Also, more and more research is being carried out into the possibility of exploiting the large amounts of manganese nodules which are available on the beds of oceans.

Metallic manganese and ferromanganese alloys are prepared from pyrolusite, as shown in Fig. 13.38.

The corrections made to the contents of the electrolytic paths requires an adjustment of the pH value which must be brought into the basic zone by the addition of ammonia because manganese is not deposited electrolytically from acidic solutions.

The uses of manganese

About 10 million tons of manganese are used annually in the form of ferroalloys and the pure metal, in the production of every type of steel, but, especially, wear-resistant and shock-resistant steels, and for supplying cast iron foundries. Many smaller, but still appreciable, quantities of manganese are made use of in the preparation of bronzes and brasses, and in order to satisfy the specifications for aluminium-, magnesium-, and titanium-based light alloys.

The dioxide, which is the most used of the manganese compounds, is employed as an oxidizing agent and as a decolorant in glass melts. Potassium permanganate,

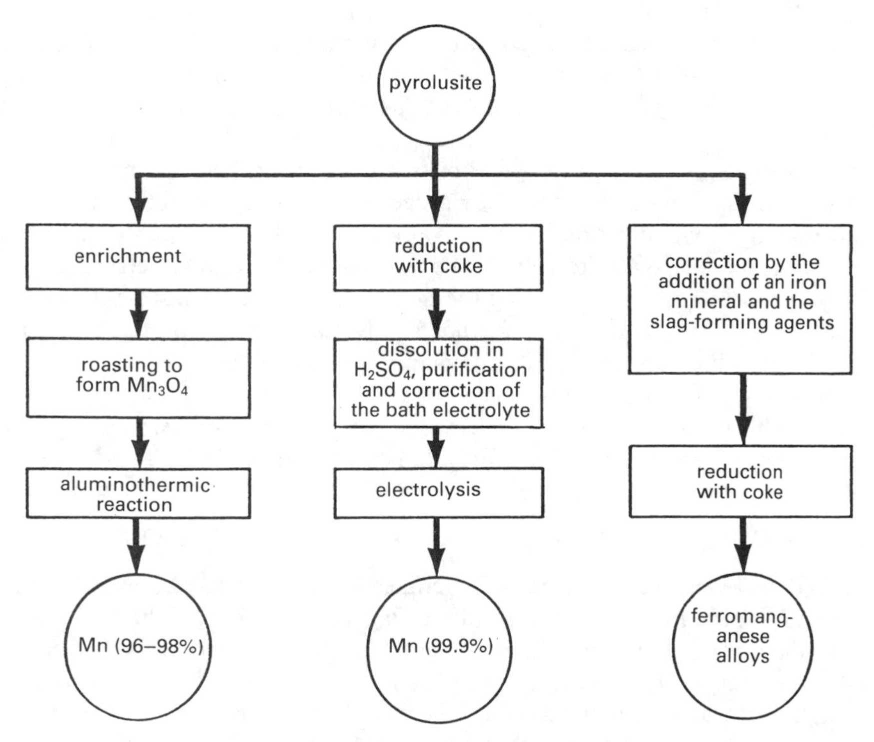

Fig. 13.38—According to how it is treated, pyrolusite is converted to more or less pure manganese metal or ferromanganese alloys.

which is used as an oxidizing agent, follows in order of importance. Considerable amounts of manganese compounds, mainly salts, are destined for use in processes of an oxidation–reduction type where either positive or negative catalysis is used.

Tin

There is a very great difference between the melting point (231.9°C) and the boiling point (2362°C) of this silvery white metal which can be passivated very readily and is already quite malleable at around 100°C.

Common tin (white tin) undergoes a phase transition and is converted to grey tin at 13.2°C. This entails a structural change from a denser crystalline form ($d = 7.28\ g/cm^3$) to a form which is appreciably lighter ($d = 5.76\ g/cm^3$) accompanied by the pulverization of the compact mass (tin pest). Fortunately, this transition is very slow in practice. It is retarded by bismuth and antimony inclusions but accelerated by traces of aluminium, zinc, manganese, and cobalt.

Tin is perfectly stable in contact with air and organic acids. When hot, it is oxidized to SnO_2, and it is dissolved by mineral acids, especially when hot, to form stannous salts, if the acids are non-oxidizing, and stannic compounds in the opposite case.

The sole mineral of tin which is of practical interest is cassiterite, SnO_2, which is rare in nature and concentrated in a few mining districts in South China, South-East Asia, the USSR, and Bolivia. Small-scale tin-mining is found in Britain and elsewhere.

The metallurgy of tin

Cassiterite is accompanied in its deposits by the sulphides of lead, zinc, iron, antimony, and copper and certain arsenides. These may assume the form of mineral veins in granitic rocks (primary deposits) or alluvial cassiteritic sands and gravels (secondary deposits).

After the rocks from the primary deposits have been broken up and crushed, the useful mineral (SnO_2) is concentrated by processes which exploit difference in specific weight in an aqueous medium, while levigation is the classical method for the concentration of alluvial cassiterites. Having completed the hydrogravimetric enrichment, one way of proceeding entails mixing the mineral with NaCl, roasting it in terraced furnaces of the Herreshoff type, and thoroughly leaching out the chlorides of the gangue metals which are formed in this manner.

The resulting SnO_2 concentrate is agglomerated into lumps which are loaded into reverbatory furnaces or electric furnaces with carbon and lime which is added in small amounts so as to obviate the formation of stannates. The following reactions occur:

$$SnO_2 + 2C \rightarrow 2CO + Sn \qquad SnO_2 + CO \rightarrow CO_2 + SnO$$

This is done by choice under partially reducing conditions to obtain a purer metal (98–99%) from the first melt. The resulting slag still contains 15–40% of the initial SnO_2 in the form of SnO (rich slag). When resmelted in reverbatory furnaces with the addition of iron, this slag furnishes 95% tin as the result of a reaction in which the tin is electrochemically displaced by the iron. When smelted for a third time, the slag from the second smelting passes from 3–5% Sn to 1% and is then discarded, while the resulting Sn–Fe alloy (hard head) is recycled to the second smelting stage.

Finally, complex methods of thermal and electrolytic refining lead to tin with a minimum purity of 99.5%.

Tin can also be recovered from tinned rejects, scrap, and clippings by treating them with chlorine gas followed by hydrolysis of the resulting $SnCl_4$ to SnO_2, or by using an electrolytic process in which the tin is anodically dissolved and cathodically redeposited.

The uses of tin

A large part of the tin produced is used in the metallic state for the electrolytic coating of sheet steel to obtain tinned strip (tinplate). Deformable lead containers and copper wires are also tin-plated.

As regards alloys, the use of tin in bronzes has fallen off considerably, but there has been a progressive increase in the amount of tin destined for use in tin antifriction alloys, soldering alloys, and special alloys.

The antifriction alloys which are abbreviated to alloy 1, 2, 3, 4, and 5, contain amounts of tin from 91% to 65%, while the antimony content, on the other hand, varies from 4.5% to 15%, and Cu and Pb are also present in various percentages. Traditionally, solders were made from lead and tin only, but nowadays they also contain Cd, Zn, Cu, and even Ag in various percentages.

Among the special alloys which contain tin, mention should be made of pewter (90–95% Sn, 1–3% Cu, and the remainder is antimony), mirror amalgam (92% Sn and 8% mercury), and the alloys which have recently been developed for use in the nuclear industry. These alloys have high mechanical strength when hot but low mechanical strength at ambient temperatures, and can be readily welded. They contain various amounts of titanium, aluminium, and zirconium. Among the compounds of tin which are used industrially, mention may be made of SnO_2 for glass and enamelling works, $SnCl_2$, a reducing agent in various organic syntheses, various stannates which are finishing aids in the textile industry, and the many organotin compounds which are widely employed as pesticides.

13.20 METALS FOR SPECIALIZED TECHNOLOGIES

Under this heading, we conventionally include those metals with rather specialized areas of use such as magnesium, which reduces the weight of alloys and acts as a deoxidant, titanium, which serves the aerospace and chemical industries, and molybdenum which finds special use in certain types of steels (stainless, constructional, and maraging steels) and in furnishing catalyst promoters.

Many of these metals (Sb, Hg, Mg, etc.) have been exploited for many years, while others, including Be, Ge, Nb, and Hf, have passed during the last few decades from being laboratory curiosities to rank among the top elements in modern technologies.

Since there is no strictly logical way of treating them, they will be examined in alphabetical order.

Antimony

m.p. = 630.5°C, b.p. = 1380°C, $d_{metallic} = 6.68$ g/cm^3

Antimony is an element with predominantly metallic characteristics. In fact, it is similar to silver when in its stable allotropic form. However, its nature as an element at the limit of the metallic state is demonstrated, inter alia, by the fact that it burns with a flame at 600–700°C.

The antimony minerals are the sulphide, stibnite, Sb_2S_3, and the oxide, senarmontite, Sb_2O_3. The metallurgy of the sulphide uses enrichment by flotation and the conversion of the enriched sulphide to the metal by the oxidation and reduction methods as indicated under (4) in Table 13.3 (page 468). On the other hand, the metallurgy of the previously enriched oxide requires only a reduction step.

Antimony is purified electrolytically by arranging anodes of impure antimony in a solution of hydrochloric and hydrofluoric acids.

The major part of the antimony produced is used in alloys. These are mainly with lead, then with tin and lead, followed by alloys with tin and copper.

Some antimony compounds are mainly used in the textile industry, in the technology for the protection of iron by burnishing, in glazes and ceramics, and as fire-prevention agents.

Beryllium

m.p. = 1315°C, b.p. = 2970°C $d = 1.86$ g/cm^3

Beryllium is a metallic element which is similar to aluminium in both its appearance and its tendency to become passivated. On the other hand, unlike aluminium, it is extremely poisonous and much harder than aluminium when cold. In fact, it is ductile and malleable only at about 1000°C.

The minerals of beryllium are extremely numerous, but the only really important one is beryl $Be_3Al_2(SiO_3)_6$ from which the metal is obtained by:

- fusing it at 700–800°C with soda and sodium fluorosilicate:

$$Be_3Al_2(SiO_3)_6 + Na_2CO_3 + 2Na_2SiF_6$$
$$\rightarrow 3Na_2BeF_4 + 8SiO_2 + Al_2O_3 + CO_2$$

- treating the sodium fluoroberyllate solution with caustic soda, and calcining the resulting hydroxide to form the oxide:

$$Na_2BeF_4 + 2NaOH \rightarrow Be(OH)_2 + 4NaF$$
$$Be(OH)_2 \rightarrow BeO + H_2O$$

- converting the oxide which has been formed into ammonium fluoroberyllate with ammonium bifluoride and thermally decomposing the fluoroberyllate to form beryllium fluoride from which the metal is recovered by the magnesiothermic reaction:

$$BeO + 2NH_4(HF_2) \rightarrow (NH_4)_2BeF_4 + H_2O$$
$$(NH_4)_2BeF_4 \rightarrow BeF_2 + 2NH_4F$$
$$BeF_2 + Mg \rightarrow MgF_2 + \mathbf{Be}$$

If mother alloys (e.g. beryllium–copper) are to be produced for use in the preparation of alloys of practical importance, the beryllium oxide is reduced in an electric arc furnace in the presence of the alloying metal.

Beryllium has the property of hardening the copper in an alloy without reducing its conductivity. It also makes steel extremely hard by case-hardening it.

Beryllium is a neutron moderator in nuclear reactors, and, when mixed with radium, acts as a neutron source.

Technologically, the most important compound of beryllium is the oxide. It is a first-rate refractory with a melting point of 2530°C.

Cadmium

m.p. = 320.9°C, b.p. = 765°C $d = 8.64\ g/cm^3$

Cadmium is a bright white metal which is softer, more malleable, and more readily passivated than zinc. It is also considerably more metallic than zinc to a point where its oxide, unlike zinc oxide, does not dissolve in alkalis.

Cadmium, which is not very abundant in nature, has as its principal mineral, greenockite, CdS, which is always associated with other sulphides, especially blende.

The metallurgy of cadmium exploits the fact that it is more volatile than zinc, and, as a result, it preferentially collects in the fumes produced during the pyrometallurgical processing of zinc. The powder which is collected when these fumes are cleansed is dissolved in sulphuric acid which precipitates any lead that is present as $PbSO_4$. 'Cadmium sponge' is first formed from the solution by the electrochemical displacement of cadmium by zinc, and this precipitate, having been separated and dried, is mixed with carbon powder and then distilled. By subsequently fusing the distillate with caustic soda, all the zinc is brought into solution, while the undissolved cadmium which remains has a purity of greater than 99.5%. Even purer cadmium can be prepared by further distillation.

The metallurgical processes for the production of cadmium, which utilize the electrochemically displaced materials obtained during the processing of zinc, are more complex, and are summarized in Fig. 13.39.

Cadmium is mainly used for the galvanic protection of iron. The covering is much more resistant than that of zinc, but far more costly. Cadmium is also used in other metallurgical fields such as low-friction and low-melting-point alloys.

On account of its high 'collision cross-section' for neutrons, cadmium is used as a neutron absorber in nuclear reactors.

Considerable amounts of cadmium are also used in the form of compounds as yellow pigments (cadmium yellows).

Cobalt

m.p. = 1495°C, b.p. = 2900°C $d = 8.90\ g/cm^3$

Cobalt is a silvery white metal which is brittle when cold but workable when hot, and is attacked to a greater (HNO_3) or lesser (H_2SO_4) degree by mineral acids.

At present, the most used mineral for the winning of cobalt is carrolite ($CuS \cdot Co_2S_3$) which, after enrichment by flotation and the addition of pyrites, is oxidized by using a Dwight–Lloyd machine and fused with coke in an electric furnace so as to obtain a Co–Fe–Cu alloy. When this alloy is treated with sulphuric acid, practically only the iron and the cobalt are dissolved. The solution obtained, when treated with H_2S, forms a precipitate containing any copper which is present, while the iron is

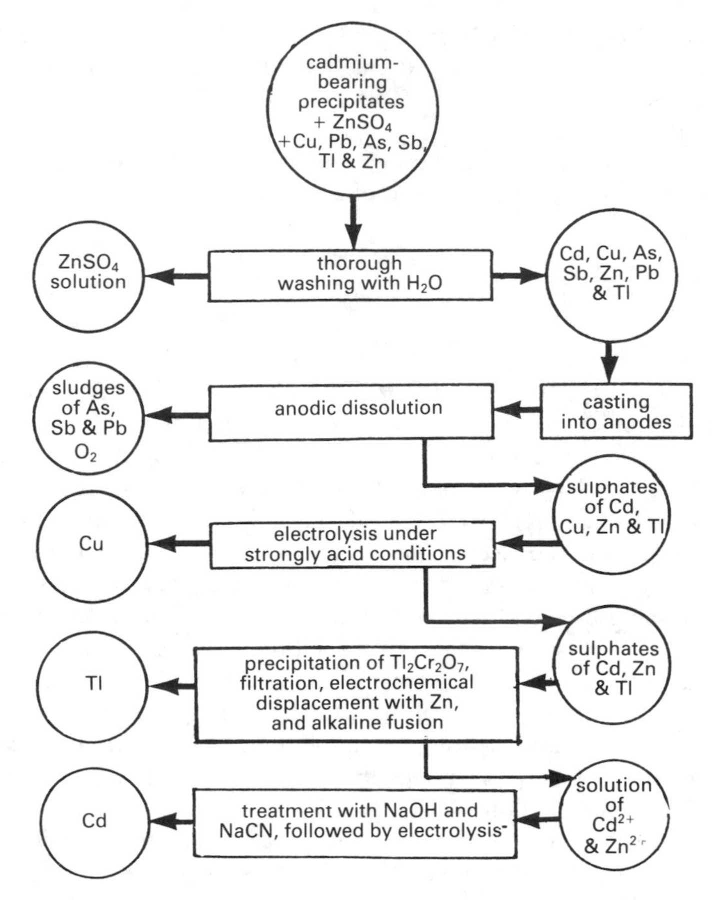

Fig. 13.39—The working up of the cadmium-bearing precipitates which are byproducts from the metallurgical treatment of blendes with the aim of obtaining cadmium and thallium (a metal similar to lead which is used for, among other things, the doping of transistors).

precipitated at a pH value of 3–4. After filtration, neutralization of the solution leads to the precipitation of $Co(OH)_3$ which, when calcined, yields the oxide. This, when reduced with carbon, yields the metal.

Cobalt is a component in high-speed and super-high-speed steels, magnetic steels, ferromagnetic alloys such as AlNiCo, hard alloys 'Stellites' with chromium and carbon, and the super-hard alloy 'widia metal' containing 10% Co and 90% W_2C which is a substitute for diamond. Other cobalt alloys, which are used in jet engines and gas turbines, are resistant to thermal shock.

Cobalt compounds are used as catalysts and as characteristic pigments for colouring ceramics, and, especially, glasses.

Finally, cobalt which has been irradiated and activated in the channels of nuclear reactors serves as a source of γ-rays with medical and technological uses.

Germanium and silicon

We shall combine the treatment of germanium and silicon on account of their well known physicochemical similarities and their common uses.

Germanium

m.p. = 937.4°C, b.p. = 2830°C $d = 5.35$ g/cm^3

Germanium is a bright grey metal which is hard and brittle and has a diamond-type crystal lattice.

While it exists in nature in many minerals, for the most part as a sulphide, there are also some typical germanium minerals such as argyrodite $4Ag_2S \cdot GeS_2$ and germanite $5CuS \cdot FeS_2 \cdot GeS_2$.

In particular, the production of germanium makes use of the ashes of certain coals which contain it in amounts of 0.5 kg/ton. It is present in these ashes as GeO_2 which, when treated with very concentrated hydrochloric acid, yields $GeCl_4$ with a boiling point of 83°C, and which therefore can be readily distilled. After distillation, the resulting $GeCl_4$ is hydrolyzed with water and yields a precipitate of pure GeO_2 from which the metal is produced by reducing the oxide with hydrogen in a stepwise manner to avoid any sublimation, since GeO_2 sublimes at 710°C. The stages in this reduction are:

$$GeO_2 + H_2 \xrightarrow{665^\circ C} GeO + H_2O$$

$$GeO + H_2 \xrightarrow{1000^\circ C} Ge + H_2O$$

In this way one obtains germanium with a purity of 99.99%. However, this grade of purity is inadequate for certain uses of germanium, and it is therefore further purified by the zone refining method which will be explained in connection with silicon.

The uses of germanium are dictated by its special properties. It is a semiconductor, that is, a material with an electrical conductivity which increases as the temperature is raised and which varies greatly as a function of the impurity content. For this reason, it serves, when suitably doped, in the construction of diodes and transistors.

Moreover, germanium is transparent to certain infrared radiations, and is therefore used in the construction of lenses and filters in apparatus intended for experiments in physical optics.

Silicon

m.p. = 1410°C, b.p. = 22355°C $d = 2.33$ g/cm^3

After oxygen, silicon is the most important component of the lithosphere. Its most pure natural form is quartz, SiO_2.

Silicon is a non-metallic semiconducting element which is practically insoluble in all acids.

The silicon used in industry is obtained by reducing silica sand with carbon in electric arc furnaces:

$$SiO_2 + 2C \rightarrow Si + 2CO$$

Various expedients are employed to ensure that it does not volatilize in appreciable quantities.

This silicon is converted into commercially pure silicon by pulverizing the ingots which are cast from the contents of the production furnace, having been allowed to solidify. The resulting material is treated with a mixture of sulphuric and dilute hydrofluoric acids because, in this way, the impurities are dissolved while the silicon is not.

For electronic uses, the silicon must be produced from its halides by reduction with hydrogen:

$$SiCl_4 + 2H_2 \rightarrow Si + 4HCl$$

The silicon tetrachloride, in its turn, is prepared by the action of chlorine on a finely subdivided mixture of silica and coke at an elevated temperature:

$$2Cl_2 + SiO_2 + 2C \rightarrow SiCl_4 + 2CO$$

Silicon produced in this manner must also be further purified by the zone refining method which is also employed in the purification of germanium.

It is well known that impurities tend to lower melting points, that is, to maintain bodies in a molten state. Therefore, if an ingot of silicon or germanium is melted and slowly solidified, the impurities tend to accumulate in the part which is still molten. At the limit, they finish up concentrated in the molten tail of the ingot which has been allowed to resolidify, and this part is therefore discarded.

Processes like that which has just been described are carried out several more times on the silicon and germanium ingots until they attain the grade of purity requested by the users. Once such a hyper-pure ingot has been prepared, a small hyper-pure single crystal (a 'seed') is placed on top of it in the atmosphere under which the zone refining is carried out. The introduction of this seed, which has the task of acting as a crystallization nucleus, is followed by small doses of the dopant element, and the zone refining is carried out once again. In this way the dopant is distributed, almost as if it were an impurity, by travelling throughout the length of the growing single crystal into the entire mass of the silicon or germanium, which is thereby doped. The doping of a semiconductor means that the atoms with one more valence electron (e_v) or one fewer valence electron than the matrix atoms are inserted here and there in the crystal lattice.

Let us now briefly indicate the simplest effects of the doping of semiconductors. For instance, germanium ($4e_v$) can be doped with, inter alia, arsenic ($5e_v$) and aluminium ($3e_v$). Since the dopants are necessarily compelled to adapt themselves to the environment of the matrix, they either give up non-bonding electrons to the crystal, thereby behaving as electron donors, or they stimulate the formation of holes, by acting as electron acceptors.

The electrons which are in excess over those required for the formation of bands and the positive holes travel under the influence of an electric field, thereby giving rise to two types of electrical conductivity, negative and positive. One therefore speaks of n-type and p-type doped semiconductors.

n-type semiconductors are prepared with P, As, and Sb, while p-type semiconductors are prepared with B, Al, Ga, In, and Tl.

Overall, doped semiconductors are neutral, but, when the two types are joined so as to constitute a 'pn-junction which forms a semiconductor diode, there is a contact zone (depletion layer) which is positively charged in the n-type material and negatively charged in the p-type material. This is because the delocalized electrons in the former diffuse across the junction to saturate the positive valence holes in the latter (Fig. 13.40). In the interior of the depletion layer of a pn-junction, a semiconductor appears not to be doped, that is, it does not act as a current conductor when a voltage is applied in a direction (reverse biasing) which maintains or enhances this situation. On the other hand, it does act as a current conductor when a voltage is applied in the opposite direction (forward biasing) which tends to destroy the depletion layer. The rectifying action of a diode semiconductor on an alternating current is based on this fact.

That which has just been described and illustrated in Fig. 13.41 is the simplest type of junction which can be constructed, and, as further information on the effects of the doping of semiconductors, we shall only add that transistor action is achieved with npn (or pnp) junctions.

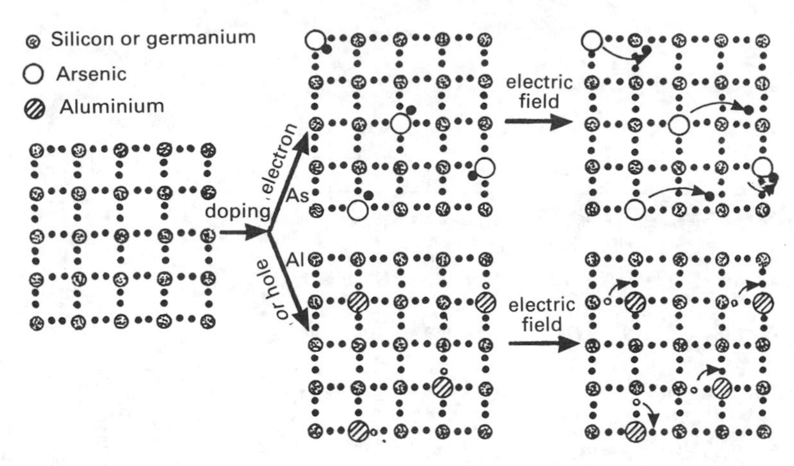

Fig. 13.40—Left. A schematic two-dimensional representation of a silicon or germanium crystal in which the valence electrons are paired between the atoms. Centre. n-type doping with As and p-type doping with Al (the open circles and the hatched in circles respectively). Right. Conduction by electronic displacements in the two types of doped systems when they are subjected to the action of an electric field. This is negative conduction when it is due to the motion of the excess electrons and positive conduction when it involves the indirect displacement of the holes.

In addition to its use in electronics, silicon is also employed in the metallurgical industry for the production of sheet silicon steel for transformer laminations, for the manufacture of steels with a high elastic modulus, in the formulation of silicon and aluminium alloys, and so on. Nowadays, a large amount of silicon is also used in the production of intermediates for various types of silicones.

Magnesium

m.p. = 650°C, b.p. = 1103°C $d = 1.74$ g/cm^3

Magnesium, a silvery white metal, has a very low reduction potential ($E^{\ominus} = +2.37$ V), and this makes it very reactive. In fact, even when cold, it undergoes extensive surface oxidation in air and develops hydrogen from acids. When hot, it displaces hydrogen from water, and, at temperatures of the order of 400°C, it ignites and burns violently, producing a magnesium lamp.

The minerals used for the preparation of magnesium are magnesite, dolomite, and the magnesium chloride in marine waters.

Fig. 13.42 shows the prominent phases in the production of magnesium by the electrolytic route, starting from all three raw materials.

The most difficult to implement is always that of converting magnesium oxide into the chloride. The reactions are:

$$MgO + C + Cl_2 \rightarrow MgCl_2 + CO + 36.2\ \text{kcal}$$

$$2MgO + C + 2Cl_2 \rightarrow 2MgCl_2 + CO_2 + 113.5\ \text{kcal}$$

The electrical resistance heating device in this furnace (Fig. 13.43) is used only to prime the reaction, as the exothermicity of the reaction subsequently covers all the energy requirements.

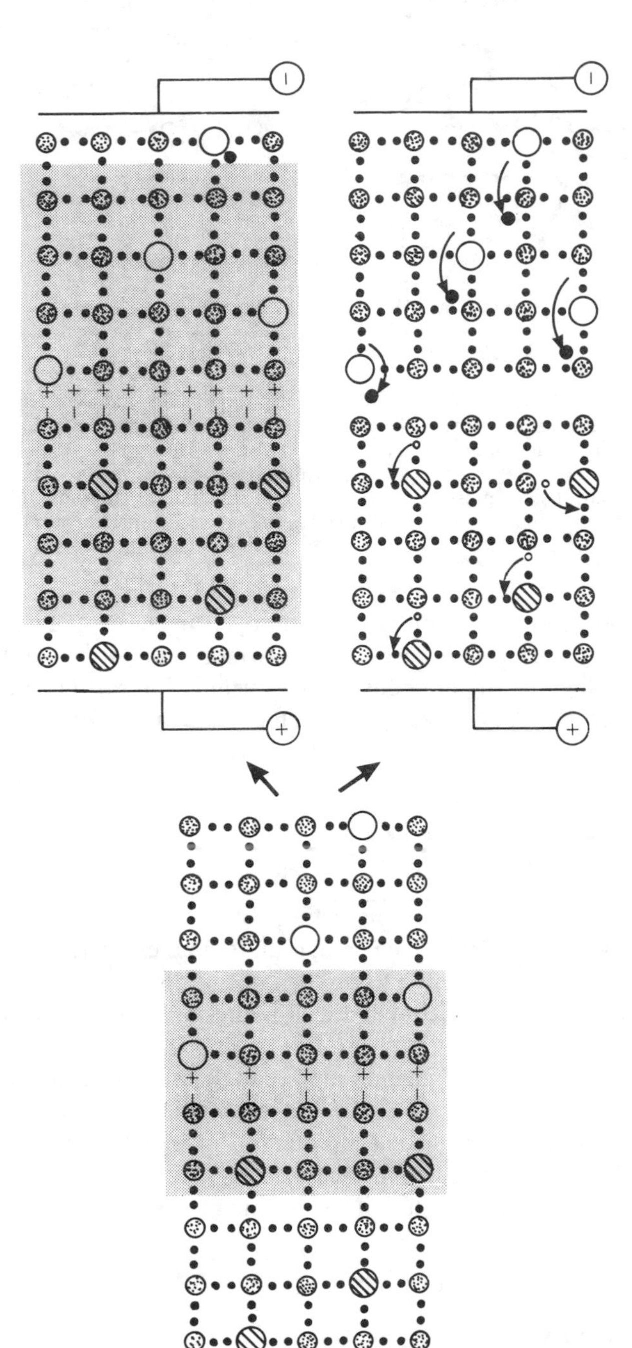

Fig. 13.41—Left: pn-Junction with a depletion layer (the shaded area), that is, a layer containing no free electrons or holes but the seat of an electrical double layer where the negative side of the layer has received electrons and the positive side has lost them. Top right: The application of a negative bias to the negative side of the depletion layer and a positive bias to the positive side of this layer (reverse biasing) leads to a widening of the depletion layer. The system is, however, non-conducting on account of the large barrier presented by the double (depletion) layer. Bottom right: Electric field applied so as to reduce the width of the depletion layer, whereupon the system becomes conducting.

Therefore, if an alternating field is applied, the current flows in only one direction, that is to say, it is rectified.

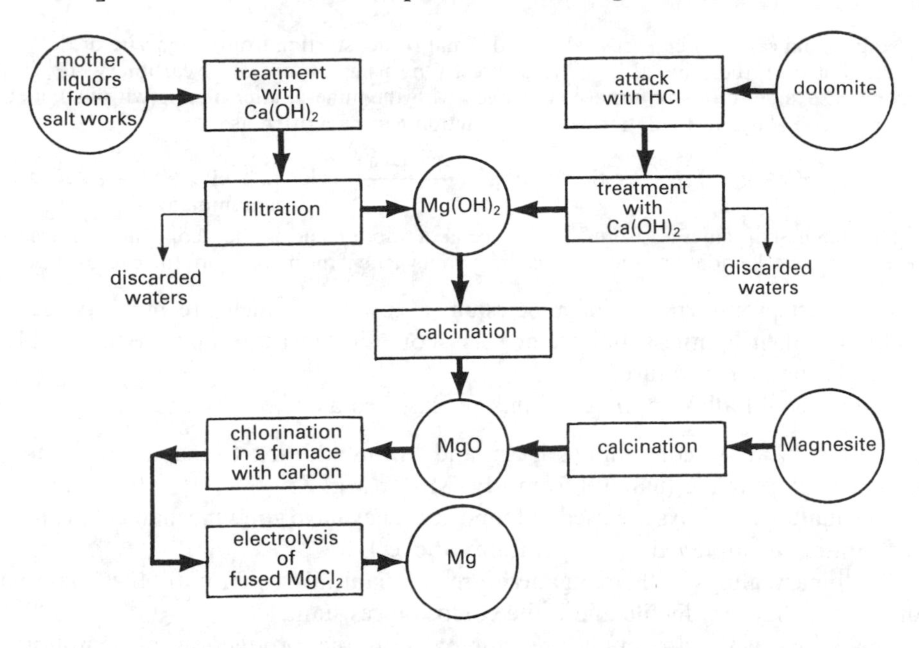

Fig. 13.42—Coordinated scheme for the production of magnesium from raw materials, that is, the mother liquors from marine salt-works, dolomite, and magnesite. The other liquors contain magnesium in the form of $MgCl_2$.

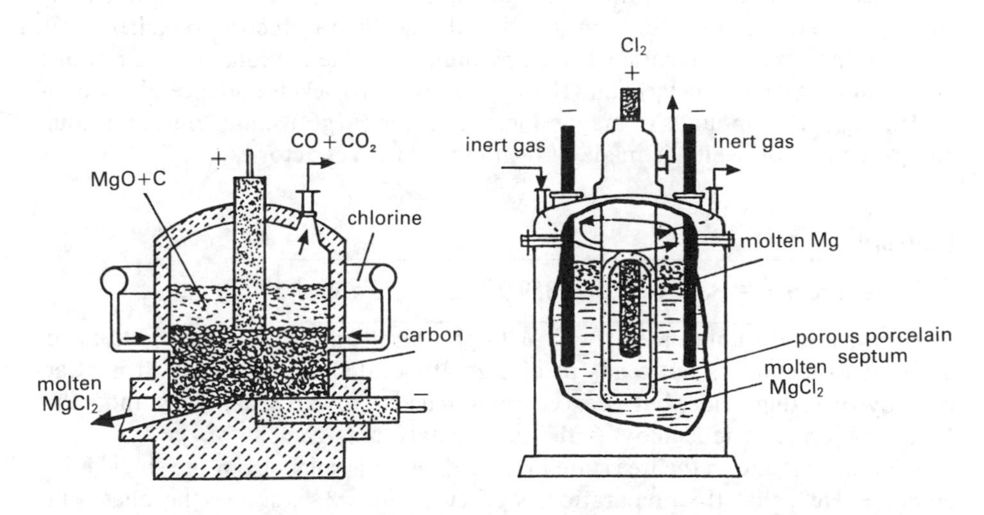

Fig. 13.43—Left. Furnace for the conversion of MgO into $MgCl_2$ which operates under the combined action of C (a reductant) and Cl_2 (a chlorinating agent) and entails exothermic reactions which are primed by the heat produced by an electric resistance. Right. Fused $MgCl_2$ electrolysis cell. A porous porcelain septum separates the anode compartment, where Cl_2 is evolved, from the cathode compartment where, upon discharge, Mg is produced which is protected from attack by oxygen, nitrogen, etc. by the circulation of a noble gas (Ar). The magnesium is continuously discharged through a small side orifice which is not shown in the figure. In this type of cell, a carbon anode (+) is used in conjunction with a carbon steel (iron) cathode (−).

Magnesium can also be prepared by a thermal route, starting from magnesite or dolomite. In the first case, the mineral is reduced directly by using an excess of carbon, while, in the second case, an electric arc furnace is charged with the mineral after mixing with bauxite and a ferrosilicon alloy. In the latter case, the reaction can be written as:

$$n\mathrm{CaMg(CO_3)_2} + \mathrm{Fe}_x\mathrm{Si}_y + z\mathrm{Al_2O_3} \xrightarrow{1100-1200^\circ\mathrm{C}} \underset{\text{silicoaluminate}}{\text{calcium}} + x\mathrm{Fe} + n\mathbf{Mg} + 2n\mathrm{CO_2}$$

The magnesium emerges from the furnace as a vapour and is then condensed. Here, it is necessary to work under an argon atmosphere or in vacuo if the metal is not to be re-oxidized.

The most prominent use of magnesium is in alloys which are highly prized on account of their lightness, but are not very corrosion resistant and are flammable in air at elevated temperatures.

The ultralight alloys of magnesium are classified as:

- quaternary alloys containing Al, Zn, and Mn in addition to magnesium;
- binary alloys made (mainly) from Mg–Mn or Mg–Zr.

The quaternary alloys are used in foundries. They have good mechanical properties but cannot be employed at temperatures above 120°C.

The binary alloys with manganese are mechanically inferior to the quaternary alloys, but they are ductile and quite corrosion resistant.

The alloys with zirconium are employed in the production of semi-finished components for the aeronautical and nuclear industries. These alloys can also operate at quite high temperatures if they also contain thorium which tends to impede their creep when hot.

Magnesium is also used as a deoxidant in metallurgy and for the reduction of chlorides (titanium and zirconium, for example). It also serves as a graphite modifier in iron and steel making for, among other things, the production of spheroidal cast irons. Minor uses of magnesium are encountered in the manufacture of ribbons for pyrotechnics and flash devices and the powders used in rockets and incendiary bombs.

Among the compounds of magnesium, the oxide (MgO) stands out on account of the particular use which is made of it in the field of refractories.

Mercury

m.p. $= -38.8$°C, b.p. $= 359.9$°C $d = 13.6$ g/cm^3

Mercury, a quite noble metal with a bright silver appearance, is stable in air at ambient temperatures but tends to oxidize at elevated temperatures. It is attacked only by oxidizing acids, but is exceedingly reactive towards halogens and sulphur. It readily forms various alloys with many metals (amalgams).

Mercury is found in the free state in nature, but it is solely from the single mineral, cinnabar HgS, that the preparation is effected. Fig. 13.44 shows the phases in the metallurgical processing of mercury.

The industrial uses of mercury are mainly as an amalgamating agent in certain metallurgical processes and in the chlorine–soda process, as an activator of the protective power of sacrificial anodes, and as a key component in the construction of a number of pieces of scientific apparatus. Some of its compounds and, in particular, the chlorides, are employed in medicine, in the construction of electrodes, and as wood preservatives and in various histological preparations.

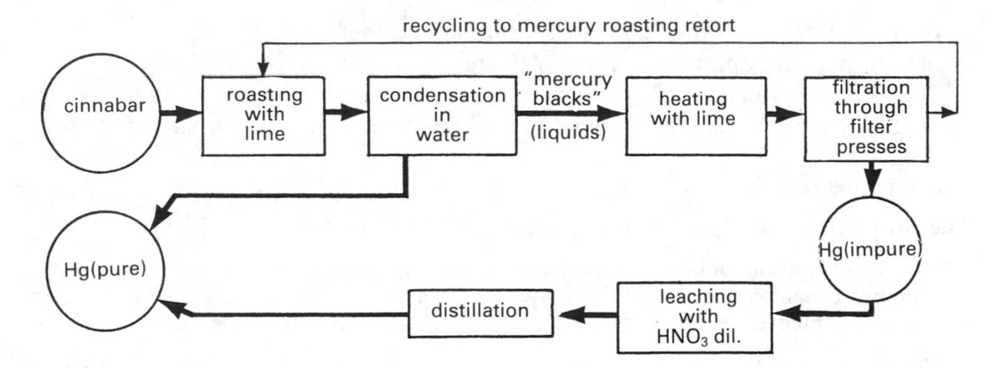

Fig. 13.44—Block diagram of the integral utilization of the cinnabar-bearing rocks for the preparation of mercury. The gangues are discharged from the furnace where the mineral is roasted with lime, and the impurities which are dissolved by nitric acid remain as residues after the distillation.

Molybdenum

m.p. = 2610°C, b.p. = 4800°C $d = 10.20$ g/cm^3

Molybdenum is a hard and brittle metal with a silvery white appearance which is a moderate conductor, stable in air at ambient temperatures, passivated by nitric acid, and is not attacked by alkalis even if they are molten or by dilute acids, but which readily dissolves in aqua regia.

The minerals of molybdenum are not abundant, but they are quite numerous. From the point of view of the preparation of the metal, only molybdenite MoS_2, powellite $CaMoO_4$, and wulfenite $PbMoO_4$ are of importance.

Starting from a molybdenite-bearing mineral, the MoS_2 is first concentrated to at least 90% by flotation and is then roasted in air, thereby inducing the reaction:

$$MoS_2 + 7/2O_2 \rightarrow MoO_3 + 2SO_2$$

The resulting oxide is then dissolved owing to complexation with an aqueous solution of ammonia:

$$MoO_3 + 2NH_4OH \rightarrow (NH_4)_2MoO_4 + H_2O$$

Molybdic acid is precipitated from the molybdate, and this acid, when heated, yields the pure trioxide which is then reduced with hydrogen:

$$(NH_4)_2MoO_4 + 2HCl \rightarrow H_2MoO_4 + 2NH_4Cl$$

$$H_2MoO_4 \xrightarrow{450^\circ C} MoO_3 + H_2O$$
$$MoO_3 + 3H_2 \rightarrow Mo + 3H_2O$$

The resulting metal is pressed into bars which are subsequently employed in certain metallurgical technologies or directly used.

In iron and steel making, molybdenum is a component in constructional, high-speed, super-high-speed, and stainless steels.

In astronautics, the parts of missiles which are exposed to high temperatures are built out of molybdenum, and it is used in the nuclear industry on account of its low neutron capture cross-section.

Wires, plates, and other accessories for lamps, thermionic valves, and various devices in physical optics and electronics are also made out of molybdenum.

Chemical compounds of molybdenum and, in particular, the oxides and molybdates, are used for the production of pigments and, even more so, in catalytic processes.

Niobium and tantalum

The properties of niobium and tantalum are summarized in Table 13.11.

The most important niobium and tantalum mineral, from which both metals are extracted, is columbite or tantalite $(Fe, Mn)(Nb, Ta)_2O_6$. As can be seen from Fig. 13.45, the processes used in the extractive metallurgy of these elements start off from this mineral.

Niobium is used in the metallurgy of stainless steels to endow them with high temperature stability, and in nuclear reactor technology on account of its low neutron capture cross-section.

Appreciable use is made of tantalum in the construction of pieces of apparatus (exchangers, stirrers, heating coils, etc.) and in the lining of others (autoclaves, pumps and valves) in the chemical industry. In electronics use is made of its ability to absorb gases in order to maintain high vacua (its 'getter' action), while, in surgery, it finds use in prostheses on account of its great compatibility with animal tissues.

Selenium and tellurium

The properties of selenium and tellurium are given in Table 13.12. These elements are rather rare in nature, where they are usually combined with noble metals (Ag, Au, Cu, etc.) in the form of selenides or tellurides.

Selenium is essentially a non-metal, while tellurium has semi-metallic properties which also tend towards the non-metallic rather than the metallic.

Table 13.11—Properties and behaviour of niobium and tantalum

Metal	*Appearance*	*Physical properties*	*Mechanical properties*	*Chemical properties*
Niobium	white-grey, bright metallic	m.p. 2468°C b.p. 4927°C *d* 8.57 g/cm^3	Hard, but sufficiently malleable and ductile	Extremely corrosion resistant. No acid other than HF attacks it. Resistant to aqua regia and (non-fused) alkalis. Forms extremely stable NbN and NbC at high temperatures. Burns in Cl_2 to form $NbCl_5$. No negative oxidation numbers.
Tantalum	grey and less bright than niobium	m.p. 2996°C b.p. 5425°C *d* 16.6 g/cm^3	Exceedingly ductile and workable in the cold and in air. When hot, must be worked in an inert gas atmosphere, if it is not to be embrittled by oxidation	Optimal resistance to acid corrosion, not even attacked by aqua regia. Resistant to halogens. Attacked by alkalis, fluorine, and hydrofluoric acid. Forms a stable carbide and nitride and, like Nb, cannot enter into combination as a negative component.

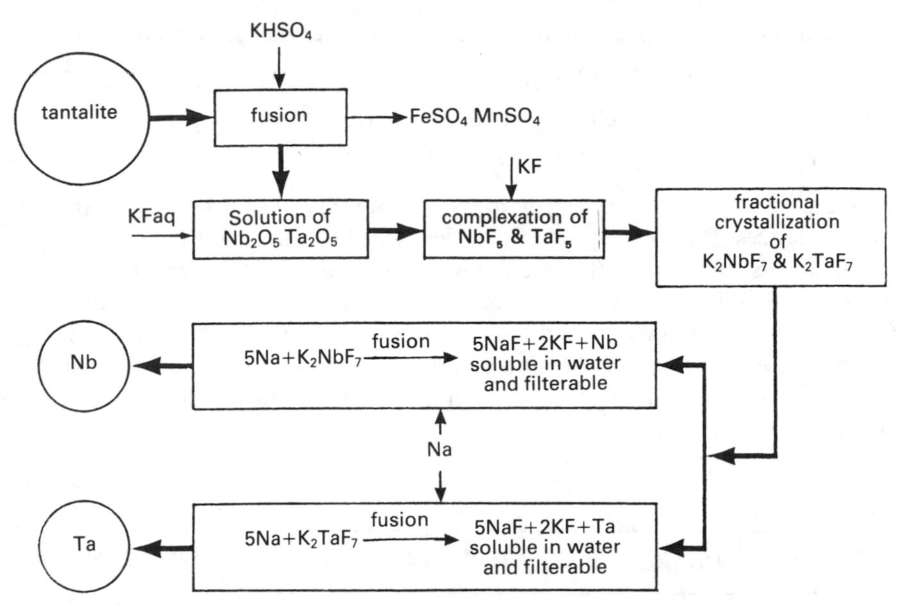

Fig. 13.45—The extraction of niobium and tantalum from tantalite by fusion with bisulphate, complexation, fractional crystallization and a further fusion with metallic sodium under an inert atmosphere.

Table 13.12—Summary of the properties of selenium and tellurium

Element	*Physical properties*	*Chemical properties*
Selenium	m.p. 220°C, b.p. 684°C, $d = 4.82$ g/cm^3 Its grey form has an electrical conductivity which increases by more than 1000-fold upon exposure to light. A thin layer, oxidized on just one side, only permits current to flow in one direction.	Burns in air to produce SeO_2, reacts with H_2 and halogens. Dissolves in HNO_3, $H_2SO_{4(conc)}$ and alkalis. The toxicity of selenium is to be considered in relation to its ability to replace sulphur in proteins.
Tellurium	m.p. 444°C, b.p. 1392°C, $d = 6.25$ g/cm^3. Very grey appearance, brittle and a very poor conductor.	Burns in air to produce TeO_2. Reacts with halogens and reacts slowly with mineral acids but more readily with hot hydrofluoric acid and hot sulphuric acid.

Selenium and tellurium are mainly produced by working up the anodic slimes obtained during copper refining processes where (for ecological reasons) it is imperative that both the fumes from the initial calcination of the slimes and from their subsequent alkaline fusion should be cleaned up. During this stage, SeO_2 and TeO_2 are recovered. These oxides are converted into Na_2SeO_3 and Na_2TeO_3.

During the course of the abovementioned alkaline fusion, the rest of the two elements which is contained in the sludges is transformed into Na_2SeO_3 and Na_2TeO_3 so that they are now conclusively and completely converted into a mixture of sodium selenite and tellurite which are soluble in water.

The subsequent stages in the metallurgical treatment consist of:

* acidification with sulphuric acid to precipitate TeO_2 while the SeO_2 remains in solution, and the two elements can therefore be separated by filtration;

* the reduction of the precipitate to Te with carbon and the addition of a borax flux:

$$TeO_2 + 2C \rightarrow 2CO + Te$$

* prolonged treatment of the solution with SO_2 to obtain Se:

$$H_2SeO_3 + 2SO_2 + H_2O \rightarrow 2H_2SO_4 + Se$$

Selenium is used in the construction of photocells, diode rectifiers, and other electronic devices. It is also added to stainless steels to improve their machinability. Selenium and its compounds are used as pigments for glasses, as vulcanizing agents for rubbers, and SeO_2 is much used as a selective oxidizing agent in organic syntheses.

Tellurium is an extremely versatile metallurgical additive because it increases the ductility of steels, hardens cast iron and lead, improves the corrosion resistance of lead itself and the mechanical properties of anti-friction alloys, renders copper more workable, and so on.

Titanium

m.p. = 1675°C, b.p. = 3260°C $d = 4.5\ g/cm^3$

Titanium metal has a steely grey appearance. It is hard but deformable (ductile) and stable in air which, like nitric acid, strongly passivates it. The great use which is made of titanium as a substitute for stainless steel in the chemical industry is based on its excellent corrosion resistance which is due to its passivation. The film of TiO_2 which covers it is more stable, both with respect to oxidizing and reducing agents, than that which passivates steels.

At a red heat, however, titanium burns both in oxygen and nitrogen with the formation of TiO_2 and TiN. On account of this, the melting, welding, and other heat treatments of titanium must be carried out in a noble gas atmosphere or under a vacuum.

The metallurgy of titanium

Titanium is extremely abundant in nature (0.6%), the only useful minerals for the winning of titanium being rutile TiO_2 and, more importantly, ilmenite $FeTiO_3$.

Starting from ilmenite, a large part of the iron is converted into a poor quality cast iron and a TiO_2-containing slag by treatment with carbon in a blown-furnace of the blast furnace type or an electric arc furnace.

From this slag, it is possible, after several stages, to arrive at metallic titanium. This is achieved by:

* treating it, after it has been formed into briquettes with coal and tar, in a furnace which is heated (by an electric resistance or by the heat released in the reaction) to about 800°C, with a stream of chlorine:

$$TiO_2 + 2C + 2Cl_2 \rightarrow TiCl_4 + 2CO$$

* reducing the $TiCl_4$, which has been formed and purified by selective condensation, with magnesium:

$$TiCl_4 + 2Mg \xrightarrow{800^\circ C} Ti + 2MgCl_2$$

so as to form 'titanium sponge' and magnesium chloride which is removed from the furnace in a fused state;

* washing the titanium sponge with dilute HCl to dissolve the magnesium and magnesium chloride which contaminate it, or distilling them off under a vacuum at about 1000°C.

As can be seen from Fig. 13.46, the components to be recycled (Mg and Cl_2) are regenerated by electrolysis of the molten $MgCl_2$, while the titanium sponge which has been freed from Mg and $MgCl_2$ is converted to the compact metal by a first melting with additives which fix the impurities in it, and then with absorbers to repartition other impurities so as to produce titanium which is at least 99.9% pure.

The uses of titanium

When added to steels, titanium also stabilizes their structures under severely corrosive conditions and mechanically demanding conditions, and enhances their weldability. On account of its properties, which include lightness, it is employed in place of stainless steels in many fields of the chemical and parachemical industries which have to deal with saline solutions, alkaline solutions, chlorine compounds and moist

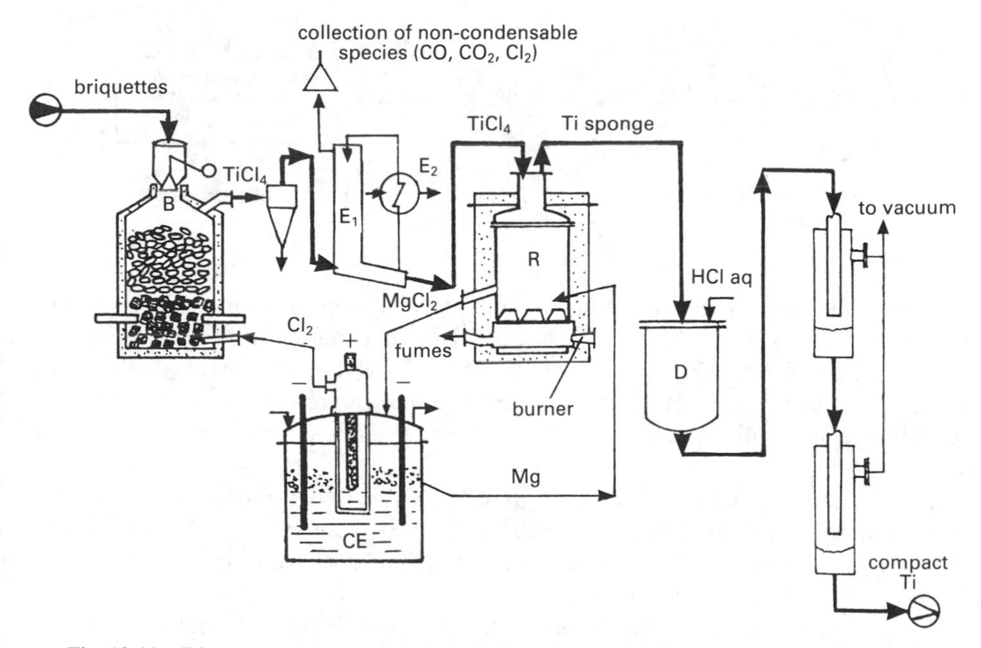

Fig. 13.46—Diagrammatic representation of the principle of the extraction metallurgy of titanium. An alternating current is supplied to the two electrodes in the electric furnace B which, gives rise to intense Joule heating, and, by circulating in the carbon briquettes, heats the charge lying above, consisting of powdered TiO_2-slag which has been formed into briquettes with coal and tar. When chlorine is fed in, this charge produces $TiCl_4$. The latter is removed from the top of the furnace, cleaned in cyclones, and selectively condensed in E_1 by being brought into contact with a recycled fraction of the condensed $TiCl_4$ which has been cooled in E_2. The tetrachloride is then passed into the externally heated reactor R where there is some magnesium which reduces it to titanium sponge. The magnesium chloride which is formed is electrolyzed in the molten state in the electrolytic cell CE which is of the type shown in Fig. 13.43. The chlorine is then recycled to the furnace B while the magnesium is recycled to the reactor R. The crude titanium sponge, which is contaminated with Mg and some $MgCl_2$, is unloaded from the latter and the contaminants are dissolved by dilute HCl in the tank with a flanged cover D. The pure titanium sponge is then pressed into bars with additives which combine with the impurities and melted under a vacuum after which it is remelted, always under a vacuum, with absorbers of other impurities. In this way, high-purity compact titanium may finally be obtained.

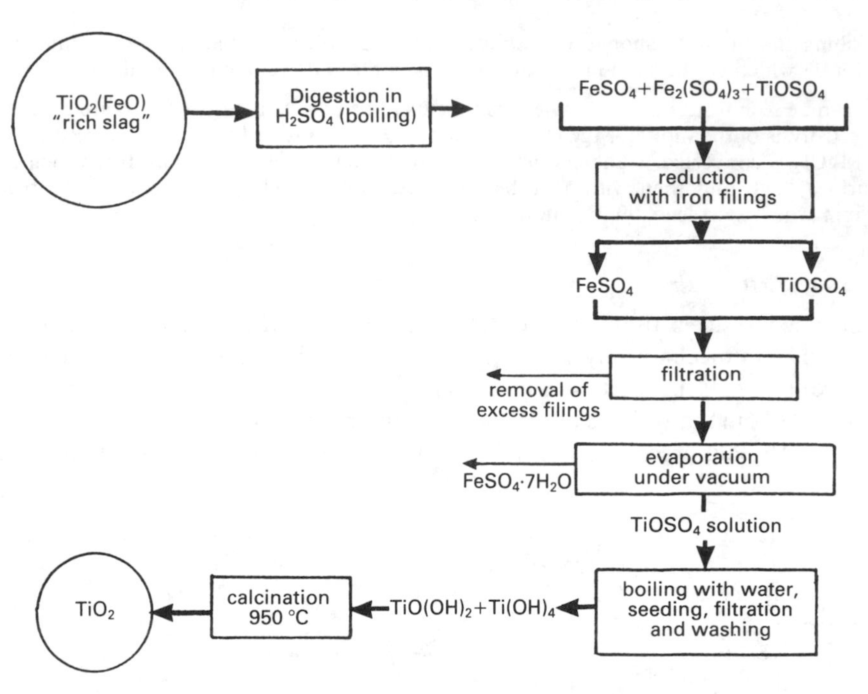

Fig. 13.47—Preparation of TiO_2 from the 'rich slag' obtained from ilmenite.

chlorine itself, certain concentrations of mineral acids, sulphonated compounds, and the majority of organic reagents which are notoriously aggressive.

Titanium also forms alloys of its own: with Mn, Cr–Fe–Mo, Al–Mn, Al–V, Al–Sn, and so on. On account of their lightness, mechanical resistance when hot, and workability, such alloys are mainly employed in the aeronautical and missile industries.

The well known pigment titanium white stands out in importance among the compounds of titanium. Chemically, this is formed from pure TiO_2 and can be prepared, in accordance with the scheme shown in Fig. 13.47, from the same TiO_2-rich slag from which titanium is produced.

The TiO_2 prepared in this manner must then be specially treated (moistened, dried, and pulverized) for it to exhibit maximum covering power as a white pigment.

Tungsten

m.p. = 3370°C, b.p. = 5927°C $d = 19.35$ g/cm^3

Tungsten, which when compact, is a bright white metal, is the element which has the second highest melting point after carbon, while its boiling point is even higher than that of carbon.

It does not dissolve in aqua regia, and, as regards other acids, it is perceptibly attacked only by a mixture of nitric and hydrofluoric acids. It is also attacked by fluorine. It is resistant towards alkalis provided that they are not hot and in an oxidizing atmosphere, since, in the latter case, soluble tungstates (WO_4^{2-}) are formed.

At a red heat, it is attacked by chlorine, while, under other conditions it is oxidized. It forms a nitride (W_2N) and carbides (WC and W_2C) which are extremely stable.

The principal minerals of tungsten are scheelite $CaWO_4$, ferberite $FeWO_4$, huebnerite $MnWO_4$, and the isomorphous mixture wolframite (Fe, Mn)WO_4. All the iron-bearing minerals are ferromagnetic. Huebnerite is a rare mineral, whereas ferberite is quite abundant, and wolframite containing a large amount of ferberite component, is even more abundant. They are always accompanied by cassiterite.

The ferberite–wolframite concentrates, which are mainly obtained by magnetic separation, are smelted with sodium carbonate in an oxidizing atmosphere:

$$2FeWO_4 + 2Na_2CO_3 + \tfrac{1}{2}O_2 \rightarrow 2Na_2WO_4 + Fe_2O_3 + 2CO_2$$

$$3MnWO_4 + 3Na_2CO_3 + \tfrac{1}{2}O_2 \rightarrow 3Na_2WO_4 + Mn_3O_4 + 3CO_2$$

Leaching of the resulting mass brings the sodium tungstate into solution which, after appropriate purification, is converted into tungstic acid and then into tungsten trioxide from which the metal is won in the form of a powder by stepwise reduction with hydrogen:

$$2WO_3 + H_2 \xrightarrow[-H_2O]{700^\circ C} W_2O_5 \xrightarrow[+H_2-H_2O]{740^\circ C} 2WO_2 \xrightarrow[+4H_2-4H_2O]{860^\circ C} W$$

Ferrotungsten for metallurgical purposes can be prepared by reducing concentrates containing a large amount of ferberite, that is, minerals which are poor in manganese, with aluminium powder:

$$3FeWO_4 + 8Al \rightarrow 3Fe + 3W + 4Al_2O_3$$

Finally, carburized tungsten powder for metallurgical use can be prepared by the reduction of tungstates at 1400–1800°C:

$$CaWO_4 + 3C \rightarrow CaO + 3CO + W$$

The powders obtained during the metallurgical processing of tungsten minerals, especially those of the pure metal, must subsequently be converted into compact structures. Powder metallurgy or sintering is employed for this purpose, as in all cases where the powders to be compacted have a melting point which is too high, and there is a danger that they will be contaminated by material from the crucibles in which they are fused and that the components of the powders themselves may decompose (as in the case of the carbides). Powder metallurgy also lends itself to the alloying of metals with widely differing melting points (e.g. AlNiCo), and, by using this technique, it is possible to manufacture finished components in one single operation with appropriately shaped moulds.

In the case of tungsten powder, the sintering process is started by putting it into steel moulds and subjecting it to a pressure of several ton/m^2. The objects which are moulded in this fashion are gradually heated over a period of one hour to temperatures of the order of 1200–1400°C while they are kept under a hydrogen atmosphere, which, by reduction activation of the surface of the powder particles, favours an increase in the dimensions of the latter, entailing recrystallization phenomena which take place between the particles in contact.

Subsequently, and always under a reducing atmosphere, the temperature is considerably increased (normally to 400–500°C below the melting point), using high-frequency alternating currents which are passed into the body in question. A final slow cooling under pressure ensures the formation of products which can be turned into sheets and extrusions in the normal manner.

The treatment which has just been described outlines the basic steps which all operations in powder metallurgy have in common.

The uses of tungsten

Pure tungsten or tungsten in the form of a ferroalloy or carburized tungsten is mainly used for the production of high-speed and super-high-speed steels and super-alloys which are extremely hard (widia metal) and corrosion and heat resistant. The remainder is employed as the pure metal in electrotechnology and electronics for plates and, especially, filaments.

Among the compounds of tungsten which are of practical use, the tungstates of the alkaline-earth elements and magnesium stand out on account of their utilization as phosphors.

Uranium

m.p. = 1133°C, b.p. = 3818°C $d = 19.05$ g/cm^3

Elemental uranium is a white metal which is similar in appearance to steel. It is ductile and malleable and feebly paramagnetic. Cold working hardens the metal, and it has a low electrical conductivity. While it is inert towards alkalis, it is more or less reactive towards all acids.

Interest in uranium grew greatly after the discovery of nuclear fission because uranium is the only naturally occurring element which is fissile.

The principal uranium minerals are uraninite (pitchblende), which is essentially U_3O_8. Other minerals which are of interest in the production of uranium or, rather, in the production of uranium-based fissile materials, are carnotite $KUO_2VO_4 \cdot 1\frac{1}{2}H_2O$ and autunite $Ca(UO_2)_3(PO_4)_2 \cdot 8H_2O$.

The metallurgy of uranium

The chemistry of the treatment of uranium minerals in order to win the element, like the remainder of uranium chemistry in general, is today perhaps the best known among the analogous sciences and techniques regarding other elements.

Here, however, we can only indicate for didactic and guidance purposes the exhaustive information on this subject resulting from the execution of specific projects, details of which are more readily available from monographs and reviews.

So far as the metallurgy of the winning of metallic uranium and the preparation of nuclear reactor fuel is concerned, it should be noted that this field is vast and complex owing to the variety of uranium-bearing minerals from which it is possible to start out, and on account of the fact that they are mutually mixed and accompanied by gangues which differ very widely both as regards the amounts of them present and their types.

In short, the metallurgical process which is, on average, the most used for the preparation of the important compound, UF_4, starting from pitchblende, comprises the following steps:

- concentration of the mineral while it is still within the confines of the mining plant, using conventional methods for the enrichment of minerals and, especially, classification and flotation;

- reduction of the concentrate, which has been converted to a suitable particle size distribution, with hydrogen at a temperature of about 600°C:

$$U_3O_8 + 2H_2 \rightarrow 3UO_2 + 2H_2O$$

- treatment with anhydrous HF at a lower temperature ($\simeq$480°C):

$$UO_2 + 4HF \rightarrow UF_4 + 2H_2O$$

The preparation of uranium tetrafluoride (green salt) in this manner is a necessary step both when metallic uranium is to be produced for particular purposes as well as when the uranium is to be enriched in the isotope U-235 which is subsequently to be used in the production of nuclear energy.

When elemental uranium is to be produced, this is achieved by the reduction of UF_4 with metallic calcium or magnesium:

$$UF_4 + 2Mg \rightarrow U + 2MgF_2$$

While processes of the reduction and fluorination type are carried out in fluid-bed reactors, autoclaves are used for the reduction of the fluoride to uranium, and, in every case, the apparatus is suitably protected by the use of appropriate refractories.

The preparation of reactor fuels

Uranium which was not particularly enriched in U-235, such as that which is obtained by using the metallurgical procedures which have just been described, was employed in the earliest reactors; but, nowadays, simple uranium metal which has not been enriched is no longer used as fissile material, mainly because it tends to deform at temperatures above 600°C.

Present-day technology is concerned with first obtaining enriched metallic uranium and then converting this into alloys or dispersions in metallic matrices or into 'ceramic fuels' (oxides, carbides, or nitrides) among which UO_2 excels.

Enriched uranium metal is obtained by first converting UF_4 into UF_6 by fluorination at about 400°C:

$$UF_4 + F_2 \rightarrow UF_6$$

The resulting product is a mixture of the molecules $^{235}UF_6$ and $^{238}UF_6$. This is introduced into a separation plant which is subdivided into two chambers by a porous membrane across which a pressure difference exists. The molecular species with the lower mass diffuses across this membrane more rapidly than the other, and the degree of separation is given by $\sqrt{M_2/M_1}$, where $M_1 < M_2$. However, given the fact that the difference in mass between two isotopes of the same element is extremely small, this system requires an enormous number of successive separation stages, and the plants are therefore very large and consume large amounts of energy in circulating the gases. The attrition of the barrier is also very large on account of the fact that the pores of the membrane are extremely small with diameters of 3 to 7×10^{-6} cm. Therefore, given the enormous size of the plants and the high energy consumption in the compression and recompression of the gases, it comes as no surprise that there has been reluctance to the general adoption of this process.

The industrial method for the separation of $^{235}UF_6$ and $^{238}UF_6$ is still ultracentrifugation which concentrates the heavier molecules towards the outside of centrifuges which operate at a very high number of revolutions per minute. The degree of separation depends on the square of the rotor velocity and on the mass difference between the molecules being separated. When this method is employed, there are well recognized mechanical problems concerned with the extremely high velocities which are employed, but, from the point of view of the amount of energy which is consumed, it is far more acceptable than the diffusion method.

Subsequent reduction of the enriched uranium hexafluoride yields uranium containing about 3% of ^{235}U as compared with the 0.7% in the starting material. This uranium is next either converted into alloys, dispersed in metallic matrices, or transformed into ceramic fuel to make it suitable for feeding into nuclear reactors in the form of small cylinders or microspheres.

However they are moulded, nuclear fuels must exhibit the lowest possible degree of chemical reactivity towards the substances which are to be found at any level and may come into contact with them within nuclear reactors. Furthermore, they must also possess the greatest possible structural rigidity up to their melting point. The small cylinders, microspheres, and other forms of the fuels must then be put into containers which are constructed of, or at least lined with, materials such as zirconium and its alloys, aluminium alloyed with nickel, steels containing high percentages of nickel and chromium, and alloys with high contents of niobium, molybdenum, tantalum and vanadium.

The uses of uranium

Apart from its very prevalent use in the production of nuclear energy, low percentages of uranium are used to reinforce the hardening of steels because of the stable carbides, UC, UC_2, and U_2C_3 which it forms.

Moreover, small amounts of uranium and its compounds (sodium diuranate, $Na_2U_2O_7 \cdot 6H_2O$, for example) are used in the colouring of ceramics, glasses, and glazes. Finally, mixtures of the cerium and barium salts are employed as dark blue pigments.

Vanadium

m.p. = 1890°C, b.p. = 3000°C $d = 5.96$ g/cm^3

Vanadium is a metal with a steel grey colour which is ductile and malleable when hot, provided that it is worked under an inert atmosphere because otherwise it oxidizes at elevated temperatures with losses in its mechanical properties. When heated in an atmosphere of chlorine, it forms VCl_4, whereas, in nitrogen, it forms VN, and, when heated to a red heat in carbon, it yields V_4C_3. When cold it is resistant to attack by mineral acids, but it is attacked by them when hot. It is attacked by molten alkalis to yield vanadates.

Vanadium is quite abundant in nature, but it is never very concentrated. Its principal minerals are vanadinite $3Pb_3(VO_4)_2 \cdot PbCl_2$ and carnotite $K(UO_2)VO_4 \cdot 1\frac{1}{2}H_2O$, so that it is usually a by-product of other metallurgical processes.

Metallurgy and uses of vanadium

The first stage in the metallurgical processing of vanadium is always the recovery of the pentoxide, V_2O_5. This compound is contained in the ashes of fuel oils and the bitumens from certain petroleums (in particular, Venezuelan petroleums), owing to the fact that it is responsible for certain biological functions in marine organisms (e.g. Holothurians) which have contributed to the formation of such petroleum deposits.

After they have been enriched, an alkaline fusion can be carried out on some minerals containing vanadium in such a way as to form alkali vanadates which are subsequently converted to V_2O_5 by acidification and calcination.

A scheme for an 'acid' metallurgical process which also leads to vanadium pentoxide is shown below.

$$\text{Vanadinite} \xrightarrow{\text{HCl(conc)}} \begin{cases} \rightarrow HVO_2 \xrightarrow[\text{air}]{NH_{3(aq)}} NH_4VO_3 \rightarrow V_2O_5 \\ \rightarrow PbCl_2 \end{cases}$$

'Vanadium bullion' can be obtained by using aluminothermic methods; a ferroalloy used in metallurgy can be prepared from V_2O_5, carbon, and silica in an electric furnace; and very pure vanadium, which is almost as soft as lead, can be prepared by the reduction of V_2O_5 with calcium in the presence of iodine:

$$V_2O_5 + 5Ca \rightarrow 5CaO + 2V$$

The principal uses of vanadium are in metallurgy, in the field of machine tool steels, especially in high-speed steels, spring steels, and low-alloy steels in general.

The salts of vanadium and V_2O_5 are much used as oxidation catalysts both in industrial inorganic and industrial organic chemistry. Small quantities of V_2O_5 are also used for rendering glasses and crystals opaque to ultraviolet rays.

Zirconium and hafnium

The properties of these two metals are summarized in Table 13.13.

Production of zirconium, hafnium, and ZrO_2

To prepare zirconium, a mineral, which is rich in zircon, and carbon are heated in an arc furnace to 2700°C. At this temperature, the greater part of the silica which is liberated volatilizes, and a mixture of carbides is formed in which ZrC predominates. This mixture, after it has been formed into briquettes, is loaded into a furnace which is similar to that used in the metallurgical processing of titanium (Fig. 13.46) where $ZrCl_4$ is formed in a current of chlorine together

Table 13.13—Properties pertinent to the uses of zirconium and hafnium

Metal	*Physical properties and natural states*	*Chemical properties*
Zirconium	A ductile silvery metal when compact. Black in powder form. m.p. 1860°C, b.p. 4750°C and $d = 6.53$ g/cm^3. Low neutron capture cross-section and insensitive to radiations. Rather abundant in nature but the only exploitable minerals are baddeleyite ZrO_2 and zircon $ZrSiO_4$.	As a powder burns readily in air. When compact, only burns at a very high temperature. Extremely readily passivated and corrosion resistant but not very resistant to HCl_{conc}, HF and moist Cl_2 (cf. titanium). Good corrosion resistance properties in the environment of nuclear reactors high temperatures and pressures, presence of CO_2, etc.).
Hafnium	Similar in appearance to Zr. m.p. 2200°C, b.p. $\sim$5400°C, and $d = 3.09$ g/cm^3. Unlike Zr, it has a high neutron capture cross-section. This means that it must be separated from Zr to be used in nuclear reactors where the insensitivity of Zr to radiations is valued. There are no specific minerals. Found in Zr minerals in average amounts of 0.1% but exceptionally up to 7%	Corrosion resistance comparable and often superior to that of zirconium, e.g. toward pure water and water vapour. An analogy is also shown in the types of compounds formed: e.g. K_2MeF_6 and $(MeCl_4)\cdot(POCl_3)_2$ where Me = Zr or Hf. Not damaged at all by radiations as regards its ductility and ultimate tensile stress.

with the other more volatile chlorides ($TiCl_4$, $SiCl_4$, etc.). The $ZrCl_4$ can be selectively condensed, redistilled, and subsequently reduced with magnesium to obtain zirconium sponge and magnesium chloride:

$$ZrCl_4 + 2Mg \rightarrow Zr + 2MgCl_2$$

The zirconium sponge is crushed and pressed into bars which are melted in a vacuum furnace to obtain compact zirconium, while the magnesium chloride, after being electrolyzed in the molten state, yields the magnesium metal used as the reducing agent.

To obtain very pure zirconium, technical zirconium sponge is first produced which, when kept at about 200°C in contact with the walls of a vessel which has been evacuated after an iodine crystal has been put into it, and is thereby full of iodine vapour, forms volatile ZrI_4 which can be readily re-decomposed when it comes into contact with a filament of pure tungsten which is placed in the centre of the reactor and heated to 1200–1400°C electrically.

The pure metallic zirconium which is formed adheres to the filament, while the iodine is recycled to the walls, thereby repeating the operation. Overall, the iodine acts as a zirconium purifying agent by transferring it from one point to another.

Metallurgically, hafnium follows zirconium throughout its processing, so that, when the latter is required for nuclear uses, the hafnium must be recovered from it. Various procedures have been developed for this purpose, the most important of which are the fractional crystallization of the alkali hexafluorides and the fractional distillation of the phosphoryl chloride complexes formed by zirconium and hafnium (see Table 13.13).

Zirconia, ZrO_2 (m.p. 2700°C), is used as a white pigment for the production of zirconium compounds and as a refractory. It is prepared from zircon or baddeleyite as shown in Fig. 13.48. While it is formally similar to baddeleyite, it differs from it in its covering power and refractory behaviour.

The uses of zirconium and hafnium

The chemical and nuclear industries are interested in the use of zirconium metal. On the one hand, this is in the sectors of chemical plant technology where acetic, phosphoric, chromic, hydrochloric, sulphuric, and nitric acids as well as ammonia and caustic soda are handled, and, on the other hand, as a getter, for the encasement of strongly radioactive nuclear fuel elements, and in the construction of the containers of reactor cores.

Zirconium alloys, which are more resistant than the metal in specific sectors where they are used and named using the abbreviation Zircaloy n (where n is the class of alloy and $n = 1, 2, 3$, etc.), are also used by the same industries. Such alloys contain more than 97% of zirconium, and then mainly Sn or Nb or Cu, depending on which class they belong to.

Pure hafnium is also used in the nuclear industry, essentially for the manufacture of neutron moderating rods and for the construction of anti-radiation linings.

Finally, appreciable amounts of zirconium compounds find various technological and parachemical outlets as pigments, refractories, tanning agents, textile finishing agents, in the structures of catalysts, and, nowadays, in cosmetics (antiperspirants and antiodour agents).

13.21 THE PRECIOUS METALS

The metals silver, gold, and platinum as well as the metals of the platinum family and especially palladium, are commonly referred to as precious metals.

They are also noble metals, par excellence, in so far as they all have a reduction potential which is considerably higher than that of hydrogen. Consequently, they are

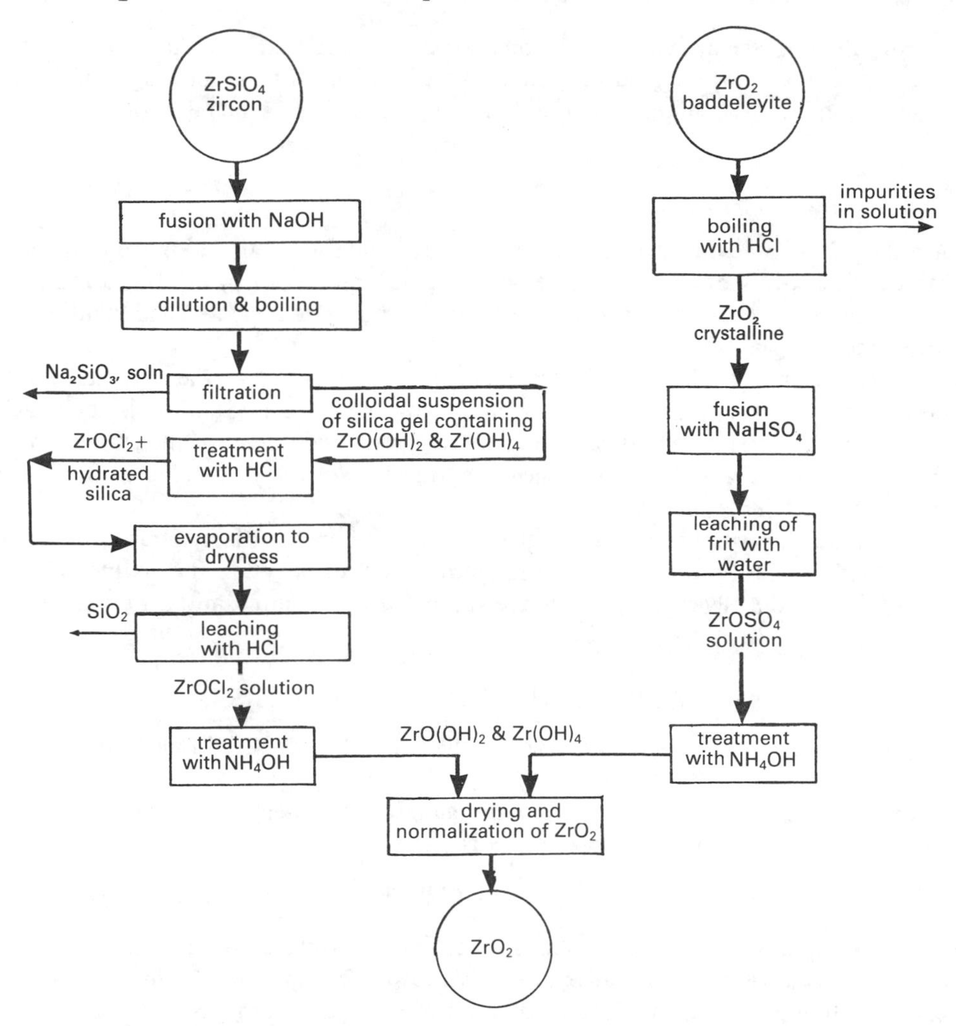

Fig. 13.48—Industrial production of zirconia from zirconium minerals.

notoriously reluctant to part with electrons and become oxidized, and, hence, to become corroded.

Apart from their specific minerals which are probably constituted of the same metals in the native state, silver, gold, platinum, and palladium are recovered as by-products of other metallurgical processes such as those used in the extraction of copper, nickel, and lead, in particular.

Silver

m.p. = 960°C, b.p. = 1950°C $d = 10.5$ g/cm^3

Silver is a bright typically 'silvery' metal which is ductile and malleable and the best conductor of electricity and heat among all the metals.

In air, silver becomes coated with a very thin layer of the sulphide which protects it from any further attack. This is something of a handicap as regards its use for electrical contacts where gold or gold alloys are preferred to it, although they have a lower conductivity. In particular, gold has a conductivity which is about 70% that of silver.

The metallurgy of silver

Among all the precious metals, silver is the one which is obtained for the most part by recovery during other processes. For this reason, the recovery of silver during the metallurgical processing of lead will be treated. First, however, we shall mention its direct metallurgy.

There are many minerals of silver. The principal ones are the sulphide Ag_2S (argentite), the sulphoantimonite Ag_3SbS_3 (pyrargyrite), and the chloride AgCl (cerargyrite). While, at one time, amalgamation was much used for the extraction of silver*, the low-grade minerals which are treated nowadays are more profitably treated by a cyanide process.

The mineral, possibly enriched, is ground and left for about forty hours in contact with a 0.5% solution of sodium cyanide into which air is systematically blown.

In this way the silver and its salts are solubilized by complexation. For example:

$$4Ag + 8NaCN + 2H_2O + O_2 \rightarrow 4NaAg(CN)_2 + 4NaOH$$

$$Ag_2S + 4NaCN \rightarrow 2NaAg(CN)_2 + Na_2S$$

Other metals (copper and gold, in particular) are also dissolved by treatment with cyanide.

The solution, after separation from the gangue, is treated with zinc so that the silver is precipitated by electrochemical displacement:

$$2NaAg(CN)_2 + Zn \rightarrow 2Ag + Na_2Zn(CN)_4$$

The precipitate, which also contains other metals displaced by the zinc as well as the excess zinc is melted in an oxidizing atmosphere and, after cooling, is treated with a solution of acid in order to dissolve the oxides of zinc, copper, etc. which have been formed.

The remaining silver, which is still contaminated by other precious metals, is then electrolytically refined.

However, a large part of the silver produced is recovered from argentiferous galenas by separating it from lead bullion using the 'Parkes process'. For this purpose, zinc is added to molten lead bullion from which the copper has been removed and which has been chemically purified so as to remove any As, Sb or Sn which was initially present. Silver is 3000 times more soluble in zinc than in lead and, as a result, an extraction is accomplished by the preferential partitioning of the silver into the zinc.

---

*Amalgamation is not directly applicable to silver sulphide and, for this reason, it must first be converted into the chloride:

$$Ag_2S + CuCl_2 \rightarrow CuS + 2AgCl$$

which is followed by:

$$2AgCl + 2nHg \rightarrow 2Ag(Hg)_{n-1} + Hg_2Cl_2$$

When this amalgam is heated to about 400°C, the amalgam and the calomel sublime to leave pure silver.

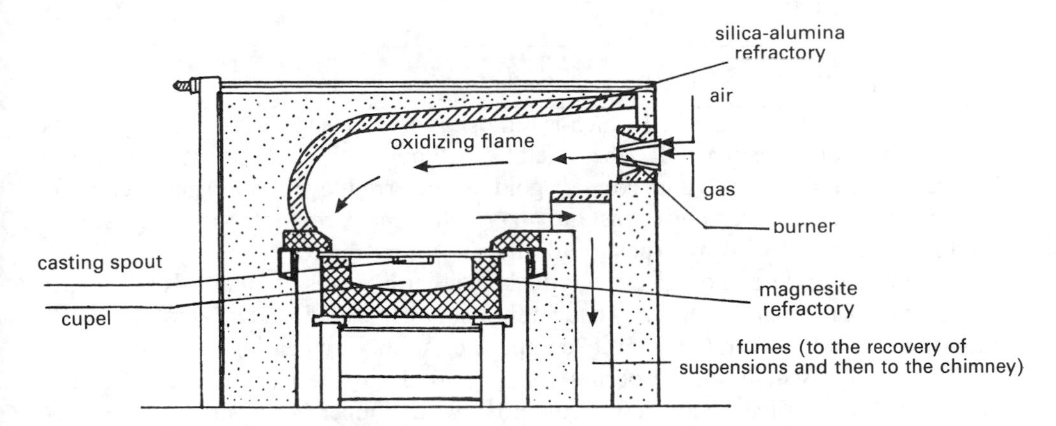

Fig. 13.49—Cupellation furnace specifically for the removal of lead from silver, but which can also serve in the separation of gold from its commonest alloying elements (copper and silver). In every case, all the oxides which emerge from the casting spout or which are carried over by the fumes are recycled for the recovery of the corresponding metals.

The solution of silver in zinc which is formed in this manner separates from the molten lead as a solid phase and can therefore be removed. As in all solvent extraction operations, the operation is repeated several times, so that practically all of the silver is finally recovered.

The subsequent separation of silver (b.p. 1950°C) from zinc (b.p. 906°C) is carried out by distillation, thereby obtaining a silver/lead alloy which can be separated into its components by 'cupellation'.

With this aim the alloy is melted in a special type of reverbatory furnace, that is, a cupel (Fig. 13.49) in which the molten alloy is superficially 'grazed' by a flame accompanied by an excess of combustion air (an 'oxidizing flame'). Having been oxidized in this way, the resulting lead oxide floats and is forced out from the casting hole of the cupel by the turbulence produced by the burning gases. When the surface appears to be only streaked with lead oxide, the mass in the cupel is made smaller and the operation is repeated.

By continuing in this manner, one arrives at silver with a completely shiny surface (a silver flash) due to the high grade of purity of the metal. It is still necessary, however, to electrolytically refine the silver.

The uses of silver

Silver is a metal with a considerable number of industrial uses apart from currency, money, and jewellery. In fact, it is used in the construction of high-performance electrical conductors, special components in aircraft, alkali resistant chemical containers, chemical catalysts, brazing alloys, and special alloys. It has also been used in the production of numerous devices used in astronautics (e.g. solar cells).

Certain compounds of silver are also of industrial importance. For instance, silver nitrate, the starting point for the production of other silver compounds, is at the foundation of all the electroplating which uses silver, and finds many uses in medicine, in the printing industry, in the manufacture of mirrors, and in chemical analysis. The halides of silver and, above all, the bromide and chloride, are important in the photographic and photocomposition fields.

Gold

m.p. = 1062°C, b.p. = 2966°C $d = 19.3$ g/cm^3

Gold is a reddish-yellow metal which is almost as soft as lead. It is extremely ductile and more malleable than all other metals. Although its electrical conductivity is inferior to that of silver and copper, gold is preferred over the latter metals for electrical contacts which must constantly have a high performance, because its surface is absolutely unaffected by exposure to air.

Both when cold and hot, gold is unattacked either by acids or alkalis. It is attacked only by aqua regia and, slowly, by hydriodic acid, selenic acid, and by aerated cyanide solutions. Gold is liberated from its compounds simply by heating.

The gold present in nature occurs almost entirely in the native state. It is often associated with silver, and, less frequently, with copper. Calaverite $AuTe_2$ and galvanite $AuGeTe_4$ are rare gold minerals.

The metallurgy of gold

Three methods for the extraction of gold have been used:

- levigation followed by selection;
- amalgamation followed by distillation of the mercury;
- treatment with cyanide under oxidizing conditions, which is carried out as in the case of silver but, with far more dilute solutions of sodium cyanide.

The reactions entailed in the oxidative cyanation are those shown in Table 13.13(1) on page 541.

The zinc is separated by fusion from the gold-bearing slime obtained by precipitation with zinc, and copper and silver can be removed from the remaining impure gold by leaching with boiling concentrated sulphuric acid (the 'quartation process'). More often, nowadays, the impure gold is remelted and chlorine gas is bubbled into the melt. This leads to the volatilization of Pb, Cu, Sb, and As as their chlorides, while the silver chloride floats on top of the melt. After the latter has been removed, the gold which remains with a purity of 99.6% can be dissolved in aqua regia and then refined electrolytically.

The uses of gold

The principal uses of gold are still the traditional ones of constituting a metallic reserve as a monetary guarantee, in providing an investment bolt-hole and/or ornamentation in the form of coins, and the extremely varied forms of jewellery.

The mechanical strength of gold is always increased by alloying. This is done, for the greater part, with copper, but also with silver, nickel, platinum, and palladium. The proportion of gold in the alloys is expressed in carats, with pure gold corresponding to 24 carats.

Nowadays, it is officially preferred that the purity of gold should be expressed as a percentage, and the percentages which frequently occur in coinage 90 (21.6 carat), 91.66 (22 carat), whereas 75 (18 carat) and 53.8 (14 carat) are common in jewellery.

There are numerous industrial uses of gold on account of its corrosion resistance, the possibility of preparing it in the form of extremely thin foils which can be handled mechanically without breaking it, and of the many alloys obtainable from it which lend themselves to a wide variety of applications.

Gold is used in printed circuits and in many other components found in radios, television sets, and computers. Gold foils are manufactured which provide protection against the very high frequency radiations encountered in astronautics, while other foils are made which, when they are included in glasses, make the glasses heat-insulating without making them opaque. Gold is also used in the construction of aero-engine turbine blades. As an example of this, about half a kilogram of gold is required in the fabrication of a turbine for a 'Jumbo' jet. Gold is also sometimes used decoratively in bookbinding.

Platinum and palladium

Table 13.14 permits a comparative study to be made of these two typical representatives of the elements of the second and third triads of Group VIII. While they are very similar in appearance, they are extremely different in their densities and their abilities to resist oxidizing acids.

The metallurgy of platinum, palladium, and other elements of the platinum group

Platinum is found in alluvial deposits, at least partially alloyed with other metals, in the form of particles of dust (nuggets). The platinum-bearing mineral, after enrichment by hydrogravimetric methods, is treated by using the basic* metallurgical processes shown in the scheme in Fig. 13.50.

Nevertheless, both platinum and almost all the palladium are mainly obtained

Table 13.14—Properties and reactions of platinum and palladium

Metal	*Physicomechanical properties*	*Chemical properties*
Platinum	Similar to silver in appearance, ductile and malleable. Black when finely subdivided and absorbs hydrogen to which sheets of red hot Pt are permeable (unlike any other gases). The physical constants are: m.p. 1773°C, b.p. 3827°C, and $d = 21.45$ g/cm^3.	At moderate temperatures resistant to all chemical reagents apart from aqua regia. At elevated temperatures, it is attacked by alkalis and alkali nitrates, cyanides, sulphur and sulphides as well as by elements with which it forms alloys (Pb, As, Sb, C, etc.). Combines with oxygen to form labile compounds at high temperatures.
Palladium	Also, a silvery-white metal, ductile and, above all, malleable. At elevated temperatures, sheets of this metal are selectively permeable to hydrogen. It has the following properties: m.p. 1552°C, b.p. 2927°C, and $d = 11.4$ g/cm^3. More abundant in nature than platinum: $\sim 10^{-6}$% as opposed to $\sim 5 \cdot 10^{-7}$%.	Soluble, when finely subdivided, in hot concentrated mineral acids and in the reagents which dissolve Pt. It is oxidized in a red heat in air but the oxide formed decomposes at a higher temperature. Has outstanding hydrogen absorption properties in the form of sheets, plates and, above all, as a powder forming labile compounds which do not modify its crystal lattice and give up hydrogen to unsaturated and organic compounds.

*The following metallurgical methodologies are not exclusive and not even absolute, because the treatments of these metals (in common with all precious metals) are largely trade secrets of the firms which treat them.

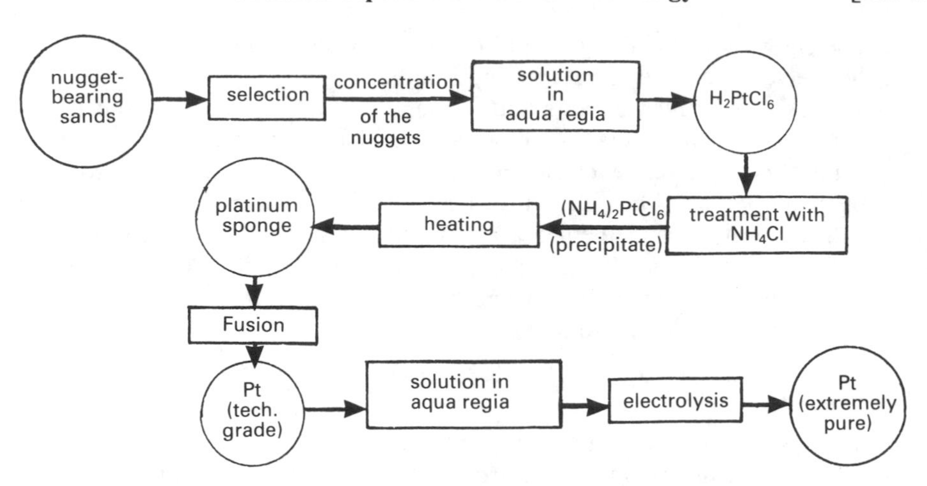

Fig. 13.50—The preparation of extremely pure platinum from sands which contain nuggets of it.

from the anodic slimes formed during the refining of nickel, from the residues which are not volatilized by CO in the Mond process, and from the anodic slimes from the refining of copper, since they occur in copper and nickel minerals, being alloyed with these metals and with other Group VIII metals (with Fe, in particular, as well as in the form of specific minerals such as sperrylite $PtAs_2$, cooperite PtS, and braggite (Pt, Pd, Ni)S.

After the contaminating elements such as Fe, Pb, Se, Te, Cu, As, Sb, and Sn have been removed, the separation from such mixtures is undertaken, in principle, in accordance with the scheme shown in Fig. 13.51.

The various fractions, which are separated, are subsequently subjected to specific treatments for the recovery of the individual components.

The uses of platinum and palladium

Platinum–palladium alloys are used for electrical contacts, while Pt–Rh alloys are also used in electrical technology and in various fields of the chemical industry. Other alloys of platinum and other Group VIII elements are employed in radar devices, electrochemical apparatus, jewellery, and odontology. Alloys of Pd and Pt with Au (mainly) and also with Ag, Ni, etc. are also used in jewellery. Palladium is employed in the construction of antimagnetic clocks, parts for balances, and so on.

Nevertheless, great use is made of platinum and palladium in the chemical industry for the construction of electrodes, extrusion nozzles reinforced by alloying, laboratory apparatus, foils for purification by the gaseous diffusion of hydrogen, and especially for the preparation of a whole range of catalysts in the form of meshes and other compact forms and as 'sponges', 'blacks', colloidal solutions, and solid and other catalytic complexes.

Platinum black and palladium black are obtained by treating solutions of the corresponding salts with reducing agents. The sponges are prepared as the residues

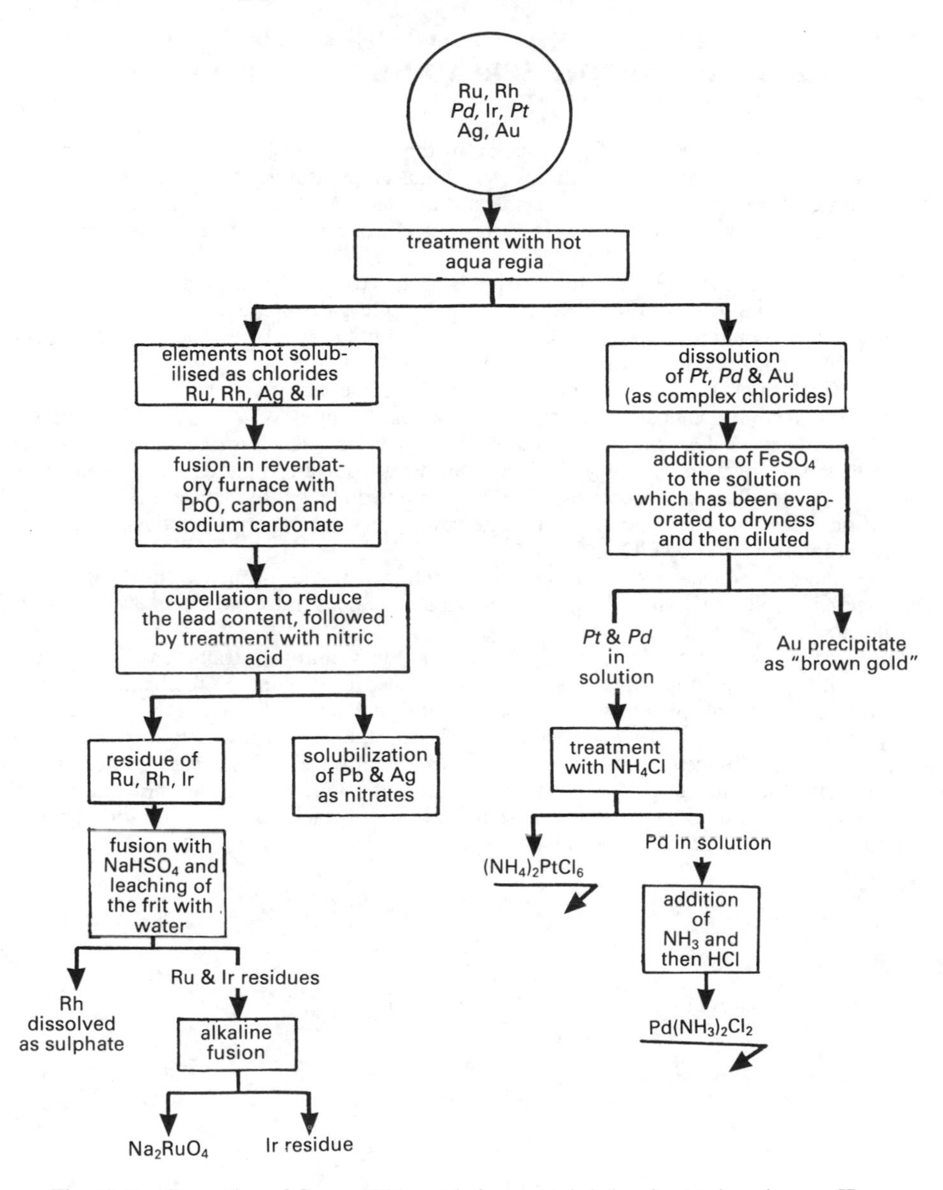

Fig. 13.51—Separation of Group VIII metals from metals belonging to the subgroup IB.

which remain after the heating of complex salts, ammonium salts, or salts of organic bases of the two metals, while the platinized or palladized asbestoses and ceramic sponges are prepared by depositing the finest metal powders from salts onto the asbestos fibres or onto extremely porous ceramic supports. Platinized platinum and palladized palladium are made by electrolytically covering cathodes of various shapes formed from the corresponding compact metals with finely subdivided metallic particles which are deposited from their salts which constitute the electrolytes.

13.22 THE DANGERS ARISING FROM METALLURGICAL PROCESSES AND THE SPREADING OF METALS IN THE ENVIRONMENT

It is obvious from the foregoing discussion that metallurgy in general and iron and steel making in particular always employ very high temperatures, the evolution of gases, the dispersion of powders, and the use of acids and alkalis.

The uses of some metals and of certain metallic compounds may also carry quite considerable dangers.

Since it is not possible here to provide a detailed treatment of all the risks to the physiology of living organisms which arise from metallurgical processes and the spreading of metals, we shall confine ourselves to some examples of pollutants from iron and steel working, of how metal powders are metabolized, and the toxic action of lead compounds which have spread in the atmosphere and cadmium compounds dispersed in waters.

Pollutants from iron and steel making are the gases (CO and SO_2, in particular) and solids (metal powders, Fe_2O_3, SiO_2, and many other oxides) which originate from furnaces, convertors, and upstream processes (selection and transportation), and intermediates in true and proper metallurgical processes. The most immediate consequences arising from the inhalation of such powders are bronchitis and respiratory difficulties. Cardiovascular troubles are also rapidly caused by CO and SO_2.

Metal powders which have been inhaled are retained in the bronchia and lungs, causing emphysema, while that part which is solubilized is carried into the circulatory system by biochemical processes, leading to poisoning.

Lead compounds (originally halides) present in the vicinity of traffic can lead, through inhalation, to the metal damaging the erythrocytes because it hinders the fixation of iron in the haemoglobin, while also harming the bones and the liver and kidneys.

Cadmium compounds dispersed by the electrochemical, pigment, and metallurgical industries are present in the sludges arising from the purification of waters. When such sludges are used as fertilizers, they introduce the metal into the food chain until it reaches man, causing malfunctioning of the kidneys, and it presents both an obstacle to bone formation as well as being responsible for malformation of the bones.

14

Silicate products and their substitutes

Silicon with an oxidation number of four in the form of 'silicate' is a very important fundamental component of industrial products which has long been a subject of study in applied chemistry. We shall consider it in this chapter which deals with ceramics, glasses, and constructional binders.

However, some alternative materials or materials with properties in common with those of the best known examples of silicate products such as organic glasses, gypsums, limes, and modern ceramic products, contain no silicon or silicates. It is clear from the title of this chapter that the treatment will include those materials which, even if they do have a different constitution from silicate products, are in some way allied to them.

Part 1 CERAMICS

14.1 DEFINITION AND CLASSIFICATION OF CERAMIC PRODUCTS

Until the 1950s the etymological definition of ceramic products had been accepted, that is, they were defined as manufactures basically consisting of clay (*keramos* = clay).

Later on, the advent of hyperthermal (and, in particular, metallurgical) technologies, nuclear reactor technology, space research technology, and the many fields which are interested in 'magnetic' memory devices rapidly brought about the development of a large number of materials with very high dielectric constants and low ablation rates, as well as thermal super-refractories, materials which are subject to nuclear fission and so on which have now also been included in the present-day inventory of ceramic products.

Consequently, any composition prepared from inorganic chemical compounds which is capable of maintaining the shape previously imparted to it after it has been suitably fired is today considered as a ceramic product.

Therefore, at the present time, ceramic products may be classified according to the following scheme:

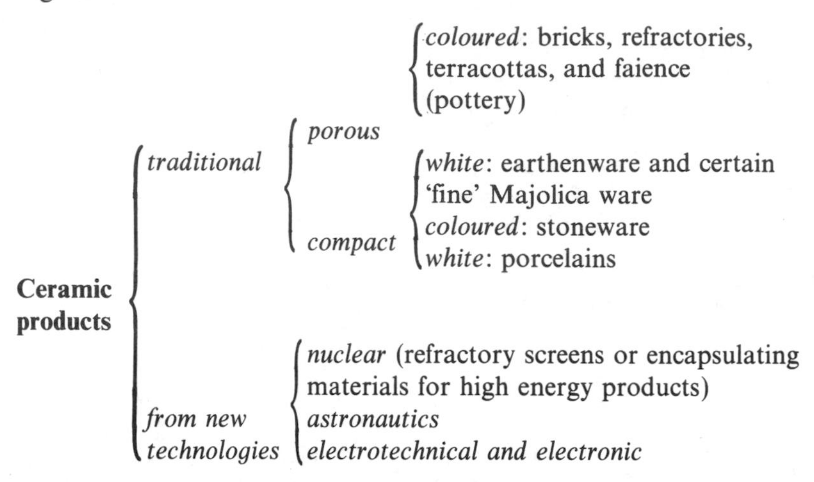

While traditional ceramics consists of a single pre-eminent raw material—clay, the ceramics developed by the new technologies do not have a single type of basic matrix which is common to all of them.

14.2 TRADITIONAL CERAMICS

The raw materials and secondary components of ceramic products

The essential components of ceramics are clays, and, frequently, also quartz and feldspars which as a whole are referred to as 'ceramic aggregates' from which the skeleton is constituted.

The secondary components ('accessories') of ceramics are the plasticizers, the fluxes, and the modifiers.

The plasticizers, such as 'sulphite lyes' and 'humic products', are materials which increase the creep and deformation properties of a ceramic product while it is still in the state of a moist mix, that is, in a 'green' state. Since plasticizers are products of an essentially organic nature, they are substantially decomposed during firing.

Binders which include starch, alginates, and cryolite are substances which give compactness to the shaped ceramic body. Small amounts of gums and urea can also act as binders. The binders are often destroyed during firing.

Fluxes (borax and soda are very well known) have the task of lowering the temperature at which the reactions (dehydration, disintegration, reformation of silicates, etc.) characterizing the firing of ceramic bodies, occur.

Modifiers, such as 'chamotte' (burnt clay) and $CaCO_3$, bring about structural changes in ceramic bodies. Physical effects such as pressure, firing temperature, and cooling conditions, and environmental agents (the type of atmosphere in the furnaces) can also act as modifiers.

When their composition lends itself to this, the clays themselves may act as plasticizers and binders. In fact, we shall see that, by starting out from particular types of clays ('bentonitic clays') it is possible to achieve a mixture with good workability and rheological properties, while the bodies produced retain their shapes well.

On the other hand, feldspars and silica may act as fluxes and modifiers respectively.

Nevertheless, it is only when they are suitably chosen and added in the appropriate amounts that ceramic aggregates are obtained which permit the production of the finest ceramic objects, as is demonstrated by the rich history of ceramic art. This art has evolved through the use of those raw materials which are still employed for ceramics today in various parts of the World, and in China in particular and the Far East in general.

The basic structure of clays

Clays are natural sedimentary rocks derived from the weathering of several types of primary rocks composed of crystalline particles of one or more minerals which are structurally (rather than chemically) different from one another and known generically as 'clay minerals'.

Regarding the origin of clays, it may be observed for example, that kaolinite, the main component of clays, is derived from the orthoclase of granites and syenites. The formal chemical process in their genesis is:

$$\underset{\textit{orthoclase}}{K_2O \cdot Al_2O_3 \cdot 6SiO_2 \cdot nH_2O} \rightarrow \underset{(H_4Al_2Si_2O_9 = \textit{kaolinite})}{Al_2O_3 \cdot 2SiO_2 \cdot 2H_2O} + K_2O + K_2O \cdot SiO_2 + 3SiO_2 + (n\text{-}2)H_2O$$

Qualitatively, clays are mainly formed from silicon, aluminium, and oxygen, but may also contain magnesium, iron, alkali, and alkaline–earth metals in smaller amounts because the clay minerals in them are accompanied by variable amounts of other materials, referred to as 'non-clay minerals'. Among the latter, quartz, calcite, witherite ($BaCO_3$), feldspars, and pyrites stand out in importance.

Given the qualitative and quantitative differences in the secondary non-clay minerals (which may also include organic substances and various salts which are soluble in water to a greater or lesser extent) accompanying clays, it is of primary importance to gain an understanding of the structures of clay minerals, that is, of the essential components which are mainly responsible for the properties of the clays.

Clay minerals can be classified as follows:

Clays minerals
- *two sheet layer silicate type*
  - kaolinite
  - halloysite
- *three sheet layer silicate type*
  - montmorillonite
  - illite

On both theoretical and practical grounds, it is necessary to look more deeply into the structures of kaolinite and montmorillonite, while a short discourse on the structural differences which exist between these and halloysite and illite respectively is sufficient.

The structure of kaolinite. The most abundant mineral forming a component of clays is kaolinite.

Physical techniques and, in particular, X-ray diffraction, have established that kaolinite is basically constructed of double sheets. The first of these consists of linked tetrahedra, while the second consists of octahedra, which are also linked, and these two sheets are joined to one another by the oxygen atoms which they share. Many such double layers are piled up to form a single stratum (lamina) of kaolinite.

Let us now look in somewhat greater detail at the actual structure of the abovementioned basic layers while emphasizing their nature.

Every layer of silicate is based on the tetrahedral laminar structure shown in Fig. 14.1. The sites marked with an A in this figure are occupied by oxygen atoms which are not shared by the silicon atoms of two tetrahedra, and therefore carry a negative charge which still has to be neutralized.

It can be seen that every unit making up the silicate layers in question has the chemical formula $Si_2O_5^{2-}$ when account is taken of the fact that, when oxygen atoms are shared by two units, only half of each atom must be attributed to each unit, These constitutive units are therefore disilicon pentoxide anions which, as a whole, confer a diffuse negative charge to the layer.

In the various types of layer silicates, there are cationic species which neutralize this charge either directly or after having coordinated oxygen and/or hydroxyl groups to an extent that the resulting complex is still positive. In this field, K^+ is often a cation which directly neutralizes a negative charge, while Mg^{2+} and Al^{3+} mainly coordinate oxygen and/or OH– groups to form cationic complexes which subsequently neutralize the charges on the silicate framework.

In the structure of layer silicate kaolinite, there is a layer consisting of a mesh of hydroxyaluminate octahedra with the structure shown in Fig. 14.2 carrying a cationic charge which neutralizes the charge on the silicate layer. The constitutive unit formed from two octahedra (a bi-octahedral unit) is picked out by a dotted line.

Hydroxyl groups occur at the positions indicated in the figure by the white circles. The oxygen atoms in these groups have one of their valencies saturated by hydrogen and the other by half an aluminium atom from one octahedron and half an aluminium from another octahedron.

The positions indicated by the black circles are occupied by the oxygen atoms marked with A in the constitutive unit of the silicate layer shown in Fig. 14.1. In fact, there is a one-to-one correspondence between each hexagonal 'star' arrangement in the silicate layer and the various arrangements, which are also hexagonal, of the hydroxyaluminate layer.

Still referring to Fig. 14.2, every bi-octahedral unit consists of two aluminium atoms (which are the centres of the corresponding octahedra and are therefore not visible), one whole

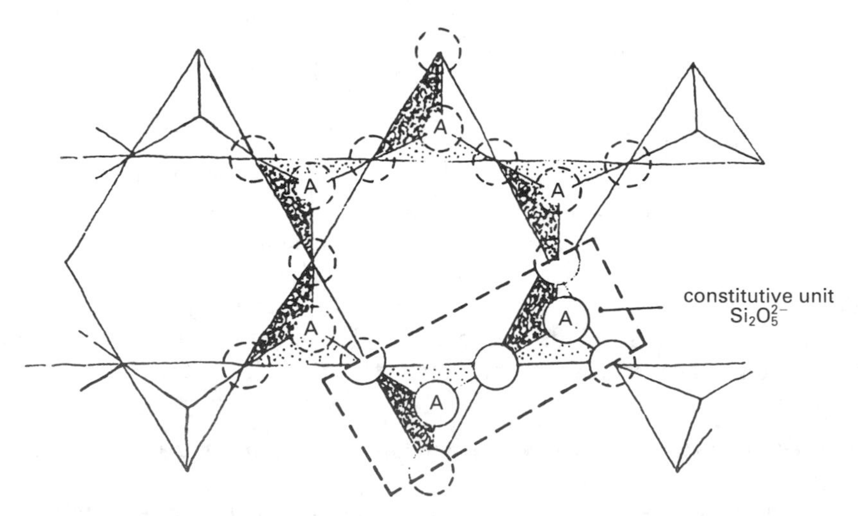

Fig. 14.1—Fragment of the first (silicate) sheet of the layer silicate (kaolinite). The disilicon pentoxide anion ($Si_2O_5^{2-}$) constitutive unit is emphasized by the rectangle.

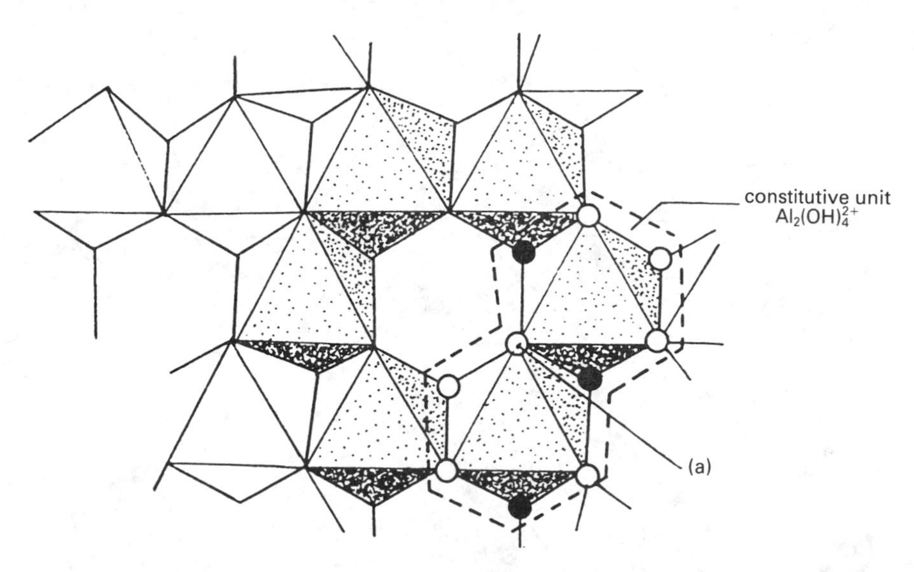

Fig. 14.2—Fragment of the second (aluminate) layer of the layer silicate kaolinite showing the cationic tetrahydroxyaluminate $[Al_2(OH)_4^{2+}]$ unit.

hydroxyl group (marked with an '*a*') and six hydroxyl groups, half of each of these groups being shared with the octahedra of other units with the result that the number of hydroxyl groups per bi-octahedral unit is equal to $3OH^-$. It can then readily be seen that the chemical formula of such a bi-octahedral unit is $Al_2(OH)_4^{2+}$.

Fig. 14.3 shows how the cationic hydroxyaluminate layers are superpositioned on the silicate layers by coming into contact with each other at the positions indicated by the black circles in Fig. 14.2. The layers are therefore mutually complementary, and the electron deficiencies on the aluminium atoms are neutralized by the surplus electrons on certain oxygen atoms of the other layer.

The two units, that is, the bi-tetrahedral and the bi-octahedral unit, which are arranged as two layers on top of one another in kaolinite constitute a larger unit which, when it is repeated a number of times, forms the laminae of kaolinite. The chemical formula of such a unit is obtained by adding the formulae for each component, that is:

$$Si_2O_5^{2-} + Al_2(OH)_4^{2+} \rightarrow Al_2(OH)_4Si_2O_5$$

Hence, the chemical formula of the double hexagonal unit consisting of hexa-tetrahedral and hexa-octahedral units respectively will be $Al_6(OH)_{12}Si_6O_{15}$.

Fig. 14.4 represents the double hexagonal unit corresponding to the formula which has just been given. The legend enables one to calculate the negative charges associated with this unit, which total 42. Hence, bearing in mind that the total number of positive charges from the six Si^{4+} ions and the $6Al^{3+}$ ions at the centres of the coordination polyhedra in this unit is equal to 42, it can be deduced that the system is electrically neutral overall.

Since it has been shown by means of electron microscopy that the thickness of the kaolinite laminae is 0.05 μm = 500 Å, and the thickness of each double layer is 7.13 Å (5.03 Å is the intrinsic thickness of the component layers, and 2.1 Å represents the distance between the two layers), each lamina of kaolinite is composed of $500/7.13 \simeq 70$ superimposed double layers.

Each lamina also has a regular hexagonal configuration with the diameter of the circumscribing circle equal to 0.7 μm = 7000 Å, while the diameter of the circle circumscribing each hexagonal tetrahedral or octahedral unit is 5.14 Å, it follows from this last each lamina surface of kaolinite has slightly fewer than $7000/5.14 \simeq 1400$ hexagonal units.

The structure of halloysite. The second clay mineral, halloysite, has a structure which is similar to that of kaolinite, but the constitutional correspondence between the hexa-tetrahedral units

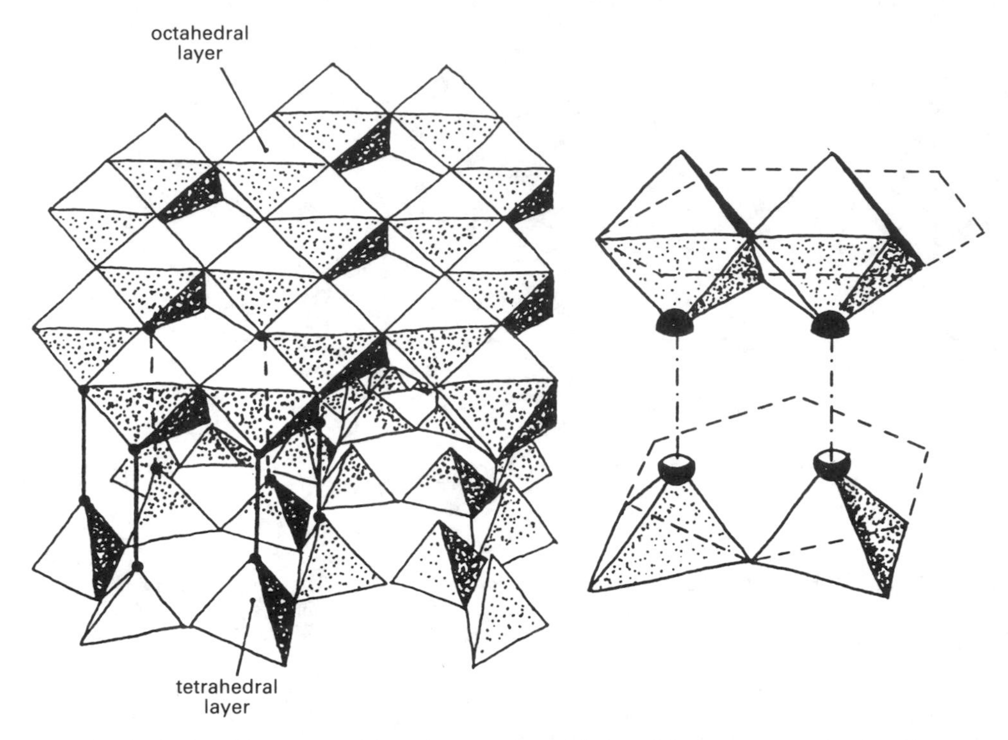

Fig. 14.3—Left: The correspondence between the polyhedra of the two hexagonal units of the elementary layers of kaolinite. Right: Detail of the correspondence between the bi-tetrahedral units (below) and the bi-octahedral units (above).

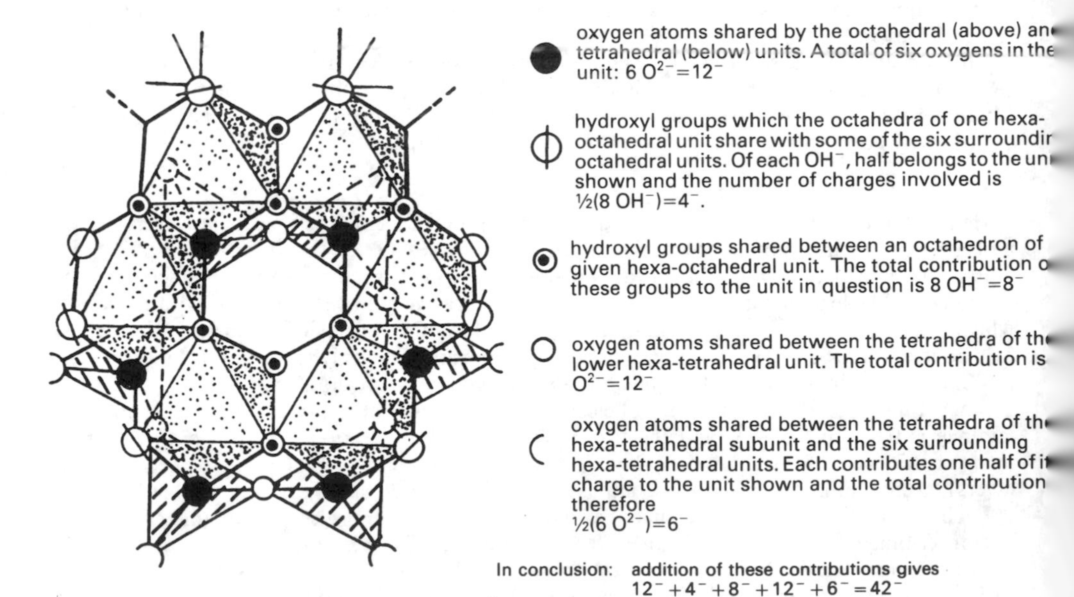

oxygen atoms shared by the octahedral (above) an
tetrahedral (below) units. A total of six oxygens in the
unit: 6 O^{2-}=12^-
hydroxyl groups which the octahedra of one hexa-
octahedral unit share with some of the six surroundir
octahedral units. Of each OH^-, half belongs to the un
shown and the number of charges involved is
½(8 OH^-)=4^-.
hydroxyl groups shared between an octahedron of
given hexa-octahedral unit. The total contribution o
these groups to the unit in question is 8 OH^-=8^-
oxygen atoms shared between the tetrahedra of th
lower hexa-tetrahedral unit. The total contribution is
O^{2-}=12^-
oxygen atoms shared between the tetrahedra of th
hexa-tetrahedral subunit and the six surrounding
hexa-tetrahedral units. Each contributes one half of i
charge to the unit shown and the total contribution
therefore
½(6 O^{2-})=6^-
In conclusion: addition of these contributions gives
$12^- + 4^- + 8^- + 12^- + 6^- = 42^-$

Fig. 14.4—Diagrammatic and descriptive demonstration that a structure with the formula $Al_2(OH)_4Si_2O_5$ is actually electrically neutral.

of the first layer and the hexa-octahedral units of the second layer is less perfect than in kaolinite. In fact, the three layers are not neutralized on a site by site basis, but only overall.

As a result, halloysite is not as tough but more plastic and easily penetrated than kaolinite. Consequently, its capacity to adsorb extraneous molecules is also greater.

Moreover, two forms of halloysite exist: a hydrated form in which a layer of water separates the layers of the mineral and a non-hydrated form which is free from this layer of water. The hydrated form has a chemical formula corresponding to a double hexagonal unit of hexa-tetrahedral and hexa-octahedral elements with water between them, $Al_6(OH)_{12}Si_6O_{15} \cdot 4H_2O$. On average, there are therefore 4 molecules of water in this unit. On the other hand, the non-hydrated form has the same chemical formula as kaolinite. Hydrated halloysite is converted into this non-hydrated form by heating it up to 400°C. A fragment of the halloysite structure is shown in Fig. 14.5.

The structure of montmorillonite. The mineral in montmorillonite clays is formed by the superpositioning of three layers to form laminae (Fig. 14.6). Of these three layers, two consist of silicate tetrahedra, while the central one consists of hydroxyaluminate groups. This holds the two layers above and below it in place and shares oxygen atoms with them; the charges on these oxygen atoms being partly neutralized by the Si^{4+} ions and partly neutralized by the Al^{3+} ions.

A small fraction (up to about 15%) of the silicon atoms in a tetrahedral layer can be substituted by aluminium atoms, while the aluminium atoms in an octahedral layer can be completely replaced by other metal cations.

Among other things, if $2Al^{3+}$ are replaced by $3Mg^{2+}$, the mineral saponite (soap stone) is formed, while if the aluminium is replaced by iron, it produces nontronite. *Sauconite* is obtained when the aluminium is replaced by zinc.

The triple layers which are joined to form the laminae of montmorillonite are weakly bound to each other so that they can accommodate layers and other polar molecules with different cross-sectional dimensions. On account of this, montmorillonite does not have a fixed distance

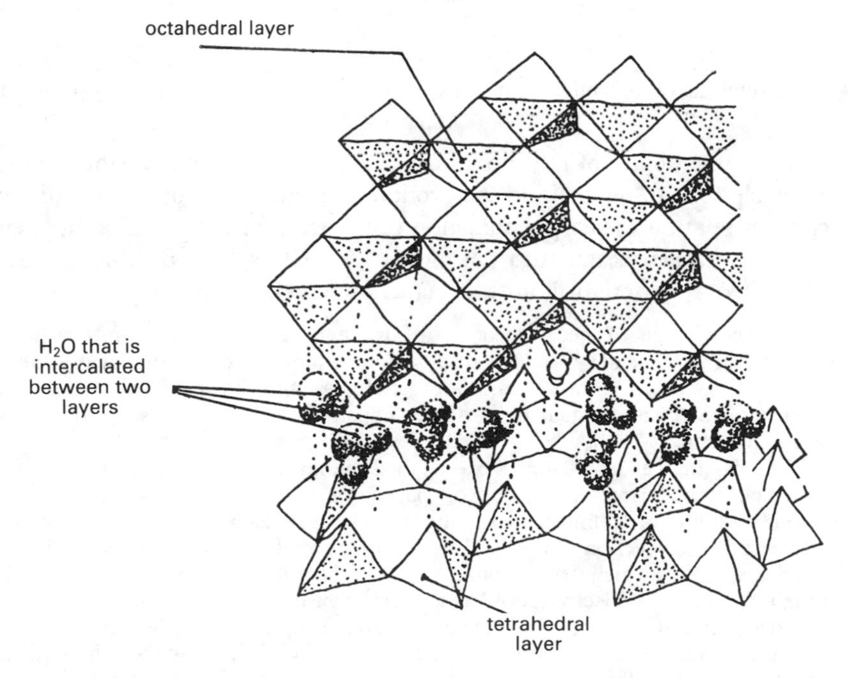

Fig. 14.5—Statistical neutralization (there is, in fact, as in kaolinite, an exact correspondence between the positive and negative charges) of the elementary polytetrahedral and polyoctahedral layers of halloysite between which water is intercalated.

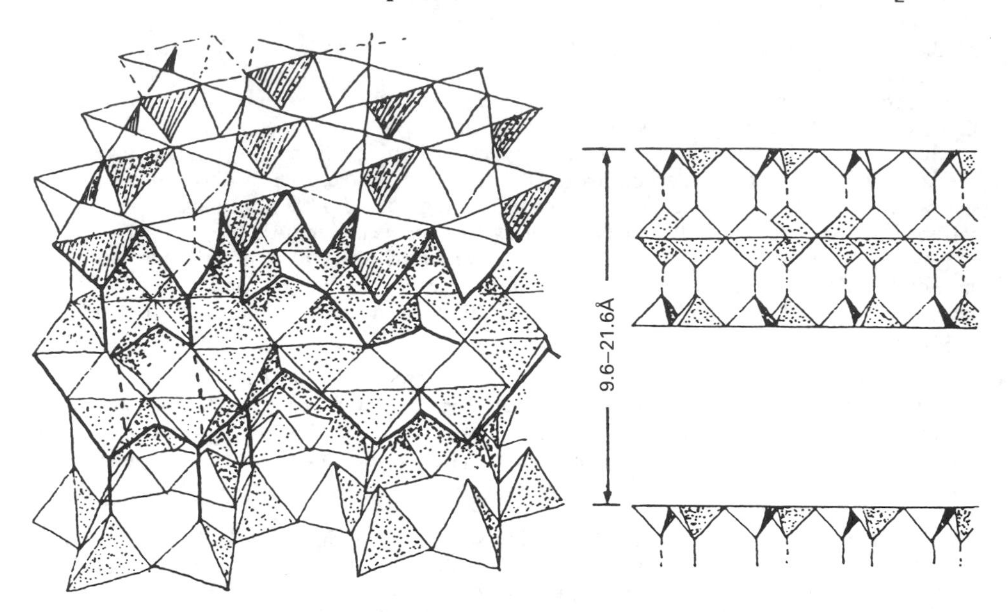

Fig. 14.6—Left. The three superpositioned layers—two (outer) layers consisting of tetrahedral units and a (centre) layer consisting of octahedral units form the layered unit (that is, a basic composite stratum) of montmorillonite. Right. There is an extremely variable distance between the corresponding heights of the successive three-layered units of montmorillonite, and, consequently, the possibility of the most widely varied molecules occupying the interstices. The likelihood of any molecules which may enter the interstices being retained between these composite layered units needs to be considered in relation to the electrical forces which are present there.

between the subunits of its laminae, and this vertical dimension may assume values between 9.6 and 21.6 Å.

The remarkable ability of the montmorillonite clays to engulf various molecular systems enables one to understand the pronounced catalytic properties exhibited by these clays in relation to the phenomenon of the transformation of components of animal and vegetable tissues into petroleum, as well as the outstanding catalytic action exhibited by montmorillonite structures in cracking processes.

The theoretical formula of the bis-hexa-tetrahedral and hexa-octahedral units of the triplet sheet is $Al_6(OH)_6Si_{12}O_{30}\cdot nH_2O$ which results from *n* molecules of water of crystallization being situated between one stratum and another. In practice, however, the mineral always differs to a lesser or greater extent from this composition owing to the substitutions which take place in its lattice.

Fig. 14.7 shows the hexagonal (bis-hexa-octahedral and hexa-tetrahedral) triplet of sheets to which the chemical formula given above, apart from the water, belongs.

The legend enables one to find the total number of negative charges in the hexagonal unit shown. There are 66, which is precisely the total number of positive charges carried by the 12 Si^{4+} and 6 Al^{3+} ions which the two outer hexa-tetrahedral and the central hexa-octahedral units contain respectively. Electroneutrality is thereby ensured.

The plasticity and absorption are therefore greater in montmorillonite than in kaolinite. The montmorillonite clays are formed from small units of indefinite shape which are dispersed in water into very thin platelets which are endowed with remarkable colloidal and plastic properties. Furthermore, the montmorillonites have characteristic physical variables depending on the amounts of water which they can accommodate, and, hence, on the extremely

● oxygen atoms shared (above and below, and hence there are 12) by the tetrahedra and octahedra. The total belonging to this unit is 12 $O^{=} \rightarrow 24^{-}$

ϕ hydroxyl groups which some of the octahedra of the unit in question share and are therefore to be considered as being half-owned by them: ½(4 OH^{-})$\rightarrow 2^{-}$

o hydroxyl groups which are shared between the octahedra of the hexa-octahedral unit. The total number belonging to the unit is therefore 4 $OH^{-} \rightarrow 4^{-}$

⊙ oxygen atoms, six of which belong to each of the outer hexa-tetrahedral units: 12 $O^{=} \rightarrow 24^{-}$

(oxygen atoms shared between the tetrahedra of the two hexa-tetrahedral units of the system under consideration and those of the twelve surrounding hexa-tetrahedral units. In this case a half share for the unit in question: ½(12 $O^{=}$)$\rightarrow 12^{-}$

In conclusion: $24^{-} + 2^{-} + 4^{-} + 24^{-} + 12^{-} = 66^{-}$

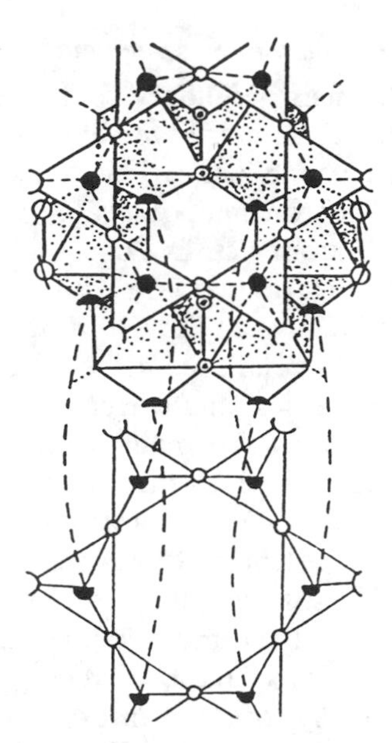

Fig. 14.7—Calculation and illustration of the negative charge distribution in a bis-hexa-tetrahedral and hexa-octahedral structural unit of montmorillonite. The calculation shows that the number and the arrangement of the oxygen atoms and the OH^{-} groups in the system required to ensure electroneutrality are attained with the 6 Al^{3+} cations and 12 Si^{4+} cations which are present in the system in question.

changeable thickness of their sheets. Finally, the montmorillonites make soils which contain them subject to crumbling and landslides by virtue of the fragility and friability which characterizes these clays.

The bentonites are special types of montmorillonite clays which are used in ceramic mixes and glazes to make them plastic. The plastic effects developed by the bentonites depend, as well as on their structure, on the fineness of the particles of which they are composed. Furthermore, more than 40% of these particles have a thickness of about 0.05 μm (5×10^{-6} cm) so that, when account is taken of the fact that the corresponding sheets have an average height of 15 Å (15×10^{-8} m), the bentonite laminae are formed on average from about thirty sheets.

The structure of illite. Fundamentally, illite has the same structure as montmorillonite, but, in this mineral, there is a greater tendency for the silicon in the tetrahedra to be replaced by aluminium. The positive charge defect which arises from such a substitution is balanced by potassium cations which are disseminated throughout the mass.

The layers of the platelets in illite are not expandable, and the thickness of a layer is almost fixed, with a value of about 10 Å.

The octahedrally coordinated cation which predominates in illite is aluminium, which, furthermore, can be replaced both by iron and magnesium.

Compared with the micas, illite has fewer atoms of silicon replaced by aluminium atoms in the tetrahedral layers, therefore it contains fewer interstitital potassium atoms. It also differs from the micas in that there is less regularity in the way in which the layers are superpositioned on one another, primarily because the other clay minerals are so intimately mixed with illite as to impede a regular correspondence between the successive layers and to prevent a precise formula being assigned to them.

The secondary components of clays

The main secondary components, which are both qualitatively and quantitatively variable in every type of clay, are the iron compounds, the silica sand, the calcium and magnesium carbonates, soluble salts, micas, and organic substances.

The iron compounds present in clays are, especially, pyrites, hydroxides, oxides, and carbonates. In a finely subdivided state, iron compounds lower the melting point of the clays, especially if it is ferrous compounds which are present. These compounds tend to give a yellow to brown coloration to ceramic mixes before firing, and a clear pink to a dark red coloration after firing.

The silica sand present in clays reduces their plasticity and their bonding power.

Calcium and magnesium carbonates (the first even more than the second) are impurities which damage the fired product by increasing the porosity of the mass while lowering its mechanical strength and refractory properties and reducing the range of temperatures over which the clays sinter.

The soluble salts present in clays are mainly sulphates, carbonates, and chlorides of the alkali and alkaline–earth metals, and of magnesium and vanadium. The frequency of the occurrence of such salts and the amounts which are present are generally low, and in exceptional circumstances they may jointly constitute up to 3% of the mass. The effect of these salts is generally that of causing efflorescences in the 'biscuit', and vanadium salts also produce yellowish-green spots there. Furthermore, sodium and potassium salts also act as fluxes by reducing the refractory properties of the clays and lowering the agglomeration point of the compressed and heated clays to slightly below the melting point (the sintering temperatures of the clays).

Micas which are generally not very abundant in clays, tend, on account of their lamellar structure and the smooth surface which they present, to reduce the cohesion of clays.

Organic impurities, consisting of humus or bitumens, impart a dark grey colour to the clay, they affect its plastic properties during working, and burn off during firing.

Silica and its role in ceramics

Types of silica. Silicon dioxide (silica) exists naturally in three main mineralogical types: rock, granular, and powdery.

Rock silica (quartz) is rarely found free from many other minerals. This reduces its non-plastic and refractory properties which rock silica in its own right should confer on ceramics. It is only after complex purification processes using flotation that this form of silica can be put to use in the production of ceramics.

The use of granular silica (silica sand) is more common. This is either initially pure or can be readily purified by simple grading and washing, and is extremely abundant.

Finally, powdery silica is not generally pure, nor is it readily purified, therefore it does not lend itself to use in the production of ceramics.

Properties of silica. Silica can exist in numerous crystalline forms, which are shown in Table 14.1.

Each form has a stable field of existence and unstable fields of existence. In the phase diagram shown in Fig. 14.8, the field of existence of the stable forms is bounded by a solid line, while the broken lines bound the range of existence of the unstable forms. It can be seen that, even

Table 14.1—Properties of the possible crystalline forms of silica

Names	*Form*	*Density*	*Field of existence*	
β-quartz	hexagonal	2.65	<575°C	stable
α-quartz	hexagonal	2.53	575–870°C	stable
			870–1250°C	unstable
			1250–1470°C	unstable
γ-tridymite	rhombic	2.27	<117°C	unstable
β-tridymite	hexagonal	2.26	117–163°C	unstable
α-tridymite	hexagonal	2.25	163–870°C	unstable
			870–1470°C	stable
			1470–1670°C	unstable
β-cristobalite	rhombic	2.33	<230°C	unstable
α-cristobalite	cubic	2.26	230–1470°C	unstable
			1470–1710°C	stable
Silica glass	amorphous	2.21	<1710°C	unstable

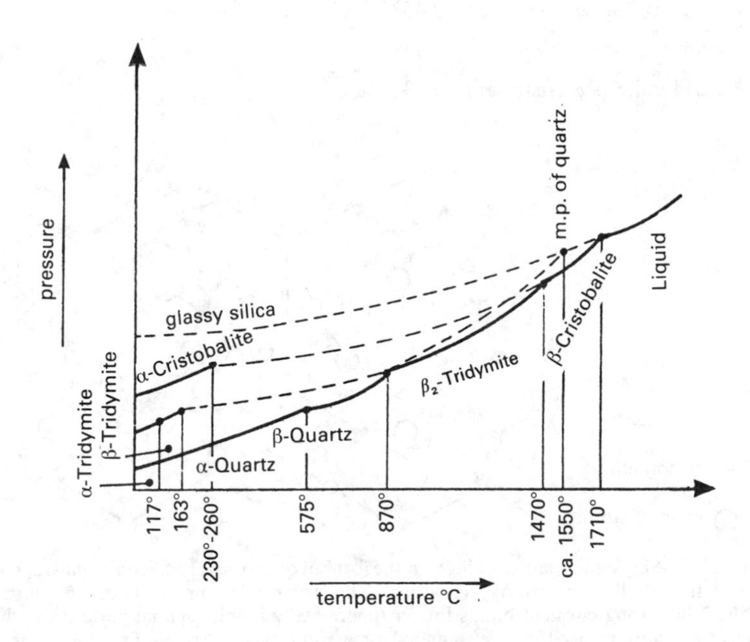

Fig. 14.8—Phase diagram of silica from which its rich polymorphism with a large number of metastable states can be seen.

under metastable conditions, each form of silica can exist at ambient temperatures. This fact is commonly met with in nature on account of the extremely variable conditions under which the deposits of the mineral are formed and the 'infinite' slowness with which the formations stabilize.

At 1710°C cristobalite melts, and amorphous silica (silica glass) is formed when the resulting melt is cooled fairly rapidly. This is an unstable phase with a very long existence since the rate at which it is transformed into a stable phase is very slow.

At the points where the lines of existence of two different phases meet, one of these phases may be transformed into the other. However, while the transformations: β-quartz $\rightleftharpoons$ α-quartz, γ-tridymite $\rightleftharpoons$ β-tridymite $\rightleftharpoons$ α-tridymite, and β-cristobalite $\rightleftharpoons$ α-cristobalite take place rapidly and with ease, this is not the case with the other transformations.

Moreover, it is possible to heat α-quartz to above 870°C without it transforming into α-tridymite, or to cool α-cristobalite to below 1470°C without it being transformed into α-tridymite, or to cool the latter to below 870°C without it transforming into α-quartz. By cooling α-cristobalite or α-tridymite it is, in fact, easier to obtain β-cristobalite and β-tridymite respectively and the γ-tridymite.

These facts, and the ease with which silica can yield numerous crystalline forms, are reflected in the arrangement of the atoms within SiO_2 crystals. In silica, as in silicates, the Si^{4+} cation is coordinated by four O^{2-} ions to form tetrahedra $(SiO_4)^{4-}$. In this very particular case, every oxygen ion is bonded to two silicon ions so that all the tetrahedra are bound to one another. Various arrangements can take place as a result of the manner of linking, and these give rise to the various crystalline forms. For example, long –Si–O–Si–O–Si–O– chains arranged in spirals can be recognized in quartz (Fig. 14.9).

In the transitions from one crystalline form to another, there are movements of the various tetrahedra leading to changes in their relative positions. Such movements are quite facile from some positions to others, but not among all the positions.

Every form of silica possesses its own density, as is shown in Table 14.1. On account of this, the transformation from one phase to another is accompanied by a change in volume which is often quite considerable. In transformations between the stable phases at the appropriate temperatures, the percentage variations in the linear dimensions are:

β-quartz	→ α-quartz	0.45%
α-quartz	→ α-tridymite	4.8%
α-tridymite	→ α-cristobalite	0.3%

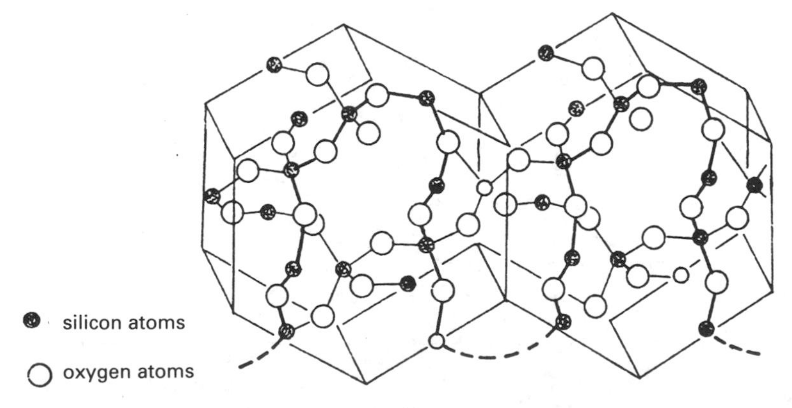

Fig. 14.9—The crystalline lattice of SiO_2 in the form of quartz. Every silicon atom is surrounded by four tetrahedrally arranged oxygen atoms. The structural components of the quartz lattice are joined by strong covalent bonds into a theoretically infinite spatial pattern in which the silicon atoms are arranged in a left-handed or a right-handed spiral. The optical activity of quartz, which is possible only in the solid state, is dependent on this (the difference in the optical activity is derived from the overall molecular asymmetry).

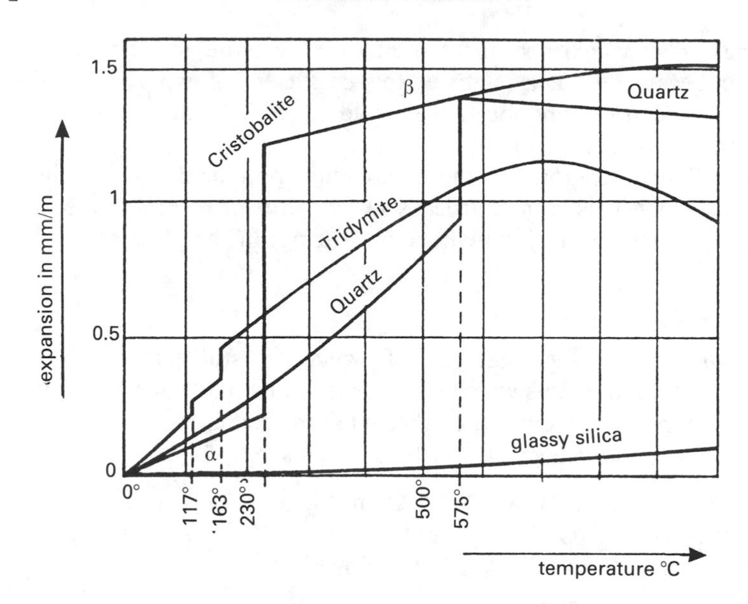

Fig. 14.10—Thermal expansion curves for the various forms of silica. It can be seen that the most dangerous of these is the expansion which takes place during the transformation of the α-form of cristobalite into the β-form.

Finally, each form of silica has its own coefficient of expansion, and there are therefore some sharp jumps in the expansion (Fig. 14.10).

This is of concern in all of the technologies in which silica is used because the raw material which is used (sand, quartzite, etc.) may be actually constituted of all the crystalline forms of silica, both on account of the processes to which such raw materials are subjected during and after their formation and the fact, which has already been mentioned, that conditions for the metastable existence of various types of silica may endure at all temperatures.

Consequently, it is good practice to discuss the type of silica (quartz, tridymite, etc.) which has to be used in the compounding of the manufactures in relation to the technological properties which the latter are required to possess.

The function of silica in a ceramic. Silica is present in ceramic mixes at variable but high percentages. Its function can be summarized in the following points:

- to make the products clear;
- to lower their plasticity during their compounding by behaving as a thinning agent;
- to reduce their contraction during the firing of the products, since they tend to expand upon heating;
- to increase the coefficient of expansion of porous mass products but to decrease it in vitrifying products. This is because crystalline silica has a coefficient of expansion which is greater than that of the clay material, while vitreous silica has a lower coefficient of expansion;
- to form a kind of framework in non-vitrified products around which the argillaceous substances and fluxes are arranged, while, in vitrified products, a part of the silica combines with the fluxes to form various silicates such as mullite ($3Al_2O_3 \cdot 2SiO_2$) and sillimanite ($Al_2O_3 \cdot SiO_2$).

The form of silica which is the most suitable for the manufacture of ceramic products is tridymite, because it has a relatively low coefficient of expansion and, especially, because they are no pronounced and sudden changes in this coefficient as the temperature is varied (Fig. 14.10).

The most dangerous form is the cristobalite modification. The sharp jump in dilatation to which it is subjected leads to fractures in ceramic objects and brings about a network of extremely fine fissures (crazing) in the glasses.

The feldspars in ceramics

The classical fluxes in the ceramic field, which are still very much used, are the feldspars. These are minerals which have the task of stimulating the formation of the vitreous phase in the ceramic materials containing them.

Three main types of feldspars may be distinguished:

- potassium feldspar: $K_2O \cdot Al_2O_3 \cdot 6SiO_2$ in the form of orthoclase or microcline;
- sodium feldspar: $Na_2O \cdot Al_2O_3 \cdot 6SiO_2$ (albite);
- calcium feldspar: $CaO \cdot Al_2O_3 \cdot 2SiO_2$ (anorthite).

Of all the feldspars, orthoclase is the most important both in absolute terms and particularly in the ceramic field even if, in practice, natural materials are always used which are mixtures of all the feldspars. For the most part, these are accompanied by quartz and by iron and magnesium compounds.

The production of ceramics

The basic phases leading to the production of ceramics are the pretreatment of the ingredients, the preparation of the mixes, the shaping and drying of the shaped objects, the firing of the manufactures and the possible finishing of the fired ceramics.

Pretreatment of the ingredients

Depending on their origin, the ingredients for the fabrication of ceramics may be subjected to various preliminary treatments. These consist of some combination of selection, washing, sedimentations, flotations, electro-osmosis, and drying.

At one time, the process of hibernation and/or estivation was a preliminary treatment of great value. This consisted of excavating the raw materials and, especially, clays at suitable times during the year, and then leaving them to be subjected to the effects of freezing during the winter and to the heat of the sun and rather high temperatures during the summer. Nowadays, the refined mechanical treatments which are possible have supplanted and improved upon the effects of natural climatic agents.

The processes concerned with the crushing and the grinding of the ingredients have always been of great importance in the ceramic field. Both of these processes are adapted to the nature (hardness, compression and tensile strength, etc.) and to the physical condition (wet, moist, etc.) of the materials being treated. Mullers, ball mills and related types, 'colloidal' mills, and 'fluid energy' mills are the most commonly used.

Colloidal mills treat aqueous suspensions of the ingredients for ceramics (slips) between superimposed alumina mill-stones which can be set at various distances apart. These produce particles in a size range from 20–50 μm and are also capable of yielding the finest systems for large display windows.

Fluid energy mills exploit the smashing action brought about when particles collide which are carried in high-velocity, counterflowing currents of compressed air or steam. When treated in this way, materials which are very hard and abrasive do not create any problems of erosion of the working parts of the milling apparatus.

The preparation of mixes

The pretreatments which have just been mentioned are carried out in every case, no matter what type of product is to be produced and regardless of whether the materials are plastic, that is, especially, clays, or hard, that is, silica sands, feldspars, and possibly carbonates and other secondary ingredients. On the other hand specific treatments are subsequently carried out which are dependent on the ceramic material which is to be produced.

The ingredients which have been ground and, possibly, reground, can either be compounded with other materials or first be classified and put into digestor silos before compounding, or be subjected to a series of preliminary treatments before compounding. The preliminary treatments may consist of sieving, mashing in water, and magnetic iron removal processes which, above all, correct the various colorations of the products.

Also, the procedures that follow blending are different:

- when it is desired to use the 'semi-dry' process, a simple moistening is carried out during the mixing procedure before the mix is sent to mature;
- when working in the 'stiff-plastic state', the mix is thoroughly wetted, mixed, and kneaded before it is sent to the maturation bins;
- when using a 'wet mud' procedure, which is typical in the manufacture of fine objects, where the most complex and highly refined preliminary treatments (such as the removal of iron compounds) are applied, it is necessary to mix, agitate, filter by using filter presses, and, after a period in the maturing bins, to pass it on to the degassing and mechanical mixing machines: that is, to multistage devices operating under a reduced pressure from which objects which are already moulded may also emerge.

In the contemporary terminology used in the ceramic field, the 'cake' unloaded from the filter presses is known as 'biscuit', and their concrete digestor silos are commonly known as 'vats'.

The shaping of ceramic objects

To give the desired form to ceramic objects (moulding), operations are carried out which take their name from the physicomechanical procedure upon which they are based.

Profile moulding. This is a method of moulding which is exploited in shaping objects which are solids of rotation (plates, cups, etc.). It is carried out by putting a pattern holder (chuck) onto the upper part of the vertical axle of a lathe (Fig. 14.11). This chuck holds the plaster pattern into or onto which the clay 'bat' to be moulded is put. As the pattern rotates together with the clay bat, a profiled piece of metal (a template) is lowered onto it and simultaneously shapes both the interior and exterior parts of the objects.

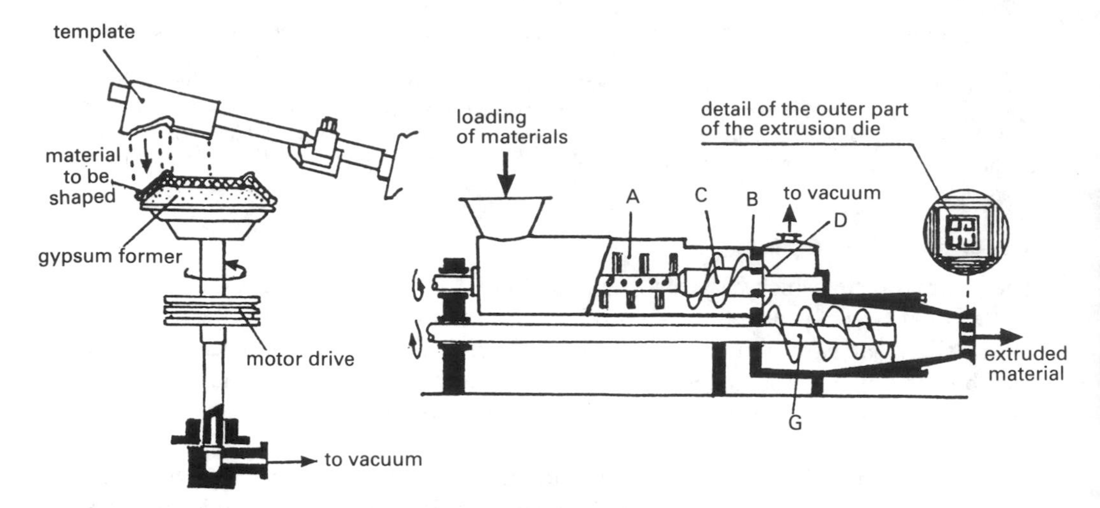

Fig. 14.11—Left: Lathe for the profile moulding of a ceramic object. Right: Mixing–degassing–extrusion machine for the final extrusion of ceramic objects: A is the mixing chamber, C is a screw conveyer, B is a perforated slab, D are knives, and G is a screw conveyer extrusion machine.

Extrusion moulding. Extrusion is used in the production of objects with a constant cross-section. It is carried out by pressing the ceramic mix through a die with a certain profile in machines known as extrusion presses (Fig. 14.11, right) which, in their simplest form, consist of a means of propulsion, a die, and a cutter.

The propulsion device compresses the mix towards the die which shapes it to the desired form, while the cutter chops the moulded section into pieces of the required length.

In such extrusion machines, two types of propulsion may be distinguished: propulsion by means of a helical screw, and hydraulic propulsion. The first type of extruders are used especially in the manufacture of bricks and refractories, and are often equipped with a degassing system. Machines with hydraulic propulsion are employed in the fabrication of technical ceramics and non-argillaceous products in general which, owing to their lack of plasticity, require (apart from the addition of organic binders to the mix) the application of high pressures in their moulding.

Moulding by simple pressure. This technique uses the mechanical compression of a wet, semi-dry, or dry mix in suitable moulds. There are a number of advantages of this technique: the high hourly production rate, cost savings, the small contraction upon drying, and the higher density of the products.

Refractories, electrical insulators, special technical ceramics, and, to a limited extent, bricks, are produced by this method (Fig. 14.12, left).

The presses which are employed attain pressures of 2000 ton/cm^2 and may be screw presses, eccentric presses, friction presses, revolver presses, or toggle presses. The operation is automatic, and the force is transmitted to the material either directly, by means of compressed air, or hydraulically.

Moulding by isostatic pressure. To impart uniform compactness and greater strength to the ceramics, and to ensure that there are no blowholes, the material to be pressed, which has been put into a rubber mould (Fig. 14.12, right) which forms an integral part of the cover or is independent of it, is pressed by pumping gas into the surroundings for a period which varies from a few seconds to several hours. Pressures of the order of 1400 kg/cm^2 are applied in order to mould some ceramic materials using this isostatic technique.

An isostatic pressure system enables one to manufacture objects with shapes which cannot be achieved with other moulding systems.

Moulding by pouring. Shapes which are not solids of revolution and are rather complex are

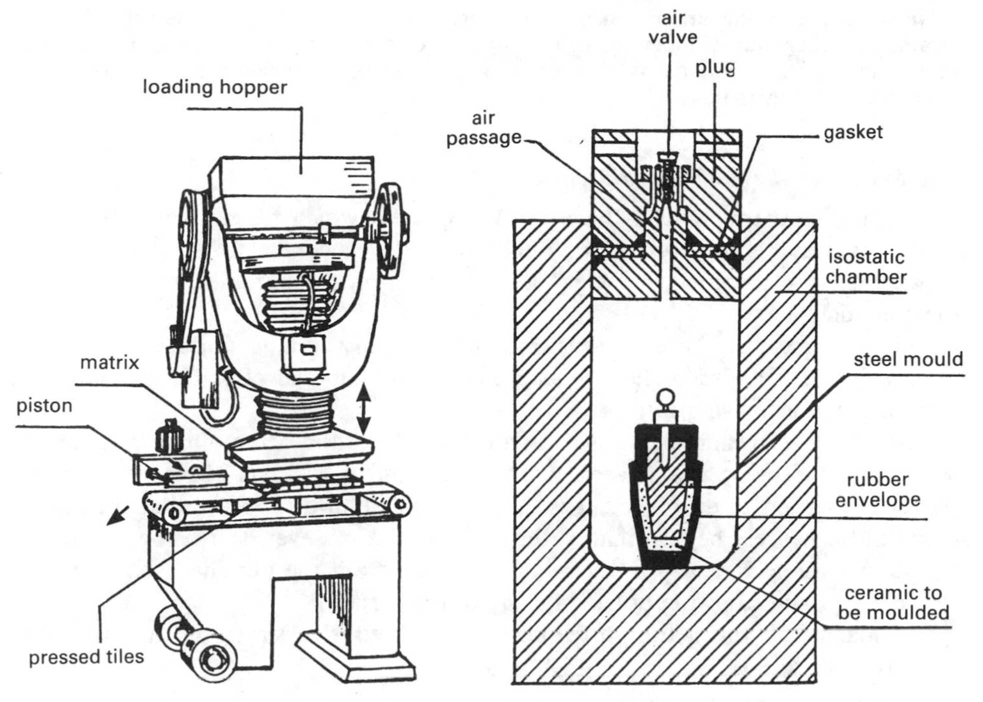

Fig. 14.12—Left: Direct high pressure (at hundreds of atmospheres) forming of mosaic lining tiles for the facings of buildings, the walls of factories, swimming pools, etc. Right: Structure of the equipment required to carry out 'isostatic pressure' moulding in which uniform compactness of the moulded products is achieved. This also offers the possibility of moulding objects with the most diverse shapes by using a wide range of different moulds.

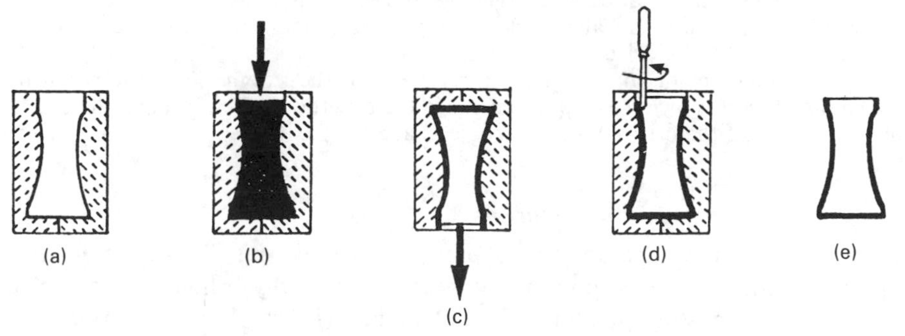

Fig. 14.13—Stages in the formation of an object from slips which are poured into plaster moulds. The technique lends itself readily to the moulding of non-plastic materials. Only the principle of the method is shown here because from time to time the moulds have very complicated shapes. The following stages are shown: (a) the mould, (b) its filling with slip, (c) the pouring off of the excess slurry, (d) the finishing of the moulded object, and (e) the shaped object before drying.

produced from slips by pouring them into suitable hollow plaster moulds (Fig. 14.13). These are filled with the suspension and then left until such time that the desired thickness has been achieved by the preferential absorption of water close to the porous plaster walls. When the desired thickness is attained the excess slip is poured out from the mould and the object is left to consolidate in contact with the plaster, while suitable finishing operations are carried out on it. Once it can be handled, the object is removed from the mould and sent to be dried.

Non-plastic materials are moulded by pouring. This is especially so in the field of special ceramics. Among other things, objects made out of the oxides of aluminium, beryllium, zirconium, thorium, uranium, and titanium as well as fused silica, spinels, nitrides, and carbides, are moulded in this manner.

The drying of the moulded objects

The moulded ceramic masses have a water content which ranges from about 5 to 30%. This may be absorbed water, adhesion water, or contraction water according to whether it is absorbed in the pores, condensed on the particles, or liberated by the colloidal particles.

This water must be almost completely removed from a ceramic object in a two-stage process before it is fired. The two stages entail the removal of water at a constant rate and at a decreasing rate.

Constant rate drying takes place initially with the evaporation of the water on the free surfaces. During this phase a continuous film of water is observed on a free transverse section of an object. It is only after this film disappears that the second phase of the process begins, that is, the drying out at an ever decreasing rate owing to the increasing difficulty in the water emerging from the moulded shape, and in the steam produced in the pores diffusing to the surface.

The maximum contraction in the body being dried is observed during the first of these two phases.

The driers may operate either intermittently or in a continuous manner, but the second solution is preferred nowadays on the grounds of cost.

Generally, the material to be dried is first loaded onto small trucks which are then transported along a channel which is traversed by a countercurrent of hot air in order to recover some of the heat from the firing kilns. The air, at a temperature of approximately 60°C, enters into the section from which the dried objects emerge to pass through the channel where it becomes progressively saturated with water until it emerges almost completely saturated from the end where the material to be dried enters.

Nowadays, however, drying with infrared rays is used in the ceramic field with ever-increasing frequency and success owing to improvements in these methods. This drying technique is used both in the case of the original moulded objects and for the drying of glazes and enamels.

The firing of ceramic manufactures

The firing of the moulded and dried ceramics generally has five stages: loading the objects into the firing environment, preheating, gradually bringing them up to the maturation temperature, cooling, and unloading. The whole of this cycle may last from a few hours to several days.

The ceramic products are fired in kilns which may be of many different types, as can be seen from Table 14.2.

The following kilns will now be described: a combustion kiln with direct intermittent heating (a down-draught kiln), a combustion furnace with direct continuous heating (a Hoffmann kiln), and a combustion kiln with indirect continuous heating (a tunnel kiln). Electric kilns will then be mentioned.

Down-draught kilns. These kilns are given this name because the flames originating from the lateral hearths or burners rise to the dome of the maximum firing chamber where their motion is inverted and the draught continues downwards to pass through apertures in the base of the chamber, after which it rises again to the upper parts of the kiln.

Table 14.2—The classificaton of kilns for the firing of ceramics

Combustion		*Electric*	
Type of environment heating	*Operation*	*Type of operation*	*Firing*
Direct	*intermittent* up-draught or or down-draught	intermittent	*chamber*
	continuous fixed fire and moving wares or moving flame and fixed *wares*	semi-intermittent	moveable hearth *chamber*
Indirect	*intermittent*	continuous	*tunnel*: on trucks or plates or grids
	continuous		*annular*: revolving hearth

In all, there are three sections (Fig. 14.14), a lower chamber (laboratory) in which the maximum firing temperature is attained, an intermediate first firing chamber (small furnace), and an upper chamber which is suitable for carrying out simpler firing procedures such as those concerned with glazes. While some wares are being totally fired in the laboratory, others are subjected to a first firing in the small furnace. At the end of such a stage, the kiln is allowed to cool, the fired ceramics are unloaded, the prefired wares are moved into the lower chamber, and new wares are loaded into the small furnace.

It can be readily understood how the intermittent steps in arriving at the correct thermal conditions, the subsequent cooling, the loading and unloading, and the changing of wares from one compartment to another lead to a loss of time, to a waste of fuel, and the need to use skilled labour. Down-draught kilns are therefore no longer used even for ceramics which are fired at high temperatures while protected, as is done with a large part of the porcelains.

Hoffmann kilns. These are ovens in which the wares remain stationary while the flames move. They are used specifically for the firing of bricks. In these kilns, there is essentially a single firing chamber which may be circular, ellipsoidal, or even zig-zag in form. The gallery, the upper part of which forms a vault, is constructed from double-walled cavity brickwork, and the cavity is filled with an insulating material. Such ovens have a width of 2–4 m, a height of 2–3 m, and a length of up to 100 m. On the inner walls appropriately positioned reliefs indicate the division of the gallery into compartments which number from 12 to 24.

There is an external communications part, which is sealed by false brickwork, corresponding to each compartment, and, at one diagonal extremity of these compartments, there is a vent tube which is connected to the fume 'collector'. The fumes are subsequently sent to the chimney. Communication between the fume collector and the compartments can be interrupted by means of a suitable gate valve, while cast iron covers off some of the apertures (from 9 to 12) arranged on the canopy of each compartment, through which the fuel is introduced. The latter consists of cheap coals and coal agglomerates.

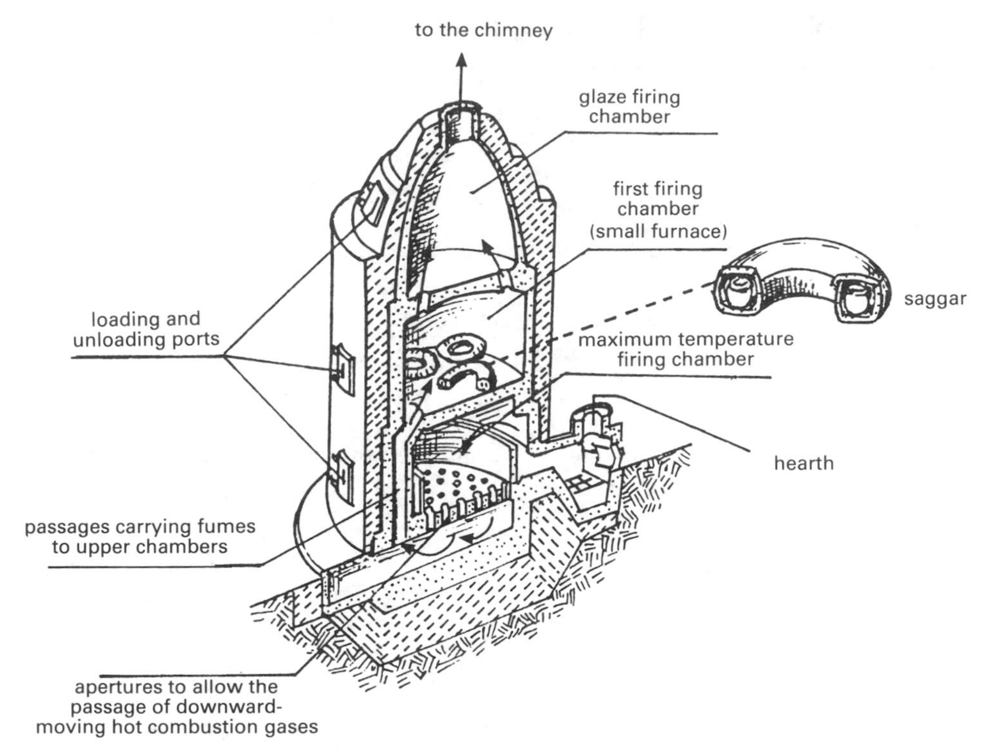

Fig. 14.14—Section of a down-draught kiln complete with accessories (the saggars which contain the products to be fired in the small furnace).

The ceramics to be fired are loaded into the compartments through the corresponding ports, taking care to leave sufficient space between individual pieces for the combustion air to circulate amongst them.

The mode of operation of a Hoffmann kiln is concisely explained in the caption to Fig. 14.15.

The atmosphere in Hoffmann kilns is generally maintained in a state such that it is neither oxidizing nor reducing, by appropriate regulation of the air flow so as to produce either an excess of oxygen or of the fuel gases (especially, CO and CH_4).

However, when the ceramics being fired contain appreciable quantities of Fe_2O_3, which tend to impart brown colorations to the fired products, a reducing atmosphere is developed towards the end of the firing in order to obtain a lighter coloured product. This reducing atmosphere converts Fe_2O_3 into FeO, which has a lighter colour.

These considerations are valid in the case of all kilns in which the firing is produced by direct contact with the combustion gas.

Tunnel kilns. These are the most commonly used kilns for the firing of ceramics on account of their high thermal efficiency, low labour costs, high production rates, and the ease with which the process is carried out.

A tunnel kiln is a straight gallery which is operationally subdivided into three zones: a preheating zone, a firing zone, and a cooling zone (Fig. 14.16).

The gases move in an opposite direction to the trolleys which pass along the kiln. The air enters into the section of the tunnel from which the trolleys emerge. It is forced along the tunnel by fans and is preheated by contact with the hot material originating from the firing zone. Suitable operation of the valves enables one to increase or decrease the cooling action of the air. In every case, this is preheated, and it is aspirated into the vicinity of the burners in order to be sent, one part to be burnt while the remainder is used to dry out freshly moulded wares.

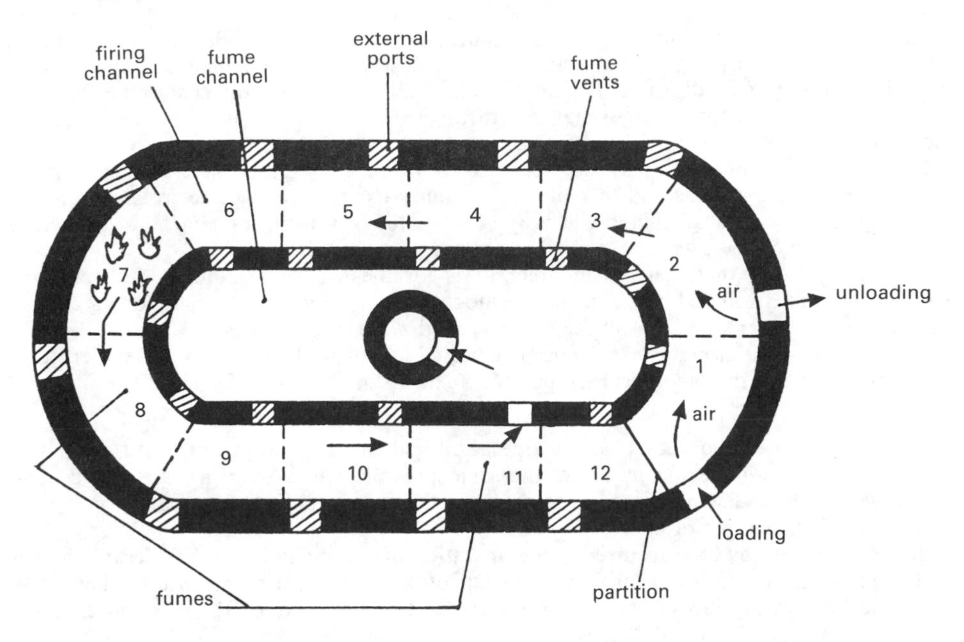

Fig. 14.15—Hoffmann kiln with all the ports sealed other than 1 and 2, and with only the vent channel from compartment 11 open. In 7, combustion and firing are taking place, while, in compartments 3, 4, 5, and 6, fired materials is in the progressive stages of cooling from 6 to 3. On the other hand, there is (green) material in 8, 9, 10, 11, and 12 which is being heated to an ever decreasing extent. After the ceramic wares in 7 have been fired, the fire next moves to compartment 8, the port leading to chamber 1 is sealed, the diaphragm between chambers 1 and 2 is raised, the old diaphragm removed, the vent from 11 is closed, and the vent from 12 is opened. In this way, chamber 2 now becomes the compartment to be loaded, while 3, after unsealing the port, becomes the compartment to be unloaded. This cycle continues in this way in an uninterrupted manner.

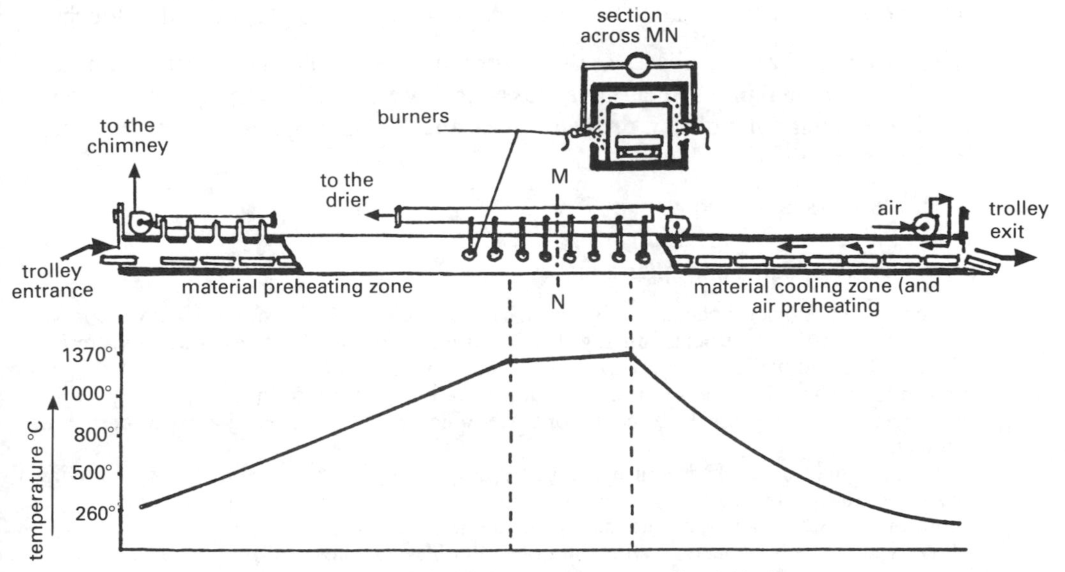

Fig. 14.16—Tunnel kiln and the temperature profile within it.

The gas stream which is directed from the central zone towards the other end of the tunnel consists of the combustion products which are cooled when they encounter the unfired wares travelling in the opposite direction. At the end of the kiln, before being directed to the chimney, the combustion gases enter heat exchangers through which other air is subsequently passed and used to dry out the ceramic wares before firing.

The length of a tunnel kiln may vary from 20 to 120 m. Nowadays, there is a tendency to construct rather short kilns and to replace the binary fixed and moveable trolley transport system by continuous metal chains which, by moving slowly, transport the metal or refractory platforms on which the ceramics to be fired are placed.

It is finally noted that, here as in other kilns for the firing of ceramics, the operating temperatures are controlled by means of thermocouples or optical pyrometers, or, in a more craftsmanlike manner, by means of the classical method of 'Seger cones'.

To ensure that the firing of a ceramic product turns out well, the recorded temperatures must correspond to the theoretical temperatures shown in Fig. 14.16 for the different zones of the kiln.

Electric kilns. Lower plant costs, ease of operation and control, savings in labour costs, and the purity and non-aggressive nature of the atmosphere in which the wares to be fired find themselves, are but a few of the advantages which electric ovens offer, either partly or in their entirety, over other types of kiln.

However, the energy costs incurred in the operation of electric kilns are considerably higher in absolute terms. It suffices to consider unit costs of the same order of magnitude when 1 kW of electrical energy produces 864 kcal, 1 kg of fuel oil produces about 10 000 kcal, and 1 kg of coal develops about 7500 kcal.

The electric kilns used in the production of ceramics are of the resistance type and are fitted with resistances made of different materials, depending on the temperature which is to be attained. Nickel–chromium alloys are used up to temperatures of about 1050°C, then 'cantal' alloys (iron–chromium–aluminium and cobalt) and, finally, carborundum rods.

The finishing of fired ceramic wares

The finishing of many fired ceramic products, such as majolicas and porcelains, includes modifying the state of their surfaces by making them, from time to time, impermeable, glossy, coloured, figured, and so on. Finishing operations are carried out on ceramics with this aim. The principal operations are glazing and colouring.

The glazing of ceramics. Glazes are thin vitreous layers which cover ceramic bodies with the aim of making their surfaces glossy and impermeable. They are classified on the basis of their composition or their appearance or their optical properties. There are, for example:

- lead, leadless, and zinc glazes;
- porcelain, glossy, and satin glazes;
- colourless, opaque, and quartz glazes.

The production and application of glazes when the required composition is known consists of weighing out the ingredients, mixing them, and wet grinding of the resulting mixture in a ball mill. The slip which is obtained, possible after the removal of iron, is passed through a vibrating sieve and then allowed to mature for a few days. This is followed by its application onto green and partly dried objects or onto ceramic bodies which have already been fired (biscuit).

The glaze may be applied with a brush, by dipping, by curtain coating, or by spraying. Nowadays, the last two methods are most common by far, as they permit the maximum degree of automation while saving on the amount of the product used.

Glazes may exhibit defects, such as pinholes, bubbles, crazing, and peeling. The causes of these are excessive firing, errors in the blending of the glaze, the presence of oil or grease on the surface being glazed, and, above all, differences between the coefficients of thermal expansion of the biscuit and the glaze, which today are avoided by modifying the formulation of the glazes and both the firing temperatures and times.

The coloration of pigments. The colours used in ceramics are classified as: pigments, underglaze colour, overglaze colours, third firing colours, reverberating colours, and lustrous colours.

Pigments may consist of oxide minerals of chromium, iron, cobalt, copper, manganese, nickel, titanium, and uranium, coloured spinels ($CoAl_2O_4$ intense blue, $CuAl_2O_4$ apple green, $MnAl_2O_4$ reddish brown, $NiAl_2O_4$ sky blue, etc.), or certain silicates, phosphates, fluorides, antimonates, and uranates. The pigments must be insoluble in the glazes, of a colour of the highest possible intensity, and of a controlled fineness to ensure colour homogeneity. Apart from being added to glazes, they are applied onto biscuit, onto an unfired glaze, or onto an already glazed lining, by various techniques. In preparing them, either in a dry or wet manner, care must be taken especially in the grinding and the firing time and temperature.

Under-glaze colours, that is, mixtures of the pigment which colour the mass and a compatible glaze which fixes the colour onto the support, are prepared by intimate wet mixing, drying, and a final pulverization. The commonest glazes are feldspars, feldspars, and kaolinite or special mixtures.

Overglaze colours are analogous to underglaze colours, but contain more fluxes to make them brighter in colour.

Third-firing colours, which are so called because they require even a third firing of the ceramic object, are made of pigments and fluxes, and are fired at temperatures below 750°C so that they may display colour tones which are more lively than underglaze colours with respect to which they are less stable (thermally labile).

Reverberating colours are those which act so that, after firing, the object appears transparent but changes its coloration depending on the angle from which it is viewed. These are mixtures of salts and oxides which are incorporated with vinegar and fired in a reducing atmosphere at 650°C.

Lustrous colours are third-firing colours initially constituted from metallic resinates which, when reduced during firing, finally take on the appearance of opaque but iridescent metallic films on bodies which have previously been sprinkled with them.

Traditional types of ceramic products

It now remains to describe the various types of both porous and compact ceramic products such as bricks, refractories, minor ceramics (terracottas, earthenware, majolica, or faience), stoneware, and porcelain, while emphasizing their specific nature and some details of the productive processes and their applications.

Bricks

According to accepted standards, bricks ard constructional materials formed from clay containing variable amounts of sand, ferric oxide, and calcium carbonate (and therefore a clay of the illitic type) which has been cleansed, milled, mixed, and formed into pieces of a suitable size and shape which are fired after drying.

Since the firing of bricks takes place at relatively low temperatures, the resulting products are always porous to a greater or lesser degree.

Bricks are classified as:

- solid brick materials which includes common bricks, bricks, paving tiles, etc.;
- perforated brick materials which includes brick with various numbers of perforations, and large building blocks;
- brick materials for covering purposes (pantiles, roofing tiles, and flat tiles).

The second and third classes are illustrated in Fig. 14.17.

On the basis of the degree of firing, an operation which is carried out either in Hoffmann kilns or tunnel kilns, bricks are subdivided into poorly fired bricks (slightly fired, clear, and with low mechanical strength), strongly fired bricks (well fired, bright red, and with a high mechanical strength), and brittle bricks (over fired, or low porosity and of a dark coloration).

Refractories

According to the accepted standards, refractories are constructional materials which are suitable for withstanding high temperatures without melting, which also have sufficiently adequate mechanical properties at elevated temperature to withstand the mechanical forces to which they may be subjected in the environment in which they are employed.

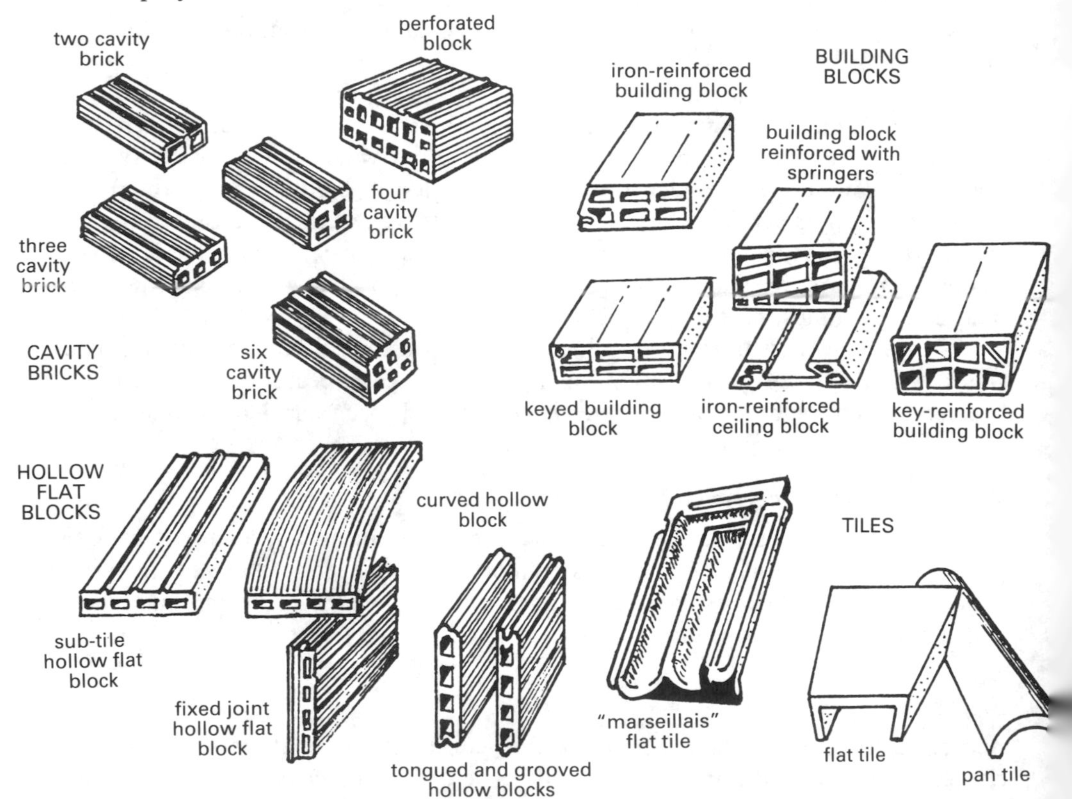

Fig. 14.17—An assortment of perforated and covering bricks showing the classes, types, and uses in accordance with the official Italian standards which control the type of such bricks on the basis of the uses to which they are to be put in ceilings, walls, and partition walls or as different types of roofing.

The raw materials for the production of refractories vary fundamentally, depending on whether they are acid, neutral, or basic refractories, and with respect to the sub-class to which they belong.

Quantitatively, acid refractories are the most used. They are based on silica, clays, particular aluminium silicates, clay-alumina mixtures, and more or less pure alumina. The neutral refractories are mainly graphite or chromite, and, in special cases, mixtures of various silicates with special neutralizing agents. Basic refractories have only oxides (or carbonates) as the raw materials, or they may also contain chromite.

Acid refractories. It is only those refractories which are based on silica which are absolutely acid, because, if they are made from clays or from particular silicates and/or alumina, they may also behave as basic refractories when they come into contact with silica, for example. Typical representatives of acid refractories are dinas, argillaceous refractories, mullite refractories, and the alumina refractories which are clearly characterized, in the order shown, by the increasing percentage of alumina which they contain.

Dinas refractories are prepared from quartzites containing more than 93% of silica which, during firing, must be transformed into α-cristobalite, the form of the lowest density which is stable at the temperatures at which it will later operate (Table 14.1). Small additions of lime water during its preparation (2–3% maximum) have a binding effect on the granules. The firing lasts for several days with a long period of maturation at between 1450 and 1500°C to allow the abovementioned transition to take place. They are mainly used as lining materials for the metallurgical and ceramic industries.

Clay (silica-alumina) *refractories* are the oldest and the most used. They are prepared from plastic (kaolinite–halloysite) clays with a high melting point, suitably thinned with chamotte (the best) which is also produced from high-melting-point clays by grinding, agglomeration, a sintering firing, and a purifying grading. The maximum stagnation temperature during the firing is between 1350 and 1380°C, and the final properties vary as a function of this stagnation period and the clay–chamotte ratio.

Mullite refractories are properly derived from sillimanate ($SiO_2 \cdot Al_2O_3$) and its polymorphic modifications, andalusite, and cyanite. Consequently, the alumina content, which does not exceed 46% in clay refractories, increases to 63% if they are made from pure sillimanite, and may be increased up to 70% by preparing them from argillaceous refractories and bauxites or from such clays and alumina. These refractories are characterized by the presence of mullite ($3Al_2O_3 \cdot 2SiO_2$) which is formed together with silica in the cristobalite form at temperature above 1500°C, and they are more resistant to thermal shock than refractories based purely on clay products.

Finally, *alumina refractories* have an alumina content of greater than 70%, and they usually contain between 80% and 90% of it. In this case, they are prepared from selected (white) bauxites, while, when they are prepared from natural corundums containing 94–98% of Al_2O_3, the alumina content exceeds 90% even when some use is made of bauxites and argillaceous refractories. Owing to the cost of the raw material, the costs of grinding it and of firing it at temperatures of 1700–1800°C, the overall costs of production of these refractories is extremely high. Nevertheless, since they are extremely abrasion resistant and perfectly refractory even when taken up to temperatures of around 2000°C, they are deemed to be suitable for some industrial applications.

Neutral refractories. Refractories produced with graphite or with amorphous carbon which graphitizes once it is put into operation, are naturally neutral. However, refractories based on chromite in which the weakly basic FeO is neutralized by the amphoteric oxide Cr_2O_3, are also neutral. Steatitic and cordieritic refractories, which we shall consider under the heading of ceramics for the new technologies (page 582) are also neutral but very expensive refractories. Therefore, when they are put into service, it is usually as high-performance electrical insulators. The graphitic refractories are prepared by the mixing of graphite or lamp black and variable amounts of high-quality argillaceous refractories. The mix is shaped and fired at temperatures of the order of those at which the material will subsequently operate (>2500°C), taking account of the fact that graphite does not have a melting point from a practical point of view but sublimes at above 4300°C. Metallurgy is the industrial sector which makes the most use of this graphitic refractory material, mainly in the form of crucibles.

Chromite refractories are produced from chromite, $FeCr_2O_4$, by taking the natural mineral, grinding it, and then agglomerating it with a little argillaceous refractory. In every case, the shapes obtained are 'bricks' with a first firing at 1000–1200°C, and then, at the second stage, the bricks are fully fired at above 1600°C. Such chromites are refractories which withstand temperatures of up to 1900–2000°C with retention of their mechanical properties, and are especially used in furnaces to separate acid refractories from basic refractories, to which they are equally indifferent.

Basic refractories. The iron and steel making industry, especially in the production of steel, are among the major concerns interested in the use of this class of refractories which all have MgO as the characteristic component.

They are prepared from magnesite, $MgCO_3$, or microcrystalline magnesium hydroxide produced by the conversion of the magnesium chloride from sea water into it, or from dolomite, $CaMg(CO_3)_2$, forsterite Mg_2SiO_4, and/or olivine $(Fe, Mg)_2SiO_4$, or, finally, from a mixture of magnesite, $MgCO_3$, and chromite $FeO \cdot Cr_2O_3$.

The magnesite, magnesium hydroxide, and dolomite are calcined and then 'dead-burned' under high pressures which favour their sintering. This is, in part, an effect of the action of the vitreous component, derived from the low melting impurities contained in the raw materials, which is present there. This component does, however, lower their operational refractory qualities, but it is a defect which can be remedied by preparing them from the purest raw materials, using a special and costly technique (electrofusion).

Forsterite and chromite produce refractories which are more or less basic when they are sintered under pressure during a firing which is preceded by calcination.

Magnesite and chromite lead to magnesia–chromite or chromite–magnesia refractories according to the component which is present in an amount greater than 50%. A conventional preparation method is employed: precalcination, mixing, and a sintering firing to obtain refractories which, on account of the chromite component, are more resistant to thermal shock than simple magnesia refractories.

Properties governing the use of refractories. The technological characteristics on the basis of which the suitability of a refractory for a certain use can be adjudged are:

- the post-contraction and additional expansion;
- resistance to thermal fracture;
- resistance to chemical attack;
- thermal conductivity, heat capacity, and thermal diffusivity;
- electrical conductivity;
- specific weight and compression strength.

Post-contraction and additional expansion. If the phases making up the structure of a refractory tend to vary in composition, to recrystallize, or continue to sinter, the material changes in volume to a greater or lesser extent. For instance, refractories rich in alumina undergo high temperature contraction due to the transformation of α-alumina ($d = 3.46\ g \cdot cm^{-3}$) into β-alumina ($d = 4\ g \cdot cm^{-3}$). On the other hand, dinas expand if they contain residual quartz ($d = 2.60\ g \cdot cm^{-3}$) which is transformed into cristobalite ($d = 2.3\ g \cdot cm^{-3}$).

Contraction causes cracking and breaking up, while expansion causes buckling and warping. These can both be avoided by firing the refractories thoroughly over a long period during the production phase.

Thermal fracture resistance. Fractures due to intermittent variations in the temperature are referred to as thermal fractures. This defect usually manifests itself in the form of fissures parallel to the heated surface which become crevices with the passage of time. Refractories which are exposed to the forces causing thermal fracture, especially during discontinuous operations, must be particularly resistant and reluctant to form fissures.

Resistance to chemical attack. There are numerous chemical agents (acids, oxidants, etc.) which are aggressive towards operating refractories, and this, in addition, may be aggravated by the severe thermal conditions. The chemical attack of a refractory not only destroys it but also contaminates the material being worked. To ensure that a refractory has chemical resistance, it is necessary to study its behaviour in relation to the attacking reagents at the working temperature.

Thermal conductivity, heat capacity, and thermal diffusivity. The thermal conductivity ($kcal/m^2 \cdot h \cdot °C$) must be low in so far as refractories must not lead to large amounts of heat being lost, bearing in mind that the thermal conductivity increases with increasing temperature. The heat capacity ($kcal/kg \cdot °C$) provides a measure of the ease with which a heated refractory increases its temperature, and this parameter must lie within quite precise limits (0.18–0.22). The thermal diffusivity ($°C/m \cdot h$) expresses the rate at which heat is transmitted through a working refractory as it is heated or cooled. If it is too low, dangerous thermal gradients are established when the refractories are subjected to thermal shocks.

Electrical conductivity. At low temperatures, refractories are dielectrics, but care must be taken to ensure that their electrical conductivity, which increases with increasing temperature, does not exceed the permissible limits for certain uses.

Specific weight and compression resistance. The values of these quantities, measured in g/cm^3 and kg/cm^2 respectively, are directly proportional to the structural homogeneity of the refractory and therefore reflect on the resistance of a refractory towards chemical attack (the higher the specific weight, the greater the chemical resistance of the refractory), and to the effects of impacts and compressions. In adjudging the suitability of the specific weight with regard to the uses to which the refractories are to be put, attention must be paid to the material constituting the refractory, while there exists a minimum value of the compression strength ($250\ kg/cm^2$) for all refractories.

Porcelains

The ceramic products, based on a vitrified compact paste, which are white and translucent, are called porcelains. These are subdivided into soft and hard porcelains

Table 14.3—The qualitative and quantitative composition of porcelains

Type of porcelain	*Kaolin*	*Silica*	*Feldspar*	$CaCO_3$	*ZnO*
Hard	50%	25%	25%	–	–
Soft	41.7%	27.6%	27.6%	1.5%	1.6%
Electrotechnical	43.7%	37.8%	18.5%	–	–

according to whether they have a relatively low or a high melting point, which, in turn, depends on their composition. The soft porcelains contain more fluxes, while the hard porcelains contain more kaolin. The compositions of the principal types of porcelain are shown in Table 14.3.

The 'structure and properties of porcelains. Research microscopic techniques have shown that the following phases are present in the porcelains:

- an isotropic glassy mixture formed from feldspar and silica and saturated to varying extends with the higher melting 'γ-alumina', (Al_2O_3);
- particles of the different crystalline modifications of silica (depending on the cooling conditions used) which are not dissolved in the glass;
- mullite crystals distributed throughout the silica–feldspar glass.

This stricture mainly arises from the chemicophysical phenomena which occur in the material during firing.

During the first stage of the firing at between 500 and 550°C, the kaolin is dehydrated:

$$Al_2O_3 \cdot 2SiO_2 \cdot 2H_2O \rightarrow Al_2O_3 \cdot 2SiO_2 + 2H_2O$$

while, at 800–900°C, the metakaolinite decomposes into the oxides:

$$Al_2O_3 \cdot 2SiO_2 \rightarrow Al_2O_3 + 2SiO_2$$

γ-Alumina crystallizes between 950 and 1000°C, while the alumina reacts with silica to form mullite at 1150–1250°C:

$$3Al_2O_3 + 2SiO_2 \rightarrow 3Al_2O_3 \cdot 2SiO_2$$

In the meantime, a feldspathic glass is formed, and a large part of the silica present in the system is united into this, also in a vitreous form. On the other hand, mullite and alumina remain dispersed in this glass.

The properties of a given porcelain are mainly dependent upon:

- the quantity and quality of the glassy mass forming it, while its translucent property is mainly due to feldspar;
- on the absolute amount of mullite present. Before its decomposition, mullite is a high-melting-point (1810–1830°C), mechanically, electrically, and chemically resistant substance which is also resistant to thermal shock on account of its low coefficient of thermal expansion;
- on the quantitative proportions between the mullite, the vitreous mass, and the

pore volume. The latter lower the value of porcelain by lowering its mechanical hardness and making it receptive towards fluids. The porosities of a good porcelain must not exceed 0.1–0.3% in terms of the amounts of water which are absorbed.

Details regarding the production of porcelains. The values for the coefficients of the properties of porcelains which have just been listed are mainly determined by the chemical composition of the raw materials, the extent to which they are milled, and the firing temperature and firing time.

Manufactured pieces may be moulded by jigs, casting, turning, or pressing. Turning and pressing are specific techniques for the fabrication of electrical insulators.

The moulded pieces are dried in tunnel driers, and this process must be carried out precisely in order to avoid fractures and cracks.

Porcelain may be subjected to one, two, or three firings ('decoration during the third firing'). For example, a single firing is sufficient for the electrical porcelains which, having been glazed when green, are fired at 1350–1400°C. On the other hand, porcelains for domestic use require two or even three firings if they are decorated.

When a double firing is carried out, the first (reviving) firing serves to dehydrate the paste and to bring about the incipient cementation of the grains to facilitate the glazing. After the glaze has been applied to the porous piece, it is subjected to a second firing at 1300°C in the case of soft porcelains and 1450°C for hard porcelains. In every case, from 1000°C upwards, the atmosphere in the furnaces has an important effect on the properties of the final products. This atmosphere should have the following characteristics:

* it must be oxidizing from 1000 to 1100°C so that any carbon particles which may have been deposited on the ceramic body may burn off;
* it must be reducing from 1100 to 1200°C to convert Fe_2O_3 to FeO which subsequently forms bluish-grey silicates with an agreeable coloration;
* it must be neutral from 1200°C upwards.

The firing in intermittent kilns (down-draught type) must last for about 60 hours, while a firing time of 40–50 hours is used in tunnel kilns. Rather rapid cooling is employed to avoid re-oxidations.

Stonewares

The opaque ceramics which are usually coloured, heavy, hard and sonorous, mechanically resistant, and fired at temperatures beween 1200 and 1300°C, are known as stonewares.

In the final analysis, stonewares differ from porcelains especially in their opacity and coloration on account of the less stringent selection of the raw materials. They may be subdivided into the following types:

- common (natural) stonewares which, having been prepared solely from iron-oxide contaminated (illitic) vitrifiable clays which melt at about 1250°C, are used in the manufacture of drainage pipes and systems, discharge conduits, and as building linings for domestic purposes;
- chemical stonewares are constitutionally similar to common stonewares but have greater chemical resistance (obtained by the addition of BaO which provides resistance towards basic agents, ZnO which provides resistance towards acidic agents, etc.), and are generally made impermeable by spraying NaCl powder onto the ceramic bodies before the final firing stage so as to provide them with surface coatings of alkali chlorosilicates and chloroaluminates;
- fine stonewares are similar to porcelains as a result of a modification of the clay which is usually employed in stoneware by the addition of plastic clays (bentonites)

and feldspars. Fine stonewares are then fired at temperatures (1350°C) which are higher than those used for other stonewares.

Chemical stonewares are used for the lining of benches, fume cupboards, walls, tanks, and chemical laboratory floor tiles, while fine stonewares are mainly used in tiles and tesserae, coloured with pigments which act as facings, mosaics, and floors in buildings (industrial building ceramics).

Minor ceramics

Under this heading are the porous paste ceramic materials which are predominantly used for the fashioning of household utensils, ornaments, and *objets d'art*, hygiene and sanitary equipment, and relatively fragile and abradable tiles and paving tiles. All of these products are characterized by the fact that they can be glazed, enamelled, and have decoration applied to their surfaces.

The most important data regarding the preparation and destinations of the various types of minor ceramics are presented schematically in Table 14.4.

A minor ceramic which has been very highly developed in recent times in the building industry is fireclay, which is similar to porcelains but with paste grains which are coarser and less homogeneous. The surfaces are treated with titanium dioxide based glazes. The characteristic use of this type of ceramic is in the covering of the external facades of buildings.

14.3 THE CERAMICS FOR THE NEW TECHNOLOGIES

The electrotechnological and electronic fields in the first place, but, nowadays, the missile and nuclear technologies, in particular, have promoted research into and the production of new ceramics (special ceramics) which have only certain characteristics and some phases of the production cycle in common with traditional ceramics. Their refractory nature, indeformability, and the absence of phenomena leading to the ageing of the structure and properties under operational conditions are the ways in which the old and new ceramics most closely resemble one another.

Known procedures for the preparation of special ceramics will be described below. However, it must be stated that the methods described cannot be generalized, owing to the fact that, in particular, the processes used in the preparation of ceramics intended for the nuclear reactor industry remain closely guarded secrets.

The production of special ceramics

In principle, the known* procedures for the preparation of these ceramics consist of three phases:

- the preparation of the pastes or 'wet muds';
- the drawing or casting of the paste or the suspensions, and the compression moulding of suitably prepared granular materials;
- the consolidation of the moulded pieces by normal firing or by sintering.

*Many of these materials, which are considered to be of strategic importance, are produced by methodologies which are kept secret; and, what is more, these technologies, even the most advanced, are always in a state of rapid evolution. The description provided here is therefore to be considered only as a guide.

Table 14.4—Characterization, production, and uses of minor ceramics

Type of ceramic	*Appearance*	*Raw materials*	*Types and firing*	*Uses*
Earthenware	Granular white paste. Dark yellow after firing. Porous if not glazed.	Kaolinite, quartz and feldspar. Possible application of pigmented glazes.	*Strong* or *hard*, *semi-hard* or *soft*. Firing from 950 to 1300°C according to type.	For sinks, hygiene-sanitary-ware, crockery and artistic products.
Majolica	'Fine' majolicas are white and have the appearance of porcelains. 'True' majolicas are yellow-red. The porosity and effervescence when in contact with acids are proportional to the lime-stone content. A greyish paste if they are *faiences*.	Kaolinite, quartz, and feldspar, for fine majolicas. Kaolinite, quartz, dolomite and limestone ('lean clays') in other majolicas. White or coloured glazing and possibly with decorations.	Like earthenware, fine majolicas may be soft or hard, i.e. fired at 1100–1300°C (hard) or 900–1000°C (soft). As a rule, they are fired twice at quite high temperatures.	Floor and wall tiles, wash basins and entire bath tubs. The *faiences* varieties are used for artistic products
Terracottas	Reddish colours or tending to such. Glazed or enamelled in the *crockery* varieties.	Typically illitic clays with an appreciable amount of montmorillonite	Generally fired at ~850–1050°C. Double firing in the case of *crockery* terracottas.	Ornamental motives and decorative china. Kitchen-ware if they are of the *crockery* type.

To prepare the slips, the raw materials are pulverized, mixed in the required proportions, and brought to the consistency of a wet mud. It is possible to carry out a further pulverization of the material in an aqueous suspension, and to reduce the particle size to the order of 1 μm without superheating.

The pastes to be drawn are unloaded as cakes from the filter presses which have been fed with the slips, while the suspensions used for casting are usually the same slips. If, however, these cakes from the filter presses undergo a reduction in the number of small particles which they contain and are then treated in granulators with the addition of binding agents (dextrins, starch, casein, and alginates) and lubricants (olein and waxes), small spheres are formed which can be dried and sieved with the recycling of those which are either too small or too large.

This is followed by the compression moulding of spheres of the correct diameter to form the ceramic components which are to be produced.

The firing of the moulded components never brings about true fusions, and is therefore always of the nature of a sintering process carried out at temperatures such that the atoms of neighbouring grains are sintered together to form a continuous whole.

Summary of the best known ceramics used in the new technologies

Fissile ceramics

Today, oxides, carbides and nitrides are the ceramic phases which are the most used in nuclear reactors as fuels in the form of small cylinders or microspheres, which are chemically inert and structurally rigid up to their melting point.

The ceramic behaviour of these fuels is very important because the environment within nuclear reactors is inherently corrosive and tends to deform and swell the solids which undergo nuclear fission reactions.

However, after emphasizing the great importance of these ceramics at the present time and, without doubt, their even greater importance in the future, we shall not dwell upon these ceramics any further, as, subsequent to their preparation, the manner in which they are put into operation and the details of the processing work carried out on them, after they have become 'poisoned' by the phenomena taking place and fission products, are secret.

Electrical and electronic ceramics

The best high-frequency and mechanically stable insulators are steatitic ceramic masses (steatites) obtained from the layer silicate, talc, and barium carbonate which forms BaO when fired at a high temperature. The raw materials are worked in the plastic state and then generally pressure moulded into the required form after granulation of the raw material. The resulting product is practically neutral because the cristobalite silica which is freed during firing is neutralized by the BaO.

* * *

In the electrotechnic technologies, especially where high resistance to thermal shock may be required, coatings in ceramics with very low coefficients of expansion are used. These ceramics are the cordierites, prepared from kaolin, talc, and magnesia which have been precalcined, pulverized, mixed in the respective ratios of 5.3:2.7:2, brought to the state of a wet mud, and then moulded and subsequently fired. During firing, cordierite ($2MgO \cdot 2Al_2O_3 \cdot 5SiO_2$) is predominantly formed. This has a structure which is crystallographically different from the corresponding natural mineral which therefore does not quite possess the refractory properties of the synthetic material.

* * *

The ferrites, which are shaped into any required form before sintering, are ceramics which have important applications in electrical technology and radiotechnology (in transformer components, aerials, and television tubes) and in electronics (calculator components). The final sintering is carried out at pressures of the order of 100 ton/cm^2 and at temperatures of approximately 1000–1250°C for more than 20 hours to obtain products in which there has been pronounced binding and homogenization of the raw materials. The pretreatment of the ingredients comprises the following steps:

- the mixing of other oxides (e.g. ZnO, NiO and PbO) and carbonates (e.g. $BaCO_3$ and $CaCO_3$) with Fe_2O_3 in predetermined ratios (generally Fe_2O_3: other oxides or carbonates = 6:1);
- calcination of the mixtures until temperatures of the order of 1300–1400°C are attained;
- the crushing of the product followed by wet grinding until particles with sizes of the order of 1 μm are obtained.

In these materials, the crystals are aligned parallel to the axes during pressing, and so, by magnetizing pieces of them which have been sintered and ground with diamond, materials with quite good magnetic properties are obtained. In fact, while certain (soft) ferrites tend to become demagnetized, others which are much more highly valued (hard ferrites) behave as permanent magnets (as magnetic memory devices) and are used as such in the abovementioned technologies.

Titanium and zirconium electrical ceramics consisting of $BaTiO_3$ and $BaZrO_3$, are prepared by starting out from $BaCO_3$ and TiO_2 or ZrO_2 with a purity of at least 99% in a production cycle which is similar to that used in the case of the ferrites up to the final sintering. They possess very good high-temperature dielectric properties and outstanding piezoelectric properties.

Superrefractories

Materials which, either inherently or by modification, are resistant to attack by the atmosphere in which they are employed at very high temperatures (on average from 2000–3000°C and even up to 3600°C), to thermal shocks, corrosion, and abrasion, are referred to as superrefractories or ultrarefractories. They are employed in nuclear reactors, jet engines, missiles, and in other areas of modern technology where extremely severe conditions are encountered.

Ceramic components are made from powders of the individual oxides, mixed oxide, boride, carbide, nitride, beryllide, silicide, and even certain sulphide powders. The conventional techniques of casting, extrusion, and compression are used for this purpose. Casting requires the use of the finest powders which are prepared by the wet route, while extrusion and compression necessitate prior granulation of the materials. The subsequent sintering heat treatment is carried out at temperatures which labilize the surfaces of the crystalline lattices of the particles which are in contact, thereby promoting, by recrystallization, the sintering of the latter to form a single crystalline matrix. In this way, the melting point of the finished product tends to be found close to the theoretical value for the pure crystallinity superrefractory.

Al_2O_3, BeO, MgO, CaO, ZrO_2, and ThO_2 are included among the superrefractory oxides, which form the most important class of ultrarefractories, because they are indifferent to oxidation.

BaB_6, Ti_2B, and TiB_2, ZrB, ZrB_2, and ZrB_{12}, TaB_2, and Ta_3B are the most important boride superrefractories. These are expensive and relatively little used. They are usually extremely resistant to oxidation. As can be seen, the same metal may form more borides than indicated by the valency rules.

The most important of the superrefractory nitrides is SiN. Others, which are of noteworthy importance, are: BN, which is structurally similar to graphite but an electrical insulator, AlN, TiN, and HfN, which melt at above 3200°C.

Of the carbide superrefractories, SiC (carborundum) is technically the most important. In a non-oxidizing environment, it can be used at temperatures up to about 2400°C.

Carborundum finds specific applications on account of its unique qualities. Among others, these are:

- a hardness which approaches that of diamond, and it is therefore used as an abrasive;
- an ability to become passivated by the formation of thin layers of Si_3N_4 and SiO_2 upon coming into contact with nitrogen and oxygen;
- electrical resistance properties which are such that it can be used as a resistance in electric furnaces;
- an optimal chemical resistance towards acids, sulphur, and sulphur compounds;
- a thermal conductivity which makes it suitable for use in the lining of systems with indirect heating.

Carborundum can be prepared by treating a mixture of finely subdivided silica sand and an excess of coke in an electric furnace at temperatures of about 2000°C. The reaction essentially consists of two steps: the reduction of SiO_2 to Si and the subsequent combination of the latter with carbon:

$$SiO_2 + 2C \rightleftharpoons Si + 2CO$$

$$Si + C \rightleftharpoons SiC$$

NaCl is also added to the reaction mixture to volatilize the impurities (mainly Fe and Al) as their chlorides. In every case, the best carborundum is that which is formed around the carbon electrodes of the production furnace. This product is also employed in the construction of high-temperature transistors.

Al_4C_3, $B_{12}C_3$, TaC, and hafnium carbide, HfC, which, among the 'ultrarefractories', holds the melting point record (>3900°C), should also be mentioned among the other carbide superrefractories.

Cermets

The word 'cermet' is derived from the initial syllables of ceramic and metal. By definition, cermets are heterogeneous mixtures (given the low mutual solubility of ceramic and metal phases at the preparation temperature) of one or more metallic materials (metals and alloys) with ceramic (refractory) phases.

The metals which, either in a pure form or as alloys, are most commonly included in cermets are chromium, cobalt, nickel, iron, molybdenum, and copper, while the ceramic phases most widely employed include, in addition to oxide refractories in general, carbides, nitrides, and borides. Cermets have been discovered during research projects with the aim of developing materials which combine the properties of metals with those of ceramic products, such as: good refractory properties, impact, and shock resistance, ductility, thermal conductivity, and mechanical strength.

Cermets are used where great resistance to high temperatures and thermal shock, excellent refractory and abrasion resistance properties, outstanding corrosion resistance, and, from time to time, other superoperational qualities, are required. These depend on the ratio between the ceramic and metallic materials from which they are formed. The nuclear reactor, aircraft jet engine, and astronautic fields in general are where they are most required.

When cermet antithermal coatings are to be applied, the work is carried out with a plasma gun in which the cermet powders are fused at thousands of degrees, and the resulting liquid is sprayed onto the body to be coated, where it cools and solidifies.

Cermets are produced by starting from the powders of the individual phases, moulding of the ceramic type, and subsequent sintering.

They may be moulded by pressing, drawing, casting, and also by 'infiltration', that is, by allowing a liquid metal (the substance which filters in) to act on a solid high-melting-point ceramic phase (the 'skeleton'). Overall, this is a question of first preparing the ceramic matrix in the conventional manner and then filling its pores with the molten metallic phase by capillarity.

Cermets which contain oxides as the ceramic phase are sintered in a controlled oxidizing atmosphere with the aim of developing thin layers of the oxide on the surface of the metal which allow it to bond readily with the ceramic phase.

A cermet derives its name from the ceramic phase which characterizes it and is referred to as the base of the cermet.

Oxide-base cermets are endowed with an elevated mechanical resistance under a load. Of these cermets, the alumina-base chromium alloyed cermets stand out in importance. These can be prepared by the aluminothermic reaction:

$$Cr_2O_3 + 2Al \rightarrow Al_2O_3 + 2Cr$$

followed by sintering.

The oxide-base cermets have good erosion resistance and are extremely resistant towards fused metals and oxidation. Additions of small quantities of metal oxides (of tungsten, molybdenum, and titanium, for example) increase certain other resistance properties such as their resistance to thermal shock.

Carbide-base cermets are less mechanically strong when under load than the oxide-base cermets, but cost less to produce because their contrary phases readily bind together. Among these cermets, those with a titanium carbide (base) and nickel, cobalt, chromium, and molybdenum (binders), tungsten carbide (base), and cobalt (binder) and those with a chromium carbide (base) and nickel or copper (binders) stand out in importance.

Boride-base cermets such as molybdenum boride (base)/chromium (binder), iron boride (base)/iron (binder), and chromium boride (base)/nickel, iron or cobalt (binders) are used where a high melting point and high-temperature erosion resistance are required.

Among other things, it may be noted that a chromium boride-base or molybdenum boride-base cermet is more heat resistance than a titanium carbide-case cermets, whilst having the same thermal shock resistance.

The most important aluminide-base cermets are those of molybdenum, which are stable and resistant to oxidation at very high temperatures, and those of nickel which have outstanding resistance to thermal shock as well as to hot oxidation.

Furthermore, the latter cermets are ductile at ambient temperature, where, however, their mechanical strength is poor.

Nitride-base cermets have properties which are similar to those of the carbide-base cermets, and the silicide-base cermets stand out on account of their resistance to high temperatures and oxidation owing to a form of passivation which is brought about by the formation of a superficial covering of vitreous silica.

Part 2 SILICATE GLASSES

14.4 THE VITREOUS STATE

All substances and mixtures which, while non-crystalline, are in a state such that they are practically undeformable* are called glasses, and their state is defined as the vitreous state.

Both organic and inorganic glasses exist, but only silica-based glasses will be considered in this chapter.

It would be a mistake to define glasses as 'bodies which do not crystallize upon solidification' because, from a physicochemical point of view, solidification is a synonym of crystallization, that is, of the assumption of symmetry and periodicity in the arrangement of the atoms or molecules or ions in the undeformable body which is formed.

A glass is therefore simply a supercooled liquid or an extremely viscous liquid which tends to devitrify with the passage of time.

As they are naturally amorphous, glasses are in an unstable state which may be compared with the state of a mixture of hydrogen and oxygen or hydrogen and nitrogen under normal conditions. In all of the cases, in fact, certain 'retardation phenomena' (molecular inertia, viscosity, freezing of the transformation equilibrium, etc.) make the evolution of glasses or of the mixtures of gases from the unstable state, in which they find themselves, into the stable state infinitely slow.

The analogy between the systems which have just been compared can be taken further by observing that, in the gaseous mixtures, the intervention of a catalyst or of a sufficiently energetic action (irradiation, electric or electromagnetic pulses, etc.) can bring them into a stable equilibrium state with their reaction products, and that, in glasses, such physico-mechanical stimuli as impacts, heating, and vibrations or the introduction of crystallization seeds can prime the crystallization (devitrification) process and bring the crystallization of the whole mass to a rapid conclusion.

It has been observed that the devitrification of glasses due to the appearance of crystallization seeds takes place, for the most part, during the period when the glasses are maturing, while the devitrification which is attributable to mechanical and physical causes constitutes the major defect in glasses after they have been put into service.

---

*Actually, a glass is not absolutely undeformable, and not only very slightly deformable but deformable to only a small extent. Inter alia, it is known that common glass, especially if it is moulded into rods, small plates, and various profiles, tends to bend and elongate under its own weight.

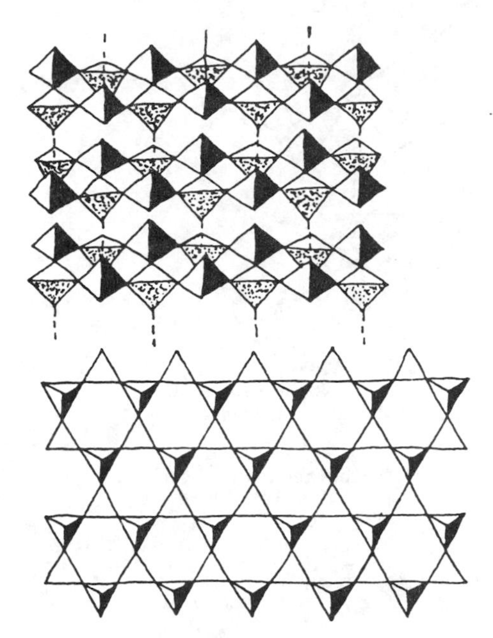
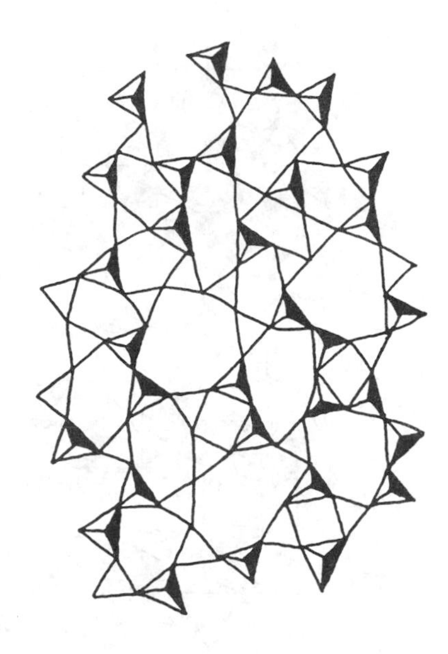

Fig. 14.18 Top left: Planar schematization of the quartz structure of Fig. 14.9 consisting of regularly arranged linked tetrahedra. Bottom left: Reduction of the spatial lattice of quartz to superimposed planes with a regular linked structure. Right: The aperiodic and asymmetric lattice of vitreous silica found in silica glasses.

14.5 THE STRUCTURE OF GLASSES

X-ray crystallographic studies of glasses have revealed the existence of an aperiodic and asymmetric, spatial atomic lattice which is far removed from the well-ordered arrangement of the atoms which may be found in quartz, no matter how it is viewed (Fig. 14.18).

Consequently, glass is isotropic and does not possess a well-defined melting point, because the bonds between the atoms, which are symmetric, are gradually broken, and melting therefore takes place over a certain temperature range. Glasses also naturally tend to devitrify because their energy content is higher than that of the crystalline state.

In addition to silica, containing the Si^{4+} ion which coordinates four O^{2-} ions to form the tetrahedral SiO_4^{4-} ion which is mutually linked to other such ions, the oxides of other elements also form glasses. Among others, Al_2O_3, B_2O_3, P_2O_5, and As_2O_5 exist in the vitreous state owing to the tendency of the corresponding base elements to form tetrahedra (PO_4^{3-}) or octahedra (AsO_6^{7-}) or both (AlO_4^{5-} and AlO_6^{9-}).

Fig. 14.19 shows how the hexaoxoarsenate octahedra link together to form regular structures.

On the other hand, Fig. 14.20 shows the discontinuous, aperiodic, and therefore vitreous structure formed by the linking of orthophosphate anions.

In the glasses formed from more than one oxide, acidic oxides, and basic oxides respectively, various cations are intercalated in the basic tetrahedral or octahedral lattices by the breaking of a certain number of the oxygen bridges (Fig. 14.21).

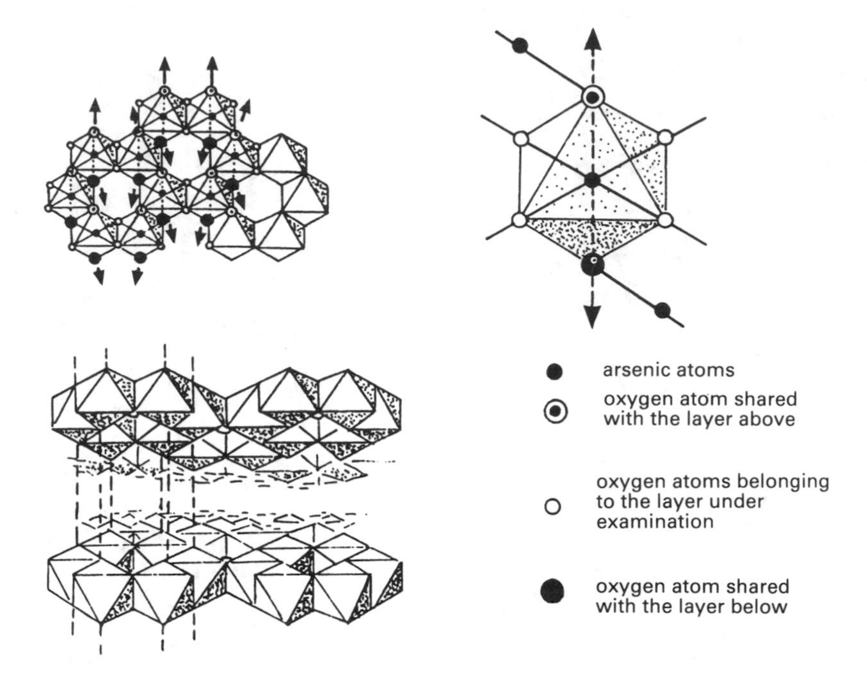

Fig. 14.19—Top left: Structure formed by the linking of AsO_6 octahedra to form six-membered rings. Bottom left: how the planes consisting of the six-membered rings of AsO_6 octahedra are packed together with the sharing of oxygen atoms. Right: Graphic detail of the manner in which atoms in a single octahedron are shared (above) and a legend which enables one to evaluate the nature and/or the role of the atoms shown in this detail (below).

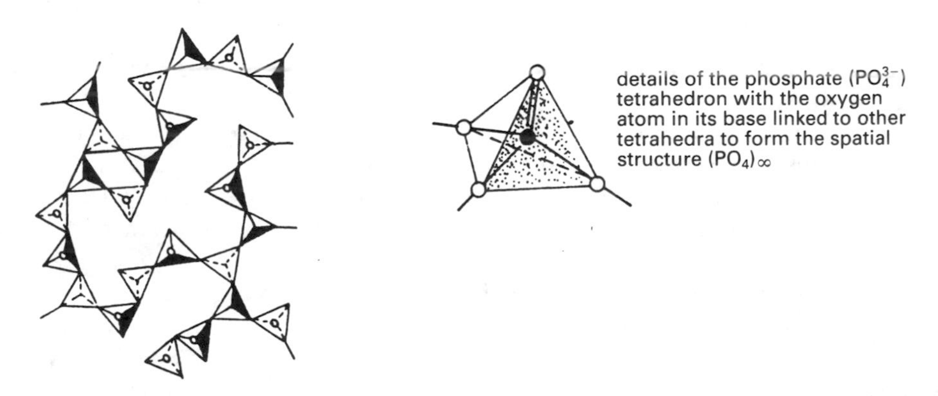

details of the phosphate (PO_4^{3-}) tetrahedron with the oxygen atom in its base linked to other tetrahedra to form the spatial structure $(PO_4)_\infty$

Fig. 14.20—Left: Irregular lattice structure arising from the random linking of tetrahedra (only the links between the tetrahedra in the plane are shown) to form 'phosphate glass', $(PO_4)_\infty$. Right: Detail of the same structure.

Of all the possible types of inorganic glasses, the one which is most common by far is based on SiO_2, that is, a glass which fundamentally exploits the aperiodic and asymmetric lattice shown in Fig. 14.18 which is randomly interrupted by the presence of metal cations which neutralize the charges on the non-bridging oxygen atoms. This can also be seen from Fig. 14.21.

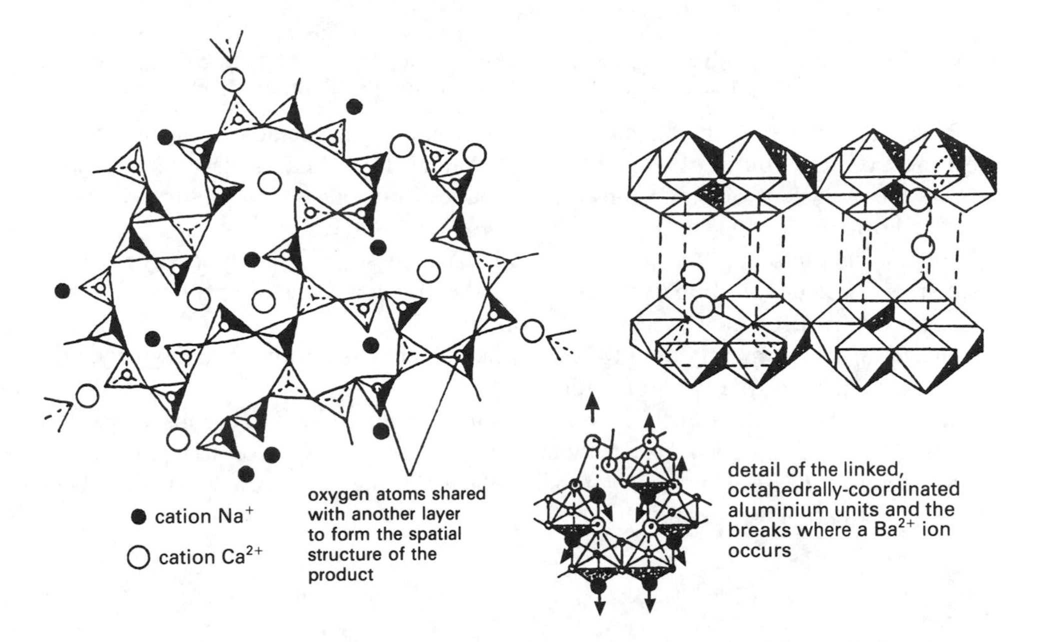

Fig. 14.21—Left: The aperiodic and asymmetric structure of common glass. This can be thought of as silica glass (Fig. 14.18) with random breaks in the spatial linking of the tetrahedra with the formation of terminal anionic —SiO^- or —SiO^- groups due to the oxygen contributed by the CaO and Na_2O components. Such terminal groups are neutralized by Na^+ and Ca^{2+} cations. Right: The structure of a glass containing alumina (Al_2O_3) and BaO. Here also, the lattice discontinuities leads to the formation of anionic terminal groups which are neutralized by the barium cations.

14.6 RAW MATERIALS FOR GLASSES OF PRACTICAL USE

Four types of raw materials are used in the industrial production of glasses: glass-forming substances, fluxes, stabilizers, and secondary components.

Glass-forming substances

All mineral substances which tend to assume indeformability of the vitreous type when melts of them are cooled, are referred to as glass-forming substances. The main substances of this type are silica, boric oxide, phosphorus pentoxide, and feldspars. Al_2O_3 is also a complementary glass-forming compound.

Silica is the glass-forming substance of the greatest interest. Sand, containing 99.1–99.7% of SiO_2, is used for this purpose.

Sometimes, natural silica is discovered which has a sufficient degree of purity to meet the requirements regarding its use in the production of glasses, but, nowadays, it is more and more often found that recourse must be made to the enrichment of relatively low-quality natural sands by flotation.

The commonest impurities occurring in natural silica are iron sesquioxide (Fe_2O_3), alumina (Al_2O_3), and calcium compounds. Ferric oxide, even in amounts as small as 0.1%, imparts an intense yellow–green coloration to glasses, and also reduces their thermal and mechanical properties. Alumina induces devitrification in the products. Calcium oxide is less harmful provided that it is compatible with the composition of the glass which is to be produced and account is taken of its presence when calculating the amounts of stabilizers to be used.

Boric oxide increases the fusibility of the mass and its refinement. Furthermore, it also increases the ability of glasses to resist changes from taking place in them and lowers the coefficient of expansion. It is introduced into glasses in the form of boric acid or borax.

Phosphorus pentoxide, which can be added to glasses as pure phosphoric acid, calcium phosphate, or bone ash, increases the transparency of the products but considerably reduces their mechanical properties.

The *feldspars* are mainly used to reduce the tendency of products to devitrify, and are mainly added in the form of orthoclase.

Alumina is not a glass-forming substance in its own right but becomes so when combined with SiO_2 and B_2O_3. Its function is to impede the hydrolysis of the silicate components in glasses, that is, to stabilize glasses in the sense of neutralizing them (since glasses are naturally alkaline).

Fluxes

Of course, feldspars, as always, also act as fluxes. Nevertheless, the fusibility and the temperature range over which the glasses can be worked are specifically increased by the use of alkali metal oxides, especially sodium oxide and potassium oxide.

Such oxides can be added either as carbonates or as sulphates, the first being preferred both because it contains less iron, attacks the refractory in the glass-making furnaces to a lesser extent, and give rise to less slag.

Stabilizers

Stabilizers are correctly defined as the components of glasses which reduce their solubility in water and the extent to which glasses are attacked by chemical agents in general, and by agents present in the atmosphere in particular. More generally, however, stabilizers are all the components which determine certain characteristics of glasses.

The most notable stabilizers and their corresponding effects are:

- calcium carbonate which makes the finished product insoluble in water;
- barium carbonate which increases the specific weight, the refractive index, and the sonority of glasses;
- lead oxide which is introduced as red lead (Pb_3O_4) or even, more commonly, as litharge (PbO). This imparts perfect transparency, brightness, and a high refractive index to the glass and facilitates its cutting;
- zinc oxide which makes the glass resistant to thermal shock, improves its mechanical and chemical properties, and increases its refractive index;
- dolomite which facilitates the fusion of the glass-forming mixture.

Aluminium oxide, as well being a glass-forming substance, can also be considered to be a stabilizer since it increases the viscosity of glasses, their physical strength, and their chemical resistance.

Secondary components

These are substances which correct defects in glass and determine its properties. They are classified, on the basis of the action which they exhibit, as refiners, decolorants, opacifiers, and colorants.

Refining agents

The secondary refining additives develop gases under the action of heat which homogenize the fused mass by carrying away with them the gas bubbles which are formed during the course of the melting.

From a chemical point of view, refining additives at first exercise a dehydrating action on the contaminating organic debris which may be present, and then carbonize any such materials, and finally oxidize it to CO and CO_2. For instance, the refining agents, sodium nitrate and sodium sulphate, liberate the corresponding anhydrides when they react with SiO_2:

$$2NaNO_3(Na_2SO_4) + SiO_2 \rightarrow Na_2SiO_3 + N_2O_5(SO_3)$$

which are notorious dehydrating and oxidizing compounds.

Decolorants

The problem of removing the coloration which is imparted to glasses by the traces of iron oxides present in the raw materials always exists to some extent. Decolorants must therefore be added to them. Manganese dioxide is mainly used for this purpose and, to a lesser extent, selenium metal and nickel oxide in that order.

When manganese dioxide is introduced into a melt in the form of its mineral pyrolusite, it first acts by evolving oxygen:

$$2MnO_2 \rightarrow Mn_2O_3 + O$$

which oxidizes the ferrous silicate to ferric silicate:

$$2FeSiO_3 + SiO_2 + O \rightarrow Fe_2(SiO_3)_3$$

The manganese (III) oxide which is first formed imparts a violet coloration to the glass which is complementary to the yellow hue of the ferric silicate, and the mass therefore appears to be colourless.

Selenium, which is much more expensive than MnO_2, confers a pinkish-red coloration on the mass, which directly extinguishes the green tint of ferrous silicate by optical compensation.

The glass treated with selenium is a colourless, more uniform, brighter glass which has improved properties even at elevated temperatures, and these properties are more persistent during the refining processes in comparison with those induced by MnO_2. This means that selenium is indispensable in certain cases.

Nickel oxide has an advantage over manganese dioxide in that its action is not hindered by the atmosphere within the furnace, but it is more expensive and, above all, suffers from the defect that it is quite a strong absorber of light.

Colorants

Characteristic colours are imparted to glasses by the addition of 'colorants' (Table 14.5).

Table 14.5—Nature and action of the principal glass colorants

Colourants	*Formulae*	*Original colours*	*Active components*	*Imparted colours*
Manganese compounds	Mn_2O_2 MnO_2 $KMnO_4$	brown powder grey black violet	Mn_2O_3 for all	various shades of violet depending on the temperature and the duration of the melting.
Oxides of iron	FeO Fe_2O_3	black powder reddish-brown powder	FeO Fe_2O_3	light green or bottle green or, finally, yellowish green
Chromium compounds	Cr_2O_3 K_2CrO_4 $K_2Cr_2O_7$	green powder yellow powder orange crystals	Cr_2O_3 However, instead of this powder, the more readily fusible chromates and dichromates are usually used.	green tending to yellow
Cobalt	Co_3O_4 Co_2O_3	grey powder blackish powder	CoO	blue
Cadmium	CdS	yellow-orange powder	CdS	yellow
Copper oxides	Cu_2O CuO	brown-red powder black powder	Cu_2O CuO	various shades of red sky blue
Metallic	Au	yellow gold powder	colloidal Au	various shades of ruby

As an addendum to this table it should be stated that the active components do not generally act as such, but, because by forming salts with silica they produce different coloured ions (e.g. the blue Co^{2+} ion).

Opacifiers

These are substances which, by dispersing in the vitreous mass, confer opacity on the products.

Possible opacifiers are: fluorite (CaF_2), cryolite (Na_3AlF_6), and sodium fluorosilicate (Na_2SiF_6). However, calcium phosphate and the strong opacifiers, tin phosphate and zirconium phosphate, are the most commonly used.

14.7 THE INDUSTRIAL PRODUCTION OF GLASSES

The cycle for the industrial fabrication of glasses is shown in the block diagram in Fig. 14.22.

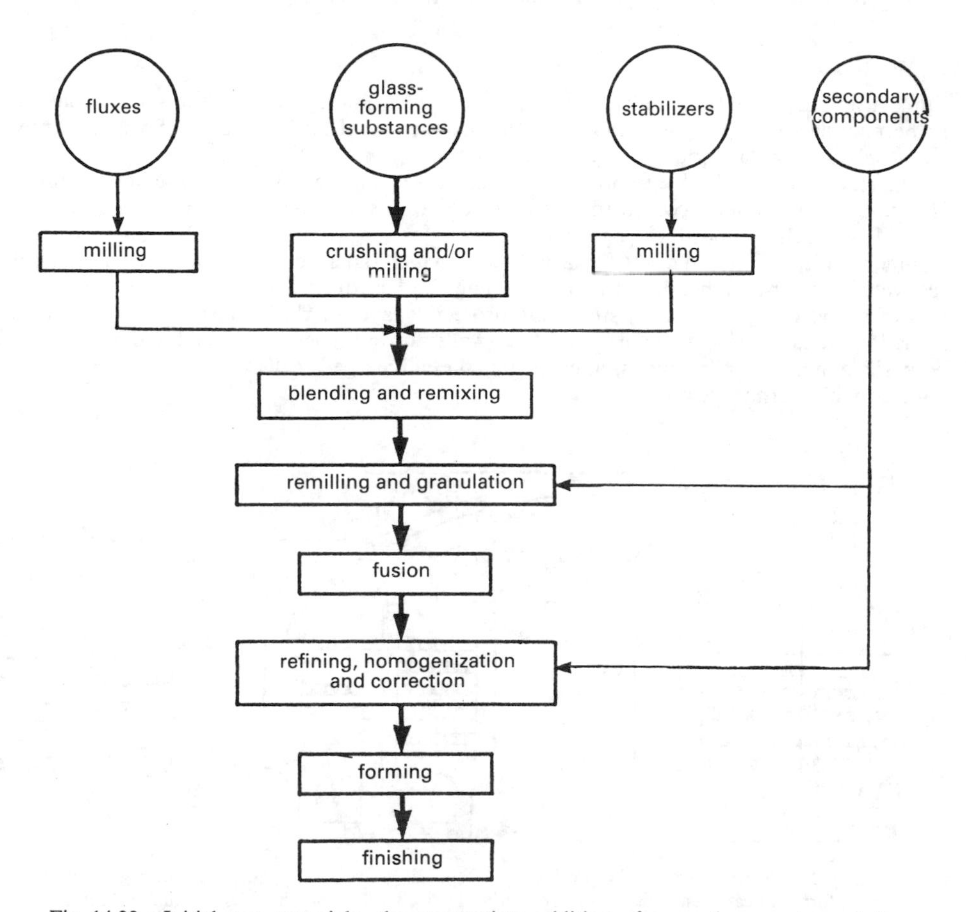

Fig. 14.22—Initial raw materials, the appropriate addition of corrective agents, and the operational phases in the preparation of glasses.

Preparation of the charge and execution of the fusion

The operations which precede fusion have the aim of preparing a good paste to be loaded into the melting furnace, and terminate with the granulation of the previously moistened material, containing not more than 4% of water in order to avoid the formation of clots. Such moistening avoids the costs incurred by the loss of ecologically damaging powders during the charging of the various types of furnaces, and it promotes the fusion process.

Although the fusion furnaces may be constructed differently, they do have common functional characteristics in that they are all heated by natural gas, by gazogene, or by fuel oils; temperatures of the order of 1400–1500°C are attained in all of them, and heat is saved in all of them by means of a system of Siemens regenerators. The furnaces are classified as: pot furnaces, tank furnaces, and well or channel furnaces.

Pot furnaces

To produce special glasses (optical glasses, technical scientific glasses, etc.), pots are used which are suitably moulded from a silica-alumina refractory. These are air dried, slowly fired, and then installed in the fusion furnace (Fig. 14.23).

Tank furnaces

Tank fusion furnaces (Fig. 14.24) are used where large quantities of several type of glass are to be manufactured, such as happens in the production of hollow glasses (bottles, carboys, flasks, phials, tubular moulded objects, caps, etc.).

In fact, to make it possible to produce several glasses simultaneously, the tank of the furnace is separated into several compartments by means of dividing walls which are not shown in Fig. 14.24. Each compartment opens at the front for the withdrawal of the glass. There are channels under the hearth of the furnace through which the air circulates with the aim of both recovering heat by preheating it and of cooling the hearth somewhat, which in the process becomes covered by a relatively sticky and protective layer of glass. The crown of the furnace is made of dinas bricks, while the part of the walls above the level of the molten glass is lined with silica–alumina refractories, and the pit, the submerged walls, and the dividers are all covered with Corhart refractory.

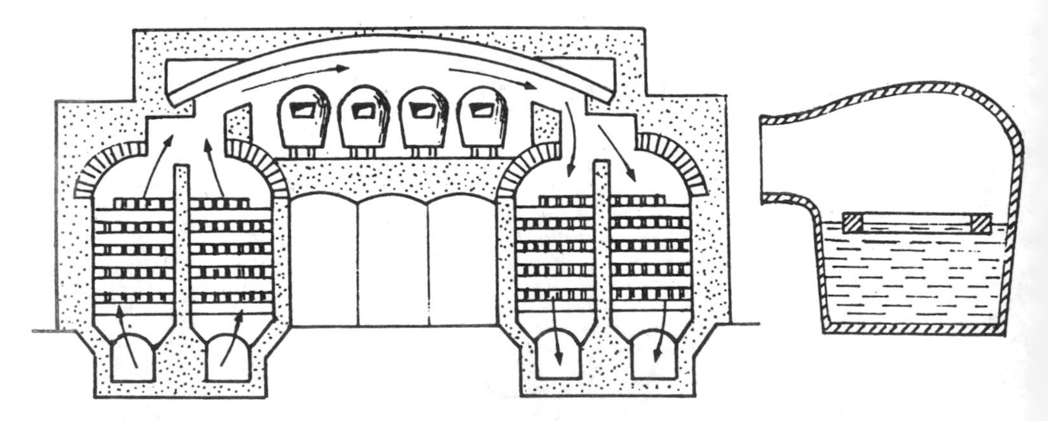

Fig. 14.23—Pot furnace for the production of special glasses equipped with heat regenerators in the form of heat exchangers with a refractory lattice work through which the fuel–air mixture and the combustion fumes alternately pass. A cross-section of a pot with the floating shuttle which is used to withdraw the glass is shown on the right.

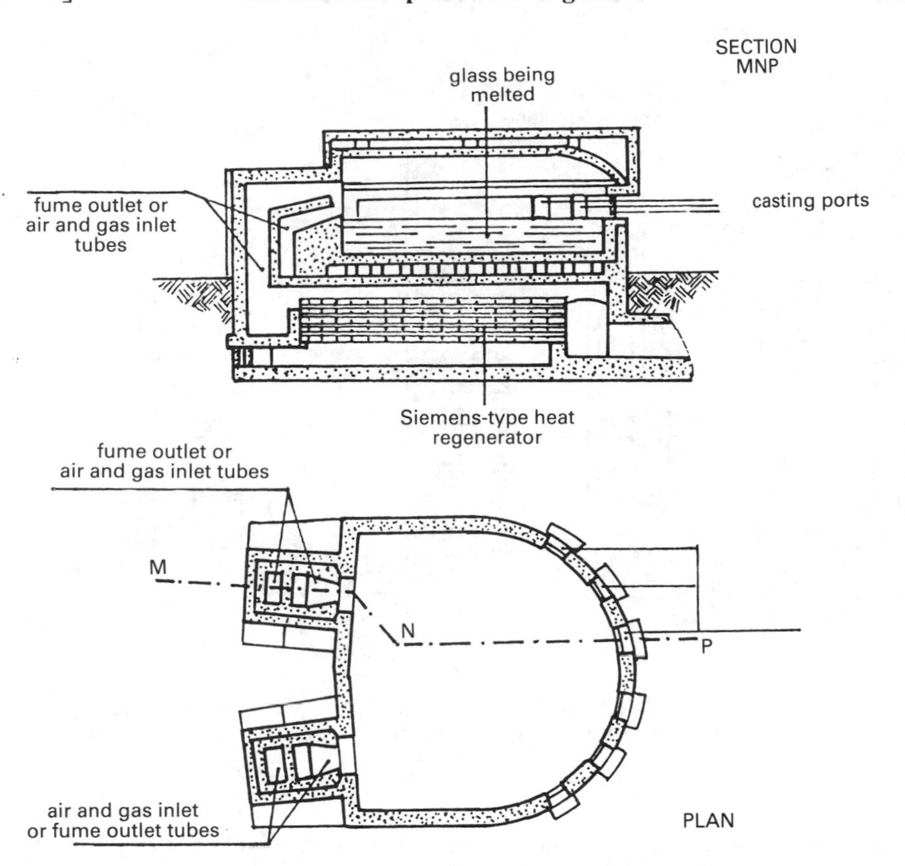

Fig. 14.24—Section and plan view of a tank furnace. The numerous casting ports with which it is equipped permit the simultaneous production of several types of glass objects which, for the most part, are hollow.

Corhart (or, more correctly, Corhart Standard Electrocast) is a refractory which has the best high-temperature resistance and resistance against corrosion by molten glass of any type of refractory. It is prepared by melting clays with a high alumina content and a composition approaching that of mullite in an electric furnace, and pouring the fused mass into sand moulds where a second firing is carried out. Extremely hard blocks are obtained which cannot be cut; they can be worked only by using sintered alumina (alundum) grindstones.

Continuous tank furnaces

When large quantities of a single type of glass, generally sheet glass, are to be produced, continuous furnaces are used. These operate in a reverbatory manner and are shaped in the form of a channel lined with the appropriate refractories in the various zones as in the case of simple tank furnaces. They are equipped with efficient heat regenerators to recover some of the heat produced by the combustion of various gases and heavy mineral oils. Such furnaces are known as continuous tank furnaces (Fig. 14.25).

As in analogous cases, when gaseous fuels are used, both the gases and the air required to burn them, or only the air when fuel oils are used, are made to pass alternately into one of two piles of refractory material located at the sides of the furnace channel. The same course is followed alternately, but in the opposite direction, by the combustion fumes. Before loading

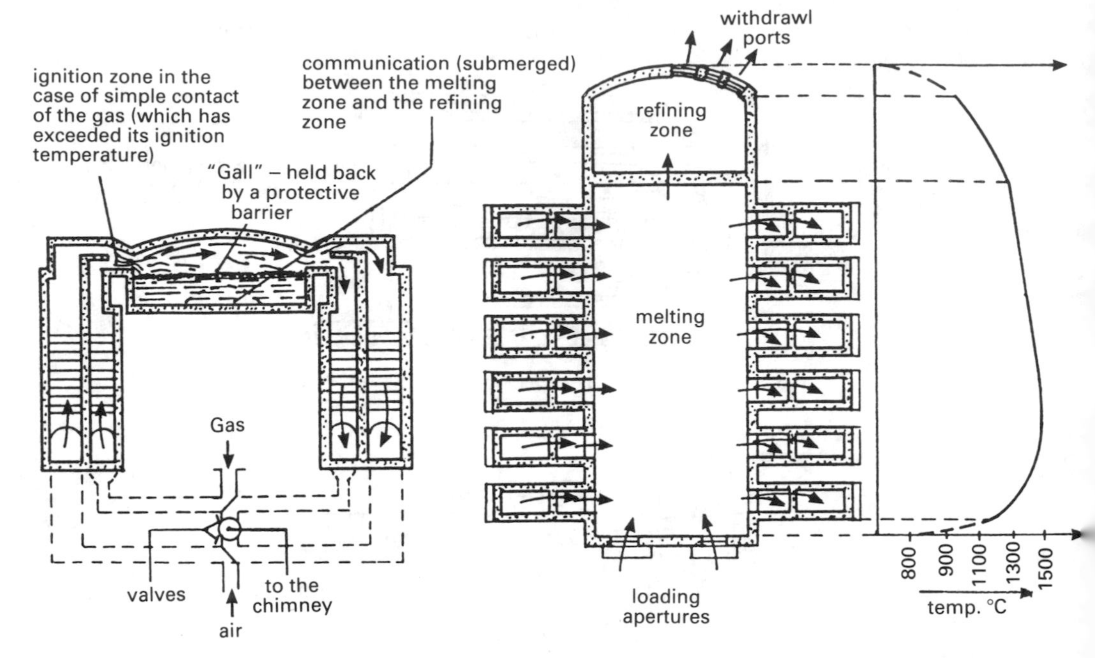

Fig. 14.25—Continuous tank melting furnace intensively supplied with gaseous fuels shown in cross-section (left) and plan view (right). The temperature profile within the furnace is reproduced on the far right.

the vitrifiable mass into a continuous tank furnace, the furnace is filled with scrap glass which, when molten, lines the refractory with a thin protective coating of glass.

The theory of fusion and the refining and correcting operations

The fusion process uses a series of chemical reactions which start to take place as soon as the temperature has increased to its maximum. These are decomposition reactions, solid state reactions, and vitrification reactions.

The decomposition reaction mainly involves the carbonates in the raw materials, e.g.

$$CaCO_3 \rightarrow CaO + CO_2$$

Solid state reactions are involved when silica reacts with oxides, for example,

$$SiO_2 + CaO \rightarrow CaSiO_3$$

or those which lead to the formation of double carbonates:

$$Na_2CO_3 + CaCO_3 \rightarrow Na_2Ca(CO_3)_2$$

while an example of a vitrifying reaction is provided by:

$$Na_2Ca(CO_3)_2 + SiO_2 \rightarrow Na_2SiO_3 + CaCO_3 + 2CO_2$$

In practice, however, not all of the mass undergoes reaction upon melting. A part of it is carried up to float on the surface and thereby form a slag in which fragments of the refractory and argillaceous residues from the glass-forming feldspars collect. If sodium sulphate is used among the fluxes (and not only as a refining agent), there is a particularly large amount of slag formed.

In this case, carbon is also added, which initiates a number of chemical reactions which lead to the total vitrification of the sulphate:

$$4Na_2SO_4 + 2C \rightarrow 2CO_2 + 4Na_2SO_3 \rightarrow 3Na_2SO_4 + Na_2S$$

$$Na_2SO_4 + Na_2S + 2SiO_2 \rightarrow 2Na_2SiO_3 + SO_2 + S$$

with the formation of sulphur and the evolution of sulphurous gases.

Silicates, however they may be formed, are the first to melt, and the excess silica is dissolved in them upon melting.

Just after the fusion has taken place, the glass has the consistency of a viscous oil at normal temperatures. The gases, CO_2, H_2O, SO_2, N_2, O_2, and others, tend to remain occluded in such a medium, and, moreover, inhomogeneous zones are established there. The homogenizing refining, which lasts for a long time, has the aim of eliminating gas bubbles and any inhomogeneities which may be present, and is carried out, as was stated above, by using refining agents which act by remixing the mass so as to entrain gas bubbles and bring them to the surface.

It can be seen from the temperature profile on the right-hand side of Fig. 14.25, showing a continuous tank furnace, that the temperatures in the refining zone are lower than those in the melting zone since it is necessary to ensure that other reactions accompanied by the evolution of gas bubbles should not take place. If such reactions occur, the bubbles remain occluded in the mass which is cooling as it passes to the discharge apertures.

Sometimes as early as in the melting phase but, more frequently, during the refining stage, it is necessary to correct the properties of the glass in the sense of increasing certain properties, decreasing them, or modifying them, e.g. decolorizing the glasses, opacifying the glasses, modifying the refractive index, etc. To do this, one proceeds for the most part, by making suitable additions. Anticrystallizing agents (e.g. Al_2O_3 and MgO) which oppose the sudden appearance of crystals in the vitreous mass as it cools, which would lead to a fall-off in the transparency and the mechanical properties of the product, should also be considered as correcting agents.

Glass forming

The working of the glass emerging from the melting furnaces has the aim of conferring upon it shapes which are suitable for the uses to which it is to be put. Various techniques are used for this purpose, principally: press-and-blow, lamination, mechanical drawing, floating, and spinning. There are also secondary procedures such as simple blowing and casting.

Press-and-blow

Nowadays, the production of hollow glass vessels is carried out industrially by using pressing-and-blowing in machines which operate at very high rates and are fed directly from melting furnaces which are mainly of the tank type.

There is a distributor sited above the moulds sketched in Fig. 14.26 which receives the molten glass from the furnace orifices and then proceeds to distribute it in suitably

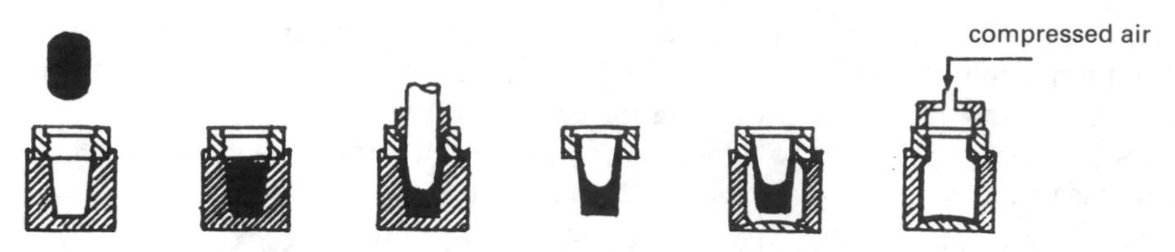

Fig. 14.26—The successive phases in the press-and-blow operation. The figure shows, successively, the blank mould into which a large drop of glass is put, the pressing in the blank mould, the transfer of the blank (the 'parison') into the final blow mould, and the blowing of the parison with compressed air so that it takes on the shape of the finished object.

sized pieces to the moulds. The course of the ensuing operation in the case of a single mould of a certain shape is illustrated in the figure.

The moulds, which are made out of steel, are kept hot during the working of the glass and are operationally synchronized so as to allow the production of hundreds of glass containers per minute when machines are used which have 40 or more press-and-blow units.

Lamination

This is the first of the ways of obtaining sheet glass. This comprises the delivery of glass from a suitable port which is opened in the refractory of a melting zone of a smelting furnace with the outflow regulated by an automatically adjustable refractory tweel (Fig. 14.27). As soon as it is outside the furnace, the molten mass is picked up by two or more pairs of laminating rollers rotating at different speeds and cooled by the circulation of water to avoid sticking.

The transporting rollers follow the forming rollers, and the glass passes to the annealing oven (lehr) where it is finished and, finally, wrapped.

Mechanical drawing

Another method of producing glass sheet is that of drawing. A number of variants of this process exist.

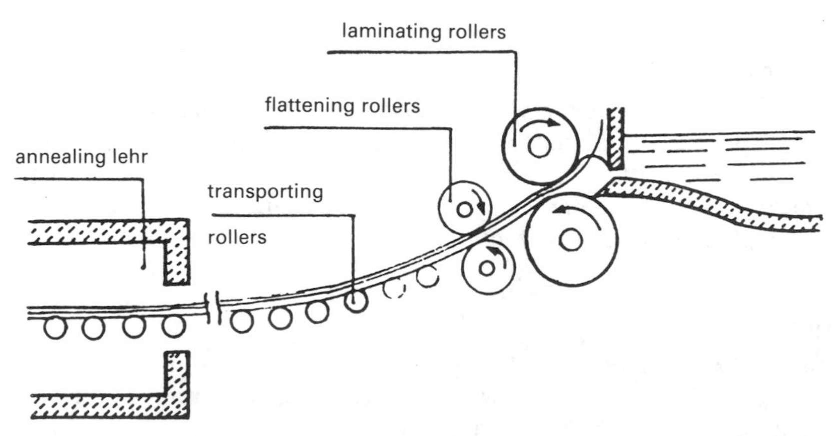

Fig. 14.27—The production of industrial crystals by lamination. Today, this method has been partly superseded because it necessitates more laborious finishing operations than the float process (Fig. 14.30), and, furthermore, induces stresses in the glasses.

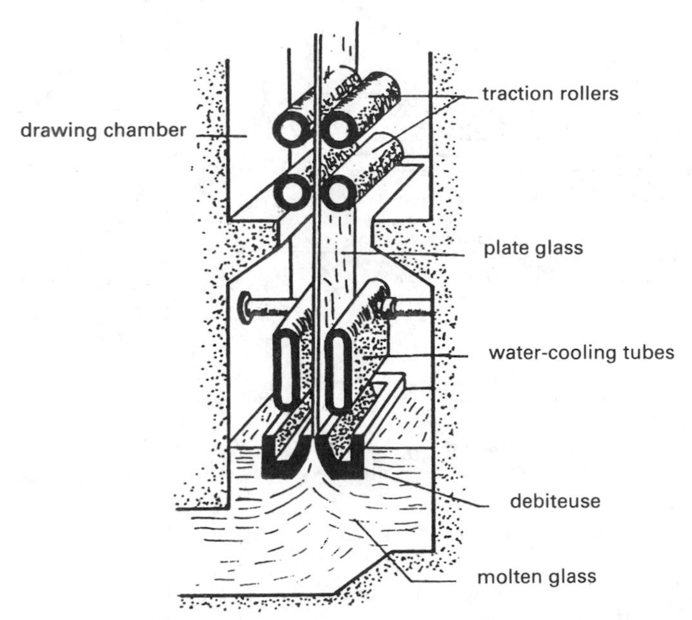

Fig. 14.28—The Fourcault method, which is still used a good deal for the fabrication of common plate glass.

One such variant is that proposed by the Belgian, Fourcault (the Fourcault method), which consists (Fig. 14.28) of the introduction of slotted pieces of refractory known as 'debiteuse(s)' into one or more of the casting ports of a melting furnace (so that it is possible to produce several sheets of glass simultaneously from a single furnace). Such a debiteuse floats on the molten glass and is kept in a given zone by tie rods.

The hemispherical protuberance of glass which is forced through the slot of the debiteuse is clasped for the whole of its length by means of a frame. This frame is then raised to higher floors or to the level of scaffolding sited much higher in the factory by means of pairs of rollers which move apart as the frame passes and then close together again to grasp the sheet firmly. The system of rollers, of which there are about fifteen pairs, constitutes the drawing machine, simply raises the plate without rolling it.

At the top of the drawing machine, operators remove the clamping frame the system at the start of each cycle, and, as the process proceeds continuously, provision is made for the sheet to be cut into suitably sized pieces.

The frame which initially clamps the glass is then used only in the case of accidental breakage or an intentional breakage of the rising ribbon of glass. This, on average, takes place after every twenty days of continuous operation of the machine.

It may also be noted that the ribbon of glass rising from the melt is immediately cooled to 650–670°C by the pair of water circulation cooling units inserted between the debiteuse and the first pair of rollers. This has the aim of preventing any pronounced drawing of the glass with a reduction in its thickness.

Finally, it should be observed that the drawing machine not only has the task of raising the large ribbon but also of cooling it gradually, so that it may be subsequently manipulated.

The American Colburn system, manufactured by the Libbey–Owens–Ford Glass Company, and the Pittsburgh system draw sheets of glass mechanically in a horizontal manner rather than vertically. They differ only in the device which forms the protuberance in the surface of the molten glass which is subsequently clamped in order to draw the glass. The operating mechanism of the Colburn–Libbey–Owens–Ford Glass Company process is schematically depicted in Fig. 14.29.

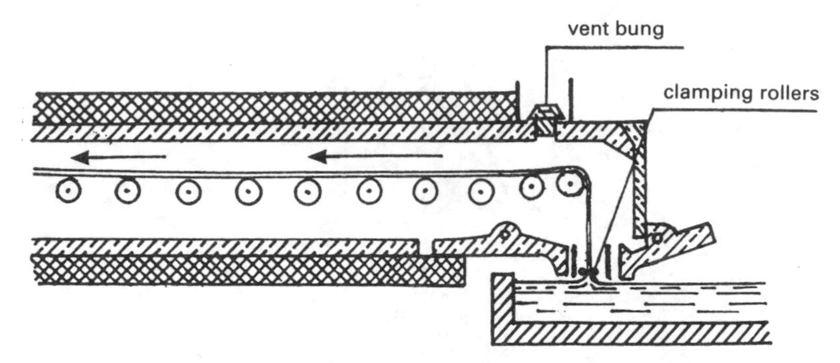

Fig. 14.29—The Colburn method (better known as the Libbey–Owens process) is an analogue of the Fourcault method for the fabrication of sheet glass but developed to operate horizontally. The clamping rollers can be manoeuvred so as to be more (during the clamping phase) or less (during the traction phase utilizing the viscosity of the product) immersed in the molten glass. The vent bung enables the temperature to be regulated in a coarse manner.

The float process

This relatively recent method, developed by the English company, Pilkington Brothers, is rapidly becoming more and more widely used owing to the optimal results to which it leads.

A ribbon of glass is first obtained (Fig. 14.30) by rolling the molten glass from a melting furnace, and this ribbon is then directed onto a bath of molten tin which is chemically indifferent towards the surface of the glass with which it is in contact but which flattens it and polishes it (Fig. 14.30).

The glass floating on the tin is kept molten by the heat produced by burners which, by controlling the environment, allow the sheet/plate glass to adapt to the shape of the absolutely flat, specular surface of the molten metal. A final adjustment to the controlled atmosphere allows the sheet/plate to leave the molten tin bath with quite constant dimensions, so that, at the end, the cut pieces can be marketed and put to use without having either to grind them or to polish them.

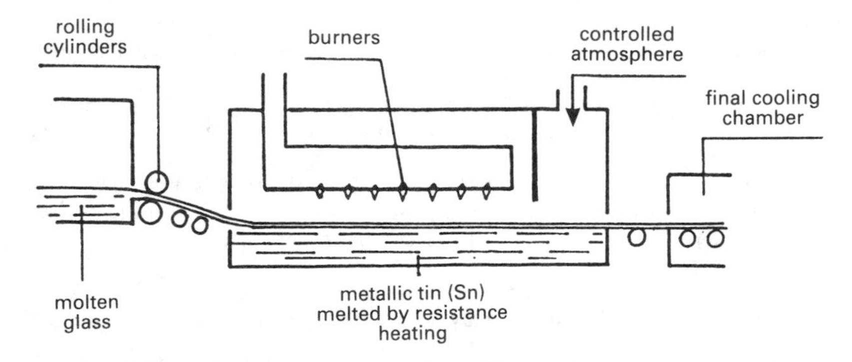

Fig. 14.30—Nowadays, the use of the float glass process for the production of flat glass in general and industrial sheet and plate glasses in particular is becoming ever more widespread as it leads to structures which are free from stresses and do not require any further finishing operations. Regulation of the temperature and the atmosphere which flows over the upper surface of the glass is essential to ensure structural homogeneity throughout the whole mass of the final manufactured product, with obvious benefits when the duration of the operation and the conditions are optimal.

Spinning

The market requires considerable quantities of long or 'textile' fibres and short ('glass wool') fibres which are prepared from the melts in glass production furnaces.

Glass wool is used in the manufacture of thermal insulation materials and in the construction of translucent glasses by interposing layers of glass wool between two plates of glass. In this way a product is obtained which, while it is not transparent, does transmit more light than ground or opal glasses.

Glass fibre, which is also prepared extremely frequently by using silicate blast furnace slags, may be either spun or woven so as to obtain manufactures which are used in the chemical industry (cloths for filters, acid-resistant garments, etc.) and the electrical technology industry for insulation systems.

Glass wool can be produced either by casting it onto notched refractory discs which rotate at a high speed and scatter it into thin threads, or by blowing liquid glass emerging from nozzles with a high-velocity gas stream. On the other hand, the textile fibres are obtained by the drawing of the molten glass through platinum bushings from the other side of which the fibres emerge. After these have been cooled, they are wound onto bobbins. The fine fibres among these have a high tensile strength of the order of ten times that of common glass. This is due to the parallel orientation of the –Si–O–Si–O–Si–O– siloxane chains.

In recent years there has been ever more research into the development of various systems for the production of double layered glass fibres (consisting of a central core and an external coating). These constitute optical fibres in which it is possible, on account of the uninterrupted total internal reflexion of the radiation passing along them, to transmit electromagnetic radiations which are modulated so as to carry information. This takes place because light which is incident on the surface of separation between a medium with a higher refractive index (the core) and a medium with a lower refractive index (the coating) is continuously refracted back into the core.

Transmission via optical fibres offers considerable advantages over transmission by means of cables. Among other things, just two glass fibres can simultaneously transmit more than 1000 signals, as opposed to the 24 which can be transmitted by a cable consisting of several strands of copper.

Pressing and casting

Certain thick glass objects (cups, ash-trays, etc.) are manufactured simply by pressing them in steel moulds covered with a carbon powder base anti-adhesive applied like a varnish with a mixture of linseed oil, pine resin, and cork powder. They are maintained at a temperature such as to avoid fracture (caused by too low a temperature) and surface roughening caused by too high a temperature).

Finally, the pressed article must be reheated to remove the flashing left where the parts of the mould join.

Casting, which is very costly and carried out by pouring the molten glass into sand moulds, is used only for the moulding of pieces of special glasses such as the mirrors of binoculars, solar energy mirrors, and certain lenses. The high cost of manufactures produced in this way arises from the need to ensure that there are no structural inhomogeneities which tend to be created in glasses for a number of reasons*, as well as the need to refinish such products by sophisticated techniques which require a large amount of time.

*Among other things, simple gaseous occlusions, thermal differences, and casting waves are responsible for the formation of inhomogeneities.

The finishing of glasses

There are four operations which may be required after the glass has been moulded in order that the products may be thought of as being finished: annealing, toughening (tempering), cleaning, and polishing.

Annealing

As the glass comes into contact with relatively cold metallic parts and with the external environment during the moulding operations, differential stresses are created between its external and internal parts which make the glass particularly sensitive to impacts and other physicomechanical actions. To eliminate these stresses, it is necessary to anneal the glass, that is, to bring the glass up to a temperature at which all of the abovementioned stresses are released, the value of which is dependent on the thickness of the piece being annealed. The rate at which the glass is subsequently cooled, which must be such as to allow the entire mass to normalize itself completely, is also dictated by the dimensions and, in particular, by the thickness of the objects which are being annealed.

Toughening

Toughening is the reverse treatment to annealing in the manner in which it is administered, and is carried out with the aim of making the glass extremely impact resistant.

Toughening is carried out by first heating the moulded glass to just below the softening point (a temperature of the order of 600°C) and then quenching it rapidly with powerful jets of air. In this way, the surfaces of the manufactured pieces stiffen while freezing their volumes, while the interiors of the objects remain soft. As a consequence of its subsequent cooling, the internal part (the heart) is stretched out by the external part as the former contracts, and is therefore subject to tensile forces. When such glass is subsequently subject to an impact, that is, a compression, this simply reduces the tensile stresses which are present, and in this way, the system is more impact resistant owing to the compensation between the tensile and compression forces.

Toughened glass cannot be worked (e.g. cut or drilled) because the forces to which its centre and shell are subjected are increased by this, which leads to the embrittlement of the object being treated.

Cleaning and polishing

To remove any particles of material which may have stuck to the glass during the forming operations and to eliminate surface irregularities, cleaning is carried out with rotating cast iron plates which are first lubricated with aqueous suspensions of sand and, subsequently, with ever more finely subdivided abrasives.

Cleaning is followed by the polishing of the glasses which is carried out by using rotating metallic plates covered with felt which is moistened with an aqueous suspension of powdered haematite (Fe_2O_3). A large amount of friction is developed in this operation, and very large amounts of energy are consumed before the worked surfaces are rendered perfectly bright and transparent in this way.

Industrial plate glasses

Industrial plate glasses are simple plates of glass with perfectly plane and parallel surfaces which are produced either simply by the float process on fused tin or by means of the normal methods for the production of sheet and plate glass, followed by levelling and polishing, using the techniques which have just been described.

14.8 DEFECTS IN GLASSES

Glass is subject to defects due to it being made incorrectly, to the wrong conditions being used for its storage and use, to the ageing of the material. Of these defects, the following are the best known:

- heterogeneity which manifests itself as threads, waves, and spots due to incomplete homogenization during fusion, to the inferior quality of the raw materials, and to insufficient refining;
- gas bubbles which are characteristic of badly refined glasses;
- stones, that is, crystalline inclusions arising from devitrification phenomena, from the incomplete reaction of the raw materials, and from fragments of refractory materials which have been eroded away;
- deformations, convexity and unequal thicknesses which are the result of poor working practices, errors in the mode of storage and transport, unbalanced stresses, etc.;
- iridescence and other chromatic defects which arise when the decoloration processes have not been carried out correctly, when there are structural inhomogeneities, and when there are other causes, present which lead to variations in the refractive index.

Such defects in glasses can be prevented by paying attention to the quality of the raw materials, the correct blending of the amounts of raw materials, by very carefully watching over the firing, refining, and finishing processes, and averting all the anomalies resulting from such defects by making every attempt to reveal them during the production stage.

Modern systems of automatic production control therefore offer secure guarantees for the prevention of defects in glasses.

Furthermore, the environments are carefully controlled, the methods are examined, and the times for the whole maturing operation are checked before starting the production of the glasses, in order to safeguard them against defects.

14.9 CLASSIFICATION AND REVIEW OF GLASSES

There is a great variety of glass products owing to the enormous number of different possible chemical compositions which differ from one another to a greater or lesser extent, and the various ways in which they can be worked.

So far as their composition is concerned, there are five types of glass: silica glasses, common glasses, potassium glasses, lead glasses and borosilicate glasses.

So far as the mode of working them is concerned, there are: ground glasses, armoured glasses, semicrystal glasses, and safety glasses.

Other classes of glass are placed within the context of the types of glasses which have just been described. For instance, coloured glasses and opacified glasses may belong to the various types of glass to which colorants (page 592) or opacifiers (page 593) have been added during their melting.

Glasses of various compositions

Silica glasses

By melting sand containing more than 99.8% of SiO_2 or quartz at temperatures above 1710°C, extremely viscous melts are obtained which can be degassed only with great difficulty. Upon cooling, either a translucent white product or, less easily, a transparent product is obtained according to the method which is used. Both of these glasses are costly, owing to the special melting techniques and the very high temperatures which have to be employed.

The translucent variety (Vitreosil) is employed in parts of plants and in vessels which are subjected to thermal shock or come into contact with hot concentrated acids, since this glass has an extremely low coefficient of thermal expansion and is absolutely resistant to acids apart from hydrofluoric acid. The electrical resistance of this type of glass is also outstanding.

The transparent variety (clear silica glass) has better mechanical properties, is less inclined to devitrify, and is less permeable to gases than Vitreosil, from which it can be formed by prolonged heating at a temperature slightly below its melting point. As this glass is perfectly transparent to ultraviolet, visible, and infrared radiation, it is mostly employed in the construction of components for scientific apparatus concerned with such electromagnetic waves.

Common glasses

Common glasses, which are obviously the most widely used, are those which are composed of the salts of SiO_2, CaO, and Na_2O with the accidental or, rarely, intentional inclusion of Al_2O_3 and Fe_2O_3. The mean compositions of these glasses are shown in Table 14.6 in relation to the uses to which they are put.

Table 14.6—Chemical composition (in %) and uses of common glasses

Uses	SiO_2	CaO	Na_2O	Al_2O_3	Fe_2O_3
Generic	72–74	10–12	12–18	1–2	traces
Green bottles	60–65	13–20	7–15	2–7	1.4
Colourless hollow ware	71–76	5–15	12–18	1–4	traces
Mirror glasses	70–73	13–15	10–15	0.5	0.1
Lamp glasses	72–75	10–12	12–15	0–1	traces
Sheet glasses	71–74	10–15	13–17	0.5	0.2
Semicrystal glasses*	70–72	14–15	12–15	0–0.4	traces

* As will be stated later, traces of the 'nucleation agents': ZrO_2, P_2O_5, ZnO, and TiO_2 are also necessary in this subtype of glasses.

Potassium glasses and lead glasses

Glasses in which the sodium silicate is largely or completely replaced by potassium silicate are potassium glasses. These glasses are more expensive, hard, and brilliant than common soda glasses.

Typical examples of potassium glasses are the Bohemian crystal glasses from which table services, chandeliers, and plates are made. These contain from 72–76% of SiO_2, 8–10% of CaO, 0–3% Na_2O, 12–15% of K_2O, and from about 1% of Al_2O_3.

A special kind of potassium glass which is mainly used in optics (crown glass) contains small amounts of B_2O_3 (about 3%) which replaces SiO_2.

Glasses containing potassium and lead silicates in a silica matrix are lead glasses which are known as lead crystal. The average composition of these lead glasses is 54% SiO_2, 14% K_2O, and 32% of PbO, that is, they are made out of raw materials which as a whole are far more expensive than those used in the production of common glasses. As well as being more expensive than the latter, they are also heavier, more fusible, and more brilliant owing to their greater refractive index and their greater ability to disperse light. Finally, they are much more opaque to X-rays and γ-rays than are normal glasses.

Lead crystal glasses are used in the production of crystal-ware, ornamental and furnishing glass-ware, and brilliants for jewellery which imitate precious stones (artificial diamonds). They are also used in the construction of insulators, electric lamps, electronic tubes, and luminescent tubes as well as in the construction of screens which are transparent but absorb high-frequency X- and γ-radiations.

Optical heavy flint glass is a special glass formed by the addition of lead (20% SiO_2 and 80% PbO). It takes its name from the fact that it has the composition of a classical flint glass which has been doped by lead to increase its weight (and not only with the normal criterion of an Abbé value < 50).

Borosilicate glasses

These are glasses which are poor or relatively poor in alkalis but, on the other hand, enriched with Al_2O_3 and containing 12% or more of B_2O_3. Borosilicate glasses, which are characterized by a low coefficient of expansion, are used in the construction of vessels, containers, and apparatus which are resistant to thermal shock.

The composition of the well known borosilicate glasses is shown in Table 14.7.

Borosilicate glasses are used in the production of chemical laboratory apparatus, thermometers, heat-resistant domestic utensils, television tubes, and other structures which are required to have constant dimensions, a high softening point, and, as has already been said, a minimum coefficient of expansion.

Table 14.7—Makes and composition (in %) of borosilicate glasses

Make	SiO_2	CaO	Na_2O	K_2O	Al_2O_3	B_2O_3	ZnO
Jena	70–72	1–2	10–12	–	5	12	–
Pyrex	80.5	–	4.5	–	8.2	12	0.8
Durax	75	–	3.5	1	5.5	15	–
Thermoglass	67–72	–	9–12	–	2.5–5	4–12	0–7

Glasses obtained by various treatments

Ground glasses and semicrystal glasses

Ground glasses are translucent, and the whole of their surface area is rough or they are partly decorated by the grinding of patterns onto a part of the surface. This is generally carried out by a sand treatment which uses either the projection of fine grains of sand onto the glass or grinding in aqueous suspensions of fine sands. However, it is also possible to make ground glass by attacking the parts of it which are not protected by either wax or paraffin, with hydrofluoric acid.

By causing a glass to assume a structure which is partly crystalline, that is, by preparing semicrystal glasses, a ceramic glass (glass-ceramic) is obtained because ceramics also have structures in which a crystalline phase is dispersed in a vitreous phase.

The processes for the conversion of a glass into a ceramic are stimulated by the introduction of nucleation catalysts (oxides of the TiO_2, ZnO, and P_2O_5) into the raw materials and then carrying out suitable heat treatments during the working of the glass after it has been moulded. The minutest crystals (crystallites) which are made to form by the nucleating agents, grow in size according to the manner in which the moulded objects are worked, without ever arriving at the complete devitrification of the mass.

The variety of possible products, which is mainly dependent on the choice of the necessary maturation temperatures, is very large because so many different degrees of crystallinity can be attained.

The common properties of semicrystal glasses, which are outstanding to a greater or lesser degree, are their non-porosity, high heat resistance, excellent mechanical strength and chemical resistance, and their excellent dielectric characteristics. These glasses are therefore used for, among other things, missile nose cones, extrusion nozzles which have been mechanically, chemically, and thermally tested in practice, and as refractories for antithermal and anti-acid linings.

Armoured glasses and safety glasses

By interposing metallic meshes between plates of glass which are in the plastic state and made plastic after putting the meshes in and then compressing and calandering the systems, very resistant glasses are obtained which, in analogy to the name given to other systems with such structures, are referred to as armoured glasses.

Greenhouses and conservatories, parapets, submerged port-holes, and the window panes for basement dwellings are constructed from such armoured glasses. By toughening the glass, which has already been described, a product is obtained which is particularly impact resistant and which, even if it is broken, fragments into minute pieces (with, however, sharp cutting edges).

Other safety glasses (shatter-proof) are obtained by the introduction of thin (slightly less than half a millimetre) layers of transparent plastic materials between perfectly polished normal sheets of glass. For the most part, the plastic foil which is used nowadays is pre-dried polyvinyl butyral.

The sandwich is heated to 180–230°C for predetermined times, and, after cooling, its edges are ground. In the event that these products are fractured, the fragments remain stuck onto the plastic foil. Such laminated glasses are often a legal requirement for cars, while the toughened glasses are used for industrial vehicles and in large-wheel transport vehicles of all kinds.

14.10 ENAMELS

By melting mixtures of feldspars, borax, silica, and sodium carbonate in rotating furnaces and quenching the fused mass, subdivided, into droplets in water, low-melting-point granular glasses are produced which, having been milled in an aqueous suspension with the addition of variable amounts of plastic clays and colouring oxides, take on the form of fluid slips which can be applied by a spray gun, a screen, or by spreading. The products obtained in this manner are referred to as enamels, and are applied to metal surfaces, generally, of cast iron or steels which, as a result of a heat treatment at 800–1200°C, finish up by bonding to them very tenaciously.

Enamel formulations are mainly based on the chemical resistance which they must exhibit, on how well they adhere to the support, and hence the nature of this support.

Special expedients, such as the application of foundation layers (counter-oxides) which are readily compatible both with the metallic material of the support and with the enamel on the metallic surfaces which are to be enamelled, enhances the adherence of the coatings and impedes the reduction of the colouring oxides by the carbon in the alloys being coated.

Part 3 BINDERS AND BUILDING AGGLOMERATES

14.11 GENERALITIES CONCERNING BINDING MATERIALS USED IN BUILDING CONSTRUCTION

The materials which are used either alone or, more frequently, as part of a mixture with suitable inert substances in order to manufacture mortars for the connection of stones and bricks, for the finishing of surfaces and joints, and the production of simple and reinforced concretes destined to be used, when cast, in the building of every kind of brickwork and masonry buildings, constitute what are known as binders for the building industry.

The term 'binders' is used to indicate the action which these materials exhibit in bonding constructional elements together in a manner which is more or less rapid and mechanically strong.

There are two basic types of binder:

- air-setting binders, where the presence of air is a prerequisite for their setting,
- hydraulic binders, where water serves to bring about the reactions which lead to hardening.

As we shall see, there also exist intermediate type binders which principally lead to hydraulic mortars, that is, to mixtures of lime and artificial and natural acidic oxides. If they are artificial, one often speaks of 'pozzolanic mortars'.

The air-setting mixed mortars manufactured from gypsum, lime, and water (and, possibly, sand) can also be considered as an intermediate type, given that the water which is added to form them also participates in the setting.

Every type of binder has its own specific field of application.

Today, the whole collection of binder-based products form an extremely versatile range of materials which have permitted structures to be erected which were quite unforeseeable even a few decades ago. Hydraulic binders have fostered the greatest innovations in this field as they have made it possible to create structures of various weight endowed with high chemical resistance and mechanical strength. They also offer appreciable aesthetic values at costs which are contained, above all, by the ease with which they can be used.

Nowadays, the production of building binders, their substitutes and additives, and also the application of mortar and concrete binders, engages whole sectors of applied and industrial chemistry which are predominantly concerned with engineering and thermotechnical activities. Our treatment will obviously tend to emphasize the chemical aspects of the problematics of binders, while not forgetting to illustrate some of the thermal and mechanical treatments used in the preparation of products destined to act as reagents in chemical processes.

14.12 AIR-SETTING BINDERS

The materials used in the manufacture of mortars which endow them with the property of setting and hardening in air are 'air-setting' binders.

It is necessary here, to agree on the meanings of the terms 'setting' and 'hardening'. A binder sets when it takes on a paste-like consistency around the bodies (pebbles, bricks, etc.) to which it mutually bonds, while it hardens when, together with the system of which it is a part, it takes on a stone-like consistency and acquires adequate mechanical strength.

Air is essential for the action of these binders, both because their setting is brought about by the evaporation of water from the paste and because the hardening is stimulated by supplying the CO_2 which is necessary for this and by providing a way in which the water formed in the chemical reaction can evaporate. The absolute necessity that they should come into contact with dry air is one of the major limitations on the use of these binders, as it is often the case that the structures are either of greater thickness or being put up in a wet atmosphere.

Technically, the most important air-setting binders are the air-setting limes and gypsum.

Air-setting limes

The materials obtained by the firing of suitable natural chalks, consisting of calcium carbonate containing small amounts of impurities, are air-setting limes.

Microcrystalline chalks containing small amounts of impurities (limestone) are the best raw materials for air-setting limes. A certain level of impurities is, however, necessary in order that the resulting lime should be in rather uniform blocks as is required in practice so as to avoid losses and handling difficulties. In fact, pure chalks lead to very fine powders which are readily dispersed.

The preparation of air-setting limes

Air-setting limes are produced by the burning of chalks in lime kilns. The pieces of chalk should be of such a size as to ensure that the decomposition reaction goes to completion during the time which they spend in the kiln, and a sufficiently fast current

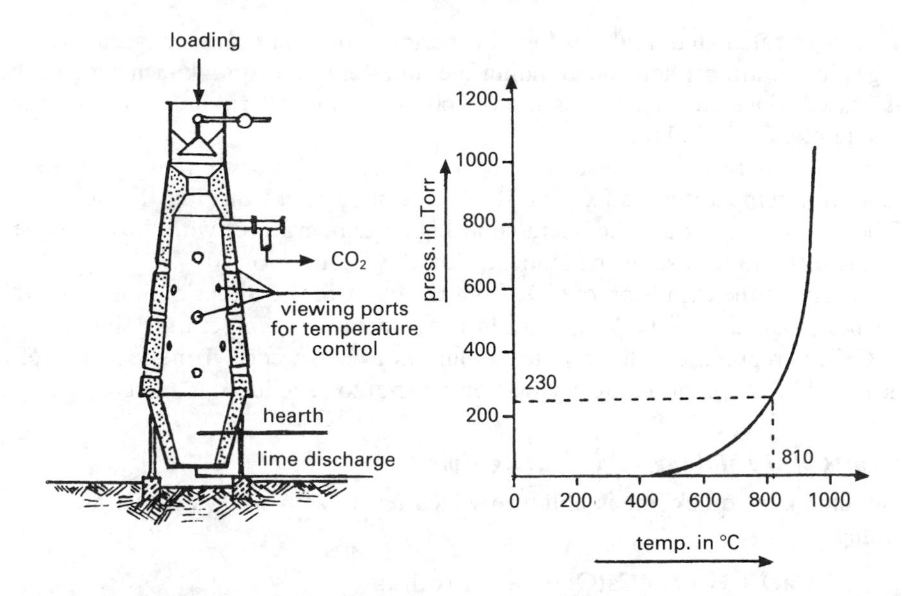

Fig. 14.31—Left: Shaft kiln for the conversion of limestone, which is loaded in from the top, into quicklime which is removed from the bottom. Right: The variation in the vapour pressure of solid $CaCO_3$ as a function of temperature. It is seen that, under the conditions for the ideal functioning of the system, the theoretical decomposition temperature of limestone in a CO_2 atmosphere is 810°C. Above this temperature, the vapour pressure of CO_2 increases rapidly, thereby allowing the reaction which transforms calcium carbonate into quicklime to take place at an enhanced rate.

of air must be passed through this kiln. In general, it is found that the size of the pieces must be inversely proportional to the visible crystalline structure of the chalk.

Lime kilns may be of various shapes, but shaft kilns, that is, kilns with the shape of the two truncated cones joined at their larger ends, are preferred. The exterior of these kilns is constructed of brick, and they are internally lined with silica–alumina refractories (Fig. 14.31).

The burning is brought about by alternately introducing suitably sized pieces of limestone and carbon (coke or lean short-flame anthracite) with a low ash and sulphur content.

Horizontal rotating kilns are also used. The limestone and the gases from the combustion of gasogene or natural gases pass through these in a countercurrent manner. In this way, not only are higher production rates achieved but contamination of the product by the fuel ash is avoided.

The reaction for the dissociation of calcium carbonate which lies at the foundation of the conversion of limestones into lime is:

$$CaCO_3 \rightleftharpoons CaO + CO_2 - 42.5\ \text{kcal/kg}$$

This is a chemical equilibrium which is, as usual, controlled by the phase rule on the basis of which, when n = the number of components = 2 and f = the number of phases = 3, it is calculated that:

$$v = n + 2 - f = 2 + 2 - 3 = 1$$

hence it is found that the system has a single degree of freedom.

Of the two physical variables acting on the system (temperature and pressure) only one of them can therefore be arbitrarily set since once the partial pressure of CO_2 (the only gaseous component) has been fixed, the kiln operating temperature is thereby determined.

It may be calculated, and it is found experimentally, that CO_2 constitutes 30% of the gas in the atmosphere surrounding the limestone in a lime kiln no matter how it is heated, since, at a total pressure of 760 torr (atmospheric pressure), the partial pressure of CO_2 is 230 torr.

On the other hand, at such a partial pressure, the minimum (theoretical) dissociation temperature is fixed at 810°C as can be seen from Fig. 14.31.

The kiln may obviously be operated at higher temperatures with the consequence that there is an increase in the tendency for CO_2 to be evolved.

To promote the expulsion of CO_2 from the cores of the pieces of limestone which have been loaded into the kiln, the kiln is operated at a temperature slightly above 900°C. This represents a limit which should not be exceeded if the tendency of the lime to yield a voluminous and plastic mortar is not to be reduced to too great a degree.

Products of the slaking of air-setting limes

From lime kilns quicklime is obtained which reacts with water exothermically with swelling:

$$CaO + H_2O \rightarrow Ca(OH)_2 + 210\ \text{kcal/kg}$$

It is, in fact, on account of this reaction that quicklime withers, by crumbling when exposed to air, owing to the moisture present in the atmosphere.

When it is suitably slaked with water, quicklime can yield two principal products: hydrated lime and chalk putty.

Hydrated lime (slaked lime) is the product obtained by the slaking of lime with the amount of water which is required to convert it into calcium hydroxide. Theoretically, 32 kg of water is required for every 100 kg of quicklime, based on the fact that every 56 g of quicklime (RMM of $CaO = 56$) requires 18 g of water. In practice, however, a greater amount of water is used in order to allow for that which evaporates.

Both quicklime and hydrated lime may be 'fat' or 'lean'. Fat limes are derived from limestones containing more than 90% of $CaCO_3$, with the products of their burning containing more than 94% of $CaO + MgO$ with an MgO content less than 5%. On the other hand, lean limestones originally contain less than 90% of $CaCO_3$ and less than 94% of $CaO + MgO$ after burning, with an MgO content exceeding 5%.

When the MgO content in limes exceeds 10%, they are unsuitable for use as constructinal binders.

Slaked lime is stored in silos where the hydration reaction goes to completion. The material is subsequently milled and bagged.

The product which is prepared by the rapid addition of a suitable amount of water to slaked lime or by the labororious direct addition of water to quicklime is called chalk putty.

To prepare this by the second (classical) method, quicklime, contained in vessels (hot fomenters) is slaked with an excess of water and the product is passed through an iron grill into large vessels with permeable walls known as lime-pits. The grill holds back any remaining granules of quicklime, while the walls absorb the excess of water and the salts which are dissolved in it.

The chalk putty remains in the lime-pits, which are covered to prevent its carbonation by air, for as long as is necessary to allow the complete hydration of the quicklime component which still remains in it. This must be done, as any remaining granules of quicklime would swell up after the material had been used, causing cracking and the breaking away of the surface of the products.

However it is prepared, chalk putty is a colloidal paste consisting of $Ca(OH)_2$ and water.

The yield of chalk putty is the volume (m^3) of the product which is obtained from one tonne of quicklime, and the corresponding quantity (1.3 metric tonne) of hydrated lime. The higher the yield of chalk putty, the greater the volume of the brickwork which can be laid or the greater the area which can be plastered. The chalk putty yield depends on the fineness to which the initial limes have been milled, and, in every case, this is smaller for lean limes than for fat limes. The lower tolerance limit for the yield is 2.5 m^3.

Air-setting mortars

The air-setting mortars are pastes consisting of slaked lime, water, and sand, or of chalk putty and sand.

The ratio in which the components are mixed in such mortars depends on the grain size distribution of the sand because, as the slaked lime which is treated with water or the chalk putty must fill all the empty spaces between the grains; the volume to be filled up depends on the shape and the dimensions of the grains. When chalk putty is used and the grain size of the sand lies within a range from 0.5 to 2 mm, it is necessary to take 2–3 volumes of sand to one of chalk putty as a rough guide.

However, when the air-setting mortars must have particular properties in relation to the uses to which they are to be put, the types of paste and the ratios between the lime, sand, and water to be adopted in their preparation may vary. For example, air-setting mortars for brickwork require a lime:sand:water ratio of 0.4:1:1.2, while those for the plastering of houses require a ratio of 1.2:24:0.9.

The name of air-setting mortars with lime-based binders derives from the fact that they can set around the constructional elements (bricks, stones, etc.) and then attain sufficient mechanical resistance (harden) solely by exposure to air.

In fact, once an air-setting mortar has been applied and exposed to air, processes of varying degrees of complexity take place in it:

- the dehydration of the chalk putty which involves the loss of water and the crystallization of $Ca(OH)_2$;
- the extremely slow carbonation of the calcium hydroxide due to the action of the CO_2 in air: $Ca(OH)_2 + CO_2 \rightarrow CaCO_3 + H_2O$;
- reactions between the slaked lime in the mortars and the oxides (Fe_2O_3, Al_2O_3, and SiO_2) of the fired clay in the bricks which exhibits the properties of artificial pozzolana. This may be represented schematically as:

$$Ca(OH)_2 + \text{fired clay} + nH_2O \rightarrow \text{hydrated calcium silicates, ferrites, and aluminates}$$

The reactions in this complex scheme are known as pozzolanic phenomena, and they require that the bricks should be soaked in order that these reactions may take place when they come into contact with the mortars. Phenomena occur here which lead to a bond being formed between the air-setting mortar and the bricks which is such that, after it has been allowed to mature, it makes it impossible to separate the constructional elements without breaking them.

Nature, reactions, and uses of gypsum

The natural rock consisting of gypsum is calcium sulphate dihydrate ($CaSO_4 \cdot 2H_2O$) and is more or less pure. It is a sedimentary rock of physicochemical origin with an average composition of 28–35% CaO, 42–45% SO_3, and 20–30% H_2O.

When calcined at between 120°C and 160°C, gypsum partly losses its water of crystallization and is transformed into the hemihydrate, $CaSO_4 \cdot \frac{1}{2}H_2O$, while, at 200°C, the hemihydrate begins to be converted into anhydrite, $CaSO_4$, which, even more slowly than the hemihydrate, also has the capacity to set by reconverting itself back into calcium sulphate dihydrate by absorbing water:

$$\left.\begin{array}{l} CaSO_4 \cdot \frac{1}{2}H_2O + 1.5H_2O \\ \text{hemihydrate} \\ \\ CaSO_4 + 2H_2O \\ \text{anhydrite} \end{array}\right\} \rightarrow CaSO_4 \cdot 2H_2O$$

Both the hemihydrate and anhydrite are more soluble in water than the calcium sulphate dihydrate. On account of this, aqueous suspensions in which the hemihydrate and anhydrite are for a large part in a state of true solution, are supersaturated with respect to the dihydrate salt, with the result that the latter crystallizes out in the form of long needle-like crystals which are interlaced with one another, in such a way as to form a compact mass, with an increase in volume. The removal of the hemihydrate and anhydrite salts from the solution takes place in this way and leads to the gradual solution of the remainder of the latter which is still in suspension, which, in turn, leads to the further precipitation of interlaced needle-like crystals of $CaSO_4 \cdot 2H_2O$ until, finally, (provided that sufficient water has geen added), it is found that all the products of calcining of the initial gypsum at temperatures below 500–600°C are transformed into a consolidated mass.

As the temperature at which natural gypsum is calcined is raised, its capacity to set falls off at above 600°C. One then has dead-burnt gypsum owing to an irreversible modification of the crystal lattice.

If, however, the calcining temperature is raised to above 1000°C, the gypsum regains its capacity to set, owing to the formation of a small amount of CaO in the interior of the superheated mass:

$$CaSO_4 \rightarrow CaO + SO_3$$

The product obtained in this manner constitutes hydraulic gypsum which, in practice, yields mechanically strong articles owing to the crystallization of basic sulphates.

The kilns used for the calcining of gypsum must be constructed in such a way that the combustion gases do not come directly into contact with them, because they tend to reduce gypsum to calcium sulphide according to the reaction: $CaSO_4 + 4CO \rightarrow CaS + 4CO_2$.

The kilns are either of the rotary or pot type (Fig. 14.32) and are heated indirectly, but may also be of the muffle tunnel type.

Putty is obtained by forming a paste from the hemihydrate with an adhesive and the addition of various oils which is used for fixing glass window panes and for making decorative finishes.

Alum gypsum or 'English cement', with which artificial stones are made, is obtained

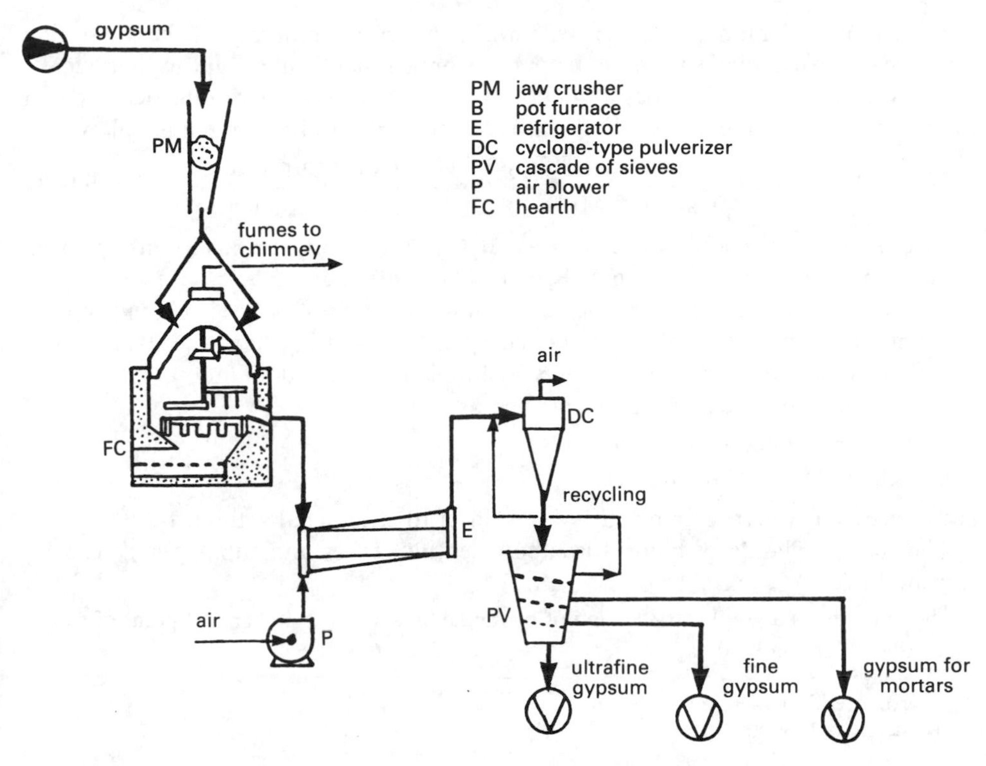

Fig. 14.32—A technological solution to the preparation of the commercial types of gypsum. Gypsum rock, which has been sorted and enriched and then crushed and subsequently calcined while being mixed and indirectly heated by the fumes emerging from the side hearth of a pot kiln, is withdrawn from the kiln and pneumatically conveyed to a cyclone which reduces it to a powder. The product formed in this cyclone is classified by sieves into a fraction to be recycled and into fractions with certain particle size distributions which are intended for various uses.

by mixing the hemihydrate of calcium sulphate with solutions of rock alum and recalcining the paste at 550°C.

Two types of mortar are then made with the gypsum:

- simple mortars, which are made by mixing gypsum which can set with water in weight ratios varying from 1:0.75 to 1:1,
- compound mortars, which are formed by blending gypsum with slaked lime (in proportions which vary, depending on the intended application), water, and, possibly, sand.

The mechanical properties of structures made of compound mortars, the setting and hardening of which entail several complex phenomena (hybrids of the reactions which take place in calcium sulphate hemihydrate, hydraulic gypsum and limes) are considerably better than those of mortars which contain only gypsum, and quite remarkable when compared with the properties of lime mortars.

14.13 HYDRAULIC BINDERS

Binders which are capable of setting under water in the absence of air are referred to as 'hydraulic' binders. The transition from air-setting binders to hydraulic binders is not sharp, and binders are met with which are predominantly of the air-setting type (weakly hydraulic) while other binders are pre-eminently hydraulic in character.

The quantity which provides a quantitative measure of the hydraulic activity of a binder is the hydraulic index (I_h) which is formulated and interpreted as follows:

$$I_h = \frac{\%SiO_2 + \%Al_2O_3 + \%Fe_2O_3}{\%CaO + \%MgO} = \frac{\%\text{ acidic components}}{\%\text{ basic components}} \quad (14.1)$$

In this formula, metal oxides such as Al_2O_3 and Fe_2O_3 are also considered as acidic components in the strongly alkaline environment.

The dependence of the hydraulic activity on the value of I_h is visible in the case of limes where those which are almost completely air-setting have $I_h = 0.16$ (max), and those which are almost completely hydraulic have $I_h = 0.52$ (max).

Hydraulic binders are officially classified as:

- standard hydraulic binders,
- non-standard hydraulic binders,

according as to whether or not they are subject to strict legal standards.

The class of 'standard' binders is subdivided into types and sub-types, as can be seen in Table 14.8.

The following is a good classification of hydraulic binders from a chemical point of view:

(1) silica–alumina binders
(2) alumina binders
(3) iron ore binders
(4) magnesia binders

Table 14.8—Official classification of hydraulic binders

Standardized			*Non-standard*	
Types	*Names*	*Sub-types*	*Types*	*Names*
(A)	normal cement	Portland, pozzolanic, and blast furnace	(A)	iron ore cements
(A)	high strength cements	Portland, pozzolanic, and blast furnace	(B)	low heat of hydration cements
(A)	high strength and rapid setting cements	Portland, pozzolanic, and blast furnace	(C)	cements for deep bores
(B)	alumina cements	same as type		
(C)	cements for retaining barrages	Portland, pozzolanic, and blast furnace	(D)	white cements
(D)	cement agglomerates	slow and fast setting	(E)	coloured cements
(E)	hydraulic limes	natural in lumps, natural or artificial as powder, predominantly hydraulic natural or artificial as powder, artificial pozzolanic as powder, artificial iron working as powder	(F)	metallurgical supersulphate cements
			(G)	magnesia cements
			(H)	other cements or whatever new hydraulic binders may be introduced

Four of the standard types of hydraulic cements ((A), (B), (C), and (D)) and the majority of the non-standardized cements belong to the chemical class (1).

The standard hydraulic binders are therefore incomparably the most important, and, among these, those of type (A) stand out. It is for this reason that the greatest amount of attention will be paid to the latter binders.

STANDARDIZED HYDRAULIC BINDERS

Portland cement

The Italian standards define Portland cement as the product which is obtained by the crushing of clinker with the addition of gypsum or anhydrite in the necessary amount in order to regularize the hydration process.

Clinker is the material which is discharged from cement kilns and essentially consists of hydraulic calcium silicates.

Portland cement is also referred to simply as 'Portland' or simply 'cement', and it is by far the most used hydraulic binder not only on account of the ready availability of the raw materials used in it but also because it forms part of the make-up of the other standardized hydraulic binders.

The composition of Portland cement

Important information on the four essential components of Portland cement is summarized in Table 14.9, and the four secondary components of the raw materials for Portland cement are described in Table 14.10.

During the calcining of Portland cement, the four essential components in the raw materials combine with one another by reacting both in the solid phase and in the liquid phase. The calcium salts which are formed in the rather complex structure are expressed in terms of the oxides which they contain, as is shown in Table 14.11.

The cementing agents start to be formed in the solid state during the calcining of the cement. However, the fluxes are formed before the cementing agents. The fluxes, on account of their melting points, liquefy as soon as they are formed, and the fused mass which is formed in this way constitutes the phase in which there is a rapid acceleration in the appearance of the cementing agents.

Table 14.9—The basic components of Portland cement

Name	*Formula*	*Symbol*	*Reactions produced*	*Origin*
Lime	CaO	C	basic	(mainly) from limestones and marls
Silica	SiO_2	S	acidic	from marls, quartzites, clay, or silica sand
Alumina	Al_2O_3	A	acidic	from marls, clays, schists, bauxites, or kaolin
Ferric oxide	Fe_2O_3	F	acidic	from marls, clays, iron minerals, or pyrites ashes

Table 14.10—The secondary components of Portland cement

Name	*Formula*	*Reactions produced*	*Origin*	*Damaging effects caused*
Magnesia	MgO	basic	from magnesia limestones, micas, micaceous clays or serpentine	if free in clinker it hydrates, increasing the volume and splitting the manufactured item†
Potassium oxide Sodium oxide	K_2O Na_2O	these alkalis are basic	from feldspars, amphiboles and pyroxenes	hinder the complete combination of the lime and cause setting anomalies
Sulphur trioxide	SO_3	acidic	from gypsum and gypsum-bearing clays	with its derivatives causes dangerous expansions in cementing work‡

† According to the Italian standards, the upper permissible limit for MgO in cements is 4%.
‡ The amount of SO_3 in finished cements must not exceed 3%.

Table 14.11—The basic composition of Portland cement

Function of groups of salts	*Names of individual salts*	*Formula of salts (in terms of the the oxides)*	*Symbols*	*Mean % of the mass*
Cementing agents	tricalcium silicate (alite)	$3CaO \cdot SiO_2$	C_3C	80%
	bicalcium silicate (belite)	$2CaO \cdot SiO_2$	C_2S	
Fluxes (celite)	tricalcium aluminate	$3CaO \cdot Al_2O_3$	C_3A	20%
	ferric phase§	$4CaO \cdot Al_2O_3 \cdot Fe_2O_3$	C_4AF	

§ Actually, the ferric phase consists of a solid solution of alumina and ferrites among which tetracalcium aluminoferrite (C_4AF), which is known as 'brownmillerite', predominates to a point where, as is indicated in the table, the entire phase is assumed to consist of it.

The moduli or 'indices' of cement

To represent useful information regarding the properties and behaviour of a cement, it is customary to make use of certain numerical indices which are known as 'moduli' and are expressed by the conventional symbols for the basic or acidic oxides present in the cement. These symbols are written here in italics to indicate that they refer to the percentages by weight which are present in the cement.

First of all, there is a modulus which is the inverse of the index of hydraulic character, the hydraulic modulus:

$$M_h = \frac{C}{S + A + F} \tag{14.2}$$

which must have values lying between 1.7 and 2.25.

Although it is widely used, M_h is not of great importance since cements containing many different cementing agents and fluxes may have the same hydraulic moduli.

Another numerical index is the silica modulus:

$$M_s = \frac{S}{A + F} \tag{14.3}$$

which expresses the ratio between the cementing agents (silicates) and the fluxes. The higher the value of M_s, the greater the difficulty in calcining the cement.

Next, there is the flux modulus:

$$M_f = \frac{A}{F} \tag{14.4}$$

which is given this name because its terms are proportional to the fluxes, with A determining the amount of C_3A and F determining the amount of C_4AF in the cement. The lower this modulus, the lower the temperature in the calcining furnaces at which the liquid phase is produced. The flux modulus is 0.64 if the amount of alumina present is so small that only C_4AF is formed. A new compound appears when its value is below 0.64. This is C_2F, called dicalcium ferrite. It is a compound which is never present in Portland cement but only in ferric cements.

Another modulus which is stated according to the Italian standards is the lime modulus:

$$M_l = \frac{C - (1.65A + 0.35F)}{S} = \frac{\text{lime combined with silica}}{\text{silica}} \tag{14.5}$$

The limiting values of this index are $1.87 \leqslant M_l \leqslant 2.8$; the practical values are $2.4 \leqslant M_l \leqslant 2.6$, and the optimal value, $M_l = 2.5$.

If there is only C_2S in the cement, the lime modulus is 1.87, while, if there is only C_3S, the value of M_l is 2.8. Although C_3S is usually desirable in cements, the M_l cannot be too high, without the danger that the basic oxides may completely combine with the acidic oxides, and, in this case, there would be free lime in the calcined material. On the other hand, M_l cannot be too low, because C_2S is much less hydraulic than C_3S.

Outside Italy and in the USA and West Germany, for example, consideration is also given to the lime saturation index which is defined by the ratio between the total lime (C) and that which can become combined in the most basic compounds (C_3S, C_3A, and C_4AF). A lime saturation index of greater than 1.0 is indicative of the presence of free lime while, on the other hand, a lime saturation index of less than 1.0 indicates that the material is in a position to be able to combine with further lime.

The optimal composition of Portland cement

Noteworthy data appertaining to the optimal composition of Portland cement, which, after the best calcining, exhibits the best results when used, are summarized in Table 14.12.

Table 14.12—Optimal values of the moduli for Portland cement and the corresponding composition

Moduli	*Optimal value*	*Components*	*% present*
M_h	2.13	S	22.14
M_l	2.50	A	5.70
M_s	2.50	F	3.16
M_f	1.80	C	66.00

However, it is not only the absolute chemical composition of the cement mixture which has an influence on the calcining but also the physical structure and the manner in which the individual components are chemically combined. For instance, a quartz silica is calcined with greater difficulty than a silica which is combined in natural silicates, and a pure limestone of the marble type mixed with kaolin is calcined with greater difficulty than a marl containing intimately mixed limestone and clay.

To ensure that a cement has the optimal composition, it is necessary to calculate the proportions of the raw materials which are to be mixed beforehand, taking account of their composition and that of the clinker which is to be produced.

Today, by using automated techniques, it is possible to blend the amounts of certain raw materials almost perfectly so as to produce a clinker with a predetermined composition intended to be used in the applications specified for it.

The microscopic appearance of clinker

The mineralogical analysis of clinker is carried out by reflection microscopy. For this purpose, a fragment of the clinker, after it has been embedded in a suitable resin, is polished with finer and finer abrasives, etched with suitable reagents, for a fixed time, and its well-revealed components are then observed in reflected light (Fig. 14.33). In this way, three principal phases are observed: alite, belite, and celite which have already been chemically defined in Table 14.11.

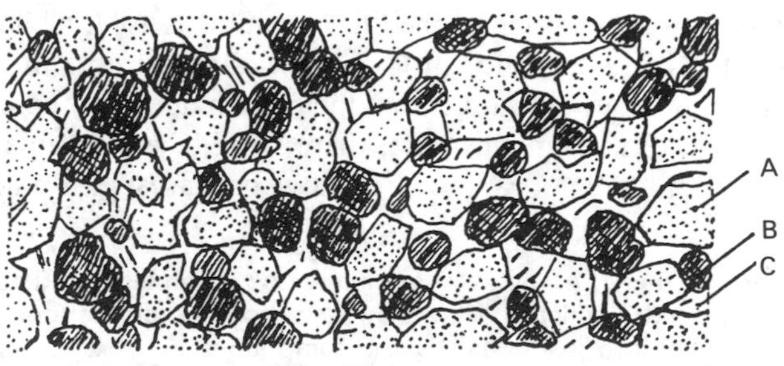

Fig. 14.33—The structured appearance of a granular clinker (that is, a clinker which has not been ground) but which has been suitably polished and etched. Alite and belite constitute about 80% of the whole observed mass, while celite, which is vitreous for the greater part and always interstitial, forms the rest. Alite is the characteristic constituent of Portland cement, and it is precisely because of the presence of this compound that Portland cement is so different from hydraulic limes. On the other hand, celite is a less essential component of the cement as regards its effects in use.

Alite is generally present in the form of grey polygons which are often banded, owing to the presence of MgO and C_3A in solid solution.

Belite has the form of dark roundish granules which are often striated. The amount of belite in a Portland cement is always lower than the amount of alite. It should be noted that the beta (β) form of belite is unstable at ordinary temperatures because, at temperatures under 520°C, a gamma-belite which is devoid of hydraulic properties is stable. When β-belite is transformed into γ-belite the clinker swells and is thereby pulverized because the gamma form is not so hard as the beta form.

Celite is vitreous if the clinker has been rapidly cooled, and, in this case, the mass contains a part of the magnesia (MgO) which is possibly present in the cement in a dissolved form.

The secondary components which are observable in the clinker under the microscope are the roundish granules of free lime and various crystalline formations produced by the compounds of the alkalis.

The setting and hardening of cement

When the granules of the finely ground cement powder come into contact with the water in the paste, two phenomena successively take place: setting, which is concluded after a period of about 25 hours, and hardening, which is a far more protracted process and is achieved after a period of about one year, but proceeds to completion only after decades or more.

From a chemical point of view, the following reactions successively take place when the cement first comes into contact with water:

$$C_3A + 6H_2O \rightarrow C_3A \cdot 6H_2O \tag{14.6}$$

$$C_4AF + nH_2O \rightarrow \underset{\substack{\text{monocalcium}\\ \text{ferrite}\\ \text{hydrate}}}{CF \cdot (n\text{-}6)H_2O} + \underset{\substack{\text{tricalcium}\\ \text{aluminate}\\ \text{hexahydrate}}}{C_3A \cdot 6H_2O} \tag{14.7}$$

The fluxes are therefore the first components to be hydrated, with the evolution of a large amount of heat as a result of reaction (14.6).

The alite reacts after a few hours of contact with water:

$$C_3S + H_2O \rightarrow C_2S + \underset{\substack{\text{calcium}\\ \text{hydroxide}}}{C \cdot H_2O} \tag{14.8}$$

Reactions (14.6), (14.7), and (14.8) are essentially the processes which are responsible for the setting of cement.

During a second stage, the chemical processes occur which bring about the hardening of the cement.

Conventionally, the point at which the setting process stops and the hardening process starts is considered to be indicated by a certain reading taken from a 'Vicat needle', the piece of apparatus shown in Fig. 14.34.

It is laid down that setting should not start until one hour has elapsed after the mixing of the water and the cement, and that it should not stop before four hours have elapsed from the beginning of the mixing.

Hardening involves both the reaction in which tetracalcium aluminate hydrate is formed:

$$C_3A \cdot 6H_2O + C \cdot H_2O + 6H_2O \rightarrow C_4A \cdot 13H_2O \tag{14.9}$$

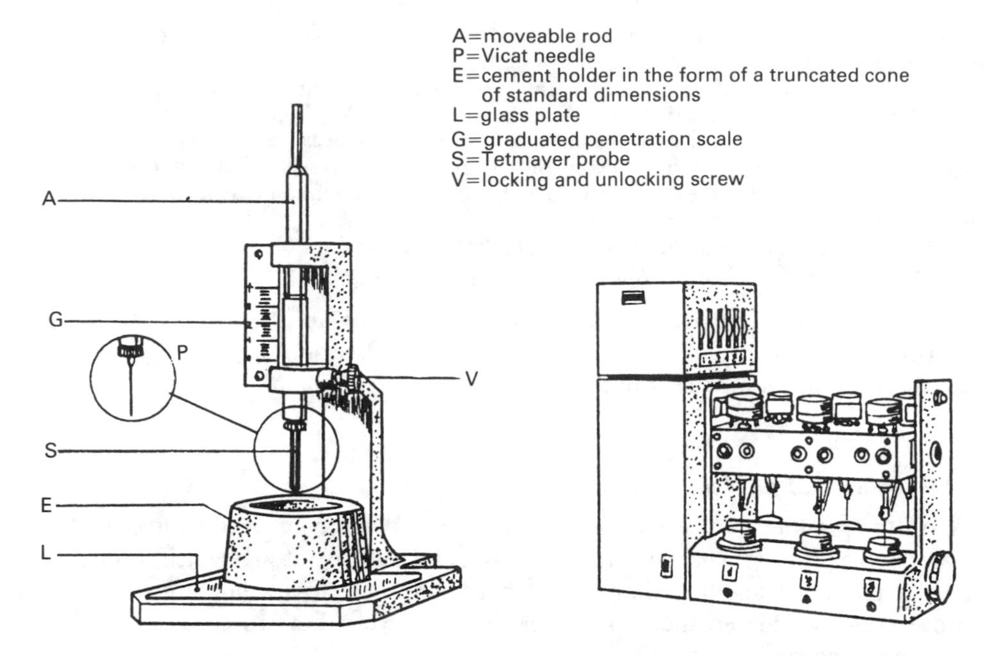

Fig. 14.34—Left: Standard apparatus (Vicat needle) for the determination of the start of the setting and the limit between the end of the setting and the beginning of hardening of the cement. Setting begins when the needle stops at a distance of 1 mm from the bottom of the cement–water paste contained in E and finishes when the Vicat needle penetrates not more than 0.5 mm into the paste. The 'normal paste test' (the test to determine the amount of water which, when mixed with the binder, furnishes a paste with specified penetrability characteristics (normal paste)) is carried out with the same apparatus, but it is fitted with a Tetmayer probe. Right: Penetration meter: a modern automated device to measure the initial and final setting times in series and for carrying out normal paste tests.

and the hydration of the original belite after the dicalcium silicate (chemically identical to belite) formed in reaction (14.8) has been hydrated:

$$C_2S + nH_2O \rightarrow C_2S \cdot nH_2O \tag{14.10}$$

So far as the dicalcium silicate formed with the belite is concerned, reaction (14.10) takes place between the setting and hardening, while the same reaction is much slower in the case of the belite present in the clinker, and components manufactured from the cement attain their definitive compactness and mechanical strength after about one year.

Therefore, while alite C_3S is indirectly (that is, through the C_2S which it produces) responsible for the early mechanical properties acquired by the cement, it is the belite C_2S which is responsible for the final mechanical properties.

If the water in the paste, owing to the way it has been prepared, is such as to lead to the excessive impoverishment of the environment in calcium hydroxide, reaction (14.9) is hindered and the dicalcium silicate is hydrolytically transformed into compounds which are poorer in lime (C_3S_2 and CS) which, when they hydrate, yield a much less consistent mass than that which is produced by C_2S.

It should also be noted that reaction (14.10) is only approximately complete within a period of one year, because the compound $C_2S \cdot nH_2O$ and the other hydrated silicates which are poorer in lime and originate from C_3S_2 and CS are of a gelatinous nature which makes them rather impervious to water.

The colloidal mass* becomes consistent and then locates itself on the periphery of the cement granules, thereby sealing the grains in a protective sheath which makes it difficult for water to reach their centres. This is why cement which has been crushed and soaked can still set and harden after a very long period.

It follows, from everything that has been said, that the water in a cement paste, which initially served to soak it and to make it fluid, is consumed by the cement at a certain rate, depending on its structural characteristics, to bring about the decomposition of the compounds (hydrolyses), to act as water of crystallization (via hydration reactions), and in the promotion of colloidal phenomena.

Gypsum and the hydration of cements

Reaction (14.9) tends to occur on its own at the expense of the free lime in the surroundings as the lime which is formed in reaction (14.8) is insufficient, at least at the start, to cater for the large amounts of calcium hydroxide ($C \cdot H_2O$) demanded by the high rate of reaction (14.6) which leads to the promoter ($C_3A \cdot 6H_2O$) of reaction (14.9) itself.

However, when there is a shortage of free lime, as stated earlier, the dicalcium silicate is decalcified and is converted into compounds which are poorer in lime (of the C_3S_2 and CS type) which, by hydrating more rapidly than C_2S, form a mass which is less consistent than that which is formed on the basis of $C_2S \cdot nH_2O$.

To avoid these serious anomalies in the setting and the hardening of cement, it is necessary to prevent the environment from becoming a large consumer of $C \cdot H_2O$ by removing it from the $C_3A \cdot 6H_2O$ which is formed by reaction (14.6).

Gypsum is a reagent which is capable of combining with tricalcium aluminate hexahydrate according to the reaction:

$$C_3A \cdot 6H_2O + 3CaSO_4 \cdot 2H_2O + 19H_2O \rightarrow C_3A \cdot 3CaSO_4 \cdot 31H_2O \quad (14.11)$$

The calcium sulphoaluminate (Candlot's salt) which is formed concurrently (or even preferentially) with $C_4A \cdot 13H_2O$ from reaction (14.9) indirectly stabilizes the C_2A (which is no longer compelled to degrade in order to supply lime to the environment), and this, furthermore, can lead, via reaction (14.10), to the mass attaining the highest possible mechanical properties.

For this reason, gypsum is considered as an essential additive with the aim of normalizing the setting and hardening of cements.

Nevertheless, the use of gypsum is subject to a criterion. In fact, if there is more than a certain amount of it present, this can lead to its action on the cement mass being prolonged, leading to dangerous cracking due to the high specific volume of Candlot's salt. In short, if the mass has not hardened, Candlot's salt finds space and then tends to split apart the already consolidated objects in which it is formed.

It should also be noted that, even if official standards permit the free use of one component or another, gypsum and not anhydrite should be added to the cement. This is because anhydrous calcium sulphate requires longer reaction times for the formation of Candlot's salt than those taken in the case of gypsum. The use of calcined gypsums ($CaSO_4 \cdot H_2O$ or $CaSO_4 \cdot \frac{1}{2}H_2O$) is

*Observations by electron microscopy show that these apparently colloidal masses are, in fact, microcrystalline (lamellar and fibrous).

therefore explicitly ruled out as they give rise to incorrect setting phenomena (false setting) which hinder the reactions involved in the correct setting of the cement.

The mono- and hemi-hydrates of calcium sulphate can be formed while grinding the clinker to which gypsum has been added on account of the friction which is associated with this process. For this reason, it is necessary to ensure that the mills used for grinding gypsum-containing clinker in cement works do not become too hot (they should not be allowed to reach temperatures greater than 110°C).

The composition of finished Portlant cement. Consequences

After setting and hardening, Portland cement is formed from a mass of substances which have grown together into a very compact texture. Some of these are of a crystalline nature, while others are of colloidal type (gels) and amorphous and therefore exhibit fluid behaviour.

The five principal components of hardened cement are $C_2S \cdot nH_2O$, $C_4A \cdot 13H_2O$, $CF \cdot (n\text{-}6)H_2O$, $C_3A \cdot 3CaSO_4 \cdot 31H_2O$, and $Ca(OH)_2$.

Such a mass has interesting properties which derive from the diffused presence of gels throughout its texture, with the result that it is quite elastic and can 'fluidify'. That is to say, it is in a position (like fluids) to adapt itself to applied forces and to distribute strains readily. The manufacture of concrete components under 'hyperstatic' conditions (with a large number of reinforcement rods) is made possible by these unusual behavioural properties.

Among the components of hardened cement, there may also be some uncombined gypsum which has led to false setting. Both this gypsum and calcium hydroxide are sensitive to humidity, and they may therefore form efflorescences and prime the washing out process which is merely anti-aesthetic or damaging to the existence of the manufactured object. In fact, Portland cements which have been subject to washing out are not resistant to prolonged contact with water, especially if it is rain water or circulating water. The soluble parts first dissolve away, and this is followed by the complete disaggregation of the cements.

$Ca(OH)_2$ is also sensitive both to the CO_2 in air, especially if it is present in polluting amounts, and to all the naturally acidic atmospheric contaminants including SO_2 and NO_2.

Owing to the effect of these gases, a type of ageing of the objects occurs, leading to the degradation of their surfaces which, among other things, impedes their bonding to newly applied cement. It is then necessary to chisel out the old cement and to then apply the new. Another type of ageing for which there is no remedy produces the irreversible dehydration of the gels forming part of the cement texture.

In environments which are not particularly corrosive, surface ageing phenomena frequently lead to a kind of passivation of the structures with the formation of a protective crust which preserves the interior from further degradation. The application of bitumens and varnishes containing products which are specifically resistant to environmental agents is obligatory for the preservation of objects made out of Portland cement in corrosive environments.

Other setting and hardening coefficients for cements

In the hydration of a cement, which is followed by it rapidly acquiring mechanical strength, the specific surface area of the cement powder* is a factor of the greatest importance. In fact, this hydration is a typical interfacial reaction which occurs at a faster rate as the surface area of contact between the phases increases.

---

*The specific surface area is the total area of one gram of the powder and is therefore measured in cm^2/g.

Consequently, cements with the highest specific surface areas (supercements) already exhibit good mechanical strength characteristics after 1–3 days, while, on the other hand, cements for petroleum wells (for example), which must remain fluid for a long time during the excavation, are based on binders with retarded rates of hydration due to the low specific surface areas presented by their powders.

As the water/cement ratio increases, the mechanical strength which is attainable in a cement paste falls off.

The amount of water which is strictly necessary for the laying of a cement is about 30% of the dry cement. Any extra water which is added serves to make the paste more workable and depends on whether the paste is to be stiff or more fluid.

One of the more reliable theories, that of Abrams, regarding the mechanical strength which is attained by objects manufactured from cement as a function of the water/cement ratio, leads to the well known Abrams water:cement ratio law:

$$R_c = \frac{A}{B^x} \tag{14.12}$$

where R_c is the compressive strength at a certain maturity and A and B are constants which depend on the materials and the test conditions.

Fig. 14.35 shows how the strength varies as a function of the water/cement ratio. It can be seen that the smaller the water/cement ratio, the greater the mechanical strength which is attained, in agreement with equation (14.12). The optimal mechanical strength is attained after the standard period (28 days) for a water/cement ratio of 0.4.

Nowadays, in order to control the conditions under which cements are applied with the view of enhancing their qualities, many additives are used which are capable

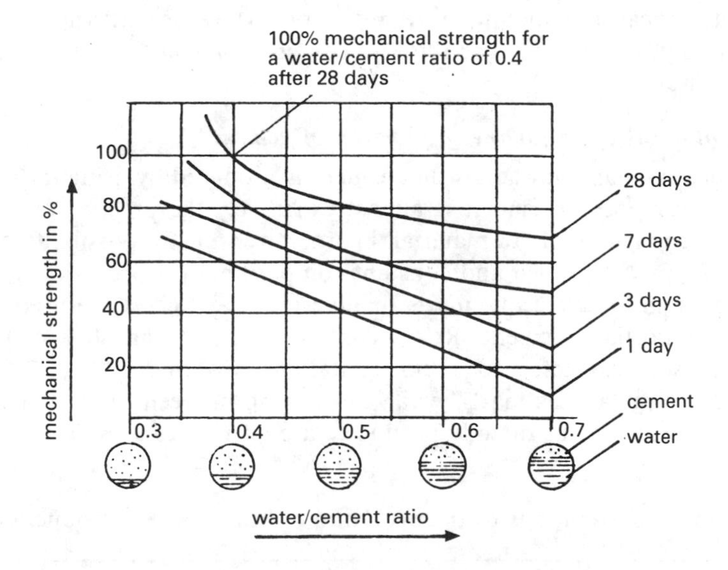

Fig. 14.35—Plots of the mechanical strength of objects made of cement against the water/cement ratio adopted in making them. The mechanical strength which the object is normally required to have 28 days after it has been made is put equal to 100%. It is seen that this is ensured when a water/cement ratio of 0.4 is adopted. Nevertheless, this is not the maximum strength which is absolutely possible, as there may be a further increase in this property over the following 28 days.

of:

- reducing the amount of water which is used, while also ensuring that all the actions to be carried out by the water are correctly undertaken;
- plasticizing the cement;
- accelerating or retarding the setting and hardening;
- ensuring greater mechanical strength by means of an increase in compactness;
- endowing the cement with water-repelling characteristics, thermal shock resistance, and other merits.

Since the additives which are used affect to some extent all of the components in the cement pastes which are being applied (and not only the binder), we shall defer their detailed treatment until cement conglomerates are dealt with.

The heat of hydration of cement

Much heat is evolved from a cement as the setting and hardening reactions take place. This can constitute a danger, especially when large amounts of cement are poured, and, for this reason, cements with a low heat of hydration are used.

It may be calculated that a Portland cement evolves about 60 kcal/kg in 7 days and 90 kcal/kg in 28 days. The heat is slowly got rid of from the centre of the cast objects which expand and then, consequently, split because of contraction upon cooling.

As can be seen from Table 14.13, it is the tricalcium aluminate which is responsible for the evolution of the greater part of the heat.

It is therefore understandable why there is sometimes a tendency to reduce and, in the limit (by using ferric cements or ferric pozzolanic cements with $M_f = 0.64$), to eliminate the tricalcium aluminate from cements. Moreover, attempts are made, in so far as it is possible, to reduce the C_3S content in favour of the C_2S content for the same reason.

The technological cycle for the production of cement

It is only exceptionally nowadays that cement is produced by using marl limestones of a proper composition, and it is almost exclusively the practice to make use of mixtures of certain natural raw materials in order to ensure the possibility of obtaining products of a predetermined and constant composition.

Limestones and argillaceous marls or limestones and clays are used as the raw materials for Portland cement. Rocks containing more than 80% of $CaCO_3$ are considered as limestones, and rocks containing less than 70% of $CaCO_3$ are considered as argillaceous marls, while those lying between these two limits, with which cement can also be directly produced, are referred to as marl limestones.

Table 14.13—Contributions to the evolution of heat by the components of cement

Components	*Heat of reaction*	*Components*	*Heat of reaction*
C_3A	207 kcal/kg	C_4AF	100 kcal/kg
C_3S	120 kcal/kg	C_2S	62 kcal/kg

In every cement plant raw materials must also be available to make possible corrections to the mixtures. Such raw materials are mainly represented by blast furnace slags, pyrites ashes, and trachytic rocks.

Cement may be made by the dry or the wet route, and these two processes differ from one another almost only in the phase in which the charge is put into the kiln.

The dry process. The materials which have already been crushed are withdrawn from the storage warehouses, blended, ground in ball mills, and then passed into silos/maturation tanks. They are subsequently withdrawn from these tanks and put into rotating panniers which, by means of centrifugal force, keep the material in contact with their walls which are fitted with water sprays. The rotational motion and the wetting of the material act jointly in forming the material into uniform small spheres which constitute the feedstock for the rotating cement kilns.

The wet process. The raw materials, having been withdrawn from the storage warehouses, are treated either individually or as a mixture with the appropriate amount of water (35–40% by weight) and then ground separately or as a mixture in ball mills. After milling, sieving, and, if necessary, remilling, the fluid pastes are sent into large homogenization vessels into which air is blown. This latter operation may also either be carried out on the individual components or on a mixture of them.

An alternative homogenization process which is becoming ever more widely used involves the partial removal of the water from the limestone in continuous rotating filters to obtain cakes which, after having been mashed with air into the fluid paste of the other raw material within a mixing tank, enter to form a material which is quite homogeneous and contains the ideal amount of water. This solution to the preparation of the charge is shown in Fig. 14.39. As in the dry process, the charge is always introduced into the highest end of the kilns where the temperature is lowest.

Kilns. Rotating cement kilns are cylinders made of thick steel plate which is internally lined with a silica–alumina refractory material.

With an average length of 60 m, but possibly up to more than 120 m, they have a diameter of the order of 3 m and a slope of 4–6%. They rest on their rotation rings and are moved and made to rotate slowly by a motor which drives them through toothed driving bands (Fig. 14.36). A moveable head closes off the furnace at the opposite end, where there is a powdered coal, naphtha, or natural gas burner. Of these, the last two types are preferred because there is no contamination of the product and they permit precise control of the amount of energy which is being put into the kiln.

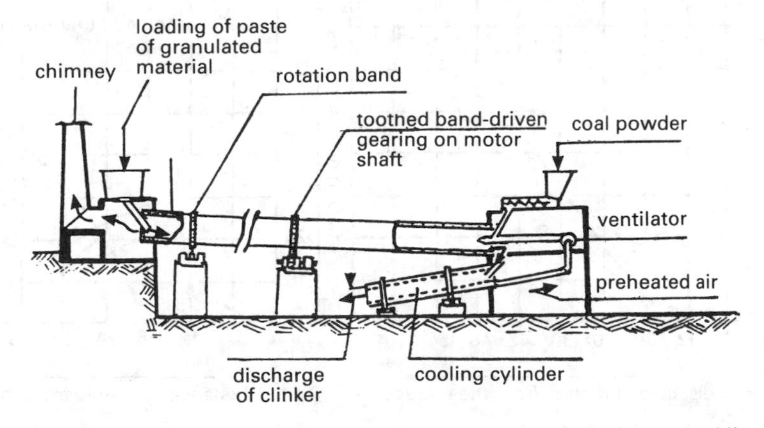

Fig. 14.36—The constructional principles and the mode of operation of a cement kiln. Many of the secondary units such as the combustion fume aspirators and the cyclones and/or electrostatic precipitators for the removal of solids from the fumes have been left out for simplicity.

Having entered the kiln, the mixture of raw materials is progressively dried out, decarbonated, and the first solid state reactions start to take place, and it arrives within range of the firing zone of the burner flame where, as the liquid phase appears, the reactions rapidly go to completion. The fusion, which is favoured by the formation of C_4AF, starts at 1335°C and is complete at 1400–1500°C.

The mass is precipitated from the lowest zone of the kiln into a rotating cooling cylinder where its temperature is lowered. It is here that the clinker is formed as a collection of spheroidal granules with diameters from 0.3 to 0.5 cm. This granulated material is actually formed by the centrifugal force to which the still semi-molten mass is subjected.

The combustion gases which are brought together by means of vents move in the opposite direction to the cement which is being calcined and pass out of the front of the loading aperture. These gases are rich in mechanically transported powders. These are calcareous powders which are ecologically extremely harmful. On account of this, it is necessary to purify these fumes by first using cyclones, and then electrostatic precipitators. It is obvious that there is a need to recycle whatever is recovered, as it is the optimal product at all the stages. The temperature profile inside a medium length kiln is shown in Fig. 14.37.

If the products from rotating cement kilns are discharged without any precautions, the energy consumption jumps up to 900 kcal/kg, while, if the heat is rationally recovered from the clinker and the fumes, it can be reduced to about 420 kcal/kg.

The most commonly used devices for recovering the heat are:

* the adoption of a curtain of chains which remixes the moist mineral charge with the fumes, or blade type exchanger/mixers in kilns which employ the dry process with granulated material being loaded into them;
* preheating of the air current for the combustion process by making it pass through the granulation cylinder in the opposite direction to the clinker;
* the use of special kilns (Lepol kilns) which are not more than 30 m in length where the charge is preheated by laying it out on a conveyor belt which advances in the opposite direction to the scorching fumes emerging from the rotating kiln;
* the installation of a series of cyclones (the Humbold system) into which the mixture to be calcined, before being put into the kiln, is passed and maintained in contact with the gases recovered from the fumes discharged from the rotating kiln.

It can be seen, from the measures which are taken to recover heat within the operating environment of cement kilns, that the maximum dispersion of heat is due to the fumes which are produced.

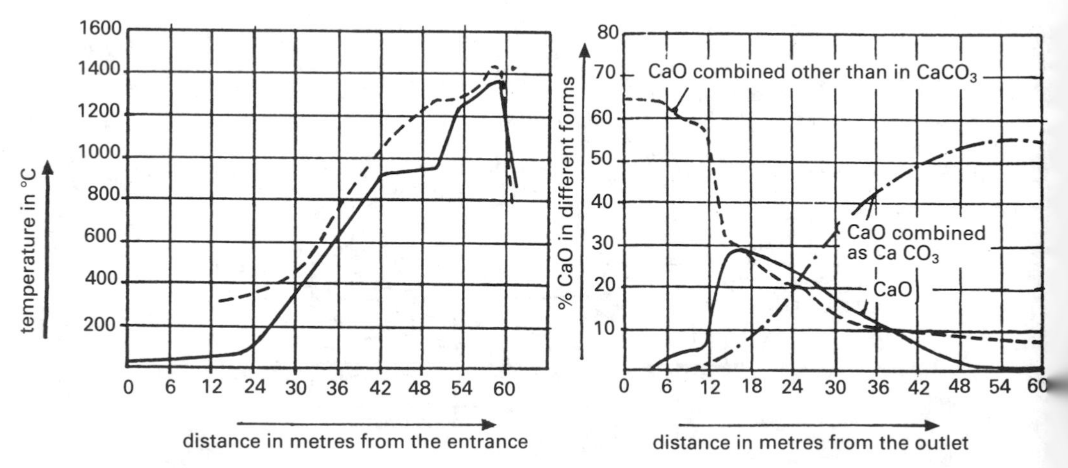

Fig. 14.37—Left: Temperature profile in a cement kiln with a hypothetical length of 60 m. The broken curve indicates what the temperature profile would be if the initial material did not have to be dried and the $CaCO_3$ dissociated at 900°C. Right: The events to which the lime is subjected in cement kilns and, in particular, its combination with SiO_2 and Al_2O_3.

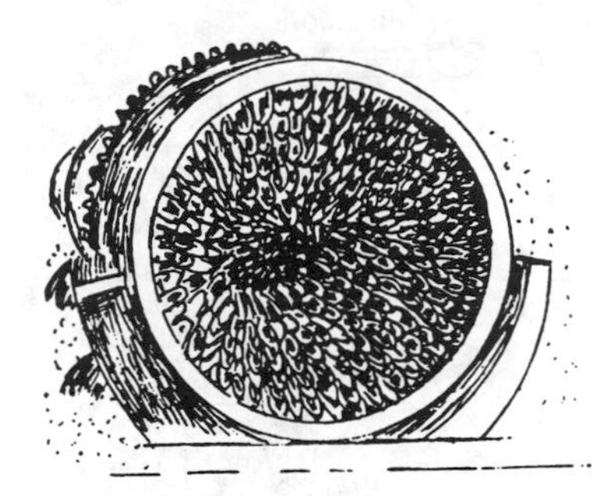

Fig. 14.38—Left: Blade-shaped exchangers set up within a cement kiln to stir the granulated material loaded into the kiln and thereby increase its contact with the combustion gases. Right: Fixed curtains of chains to prevent the agglomeration of 'moist' charges in cement kilns. Such curtains increase the contact between the particles in such charges and the combustion fumes which they encounter as they pass through the kiln.

The finishing and marketing of the clinker. After the larger pieces of material which it contains have been broken up, the clinker is sent to the storage silos. These are large containers which are widely used in factories producing granular and powdered materials on a large scale. Their function is to act as temporary warehouses for such products and to permit their further finishing treatments at programmed times. The clinker passes from these storage silos to the primary milling unit, and then, after mixing it with gypsum according to the specification, on to the second milling stage. When there is a need to adjust the indices of the cement which has been produced, further auxiliary raw materials (pyrites ashes, trachytic powders, etc.) are also added. The fraction of the resulting material which is not sufficiently fine is then recycled and stored in silos ready to be used.

A diagram of the complete technological cycle for the production of cement is shown in Fig. 14.39.

A way of marketing cement which is now very much in fashion is that of carrying it in mixer lorries which deliver ready-made concrete.

Finally, the cement can also be withdrawn from the silos and packed into sacks made out of Kraft paper. This paper consists of long fibres which have been calendered and made water-resistant by treating them with urea glues.

Table 14.14—Origin and composition of the best-known European pozzolanas

Location of excavation	*f.l.*	*S*	*A*	*F*	*C*
Rhenish Trass	10.1	54.6	16.4	3.8	3.8
French Gaize	9.2	56.4	15.3	3.6	3.9
Santorin Earth (Egeo Island)	4.9	63.2	13.2	4.9	3.2
Laziale Pozzolana, Segni (Rome)	9.6	44.1	17.3	10.7	7.5
Flegrean Pozzolana, Bacoli (Naples)	4.8	55.7	19.0	4.6	5.2

Note: The symbols at the top respectively indicate: f.l. = firing loss $S = SiO_2$, $A = Al_2O_3$, $F = Fe_2O_3$, and $C = CaO$. The %'s shown sum to 100% with the TiO_2, MgO, alkali, and SO_3 contents which are also present.

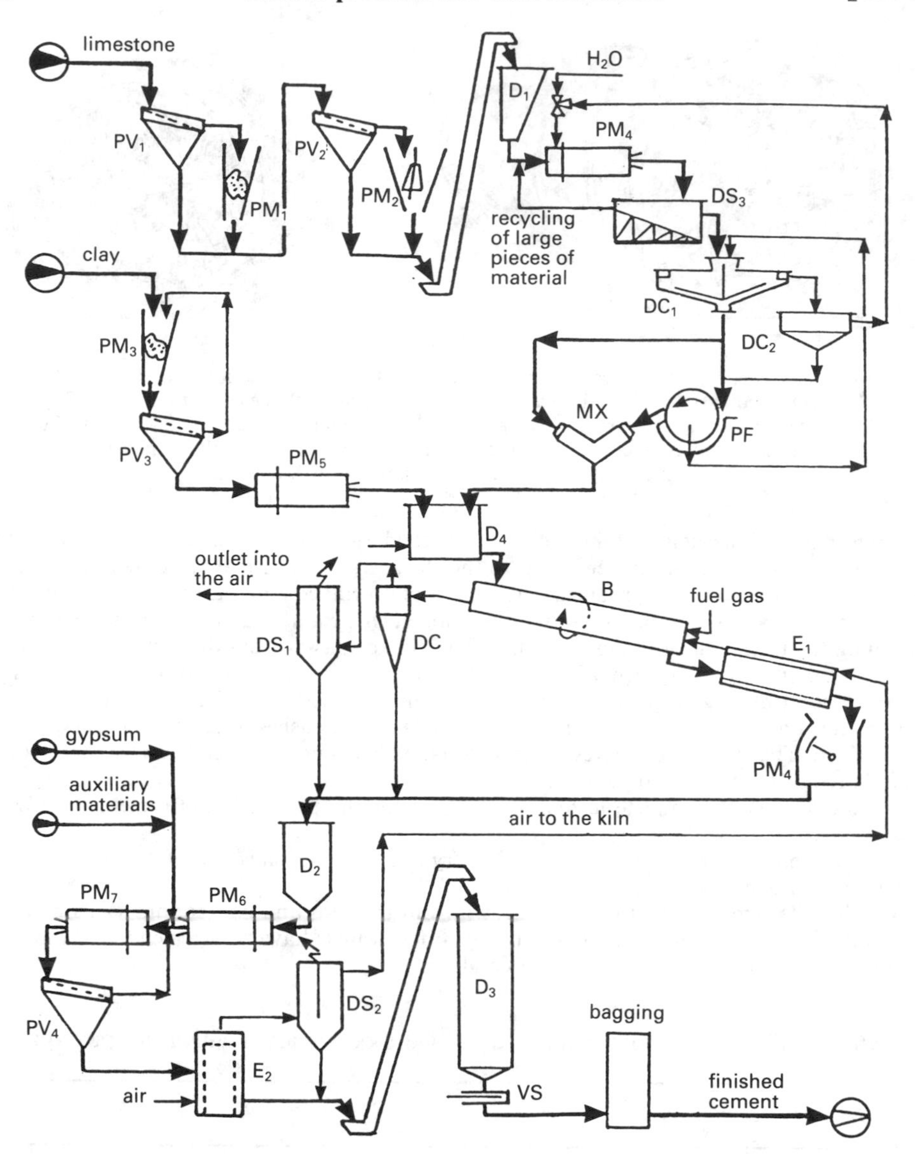

Fig. 14.39—Panoramic scheme showing how cement is produced, starting from limestone and clay with mixing of the limestone in a partially dried suspension and the milled clay.

Pozzolanic binders

Pozzolanas are eruptive rocks of the effusive type which are still quite widespread in the Campano–Laziale region of Italy.

They are materials with an average density of 1000–1100 kg/m^3 and with 'active oxide' compositions shown in Table 14.14 for typical examples. It can be seen from this table that the composition of these materials varies widely and depends on the area from which they originate.

It is seen that pozzolonas have a distinctly 'acidic' composition due to the prevalence of SiO_2 and Al_2O_3 in them. They are therefore capable of reacting with calcium hydroxide in mortars which can set and harden under water.

Good pozzolanas must originate from strata which are devoid of outcrops, they must be fine, heterogeneous materials must be absent, and they must not contain 'inert' substances such as limestones or clays.

After the silica and alumina of pozzolanas, which are naturally vitreous or become so when they are fired and tempered, have been mixed with lime or with chalk putty, they are in a position to set and harden like the analogous components of Portland cement.

The sieved pozzolana-milled lime (or slaked lime putty)–sand–water pastes constitute the pozzolanic mortars. These are subdivided into walling pastes (4 parts of pozzolana, 3 parts of sand, and a part of slaked lime putty) and normal pozzolanic mortar (3 parts of pozzolana, 3 parts of sand, and 1 part of slaked lime).

Similar pastes, but made with gypsum rather than sand, constitute the gypsum-pozzolanic mortars.

Portland-pozzolanic (or pozzolanic) cements are also produced by milling mixtures of Portlant cement and pozzolanas. These binders exhibit a chemical resistance which is much superior to that of Portlant cement because the pozzolanic silica neutralizes both the free lime and that which is formed by hydrolysis, as has been noted, during the setting of Portland cement. As a result of this, the washing out of the lime, the reactions leading to the formation of hydrated sulphoaluminates upon contact with sulphate-containing waters, and the processes involved in the combination of lime with CO_2 and other acidic atmospheric contaminants, are hindered.

Pozzolanic cements are also advantageous on account of the very low heat of hydration which they have in comparison with Portlant cement. The iron ore-pozzolanic cements are still better than the pozzolanic cements both with respect to the lowest heat of reaction which they exhibit and their chemical resistance. These are obtained by mixing iron ore cements, which are discussed later, with pozzolanas.

Blast furnace binders

Blast furnace slags are chemically basic and can readily act as raw materials for hydraulic binders. If they are to manifest hydraulic properties such slags must be quenched in water when they emerge from a blast furnace. When this is done they become vitreous, porous, and chemically quite reactive.

A blast furnace slag which is destined for use in cements must not contain more than 5% of MnO, a hydraulically damaging compound, and the useful components must be present within the following percentage limits:

$$S = 28\text{–}38\%,\ A = 8\text{–}18\%,\ \text{and}\ C = 35\text{–}45\%$$

The amounts of MgO in blast furnace slags are also, generally, quite considerable. However, since these are uniformly distributed throughout the mass, they do not lead to localized expansions during setting. Other minor components of the slags are CaS (which, by hydrolyzing and forming H_2S, is responsible for the nauseating odour of blast furnace cements), and FeO as well as iron and alkalis.

When mixed with lime, quenched blast furnace slags have hydraulic properties due to the formation of $C_2S \cdot aq$ and $C_3A \cdot 6H_2O$, that is, the same compounds which characterize Portland cement after it has set. Slag–lime mixtures and sand constitute the metallurgical hydraulic mortars.

When milled with Portland cement, blast furnace slags produce slag cements (or metallurgical cements).

Slag cements may be alternatively produced by the firing of slag and limestone, and, furthermore, for certain purposes, these cements may even be made by mixing ground slag with Portland cement at the moment of use.

The setting and hardening of slag cements are slower than those of Portland cements, but the final mechanical properties are of the same order of magnitude. The setting and hardening of these cements first involve the fixation of the lime formed by the hydrolysis of the tricalcium silicate in the Portland cement by the slags, and, secondly, the parallel hydration of the slags and the Portland cement. The disappearance of the lime formed by hydrolysis renders the final product resistant to aggressive waters.

The heat of hydration of slag cements is much lower than that of Portland cement, and these cements, in practice, therefore undergo only small contractions and lend themselves well to the casting of large sections.

Finally, on account of their heat resistance and the fact that they also remain unaltered up to 400–500°C, slag cements are suitable for the laying of bricks in heat-deflecting screens.

Alumina cements

According to the Italian Standards, alumina cements are the products of the milling of a clinker based on the hydraulic aluminates of calcium.

The cements have the following composition:

$$SiO_2\ 3\text{–}11\%,\ Al_2O_3\ 35\text{–}44\%,\ Fe_2O_3\ 4\text{–}12\%,\ CaO\ 35\text{–}44\%,\ \text{and}\ FeO\ 0\text{–}10\%$$

A large variety of aluminates are found among the 'salt' components of these cements. Actually, even if CA does predominate, C_5A_3, CA_2, $C_{12}A_7$, and many others are also present.

Alumina cements are prepared from finely milled and well mixed bauxites and limestones. The firing is carried out at 1500–1600°C in reverbatory, shaft, or electric furnaces. The process does permit the formation of clinker since the product passes from a liquid to a solid over a narrow thermal range. For this reason, alumina cements are discharged as liquids from the firing furnaces and are known as 'molten cements'. Considerable care is paid to the manner in which the molten cement is cooled, since the properties of the cements produced are to some extent dependent on this.

During the firing, the iron oxide, Fe_2O_3, is partially reduced to iron which can be separated magnetically before the final milling.

So far as the setting of alumina cements is concerned, it should be noted that, in view of the high rate of the reaction between the starting aluminates (including dicalcium aluminate) and water, hydrated alumina is liberated:

$$2CA_2 + 11H_2O \rightarrow C_2A\cdot 8H_2O + 3A\cdot 3H_2O \qquad (14.13)$$

It is precisely on account of the high rate of the reactions occurring as soon as the cement has been mixed that, after only 24 hours, objects made from alumina cements have attained a mechanical strength which is about half of that which is attained after 28 days.

These high hydration rates just after the cement has been mixed lead to the intense evolution of heat. This heat must be allowed to escape when casting thin sections, or it must be eliminated by bathing the manufactured items and exploiting the climate at the site where the sections are cast.

The mechanical strength of alumina cements starts to fall off at above 25°C owing to the decomposition of the dicalcium aluminate produced in (14.13):

$$3C_2A \cdot 8H_2O \rightarrow 2C_3A \cdot 6H_2O + 9H_2O + A \cdot 3H_2O$$

which is followed by a contraction in the volume because dicalcium aluminate octahydrate has a specific volume which is much greater than that of the tricalcium aluminate hexahydrate and the alumina trihydrate which are formed from it.

The values of the cements lie in the manner in which they lend themselves to fulfilling urgent operations requiring the fast production of components subjected to heavy loads combined with optimal chemical resistance both to pure waters and sulphate-bearing waters. These properties follow, on the one hand, from the fact that these cements do not contain soluble lime, and, on the other hand, from the impediment which the thin coating layer of colloidal alumina presents to the action of both gypsum and other solutes in the waters. In cold climates, the resistance of these cements towards sea water is also sharply increased.

Cements for retaining barrages

These cements are intended for the construction of embankments, for the isolation and protection of dry docks, certain mining works, and to stop the flow of waters in dykes and dams, and so on. The cements for retaining barrages must:

- have a final compression strength which is much greater than that of Portland cement;
- have a low heat of hydration.

Since they are produced in such a way as to eliminate the greatest possible amount of tricalcium aluminate and to increase the amount of belite (C_2S), the setting and hardening of these cements are slow to a point where, after 90 days, they have attained about $\frac{2}{3}$ of their final mechanical strength as compared with the period of 28 days which is required for Portland cements.

Cementitious agglomerates

The hydraulic binders which exhibit physicomechanical properties which are considerably lower than those which have been established for Portland cement to a point where their use is not permitted in the manufacture of reinforced cements, are simply referred to as cementitious agglomerates.

Cementitious agglomerates are subdivided into fast-setting agglomerates and slow-setting agglomerates. The first type, which is produced from upgraded raw materials (containing a lot of 'inert' substances) are, apart from in low-duty structures, used like Portland cements. The other type is obtained by the calcining of marl limestones containing about 30% clay, and their final appearance therefore tends to be yellow.

Fast-setting cementitious agglomerates are more widely used than the slow-setting variety, which, when all things are taken into consideration, are less convenient than

Portland cements; and the lower cost is not sufficient to compensate for their shortcomings in the areas where they are employed. After grinding, which must not be so extensive as to accelerate their setting any further, the 'rapid' agglomerates are made into a mortar paste with a little sand and are used in the construction of objects which are not very stable and have low mechanical properties: in brief, objects for emergency and temporary uses.

Hydraulic limes and their derivatives

Limes originating from argillaceous materials can set under water, and are called 'hydraulic limes'. For the most part, limes containing 10–22% of clay are used in their preparation. If they contain more than 22% of clay, the resulting products belong to the field of cements.

They are therefore substances which lie within the transition between air-setting limes and cements.

The ability of hydraulic limes to set under water is adjudged on the basis of their hydraulic indices, which traditionally leads to them being classified as in Table 14.15*.

The production of hydraulic limes generally starts off from raw materials of a suitable composition; these are calcined in 'lime kiln' type furnaces at 1100°C, consuming slightly more than 15 kg of coal for every 100 kg of lime produced.

During the calcining, not only is the limestone decarbonated, but compounds are formed which confer the hydraulic power on the lime which is produced.

Weakly hydraulic limes are marketed directly in the form of blocks which are to be hydrated in the factories, while the other types of lime must be hydrated and pulverized before sale. The hydration operation must be carried out in such a way that it affects only the CaO. Otherwise, the material loses its value owing to the hydration of the hydraulic components and, in the limit, becomes drowned lime.

Suitably hydrated hydraulic limes constitute the slaked hydraulic limes from which it is possible to obtain various derivatives, as is shown in the scheme in Fig. 14.40.

The uncalcined and over-calcined materials are, for the most part, used in the production of bricks. 'Flowers of lime' mainly, but also heavy lime in the case of certain uses, are, when mixed with sand, the raw materials for the preparation of hydraulic mortars.

Table 14.15—Practical classification and properties of hydraulic limes

Category	I_h	*Clay in the limestone*	*Setting time (days)*
Weakly hydraulic	0.1–0.16	5.3–8.2	30–16
Moderately hydraulic	0.16–0.31	8.2–14.8	15–10
Hydraulic	0.31–0.42	14.8–19.1	9–5
Very hydraulic	0.42–0.52	19.1–21.8	4–2

*Today, the Official (Italian) Standards anticipate that a more rational classification will be provided for these substances, based on Law No. 565 of the 26.5.1965 which refers to it.

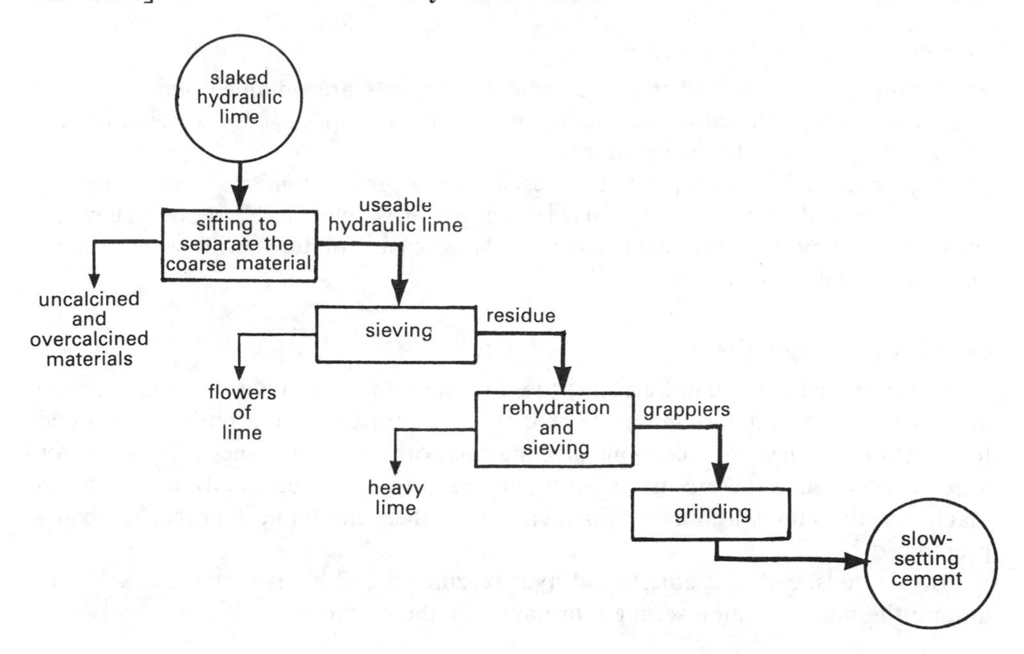

Fig. 14.40—Scheme for the coordinated preparation of the derivatives of hydraulic limes.

Slow-setting cements when mixed in suitable proportions with sand and water yield the cement mortars used in the construction of walling and as plasters. The setting and hardening of objects which are bound by using these cements involve reactions which are analogous to those taking place in the setting and hardening of Portland cement.

NON-STANDARDIZED HYDRAULIC BINDERS

Special cements

There are many variations in the formulation of Portland cement which are not included in the Official Standards. Such non-standardized hydraulic binders are known by the common name of Special Non-Normalized Cements (SNNC). Here, we shall summarize the best known types and their properties.

Iron ore cements

By making cements from limestones, clays, and pyrites ashes, hydraulic binders are obtained with a flux modulus, M_f, below 0.64. This results in C_3A being almost completely absent, which eliminates the danger of the reaction between tricalcium aluminate and the calcium sulphate present in the environment.

On account of their chemical resistance towards most substances, these cements serve in the construction of discharge channels, piers, quays and wharfs, and installations which are in continuous contact with sulphate waters, such as motorway tunnels through gypsum rocks.

Low heat of hydration cements

By mixing pozzolanas and iron ore cements, cements are obtained which are not only poor in C_3A but also in C_3S. Given their composition, these cements have a short setting phase after being poured.

They have the following properties: resistance towards sulphate waters, extremely low heat of hydration, and slow hardening which is due to C_2S alone. They are typically employed in the manufacture of large cast structures with wide-ranging degrees of reinforcement.

Cements for deep holes

As compared with Portland cement, these cements have a smaller specific surface area, a low C_3A content, and an appreciable amount of 'setting retardants' added to the clinker. These modifications prolong the normal setting times to a greater or lesser extent, since these cements must remain fluid for a sufficiently long time to reach certain sites without the final values of their mechanical properties being prejudiced.

Polyalcohols, cellulose ethers, and hydroxyzincate and hydroxystannate salts are among the most common setting retardants for the cements.

White cements and their coloured derivatives

The 'white' SNNC are obtained from extremely pure raw materials, that is, from limestones containing no iron oxide, kaolin, and/or pure sands.

The calcining of white cements is made more difficult by the lack of Fe_2O_3, which means that the flux C_4FA is absent. This difficulty can be partly overcome by the addition of CaF_2, which is well known as a flux. However, when fluorite is present, the cement clinker must be rapidly cooled as, otherwise, the fluorite catalyzes the transformation of dicalcium silicate from the hydraulic (β) form to the non-hydraulic (γ) form.

Coloured special non-standardized cements (SNNC) are prepared by the addition of coloured mineral pigments. This is generally done during the grinding of the clinker. Such pigments are English red, Siena earth, etc.

Metallurgical supersulphate cements

Metallurgical or supersulphate cements are obtained by mixing gypsum (about 15%) with blast furnace slag. These cements are relatively light and have a low heat of hydration. Small amounts (about 2%) of Portland cement are also added to them to speed up their initial hydration.

Magnesite cements or 'Sorel' cements

By forming a paste, immediately before it is to be used, by treating magnesium oxide with a solution of magnesium chloride or magnesium sulphate of a certain concentration, cements are obtained which are based on magnesium compounds (and, hence, magnesite cements) which are known as 'Sorel' cements. These cements have nothing in common with Portland cements either constitutionally or with regard to the phenomena which lead to their setting and hardening.

The setting and hardening of these mixtures are both very rapid as a consequence of the formation of compact masses of magnesium oxychlorides and basic magnesium sulphates. They also set to act as a wood binder.

Non-shrinking cements and expanding cements

Numerous kinds of cements which vary from those which consolidate without shrinking to others which actually expand can be prepared by mixing Portland cement, appreciable amounts of gypsum and blast furnace slags or metallurgical supersulphate cements and Portland cements, provided that the proportions of the components are suitably controlled. The expansion effect is associated with the formation of Candlot's salt.

Non-shrinking cements allow voids and crack formations to be avoided, that is, they have an impermeabilizing effect, while the weakly and strongly expanding cements (up to 12–15 mm/m) create the same effects in the manufactured objects as would arise from the precompression of a normal cement without the use of the complicated apparatus required by this technique, and which cannot be applied to large structures.

Anti-acid cements

By adding certain substances to cements and, in particular, to Portland cement, it is possible to produce binders which are resistant towards acids and/or other aggressive chemical reagents. In other words, their name is to be interpreted in a wide sense.

As examples of this, it may be recalled that cements containing pumice and sodium silicate are resistant towards sulphuric acid solutions; cements to which asbestos fibres, pumice, and alkaline silicates have been added are resistant towards nitric acid solutions; while cements to which barium sulphate and sodium silicate powders have been added are unaffected by chlorine.

These cements are employed in the construction of large vessels in chemical factories, channels and tubes for the delivery and discharge of materials, absorption vessels, and so on. However, all of these structures must generally be finished with bitumen-based coatings or other products which are selected on the basis of the type and concentration of the reagents to be handled.

14.14 ARTIFICIAL BUILDING CONGLOMERATES

Natural conglomerates (brecchias, pudding-stones, etc.) exist which are used in the construction of buildings. For the most part, these are large fragments of quartz rocks which have been variously cemented in a natural manner.

However, artificial building conglomerates are considerably more important on account of their widespread availability, uses, and properties. The artificial building conglomerates are water pastes of hydraulic binders and various materials which, in the vast majority of cases, consist of stony materials called aggregates or 'inert materials'.

Only these conglomerates are of interest in our study, and, throughout the following pages, these will simply be referred to as artificial conglomerates.

The following scheme presents a technical–practical classification of the artificial conglomerates which are used nowadays in building construction.

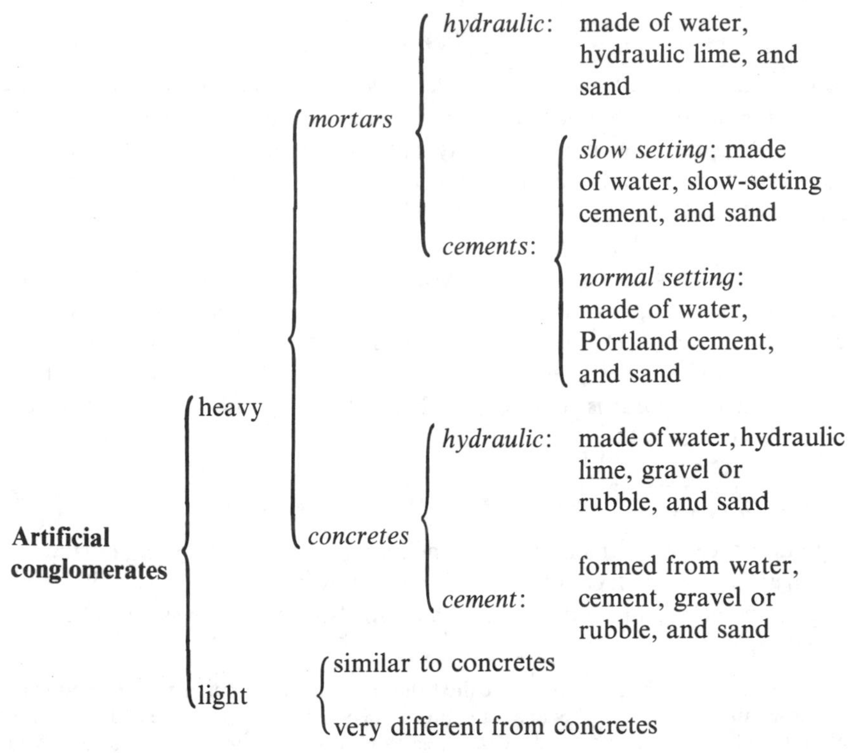

It can be seen that the component which distinguishes concretes from mortars is the gravel or the rubble. That is, mortars are made with a single aggregate (sand) while concretes are made by using two aggregates (gravel or stone rubble and sand).

It is therefore obvious that the chemical behaviour of mortars and concretes, where the binder is the same, will be similar, and the same possibilities exist for the modification of their behaviour with aerating agents, fluidifying agents, plasticizers, and so on.

On account of this, a detailed treatment of concretes only will be presented.

The basic components of concretes and concrete additives

Concrete is an artificial conglomerate. The essential raw materials for this are cement, water, and aggregates, while the secondary components are specific additives and curing agents.

Cement has been specifically considered in the preceding section, and we shall now pass on to review all the other components of concrete.

Waters for constructional purposes

The term 'waters for constructional purposes' is used to indicate the waters which may be mixed with binders to form pastes which are used in building construction.

In practice, it is easy to procure water for constructional purposes since, as well as drinking waters, spring waters, waters from non-stagnant wells, and waters from rivers and lakes can be used.

In the following paragraphs we shall review waters which are not suitable for constructional purposes, indicating only why this is so since methods for treating such waters and correcting them have been illustrated earlier in this volume. Finally, we shall point out the conditions for the use of waters in pastes for conglomerates.

Salt waters and sulphate waters. The chlorides in salty waters react with the hydrated lime to liberate alkali hydroxides which, for the most part, consist of sodium hydroxide. The following reaction then takes place when these hydroxides come into contact with CO_2:

$$2NaOH + CO_2 + 9H_2O \rightarrow Na_2CO_3 \cdot 10H_2O$$

which leads to the formation of sodium carbonate decahydrate which is both voluminous and efflorescent.

Furthermore, certain chlorides ($CaCl_2$, for example) are deliquescent. They pick up moisture and hinder the drying out of the manufactured component. Finally, all chlorides are harmful because they promote the formation of chloroaluminates* in the conglomerates, and these compounds are very effusive.

Sulphate waters are rich in $CaSO_4$ and/or $MgSO_4$. Both of these sulphates are dangerous because of the swelling up which they either directly ($CaSO_4$) or indirectly ($MgSO_4$) provoke in conglomerates owing to the formation of Candlot's salt in them. There is, moreover, a threshold above which sulphates become dangerous (about 0.3 g/l). The sulphate content in drinking waters lies below this threshold.

Magnesium sulphate acts indirectly because it must first yield calcium sulphate:

$$MgSO_4 + Ca(OH)_2 \rightarrow CaSO_4 + Mg(OH)_2$$

On account of the insolubility of the magnesium hydroxide which is formed in this way, the alkalinity of the environment falls off extremely sharply after this reaction, and the major factors determining the mechanical strength of manufactured components, that is, the silicates of calcium, are thereby rendered unstable. Consequently, magnesium sulphate is more dangerous than calcium sulphate.

Rain waters, snow field waters, and glacial waters. If they are collected in centres of habitation or industrial zones, rain waters are turbid waters and, possibly, polluted; that is, they belong to the categories which we shall consider later. In other cases, rain waters, like the waters from snow fields and glaciers, contain no salts or gases. They are therefore waters with high solvent and diluent powers in which the solubility of calcium hydroxide increases, possibly, without attaining saturation, and thereby leads to the instability of the compounds which influence the hardening of the binders and, in particular, of belite.

These waters can be converted into waters for constructional purposes by putting them into contact with non-marl limestones and blowing in CO_2, or making them percolate through suitable earths, or, again, by the addition of suitable compounds to them (containing limes or bicarbonates, etc.).

On the other hand, building structures are protected from attack by waters which contain no salts by the use of protective coatings.

*Chloroaluminates are salts with the formula: $Me_3[AlX_6]_2$, where Me indicates calcium or magnesium and X indicates chloride and/or hydroxide. They are generally basic salts.

Turbid, stagnant, and waste waters. Turbid waters contain clays and organic substances which precipitate the aluminium and calcium in binders by various complexation reactions. Stagnant waters mainly contain gases such as CO_2, H_2S, CH_4, and other hydrocarbons. The first two of these gases (CO_2 and H_2S), by forming salts, remove calcium hydroxide and thereby lead to the well known and deleterious consequences. Hydrocarbons are chemical inhibitors of the setting and hardening processes.

Waste waters contains salts, organic substances, and every kind of gas, and are therefore quite inadmissible for use in the preparation of binder pastes.

The temperature and quantity of water in conglomerate pastes. In order that the binders should set correctly, the water used in the mixing of the paste should be at a temperature of between 10 and 40°C. Higher temperatures make it difficult to dissipate the heat evolved during setting and thereby hinder its course, while temperatures below 10°C lower the reaction rates to a greater or lesser extent, and, at about 0°C, setting no longer occurs. Below zero degrees, freezing occurs, followed by the shattering of the object due to the expansion which takes place when ice is formed. The addition of certain salts (in particular, alkali nitrates) impedes freezing without hindering or modifying the reactions.

As regards the amount of water to be used in the preparation of binder pastes, it is necessary to rely on common practice or empirical tests or, better still, the special tests described in the Official Standards, Since every type of binder is characterized as being of the 'normal paste' type after the addition of the optimal amount of water, the cement and lime producers themselves implicitly indicate the appropriate amount of water by specifying the properties of the normal marketed paste on the bags (Fig. 14.41).

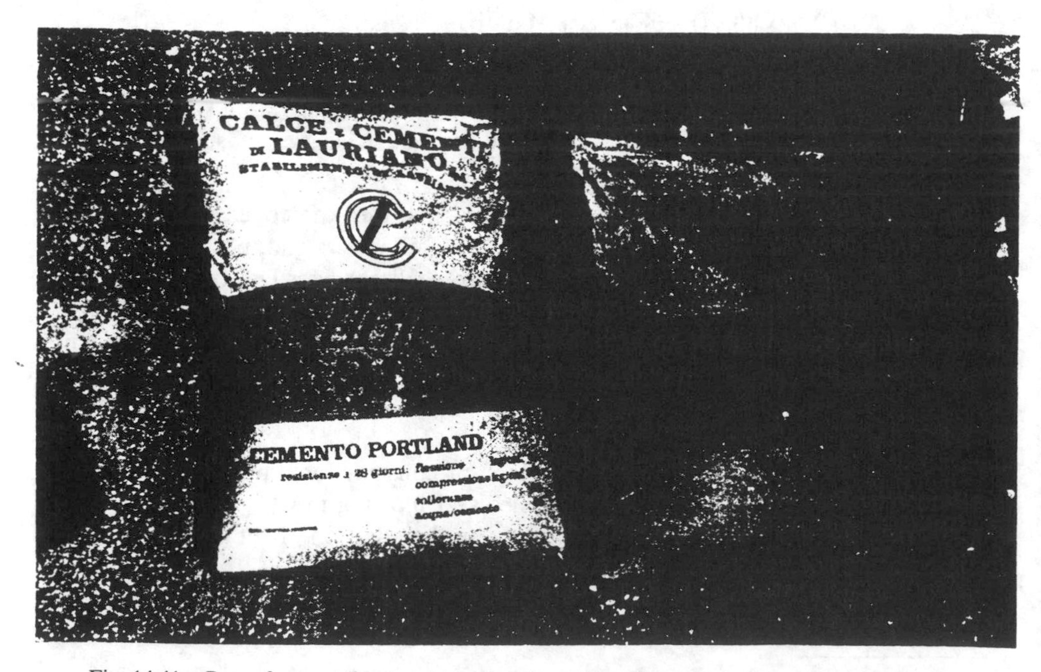

Fig. 14.41—Bags of cement (front and rear views) with the official grade and properties marked on both sides as required by law.

Concrete aggregates

Concrete aggregates such as gravel (with an upper size limit corresponding to shingle and a lower size limit corresponding to fine gravel aggregate) and sands are 'elastic products'; that is, they are formed by natural phenomena involving the breaking, flaking, crushing, abrasion, and crumbling of stones.

The aggregates used in concretes are illustrated in Table 14.16.

Natural rock aggregates are used when suitable gravels are not available. These are produced by the crushing of hard, strong, compact, and non-marl rocks, and the crushed material is classified into different sizes by washing the material with jets of water. A good natural rock aggregate for concrete must weigh about 1450 kg/m^3, when dry.

Gravels to be used in concretes must be free from saltiness, organic matter and earth materials. When necessary, they are conveniently classified using water.

Sands for conglomerates must be sharp to the touch and free from soluble salts, marls, clays, and organic matter. When a good sand is washed, it does not make the water turbid, and, when clasped in the palm of the hand, it creaks and then runs off without leaving any traces. Any shortcomings in a sand for conglomerates can therefore be corrected by washing it with water while stirring it and keeping its surfaces rubbed during the washing.

The grain size distribution of a sand to be used in conglomerates must be established on the basis of the uses to which it is to be put. The criteria are:

* sand for concretes must pass through a sieve with a mesh consisting of 5 mm circular holes;
* sand for mortars to be used with natural stone must pass through a mesh with a size of 3 mm;
* sand for mortars to be used in the laying and rendering of brickwork must pass through sieves with a mesh size of 1 mm.

On the basis of where they are found, sands may be classified as dry quarry, subaqua (or dredged), river, lacustrine, sea, or volcanic. The best quality sands are those dredged from under water and river sands. Marine sands are generally also very good after they have been thoroughly washed with fresh water.

In the case of the non-availability of silica sands, sands can be prepared for conglomerates by crushing granites or very rich limestones (practically marbles). The resulting sands are, however, expensive owing to the energy consumed in making them and the price of the raw materials.

Table 14.16—Names and properties of clastic concrete aggregates

Name	*Diameter (cm)*	*Name*	*Diameter (cm)*	*Structure*
Shingle	8–6	Coarse sands	0.5–0.25	Shingle and gravels, at least externally, are made out of silica or hard silicates. Sands must consist of 90% min. of SiO_2 with the remainder consisting of feldspars ('granitic' sands), $CaCO_3$ ('calcareous' sands), Fe_3O_4 ('magnetitic' sands), or $FeTiO_3$ (ilmenitic' sands).
Coarse gravels	6–5	Medium sands	0.25–0.1	
Medium gravels	5–3	Fine sands	0.1–0.05	
Fine gravel	3–0.5	Very fine sands	0.05–0.005	

The weight of a good sand for conglomerates must lie within the range from 1300–1600 kg/m^3.

Specific concrete additives

Concrete additives are mainly used to modify, promote the development of, or enhance the manner in which the binders acts, as has already been noted in the treatment of cement. Some, however, such as waterproofing and impermeabilizing agents, clearly act on the whole of the conglomerate.

Below, the principal types of such additives are summarized and their effects are indicated.

Vinylic products. Aqueous suspensions of vinyl acetate act, in principle, as hydric regulators and enhance the adhesion and cohesion between the aggregates. More specifically, the vinyl additives:

* remove the need to continuously soak the conglomerates during the setting phase because the highly hydrophilic particles of the polymer slow down the evaporation of the water which is present in the paste;
* facilitate the dispersion of the cement in the pastes, thereby improving their workability and permitting a reduction in the water/cement ratio;
* increase the cohesion of the paste, thereby making the final conglomerates more resistant to corrosion and erosion.

Accelerating agents. When salts such as Na_2SO_4, K_2SO_4, and $CaCl_2$ are added in suitable amounts to the conglomerate pastes, they accelerate the setting and hardening phenomena by controlling the hydration of the binder. In particular, calcium chloride, which is not harmful when present in amounts not exceeding 2%, allows the paste to attain high mechanical properties after the elapse of short periods.

Retarding agents. To retard the setting of the binder with the aim of increasing the period over which the paste can be handled (to achieve the improved dispersion of the heat of reaction, for example), sugars, glycerine, borax, carboxymethylcellulose, or salts of zinc, tin, and lead may be introduced, depending on one's choice, into the pastes.

Aerating compounds. Compounds, of fairly constant dimensions, which promote the occlusion of air bubbles and which are uniformly distributed throughout the mass are known as 'aerating compounds'. Such aeration is carried out with the aim of making the concretes resistant to freezing and thawing, making them more workable, homogenizing the paste, reducing the exudation of liquid and the tendency of the components to separate from one another. The materials used as such aerating agents are natural resins or synthetic products of the alkylsulfonic derivative type. The amounts used are very variable and cover a range from 0.005–0.08% of the binder.

Plasticizers. Plasticizers increase the workability of the paste and facilitate the laying operations. Hydraulic lime itself acts as a plasticizer, and other specific plasticizers are the bentonites, Kieselguhr, talc, and very fine sand. In general, the bentonites and sands must never exceed 5% of the concrete by weight, a limit which can be exceeded by other plasticizers.

Thinning agents. Hydrophilic, surface active substances (thinning or fluxing agents) facilitate the intimate contact between the components of the pastes and, in particular, the soaking of the various phases, thereby making it possible to reduce the amounts of water which are used to close to the theoretical minimum. As a consequence, thinning agents promote the workability of the paste and lead to an increase in the mechanical strength of the manufactured sections. Polyvinyl alcohols are the best thinning agents. They are added in a ratio of about 2–3 g per kilogram of binder.

Anti-degradation agents. To prevent the degradation of cement-based materials, it is necessary to prevent an aggressive atmosphere from coming into contact with the cement. This can be done by making the conglomerates impermeable so that the air can no longer penetrate into them.

The best way of doing this is to mix in about 5% of non-emulsified epoxy resins with the cement at the point when the concrete is being prepared. When this is done, the concretes not only acquire an optimal resistance to polluted atmospheres but also to aggressive solutions such as sulphate solutions, chlorite solutions (swimming pools), and salt water.

Manufactures made by using such additives in the pastes are lighter than the normal products. They have a greater flexural strength and the same compressive strengths. Concretes and mortars which are both impermeable and resistant to aggressive chemicals may also be obtained by mixing in 1–3 kg of waterproofing agents, which are available as a paste or a powder, with each 100 kg of cement used during the preparation of the mortar or agglomerate pastes.

The components of these additives, which consist of plastic materials (resins + plasticizers), etc, act:

* by increasing the workability of the paste without increasing the water/cement ratio;
* as anti-shrinkage agents so as to prevent dangerous cracking of the mass;
* as substances which render insoluble any hydrated lime which has not participated in the reaction leading to the hardening of the cement.

Other specific actions are possible, depending on the nature and the state of the resin used in their manufacture.

Anti-detachment additives. One of the reasons for the breakdown of reinforced concrete is the progressive detachment of the reinforcement from the conglomerate. To ensure the perfect and lasting adhesion between the components of the reinforced cement, synthetic adhesives may be added to concretes in the form of emulsions which subsequently harden owing to polymerization reactions and form macromolecules which are insoluble in water, benzene, and oils, but readily bond to both the metal component and the concrete.

Curing agents

Both during the delicate maturing phase when, among other things, a concrete may undergo degradation due to the evaporation of the water which has not already reacted, and when the structure is actually put up in its final state and it is necessary to protect it, treatments on the surfaces of cast sections of it or items manufactured from it are carried out by using chemical products known as 'curing agents'.

While the concrete is maturing, it is protected, for example, with films of an organic nature which, when sprayed on the surfaces, retain the water there for a long time. Examples of curing agents for cast conglomerates are:

* protective coating epoxy resins which, by strongly adhering to the surfaces and rendering them absolutely impermeable, thereby prevent osmosis from occurring which would lead to their swelling if they come into contact with fresh or slightly salty water, or their wrinkling and puckering, if they come into contact with brines;
* silicone solutions in suitable solvents. These have the ability to penetrate so deeply as to be able to compete with the effects achieved by techniques using the addition of resins during the making of the manufactured objects. They render the structure completely inaccessible to aqueous solutions, moisture, dirt, and grime and corrosive agents (anti-soiling and anti-corrosive actions).

Some metallic stearates, various resins, and bituminous mixtures also act as waterproofing and impermeabilizing curing agents. These are mainly used for walling and rendering made out of every kind of concrete and mortar. Among these agents, the epoxy–bituminous and the analogous epoxy–tar agents stand out in importance. By copolymerizing and linking themselves into compact masses, these agents form hard and tenaciously adhering coatings. They are particularly suitable for the protection of concretes which are exposed to highly aggressive industrial atmospheres or to chemical attack combined with the continual presence of water at temperatures up to 75°C.

The grain size of concrete aggregates

Structurally, concrete resembles natural agglomerates of the sandstone and breccia type. On the one hand, however, it must have superior qualitites of compactness, mechanical strength, and chemical resistance, while, on the other hand, it must lend itself to being reinforced. In addition to the correct ratio between the main components which has already been mentioned, it is also required that the aggregates would have a suitable grain size distribution.

Well-proved experimental tests, as well as many profound theoretical studies, have demonstrated that the particle size distribution of the aggregates is statistically optimal if their dimensions are such that, when screened, they correspond to the ideal Füller curve, and it is good in the case of statistical distributions lying in the particle size distribution envelope (the hatched area in Fig. 14.42), containing the ideal curve.

In brief, a concrete has a correct particle size distribution of the aggregates if these, when screened with sieves having predetermined hole sizes, turn out to have a size distribution, regarding the percentages which pass through the calibrated holes in the various screens, so as to yield values which lie within the particle size distribution envelope and, at the limit, correspond to the ordinate values of the ideal Füller curve.

In practice, to establish that the particle size distribution of the aggregates in a concrete is correct, it is checked that at least three* fractions of the aggregates fall within the particle size distribution envelope in the Füller diagram. For this purpose, a preweighed amount of the aggregate is put into the highest of a series of sieves with ever decreasing hole sizes, and, by shaking, is distributed among the lower sieves. The various fractions are then weighed, divided by the total weight, and multiplied by 100. The resulting values are then checked against the particle size distribution envelope in the Füller diagram at the abscissa values corresponding to each of the hole sizes in the sieves through which they have been graded. They must lie within this envelope if the size distribution of the aggregates in the particular preparation in question is to be considered as acceptable (see Fig. 14.43).

Reinforced concrete

The term 'reinforced concrete' indicates a structure consisting of a metal frame embedded into the concrete.

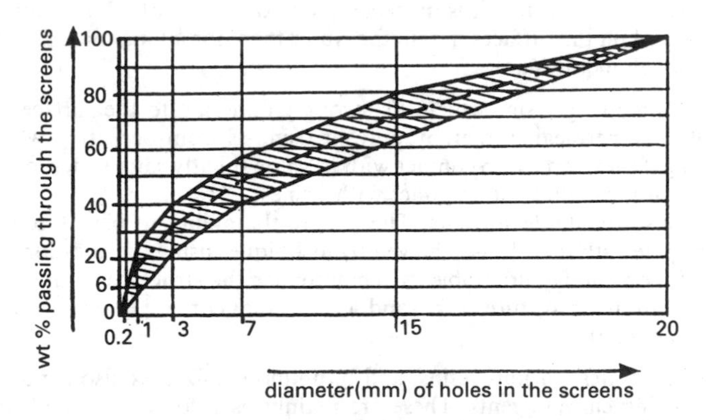

Fig. 14.42—Füller diagram. The broken curve within the hatched area shows the ideal grain size distribution of the aggregates for the production of a concrete with optimal strength. The hatched area shows the limits of variation.

*This is equivalent to checking that a formula corresponds to a curve, using several pairs of parametric values.

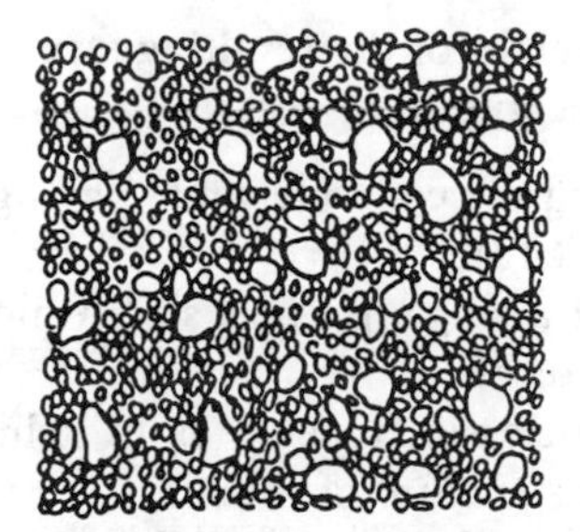
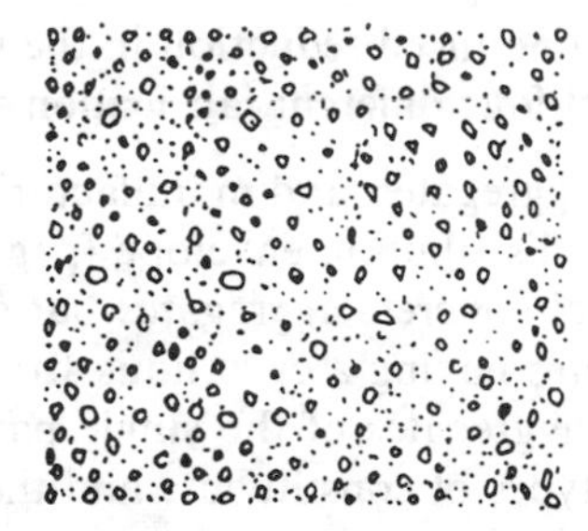

Fig. 14.43—Left: Structure of concrete produced with aggregates with sizes which are clearly too small with the result that the concrete has poor mechanical properties. Right: Concrete made with aggregates with the correct particle size distribution which endows it with high mechanical strength. In both cases, the voids between the aggregate particles must be looked upon as being occupied by the binder.

The metallic reinforcement is generally made of ordinary steel with a tensile strength of at least 45 kg/mm^2, and the thickness of the concrete around the steel must be at least 3 cm.

The metal rods must be thoroughly cleaned before being used in order to facilitate their adherence to the concrete. In principle, concrete and iron should adhere extremely well to one another on account of both the forces of an electrostatic nature which are operative between them and because they are in a position to react chemically to form ferrites, which, by interpenetrating the surfaces in contact, consolidate the bond between them.

The bonding between the steel and the concrete ensures that the resulting structure has a high tensile strength and flexural strength on account of the reinforcing which contributes the toughness of metals to the system. Another great merit of reinforced concrete is the fact that its mechanical properties do not fall off because of the effect of heat or thermal gradients, since its two components possess coefficients of thermal expansion (α) which are almost equal $\alpha_{steel} = 1.3 \times 10^{-5}$ cm/m·°C and $\alpha_{concrete} = 1.2 \times 10^{-5}$ cm/m·°C.

Nevertheless, physicochemical phenomena do exist which can ruin structures made of reinforced cement. In fact, if the iron is not well protected, corrosion may occur because of differential aeration, and the stray electric currents cause the iron to rust and to be pulled away from the cement.

Reinforced concretes with enhanced mechanical properties are attained by producing precompressed reinforced cement, that is, by putting the metallic reinforcing rods under tension during the casting and hardening of the cement. In this way, the possibility of damage occurring owing to tensile forces arising when loads are applied after the hardening of the castings is eliminated beforehand.

A particular type of reinforced concrete is 'steel concrete' which is made by the incorporation of iron alloy powders, turnings, or granules into Portland cement.

Steel concrete is used for the paving of military establishments, workshops used in the production and working of ferrous alloys, cement and glass works, and ceramic factories. In addition, it is used for the internal lining of silos which come into contact with abrasive materials and for the stone paving of the tracks used for the testing of heavy tracked vehicles and machines such as various types of excavators and armoured cars.

Light conglomerates similar to concretes

Nowadays, a numerous group of articial conglomerates exist which are far lighter than normal concrete but to which they bear a resemblance because they are also the products of a cement paste, water, and inert granules.

Whereas concretes normally have specific weights within the range from 2400–2600 kg/m^3, the conglomerates have specific weights which range from 200–1800 kg/m^3.

The very pronounced reduction in the weight of the resulting structures is arrived at by means of four different fabrication techniques:

- using light aggregates, and manufacturing the concrete while paying regard to the particle size distribution envelope (page 642);
- using normal concrete aggregates but of a single particle size, which leads to the conglomerates having a cavernous structure;
- using light aggregates of the single particle size type so as to achieve lightening due to the types of constituents and the structure of the mass;
- promoting the formation of 'concrete foams', using special swelling or aeration agents, which subsequently set in the form of products with a cellular structure.

The first three methods can be realized in building factories, while the fourth method can be carried out within the framework of the industrial technologies directed towards the preparation of artificial stones and prefabricated panels which, nowadays, frequently make use of the burnt sludges obtained from the products of water purification processes.

Expanded clays, pumice-stone, blast furnace slags, vermiculite, and expanded polystyrene are light aggregates for conglomerates. More recently, the ashes obtained from the combustion of coal (fly ash) in large thermal power stations have been used for this purpose.

Both the particle size distribution of the aggregates and the percentage that forms part of the conglomerate are specified, as is also the weight of cement per cubic metre of the conglomerate produced.

Concretes of this type are applied in two main areas, depending on the mechanical compression strength which they attain after 28 days. More precisely:

* those with a greater strength serve for the casting of light structures which have already been thoroughly checked for uses such as penthouses, lean-to buildings, canopies over railway stations, arches, and roofs;
* those with medium to low mechanical strength are used in the manufacture of non-load carrying artificial stones in the form of blocks and slabs and in the preparation of filling materials with high insulating properties such as prefabricated panels.

By forming a paste from a gravel, where all the particles have the same size, water, and cement which is added in amounts of 150–250 kg/m^3 at a water:binder ratio of 0.4:0.5, cavernous honeycomb concretes are obtained with a specific weight of about 1700 kg/m^3.

These concretes are suitable for the casting of quite massive and highly-insulating walls. This is ensured by the weight of the aggregate included in them and the avoids which are created as a consequence of the single particle size of the aggregate respectively.

Concretes with excellent insulating qualities but low mechanical properties are manufactured with vermiculite and expanded clays, both of which have a uniform particle size.

The principal use of these types of concrete is in artificial stones to be used in structures and, for the most part, dividing walls and partitions which are either not subjected or only slightly subjected to mechanical loads where a high degree of both thermal and acoustic insulation is required.

By forming a paste from hydraulic binders, water, and fine aggregates with the addition of substances which bring about the formation of quite uniform bubbles which are regularly distributed throughout the mass, solid foams are obtained which are so light that they float on water but have excellent insulating qualities and good mechanical properties.

These products, known as 'aerated concretes', can be subsequently worked with great ease by sawing, nailing, and chiselling.

Their mechanical strengths vary, depending on the density and the type of maturing procedure used, from 40 kg/cm^2 to about 300 kg/cm^2, and these concretes are most commonly used in the production of artificial stones and prefabricated panels.

The principle method for the production of aerated concretes is the Gasbeton method which entails the addition of gas-producing additives such as aluminium powder which reacts according to the equation:

$$2Al + (Ca(OH)_2 + 6H_2O \rightarrow Ca[Al(OH)_4]_2 + 3H_2$$

or hydrogen peroxide which reacts according to the equation:

$$2H_2O_2 \rightarrow 2H_2O + O_2$$

The normal ways for preparing these concretes (cellular concretes) are as follows:

* blast furnace slags + cement + single-sized sand + aluminium powder + water, maturation for 18 hours under a steam pressure of 15 atmospheres;
* volatile ashes + cement + single-sized sand + aluminium powder + dispersing additives + water, maturation after maximum aeration in an autoclave.

The Schaumbeton method is another method for the production of aerated concretes which uses the introduction of an air-filled foam which is made to form by foaming agents which are resistant to an alkaline and lime-containing environment.

However, aerated concretes are not always directly produced, and it is sometimes necessary to resort to preparative techniques which exploit preaerated materials.

For instance, when cement is mixed with finely ground silica and water and small polystyrene spheres, which are coated with an epoxide resin to improve the adhesion between the mortar and the plastic, a concrete is obtained with properties and uses analogous to those of normal aerated concretes.

The maturation of the cast object takes place automatically (auto-maturation) owing to the heat of reaction which the mass retains because of its extremely high insulating characteristics.

Light conglomerates which differ from concretes

By using fibrous aggregates instead of granular aggregates, and making use of cements other than Portland cement, conglomerates are obtained which have their own fields of use.

Conglomerates based on vegetable fibres

By forming a paste from the twigs of aquatic plants or long and pliable wood shavings and chips, especially those of poplar trees, which are soaked in solutions of magnesium chloride or sulphate, with a magnesite cement or a mixed magnesite–Portland cement, conglomerates are obtained which have low mechanical properties but extremely good insulating qualities to an extent where a 3–5 cm thick layer of this material has insulating characteristics which are equal to those of a normally constructed wall with a thickness of 30 cm.

These materials are most commonly used for thermal and acoustic insulation purposes, in the construction of light insulating walls, attics, and partition walls of various types and purposes.

Here are some specific materials which, in their properties, correspond to these conglomerates:

Eraclit: marine plants formed into a paste with magnesite cement in solutions of magnesium salts with subsequent moulding into sheets and maturation in an oven;
Populit: Poplar wood shavings or chips are soaked in magnesium salt solutions and formed into a paste with magnesite cement. They are subsequently moulded into simple slabs or more complex forms;
Durasol: obtained in the same manner as Populit but using the fibres of various woods. Also made into sheets, slabs, and more complex shapes;
Carpilite: differs from Durasol only in the aggregate used which, in this case, is wood wool, that is, fine woolly wood fibres.

Cement–asbestos concretes (Eternit)

For some time fibrous concretes have been produced with asbestos fibres which are formed into a paste with cement and water. All types of asbestos (chrysotile, tremolite, anthophyllite, etc.) are used for this purpose, so that it would be more correct to speak of fibrous silicate–cement concretes.

To understand some of the merits of cement–asbestos conglomerates, it must be remembered that, on the one hand, their mineral fibres components are unalterable, like stone, and as strong as steel of the same cross-sectional area, while, on the other hand, the binder is the same as that used in a reinforced concrete. As a result of this, the product can readily be used in the form of thin, relatively light sheets or slabs which are required in modern building construction.

The techniques used to obtain these concretes are similar to that used in the production of paper. It starts by taking successive aliquots from a bath containing an aqueous suspension of fibrous silicates and cement from which water is removed when they are put onto a rotating absorbent cylinder which moulds their moist fibrous layers. These layers are deposited on another cylinder until they form a panel of the required thickness. This is followed by washing and levelling so as to obtain a type of cardboard which is extremely strong and can be readily worked with mechanical tools.

Objects made out of asbestos–cement mixtures which are very widespread in every building sector exhibit the outstanding merits of lightness, impermeability, and high temperature resistance. They also have a high chemical resistance towards aggressive and filthy waters (which can be enhanced by epoxy-tar coatings) as well as fairly high tensile and compression strengths*. Finally, it should be mentioned that these materials can be coloured with mineral pigment-based paints which can be baked.

*The (average) tensile, strengths are about 350 kg/cm^2 and the average compression strengths are 700 kg/cm^2. Furthermore, it may be recalled that tubes of Eternit and similar concretes can withstand pressures of up to 50 atmospheres.

15

Bitumens, tars, and derivatives

15.1 BITUMENS: NATURE, PROPERTIES, AND DERIVATIVES

Chemically, bitumens, for the greater part, consist of mixtures of very high molecular weight hydrocarbons and acids which structurally include all possible combinations among alkanes, cycloalkanes, and aromatic hydrocarbons with or without carboxyl groups on the open chains or rings.

There are also nitrogenous compounds and, especially, sulphur compounds in bitumens, and these are responsible for the optico-olfactory properties which are so characteristic of these substances.

In general, and especially when they have just been recovered, bitumens are black or dark brown materials which are bright to a greater or lesser degree and have a low softening point.

According to Italian standards, bitumens are extremely complex mixtures of higher hydrocarbons and the products of their partial oxidative transformation endowed with a binding capacity and which dissolve in carbon disulphide to leave an insoluble residue which is not greater than 2% of the material taken.

The most recent and reliable technical and scientific research on bitumens suggests that they generally consist of colloidal dispersions of asphaltenes, maltenes, and other substances which are difficult to define from the point of view of their true chemical nature.

Their colloidal nature endows the bitumens with internal elasticity and thixotropic properties which leads to a reduction in their viscosity as a result of mechanical mixing. However, when they are allowed to rest, the original viscosity is re-established, or there is even an increase over this value.

The asphaltene micelles are composed of aromatic compounds (relative molecular masses from 600 to 2000) including carboxylated compounds with high melting points. On the other hand, the maltene micelles are predominantly paraffinic/aromatic or cycloalkane compounds, and carboxylated compounds are also quite abundant.

The compounds which are of an uncertain nature have a high sulphur content and have a high intrinsic carbon/hydrogen ratio.

The properties which are legally required of bitumens are shown in Table 15.1, and some of the pieces of apparatus which are recommended in the official standards for the determination of some of these properties are depicted in Figs. 15.1, 15.2, and 15.3.

Bitumens may be of either natural or artificial origin, with a great prevalence of the latter at the present time.

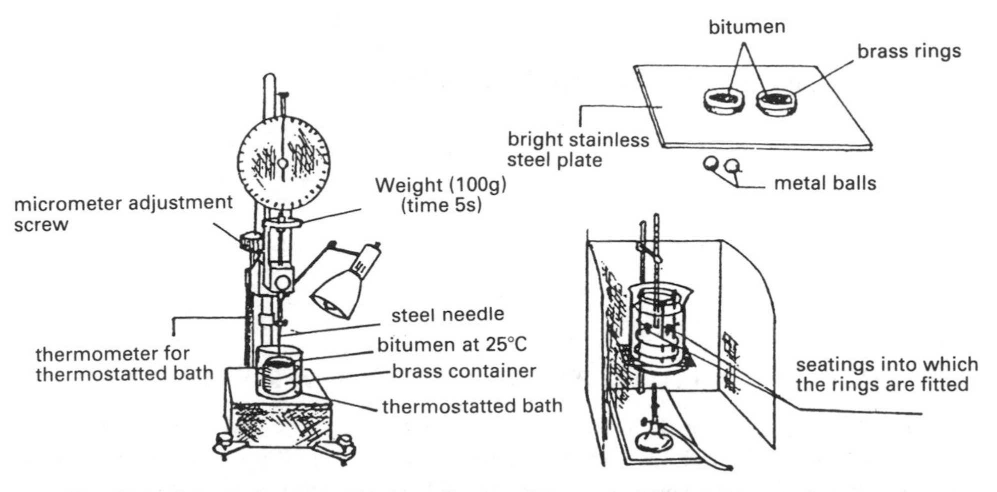

Fig. 15.1—Left: Apparatus for the determination of the penetrability of a bitumen by measuring how far a steel needle, under constant load and at 25°C, sinks into it. Right: 'Ball and ring' device for the determination of the softening point of bitumens. These are pressed by metal balls into the rings, which have been put into a heated chamber in which they have been poured and cooled, using a technique which ensures that they have a smooth surface.

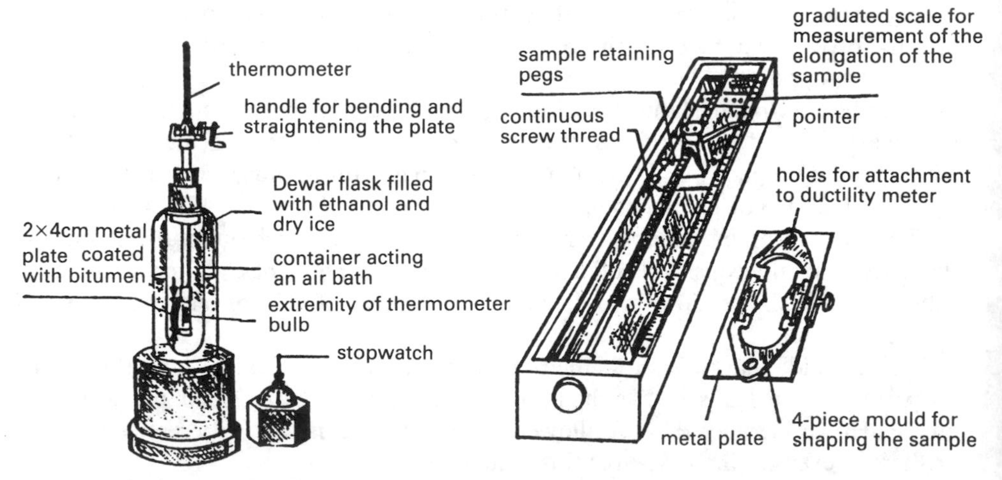

Fig. 15.2—Left: Frass apparatus for determining the 'fracture point' of a bitumen by noting when the bitumen, which has been spread out on a thin plate which is repeatedly bent and straightened, finally cracks. Right: Ductility meter which measures the ductility of a bitumen sample of a certain shape made by using a mould and subjected to traction at a certain velocity until it breaks.

Table 15.1—Values required of bitumens in order for them to be acceptable

Properties	*B*† 20/30	*B* 30/40	*B* 40/50	*B* 50/60	*B* 60/80	*B* 80/100	*B* 130/150	*B* 180/200
Penetration at 25°C- dmm (in tenths of a millimetre)	>20 to 30	>30 to 40	>40 to 50	>50 to 60	>60 to 80	>80 to 100	from 130 to 150	from 180 to 180
Softening point (*ball and ring*) in °C	56/68	45/64	51/60	48/56	45/54	44/49	40/45	37/42
Breaking point, max.: in °C	−2	−4	−6	−7	−8	−10	−12	−14
Ductility at 25°C, min.: in cm	25	50	70	80	90	100	100	100
Solubility in CS_2/min. %	99	99	99	99	99	99	99	99
Maximum volatility at 163°C	–	–	–	–	–	0.5	1	1
Maximum volatility at 200°C	0.5	0.5	0.5	0.5	0.5	–	–	–
Penetration at 25°C of the residues from volatility test: minimum value expressed as a percentage of that of the original bitumen	60	60	60	60	60	60	60	60
Maximum breaking point of the residue from the volatility test: in °C	0	−2	−4	−5	−6	−7	−9	−11
Maximum wt.% of paraffin	2.5	2.5	2.5	2.5	2.5	2.5	2.5	2.5
Minimum adhesion to:								
San Fedelino granite, *dried samples* kg/cm^2	7.50	7.00	6.50	6.00	5.50	5.00	3.50	3.00
San Fedelino granite, *soaked samples* kg/cm^2	3.60	3.00	2.50	2.25	2.00	1.75	1.50	1.25
Standard Carrara marble, dried samples, kg/cm^2	7.00	6.50	6.00	5.50	5.00	4.50	3.00	2.60

† The letter B, followed by the penetration range characterizing the bitumen, distinguishes between the various types of bitumens.

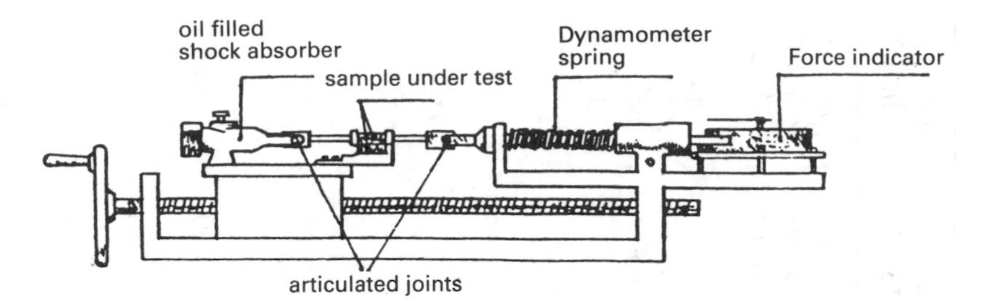

Fig. 15.3—Machine for testing the adhesion of bitumen to stone. This determines the force required to pull two standard stones (San Fedelino granite and Carrara marble) apart when they are bonded by a thin film of the bitumen under examination. The stones are shaped in the form of prisms. One is firmly clamped while the other is encouraged to move under a load increasing at a rate of 5–10 kg/s. The adhesion is proportional to the maximum load at which the bond is broken.

Natural bitumens fill various types of terrestrial spaces (pockets, fractures, or lacustrine passages) which are mainly located in South America (the island of Trinidad) and Central America. However, bituminous asphalts or the rock asphalts from which bitumens can be recovered by fusion under pressure are more frequent in nature.

The best known European and Eurasian bitumen deposits occur in Transylvania, Albania, in the Caucasus, and, especially, in the Dead Sea region (Jewish bitumen).

Notable deposits of rock asphalts occur in Italy in the Ragusa district of Sicily and the Pescara region in Abbruzzo. These materials are readily exploited because they are easily crushed.

Bitumen products which are of industrial interest are characterized on the basis of the percentage of the product formed by the polymerization and oxidation of petroleums (bitumen) with respect to the inert minerals which they contain. There are therefore:

- the correctly named bitumens which must contain at least 98% of true bitumen, with the remainder consisting of lithoid (inert) inorganic substances;
- bituminous asphalts containing from 98 to 15% of bitumen, while the remainder consists of inert substances;
- rock asphalts, which are formed from 15–6% of bitumen and 85–94% of inert material.

When there is less than 6% bitumen in the material, it is simply considered as a bitumen contaminated rock. These are further subdivided into rocks containing from 2–4% of bitumen and rocks containing 4–6% of bitumen.

Artificial bitumens are obtained as the final 'vacuum' residue in petroleum refineries. Stabilizing blowing is carried out on them to bring their properties close to those of natural bitumens. Air is blown over a period of several hours into the mass which is maintained in a semifluid state. In every case, artificial bitumens are much poorer in sulphur than natural bitumens because the sulphur-containing compounds have already been distilled over and passed into the petrols, gas oils, and heating oils during the topping distillation stage which precedes the vacuum stage.

15.2 TARS

When they are formed, tars are oily, dense, viscous, brown or blackish liquids which are obtained by the distillation of anthracite grade coals in the absence of air. It is for this reason that one speaks of coal tars.

Lower quality tars are a dark chestnut brown in the form in which they are formed, but can be improved with the aim of using them practically, both by separating off the hydrocarbon components and the alcohols and lower acids which they contain from them, or by treating them with oxidation–polymerization agents. They are obtained by the distillation of lignites. Tars are produced from lignites on a large scale in West Germany.

The original tars must first be treated before being used as industrial tars. This treatment comprises:

- decantation to separate the waters with dissolved inorganic compounds (ammonium salts);
- heating to remove further water and benzol;
- distillation under vacuum in boilers.

Consequently, industrial tars are the non-distillable constituents (with boiling points above 400°C) of coal tars which are often commonly known as pitches.

The properties of tars which are obtained directly by the distillation of coal are shown in Table 15.2.

Usually*, industrial tars are black translucent solids at normal temperatures with a phenolic smell. They are predominantly formed from polynuclear aromatic hydrocarbons characterized by the presence of diffused hydroxyl functional groups. The remainder consists of complex hydrocarbons of the type found in bitumens.

15.3 ASPHALTS AND THEIR CLASSICAL FORMS OF USE

Asphalts are intimate mixtures of bitumens and tars, which may be of natural or artificial origin, and inert inorganic materials.

Natural asphalts are prepared for use by crushing asphaltic rocks, that is, rocks which contain from just over 6% to about 15% of bitumens. These rocks are said to be poor if they contain between 6 and 10% of bitumen, and rich if they contain more than 10% of bitumen.

Artificial asphalts are products which are obtained by mixing:

- impure bituminous rocks (2–6% of bitumen), which have been suitably crushed to the size of fine gravel, with asphaltic bitumens;
- impure bituminous rocks which have been reduced to the size of fine gravel with artificial bitumens or tars;
- various mineral aggregates (sands, gravels, and limestones) crushed to the size of fine gravel if necessary, with artificial bitumens or with tars.

It is obvious that all the mixtures of stony aggregates with bitumens or tars which have just been listed, correspond to the definition of asphalts presented above.

*** *Usually*, because, depending on how much is pushed down, on how the vacuum distillation from which they originate is carried out, and the nature of the original fossil carbon (rich or lean anthracites), pitches can turn out to be rich, semi-rich, or hard, with softening points increasing in that order.**

Table 15.2—Acceptance properties of tars

Properties	*C* 10/40†	*C* 40/125	*C* 125/500
Viscosity at 30°C: in seconds	10–40	>40–125	>125–500
Specific weight at 15°C	1.14–1.22	1.15–1.24	1.16–1.25
Maximum weight % of water	0.5	0.5	0.5
Maximum weight % of distillate (excluding water) up to 200°C	1	1	1
Weight of % of distillate between 200°C and 270°C	9–17	5–14	4–10
Weight % of distillate between 270°C and 300°C	4–12	4–12	2–10
Minimum weight % of distillate between 300°C and 350°C	15	16	16
Softening point (ball and ring) of residue from distillation at 350°C, maximum in °C	85	85	85
Maximum percentage of phenols (cm^3 per 100 g of tar)	3	3	3
Maximum weight % of naphthalene	4	3	3
Weight % insoluble residue	4–18	5–20	6–22
Minimum adhesion to:			
San Fedelino granite (dry samples: kg/cm^2)	3.00	5.00	6.00
San Fedelino granite (wetted samples: kg/cm^2)	1.25	1.75	2.00
Standard Carrara marble, dry	3.00	5.00	6.00

Note: Purchasers have the authority to request tars with viscosities lying between limits which are closer than those indicated in this table for each type of tar.
† The letter C, followed by the viscosity range which characterizes the material, is used to distinguish between the various types of *tars*.

From a chemical point of view, the good adherence between the mineral support and the plastic organic component in the asphalts is specifically guaranteed by the compatibility which exists between the naphthenic acids binding the bitumens and tars and the silicates and carbonates of the lithoid (stony) additives.

Asphalt flours are obtained by finely crushing asphalts. These can be laid, when molten, on concrete road beds and then pressed with rollers to obtain compressed asphalts. Alternatively, they can be heated up to about 150°C in boilers and mixed in amounts up to 6% with bitumen, whereupon, after complete homogenization, asphalt mastics are obtained.

Asphalt mastics, when mixed with bitumen (or tar) and sand containing a large fraction of minute particles, yield the cast asphalts which, after fusing at about 200°C, are spread out in a highly plastic state on road surfaces.

Finally, rolled asphalts are obtained by the hot laying of artificial asphalts prepared by using various aggregates on road surfaces and then consolidating the layer of asphalt by heavy rolling.

15.4 BITUMINOUS CONGLOMERATES

Materials exist which have binding properties similar to those of a binary water–cement combination, these properties being ensured by bitumen. They are used in various sectors of the constructional and, especially, road building technologies, and such materials are defined as bituminous conglomerates. The mechanical strength of such agglomerates is ensured by the mineral aggregates which they contain. According to the uses to which they are to be put, various levels of compression strength, wear resistance, and fatigue resistance are required. Furthermore, they are found to be impermeable to water and adhesives.

Finally, the ease with which all bituminous conglomerates can be prepared and applied is greatly valued.

According to the percentage of voids which remain after compaction, normal bituminous conglomerates are subdivided into:

- closed conglomerates with a percentage of voids not exceeding 7%;
- semi-open conglomerates with a percentage of voids below 12%;
- open conglomerates with more than 12% of voids.

Closed bituminous conglomerates

For the construction of road surfaces which carry high volumes of traffic, carriage-ways, motorways, and airport runways, appropriate load-bearing structures must be employed for the top layer of the coating. Closed bituminous conglomerates lend themselves to this, as they exhibit the highest values of properties which, in general, characterize bitumen bound conglomerates.

There are two types of closed bituminous conglomerates: bituminous mortars and bituminous concretes.

The bituminous mortars are formed from sands (the aggregate), additives, and bitumen, while the bituminous concretes are formed from crushed stone, pebbles, and sands (the aggregates), additives, and bitumens.

The bitumens used for closed aggregates must have melting points which are higher, the higher the average temperature of the site where they are employed; and they must constitute from 6.5–8.5% of the weight of the inert material.

The additives (fillters) are, for the most part, hydrated lime, silica* powders, or asbestos and cements.

The aggregates are subdivided into fine and coarse aggregates according to whether one is dealing with sands, or pebbles and crushed stone respectively.

*The silicas are, for example, quartziferous porphyries, quartzites, granites, and granodiorites.

Semi-open bituminous conglomerates

Wherever possible, closed bituminous conglomerates are replaced by products which are cheaper because of the reduced amount of care required in their manufacture and the savings made in valuable raw materials (the bitumen and the selected fillers).

The semi-open conglomerates are mainly used in the preparation of the binding strata between the layers in direct contact with the earth (the base layers) and the top layer which is subject to the wear and tear. They are also employed in the reprofiling of surfaces and as the top layers on roads which carry only light traffic.

Depending on their use, the composition of these conglomerates more or less approaches that of the closed agglomerates, while in those intended for the commonest uses, no additives are employed, but 4–4.5% of bitumen with respect to the inert material is then included. The climate of the region in which they are employed must also be taken into account when establishing their composition.

Open bituminous conglomerates

For roads and pavements which are used only by pedestrians, roads with light traffic in areas with a cold but not freezing climate, and, especially, temporary coatings and patching work, use is made of bituminous conglomerates formed from bitumen and gravel or crushed stone, which are commonly referred to as bitumenized gravel or bitumenized crushed stone.

Since the particles in these aggregates are almost all of the same size, the percentage of voids in these materials increases to above 12%, and one therefore speaks of open bituminous conglomerates. These materials are very cheap on account of the ease with which they can be produced, the low grade of the bitumens which can be used in their preparation, and because these materials require only up to 4.5% (and even 3%) of tar.

Non-standardized bituminous conglomerates

The term base layers has already been mentioned. It indicates the foundations laid on the earth which support the bonding layer and the finishing layers of road surfaces. A suitable type of bituminous conglomerate has been produced which is intended solely for the construction of such foundation layers, and this is but one of the many non-standardized conglomerates which have been developed in the last thirty years for specific purposes. They are described in the appropriate manuals on the theory of the action and the practice of their manufacture.

Referred to as bitumenized compounds, this class of conglomerates is formed from gravel and river sand mixed with bitumen, or from clean crushed stone and pit sand mixed with bitumen.

These are economical mixtures, as the components of the correct grade are readily found in the requisite amounts, and they are very simple to prepare and apply.

Either bitumens separated from natural asphalts or bitumens produced by carrying out artificial air-induced oxidative polymerization processes on the heavier fractions of petroleum distillates are used in the manufacture of bituminous conglomerates which may be adapted to fulfil certain requirements and are therefore not constrained to have any preordinated compositions.

Although the ratios between the percentages of the inert materials and those of the binder may vary widely so as to yield a whole range of bituminous compounds, the most commonly used types contain from 3.5–5% with respect to the inert materials of the kind of medium soft bitumen which is used in the manufacture of semi-open bituminous conglomerates for foundation layers.

15.5 OTHER MODES OF USE OF BITUMENS AND TARS

Bitumens obtained from the petroleum industry are almost exclusively used in conglomerates for the surfacing of roads, and the same can be said regarding the origin of the bitumens destined for use in other products. It is, however, to be anticipated that tars will become of greater importance with the tendency to turn to the development of the use of coals and the possible modification of the formulation of the various products which will contain tars rather than bitumens as binders.

The following review provides some idea as to the extent to which bitumens can be replaced by tars.

Bituminous and tar emulsions

Aqueous suspensions of bitumens or tars (emulsions) stabilized by soaps or with other artificial synthetic surfactants which are never present in amounts exceeding 1.5%, are very commonly used nowadays for road works and waterproofing. They can be carried out in wet weather, but not when it is ‘beating down’.

These are emulsions of water in bitumens or tars because the percentages of the non-aqueous component which vary from 60–70% in suspensions for road works and from 70–80% in waterproofing suspensions are greater than the percentages of water. The emulsions are prepared by ‘whipping’ the components together at temperatures close to 100°C, thereby mixing molten bitumens or tars with preheated water*.

The resulting emulsions formed at the abovementioned temperatures do not separate into two phases when cold, but, when the water evaporates after they have been applied to the inert materials to be consolidated, the bitumen (or the tar) forms a film which adheres strongly to the supports and to the inert materials onto which it is spread. At the end of the setting process, the waterproof nature of the films is ensured by the fact that the particles of the inert materials are bonded to one another, while the mechanical strength arises from the bonding of the film to the support.

The operational efficiency, which is understood as the intrinsic tenacity and adherence to supports and to mineral coatings in contact with them of the films formed by bitumens and tars which have been pre-emulsified with water, is attained as the emulsifying agent is slowly released from the mass. This provides an opportunity for the film-forming components to orient themselves in the most favourable manner and to promote and develop the most stable form of bonding both with respect to their own interior and where they are coming into contact with the other materials.

Fluxed (or ‘liquid’) bitumens or tars

By forming highly-dispersed suspensions of bitumens and tars in solvents of the

*Although the components to be emulsified are molten in the case of these preparations, the water must also already be well heated if the whole system is to be brought up to 100°C, because the specific heat of water is considerably higher than the specific heats of bitumens and tars.

benzene type or oils obtained from the distillation of coal tar, preparations are obtained which can be applied when cold in order to waterproof or, at least, to consolidate sites after excavation, patchwork, and so on. These products behave as liquids when applied (hence the name 'fluxed'), and thereby ensure both economy in use and a uniform penetration in the soils which are treated.

A guide to the relationship between the characteristic values of the properties and the behaviour of the materials also exists for the (fluxed) liquid bitumens, as shown in Table 15.3.

Bituminous paints

Like all paints, bituminous paints primarily contain the two essential components: the film-forming component and the solvent which are entrusted with the respective tasks of forming thin but strong and elastic films and of dissolving the film-forming component(s) in the phase in which the paint is applied.

Table 15.3—Properties for the acceptability of liquid bitumens (LB)

Properties	*LB* 0–1	*LB* 5–15	*LB* 25–75	*LB* 150–300	*LB* 350–700
Redwood viscosity at 25°C:					
10 mm hole	0–1	5–15	25–75	150–300	350–700
4 mm hole	max. 30				
Flash point in an open vessel, minimum in °C:	65	70	75	85	90
Distillation (% of the the total volume of the distillate up to 360°C):					
water, maximum	0.5	0.5	0.5	0.5	0.5
up to 225°C	max. 15	max. 10	max. 6		
up to 260°C	max. 60	10–50	5–40		
up to 315°C	75–80	min. 70	min. 65	min. 60	min.50
Distillation residue at 360°*C*:					
vol. %, minimum:	60	70	75	80	85
solubility in CS_2, minimum:	99.5	99.5	99.5	99.5	99.5
softening point:	min. 25°C	–	–	–	–
penetration:		80–300	80–300	80–300	80–250
Adhesion to San Fedelino granite, minimum:					
(a) dry samples, kg/cm^2	2.5	2.75	2.75	2.75	3.00
(b) wet samples, kg/cm^2	1.1	1.25	1.25	1.25	1.25
Adhesion to statuary Carrara marble:					
dry samples: kg/cm^2	2.3	2.5	2.50	2.50	2.75

If simple bituminous paints are being considered, there is a single film-forming agent: bitumen or tar (even if it is somewhat unusual to speak of tar paints). On the other hand, if, as is becoming ever more common nowadays, auxiliary film-forming agents such as urea or epoxy resins are included in these paints in addition to the main film-forming component, then they are complex bituminous paints, among which the best known are the epoxybituminous paints and the epoxytar paints. The auxiliary film-forming agents are added to enhance the adhesion, the hydrophobicity, and the waterproofing characteristics of the resulting coatings.

The solvents used in bituminous paints are either aromatic (benzene or xylenes) or of the terpenic (turpentine) type. Both of these types of solvent act as a complementary component (diluent) for the paints.

The only solid complementary component of bituminous paints which is really effective is the pigment lamp black, which is inherently present in bitumens and tars where it originates from the carbonization of organic substances.

Waterproofing treatments, the protection of drainage systems, and simple discharge systems, and the provision of protection against corrosion and moisture, are among the most typical services rendered by these paints.

Bitumenized or tarred cardboards and sheathings

For some time cardboards impregnated with bitumens or tars mixed with rubber adhesives have been used for the protection and waterproofing of channels and concrete and masonry structures.

While cardboards were initially employed as the support materials, fibres began to be used, and their manufacture now also includes the use of foils or meshes of plastic materials such as polyethylene, polypropylene, or polyesters.

Having been made available in wound rolls which, when unwound, are self-adhesive, these products find many contemporary uses as linings and sealants.

Thick leaves of polyethylene or polypropylene meshes are bonded with special mixtures of rubbers and bitumen, thereby obtaining self-adhesive, waterproofing sheathings which can be hung up simply by pressing them onto the supporting materials. The bonding of two different but complementary materials, although they have different physical properties, ensures a high resistance to abrasion or tearing combined with perfect sealing.

Flexible and elastic, rubberized bitumen supported by a resin can be adapted to any shape, is pliable, and it may be cut and formed with great ease. Moreover, it does not split when structural movement or settling occurs.

Tarred and bituminized sheathings are also bonded to metallic layers, mainly of aluminium, and the resulting composites may be used to replace wholly metal structures to provide greater corrosion resistance. Some of the formulations including epoxy resins also exhibit waterproofing action and possess enhanced heat resistance and a high chemical resistance.

Mixtures of bitumen and/or tar with epoxide reagents to which suitable catalysts have been added, bringing about their cross-linking until they are completely hardened on the supports to which they are anchored, are also widely used. Using spray guns, they are rapidly applied to metallic supports or structures made of conglomerates. When they have set, they ensure the absolute protection of the underlying structures under both very testing stagnant and dynamic conditions.

Appendices

A.1—The elements, with atomic numbers and atomic masses, $A_r(^{12}C) = 12$

Name	*Symbol*	*Atomic number*	*Rel. atomic mass*†	*Name*	*Symbol*	*Atomic number*	*Rel. atomic mass*
Actinium	Ac	89	[277]	Molybdenum	Mo	42	95.94
Aluminium	Al	13	26.9815	Neodymium	Nd	60	144.24
Americum	Am	95	[243]	Neon	Ne	10	20.183
Antimony	Sb	51	121.75	Neptunium	Np	93	[237]
Argon	Ar	18	39.948	Nickel	Ni	28	58.71
Arsenic	As	33	74.9216	Niobium	Nb	41	92.906
Astatine	At	85	[210]	Nitrogen	N	7	14.0067
Barium	Ba	56	137.34	Nobelium	No	102	[254]
Berkelium	Bk	97	[249]	Osmium	Os	76	190.2
Beryllium	Be	4	9.0122	Oxygen	O	8	15.9994
Bismuth	Bi	83	208.980	Palladium	Pd	46	106.4
Boron	B	5	10.811	Phosphorus	P	15	30.9738
Bromine	Br	35	79.909	Platinum	Pt	78	195.09
Cadmium	Cd	48	112.40	Plutonium	Pu	94	[242]
Caesium	Cs	55	132.905	Polonium	Po	84	[210]
Calcium	Ca	20	40.08	Potassium	K	19	39.102
Californium	Cf	98	[251]	Praseodymium	Pr	59	140.907
Carbon	C	6	12.01115	Promethium	Pm	61	[147]
Cerium	Ce	58	140.12	Protactinium	Pa	91	[231]
Chlorine	Cl	17	35.453	Radium	Ra	88	[226]
Chromium	Cr	24	51.996	Radon	Rn	86	[222]
Cobalt	Co	27	58.9332	Rhenium	Re	75	186.2
Copper	Cu	29	63.54	Rhodium	Rh	45	102.905
Curium	Cm	96	[247]	Rubidium	Rb	37	85.47
Dysprosium	Dy	66	162.50	Ruthenium	Ru	44	101.07

Element	Symbol	Atomic number	Atomic weight	Element	Symbol	Atomic number	Atomic weight
Einsteinium	Es	99	[254]	Samarium	Sm	62	150.35
Erbium	Er	68	167.26	Scandium	Sc	21	44.956
Europium	Eu	63	151.96	Selenium	Se	34	78.96
Fermium	Fm	100	[253]	Silicon	Si	14	28.086
Fluorine	F	9	18.9984	Silver	Ag	47	107.870
Francium	Fr	87	[223]	Sodium	Na	11	22.9898
Gadolinium	Gd	64	157.25	Strontium	Sr	38	87.62
Gallium	Ga	31	69.72	Sulphur	S	16	32.064
Germanium	Ge	32	72.59	Tantalum	Ta	73	180.948
Gold	Au	79	196.967	Technetium	Tc	43	[99]
Hafnium	Hf	72	178.49	Tellurium	Te	52	127.60
Helium	He	2	4.0026	Terbium	Tb	65	158.924
Holmium	Ho	67	164.930	Thallium	Tl	81	204.37
Hydrogen	H	1	1.00797	Thorium	Th	90	232.038
Indium	In	49	114.82	Thulium	Tm	69	168.934
Iodine	I	53	126.9044	Tin	Sn	50	118.69
Iridium	Ir	77	192.2	Titanium	Ti	22	47.90
Iron	Fe	26	55.847	Tungsten	W	74	183.85
Krypton	Kr	36	83.80	Uranium	U	92	238.03
Lanthanum	La	57	138.91	Vanadium	V	23	50.942
Lawrencium	Lr	103	[257]	Xenon	Xe	54	131.30
Lead	Pb	82	207.19	Ytterbium	Yb	70	173.04
Lithium	Li	3	6.939	Yttrium	Y	39	88.905
Lutetium	Lu	71	174.97	Zinc	Zn	30	65.37
Magnesium	Mg	12	24.312	Zirconium	Zr	40	91.22
Manganese	Mn	25	54.9380				
Mendelevium	Md	101	[256]				
Mercury	Hg	80	200.59				

† The values in parenthesis indicate the mass number of the longest lived isotope or the best known isotope of the element.

A.2—The series of standard electrode potentials† (at 25°C, $E^{\ominus}$ in volt)

Half reactions	$E^{\ominus}$	*Half reactions*	$E^{\ominus}$
$Li^+ + e^- \rightarrow Li$	−3.05	$IO^- + H_2O + 2e^- \rightarrow I^- + 2OH^-$	+0.49
$K^+ + e^- \rightarrow K$	−2.92	$NiO_2 + 2H_2O + 2e^- \rightarrow Ni(OH)_2 + 2OH^-$	+0.49
$Ba^{2+} + 2e^- \rightarrow Ba$	−2.90	$Cu^+ + e^- \rightarrow Cu$	+0.52
$Sr^{2+} + 2e^- \rightarrow Sr$	−2.89	$I_2 + 2e^- \rightarrow 2I^-$	+0.54
$Ca^{2+} + 2e^- \rightarrow Ca$	−2.81	$MnO_4^- + e^- \rightarrow MnO_4^{2-}$	+0.55
$Na^+ + e^- \rightarrow Na$	−2.71	$MnO_4^{2-} + 2H_2O + 3e^- \rightarrow MnO_2 + 4OH^-$	+0.58
$Mg(OH)_2 + 2e^- \rightarrow Mg + 2OH^-$	−2.67	$BrO_3^- + 3H_2O + 6e^- \rightarrow Br^- + 6OH^-$	+0.61
$Mg^{2+} + 2e^- \rightarrow Mg$	−2.37	$ClO_3^- + 3H_2O + 6e^- \rightarrow Cl^- + 6OH^-$	+0.62
$H_2AlO_3^- + H_2O + 3e^- \rightarrow Al + 4OH^-$	−2.35	$O_2 + 2H^+ + 2e^- \rightarrow H_2O_2$	+0.68
$Ti^{3+} + 3e^- \rightarrow Ti$	−2.00	$Fe^{3+} + e^- \rightarrow Fe^{2+}$	+0.77
$Al^{3+} + 3e^- \rightarrow Al$	−1.67	$Hg_2^{2+} + 2e^- \rightarrow 2Hg$	+0.79
$ZnO_2 + 2H_2O + 2e^- \rightarrow Zn + 4OH^-$	−1.22	$Ag^+ + e^- \rightarrow Ag$	+0.80
$Mn^{2+} + 2e^- \rightarrow Mn$	−1.03	$Pd^{2+} + 2e^- \rightarrow Pd$	+0.83
$Sn(OH)_6^{2-} + 2e^- \rightarrow HSnO_2^- + 3OH^- +$		$Hg^{2+} + 2e^- \rightarrow Hg$	+0.85
$+ H_2O$	−0.96	$TiO_2 + 4H^+ + 4e^- \rightarrow Ti + 2H_2O$	+0.86
$SO_4^{2-} + H_2O + 2e^- \rightarrow SO_3^{2-} + 2OH^-$	−0.92	$ClO^- + H_2O + 2e^- \rightarrow Cl^- + 2OH^-$	+0.90
$2H_2O + 2e^- \rightarrow H_2 + 2OH^-$	−0.83	$2Hg^{2+} + 2e^- \rightarrow Hg_2^{2+}$	+0.90
$Zn^{2+} + 2e^- \rightarrow Zn$	−0.76	$NO_3^- + 3H^+ + 2e^- \rightarrow HNO_2 + H_2O$	+0.94
$Cr^{3+} + 3e^- \rightarrow Cr$	−0.74	$NO_3^- + 4H^+ + 3e^- \rightarrow NO + 2H_2O$	+0.96
$AsO_4^{3-} + 2H_2O + 2e^- \rightarrow AsO_2^- + 4OH^-$	−0.71	$HNO_2 + H^+ + e^- \rightarrow NO + H_2O$	+0.99
$PbO + H_2O + 2e^- \rightarrow Pb + 2OH^-$	−0.57	$Br_2 + 2e^- \rightarrow 2Br^-$	+1.09
$Fe(OH)_3 + e^- \rightarrow Fe(OH)_2 + OH^-$	−0.56	$ClO_4^- + 2H^+ + 2e^- \rightarrow ClO_3^- + H_2O$	+1.19
$Cr^{2+} + 2e^- \rightarrow Cr$	−0.56	$2IO_3^- + 12H_+^2 + 10e^- \rightarrow I_2 + 6H_2O$	+1.19

$S + 2e^- \rightarrow S^2$	−0.51	$Pt^{2+} + 2e^- \rightarrow Pt$	+1.20
$H_2SO_3 + 4H^+ + 4e^- \rightarrow S + 3H_2O$	−0.45	$MnO_2 + 4H^+ + 2e^- \rightarrow Mn^{2+} + 2H_2O$	+1.21
$Fe^{2+} + 2e^- \rightarrow Fe$	−0.44	$O_2 + 4H^+ + 4e^- \rightarrow 2H_2O$	+1.23
$Cr^{3+} + e^- \rightarrow Cr^{2+}$	−0.41	$O_3 + H_2O + 2e^- \rightarrow O_2 + 2OH^-$	+1.24
$Cd^{2+} + 2e^- \rightarrow Cd$	−0.40	$Au^{3+} + 2e^- \rightarrow Au^+$	+1.29
$Co^{2+} + 2e^- \rightarrow Co$	−0.27	$Cr_2O_7^{2-} + 14H^+ + 6e^- \rightarrow 2Cr^{3+} + 7H_2O$	+1.33
$Ni^{2+} + 2e^- \rightarrow Ni$	−0.25	$ClO_4^- + 8H^+ + 7e^- \rightarrow 1/2Cl_2 + 4H_2O$	+1.34
$O_2 + 2H_2O + 2e^- \rightarrow H_2O_2 + 2OH^-$	−0.146	$Cl_2 + 2e^- \rightarrow 2Cl^-$	+1.36
$Sn^{2+} + 2e^- \rightarrow Sn$	−0.14	$ClO_4^- + 8H^+ + 8e^- \rightarrow Cl^- + 4H_2O$	+1.37
$Pb^{2+} + 2e^- \rightarrow Pb$	−0.13	$Au^{3+} + 3e^- \rightarrow Au$	+1.42
$CrO_4^{2-} + 4H_2O + 3e^- \rightarrow Cr(OH)_3 +$		$BrO_3^- + 6H^+ + 6e^- \rightarrow Br^- + 3H_2O$	+1.44
$+ 5OH^-$	−0.12	$Ce^{4+} + e^- \rightarrow Ce^{3+}$	+1.44
$Fe^{3+} + 3e^- \rightarrow Fe$	−0.04	$ClO_3^- + 6H^+ + 6e^- \rightarrow Cl^- + 3H_2O$	+1.45
$2H^+ + 2e^- \rightarrow H_2$	**0.000**	$PbO_2 + 4H^+ + 2e^- \rightarrow Pb^{2+} + 2H_2O$	+1.46
$NO_3^- + H_2O + 2e^- \rightarrow NO_2^- + 2OH^-$	+0.01	$ClO_3^- + 6H^+ + 5e^- \rightarrow 1/2Cl_2 + 3H_2O$	+1.47
$S + 2H^+ + 2e^- \rightarrow H_2 + S_{aq}$	+0.14	$HClO + H^+ + 2e^- \rightarrow Cl^- + H_2O$	+1.49
$Sn^{4+} + 2e^- \rightarrow Sn^{2+}$	+0.15	$MnO_4^- + 8H^+ + 5e^- \rightarrow Mn^{2+} + 4H_2O$	+1.50
$Cu^{2+} + e^- \rightarrow Cu^+$	+0.16	$BrO_3^- + 6H^+ + 5e^- \rightarrow 1/2Br_2 + 3H_2O$	+1.52
$SO_4^{2-} + 4H^+ + 2e^- \rightarrow H_2SO_3 + H_2O$	+0.20	$HBrO + H^+ + e^- \rightarrow 1/2Br_2 + H_2O$	+1.59
$AgCl + e^- \rightarrow Ag + Cl^-$	+0.22	$HClO + H^+ + e^- \rightarrow 1/2Cl_2 + H_2O$	+1.63
$IO_3^- + 3H_2O + 6e^- \rightarrow I^- + 6OH^-$	+0.26	$H_2O_2 + 2H^+ + 2e^- \rightarrow 2H_2O$	+1.78
$Cu^{2+} + 2e^- \rightarrow Cu$	+0.34	$O_3 + 2H^+ + 2e^- \rightarrow O_2 + H_2O$	+1.78
$O_2 + 2H_2O + 4e^- \rightarrow 4OH^-$	+0.40	$S_2O_8^{2-} + 2e^- \rightarrow 2SO_4^{2-}$	+2.07
$Fe(CN)_6^{3-} + e^- \rightarrow Fe(CN)_6^{4-}$	+0.46	$F_2 + 2e^- \rightarrow 2F^-$	+2.87

† The values of the standard oxidation potentials are obtained by changing the sign.

A.3—Conversions from °C to °F and vice versa†

in °C	Temperature to be converted	in °F	in °C	Temperature to be converted	in °F	in °C	Temperature to be converted	in °F
−128.9	**−200**	−328.40	2.2	**36**	96.8	27.2	**81**	177.4
−73.3	**−100**	−148.0	2.8	**37**	98.6	27.8	**82**	179.6
−62.2	**−80**	−112.0	3.3	**38**	100.4	28.3	**83**	181.4
−51.1	**−60**	−76.0	3.9	**39**	102.2	28.9	**84**	183.2
−45.6	**−50**	−58.0	4.4	**40**	104.0	29.4	**85**	185.0
−40.0	**−40**	−40.0				30.0	**86**	186.8
−34.4	**−30**	−22.0	5.0	**41**	105.8	30.6	**87**	188.6
−28.9	**−20**	−4.0	5.6	**42**	107.6	31.1	**88**	190.4
−23.3	**−10**	14.0	6.1	**43**	109.4	31.7	**89**	192.2
−17.8	**0**	32.0	6.7	**44**	111.2	32.2	**90**	194.0
			72.	**45**	113.0			
−17.2	**1**	33.8	7.8	**46**	114.8	32.8	**91**	195.8
−16.7	**2**	35.6	8.3	**47**	116.6	33.3	**92**	197.6
−16.1	**3**	37.4	8.9	**48**	118.4	33.9	**93**	199.4
−15.6	**4**	39.2	9.4	**49**	120.2	34.4	**94**	201.2
−15.0	**5**	41.0	10.0	**50**	122.0	35.0	**95**	203.0
−14.4	**6**	42.8				35.6	**96**	204.8
−13.9	**7**	44.6	10.6	**51**	123.8	36.1	**97**	206.6
−13.3	**8**	46.4	11.1	**52**	1256.	36.7	**98**	208.4
−12.8	**9**	48.2	11.7	**53**	127.4	37.2	**99**	210.2
−12.2	**10**	50.0	12.2	**54**	129.2	37.8	**100**	212.0
			12.8	**55**	131.0			
−11.7	**11**	51.8	13.3	**56**	132.8	48.9	**120**	248.0
−11.1	**12**	53.6	13.9	**57**	134.6	60.0	**140**	284.0
−10.6	**13**	55.4	14.4	**58**	136.4	71.1	**160**	320.0
−10.0	**14**	57.2	15.0	**59**	138.2	93.3	**200**	392.0

−9.4	**15**	59.0	15.6	**60**	140.0	100.0	**212**	413.6
−8.9	**16**	60.8				115.6	**240**	464.0
−8.3	**17**	62.6	16.1	**61**	141.8	137.8	**280**	536.0
−7.8	**18**	64.4	16.7	**62**	143.6	160.0	**320**	608.0
−7.2	**19**	16.2	17.2	**63**	145.4	182.2	**360**	680.0
−6.7	**20**	68.0	17.8	**64**	147.2	204.4	**400**	752.0
			18.3	**65**	149.0			
−6.1	**21**	69.8	18.9	**66**	150.8	232.0	**450**	842.0
−5.6	**22**	71.6	19.4	**67**	152.6	260.0	**500**	932.0
−5.0	**23**	73.4	20.0	**68**	154.4	288.0	**550**	1022.0
−4.4	**24**	75.2	20.6	**69**	156.2	316.0	**600**	1112.0
−3.9	**25**	77.0	21.1	**70**	158.0	343.0	**650**	1202.0
−3.3	**26**	78.8				371.0	**700**	1292.0
−2.8	**27**	80.6	21.7	**71**	159.8	399.0	**750**	1382.0
−2.2	**28**	82.4	22.2	**72**	161.6	427.0	**800**	1472.0
−1.7	**29**	84.2	22.8	**73**	163.4	482.0	**900**	1652.0
−1.1	**30**	86.0	23.3	**74**	165.2	538.0	**1000**	1832.0
			23.9	**75**	167.0			
−0.6	**31**	87.8	24.4	**76**	168.8	649.0	**1200**	2192.0
0.0	**32**	89.6	25.0	**77**	170.6	816.0	**1500**	2732.0
0.6	**53**	91.4	25.6	**78**	172.4	1093.0	**2000**	3632.0
1.1	**34**	93.2	26.1	**79**	174.2	1371.0	**2500**	4532.0
1.7	**35**	95.0	26.7	**80**	176.0	1649.0	**3000**	5432.0

† The conversion formulae are: $°C = \frac{5}{9}(°F - 32)$ and $°F = \frac{9}{5}°C + 32$. The **bold** numbers in the central column of the sets of three columns are the 'neutral' values of the temperature in the sense that they can be understood as being either in °C or °F. In the first case, the corresponding (*conversion*) value in °F is found to the right of the 'neutral' value, whereas, in the second case, the corresponding value in °C is found to the left of the neutral value. Thus:

$10°F = -12.2°$
$10°C = 50.0°F$
$-40°F = -40°C$
$-40°C = -40°F$

A.4—Formulae, hardness on Mohs scale (H_M), and specific weight (W_{sp}) of minerals

Name	*Formula*	H_M	W_{sp}	*Name*	*Formula*	H_M	W_{sp}
A				**D**			
Agate	SiO_2	6	2.6	Diamond	C	10	3.5
Albite	$NaAlSi_3O_8$	6–7	2.6	Dolomite	(Ca, Mg)CO_3	3.5–4	2.8
Amber	–	2–2.5	1–1.1				
Amethyst	SiO_2	7	2.6	**E**			
Anatase	TiO_2	5.5–6	3.9				
Andalusite	Al_2SiO_5	7.5	3.2	Erythrite	$Co_3As_2O_8 \cdot 8H_2O$	2	3
Anglesite	$PbSO_4$	3	6.2				
Anhydrite	$CaSO_4$	3–3.5	3	**F**			
Anorthite	$CaAl_2Si_2O_8$	6–7	2.8				
Anthracite	C 90%	2–2.5	1.1–1.7	Feldspar	$KAlSi_3O_8$	6	2.5
Apatite	$3Ca_3(PO_4)_2 \cdot CaF_2$	5	3.3	Fluorite	CaF_2	4	3.2
Aragonite	$CaCO_3$	3.5–4	3	Fluorspar	CaF_2	4	3.2
Argentite	Ag_2S	2–2.5	7.2				
Arsenolite	As_2O_3	1.5	3.7	**G**			
Arsenopyrite	FeAsS	5.5–6	6				
Asbestos	Mg(Ca)SiO_3	5	2.9–3.2	Galena	PbS	2–3	7.4
Autunite	$Ca(UO_2)_2O_8 \cdot 12H_2O$	2–2.5	3	Graphite	C	1–2	2.2
Azzurrite	$CuCO_3 \cdot Cu(OH)_2$	3.5–4.3	3.8	Gypsum	$CaSO_4 \cdot 2H_2O$	1.5–2	2.3
B				**H**			
Barytes	$BaSO_4$	3–3.5	4.5	Haematite	Fe_2O_3	5.5–6.5	5.2
Bauxite	$Al_2O_3 \cdot 2H_2O$	1–3	2.5	Hausmannite	Mn_3O_4	5–5.5	4.8
Bentonite	$Al_2Si_4O_{10}(OH)_2$	1–2	2.8–3	Hemimorphite	$Zn_4Si_2O_7(OH)_2 \cdot H_2O$	5	3.5
Beryl	$Be_3Al_2Si_6O_{18}$	7–8	2.7				

Biotite	$K(Mg, Fe)_3AlSi_3O_{10}\cdot$		
(black mica)	$\cdot(F, OH)_2$	2.5–3	2.8
Bismuthinite	$Bi_2O_3\cdot CO_2\cdot H_2O$	2	6.5
Blende	ZnS	3.4–4	4
Borax	$Na_2B_4O_7\cdot 10H_2O$	2.3	1.7
Bornite	Cu_3FeS_3	3	5.2
Brookite	TiO_2	5.5–6	3.8
C			
Calamine	$Zn_4Si_2O_7(OH)_2\cdot H_2O$	4.5–5	3.5
Calcite	$CaCO_3$	~3	2.7
Carnallite	$KMgCl_3\cdot 6H_2O$	1	1.6
Cassiterite	SnO_2	6–7	7
Celestine	$SrSO_4$	3–3.5	4
Cerussite	$PbCO_3$	3–5	6.5
Chalcedony	SiO_2	7	2.6
Chalcocite	Cu_2S	2.5–3	5.6
Chalcopyrite	$CuFeS_2$	3.5–4	4.2
Chromite	(Fe, Mg)		
	$(Cr, Al, Fe)_2O_4$		
Chrysoberyl	$BeAlO_4$	8.5	3.8
Cinnabar	HgS	2–2.5	8–8.2
Cobaltite	$CoAsS$	5.5	6.3
Corundum	Al_2O_3	9	4
Cristobalite	SiO_2	6–7	2.5
Cryolite	Na_3AlF_6	2.5	3
Crocoite	$PbCrO_4$	2.5–3	6
Cuprite	Cu_2O	3.5–4	6
I			
Ilmenite	$FeTiO_3$	5–6	4.7
J			
Jadeite	$NaAlSi_2O_6$	6.5–7	3.3
Jasper	$AlO(OH)$	6.5–7	3.5
K			
Kainite	$MgSO_4\cdot KCl\cdot 3H_2O$	3	2.2
Kaolin	$Al_2Si_2O_5(OH)_4$	1	2.6
L			
Lapis lazuli	$3Na_2O\cdot 3Al_2O_3\cdot$		
	$2Na_2S\cdot 6SiO_2$	5.5	2.4
Leucite	$KAlSi_2O_6$	5.5–6	2.5
Limonite	$FeO(OH)\cdot nH_2O$	5–5.5	3.5–4
Litharge	PbO	2	9.2
M			
Magnesite	$MgCO_3$	3.5–4.5	3
Magnetite	Fe_3O_4	5.5–6.5	5.2
Malachite	$CuCO_3\cdot Cu(OH)_2$	3.4–4	4
Manganite	$Mn_2O_3\cdot H_2O$	4	4.5
Marcassite	FeS_2	6–6.5	4.9

A.4—*Continued*

Name	*Formula*	H_M	W_{sp}
M			
Millerite	NiS	3–3.5	5.2–5.9
Molybdenite	MoS_2	1–1.5	4.8–5
Monazite	$(Ce, La, Y, Nd) \cdot PO_4$	5.2	5
Muscovite (white mica)	$KAl_3Si_3O_{10}(F, OH)_2$	2–2.5	2.7–3.1
N			
Nepheline	$Na_3KAl_4SiO_4O_{16}$	5.5–6	2.6
Niccolite	$NiAs$	5–5.5	7.5
O			
Olivine	$(Mg, Fe)SiO_4$	6–7	3.4
Opal	$SiO_2 \cdot nH_2O$	5.5–6.5	2.3
Orpiment	As_2S_3	1.5–2	3.5
Orthoclase	$KAlSi_3O_8$	6	2.5
Ozocherite	Higher paraffins	1	0.9
P			
Perovskite	$CaTiO_3$	5.5	4
Pyrites	FeS_2	6–6.5	5
Pyrolusite	MnO_2	2–2.5	4.8
Senarmontite	Sb_2O_3	2–2.25	5.2
Serpentine	$Mg_3Si_2O_5(OH)_4$	3–4	2.5
Siderite	$FeCO_3$	3.5–4	3.8
Silica	SiO_2	7	2.6
Silvite	KCl	2	2
Smithsonite	$ZnCO_3$	5.5	4.4
Sphalerite	ZnS	3.4–4	4
Spinel	$MgAl_2O_4$	8	3.5–4
Stibine	Sb_2S_3	2	4.6
Strontianite	$SrCO_3$	3.5–4	3.7
T			
Talc	$Mg_3Si_4O_{10}(OH)_2$	1	2.6
Thorite	$ThSiO_4$	4.5–5	5.3
Titanite	$CaTiSiO_5$	5–5.5	3.5
Topaz	$Al_2SiO_4(OH, F)_2$	8	3.5
Tourmaline	$Na(Mg, Fe, Mn) \cdot Al_6Si_6O_{18}(BO_3)_3$	7–7.5	3.2
Tridymite	SiO_2	7	2.3
Turquoise	$Cu(Al, Fe)_6(PO_4)_4 \cdot (OH)_8$	5–6	3
U			
Ulmannite	$NiSbS$	5–5.5	6.6

Pyroxene	$NaAlSi_2O_6$	5.5–6.5	3.3
Pyrrhotite	from Fe_5S_6 to $Fe_{16}S_{17}$	3.5–4.5	4.6
Q			
Quartz	SiO_2	7	2.6
R			
Realgar	As_2S_2	1.5–2	3.5
Red lead	Pb_3O_4	2–3	4.7
Rhodochrosite	$MnCO_3$	3.5–4.5	3.7
Rock salt	NaCl	2–2.5	2.2
Ruby	Al_2O_3	9	4
Rutile	TiO_2	6–6.5	4.4
S			
Saltpetre	KNO_3	2	2.3
Sapphire	Al_2O_3	9	4
Scheelite	$CaWO_4$	4.5–5	6
Uraninite	UO_2	5.5	10
V			
Valentinite	Sb_2O_3	2.5–3	5.8
Vermiculite	$(Mg, Ca)(Fe, Al)_6 \cdot Si_8O_{20}(OH)_4$	1.5	2.4
Vesuvianite	$Ca_{10}(Mg, Fe)_2Al_4 \cdot (Si_2O_7)_2(SiO_4)_5 \cdot (OH, F)$	6.5	3.4
W			
Willemite	Zn_2SiO_4	5–5.5	4
Wolframite	$(Fe, Mn)WO_4$	5–5.5	7.5
Wollastonite	$CaSiO_3$	4.5–5	2.7–2.9
Wurtzite	ZnS	3.5–4	4
Z			
Zircon	$ZrSiO_4$	7.5	4.5

A.5—Corresponding values between the density in degrees Baumé (Bé) and the relative density ($d^{15°C}$) at a temperature of 15°C

	$d^{15°C}$			$d^{15°C}$	
Bé	L_{lg}	L_{hv}	*Bé*	L_{lg}	L_{hv}
1	1.0670	1.0070	**34**	0.8574	1.3080
2	1.0590	1.0140	**35**	0.8524	1.3200
3	1.0510	1.0210	**36**	0.8473	1.3320
4	1.0430	1.0290	**37**	0.8424	1.3450
5	1.0360	1.0360	**38**	0.8375	1.3570
6	1.0290	1.0430	**39**	0.8327	1.3700
7	1.0210	1.0510	**40**	0.8279	1.3830
8	1.0114	1.0590	**41**	0.8232	1.3970
9	1.0070	1.0670	**42**	0.8185	1.4100
10	1.0000	0.0740	**43**	0.8139	1.4200
11	0.9931	1.0830	**44**	0.8093	1.4390
12	0.9863	1.0910	**45**	0.8048	1.4530
13	0.9796	1.0990	**46**	0.8004	1.4680
14	0.9730	1.1070	**47**	0.7959	1.4830
15	0.9665	1.1160	**48**	0.7916	1.4980
16	0.9601	1.1250	**49**	0.7873	1.5140
17	0.9537	1.1340	**50**	0.7830	1.5300
18	0.9475	1.1420	**51**	0.7788	1.5470
19	0.9413	1.1520	**52**	0.7746	1.5630
20	0.9352	1.1610	**53**	0.7704	1.5800
21	0.9292	1.1700	**54**	0.7664	1.5980
22	0.9232	1.1800	**55**	0.7623	1.6160
23	0.9174	1.1900	**56**	0.7583	1.6340
24	0.9116	1.1990	**57**	0.7543	1.6530
25	0.9058	1.2100	**58**	0.7504	1.6720
26	0.9002	1.2200	**59**	0.7465	1.6920
27	0.8946	1.2300	**60**	0.7427	1.7120
28	0.8891	1.2410	**61**	0.7389	1.7320
29	0.8837	1.2510	**62**	0.7351	1.7530
30	0.8783	1.2620	**63**	0.7314	1.7750
31	0.8730	1.2740	**64**	0.7277	1.7970
32	0.8677	1.2850	**65**	0.7241	1.8190
33	0.8625	1.2960	**66**	0.7204	1.8430

The formulae which allow the calculation of these values for liquids which are lighter than water (L_{lg}) and liquids which are havier than water (L_{hv}) are respectively:

$$d_{L_{lg}}^{15°C} = \frac{144.32}{134.32 + Bé} \Rightarrow Bé = \frac{144.32}{d_{L_{lg}}^{15°C}} - 134.32$$

$$d_{L_{hv}}^{15°C} = \frac{144.32}{144.32 - Bé} \Rightarrow Bé = 144.32 - \frac{144.32}{d_{L_{hv}}^{15°C}}$$

For example, if $Bé = 0$: $d_{L_{lg}}^{15°C} = 1.0722$ and $d_{L_{hv}}^{15°C} = 1.0000$.

A.6—Percentage composition (%) and content in weight per dm^3 of sulphuric acid solutions

% H_2SO_4	g/l H_2SO_4	% H_2SO_4	g/l H_2SO_4	% H_2SO_4	g/l H_2SO_4
0.09	1	28.58	346	52.15	740
0.95	9	29.21	355	52.63	750
1.57	16	29.84	364	53.11	759
2.30	23	30.48	373	53.59	769
3.03	31	31.11	382	54.07	779
3.76	39	31.70	391	54.55	789
4.49	46	32.28	400	55.03	798
5.23	54	32.86	409	55.50	808
5.96	62	33.43	418	55.97	817
6.67	70	34.00	426	56.43	827
7.37	77	34.57	435	56.90	837
8.07	85	35.14	444	57.37	846
8.77	93	35.71	453	57.83	856
9.48	101	36.29	462	58.29	866
10.19	109	36.87	472	58.75	876
10.90	117	37.45	481	59.22	886
11.60	125	38.03	490	59.70	896
12.30	133	38.61	500	60.18	906
12.99	142	39.19	410	60.65	916
13.67	150	39.77	519	61.12	926
14.35	158	40.35	529	61.59	936
15.03	166	40.93	538	62.06	946
15.71	175	41.50	548	62.53	957
16.36	183	42.08	557	63.00	967
17.01	191	42.66	567	63.43	977
17.66	199	43.20	577	63.85	987
18.31	207	43.74	586	64.26	997
18.96	215	44.28	596	64.67	1007
19.61	223	44.82	605	65.20	1017
20.26	231	45.35	614	65.65	1027
20.91	239	45.88	624	66.09	1038
21.55	248	46.41	633	66.53	1048
22.19	257	46.94	643	66.95	1058
22.83	266	47.47	653	67.40	1068
22.47	275	48.00	662	67.83	1078
24.12	283	48.53	672	68.26	1089
24.76	292	49.06	682	68.70	1099
25.40	301	40.59	692	69.13	1110
26.04	310	50.11	702	69.56	1120
26.68	319	50.63	711	70.00	1131
27.32	328	51.15	721	70.42	1141
27.95	337	51.66	730	70.85	1151

A.6—*Continued*

% H_2SO_4	g/l H_2SO_4	% H_2SO_4	g/l H_2SO_4	% H_2SO_4	g/l H_2SO_4
71.27	1162	78.04	1334	85.70	1534
71.70	1172	78.48	1346	86.30	1549
72.12	1182	78.92	1357	86.92	1564
72.55	1193	79.36	1369	87.60	1581
72.97	1204	79.80	1381	88.30	1598
73.40	1215	80.24	1392	89.16	1618
73.82	1225	80.68	1404	90.05	1639
74.24	1236	81.12	1416	91.00	1661
74.66	1246	81.56	1427	92.10	1685
75.08	1257	82.00	1439	93.56	1717
75.50	1268	82.44	1452	95.60	1759
75.94	1278	83.01	1465	95.95	1765
76.38	1289	83.51	1478	96.38	1774
76.76	1310	84.02	1491	97.35	1792
77.17	1312	84.50	1504	98.30	1808
77.60	1323	85.10	1519	98.52	1814

A.7—Percentage composition (%) and content in weight per dm^3 of hydrochloric and nitric acid solutions

Hydrochloric acid		*Nitric acid*			
Wt. % HCl	g/l HCl	Wt. % HNO_3	g/l HNO_3	Wt. % HNO_3	g/l HNO_3
0.16	1.6	0.10	1	14.74	160
1.15	12	1.00	10	15.53	169
2.14	22	1.90	19	16.32	179
3.12	32	2.80	28	17.11	188
4.13	42	3.70	38	17.89	198
5.15	53	4.60	47	18.67	207
6.15	63	5.50	57	19.45	217
7.15	74	6.38	66	20.23	227
8.16	85	7.26	75	21.00	236
9.16	96	8.13	85	21.77	246
10.17	107	8.99	94	22.54	256
11.18	118	9.84	104	23.31	266
12.19	129	10.68	113	24.08	276
13.19	140	11.51	123	24.84	286
14.17	152	12.33	132	25.60	296
15.16	163	13.15	141	26.36	306
16.15	174	13.95	151	27.12	316

A.7—*Continued*

Hydrochloric acid		*Nitric acid*			
Wt. % HCl	g/l HCl	Wt. % HNO_3	g/l HNO_3	Wt. % HNO_3	g/l HNO_3
17.13	186	27.88	326	59.39	814
18.11	197	28.63	336	60.30	829
19.06	209	29.38	347	61.27	846
20.01	220	30.13	357	61.92	857
20.97	232	30.88	367	62.24	862
21.92	243	31.62	378	63.23	879
22.86	255	32.36	388	64.25	896
23.82	267	33.09	399	64.30	914
24.78	279	33.82	409	66.40	933
25.74	291	34.55	420	67.50	952
26.70	302	35.28	430	68.63	971
27.66	315	36.03	441	69.80	991
28.14	321	36.78	452	70.98	1011
28.61	328	37.53	463	72.17	1032
29.57	340	38.29	475	73.39	1053
29.95	345	39.05	486	74.68	1075
30.55	353	39.82	498	75.98	1098
31.52	366	40.58	509	77.28	1121
32.10	374	41.34	521	78.60	1144
32.49	379	42.10	533	79.98	1168
33.46	391	42.87	544	81.42	1193
33.65	394	43.64	556	82.90	1219
34.42	404	44.41	568	84.45	1246
35.39	418	45.18	581	86.05	1274
36.31	430	45.95	593	87.70	1302
37.23	443	46.72	605	89.60	1335
38.16	456	47.49	617	91.60	1369
39.11	469	48.26	630	94.09	1411
		49.07	643	95.08	1428
		49.89	656	96.00	1444
		50.71	669	96.39	1451
		51.53	683	96.76	1457
		52.37	697	97.50	1470
		52.80	704	98.10	1481
		53.22	711	98.53	1490
		54.07	725	98.90	1497
		54.93	739	99.07	1501
		55.79	753	99.21	1504
		56.66	768	99.46	1510
		57.57	783	99.67	1515
		58.48	789		

A.8—Percentage composition (%) and content in weight per dm^3 of solutions of inorganic bases

Ammonia		*Sodium hydroxide*		*Potassium hydroxide*	
% NH_3	g/l NH_3	% NaOH	g/l NaOH	% KOH	g/l KOH
0.00	0.0	0.08	0.78	0.10	0.98
0.45	4.5	0.95	9.57	1.18	11.97
0.91	9.1	1.83	18.69	2.27	23.17
1.37	13.6	2.72	27.97	3.36	34.60
1.84	18.2	3.61	37.51	4.44	46.20
2.31	22.8	4.51	47.30	5.52	57.98
3.30	32.5	5.41	57.29	6.60	69.99
4.30	42.2	6.31	67.47	7.68	82.14
4.80	47.0	7.21	77.81	8.75	94.45
6.30	61.3	8.10	88.28	9.79	106.73
7.31	70.8	9.00	99.00	10.86	119.50
7.82	75.6	9.90	109.91	11.22	132.27
8.33	80.4	10.80	120.98	12.97	145.25
8.84	85.1	11.70	132.24	14.03	158.54
9.37	90.1	12.60	143.67	15.07	171.80
9.91	95.1	13.50	155.29	16.09	185.03
10.47	100.2	14.40	167.09	17.10	198.38
11.60	110.6	15.32	179.20	18.12	212.03
12.17	115.8	16.22	191.35	19.14	225.89
12.74	120.9	17.12	203.69	20.16	239.90
13.31	126.1	18.02	216.24	21.16	253.94
13.88	131.2	18.93	229.03	22.17	268.24
14.46	136.4	19.83	241.90	23.16	282.53
15.04	141.4	20.73	254.98	24.15	297.01
15.63	146.8	21.64	268.99	25.13	311.58
16.22	152.0	22.55	281.81	26.11	326.34
16.82	157.3	23.46	295.53	27.10	341.46
17.42	162.6	24.36	309.42	28.07	356.49
18.03	167.9	25.28	323.52	29.00	371.20
18.64	173.2	26.19	337.89	29.95	386.39
19.87	183.8	27.11	352.44	30.90	401.65
21.12	194.6	28.03	367.19	31.84	417.09
21.75	199.9	28.96	382.32	32.78	432.64
23.68	216.2	29.90	397.63	33.71	448.34
24.99	227.2	30.84	413.24	34.63	464.08
25.66	232.7	31.79	429.18	35.55	479.94
26.31	238.2	32.74	445.30	36.46	495.91
26.98	243.7	33.70	461.62	37.37	512.01
27.65	249.2	34.66	478.31	38.28	528.29
28.33	254.7	35.64	495.35	39.19	544.71

A.8—*Continued*

Ammonia		*Sodium hydroxide*		*Potassium hydroxide*	
% NH_3	g/l NH_3	% NaOH	g/l NaOH	% KOH	g/l KOH
29.01	260.3	36.62	512.64	40.08	561.12
30.37	271.3	37.60	530.22	40.97	577.73
31.05	276.7	38.60	548.12	41.87	594.50
31.75	282.3	39.61	566.37	42.75	611.27
32.50	288.3	40.62	584.97	43.62	628.17
33.25	294.3	41.65	603.91	44.50	645.19
34.10	301.2	42.69	623.24	45.37	662.33
34.95	308.0	43.73	642.82	46.23	679.59
45	382.1	44.77	662.63	47.09	696.86
50	416.0	45.81	682.61	47.94	714.31
55	448.3	46.86	702.96	48.79	731.82
60	477.6	47.92	723.53	49.64	749.50
65	504.4	48.97	744.33		
70	528.5				
75	549.8				
80	568.8				
85	584.8				
90	598.5				
95	609.9				
100	618.0				

A.9—Formulae and physical properties of inorganic substances

Substance	*Formula*	*Relative molecular mass*	*Colour*	*Solubility*†
		Aluminium		
Bromide	$AlBr_3$	266.69	white	465 g/l at 18°C
Carbide	Al_4C_3	143.96	yellow-green	decomposes → CH_4
Chloride	$AlCl_3$	133.34	white	720 g/l at 18°C
Cryolite	Na_3AlF_6	209.94	°C	slightly soluble
Fluoride	AlF_3	83.98	white	5.6 g/l
Hydroxide	$Al(OH)_3$	78.00	white	insoluble
Nitrate	$Al(NO_3)_3 \cdot 9H_2O$	375.13	white	653 g/l
Oxide	Al_2O_3	101.96	white	insoluble
*ortho*Phosphate	$AlPO_4$	121.95	white	insoluble
Potassium sulphate	$KAl(SO_4)_2 \cdot 12H_2O$	474.38	white	110 g/l at 18°C
Sulphate	$Al_2(SO_4)_3$	342.13	white	267 g/l at 18°C
Sulphide	Al_2S_3	150.14	white	decomposes → H_2S
		Ammonium		
Bicarbonate	NH_4HCO_3	79.06	colourless	very soluble
Bisulphate	$(NH_4)HSO_4$	115.11	colourless	very soluble
Bromide	NH_4Br	97.94	colourless	970 g/l
Carbamate	NH_2COONH_4	78.07	colourless	very soluble: decomposes
Chloride	NH_4Cl	53.49	white	300 g/l at 0°C
Dichromate	$(NH_4)_2Cr_2O_7$	252.07	orange	310 g/l
Hydrogen	$(NH_4)_2HPO_4$	132.06	colourless	600 g/l at 0°C
Hydroxide	NH_4OH	35.05	colourless sol.	very soluble
Iodide	NH_4I	144.94	colourless	very soluble

Molybdate	$(NH_4)_6 \cdot Mo_7O_{24} \cdot 4H_2O$	1235.86	colourless	very soluble
Phosphomolybdate	$(NH_4)_3PO_4 \cdot 12MoO_3$	1876.35	yellow	slightly soluble
Nitrate	NH_4NO_3	80.04	white-yellow	very soluble
Nitrite	NH_4NO_2	64.04	yellowish	very soluble
Oxalate	$(NH_4)_2(COO)_2 \cdot H_2O$	142.11	colourless	40 g/l
Perchlorate	NH_4ClO_4	117.49	colourless	106 g/l at 0°C
Peroxydisulphate	$(NH_4)_2S_2O_8$	228.20	colourless	580 g/l at 0°C
Reineckate	$NH_4[Cr(NH_3)_2(SCN)_4] \cdot H_2O$	354.43	violet-red	slightly soluble
Sulfamate	$NH_4SO_3NH_2$	114.12	white	very soluble
Sulphate	$(NH_4)_2SO_4$	132.14	colourless	705 g/l at 0°C
(Fe^{II}) Sulphate	$(NH_4)_2Fe(SO_4)_2 \cdot 6H_2O$	392.13	green	275 g/l
Sulphide	$(NH_4)_2S$	68.14	colourless	very soluble
Sulphite	$(NH_4)_2SO_3 \cdot H_2O$	134.15	colourless	400 g/l
Thiocyanate	NH_4SCN	76.12	colourless	very soluble
Vanadate	NH_4VO_3	116.98	colourless	very soluble
		Antimony		
Chloride	$SbCl_3$	228.11	white	very soluble
Oxide(II)	Sb_2O_3	291.50	white	insoluble
Oxide(V)	Sb_2O_5	323.50	yellowish	slightly soluble
Oxychloride	$SbOCl$	173.20	white	insoluble
Sulphate	$Sb_2(SO_4)_3$	531.67	white	slightly soluble
Sulphide(III)	Sb_2S_3	339.68	orange	insoluble
Sulphide(V)	Sb_2S_5	403.80	orange	insoluble
		Arsenic		
*ortho*Arsenic Acid	$H_3AsO_4 \cdot 0.5H_2O$	150.94	colourless	very soluble
Arsine	AsH_3	77.94	colourless gas	slightly soluble
Bromide	$AsBr_3$	314.63	colourless	decomposes
Chloride	$AsCl_3$	181.28	colourless liq.	decomposes

† At 25°C in water, if nothing is stated to the contrary.

A.9—*Continued*

Substance	*Formula*	*Relative molecular mass*	*Colour*	*Solubility*†
		Arsenic (*cont.*)		
Iodide	AsI_3	455.63	red	60 g/l
Oxide(III)	As_2O_3	197.84	white	slightly soluble
Oxide(V)	As_2O_5	229.84	white	very soluble
Sulphide(III)	As_2S_3	246.02	yellow	insoluble
Sulphide(V)	As_2S_5	310.14	yellow	insoluble
		Barium		
Acetate	$Ba(CH_3COO)_2 \cdot 1H_2O$	255.42	white	60 g/l at 18°C
*ortho*Arsenate	$Ba_3(AsO_4)_2$	689.83	white	0.6 g/l
Bromide	$BaBr_2$	297.14	white	1065 g/l
Carbonate	$BaCO_3$	197.34	white	insoluble
Chlorate	$Ba(ClO_3)_2 \cdot H_2O$	332.25	white	280 g/l at 18°C
Chloride	$BaCl_2$	208.24	white	370 g/l
Chromate	$BaCrO_4$	253.32	yellow	insoluble
Fluorine	BaF_2	175.33	white	1.2 g/l
Fluosilicate	$BaSiF_6$	279.41	white	slightly soluble
Hydroxide	$Ba(OH)_2 \cdot 8H_2O$	315.47	white	60 g/l at 18°C
Iodide	BaI_2	391.14	white	1950 g/l at 18°C
Nitrate	$Ba(NO_3)_2$	261.34	white	95 g/l
Oxalate	$Ba(COO)_2$	225.35	white	slightly soluble
Oxide	BaO	153.33	white	40 g/l as $Ba(OH)_2$
Perchlorate	$Ba(ClO_4)_2$	336.23	colourless	very soluble
Permanganate	$Ba(MnO_4)_2$	375.20	violet	750 g/l

Peroxide	BaO_2	169.33	white	slightly soluble
*ortho*Phosphate	$Ba_3(PO_4)_2$	601.93	white	insoluble
Sulphate	$BaSO_4$	233.39	white	insoluble
Sulphide	BaS	169.39	white	decomposes
Sulphite	$BaSO_4$	217.39	white	0.2 g/l
		Beryllium		
Chloride	$BeCl_2$	79.92	colourless	very soluble
Nitrate	$Be(NO_3)_2 \cdot 3H_2O$	187.07	white-yellow	very soluble
Oxide	BeO	25.01	white	insoluble
		Bismuth		
Bromide	$BiBr_3$	448.59	yellow	decomposes → BiOBr
(*basic*) Carbonate	$(BiO)_2CO_3$	509.97	white	insoluble
Chloride	$BiCl_3$	315.34	white	decomposes → BiOCl
Hydroxide	$Bi(OH)_3$	260.00	white	slightly soluble
Iodate	$Bi(IO_3)_3$	733.69	white	insoluble
Iodide	BiI_3	589.69	reddish-black	insoluble
Nitrate	$Bi(NO_3)_3 \cdot 5H_2O$	485.07	colourless	decomposes
(*basic*) Nitrate	$(BiO)NO_3 \cdot H_2O$	305.00	white	insoluble
Oxychloride	$BiOCl$	260.43	white	insoluble
Oxide	Bi_2O_3	465.96	yellowish	insoluble
*ortho*Phosphate	$BiPO_4$	303.95	white	insoluble
Sulphide	Bi_2S_3	514.14	brown-black	insoluble
		Boron		
*meta*Boric acid	HBO_2	43.82	white	slightly soluble
*ortho*Boric acid	H_3BO_3	61.83	white	60 g/l
Fluoride	BF_3	67.81	colourless gas	1050 g/l at 760 mm Hg
Oxide	B_2O_2	69.62	white	slightly soluble

† At 25°C in water, if nothing is stated to the contrary.

A.9—*Continued*

Substance	*Formula*	*Relative molecular mass*	*Colour*	*Solubility*†
		Bromine		
Bromic acid	$HBrO_3$	128.91	colourless soln.	very soluble
Cyanide	$BrCN$	105.92	white	soluble
		Cadmium		
Acetate	$Cd(CH_3COO)_2$	230.50	colourless	very soluble
Bromide	$CdBr_2$	272.22	yellowish	755 g/l at 18°C
Carbonate	$CdCO_3$	172.42	white	insoluble
Chloride	$CdCl_2$	183.32	white	1405 g/l
Cyanide	$Cd(CN)_2$	164.45	white	17 g/l at 15°C
Ferrocyanide(II)	$Cd_2[Fe(CN)_6]\cdot nH_2O$	436.78(an)	white	insoluble
Hydroxide	$Cd(OH)_2$	146.42	white	slightly soluble
Iodate	$Cd(IO_3)_2$	462.22	white	soluble
Iodide	CdI_2	366.22	yellow-green	862 g/l
Nitrate	$Cd(NO_3)_2$	236.42	colourless	1090 g/l at 0°C
Oxalate	$Cd(COO)_2$	200.43	colourless	insoluble
Oxide	CdO	128.41	dark brown	insoluble
*ortho*Phosphate	$Cd_3(PO_4)_2$	527.17	white	insoluble
Sulphate	$CdSO_4$	208.47	white	770 g/l
Sulphide	CdS	144.47	yellow-range	insoluble
Sulphite	$CdSO_3$	192.47	white	slightly soluble

Calcium

Acetate	$Ca(CH_3COO)_2 \cdot 2H_2O$	194.20	white	324 g/l
Arsenate	$Ca_3(AsO_4)_2$	398.08	white	insoluble
*tetra*Borate	CaB_4O_7	195.32	white	slightly soluble
Bromate	$Ca(BrO_3)_2 \cdot H_2O$	313.91	white	very soluble
Bromide	$CaBr_2$	199.89	white	1530 g/l
Carbide	CaC_2	64.10	white	decomposes → C_2H_2
Carbonate	$CaCO_3$	100.09	white	insoluble
Citrate	$Ca_3(C_6H_5O_7)_2$	498.44	white	insoluble
Chlorate	$Ca(ClO_3)_2$	206.99	white	very soluble
Chloride	$CaCl_2$	110.99	white	745 g/l at 20°C
Chromate	$CaCrO_4 \cdot 2H_2O$	192.09	yellow	163 g/l at 20°C
Cyanamide	$CaCN_2$	80.10	white	decomposes → NH_3
Cyanide	$Ca(CN)_2$	92.12	white	decomposes
Ferrocyanide	$Ca_2[Fe(CN)_6] \cdot 12H_2O$	508.24	yellowish	868 g/l
Fluoride	CaF_2	78.08	white	insoluble
Formate	$Ca(HCOO)_2$	130.12	colourless	167 g/l
Hydroxide	$Ca(OH)_2$	74.09	white	slightly soluble
Hypochlorite	$Ca(ClO)_2$	142.98	white	soluble
Iodide	CaI_2	293.89	yellowish	soluble
Nitrate	$Ca(NO_3)_2$	164.09	white	very soluble
Oxalate	$Ca(COO)_2$	128.10	white	insoluble
Oxalate (hydrated)	$Ca(COO)_2 \cdot H_2O$	146.12	white	insoluble
Oxide	CaO	56.08	white	→ $Ca(OH)_2$
*ortho*Phosphate	$Ca_3(PO_4)_2$	310.18	white	insoluble
Phosphide	Ca_3P_2	182.10	greyish	decomposes → PH_3
Pyrophosphate	$Ca_2P_2O_7$	254.10	white	insoluble
Sulphate	$CaSO_4$	136.14	white	slightly soluble
Sulphide	CaS	72.14	white	0.2 g/l
Thiocyanate	$Ca(SCN)_2 \cdot 3H_2O$	210.29	white	very soluble

† At 25°C in water, if nothing is stated to the contrary.

A.9—*Continued*

Substance	*Formula*	*Relative molecular mass*	*Colour*	*Solubility*†
		Carbon		
Dioxide	CO_2	44.01	colourless gas	very soluble
Fluoride	CF_4	87.99	colourless gas	slightly soluble
Oxide	CO	28.01	colourless gas	3.5 mm at 0°C
Sulphide	CS_2	76.14	colourless liq.	insoluble
Tetrachloride	CCl_4	153.81	colourless liq.	insoluble
		Cerium		
Ammonium nitrate(IV)	$Ce(NH_4)_2(NO_3)_6$	548.25	yellow	very soluble
Ammonium sulphate(IV)	$Ce(NH_4)_4(SO_4)_4 \cdot 2H_2O$	632.56	yellow	very soluble
Carbonate	$Ce_2(CO_3)_3 \cdot 5H_2O$	550.34	white	insoluble
Fluoride	CeF_3	197.12	white	insoluble
Hydroxide	$Ce(OH)_3$	191.14	white	slightly soluble
Nitrate	$Ce(NO_3)_3 \cdot 6H_2O$	434.23	white	very soluble
Oxalate	$Ce_2(C_2O_4)_3 \cdot 9H_2O$	706.44	yellow	slightly soluble
*ortho*Phosphate	$CePO_4$	235.09	red	insoluble
Sulphate(IV)	$Ce(SO_4)_2$	332.24	yellow	very soluble
Sulphate(IV)(*hydr.*)	$Ce(SO_4)_2 \cdot 4H_2O$	404.31	yellow	very soluble

		Chlorine		
Chloric acid	$HClO_3$	84.46	colourless sol.	very soluble
Hypochloric acid	$HClO$	52.46	yellow/green soln.	very soluble: unstable
Oxide(I)	Cl_2O	86.91	yellow-brown gas	$\rightarrow HClO$
Oxide(IV)	ClO_2	67.45	yellowish gas	partly $\rightarrow HClO_2 + HClO_3$
Oxide(VII)	Cl_2O_7	182.90	colourless gas	$\rightarrow HClO$
		Chromium		
Acetate(II)	$Cr(CH_3COO)_2$	170.09	red	decomposes
Acetate(III)	$Cr(CH_3COO)_3 \cdot H_2O$	147.15	green or violet	soluble
Bromide(II)	$CrBr_2$	211.81	white	soluble
Bromide(III)	$CrBr_3$	291.71	olive (green/blue)	soluble
Chloride(II)	$CrCl_2$	122.90	white	oxidizes
Chloride	$CrCl_3$	158.35	violet	soluble
Hydroxide	$Cr(OH)_2$	86.01	yellowish	oxidizes
Hydroxide	$Cr(OH)_3 \cdot 3H_2O$	157.06	green-blue	insoluble
Nitrate(III)	$Cr(NO_3)_2 \cdot 9H_2O$	400.15	purple	soluble
Oxalate(III)	$Cr_2(C_2O_4)_3 \cdot 6H_2O$	476.14	red	soluble
Oxide(II)	CrO	68.00	black	insoluble
Oxide(III)	Cr_2O_3	151.99	green	insoluble
*ortho*Phosphate	$CrPO_4 \cdot 2H_2O$	183.00	violet	slightly soluble
		Cobalt		
Acetate	$Co(CH_3COO)_2$	177.03	pink (red)	soluble
Acetate	$Co(CH_3COO)_3$	236.06	green	decomposes
Bromide(II)	$CoBr_2$	218.75	green	650 g/l at 50°C
Bromide(II)	$CoBr_3 \cdot 6H_2O$	326.84	red-violet	very soluble
Carbonate(II)	$CoCO_3$	118.94	red	insoluble
Chloride(II)	$CoCl_2$	129.84	blue (pink)	460 g/l at 15°C

† At 25°C in water, if nothing is stated to the contrary.

A.9—*Continued*

Substance	*Formula*	*Relative molecular mass*	*Colour*	*Solubility*†
		Cobalt (*cont.*)		
Chloride(III)	$CoCl_3$	165.29	red	decomposes
Chromate(II)	$CoCrO_4$	174.93	dark grey	insoluble
Cyanide(II)	$Co(CN)_2 \cdot 2H_2O$	147.00	blue-violet	slightly soluble
Nitrate(II)	$Co(NO_3)_2 \cdot 6H_2O$	291.04	red	1340 g/l at 0°C
Sulphate(II)	$CoSO_4$	155.00	black (pink)	370 g/l
Sulphide(II)	CoS	91.00	black	insoluble
Thiocyanate(II)	$Co(CNS)_2 \cdot 3H_2O$	299.14	violet	soluble
		Copper		
Acetate(II)	$Cu(CH_3COO)_2 \cdot H_2O$	199.66	dark green	75 g/l
Arsenate	$Cu_3(AsO_4)_2 \cdot 4H_2O$	540.54	blue	insoluble
Azzurrite	$2CuCO_3 \cdot Cu(OH)_2$	344.67	blue	insoluble
Bromide(I)	$CuBr$	143.45	white	insoluble
Bromide(II)	$CuBr_2$	223.36	dark green	very soluble
Carbonate(I)	Cu_2CO_3	187.10	yellow	insoluble
Chlorate(II)	$Cu(ClO_3)_2 \cdot 6H_2O$	338.54	green	very soluble
Chloride(II)	$CuCl_2$	134.45	brown (blue)	700 g/l at 0°C
Chromate basic(II)	$CuCrO_4 \cdot 2CuO \cdot 2H_2O$	374.67	yellow	insoluble
Cyanide(I)	$CuCN$	89.57	white	insoluble
Dichromate(II)	$CuCr_2O_7 \cdot 2H_2O$	315.57	black	very soluble
Ferrocyanide	$Cu_2[Fe(CN)_6]$	339.04	brown	insoluble
Fluoride(II)	CuF_2	101.55	white (blue)	50 g/l
Hydroxide(II)	$Cu(OH)_2$	97.56	blue	insoluble

Iodide(I)	CuI	190.45	dirty white	insoluble
Nitrate(II)	$Cu(NO_3)_2 \cdot 6H_2O$	295.04	blue	very soluble
Oxide(I)	Cu_2O	143.09	red	insoluble
Oxide(II)	CuO	79.55	black	insoluble
Sulphate(I)	Cu_2SO_4	223.15	grey	decomposes
Sulphate(II)	$CuSO_4 \cdot 5H_2O$	249.69	blue	335 g/l
Sulphide(II)	CuS	95.61	black	insoluble
Thiocyanate(I)	$CuSCN$	121.63	white	insoluble
		Fluorine		
Nitrate	FNO_3	81.01	colourless gas	$\rightarrow F_2O + O_2 + HF + HNO_2$
Oxide	F_2O	54.00	colourless gas	soluble
		Germanium		
Chloride(IV)	$GeCl_4$	214.41	colourless liq.	decomposes
Fluoride(IV)	GeF_4	148.58	colourless liq. or gas	decomposes
Oxide(IV)	GeO_2	104.59	colourless	insoluble
		Gold		
Bromide(I)	$AuBr$	276.88	yellowish	insoluble
Bromide(III)	$AuBr_3$	436.69	grey	slightly soluble
Chloride(I)	$AuCl$	232.42	yellow	insoluble
Chloride(III)	$AuCl_3$	303.33	reddish	680 g/l
Iodide(I)	AuI	323.87	green-yellow	insoluble
Iodide(II)	AuI_3	577.68	dark green	insoluble
Sulphide(I)	Au_2S	426.00	brown-black	insoluble
Sulphide(III)	Au_2S_3	490.13	brown-black	insoluble

† At 25°C in water, if nothing is stated to the contrary.

A.9—*Continued*

Substance	*Formula*	*Relative molecular mass*	*Colour*	*Solubility*†
		Hydrogen		
Hydrobromic acid	HBr	80.91	colourless gas	very soluble
Hydrochloric acid	H_2F_2	36.46	colourless gas	very soluble
Hydrocyanic acid	HCN	27.03	colourless gas or liquid	miscible
Hydrofluoric acid	H_2F_2	40.01	colourless gas	very soluble
Hydrogen peroxide	H_2O_2	34.02	colourless liquid	miscible
Hydrogen sulphide	H_2S	34.08	colourless gas	very soluble
Hydriodic acid	HI	127.91	colourless gas	very soluble
Water	H_2O	18.02	colourless liquid	
		Iodine		
Bromide(I)	IBr	206.81	black	soluble
Chloride(I)	ICl	162.36	cr. or liq., brown	soluble
Chloride(III)	ICl_3	233.26	yellow-brown	sol. HCl, CCl_4
Iodic acid	HIO_3	175.91	colourless	2700 g/l
Oxide(V)	I_2O_5	333.81	colourless	very soluble
Periodic acid	HIO_4	191.91	colourless	very soluble
		Iron		
Acetylacetonate	$Fe(C_4H_7O_2)_3$	353.18	red	soluble
Ammonium sulphate	$Fe(NH_4)_2(SO_4)_2 \cdot 6H_2O$	392.14	green	275 g/l

Bromide(II)	$FeBr_2$ (also $6H_2O$)	215.66	yellow-green	1100 g/l at 15°C
Bromide(III)	$FeBr_3$ (also $6H_2O$)	295.56	dark red	soluble
Carbonate	$FeCO_3$	115.86	greyish	insoluble
Carbonate (*hydr.*)	$FeCO_3 \cdot H_2O$	133.87	greyish	slightly soluble
(*penta*)Carbonyl	$Fe(CO)_5$	195.90	yellow liquid	insoluble
Chloride(II)	$FeCl_2 \cdot 4H_2O$ (also $2H_2O$)	198.81	green	1600 g/l at 15°C
Chloride(III)	$FeCl_3$ (also $6H_2O$)	162.20	brown	990 g/l
Disulphide	FeS_2	119.97	bright gold	insoluble
Fe^{III} ferricyanide	$Fe[Fe(CN)_6]$	267.81	green	insoluble
Fe^{II} ferrocyanide	$Fe_2[Fe(CN)_6]$	323.65	white → blue	insoluble
Fe^{III} ferrocyanide	$Fe_4[Fe(CN)_6]_3$	859.26	blue	insoluble
Iodide(II)	FeI_2	309.66	grey	soluble
Nitrate(III)	$Fe(NO_3)_3 \cdot 6H_2O$	349.95	colourless	1500 g/l at 0°C
Oxide(II)	FeO	71.85	black	insoluble
Oxide(III)	Fe_2O_3	159.69	reddish brown	insoluble
Oxide(II,III)	Fe_3O_4	231.54	black	insoluble
*ortho*Phosphate	$FePO_4 \cdot 2H_2O$	186.85	greyish	insoluble
Sulphate(II)	$FeSO_4 \cdot 7H_2O$	278.01	bright green	475 g/l
Sulphate(III)	$Fe_2(SO_4)_3$	399.86	yellow	slowly dissolves
Sulphide	FeS	87.91	black	insoluble
Sulphite	$FeSO_3 \cdot 3H_2O$	189.96	greenish	slowly dissolves
Thiocyanate	$Fe(SCN)_3$	230.08	dark red	very soluble

Lanthanum

Chloride	$LaCl_3$	245.27	white	very soluble
Chloride (*hydr.*)	$LaCl_3 \cdot 7H_2O$	371.38	white	very soluble
Hydroxide	$La(OH)_3$	189.93	white	insoluble
Nitrate	$La(NO_3)_3 \cdot 6H_2O$	433.02	white	1510 g/l
Sulphate	$La_2(SO_4)_3$	566.00	white	30 g/l

† At 25°C in water, if nothing is stated to the contrary.

A.9—*Continued*

Substance	*Formula*	*Relative molecular mass*	*Colour*	*Solubility*†
		Lead		
Acetate(II)	$Pb(CH_3COO)_2$	325.29	white	450 g/l
Acetate(IV)	$Pb(CH_3COO)_4$	443.38	white	decomposes
(*basic*)Acetate	$Pb(CH_3COO)_2 \cdot Pb(OH)_2 \cdot H_2O$	584.52	white	very soluble
Bromide	$PbBr_2$	367.01	white	insoluble
Carbonate	$PbCO_3$	267.21	white	9.75 g/l
(*basic*)Carbonate	$2PbCO_3 \cdot Pb(OH)_2$	775.64	white	insoluble
Chloride	$PbCl_2$	278.11	white	insoluble
Chromate	$PbCrO_4$	323.19	yellow	10.8 g/l
Dichromate	$PbCr_2O_7$	423.19	red	insoluble
Ferrocyanide	$Pb_2[Fe(CN)_6] \cdot 6H_2O$	680.41	yellowish	insoluble
Hydroxide	$Pb(OH)_2$	241.21	white	slightly soluble
Iodide	PbI_2	461.01	yellow	decomposes
Nitrate	$Pb(NO_3)_2$	331.21	colourless	580 g/l
Oxalate	$Pb(COO)_2$	295.22	white	slightly soluble
Oxide	PbO	223.20	yellow	insoluble
Oxide(II,IV)	Pb_3O_4	685.60	red	insoluble
*meta*Phosphate	$Pb(PO_3)_2$	365.14	colourless	insoluble
*ortho*Phosphate	$Pb_3(PO_4)_2$	811.54	colourless	insoluble
Sulphate	$PbSO_4$	303.26	white	insoluble
Sulphide	PbS	239.26	black	insoluble
Thiocyanate	$Pb(SCN)_2$	323.36	white	0.5 g/l

		Lithium		
Aluminium hydride	$LiAlH_4$	37.95	white	decomposes
Bicarbonate	$LiHCO_3$	67.96	white	60 g/l at 15°C
Bromide	LiBr	86.85	white	1460 g/l at 5°C
Carbonate	Li_2CO_3	73.89	white	15.4 g/l at 0°C
Chlorate	$LiClO_3$	90.39	white	4480 g/l
Chloride	LiCl	42.39	white	640 g/l at 0°C
Dichromate	$Li_2Cr_2O_7 \cdot 2H_2O$	265.90	red-orange	1850 g/l
Fluoride	LiF	25.94	white	2.7 g/l at 18°C
Hydroxide	LiOH	23.95	white	129 g/l
Iodate	$LiIO_3$	181.84	white	810 g/l at 20°C
Iodide	LiI	133.84	white	1650 g/l
Nitrate	$LiNO_3$	68.94	white	890 g/l
Oxalate	$Li_2(COO)_2$	101.90	white	80 g/l at 20°C
Oxide	Li_2O	29.88	white	$\rightarrow$ LiOH
Perchlorate	$LiClO_4$	106.39	white	600 g/l
*ortho*Phosphate	Li_3PO_4	115.79	white	slightly soluble
*meta*Silicate	Li_2SiO_3	89.96	white	insoluble
Sulphate	Li_2SO_4	109.94	white	255 g/l
Sulphide	Li_2S	45.94	white	very soluble
		Magnesium		
Acetate	$Mg(CH_3COO)_2$	142.40	white	very soluble
Ammonium arsenate	$MgNH_4AsO_4 \cdot 6H_2O$	289.36	white	slightly soluble
Ammonium phosphate	$MgNH_4PO_4$	137.31	white	slightly soluble
*ortho*Borate	$Mg_3(BO_2)_2$	109.53	white	insoluble
Bromide	$MgBr_2$	184.11	white	1030 g/l
Carbonate	$MgCO_3$	84.31	white	slightly soluble
Chlorate	$Mg(ClO_3)_2 \cdot 6H_2O$	299.30	white	1300 g/l
Chloride	$MgCl_2$	95.21	white	570 g/l

† At 25°C in water, if nothing is stated to the contrary.

A.9—*Continued*

Substance	*Formula*	*Relative molecular mass*	*Colour*	*Solubility*†
		Magnesium (*cont.*)		
Chromate	$MgCrO_4 \cdot 7H_2O$	266.40	yellow	very soluble
Cyanide	$Mg(CN)_2$	76.34	white	soluble
Hydroxide	$Mg(OH)_2$	58.32	white	insoluble
Iodide	MgI_2	278.12	white	1500 g/l
Nitrate	$Mg(NO_3)_2 \cdot 6H_2O$	256.41	white	1250 g/l
Oxalate	$Mg(COO)_2 \cdot 2H_2O$	148.36	white	slightly soluble
Oxide	MgO	40.30	white	slightly soluble
Perchlorate	$Mg(ClO_4)_2$	223.21	white	soluble
*ortho*Phosphate	$Mg_3(PO_4)_2$	262.85	white	insoluble
Pyrophosphate	$Mg_2P_2O_7$	222.55	white	insoluble
Sulphate	$MgSO_4 \cdot 7H_2O$	246.47	white	720 g/l
Sulphide	MgS	56.36	yellowish	decomposes
		Manganese		
Acetate(II)	$Mn(CH_3COO)_2$	173.03	pink	soluble
Bromide(II)	$MnBr_2$	214.75	pink	very soluble
Carbonate(II)	$MnCO_3$	114.95	pink	insoluble
Chloride(II)	$MnCl_2$	125.85	pink	720 g/l
Hydroxide(II)	$Mn(OH)_2$	88.95	white → brown	insoluble
Iodide(II)	MnI_2	308.75	brown	soluble
Nitrate(II)	$Mn(NO_3)_2 \cdot 4H_2O$	251.01	pink	very soluble
Oxide(III)	Mn_2O_3	157.88	black	insoluble
Oxide(IV)	MnO_2	86.94	black	insoluble

Sulphate(II)	$MnSO_4 \cdot 4H_2O$	223.06	pink	650 g/l
Sulphate(III)	$Mn_2(SO_4)_3$	398.05	green	decomposes
Sulphide(II)	MnS	87.00	pink	insoluble
		Mercury		
Acetate(II)	$Hg(CH_3COO)_2$	318.68	white	400 g/l
Bromide(I)	Hg_2Br_2	560.99	white	insoluble
Bromide(II)	$HgBr_2$	360.40	white	insoluble
Carbonate	Hg_2CO_3	461.19	yellowish	insoluble
Chlorate(I)	$Hg_2(ClO_3)_2$	568.08	white	decomposes
Chlorate(II)	$Hg(ClO_3)_2$	367.49	white	decomposes
Chloride(I)	Hg_2Cl_2	472.09	white	insoluble
Chloride(II)	$HgCl_2$	271.50	white	70 g/l
Chromate(II)	$HgCrO_4$	316.58	red	insoluble
Iodide(I)	Hg_2I_2	654.99	yellow	insoluble
Iodide(II)	HgI_2	454.40	red	insoluble
Nitrate(I)	$Hg_2(NO_3)_2 \cdot 2H_2O$	561.22	white	decomposes
Nitrate(II)	$HgNO_3 \cdot \frac{1}{2}H_2O$	333.61	white-yellow	very soluble
Oxide(II)	HgO	216.59	yellow-red	insoluble
Sulphate(II)	$HgSO_4$	296.65	white	decomposes
Sulphide(I)	Hg_2S	433.24	black	insoluble
Sulphide(II)	HgS	232.65	red or black	insoluble
		Molybdenum		
Chloride(II)	$MoCl_2$	166.85	yellow	insoluble
Chloride(III)	$MoCl_3$	202.30	dark red	insoluble
Chloride(V)	$MoCl_5$	273.21	green-black	decomposes
Iodide(II)	MoI_2	349.75	brown	insoluble
Sulphide(IV)	MoS_2	160.06	black	insoluble
Trioxide	MoO_3	143.94	white-yellow	insoluble

† At 25°C in water, if nothing is stated to the contrary.

A.9—*Continued*

Substance	*Formula*	*Relative molecular mass*	*Colour*	*Solubility*†
		Nickel		
Acetate	$Ni(CH_3COO)_2$	176.78	green	113 g/l at 0°C
Bromide	$NiBr_2$	218.50	yellowish	very soluble
Bromide (*hydr.*)	$NiBr_2 \cdot 3H_2O$	272.55	green-yellow	insoluble
Carbonate	$NiCO_3$	118.70	green	insoluble
*tetra*Carbonyl	$Ni(CO)_4$	179.73	colourless liq.	insoluble
Chlorate	$Ni(ClO_3)_2 \cdot 6H_2O$	333.70	dark red	650 g/l
Chloride	$NiCl_2$	129.60	yellow	insoluble
Cyanide	$Ni(CN)_2$	110.73	yellowish	9 g/l
Dimethylglyoximate	$Ni(CH_3CNO{-}CH_3CH \cdot OH)_2$	288.92	red	insoluble
Ferrocyanide(II)	$Ni_2[Fe(CN)_6]$	329.33	bright green	insoluble
Iodide	NiI_2	312.50	black	very soluble
Nitrate	$Ni(NO_3)_2 \cdot 6H_2O$	290.79	green	2400 g/l at 5°C
Oxide	NiO	74.69	black	insoluble
*ortho*Phosphate	$Ni_3(PO_4)_2 \cdot 8H_2O$	510.13	green	insoluble
Sulphate	$NiSO_4$	154.75	greenish	300 g/l at 10°C
Sulphide	NiS	90.75	black	insoluble
		Nitrogen		
Ammonia	NH_3	17.03	colourless gas	very soluble
Hydrazine	N_2H_4	32.05	colourless liq.	miscible
Hydrazine hydrate	$N_2H_4 \cdot H_2O$	50.06	colourless liq.	miscible

Hydrazine sulphate	$N_2H_4 \cdot H_2SO_4$	130.12	white	soluble
Hydroxylamine hydrochloride	$NH_2OH \cdot HCl$	69.49	white	very soluble
Hydroxylamine	NH_2OH	33.03	white	very soluble
Nitric acid	HNO_3	63.01	colourless liq.	miscible
Nitric anhydride	N_2O_5	108.01	white	$\rightarrow HNO_3$
Nitric oxide	NO	30.01	colourless gas	soluble
Nitrogen dioxide	NO_2	44.01	red-brown gas	$\rightarrow HNO_2 + HNO_3$ etc.
*di*Nitrogen-tetroxide	N_2O_4	92.01	colourless liq.	$\rightarrow HNO_2 + HNO_3$ etc.
Nitrous acid	HNO_2	47.02	azure solution	decomposes
Nitrous anhydride	N_2O_3	76.01	blue liq. at 2°C	$\rightarrow HNO_3$
Nitroux oxide	N_2O	44.01	colourless gas	very soluble
		Palladium		
Chloride	$PdCl_2$	177.33	dark red	soluble
Oxide	PdO	122.42	greenish	insoluble
Sulphide	PdS	138.48	brown	insoluble
		Phosphorus		
Chloride(III)	PCl_3	137.33	colourless fuming liquid	decomposes
Chloride(V)	PCl_5	208.24	white	decomposes
Iodide	PI_3	411.68	red	decomposes
Oxychloride	$POCl_3$	143.33	colourless fuming liquid	decomposes
Phosphine	PH_3	34.00	colourless gas	slightly soluble
Phosphoric acid	H_3PO_4	98.00	cryst. or liq.	very soluble
Phosphoric anhydride	P_2O_5	141.95	white	$\rightarrow H_3PO_4$
Phosphorous acid	H_3PO_3	82.00	colourless	very soluble
Phosphorous anhydride	P_2O_3	109.95	white	$\rightarrow H_3PO_3$

† At 25°C in water, if nothing is stated to the contrary.

A.9—*Continued*

Substance	*Formula*	*Relative molecular mass*	*Colour*	*Solubility*†
		Platinum		
Chloride(II)	$PtCl_2$	265.99	olive green	soluble
Chloride(IV)	H_2PtCl_6	409.81	yellowish brown	very soluble
Oxide(IV)	PtO_2	227.08	black	insoluble
		Potassium		
Acetate	CH_3COOK	98.14	colourless	2550 g/l
Arsenate	K_3AsO_4	256.21	colourless	190 g/l
Bicarbonate	$KHCO_3$	100.12	white	360 g/l
Bisulphate	$KHSO_4$	136.16	white	520 g/l
Bitartrate	$C_4H_5O_6K$	188.18	white	60 g/l
*tetra*Borate	$K_2B_2O_7 \cdot 7H_2O$	377.56	colourless	260 g/l
Bromate	$KBrO_3$	167.00	colourless	80 g/l
Bromide	KBr	119.00	colourless	680 g/l
Carbonate	K_2CO_3	138.21	colourless	1130 g/l
Chlorate	$KClO_3$	122.55	colourless	72 g/l
Chloride	KCl	74.55	colourless	355 g/l
Chromate	K_2CrO_4	194.19	yellow	650 g/l
Cyanate	$KOCN$	81.12	colourless	very soluble
Cyanide	KCN	65.12	colourless	510 g/l
Cyanoaurate	$K[Au(CN)_2]$	288.10	colourless	143 g/l
Dichromate	$K_2Cr_2O_7$	294.19	orange	160 g/l
Ferricyanide	$K_3[Fe(CN)_6]$	329.25	orange	488 g/l

Ferrocyanide	$K_4[Fe(CN)_6] \cdot 3H_2O$	442.40	yellow	280 g/l
Fluoride	KF	58.10	colourless	930 g/l
Hydroxide	KOH	56.11	white	1100 g/l
Iodate	KIO_3	214.00	white	88 g/l
Iodide	KI	166.00	white	1480 g/l
Manganate	K_2MnO_4	197.14	green	slowly decomposes
Metabisulphite	$K_2S_2O_5$	222.32	white	very soluble
Nitrate	KNO_3	101.10	white	373 g/l
Oxalate	$K_2C_2O_4 \cdot H_2O$	184.24	white	335 g/l
Perchlorate	$KClO_4$	138.55	white	slightly soluble
Periodate	KIO_4	230.00	white	43 g/l
Permanganate	$KMnO_4$	157.04	violet	77 g/l
Peroxydisulphate	$K_2S_2O_8$	270.32	white	17.5 g/l at 0°C
*ortho*Phosphate	K_3PO_4	212.27	colourless	900 g/l at 20°C
Pot. dihydrogen phosphate	KH_2PO_4	136.09	white	220 g/l
Pot. hydrogen phosphate	K_2HPO_4	174.18	white	very soluble
Pyroantimonate	$KSb(OH)_6 \cdot \frac{1}{2}H_2O$	271.90	white	28 g/l at 20°C
Pyrosulphate	$K_2S_2O_7$	254.32	white	soluble
Sodium tartrate	$C_4H_4O_6KNa \cdot 4H_2O$	282.22	white	1100 g/l
Sulphate	K_2SO_4	174.26	white	120 g/l
Sulphide	K_2S (also $5H_2O$)	110.26	yellowish	very soluble
Sulphite	$K_2SO_3 \cdot 2H_2O$	194.29	white	1100 g/l
Tellurite	K_2TeO_3	253.80	white	soluble
Thiocyanate	KSCN	97.18	colourless	very soluble
		Ruthenium		
Chloride(III)	$RuCl_3$	207.48	black	soluble
Oxide(VIII)	RuO_4	165.10	yellow	20 g/l

† At 25°C in water, if nothing is stated to the contrary.

A.9—*Continued*

Substance	*Formula*	*Relative molecular mass*	*Colour*	*Solubility†*
		Selenium		
Dioxide	SeO_2	110.96	white	very soluble
Oxychloride	$SeOCl_2$	165.87	pale yellow	decomposes
Sulphide	SeS	111.02	orange	insoluble
		Silicon		
Carbide	SiC	40.10	grey	insoluble
Chloride	$SiCl_4$	169.90	colourless liq.	decomposes
Dioxide	SiO_2	60.09	white	insoluble
Fluoride	SiF_4	104.08	colourless gas	decomposes
		Silver		
Acetate	CH_3COOAg	166.91	white	12 g/l
Arsenate	Ag_3AsO_4	462.52	red-brown	insoluble
Arsenite	Ag_3AsO_3	446.52	yellow	insoluble
Bromate	$AgBrO_3$	235.77	white	2 g/l
Bromide	$AgBr$	187.77	bright yellow	insoluble
Carbonate	Ag_2CO_3	275.75	white	slightly soluble
Chlorate	$AgClO_3$	191.32	white	10 g/l at 18°C
Chloride	$AgCl$	143.32	white	insoluble
Chromate	Ag_2CrO_4	331.73	red	insoluble
Dichromate	$Ag_2Cr_2O_7$	431.73	yellow	insoluble

Fluoride	AgF	126.87	white-yellow	1830 g/l at 18°C
Iodate	$AgIO_3$	282.75	colourless	slightly soluble
Iodide	AgI	234.77	yellow	insoluble
Nitrate	$AgNO_3$	169.87	colourless	1220 g/l at 0°C
Nitrite	$AgNO_2$	153.87	bright yellow	slightly soluble
Oxide	Ag_2O	231.74	dark brown	insoluble
Perchlorate	$AgClO_4$	207.32	white	very soluble
*ortho*Phosphate	Ag_3PO_4	418.58	yellow	insoluble
Sulphate	Ag_2SO_4	311.80	white	7.75 g/l at 18°C
Sulphide	Ag_2S	247.80	black	insoluble
		Sodium		
Acetate	CH_3COONa	82.03	white	very soluble
*meta*Aluminate	$NaAlO_2$	81.97	white	soluble
Amide	$NaNH_2$	39.01	white	decomposes
Arsenate	$Na_3AsO_4 \cdot 12H_2O$	424.70	white	390 g/l at 16°C
Azide	NaN_3	65.02	white	420 g/l
Bicarbonate	$NaHCO_3$	84.01	white	103 g/l
Bisulphate	$NaHSO_4$	120.06	white	very soluble
Bisulphide	$NaHS$	56.06	colourless	very soluble
Bisulphite	$NaHSO_3$	104.06	white	very soluble
*meta*Borate	$NaBO_2$	65.80	white	265 g/l
*tetra*Borate	$Na_2B_4O_7 \cdot 10H_2O$	381.37	white	31.5 g/l
Borohydride	$NaBH_4$	37.38	white	550 g/l
Bromate	$NaBrO_3$	150.89	white	300 g/l
Bromide	$NaBr$	102.89	white	900 g/l
Carbonate	Na_2CO_3	105.99	white	195 g/l
Citrate	$Na_3C_6H_5O_7 \cdot 2H_2O$	294.10	white	770 g/l
Chlorate	$NaClO_3$	106.44	white	990 g/l
Chlorite	$NaClO_2$	90.44	white	400 g/l

† At 25°C in water, if nothing is stated to the contrary.

A.9—*Continued*

Substance	*Formula*	*Relative molecular mass*	*Colour*	*Solubility*†
Chromate	Na_2CrO_4	161.97	yellow	825 g/l
Cyanide	$NaCN$	49.01	white	500 g/l
Dichromate	$Na_2Cr_2O_7 \cdot 2H_2O$	298.00	orange	2380 g/l at 0°C
Disodium arsenate	$Na_2HAsO_4 \cdot 12H_2O$	402.09	colourless	very soluble
Dithionite	$Na_2S_2O_4$	174.10	white	very soluble
Fluoride	NaF	41.99	white	43 g/l
Fluorosilicate	Na_2SiF_6	188.06	colourless	7 g/l
Formate	$HCOONa$	68.01	white	440 g/l at 0°C
Hydroxide	$NaOH$	40.00	white	1150 g/l
Hypochlorite	$NaClO$	74.44	colourless soln.	very soluble
Hypophosphite	$NaH_2PO_2 \cdot H_2O$	105.99	white	80 g/l
Iodate	$NaIO_3$	197.89	white	95 g/l
Iodide	NaI	149.89	white	1840 g/l
Metabisulphite	$Na_2S_2O_5$	190.10	white	very soluble
Molybdate	$Na_2MoO_4 \cdot 2H_2O$	241.95	white	very soluble
Nitrate	$NaNO_3$	85.00	white	920 g/l
Nitrite	$NaNO_2$	69.00	white-yellow	825 g/l
Nitroprusside	$Na_2[Fe(CN)_5NO] \cdot 2H_2O$	297.96	dark red	very soluble
Oxalate	$Na_2C_2O_4$	134.00	white	40 g/l
Oxide	Na_2O	61.98	white	$\rightarrow NaOH$
Perborate	$NaBO_3 \cdot 4H_2O$	153.85	white	25 g/l
Perchlorate	$NaClO_4$	122.44	white	very soluble
Peroxide	Na_2O_2	77/98	yellowish	decomposes

Peroxydispulphate	$Na_2S_2O_8$	238.10	white	soluble
*ortho*Phosphate	$Na_3PO_4 \cdot 12H_2O$	380.12	colourless	290 g/l
Phosphate disodium	$Na_2HPO_4 \cdot 2H_2O$	178.00	white	152 g/l
Phosphate monosodium	NaH_2PO_4	119.98	white	very soluble
Phosphite disodium	$Na_2HPO_3 \cdot 5H_2O$	216.04	white	very soluble
Polyphosphate	$(NaPO_3)_x$	x.101.96	colourless	very soluble
Pyrophosphate	$Na_4P_2O_7 \cdot 10H_2O$	446.05	white	80 g/l
Sulphate	Na_2SO_4	142.04	white	170 g/l
Sulphate (*hydr.*)	$Na_2SO_4 \cdot 10H_2O$	322.19	colourless	490 g/l
Sulphide	$Na_2S \cdot 9H_2O$	240.18	colourless	very soluble
Sulphite	Na_2SO_3	126.04	white	240 g/l
Sulphite (*hydr.*)	$Na_2SO_3 \cdot 7H_2O$	252.15	colourless	620 g/l
Thiosulphate	$Na_2S_2O_3 \cdot 5H_2O$	248.17	colourless	780 g/l
Tungstate	$Na_2WO_4 \cdot 2H_2O$	329.86	colourless	very soluble
*meta*Vanadate	$NaVO_3 \cdot 4H_2O$	193.99	white-yellow	slightly soluble
		Sulphur		
Chloride	S_2Cl_2	135.03	yellow liq.	decomposes
Chlorosulphonic acid	HSO_3Cl	116.52	colourless liq.	decomposes
Dichloride	SCl_2	102.97	red liq.	decomposes
Peroxydisulphuric acid	$H_2S_2O_8$	194.14	colourless	decomposes
Peroxymonosulphuric acid	H_2SO_5	114.08	colourless	decomposes
Pyrosulphuric acid	$H_2S_2O_7$	178.14	colourless	decomposes
Sulphuric acid	H_2SO_4	98.08	colourless liq.	miscible
Sulphurous acid	H_2SO_3	82.08	colourless soln.	soluble
Sulphur dioxide	SO_2	64.06	colourless gas	30 cm^3/l at 0°C
Sulphur trioxide	SO_3	80.06	white	$\rightarrow H_2SO_4$
Sulphuryl chloride	SO_2Cl_2	134.97	colourless liq.	decomposes
Thionyl chloride	$SOCl_2$	118.97	colourless liq.	decomposes

† At 25°C in water, if nothing is stated to the contrary.

A.9—*Continued*

Substance	*Formula*	*Relative molecular mass*	*Colour*	*Solubility*†
		Thallium		
Chloride	$TlCl$	239.84	white	slightly soluble
Nitrate(I)	$TlNO_3$	266.39	white	soluble
Sulphate(I)	Tl_2SO_4	504.82	white	slightly soluble
		Tin		
Chloride(II)	$SnCl_2$ (also $2H_2O$)	189.60	colourless	soluble
Chloride(IV)	$SnCl_4$ (also $5H_2O$)	260.50	colourless liq.	soluble
Nitrate(IV)	$Sn(NO_3)_4$	366.70	colourless	decomposes
Oxide(II)	SnO	134.69	black	insoluble
Oxide(IV)	SnO_2	150.69	white	insoluble
Sulphate(II)	$SnSO_4$	214.75	yellowish	320 g/l
Sulphide	SnS	150.75	brown	insoluble
Sulphide(IV)	SnS_2	182.82	yellow	insoluble
		Titanium		
Chloride(II)	$TiCl_2$	118.79	brown-black	decomposes
Chloride(III)	$TiCl_3$	154.24	violet	soluble
Chloride(IV)	$TiCl_4$	189.69	bright yellow liq.	decomposes when hot
Dioxide	TiO_2	79.88	white	insoluble

		Uranium		
Acetate	$UO_2(CH_3COO)_2 \cdot 2H_2O$	424.15	yellow	100 g/l
Dioxide	UO_2	270.03	brown-black	insoluble
Nitrate	$(UO_2)(NO_3)_2 \cdot 6H_2O$	502.13	yellow	very soluble
Pitchblende	U_3O_8	842.09	olive green	insoluble
		Vanadium		
Oxide(V)	V_2O_5	181.88	yellow-red	slightly soluble
Sulphide	V_2S_3	198.06	dark green	insoluble
		Zinc		
Acetate	$Zn(CH_3COO)_2$	183.47	colourless	350 g/l
Bromide	$ZnBr_2$	252.19	colourless	very soluble
Carbonate	$ZnCO_3$	125.39	white	insoluble
Chloride	$ZnCl_2$	136.29	colourless	very soluble
Hydroxide	$Zn(OH)_2$	99.39	white	insoluble
Iodide	ZnI_2	319.19	colourless	very soluble
Nitrate	$Zn(NO_3)_2 \cdot 3H_2O$	243.44	colourless	very soluble
Oxide	ZnO	81.38	white	insoluble
Sulphate	$ZnSO_4 \cdot 6H_2O$	269.53	white	very soluble
Sulphide	ZnS	97.44	white	very soluble
		Zirconium		
Nitrate	$Zr(NO_3)_4$	339.25	white	very soluble
Oxide	ZrO_2	123.22	white	insoluble
Oxychloride	$ZrOCl_2$	178.13	white	very soluble

† At 25°C in water, if nothing is stated to the contrary.

Index